AN INTRODUCTION TO OPERATIONAL AMPLIFIERS

WITH LINEAR IC APPLICATIONS

SECOND EDITION

ELECTRONIC TECHNOLOGY SERIES

Santokh S. Basi
SEMICONDUCTOR PULSE AND SWITCHING CIRCUITS (1980)

Theodore Bogart
LAPLACE TRANSFORMS AND CONTROL SYSTEMS THEORY FOR TECHNOLOGY (1982)

Rodney B. Faber
APPLIED ELECTRICITY AND ELECTRONICS FOR TECHNOLOGY, 2nd edition (1982)

Luces M. Faulkenberry
AN INTRODUCTION TO OPERATIONAL AMPLIFIERS, 2nd edition (1982)

Joseph D. Greenfield
PRACTICAL DIGITAL DESIGN USING ICs (1977)

Joseph D. Greenfield and William C. Wray
USING MICROPROCESSORS AND MICROCOMPUTERS: THE 6800 FAMILY (1981)

Curtis Johnson
PROCESS CONTROL INSTRUMENTATION TECHNOLOGY, 2nd edition (1982)

Irving L. Kosow
STUDY GUIDE IN DIRECT CURRENT CIRCUITS: A PERSONALIZED SYSTEM OF INSTRUCTION (1977)

Irving L. Kosow
STUDY GUIDE IN ALTERNATING CURRENT CIRCUITS: A PERSONALIZED SYSTEM OF INSTRUCTION (1977)

Sol Lapatine
ELECTRONICS IN COMMUNICATION (1978)

Ulises M. Lopez and George E. Warrin
ELECTRONIC DRAWING AND TECHNOLOGY (1978)

Dan I. Porat and Arpad Barna
INTRODUCTION TO DIGITAL TECHNIQUES (1979)

Herbert W. Richter
ELECTRICAL AND ELECTRONIC DRAFTING (1977)

Thomas Young
LINEAR INTEGRATED CIRCUITS (1981)

AN INTRODUCTION TO OPERATIONAL AMPLIFIERS

WITH LINEAR IC APPLICATIONS

SECOND EDITION

Luces M. Faulkenberry
TEXAS STATE TECHNICAL INSTITUTE

PRENTICE HALL, Englewood Cliffs, New Jersey 07632

Library of Congress Cataloging in Publication Data:

Faulkenberry, Luces M.
An introduction to operational amplifiers, with linear IC applications.

(Electronic technology series ; ISSN 0422-910X)
Includes index.
1. Operational amplifiers. I. Title. II. Series.

TK7871.58.06F38 1982 621.3815'35 81-13043
AACR2

Printed in the United States of America

10 9 8 7 6 5 4 3 2 1

ISBN 0-13-497090-X

PREFACE

The second edition of this textbook is designed to introduce the student of electronics to operational amplifier operation, parameters, parameter measurement, and basic operational amplifier circuitry in a clear and easily understandable way. In addition, some special purpose linear circuits and amplifiers are presented. All material is presented with a minimum of mathematics consistent with reasonable accuracy. Practical considerations for component calculation and circuit set-up are emphasized throughout the text and laboratories, and important calculations are illustrated by worked examples.

The book is intended as a text for a one-semester course in operational amplifier theory and applications in community colleges, technical institutes, and two- and four-year technology programs. It is also intended for the working electronic technician who needs more information about operational amplifiers.

Each chapter consists of an introduction, objectives, text material and examples, a brief summary, self-test questions, and a laboratory. The laboratories are constructed to use commonly available test equipment that most schools have. The operational amplifiers specified are in widespread use. No special, hard to obtain components or supplies are required. Laboratory experiments are included from which instructors will select those parts of a laboratory exercise of greatest benefit for their particular classes.

Chapters 8, 10, and 11 are new in the second edition. Chapter 8 presents several types of active filters with an absolute minimum of mathematics. The student will, however, be able to construct high performance active filters on completing the chapter. Chapter 10 presents linear and switching power supply voltage regulators.

Both general principles and IC regulators are included. Chapter 11 introduces some special purpose IC amplifiers including comparators and some comparator applications, current differencing amplifiers, instrumentation amplifiers, and isolation amplifiers. 555 timers are also covered in Chapter 11. The rest of the changes in the book are things the author wishes he had done the first time. These primarily include clearer explanations and more examples.

The author does not feel that the sequencing of the chapters is sacred and expects an instructor to sequence the chapters in a manner that suits that instructor's course, as the author did. Some teachers prefer moving from Chapter 1 to Chapter 5 and the latter portion of the text, returning to the earlier chapters (2, 3, and 4) later. The text is intended to function well in this fashion.

The coverage of the book is as follows: Chapter 1 introduces the operational amplifier, its operation, and the basic amplifier configurations. Chapter 2 discusses negative feedback more generally, introduces some sources of error resulting from nonideal amplifiers, and explains external offset compensation. Chapter 3 covers bias current, CMRR, and their measurement, error due to temperature, and chopper stabilization. Chapter 4 covers frequency response, methods of phase compensation, and slew rate. Chapter 5 covers summing circuits. Chapter 6 covers integrators and differentiators. Chapter 7 presents logarithmic circuits and Chapter 8 active filters. Chapter 9 is a collection of circuits that illustrate common operational amplifier applications. Chapters 10 and 11 are on voltage regulators and special purpose linear ICs, respectively. Chapter 12 discusses noise and grounding. Appendix A is a differential amplifier review, and Appendix B is a discussion of the 741 internal operation. Appendix C is a collection of specifications. Appendixes D, E, and F are derivations too long for the text material. Appendix G is the answers to the self-test questions. A glossary of symbols and terms is included at the end of the book.

The student will need a basic understanding of transistors and algebra to use this text successfully. Chapter 7 uses some elementary differentiation and integration, but this small amount of calculus can be presented during the course, or read around.

My thanks to all who have provided guidance and support in the preparation of this text, especially to the curious and interested students whose questions forced the author to find simple and *clear* explanations.

Luces M. Faulkenberry

CONTENTS

CHAPTER 1

THE BASIC OP-AMP

To use operational amplifiers (op-amps), we must know what they are. This chapter discusses construction, characteristics, and some important specifications of op-amps. The major amplifier configurations using op-amps are also discussed.

OBJECTIVES

After completing the study of this chapter and the self-test questions, the student should be able to:

1. List and state the operation of the major parts of an op-amp.
2. State the name and purpose of the op-amp terminals.
3. State the name and definition of the following op-amp specs: A_{ol}, V_{os}, I_B, I_{os}, R_{in}, R_{out}.
4. Calculate the components of the feedback loop for a given closed loop gain and draw from memory a voltage follower, noninverting amplifier, inverting amplifier, and differential input amplifier.
5. Given the feedback components, calculate the closed loop gain for the circuits listed in objective 4.
6. Perform the laboratory for Chapter 1.

1-1 WHAT IS AN OP-AMP?

An operational amplifier is a modular, multistage amplifier with a differential input that approximates most of the characteristics of the mythical "ideal amplifier." The properties associated with an ideal amplifier are:

1. Infinite voltage gain ($A_v \rightarrow \infty$).
2. Infinite input impedance ($Z_{in} \rightarrow \infty$).
3. Zero output impedance ($Z_{out} \rightarrow 0$).
4. Output voltage $V_{out} = 0$ when input voltages $V_1 = V_2$.
5. Infinite bandwidth (no delay of the signal through the amplifier).

In practice, none of these properties can be achieved, but they can be approximated closely enough for many applications. For example, if feedback is used to limit the amplifier circuit gain to 10, an amplifier gain (without feedback) of 1000 is close enough to infinity for practical purposes.

The first stage of an operational amplifier is a differential amplifier. The differential amplifier provides a high gain to difference signals (i.e., $V_2 - V_1$ in Figures

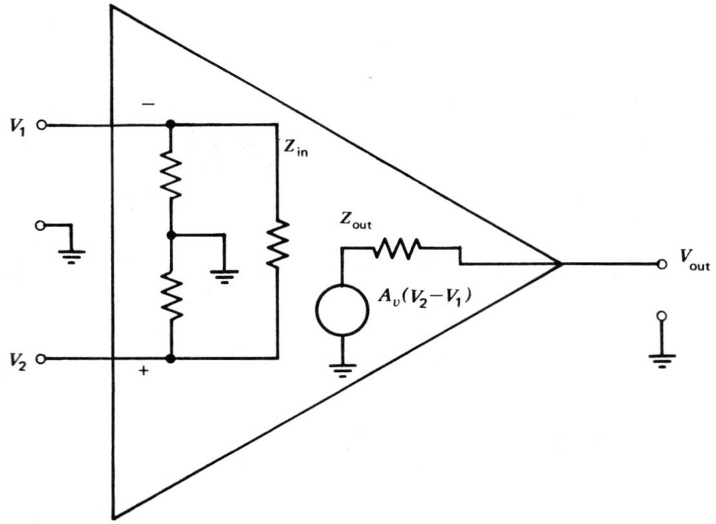

FIGURE 1.1 Amplifier representation.

1.1 and 1.2) and a low gain to signals applied to both inputs simultaneously (common-mode signals).[1]

The differential amplifier also provides a high impedance to any input signal applied to it. The input stage of an operational amplifier is the most critical since that is where input impedance is set and the common-mode response and offset voltages are minimized.[2]

One or more intermediate stages follow the input stage, as shown in Figure 1.3, to shift the quiescent voltage level to zero at the output and provide both voltage and current gain. The added voltage gain is needed for a high overall voltage gain, and the current gain is needed to supply drive current to the output stage without loading the input stage. Both single-ended and differential configurations are used for the intermediate amplification stages.

The output stage must supply a low output impedance and enough current to drive the expected load. It should also have a high enough input impedance that it does not load the last intermediate amplification stage. The output stage is usually an emitter follower or a complementary configuration.

Figure 1.4 is a schematic of a simple operational amplifier. There are a few things we should note about the input circuit. The emitter resistors of Q_1 and Q_2 increase the input impedance of the input stage. The collector currents in the input

[1] Common-mode signals are those of the same phase and amplitude applied to both inputs simultaneously.

[2] Offset voltages are small, undesired signals internally generated by the amplifier that cause some output voltage with zero volts applied to both inputs. They are caused by imperfect matching of the emitter-base voltages of the input transistors.

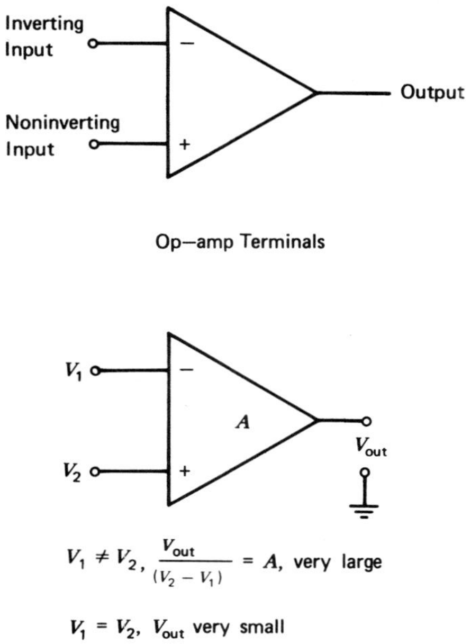

FIGURE 1.2 Common-mode signal.

stage are usually low so that the ac resistance of the emitter-base diode of the input transistors is high and the circuit can be driven with low input currents. This causes a loss in voltage gain of the first stage that is made up in the intermediate stages. A constant current source is used to provide emitter current for the first stage to reduce the circuit's sensitivity to common-mode signals. Since the internal resistance of a constant current source (r_{ac}) is high, the gain of the differential amplifier to common-mode signals (A_{cm}) is very low.[3]

To decrease the amount of input current required to drive the differential amplifier and raise input resistance, the first-stage Q_1 and Q_2 may be Darlington pairs

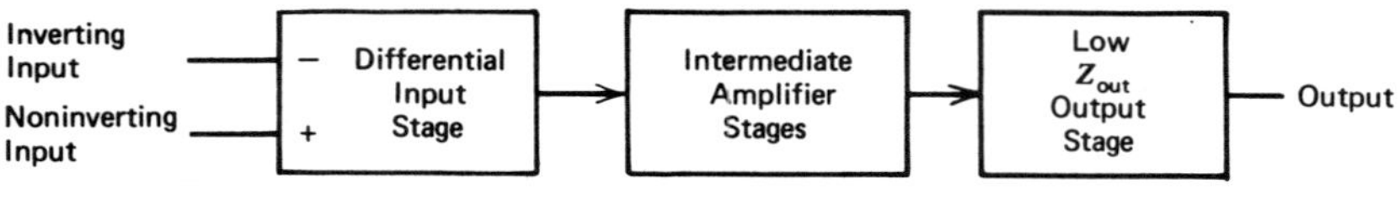

FIGURE 1.3 Block diagram of an operational amplifier.

or field-effect transistors. The use of FETs (junction or MOSFET) for Q_1 and Q_2 allows very high input resistance. The input offset voltage (V_{os}) and change in V_{os} with temperature are greater with FET differential amplifiers than with bipolar transistors, but this can be minimized with various feedback arrangements inside

[3] See Appendix A for A_{cm} equation and its cause.

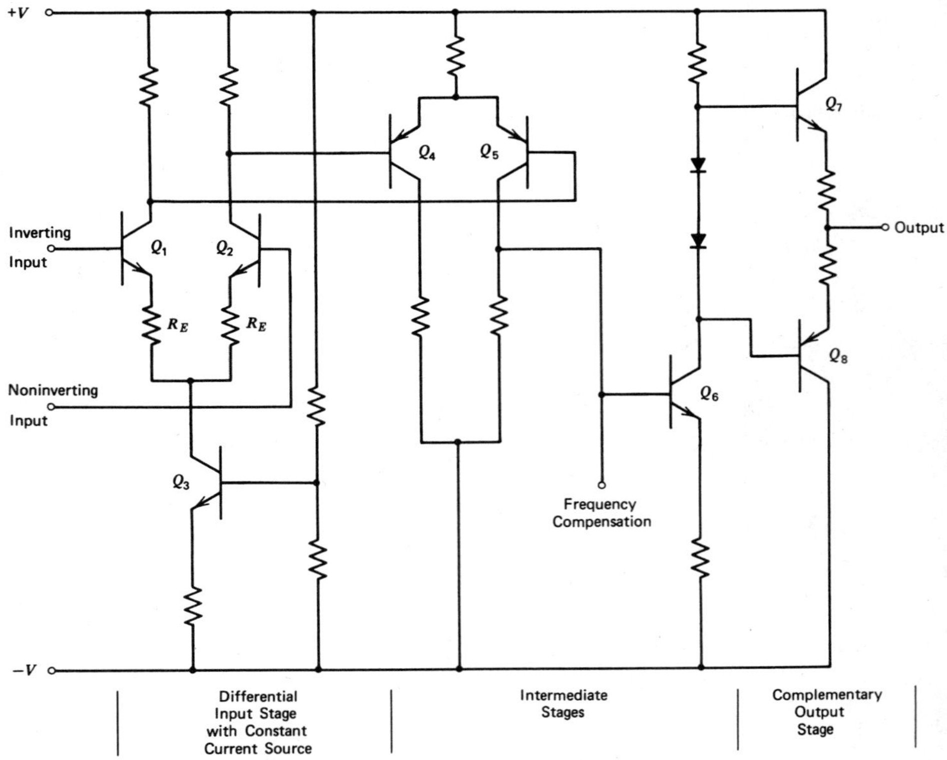

FIGURE 1.4 A simple operational amplifier circuit.

the amplifier. Integrated circuit (IC) amplifiers are available that have FET input transistors for high input impedance with bipolar transistors used for the rest of the op-amp circuitry. The use of Darlington pairs for Q_1 and Q_2 also results in increased V_{os} and change in V_{os} with temperature.

If the voltage gain of the first stage is 10 ($A_{v_1} = 10$), the second stage is 100 ($A_{v_2} = 100$) and the third stage is 20 ($A_{v_3} = 20$), the total gain (A_t) is the product of the individual stage gains so that

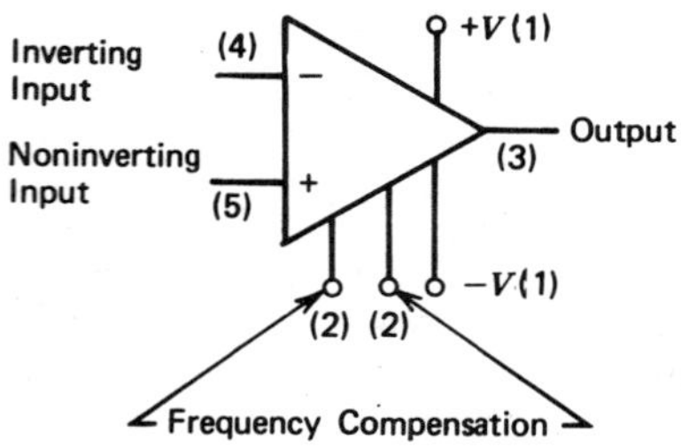

FIGURE 1.5 Terminals of an operational amplifier.

$$A_t = A_1A_2A_3 = (10)(100)(20) = 200{,}000$$

which is rather high.

Appendix B explains the operation of the Fairchild μA 741, a popular IC op-amp.

1-2 TERMINALS OF AN OPERATIONAL AMPLIFIER

Figure 1.5 shows the external terminals of an op-amp. These are:

1. +V, −V: The terminals for the power supply voltages.
2. Frequency compensation: These terminals (sometimes called lead, lag, or roll off) are used to prevent oscillation of the operational amplifier circuit when there is no compensation internally in the amplifier. More will be said about this in a later section.
3. Output: Where the amplified voltage is found.
4. Inverting input: If the noninverting input is grounded and a signal applied to the inverting input, the output will be 180° out of phase with the input signal.
5. Noninverting input: If the inverting input is grounded and a signal applied to the noninverting input, the output will be in phase with the input signal.

1-3 SOME SPECIFICATIONS

1. Open loop gain (A_{ol}): The gain of the amplifier with no feedback. Usually several thousand. Also called large signal voltage gain.
2. Input offset voltage (V_{os}): Small, undesired voltages internally generated by the amplifier, that cause an output voltage to be present when both inputs are connected to zero volts. Caused by the imperfect matching of the emitter base voltages of the input transistors. V_{os} is usually a few millivolts.
3. Bias current (I_B): The current required to drive the input stage of the operational amplifier; the base current that must be supplied to the input transistor.

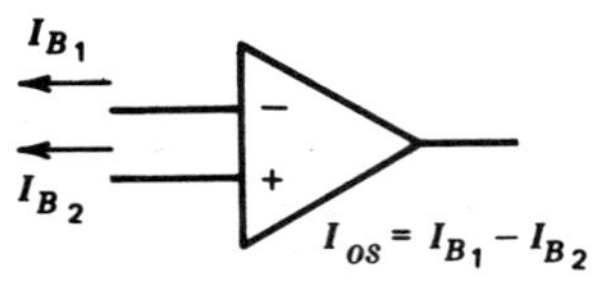

FIGURE 1.6 Defining input offset current.

4. Input offset current (I_{os}): The difference in the bias current required by the two input transistors of the operational amplifier. Caused by the imperfect matching of the betas (β) of the input transistors. In Figure 1.6, if I_{B_1} is the current required to drive the inverting input transistor, and I_{B_2} is the current required to drive the noninverting input transistor, $I_{os} = I_{B_1} - I_{B_2}$. The bias current for an input changes somewhat as the input voltage is changed, so the offset current will also change. I_{os} is usually between a few and several hundred nanoamps.

5. Input resistance: The resistance of the amplifier to an input signal (R_{in}). R_{in} is typically greater than one megohm, but it can be as high as several hundred megohms. The input resistance can be differential, between the two input terminals or common mode, from the two inputs to ground. The specification sheet will usually not specify which is which, just R_{in}.

6. Output resistance: The resistance internal to the amplifier that would be seen by a voltage applied to the output of the amplifier. R_{out} is usually less than a few hundred ohms.

7. Common-mode rejection ratio: The ability to reject (not amplify) signals applied to both inputs simultaneously. This topic is discussed in detail in Chapter 3.

8. Supply-voltage rejection ratio: The output voltage change for a power supply change (+V and −V together) of 1 V. Generally given in microvolts per volt.

9. Input capacitance (C_{in}): Capacitance from the input terminals to ground.

10. Supply current: Quiescent current the op-amp draws.

11. Power consumption: Quiescent power dissipated by the op-amp.

12. Slew rate (S): The maximum rate of output voltage change given in volts per microsecond.

13. Transient response: The response of an op-amp to a step input voltage. The rise time and output voltage overshoot are given for a specified input voltage change.

14. Absolute maximum ratings: These include such ratings as:
   (a) Maximum power dissipation.
   (b) Operating temperature range.
   (c) Maximum supply voltage.
   (d) Maximum differential input voltage (between the inverting and noninverting terminals).
   (e) Maximum common-mode input voltage.
   (f) Storage temperature range.

If these maximum ratings are exceeded, damage to the op-amp will occur.

Many manufacturers of op-amps include curves of many amplifier parameters in their specifications. These may include V_{out} (max) versus R_L, V_{out} (max) versus supply voltage, V_{os} versus temperature, and I_B versus temperature. A thorough reading of op-amp specifications is a must for using them successfully.

Critical parameters such as V_{os} and A_{ol} will usually be given at the maximum and minimum operating temperatures as well as at room temperature. More specifications will be discussed as needed in later sections.

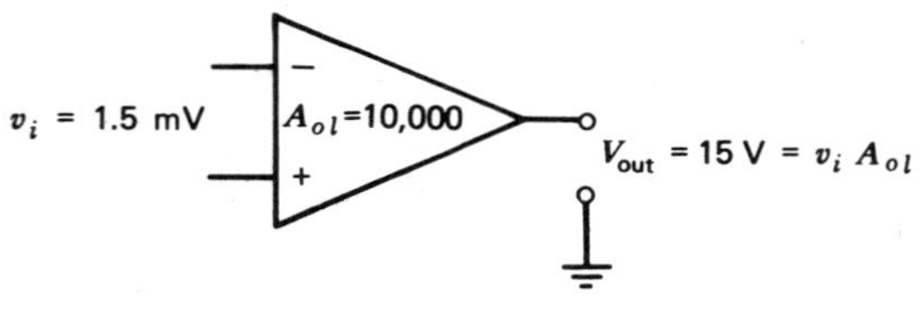

FIGURE 1.7 Restraint of differential input voltage.

1-4 SUMMING-POINT RESTRAINTS

For reasons to be discussed later, the two inputs of the amplifier are often referred to as the *summing point*. The restraints on these are that the bias current is low and that the voltage between the two input terminals is very low compared to any other circuit voltage values when the amplifier is used with feedback. The second restraint, or condition, is assured by the high open loop gain of the amplifier. For example, if the output voltage is 15 V and the open loop voltage gain is 10,000, the voltage appearing between the input terminals (v_i in Figure 1.7) must be V_{out}/A_{ol} or 15 V/10,000 or 1.5 mV. It is important to note that the output voltage is caused only by the small voltage appearing between the input terminals, and no other. Since the open loop gain is very large, this input voltage is very low.

1-5 VOLTAGE FOLLOWER

In Figure 1.8, V_{out} is fed directly back to the inverting input. If we recall that the voltage between the inputs (v_i) is the voltage the amplifier gain (A_{ol}) acts on, we see that if a voltage is put on the noninverting input, the amplifier will be driven until $v_i = V_{out}/A_{ol}$ and the output will remain at this value until the input is changed. Since the amplifier gain is very high, v_i is very small, so V_{out} will be approximately the same as V_{in}.

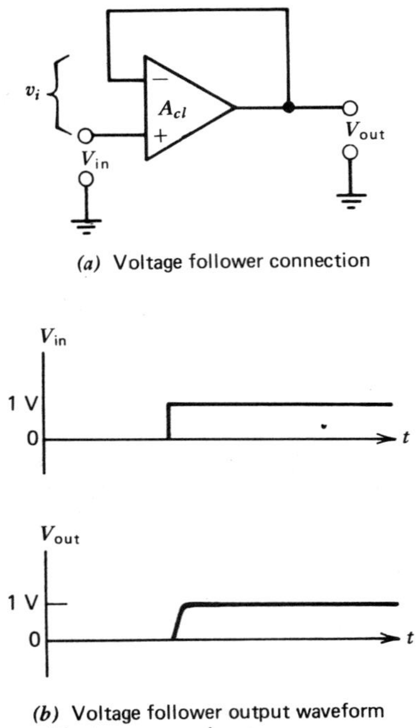

(a) Voltage follower connection

(b) Voltage follower output waveform with step response input

FIGURE 1.8 Voltage follower.

For example, if an input voltage of 1 V is applied to the voltage follower of Figure 1.8, the output will start to go positive, since +1 V is applied to the noninverting input. The output will rise until $V_{out} = V_{in}$, or $v_i \cong 0$. If the op-amp $A_{ol} = 10{,}000$, the output will stop going positive when $v_i = 1\ V/10{,}000 = 0.1$ mV. With respect to 1 V, 0.1 mV is negligible or approximately zero. If the output voltage tried to rise above 1 V, the polarity of v_i would reverse and $V_{in} - V_{out}$ would no longer be approximately zero, so the output would be driven back down to 1 V.

Applying Kirchoff's law, we see that

$$V_{in} + v_i = V_{out}$$

Since

$$V_{out} = A_{ol}v_i$$

we see that

$$v_i = V_{out}/A_{ol}$$

Hence

$$V_{in} + \frac{V_{out}}{A_{ol}} = V_{out}$$

As A_{ol} approaches infinity, the V_{out}/A_{ol} term approaches zero and we are left with

$$V_{in} = V_{out}$$

Since the input is applied to the noninverting input, the output voltage is the same phase and magnitude as the input.

The only path to ground for the input voltage is through the amplifier input resistance, which is very high, so the voltage follower makes a good buffer stage.

1-6 NONINVERTING AMPLIFIER

The noninverting amplifier configuration (Figure 1.9) allows the operational amplifier to be used as a high input impedance, noninverting amplifier in which the voltage gain of the circuit can be set within close limits with resistors R_1 and R_f. The input impedance of the circuit is high because the only path to ground for the input current is through the high input impedance of the amplifier.

R_1 and R_f act as a voltage divider with a very light load, since the current needed to drive the amplifier is very low ($I_B \cong 0$). Thus the current through R_1 and R_f is the same, and the voltage applied to the inverting input is

$$V_{out}\frac{R_1}{R_f + R_1}$$

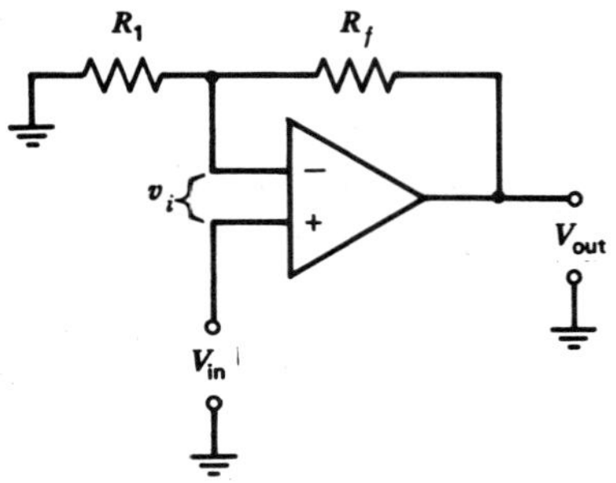

FIGURE 1.9 Noninverting amplifier.

Suppose $V_{in} = 1$ V, the amplifier will react to v_i being greater than V_{out}/A_{ol} and drive until the voltage at the inverting input is equal to the voltage at the noninverting input (i.e., $v_i = V_{out}/A_{ol} \cong 0$). If $R_1 = 10\ k\Omega$ and $R_f = 100\ k\Omega$, V_{out} will have to be 11 V for v_i to become low enough to stop driving the amplifier. The output voltage will then remain at 11 V until the input is changed.

To develop a gain equation for the circuit, recall that

$$I_{R_1} = I_{R_f}$$

Since

$$R_{in} \to \infty$$

Now

$$I_{R_1} = V_{R_1}/R_1 \qquad \text{and} \qquad I_{R_f} = V_{R_f}/R_f$$

The voltage at the inverting amplifier terminal is $V_{in} + v_i$, so

$$I_{R1} = \frac{V_{in} + v_i}{R_1} \qquad \text{and} \qquad I_{Rf} = \frac{V_{out} - (V_{in} + v_i)}{R_f}$$

Hence

$$\frac{V_{in} + v_i}{R_1} = \frac{V_{out} - (V_{in} + v_i)}{R_f}$$

Since

$$V_{out} = A_{ol} v_i$$

$$v_i = \frac{V_{out}}{A_{ol}}$$

If $A_{ol} \to \infty$, as we have assumed, $v_i \cong 0$ and we can write

$$\frac{V_{in}}{R_1} = \frac{V_{out} - V_{in}}{R_f}$$

Now we can solve for the circuit gain, V_{out}/V_{in}, which is usually called the *closed loop* (A_{cl}) or *feedback gain* (A_{fb}). Solving

$$V_{in}R_f = R_1V_{out} - R_1V_{in}$$

$$V_{in}(R_f + R_1) = R_1V_{out}$$

$$\frac{R_f + R_1}{R_1} = \frac{V_{out}}{V_{in}} = A_{fb}$$

Thus the values of R_f and R_1 set the voltage gain of the circuit. The closed loop gain formula for the noninverting amplifier is

$$A_{fb} = \frac{R_f + R_1}{R_1} = \frac{R_f}{R_1} + 1 \qquad (1\text{-}1)$$

if $A_{ol} >> A_{fb}$. The second form of the gain equation, $A_{fb} = R_f/R_1 + 1$, is easier to use for problem solving.

In the previous example where $R_1 = 10\ \text{k}\Omega$ and $R_f = 100\ \text{k}\Omega$

$$A_{fb} = \frac{100\ \text{k}\Omega + 10\ \text{k}\Omega}{10\ \text{k}\Omega} = 11$$

The use of resistors R_1 and R_f to apply a part of the output voltage to an input terminal, as is done in the noninverting amplifier, is called *feedback*. This is an important concept. $R_1 + R_f$ must be picked so that the amplifier can supply enough driving current for the feedback resistors and the load.

If we want to set the closed loop gain of a noninverting amplifier with R_1 picked, we solve the closed loop gain equation for R_f.

$$A_{fb} = \frac{R_f}{R_1} + 1$$

$$A_{fb} - 1 = \frac{R_f}{R_1}$$

so

$$R_f = R_1(A_{fb} - 1)$$

EXAMPLE 1-1

For $R_1 = 10\ \text{k}\Omega$ and $A_{fb} = 20$, $R_f = (20 - 1)(10\ \text{k}\Omega) = 190\ \text{k}\Omega$. If we pick R_f and A_{fb}, we solve the A_{fb} equation for R_1.

$$A_{fb} = \frac{R_f}{R_1} + 1$$

$$A_{fb} - 1 = \frac{R_f}{R_1}$$

$$R_1 = \frac{R_f}{A_{fb} - 1}$$

If $A_{fb} = 20$ and $R_f = 200\ \text{k}\Omega$

$$R_1 = \frac{200\ \text{k}\Omega}{19} = 10.5\ \text{k}\Omega$$

The maximum value of $R_1 + R_f$ is set by the bias current. A reasonable way to calculate $R_1 + R_f$ maximum is to allow $I_{R_F} = 20\ I_B$ when $V_{out} = +V/2$.

$$(R_1 + R_2)_{max} = \frac{(+V/2)}{20\ I_B}$$

For a μA 741C op-amp, I_B (max) $= 500$ nA. So if $+V = -V = 15$ V

$$(R_1 + R_2)_{max} = \frac{7.5\ \text{V}}{10\ \mu\text{A}} = 750\ \text{k}\Omega$$

In most instances one would use a lower value of $R_1 + R_f$ to minimize noise (see Chapter 12, Section 12-1 on thermal noise). The minimum value of $R_1 + R_f$ is limited by the op-amp output current. This is about 2 kΩ for a μA 741C. This low a value is seldom used, otherwise no output current is left to drive the load. Typical $R_1 + R_f$ values fall between 50 kΩ and 1MΩ.

1-7 INVERTING AMPLIFIER

We will now find a gain formula for the inverting amplifier. As implied by the name, the inverting amplifier input and output are 180° out of phase. As with the noninverting amplifier, the open loop gain of the amplifier is so high that only a small v_i is required to drive the amplifier output voltage to its limits. (Usually $V_{out_{max}}$ is slightly less than the supply voltage.) If a positive V_{in} is applied to the circuit, v_i will become greater than zero, driving the output voltage negative (since the input is applied to the inverting input of the amplifier). The output will continue going negative until the voltage at the inverting input terminal (point A of Figure 1.10) is almost zero, $v_i = V_{out}/A_{ol} \cong 0$). Thus R_1 and R_f act as a voltage divider between V_{out} and V_{in}, and the ratio between V_{out} and V_{in} is that of R_f to R_1. Point A is often called a *virtual ground* because it is almost at ground potential since v_i is normally very small.

To develop a closed loop gain equation, recall that, since R_{in} of the amplifier is very high, $I_{R_1} = I_{R_f}$.

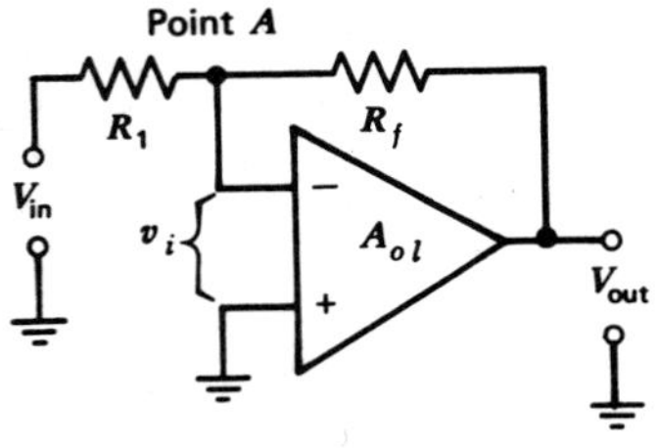

FIGURE 1.10 Inverting amplifier.

Since

$$I_{R_1} = \frac{V_{in} - v_i}{R_1} \quad \text{and} \quad I_{Rf} = -\frac{(V_{out} - v_i)}{R_f}$$

we can write

$$\frac{V_{in} - v_i}{R_1} = -\frac{(V_{out} - v_i)}{R_f}$$

The minus sign in front of the right-hand term denotes the inverted output. Since $v_i = 0$ (because $A_{ol} \to \infty$)

$$\frac{V_{in}}{R_1} = -\frac{V_{out}}{R_f}$$

The closed loop gain is

$$A_{fb} = \frac{V_{out}}{V_{in}} = -\frac{R_f}{R_1} \tag{1-2}$$

EXAMPLE 1-2

Referring to Figure 1.10

(a) Calculate A_{fb} if $R_1 = 20\ \text{k}\Omega$ and $R_f = 400\ \text{k}\Omega$.

$$A_{fb} = -\frac{R_f}{R_1} = -\frac{400\ \text{k}\Omega}{20\ \text{k}\Omega} = -20$$

Remember the minus sign merely indicates that the amplifier is an inverter.

(b) Calculate R_f if $R_1 = 10\ \text{k}\Omega$ and $A_{fb} = -15$.

$$A_{fb} = -\frac{R_f}{R_1}$$

so

$$R_f = A_{fb}R_1 = 15(10\ \text{k}\Omega) = 150\ \text{k}\Omega$$

(c) Calculate R_1 if $R_f = 1\ \text{M}\Omega$ and $A_{fb} = 50$.

$$R_1 = \frac{R_f}{A_{fb}} = \frac{1\ \text{M}\Omega}{50} = 20\ \text{k}\Omega$$

The input resistance of the inverting amplifier circuit configuration is just R_1 since point A of Figure 1.10 remains at approximately zero volts due to the feedback. R_1 must be picked so that it does not load the V_{in} source, and of course R_f must be large enough so that the operational amplifier is not excessively loaded.

1-8 DIFFERENTIAL INPUT AMPLIFIER

To consider this amplifier configuration (see Figure 1.11), we will note that the voltage difference between the inverting input and noninverting input is very small (usually less than 1 mV) since V_{out}/A_{ol} is very small. We will therefore consider the inverting and noninverting input to be at the same voltage, V_f, throughout this discussion.

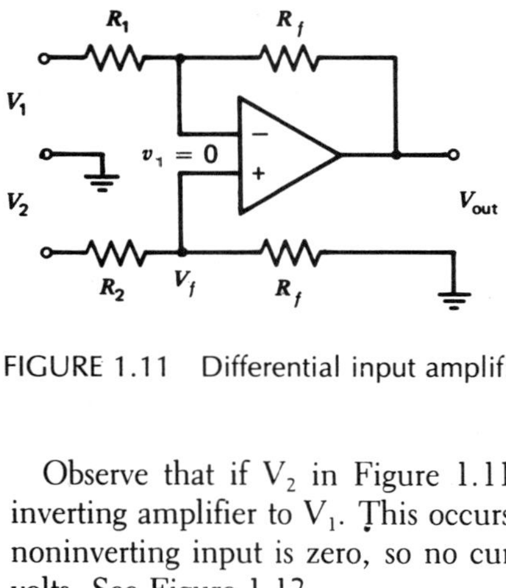

FIGURE 1.11 Differential input amplifier.

Observe that if V_2 in Figure 1.11 is zero volts, the amplifier will act as an inverting amplifier to V_1. This occurs because the input current of the amplifier's noninverting input is zero, so no current flows through R_2 or R_f', and V_f is zero volts. See Figure 1.12.

Now if we set V_1 at zero volts and use V_2 as an input signal as shown in Figure 1.13, the amplifier will act like a noninverting amplifier with a voltage divider (R_2 and R_f') supplying the input voltage, V_f, to the noninverting amplifier configuration.

When both V_1 and V_2 are applied simultaneously, the inverting input causes the output to drive to a voltage that will cause the voltage at the junction of R_1 and R_f to be V_f, where $V_f = V_2\,[R_f'/(R_2 + R_f')]$ instead of zero, as would occur in a normal inverting amplifier.

We will now develop an equation for the output voltage. Since the input resistance of the amplifier is very high, we know

$$I_{R_1} = I_{R_f} \quad \text{and} \quad I_{R_2} = I_{R_f'}$$

$$I_{R_1} = \frac{V_1 - V_f}{R_1} = I_{R_f} = \frac{V_f - V_{out}}{R_f}$$

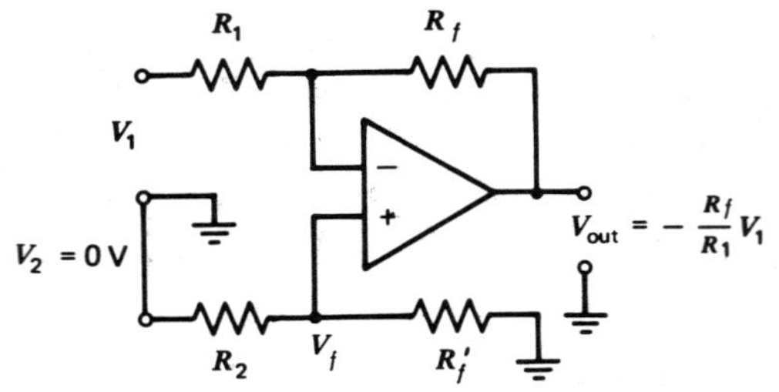

FIGURE 1.12 Differential input amplifier when $V_2 = 0$.

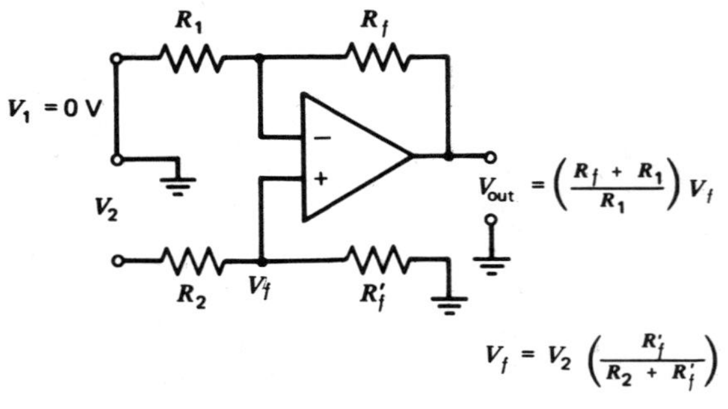

FIGURE 1.13 Differential input configuration when $V_1 = 0$.

Solving the second and fourth terms for V_{out}

$$\frac{V_1 - V_f}{R_1} = \frac{V_f - V_{out}}{R_f}$$

$$R_fV_1 - R_fV_f = R_1V_f - R_1V_{out}$$

$$R_1V_{out} = R_1V_f + R_fV_f - R_fV_1$$

$$R_1V_{out} = V_f(R_1 + R_f) - R_fV_1$$

$$V_{out} = V_f\left(\frac{R_1 + R_f}{R_1}\right) - \frac{R_f}{R_1}V_1$$

The equation for V_{out} is the equation for V_{out} of a noninverting amplifier with V_f as the input plus the equation for V_{out} of an inverting amplifier, as we would expect from our previous discussion. V_f is the voltage at the junction of the voltage divider R_2 and R_f with V_2 applied to R_2, so

$$V_f = V_2 \frac{R_f'}{R_2 + R_f'}$$

Substituting this value of V_f into the V_{out} equation, we have

$$V_{out} = V_2\left(\frac{R_f'}{R_2 + R_f'}\right)\left(\frac{R_1 + R_f}{R_1}\right) - \frac{R_f}{R_1}V_1$$

which is the general equation for V_{out}.

If we set $R_1 = R_2$ and $R_f = R_f'$ (a not unusual situation)

$$V_{out} = V_2\left(\frac{R_f}{R_1 + R_f}\right)\left(\frac{R_1 + R_f}{R_1}\right) - \frac{R_f}{R_1}V_1$$

and

$$V_{out} = \frac{V_2R_f}{R_1} - \frac{R_f}{R_1}V_1$$

so

$$V_{out} = \frac{R_f}{R_1}(V_2 - V_1) \tag{1-3}$$

In this situation ($R_1 = R_2$ and $R_f = R_f'$) the output polarity is determined by the larger of V_1 or V_2.

The same considerations exist in choosing the resistor values in this circuit that exist in choosing the values for the inverting and noninverting amplifiers.

EXAMPLE 1-3

Referring to Figure 1.11, let

$$V_1 = 0.1\text{ V}$$
$$V_2 = -0.2\text{ V}$$
$$R_f = R_f' = 100\text{ k}\Omega$$
$$R_1 = R_2 = 20\text{ k}\Omega$$

Find V_{out}.

SOLUTION

$$V_{out} = \frac{R_f}{R_1}(V_2 - V_1)$$
$$= \frac{100\text{ k}\Omega}{20\text{ k}\Omega}(-0.2\text{ V} - 0.1\text{ V}) = 5(-0.3\text{ V}) = -1.5\text{ V}$$

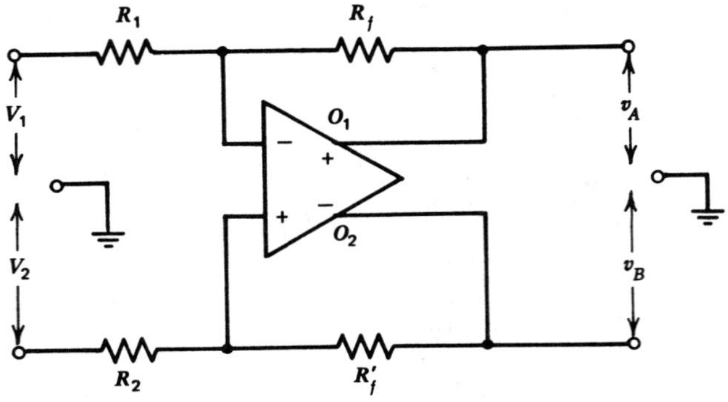

FIGURE 1.14 Differential output amplifier.

1-9 DIFFERENTIAL OUTPUT AMPLIFIER (BALANCED)

Let us assume at the outset that $R_1 = R_2$, $R_f = R'_f$, and that the voltage on the inverting and noninverting input terminals of the amplifier in Figure 1.14 are the same voltage, V_f. The last statement is assured by assuming the open loop gain of the amplifier, A_{ol}, is very high, as in our discussion of the differential input configuration.

The plus sign at terminal O_1 indicates that v_A is the opposite polarity of V_1, and the minus sign at terminal O_2 indicates the polarity of v_B is opposite that of V_2. The total output voltage is the algebraic sum of v_A and v_B. We see that if $V_1 = 0$ and V_2 is the input, the amplifier will act as an inverter with $v_A = 0$ and $v_B = -(R'_f/R_2)V_2$. In a similar manner, if $V_2 = 0$, $v_B = 0$, and $v_A = (R_f/R_1)V_1$, we have essentially two inverters in tandem.

Now let us develop an equation for V_{out}. We know that $I_{R_1} = I_{R_f}$ and $I_{R_2} = I'_{R_f}$, so

$$I_{R_1} = \frac{V_1 - V_f}{R_1} = \frac{V_f - v_A}{R_f}$$

and

$$I_{R_2} = \frac{V_2 - V_f}{R_2} = \frac{V_f - v_B}{R'_f}$$

Subtracting these two equations, where $R_1 = R_2$ and $R_f = R'_f$

$$\frac{V_2 - V_f}{R_1} - \frac{V_1 - V_f}{R_1} = \frac{V_f - v_B}{R_f} - \frac{V_f - v_A}{R_f}$$

$$\frac{V_2 - V_1}{R_1} = \frac{v_A - v_B}{R_f}$$

Since $|v_A - v_B| = V_{out}$

$$\frac{(V_2 - V_1)}{R_1} = \frac{V_{out}}{R_f}$$

$$V_{out} = \frac{R_f}{R_1}(V_2 - V_1) \tag{1-4}$$

SUMMARY

1. An operational amplifier is a multistage amplifier with a differential input that approximates the characteristics of an ideal amplifier. An op-amp will have high voltage gain, high input resistance, and low output resistance.
2. The pin connections and specifications of an operational amplifier are important to the user, who must be familiar with both.
3. The high open loop gain and high input resistance of an op-amp allow us to feed back part of the output to the inverting input through a resistive voltage divider. By using negative feedback, we can set the voltage gain of an op-amp with the resistive feedback network. We can set the gain with precision as long as the open loop gain of the op-amp is much greater than the gain set by the resistive network.
4. The voltage follower has a voltage gain of 1 and can be used either as a current amplifier or as a buffer stage.
5. The noninverting amplifier has no phase inversion and a very high input resistance. The voltage gain equation for a noninverting amplifier is

$$A_{fb} = \frac{R_f}{R_1} + 1 \tag{1-1}$$

6. The inverting amplifier has an input resistance equal to R_1 and a phase inversion. Its voltage gain equation is

$$A_{fb} = -\frac{R_f}{R_1} \tag{1-2}$$

7. The differential input amplifier output is proportional to the difference between two input signals. If $R_1 = R_2$ and $R_f = R_f'$, the output voltage of the differential input amplifier is

$$V_{out} = \frac{R_f}{R_1}(V_2 - V_1) \tag{1-3}$$

8. The differential output amplifier has two outputs as well as two inputs. The amplifier acts as two paralleled inverters operating oppositely phased. The output voltage of the differential output amplifier is the algebraic sum of the individual outputs and is given by the expression

$$|V_a - V_b| = \frac{R_f}{R_1}(V_2 - V_1) \tag{1-4}$$

if

$$R_f = R_f' \qquad \text{and} \qquad R_1 = R_2$$

SELF-TEST QUESTIONS

1-1 State the characteristics of the ideal amplifier.
1-2 Define offset voltage.
1-3 State the major cause of V_{os} and I_{os} in a bipolar transistor input op-amp.
1-4 State the major differences between an FET input op-amp and a bipolar transistor input op-amp.
1-5 A voltage follower makes a good buffer stage. State the reason.
1-6 Calculate the unknown value of A_{fb}, R_1, or R_f for an inverting amplifier, given the following values:

(a) $A_{fb} = 30$, $R_1 = 10\ k\Omega$, $R_f =$ ________.
(b) $R_1 = 10\ k\Omega$, $R_f = 1.5\ M\Omega$, $A_{fb} =$ ________.
(c) $A_{fb} = 20$, $R_f = 1.8\ M\Omega$, $R_1 =$ ________.

1-7 Calculate the unknown value of A_{fb}, R_1, or R_f for a noninverting amplifier, given the following values:

(a) $R_1 = 10\ k\Omega$, $R_f = 200\ k\Omega$, $A_{fb} =$ ________.
(b) $R_1 = 20\ k\Omega$, $A_{fb} = 20$, $R_f =$ ________.
(c) $R_f = 2\ M\Omega$, $A_{fb} = 10$, $R_1 =$ ________.

1-8 A differential input amplifier with $R_1 = R_2 = 10\ k\Omega$ and $R_f = R_f' = 200\ k\Omega$ has inputs $V_1 = +0.3$ V and $V_2 = +0.5$ V. Calculate V_{out}.
1-9 The amplifier of question 1-8 has inputs $V_1 = +0.4$ V and $V_2 = -0.2$ V. Calculate V_{out}.
1-10 An inverting amplifier has $R_1 = 10\ k\Omega$, $R_f = 120\ k\Omega$, $V_{in} = 0.2$ V, and $A_{ol} = 10{,}000$. Calculate V_{out} and v_i.
1-11 State the purpose of Q_3 in the operational amplifier circuit in Figure 1.4.
1-12 Draw from memory the following op-amp circuits:

(a) Voltage follower.
(b) Noninverting amplifier.
(c) Inverting amplifier.
(d) Differential input amplifier.

If you cannot answer certain questions, place a check next to them and review appropriate parts of the text to find the answers.

LABORATORY EXERCISE

Objective

The student should be able to calculate, breadboard, and measure the ideal gain for the noninverting, inverting, and differential input amplifiers and voltage follower on completing this lab. The student should be able to compensate the offset voltage of a μA 741 operational amplifier (or equivalent).

Equipment

1. Fairchild μA741 operational amplifier or equivalent and manufacturer's specifications.
2. 2% tolerance resistor assortment.
3. Power supply, ±15 V dc.

4. Voltmeter or oscilloscope capable of measuring as low as 5 mV dc.
5. Signal generator.
6. 10 kΩ potentiometer.
7. Breadboard such as EL Instruments SK-10 mounted on a piece of 5 in. by 8½ in. vectorboard.

Theory of Compensating Offset Voltage in a μA 741 Op-Amp

To compensate for the effect of the internally generated offset voltage of an operational amplifier, an external voltage that cancels the internal offset voltage must be added to the circuit. There are several ways to accomplish this. These will be discussed in Section 2-6.

The compensation for offset voltage (V_{os}) in the μA741 operational amplifier is based on the fact that V_{BE} increases as I_C increases. The principle is illustrated in Figure 1.15. Q_3 and Q_4 are constant current generators supplying emitter current for input transistors Q_1 and Q_2 respectively. The difference in V_{BE} of Q_1 and Q_2 is compensated by adjusting the current supplied by the constant current generators Q_3 and Q_4. This adjustment is made in the μA741 by connecting a 10 kΩ potentiometer between the offset null connections and connecting the potentiometer wiper to $-V$, as shown in Figure 1.16.

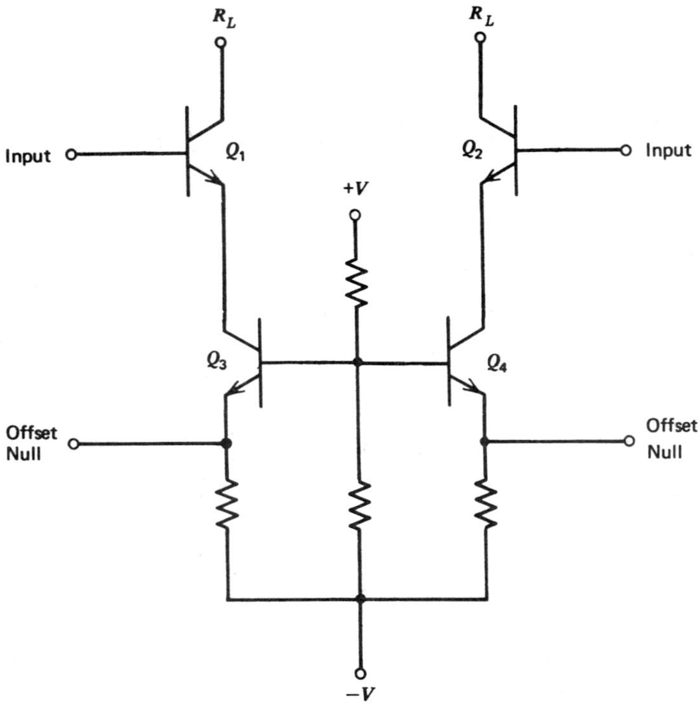

FIGURE 1.15 Compensation of V_{os} in μA 741 op-amp.

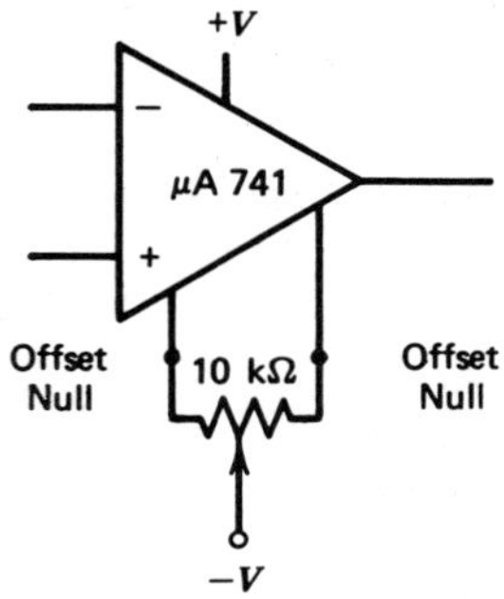

FIGURE 1.16 External potentiometer connection for offset voltage compensation.

By varying the potentiometer wiper position, we vary the effective emitter resistance of the constant current generators, which varies their current. The potentiometer is adjusted until the V_{BE} of Q_1 and Q_2 are the same, at which time the offset voltage is compensated. Many integrated circuit operational amplifiers use this method.

Note: The *circuit* input is grounded for compensating the offset, not the op-amp inputs. Of course, after compensation, the circuit input ground is removed before applying an input voltage.

Procedure

1. Compensating offset.
   (a) Connect the μA741 as shown in Figure 1.17. Be sure to check the pin connections in the manufacturer's specifications before connecting the operational amplifier. Note the output voltage and record this value.
   (b) Adjust the 10 kΩ potentiometer until the output voltage is zero.
   (c) Calculate the closed loop gain (A_{fb}) of the circuit of Figure 1.17 using

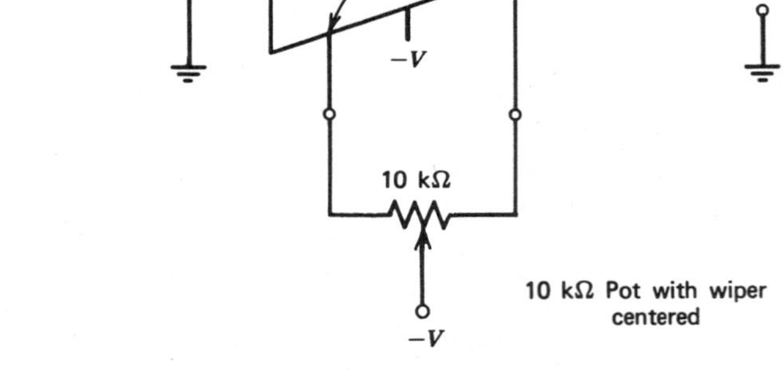

FIGURE 1.17 Connection for offset compensation.

$$A_{fb} = \frac{R_f}{R_1} + 1 \tag{1-1}$$

(d) Calculate the offset voltage before the circuit was compensated. Disconnect the wiper of the offset null potentiometer, measure V_{out} due to V_{os}, then reconnect.

$$V_{os} = \frac{V_{out}}{A_{fb}}$$

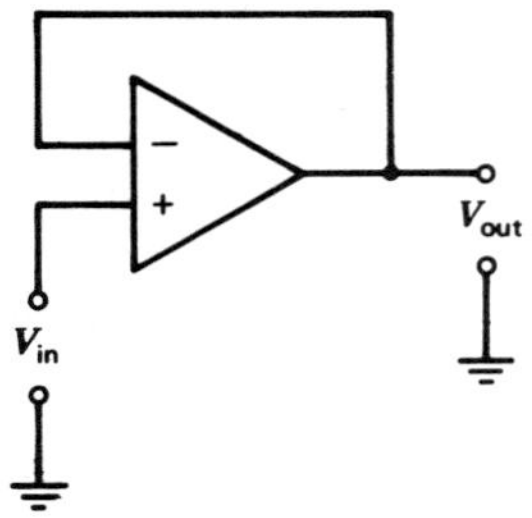

FIGURE 1.18 Voltage follower.

2. Noninverting amplifier gain.
(a) With the circuit set up as in part 1, find the value of R_f needed for an amplifier closed loop gain of first 10 and then 100.
(b) Put these values in the circuit, remove the ground from the noninverting input, and measure the ac closed loop gain.
(c) For a value of R_f = 500 kΩ, calculate the value of R_1 needed for a closed loop gain of 11.
(d) Measure the ac closed loop gain using this value of R_1.
(e) With the gain (A_{fb}) set at 11, calculate and measure the output voltage with an input voltage of 0.5 V dc. Note the polarity of the output.

3. Voltage follower.
(a) Connect the circuit as shown in Figure 1.18, but do not remove the offset voltage compensation.
(b) Apply a voltage signal (ac then dc) to the input and measure and record the input and output voltages. How do the ratios of the two signals V_{out}/V_{in} compare?

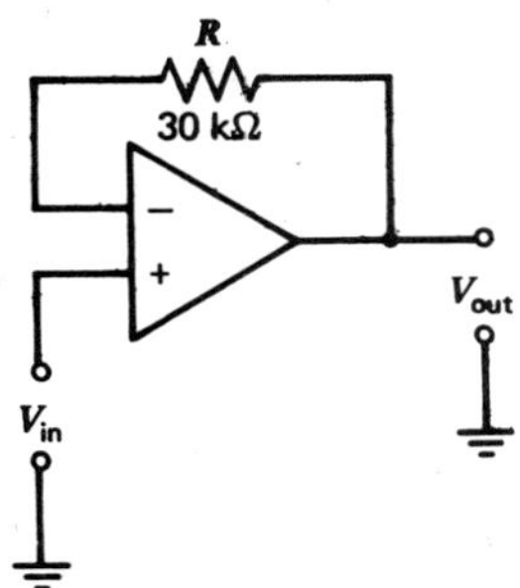

FIGURE 1.19 Effect of feedback resistor on voltage follower.

(c) Connect a resistor between the output and inverting input as shown in Figure 1.19 and repeat step (b). Does the closed loop voltage gain change? Why or why not?

4. Inverting amplifier.
(a) Connect the μA 741 in the configuration in Figure 1.20 (with the V_{os} compensation in the circuit).
(b) Calculate and measure the gain (A_{fb}) of the circuit. *Note:* $A_{fb} = -R_f/R_1$. How does this gain compare with the same values of R_1 and R_f used in the noninverting amplifier?
(c) Calculate the value of R_1 necessary to set the gain at 20 with $R_f = 200$ kΩ. Calculate and measure A_{fb} with these values of R_1 and R_f.
(d) Set $R_f = 100$ kΩ and $R_1 = 10$ kΩ. Note the value and polarity of the output voltage with 0.5 V dc as the input.

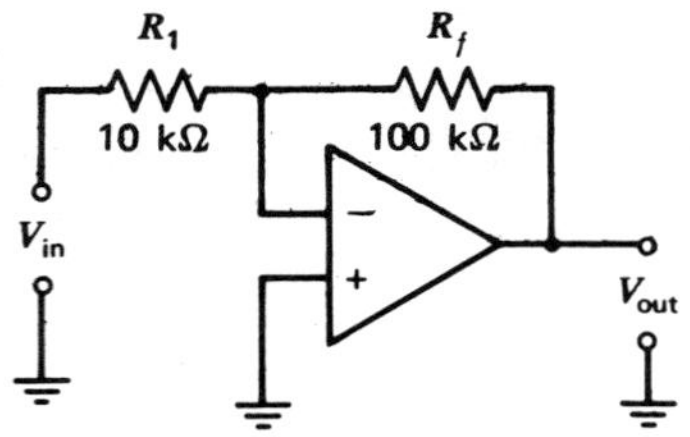

FIGURE 1.20 Inverting amplifier.

5. Differential input amplifier.
(a) Connect the circuit shown in Figure 1.21. (Be sure V_{os} is compensated.) *Note:*

$$V_{out} = \frac{R_f}{R_1}(V_2 - V_1)$$

(b) Calculate and measure the output voltage with $V_1 = +0.2$ V dc and $V_2 = -0.3$ V dc. A voltage divider or potentiometer can be used to set V_1 and V_2.
(c) Calculate and measure the output voltage with $V_1 = +0.2$ V dc and $V_2 = +0.3$ V dc.
(d) What must R_f and R_f' be for $V_{out} = 20\ (V_2 - V_1)$ if $R_2 = R_1 = 20$ kΩ. Connect these values and measure V_{out} with $V_1 = +0.1$ V dc and $V_2 = -0.1$ V dc.

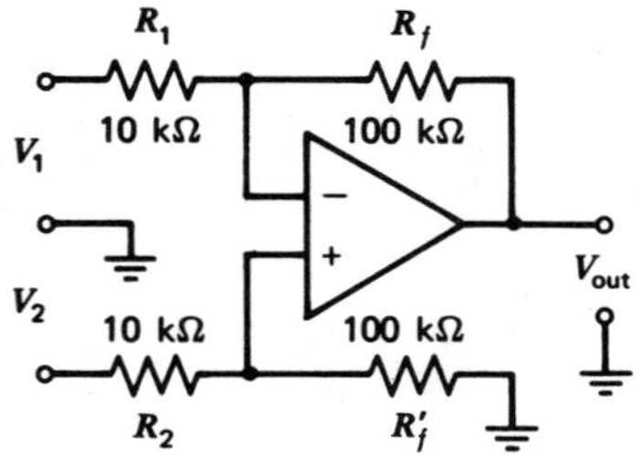

FIGURE 1.21 Differential input amplifier.

(e) What values would you pick for R_f and R_f' with $R_1 = R_2 = 10\ k\Omega$ if you wish the output to be twice as large for V_1 as it is for V_2 if $V_1 = V_2$? *Recall:*

$$V_{out} = \left(\frac{R_f'}{R_2 + R_f'}\right)\left(1 + \frac{R_f}{R_1}\right)V_2 - \frac{R_f}{R_1}V_1$$

Put these values in the circuit and verify your answer by measurement.

Discussion

For each section compare the measured data with the results predicted from theory. How well do the experimental results conform to theoretical results? Discuss any discrepancies between theory and practice and account for them.

CHAPTER

2

NEGATIVE FEEDBACK AND EXTERNAL OFFSET COMPENSATION

In the first chapter we assumed an ideal amplifier for all of our discussions. In the next two chapters we will look at some consequences of our amplifiers not being as perfect as we would like. To begin, let us look at some general consequences of negative feedback.

OBJECTIVES

After completing the study of this chapter and the self-test questions, the student should be able to:

1. State and calculate the effect of negative feedback and finite open loop gain on closed loop gain, input resistance, and output resistance.
2. State the theory of operation, calculate the components, and draw circuits for the external compensation of offset voltage in the inverting, noninverting, and differential input amplifier.
3. Perform the laboratory for Chapter 2.

2-1 EFFECT OF NEGATIVE FEEDBACK ON GAIN

Negative feedback decreases circuit gain. To see this we construct the equivalent circuit of Figure 2.1. In this circuit the block labeled A is an amplifier with open loop gain A. The block labeled β is the feedback network. β is the *feedback factor* and can be thought of as the fraction of the output fed back to the input. The

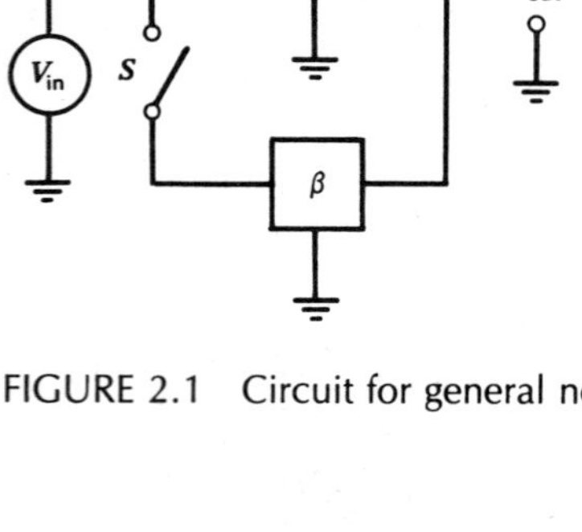

FIGURE 2.1 Circuit for general negative feedback.

symbol ⊗ denotes the summing point where the negative feedback and the input V_{in} are applied.

Before the switch S is closed, the output is just

$$V_{out} = A\,V_{in}$$

The moment the switch is closed, the input to the amplifier (V_S) becomes

$$V_S = V_{in} - \beta\,V_{out}$$

The minus sign is a consequence of the negative feedback (out of phase with the input). Now

$$V_{out} = A\,V_S$$

and

$$V_{out} = A\,(V_{in} - \beta\,V_{out})$$

Solving

$$V_{out} = A\,V_{in} - A\beta\,V_{out}$$

$$V_{out}(1 + A\beta) = A\,V_{in}$$

$$\frac{V_{out}}{V_{in}} = \frac{A}{1 + A\beta} = A_{fb} \qquad (2\text{-}1)$$

where A_{fb} is the voltage gain with feedback applied; all other terms have been defined previously.

Equation 2-1 is the general form of the feedback voltage gain equation. If we divide the numerator and denominator by $A\beta$, we get another form that is often used.

$$A_{fb} = \frac{A}{1 + A\beta} = \frac{1/\beta}{1 + 1/A\beta} \qquad (2\text{-}2)$$

Note in Equation 2-2 that if $A\beta >> 1$,

$$A_{fb} = \frac{1/\beta}{1 + 1/A\beta} = \frac{1/\beta}{1 + 0}$$

therefore

$$A_{fb} = \frac{1}{\beta}$$

if

$$A\beta >> 1$$

Normally, for an op-amp connected as an amplifier, $A_{ol}\,\beta >> 1$, so $A_{fb} = 1/\beta$.

The quantity Aβ is called the *loop gain* and must be positive if the circuit is an amplifier. Since the feedback is negative and subtracts from the input, the loop gain is positive.

2-2 EFFECT OF NEGATIVE FEEDBACK ON OUTPUT RESISTANCE

Negative feedback decreases the effective output resistance of an amplifier. To understand why, refer to Figure 2.2.

If we consider i_{fb} negligible, then for some particular values of R_1 and R_f we will have a particular output voltage

$$V_{out} = -V_{in}\frac{R_f}{R_1} = v_i A_{ol}$$

If we close S, applying a load R_L, the current the amplifier supplies to the load (i_L) will cause a voltage drop across the amplifier internal output resistance, causing the voltage at the output terminal to drop, thus becoming

$$V_{out} = -V_{in}\frac{R_f}{R_1} - i_L\ R_{out} = v_i A_{ol} - R_{out}\ i_L$$

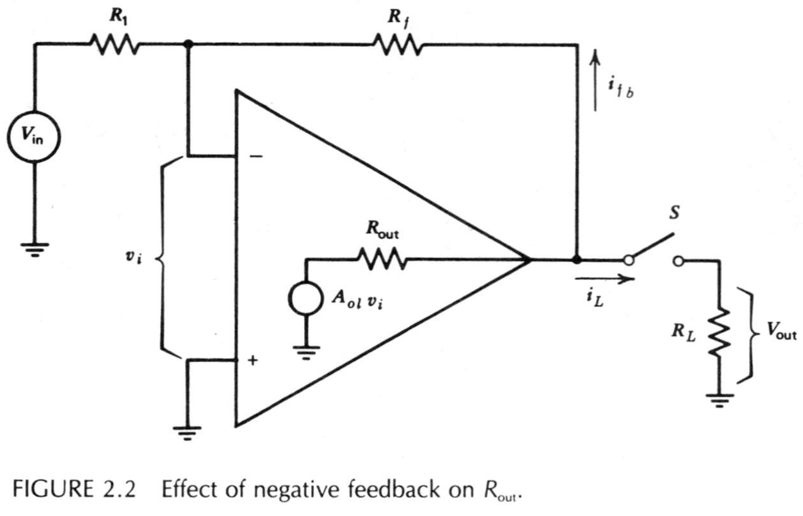

FIGURE 2.2 Effect of negative feedback on R_{out}.

This change is fed back to the inverting input by the voltage divider R_1 and R_f, causing v_i to increase. This increase of v_i causes the amplifier to drive until the output voltage reaches the value it was before we applied the load. Since the output voltage drops less than predicted when the load is applied, we say that negative feedback decreases the effective output resistance of the amplifier.

To find the magnitude of this decrease, we will use a diagram similar to Figure 2.1, but we will add an internal output resistance to the amplifier (see Figure 2.3). Note that the current through the feedback loop is small compared to i_{out}.

For the purpose of this discussion we will consider $V_{in} << A\beta\, V_{out}$. Note that we can express the output current in terms of the voltage across R_{out}

$$i_{out} = \frac{V_{R_{out}}}{R_{out}} = \frac{V_{out} - A\,V_S}{R_{out}}$$

but

$$V_S = -\beta\, V_{out}$$

so

$$i_{out} = \frac{V_{out} - (-A\beta\, V_{out})}{R_{out}} = \frac{V_{out}\,(1 + A\beta)}{R_{out}}$$

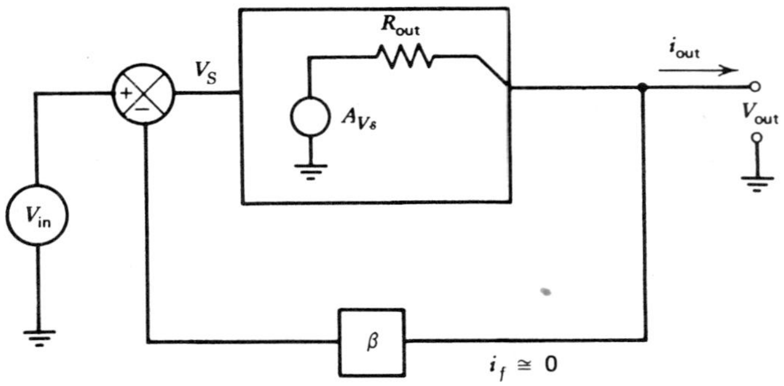

FIGURE 2.3 Feedback diagram of effect of negative feedback on R_{out}.

Solving now for V_{out}/i_{out} we find

$$\frac{V_{out}}{i_{out}} = \frac{R_{out}}{1 + A\beta}$$

V_{out}/i_{out} is an output resistance term, so we call this $R_{out\ fb}$ (output resistance with feedback).

$$R_{out\ fb} = \frac{R_{out}}{1 + A\beta} \tag{2-3}$$

Example 2-1

If A = 10,000, β = 0.01, and the manufacturer specifies R_{out} at 3 kΩ, with feedback the effective output resistance will be

$$R_{out\ fb} = \frac{R_{out}}{1 + A\beta}$$

$$= \frac{3\text{ k}\Omega}{1 + 10^4 \times 10^{-2}} = 30\ \Omega$$

2-3 EFFECT OF NEGATIVE FEEDBACK ON INPUT RESISTANCE

Negative feedback in an operational amplifier circuit causes the effective input resistance to increase. To get an intuitive idea why this happens, we apply an input signal to the amplifier in Figure 2.4 with the switch in the position shown. If we consider the constant current source as having infinite impedance, the input resistance is

$$R_{in} \simeq 2(\beta + 1)(r_D + r_E)$$

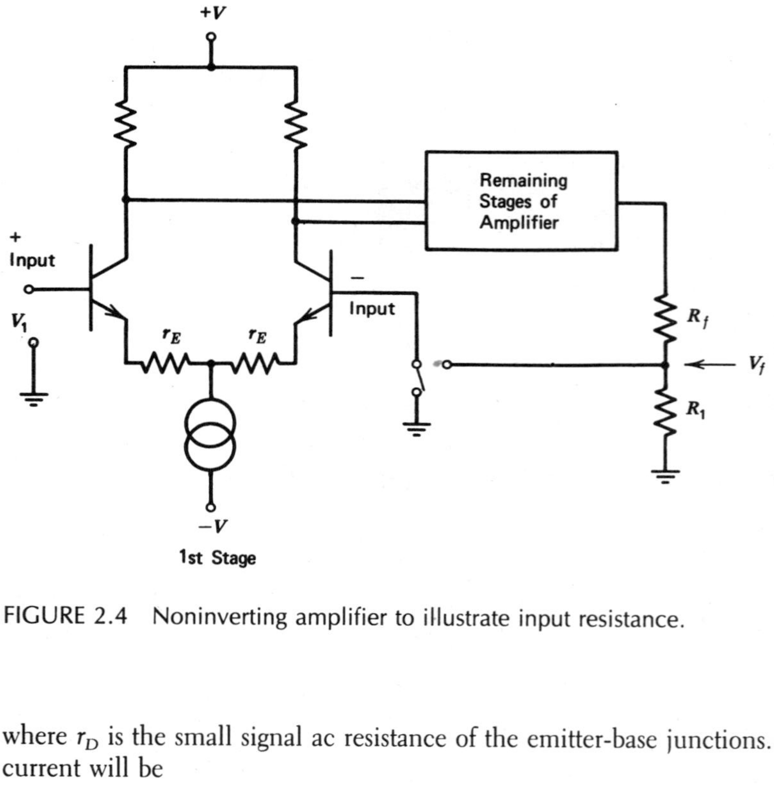

FIGURE 2.4 Noninverting amplifier to illustrate input resistance.

where r_D is the small signal ac resistance of the emitter-base junctions. The input current will be

$$i_{in} \simeq V_1/R_{in}$$

Now we change the position of S so that the feedback resistors R_f and R_1 are in the circuit. The amplifier will now drive until V_f at the junction of R_1 and R_f is almost the same as the input voltage, V_1. This means the input current will decrease since the voltage at the + and − terminals is almost the same, causing the effective input resistance to increase.

To find out how much the input impedance is increased, refer to Figure 2.5.

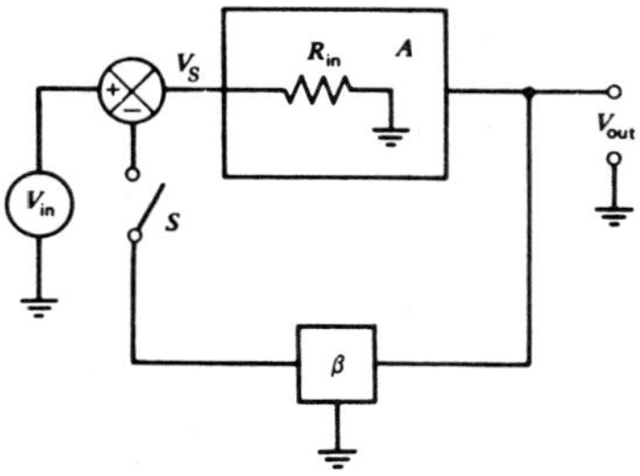

FIGURE 2.5 Effect of negative feedback on R_{in}.

With S open, no feedback, we see that R_{in} is just V_{in}/i_{in} since $V_{in} = V_S$. When we close S we feed back part of the output, $-\beta V_{out}$, so that

$$V_S = V_{in} - V_f$$

but

$$V_f = -\beta V_{out}$$

Therefore

$$V_s = V_{in} + BV_{out}$$
$$V_{out} = -AV_s$$
$$V_s = V_{in} - ABV_s$$
$$V_{in} = V_s + ABV_s$$
$$V_s = V_{in}(1 + \beta A)$$
$$V_s = V_{in}$$
$$V_{in}$$
$$V_s$$
$$V_{in} = V_S(1 + \beta A)$$

and

$$\frac{V_{in}}{i_{in}} = \frac{V_S}{i_{in}}(1 + \beta A) = R_{in}(1 + \beta A)$$

The input resistance with feedback (R_{infb}) is defined to be V_{in}/i_{in}, and the input resistance of the amplifier is V_S/i_{in}, so

$$R_{infb} = R_{in}(1 + \beta A) \tag{2-4}$$

Note. Equation 2-4 refers only to the input impedance between the op-amp terminals and not to the common-mode resistance from both input terminals to ground. The equation is good for the differential input resistance of a noninverting amplifier, but not for the input resistance of an inverting amplifier.

In an inverting amplifier the Miller effect causes R_f to look smaller to an input signal because V_{out} causes more current through R_f than v_i could alone. Thus, at the inverting terminal of an inverting amplifier

$$R_f\,(\text{eff}) = \frac{R_f}{1 + A_{ol}}$$

This is a small value. The input resistance of an inverting amplifier is

$$R_{in}\text{ (inv)} = R_1 + \frac{R_f}{1 + A_{ol}} \,\Big|\Big|\, R_{in}\text{ (op-amp)}$$

so

$$R_{in}\text{ (inv)} \cong R_1$$

2-4 NONIDEAL NONINVERTING AMPLIFIER

Using the general negative feedback gain equation, all we have to find is the feedback factor and the open loop gain to know how much our ideal A_{fb} equation will vary from the actual closed loop gain.

Referring to Figure 2.6, note that $V_{out} = A_{ol}v_i$, but $v_i = V_{in} - V_f$ (V_{in} must be larger than V_f to get a noninverted output). We can now write

$$V_{out} = A_{ol}(V_{in} - V_f) = A_{ol}V_{in} - A_{ol}V_f$$

but

$$V_f = V_{out}\frac{R_1}{R_1 + R_f}$$

so

$$V_{out} = A_{ol}V_{in} - \frac{A_{ol}R_1V_{out}}{R_1 + R_f}$$

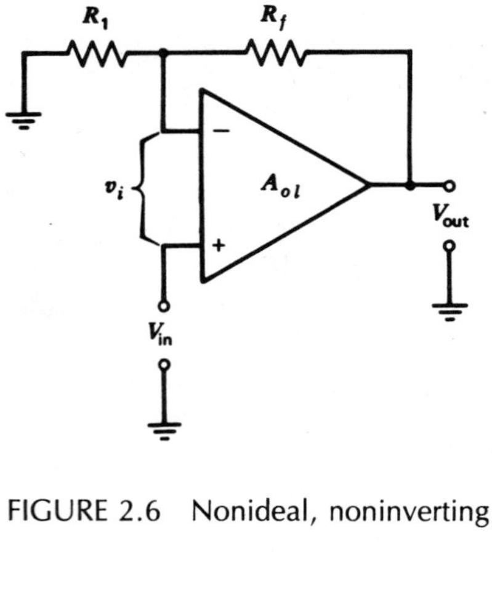

FIGURE 2.6 Nonideal, noninverting amplifier.

Solving

$$A_{ol}V_{in} = V_{out}\left(1 + \frac{A_{ol}R_1}{R_1 + R_f}\right)$$

$$\frac{V_{out}}{V_{in}} = A_{ol}\bigg/\left(1 + \frac{A_{ol}R_1}{R_1 + R_f}\right)$$

Referring to the general gain equation (2-1)

$$A_{fb} = \frac{A}{1 + A\beta}$$

we see that β must be

$$\beta = \frac{R_1}{R_1 + R_f} \tag{2-5}$$

Note that $1/\beta$ is the ideal closed loop gain since

$$A_{fb} = \frac{R_1 + R_f}{R_1} = 1/\beta \tag{2-6}$$

The product of the open loop gain A and the feedback factor β or ($A\beta$) is called the *loop gain*. We will see more of the loop gain in Chapter 4 on frequency response.

EXAMPLE 2-2

If the open loop gain is 1200 and we want a closed loop gain of 101 ($R_1 = 10\ k\Omega$ and $R_f = 1\ M\Omega$ will do this), we find the closed loop gain is

$$A_{fb} = \frac{1200}{1 + 1200(1/101)} = \frac{1200}{12.9} = 93$$

which is about a 7.9% error. If this is too much error, we must use an amplifier with a higher open loop gain.

Note as A_{fb} decreases so does the error due to the operational amplifier's finite open loop gain. For example, to set $A_{fb} = 11$ ($R_1 = 10\ k\Omega$ and $R_f = 100\ k\Omega$) with $A_{ol} = 1200$,

$$A_{fb} = \frac{1200}{1 + 1200(1/11)} = 10.9$$

a 0.91% error.

Suppose in the first part of this example we need $A_{fb} = 100 \pm 1\%$. The general gain equation can be used to find the minimum open loop gain the op-amp must have. From

$$A_{fb} = \frac{A_{ol}}{1 + A_{ol}\beta}$$

we solve for A_{ol}

$$A_{fb}(1 + A_{ol}\beta) = A_{ol}$$

$$A_{fb} + A_{fb}A_{ol}\beta = A_{ol}$$

$$A_{ol} - A_{fb}A_{ol}\beta = A_{fb}$$

$$A_{ol}(1 - A_{fb}\beta) = A_{fb}$$

thus

$$A_{ol} = \frac{A_{fb}}{1 - A_{fb}\beta}$$

To solve, use A_{fb} minimum and β should be 1/(ideal closed loop gain). Thus for the first part of this example

$$A_{ol} = \frac{99}{1 - 99\,(0.01)} = 9900$$

So to obtain $A_{fb} = 100 \pm 1\%$ the op-amp must have $A_{ol} \geq 9900$.

2-5 NONIDEAL INVERTING AMPLIFIER

We develop the equation for an inverting amplifier with finite gain in the same way we developed the ideal closed loop gain, but we do not let any of the terms go to zero. The nonideal equation is still an approximation, since we assume in Figure 2.7 that $I_B = 0$. With that assumption, we note that

$$I_{R1} = -I_{Rf}$$

so

$$\frac{V_{in} - v_i}{R_1} = -\frac{V_{out} - v_i}{R_f}$$

Since $V_{out} = -A_{ol}\,v_i$, $v_i = -V_{out}/A_{ol}$. Substitute this into the equation for v_i to obtain

$$\frac{V_{in}}{R_1} + \frac{V_{out}}{A_{ol}\,R_1} + \frac{V_{out}}{R_f} + \frac{V_{out}}{A_{ol}\,R_f} = 0$$

Solving for V_{out}/V_{in}

$$\frac{V_{in}}{R_1} = -V_{out}\left(\frac{1}{A_{ol}\,R_1} + \frac{1}{R_f} + \frac{1}{A_{ol}\,R_f}\right)$$

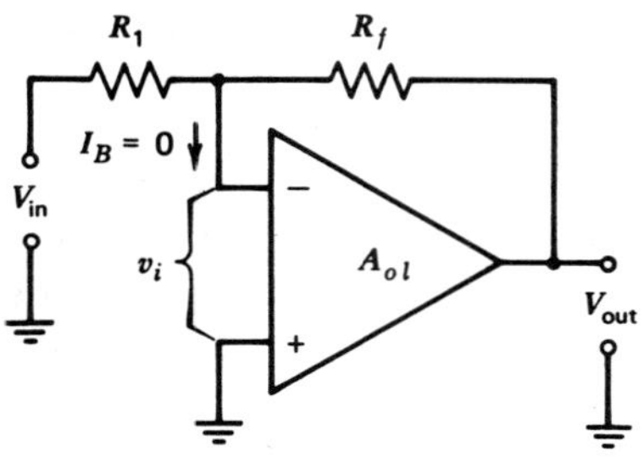

FIGURE 2.7 Nonideal inverting amplifier.

Multiplying both sides of the equation by A_{ol}, R_1, and R_f obtains

$$V_{in} A_{ol} R_f = -V_{out} (R_f + A_{ol} R_1 + R_1)$$

Therefore

$$A_{fb} = \frac{V_{out}}{V_{in}} = -\frac{A_{ol} R_f}{R_f + R_1 + A_{ol} R_1}$$

To obtain an equation that fits the general feedback equation

$$A_{fb} = \frac{A}{1 + A\beta}$$

where $\beta = 1/A_{fb}$ (ideal), divide the A_{fb} equation by $R_1 + R_f$ and multiply the right-hand term in the denominator by one, in the form R_f/R_f, to obtain

$$A_{fb} = -\frac{A_{ol} R_f/(R_1 + R_f)}{1 + \dfrac{A_{ol} R_1}{R_1 + R_f}\left(\dfrac{R_f}{R_f}\right)} = -\frac{A_{ol} R_f/(R_1 + R_f)}{1 + \dfrac{A_{ol} R_f}{R_1 + R_f}\left(\dfrac{R_1}{R_f}\right)}$$

Now let $A_{ol} R_f/(R_1 + R_f) = A_{oe}$, the effective open loop gain of an inverter, then the actual closed loop gain equation for the inverter is

$$A_{fb} = -\frac{A_{oe}}{1 + A_{oe}\beta} \tag{2–7}$$

where

$$\beta = \frac{R_1}{R_f}$$

$$A_{oe} = A_{ol}\frac{R_f}{R_1 + R_f} \tag{2–8}$$

and

$$A_{ol} = \text{open loop gain}$$

Note that $R_f/(R_1 + R_f)$ is the voltage divider ratio of R_f and R_1, the inverter's feedback resistors, to V_{in}. Referring to Figure 2.7, if $V_{out} = 0$, then the input voltage at the summing point, v_i, is equal to $V_{in}\, R_f/(R_1 + R_f)$. This voltage dividing effect to V_{in} occurs even when $V_{out} \neq 0$ and causes an effective decrease in v_i proportional to the voltage divider ratio of $R_f/(R_1 + R_f)$. This is more easily expressed in the actual closed loop gain formula as a reduction in A_{ol} to A_{oe} by the ratio just expressed:

$$A_{oe} = A_{ol}\frac{R_f}{R_1 + R_f}$$

A_{ol} is the open loop gain we would find in the manufacturer's specifications (often specified as large signal voltage gain).

EXAMPLE 2-3

The inverter of Figure 2.7 has:

$$R_f = 1\ \text{M}\Omega$$
$$R_1 = 20\ \text{k}\Omega$$
$$A_{ol} = 50{,}000$$

Find the actual closed loop gain.

SOLUTION

$$\beta = \frac{R_1}{R_f} = \frac{20\ \text{k}\Omega}{1\ \text{M}\Omega} = 0.02$$

$$A_{oe} = A_{ol}\left(\frac{R_f}{R_1 + R_f}\right) = 50{,}000\left(\frac{1\ \text{M}\Omega}{1\ \text{M}\Omega + 20\ \text{k}\Omega}\right) = 49{,}019$$

$$A_{fb} = \frac{-A_{oe}}{1 + A_{oe}\beta} = -49.95$$

Note that this is very close to $\dfrac{R_f}{R_1} = 50$, since $A_{oe} >> A_{fb}$.

The actual closed loop gain equation can be used to find the minimum A_{ol} necessary for a given closed loop gain error, as shown in Figure 2-4.

EXAMPLE 2-4

An inverter is to be used as an amplifier with $A_{fb} = 100 \pm 1\%$, with $R_1 = 10\ k\Omega$ and $R_f = 1\ M\Omega$. Calculate the minimum A_{ol} we must use to obtain this small an error.

$$R_f/(R_1 + R_f) = 1\ M\Omega/1.01\ M\Omega = 0.99$$

$$A_{fb}(\min) = 100 - 1\% = 99$$

$$1/\beta = 1/100 = 0.01$$

Just set up the general A_{fb} equation where $A_{fb} = A_{fb}(\min)$ and solve for A_{ol}.

$$A_{fb} = -\frac{A_{oe}}{1 + A_{oe}\beta} = -\frac{A_{ol}R_f/(R_1 + R_f)}{1 + A_{ol}\left(\dfrac{R_f}{R_1 + R_f}\right)\beta} \tag{2–9}$$

Substituting above values into Equation 2–9 yields

$$99 = -\frac{A_{ol}\,(0.99)}{1 + A_{ol}\,(0.99)\,(0.01)}$$

Solving for A_{ol} yields

$$99\,[1 + A_{ol}\,(0.99)\,(0.01)] = -A_{ol}\,(0.99)$$

$$99 + 0.98\,A_{ol} = -99\,A_{ol}$$

$$A_{ol} = -99/0.01 = -9900$$

We must therefore buy an amplifier with a minimum open loop gain of 9900 to obtain our desired accuracy. In practice we would buy an op-amp with a much higher A_{ol} in order to allow for resistor tolerances and other sources of error to be discussed in Chapter 3.

The procedure for finding $A_{ol}(\min)$ for a given A_{fb} error is the same for a non-inverting amplifier except that we use A_{ol} instead of $A_{ol}\,[R_f/(R_1 + R_f)]$ or A_{oe}.

2-6 EXTERNAL OFFSET COMPENSATION

Some hybrid and discrete component amplifiers have a screwdriver adjust to null offset internally. In amplifiers that have no internal means of correcting for V_{os} we must add an external resistor network to compensate for offset voltage. The Texas Instruments TL082 and many hybrid operational amplifier modules are examples of amplifiers compensated in this manner.

Before we discuss trimming out the offset voltage, we should discuss the effect of the bias current (I_B) on offset voltage. Even though I_B is small, it does exist, and,

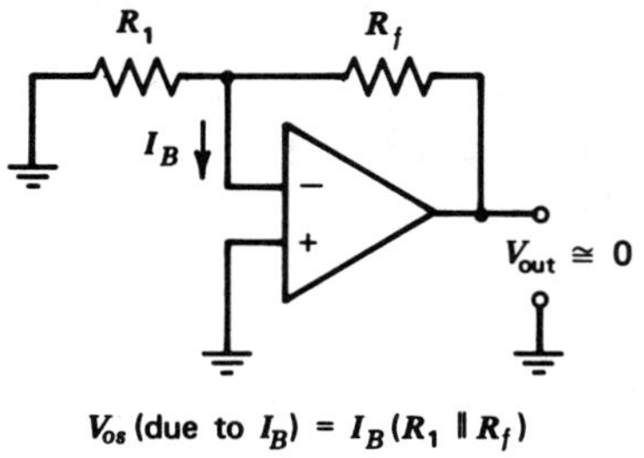

FIGURE 2.8 Offset voltage due to bias current.

referring to Figure 2.8, we see that if V_{os} (due to V_{BE}) is zero, I_B flows through the parallel combination of R_1 and R_f causing a voltage V_{os} (due to I_B) equal to I_B $(R_1 \|R_f)$.

Since there is a bias current flowing into (or out of) the noninverting terminal (I_{B_2}) in Figure 2.9 approximately equal to the bias current flowing into (or out of) the inverting terminal (I_{B_1}), if we place a resistor in the noninverting terminal (R_S) that is equal $R_1\|R_f$, then the voltage generated across R_S will be equal (approximately) to the offset voltage at the inverting terminal due to I_{B_1} $(R_1\|R_f)$.

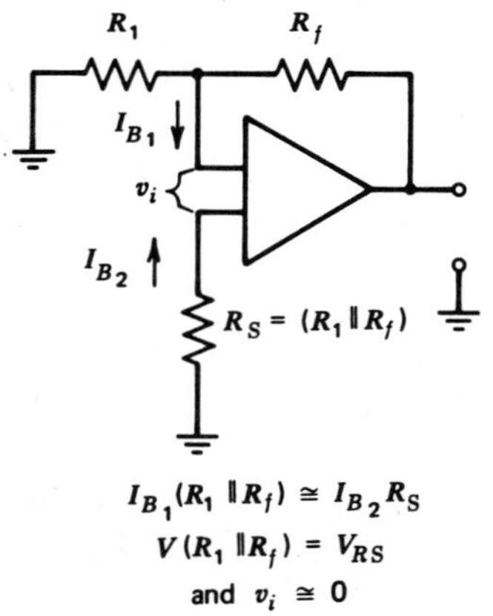

FIGURE 2.9 Bias current compensation.

To compensate for V_{os} due to mismatched V_{BE}, we will set up a voltage divider with which Vos_{max} can be canceled without disturbing the feedback network. Usually we will try to compensate for offset voltage from any source at one time. The offset voltage adjust (P_1) for an inverting circuit is shown in Figure 2.10. Note that $R_A + R_2 = R_S$ is our compensation for offset voltage due to bias current. R_4 is picked so that the parallel combination of R_A and R_4 is approximately equal to R_A; this means keeping R_A small and R_4 large. The offset voltage adjustment range is

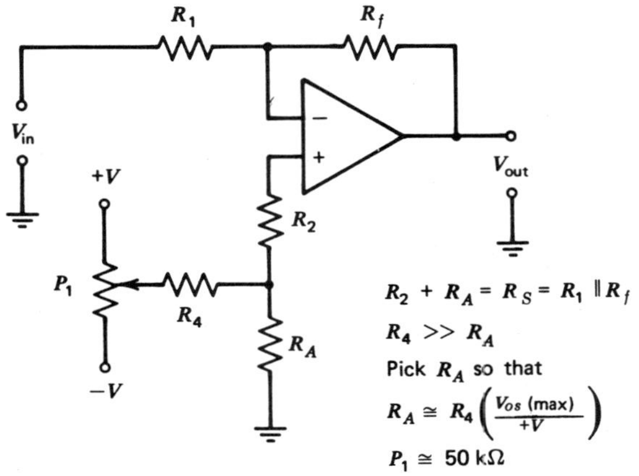

FIGURE 2.10 Offset voltage compensation, inverting amplifier.

approximately $\pm V\,(R_A/R_4)$ since $R_4 >> R_A$. The potentiometer P_1 should be large enough that the supply is not unnecessarily loaded, but the current through the potentiometer should be at least 20 to 40 times I_B. In like manner the current through R_4 should be larger than I_B, so that R_A and R_4 act like a voltage divider.

EXAMPLE 2-5

Suppose in Figure 2.10 $R_1 = 20$ kΩ and $R_f = 200$ kΩ. We would set $R_S \simeq 18.2$ kΩ. If $V = \pm 15$ V, $I_B = 0.8$ μA, and $V_{os}(\text{max}) = 20$ mV, we pick $R_4 \simeq \pm V/20\,I_B = 15\text{ V}/16\ \mu\text{A} = 800$ kΩ. In practice we use a value smaller than this if we can, to keep R_A small, so we pick $R_4 = 400$ kΩ. Now we find R_A

$$\frac{R_A}{R_4} = \frac{V_{os}(\text{max})}{+\,V}$$

therefore

$$R_A = \frac{V_{os}(\text{max})}{+\,V}\,R_4$$

$$R_A = 400\text{ k}\Omega\,\frac{(20\text{ mV})}{(15\text{ V})} = 540\ \Omega$$

Now, $R_2 = R_S - R_A = 18.2\text{ k}\Omega - 540\ \Omega \approx 17.66$ kΩ.

Compensating V_{os} in the noninverting amplifier is similar; however, the voltage divider is in the feedback network, so it is very important that R_4 be much larger than R_A (see Figure 2.11). Note that $R_1 = R_A + R_B$ and this sum is used in the closed loop gain equation. P_1 and R_4 are selected in the same manner as for the inverting amplifier.

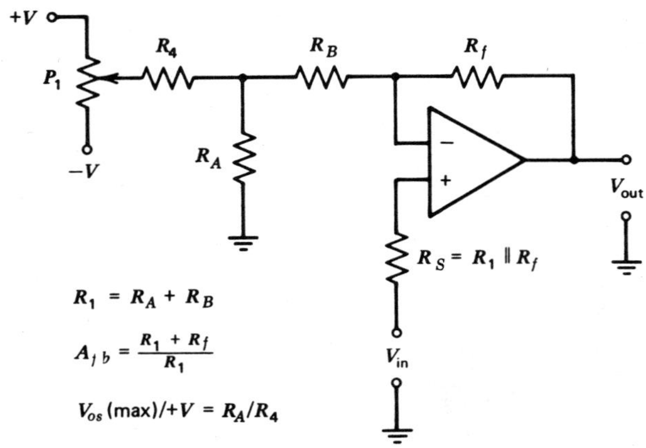

FIGURE 2.11 Offset voltage compensation, noninverting amplifier.

The differential input circuit is compensated similarly, but the voltage divider circuit is made part of R_f' as in Figure 2.12. In this circuit $R_f' = R_B + R_A$ and this sum is used with the output voltage equation. Note that R_4 must be much larger than both $R_2 + R_B$ and R_A. Otherwise the components are picked in the same manner as the preceding two circuits. R_S is not in this circuit since usually $(R_2 \parallel R_f') \simeq (R_1 \parallel R_f)$ so that offset due to bias current is compensated.

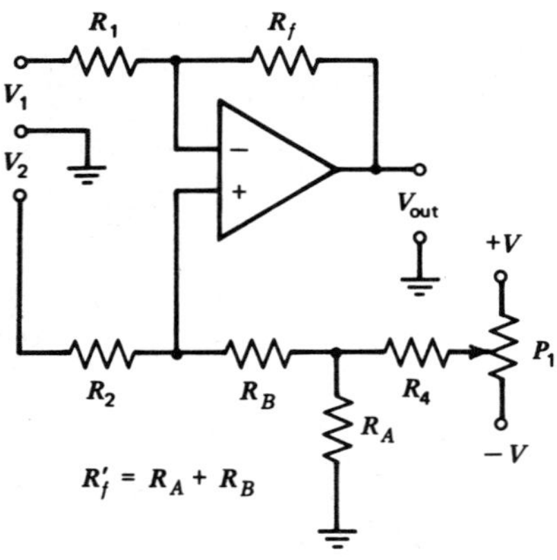

FIGURE 2.12 Offset voltage compensation; differential input amplifier.

SUMMARY

1. The addition of negative feedback causes amplifier gain to decrease from its open loop value to that set by the feedback. The general closed loop gain (A_{fb}) expression is

$$A_{fb} = \frac{A}{1 + A\beta} \tag{2-1}$$

where

$$A = \text{amplifier gain}$$
$$\beta = \text{feedback factor}$$

2. Negative feedback causes the effective input resistance of an amplifier to increase. The input resistance with feedback (R_{infb}) is found with this equation:

$$R_{infb} = R_{in}(1 + A\beta) \tag{2-4}$$

where R_{in} = input resistance without feedback.

3. Negative feedback causes the effective output resistance of an amplifier to decrease. The output resistance with feedback (R_{outfb}) is

$$R_{outfb} = \frac{R_{out}}{1 + A\beta} \tag{2-3}$$

where R_{out} = output resistance without feedback.

4. Since open loop gain is finite, the actual A_{fb} will always be less than the ideal $A_{fb} = 1/\beta$. By expressing the closed loop gain of an inverter and noninverting amplifier in the form of the general A_{fb} equation, we can find out how much error we obtain from A_{ol} being less than infinite. More importantly, we can calculate the minimum open loop gain we must have for a given amount of error.

The general A_{fb} equations are the following:

(a) For an inverter

$$A_{fb} = -\frac{A_{oe}}{1 + A_{oe}\beta} \tag{2-7}$$

Where

$$\beta = R_1/R_f$$
$$A_{oe} = A_{ol}R_f/(R_f + R_1) \tag{2-8}$$

(b) For a noninverting amplifier

$$A_{fb} = \frac{A_{ol}}{1 + A_{ol}\beta}$$

Where

$$\beta = R_1/(R_f + R_1)$$
$$A_{ol} = \text{open loop gain}$$

5. For accurate amplification of dc voltages, input offset voltages and currents must be nulled. The principle of nulling V_{os} is to apply a voltage equal, but opposite in polarity, to the input offset voltage—thereby canceling it. Various voltage dividers are used to accomplish this.

SELF-TEST QUESTIONS

2-1 A noninverting amplifier with $R_1 = 20\ \text{k}\Omega$ and $R_f = 2\ \text{M}\Omega$ has the following specifications:

$$A_{ol} = 20{,}000$$
$$R_{in} = 200\ \text{k}\Omega$$
$$R_{out} = 1\ \text{k}\Omega$$

Calculate β, A_{fb}, R_{infb}, R_{outfb}.

2-2 An inverting amplifier with $R_1 = 10\ \text{k}\Omega$ and $R_f = 1\ \text{M}\Omega$ has the following specifications:

$$A_{ol} = 30{,}000$$
$$R_{in} = 300\ \text{K}\Omega$$
$$R_{out} = 500\ \Omega$$

Calculate β, A_{fb}, A_{oe}, R_{infb}, R_{outfb}.

2-3 State what happens to A_{fb}, R_{infb}, and R_{outfb} as loop gain increases.

2-4 State the cause of V_{os} due to I_B.

2-5 State the principle of external offset compensation.

2-6 The amplifier of Figure 2.10 has supply voltages of ± 15 V, $V_{os}(\text{max}) = 10$ mV, $R_1 = 10\ \text{k}\Omega$, and $R_f = 100\ \text{k}\Omega$. If $R_4 = 200\ \text{k}\Omega$, calculate R_2 and R_A.

2-7 The amplifier of Figure 2.11 has supply voltages of ± 15 V, $V_{os}(\text{max}) = 7$ mV, $R_1 = 10\ \text{k}\Omega$, and $R_f = 200\ \text{k}\Omega$. Let $R_4 = 200\ \text{k}\Omega$ and calculate R_S, R_A and R_B.

2-8 State how R_4 is chosen.

2-9 State the condition necessary for the ideal closed loop gain calculation to be accurate.

2-10 Calculate the open loop gain necessary for a noninverting amplifier that needs $A_{fb} = 50 \pm 0.1\%$.

If you cannot answer certain questions, place a check next to them and review appropriate parts of the text to find the answers.

LABORATORY EXERCISE

Objective

The student should be able to compensate externally for offset voltage, measure open loop gain, and compare actual and ideal closed loop gains on completing this laboratory.

Equipment

1. Fairchild μA 741 operational amplifier or equivalent and manufacturer's specifications.
2. Assortment of 2% tolerance resistors.
3. ± 15 V dc power supply.
4. Voltmeter or oscilloscope capable of measuring 5 mV dc and ac.

5. Signal generator.
6. Potentiometer between 10 kΩ and 50 kΩ.
7. Breadboard such as EL Instruments SK-10 mounted on vectorboard.

Procedure 1.

(a) Connect the circuit shown in Figure 2.13 to measure open loop gain. V_{in} should be less than a 10 Hz ac signal and large in amplitude (about 15 V peak to peak). The closed loop gain is 1, so offset is not a problem in the open loop gain measuring circuit. The 99 kΩ and 1 kΩ resistors act as a voltage divider. We will measure V_S, but the amplifier input voltage, v_i, will be $V_S/100$. The open loop gain is then

$$A_{ol} = \frac{V_{out}}{V_S/100} = V_{out}/v_i$$

Measure V_{out} and V_S carefully, compute the open loop gain, and record the value.
(b) Connect the inverting amplifier shown in Figure 2.14. Set R_f = 1 MΩ and calculate R_1 for A_{fb} = 1000. Calculate R_2 and R_3 for R_4 = 200 kΩ and P_1 = 50 kΩ.
Note. We are ignoring the offset null pins of the μA741 so that we can practice nulling the offset externally.
(c) Ground the circuit input and null the offset.
(d) With V_{in} = 0.01 V at a frequency of about 5 Hz, measure the closed loop gain.
(e) Compute the closed loop gain of the circuit in step (d) using these equations:

$$A_{fb} = \frac{A_{oe}}{1 + A_{oe}\beta}$$

where

$$\beta = R_1/R_f$$

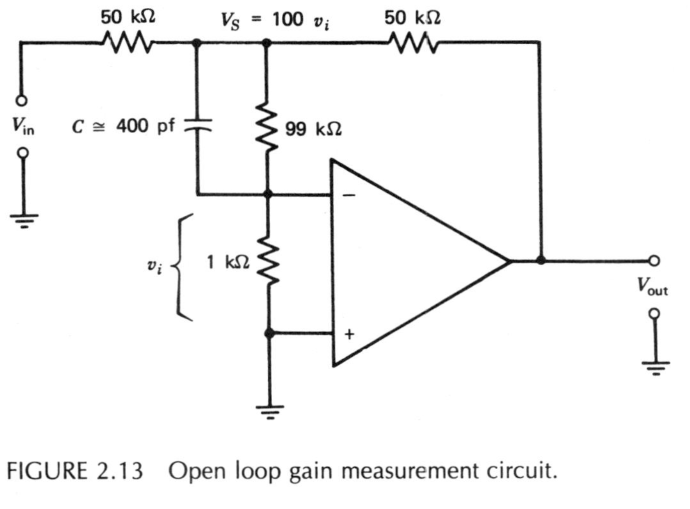

FIGURE 2.13 Open loop gain measurement circuit.

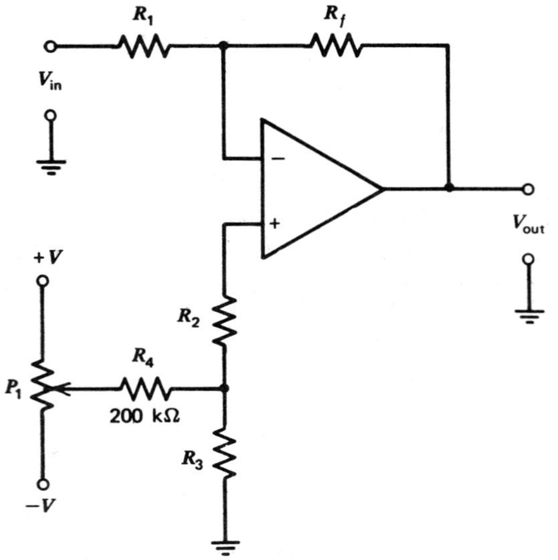

FIGURE 2.14 Offset null connection, inverting amplifier.

and

$$A_{oe} = \frac{A_{ol}R_f}{R_1 + R_f}$$

How does this calculated value agree with your measured value?

2.

(a) Connect the circuit in Figure 2.15 and compensate the input offset voltage of the amplifier.

(b) Calculate the ideal closed loop gain using the equation

$$A_{fb} = \frac{R_f + (R_A + R_B)}{(R_A + R_B)}$$

(c) Calculate the closed loop gain using the nonideal equation

$$A_{fb} = \frac{A_{ol}}{1 + A_{ol}\beta}$$

where

$$\beta = \frac{R_1}{R_1 + R_f} \qquad \text{(recall } R_1 = R_A + R_B\text{)}$$

(d) Measure the closed loop gain. How do steps (b), (c), and (d) agree?

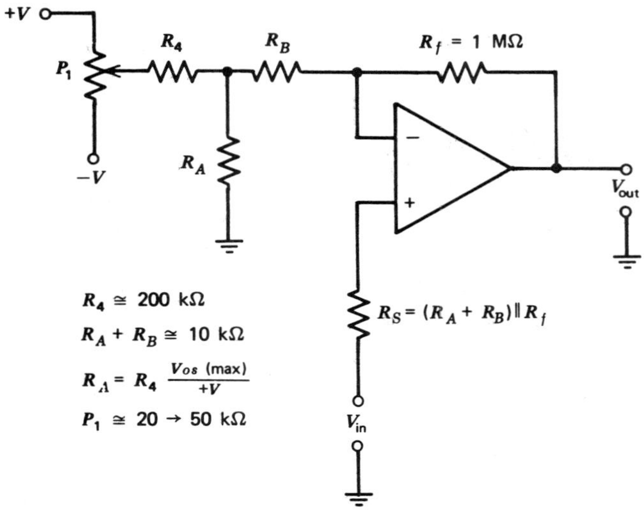

FIGURE 2.15 Offset null connection, noninverting amplifier.

Discussion

State the relationship you observed between the values of the ideal closed loop gain and the measured closed loop gain. Discuss the causes for any differences you observed in the calculated and observed closed loop gain.

CHAPTER

3

BIAS CURRENT, CMRR, TEMPERATURE DRIFT, AND CHOPPER STABILIZATION

To choose the values of our op-amp feedback resistors, we must know how much current is necessary to operate the amplifier. For accurate amplification of differential signals, we must know the effect a common-mode signal will have on our system. This chapter discusses these two measurements. The offset voltage and current can be completely compensated at one temperature. Input offset voltage and current change with temperature. Since it is often impractical to compensate for the drift in V_{os} and I_{os} with temperature, we must be able to predict the error from the drift to know if it is acceptable over our circuit's operating temperature range.

OBJECTIVES

After completing the study of this chapter and the self-test questions, the student should be able to:

1. State the principle behind, choose the components for, and state the steps involved in measuring I_B and I_{os}.

2. Define and calculate the error caused by CMRR.

3. Draw a circuit for and state the method of measuring CMRR.

4. State the two major factors contributing to temperature drift.

5. Given $\Delta I_{os}/\Delta T$, $\Delta V_{os}/\Delta T$, and ΔT, calculate the error in output voltage caused by temperature change.

6. Briefly explain the operation of a chopper-stabilized and a varactor diode carrier operational amplifier.

7. Perform the laboratory for Chapter 3.

3-1 BIAS CURRENT AND MEASUREMENT

To measure the bias current, the circuit shown in Figure 3.1 is assembled. Note that this circuit resembles a follower with the input grounded if S_1 and S_2 are closed.

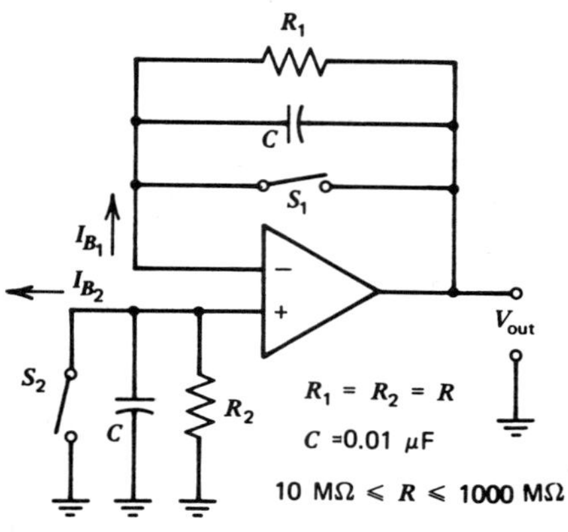

FIGURE 3.1 I_B and I_{os} measuring circuit.

If there is an internal offset voltage adjust, V_{os} is nulled as well as possible with S_1 and S_2 closed. The capacitors are used to avoid frequency instability or oscillation. If V_{os} cannot be adjusted out, it should be noted. Since the circuit is a follower, V_{out} with S_1 and S_2 closed is just V_{os}. Note that when S_1 is open and S_2 is closed there will be a voltage developed across R_1 from I_{B_1} flowing through it. (R_1 and R_2 are large since I_B is small; in this way we get a readable voltage.) Since the circuit is a follower, the output voltage is just

$$V_{out} = I_{B_1}R_1 \qquad \text{if} \qquad V_{os} << V_{out}$$

Therefore:

$$I_{B_1} = \frac{V_{out}}{R_1} \qquad \text{if} \qquad V_{out} >> V_{os} \tag{3-1}$$

If V_{os} is not negligible with respect to V_{out} with S_1 open, then

$$I_{B_1} = \frac{V_{out} - V_{os}}{R_1} \tag{3-2}$$

EXAMPLE 3-1

If $V_{os} = 10$ mV and V_{out} is -15 mV, where $R_1 = 10\ M\Omega$,

$$I_{B_1} = \frac{-15\text{ mV} - 10\text{ mV}}{10\ M\Omega} = \frac{-25\text{ mV}}{10\ M\Omega} = -2.5\text{ nA}$$

If $R_1 = 10\ M\Omega$, $V_{os} = -5$ mV, and $V_{out} = -10$ mV

$$I_{B_1} = \frac{-10\text{ mV} - (-5\text{ mV})}{10\ M\Omega} = \frac{-5\text{ mV}}{10\ M\Omega} = -0.5\text{ nA}$$

If we now close S_1 and open S_2

$$I_{B_2} = \frac{V_{out} - V_{os}}{R_2} \tag{3-3}$$

Since $R_1 = R_2$ if we open S_1 and S_2 at the same time

$$V_{out} = I_{B_1}R - I_{B_2}R = R(I_{B_1} - I_{B_2})$$

But $I_{B_1} - I_{B_2}$ is the input offset current, I_{os}. So if $V_{os} << V_{out}$,

$$I_{os} = \frac{V_{out}}{R} \tag{3-4}$$

with S_1 and S_2 open

If V_{os} is not much smaller than V_{out}, then

$$I_{os} = \frac{V_{out} - V_{os}}{R} \tag{3-5}$$

where V_{os} is measured with S_1 and S_2 closed.

Table 3.1 summarizes our results.

TABLE 3.1

BIAS AND OFFSET CURRENT MEASUREMENT EQUATIONS

Switches OPEN	I_B	Equation
S_1	$I_{B_1} = \frac{V_{out} - V_{os}}{R}$	(3-2)
S_2	$I_{B_2} = \frac{V_{out} - V_{os}}{R}$	(3-3)
S_1 and S_2	$I_{os} = \frac{V_{out} - V_{os}}{R}$	(3-5)

Note. With S_1 and S_2 closed, $V_{out} = V_{os}$. $R_1 = R_2 = R$

3-2 COMMON-MODE REJECTION RATIO (CMRR)

To discuss common-mode rejection ratio, we must first discuss common-mode gain. Ideally, if two voltages of equal amplitude are applied to the inputs of an operational amplifier, the output of the amplifier is zero (Figure 3.2). This is seldom the case with a realizable amplifier,[1] when $V_1 = V_2$ there is always some output,

[1] For the reason see Appendix A.

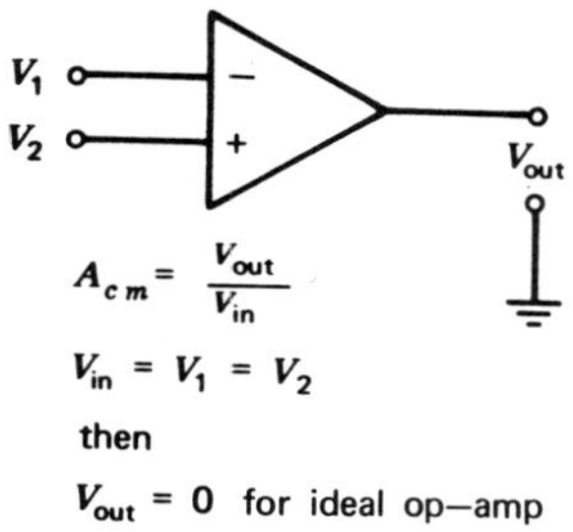

FIGURE 3.2 Defining CMRR.

however small. Common-mode gain is V_{out}/V_{in} when $V_1 = V_2 = V_{in}$. Usually common-mode gain, A_{cm}, is much less than one. $A_{cm} = 0.01$ is a typical value for an operational amplifier.

Another method of expressing the common-mode sensitivity is the common-mode rejection ratio (CMRR). The CMRR is defined to be:

$$\text{CMRR} = \frac{A_D}{A_{cm}} = \frac{\text{differential gain}}{\text{common-mode gain}}$$

The differential gain of an op-amp without feedback is just its open loop gain. Thus most specification sheets give CMRR as:

$$\text{CMRR} = \frac{A_{ol}}{A_{cm}} = \frac{\text{open loop gain}}{\text{common-mode gain}}$$

Values from 1000 to 10,000 are typical for CMRR, with higher values preferable. Often CMRR is expressed in decibels so that

$$\text{CMRR (dB)} = 20 \log_{10} \frac{A_{ol}}{A_{cm}}$$

$$\text{CMRR (dB)} = 20 \log_{10} \text{CMRR}$$

Let us look at the error caused by CMRR. Referring to Figure 3.3 observe that

$$V_{out} = -A_{ol}v_i + A_{cm}V_2$$

since $V_2 \simeq V_1$. Also observe that

$$v_i = V_1 - V_2 = V_{out}\left(\frac{R_1}{R_1 + R_f}\right) - V_2$$

We substitute this for v_i in the first equation:

$$V_{out} = -A_{ol}\left(\frac{R_1}{R_1 + R_f}\right) V_{out} + A_{ol}V_2 + A_{cm}V_2$$

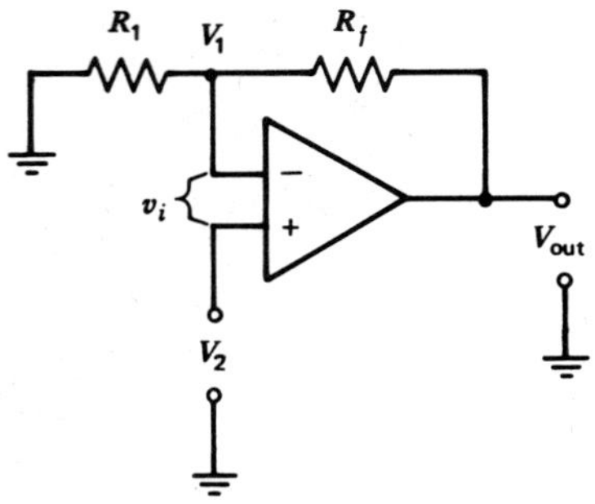

FIGURE 3.3 Noninverting amplifier for demonstrating error caused by CMRR.

Solving for V_{out} we find

$$V_{out}\left[1 + A_{ol}\left(\frac{R_1}{R_1 + R_f}\right)\right] = V_2(A_{ol} + A_{cm})$$

and

$$V_{out} = \frac{V_2(A_{ol} + A_{cm})}{1 + A_{ol}\left(\frac{R_1}{R_1 + R_f}\right)}$$

We now solve for V_{out}/V_2, which is just A_{fb}, and find

$$\frac{V_{out}}{V_2} = A_{fb} = \frac{A_{ol} + A_{cm}}{1 + A_{ol}\left(\frac{R_1}{R_1 + R_f}\right)}$$

Observe that $R_1/(R_1 + R_f) = \beta$ for the noninverting amplifier, so

$$A_{fb} = \frac{A_{ol} + A_{cm}}{1 + A_{ol}\beta} = \frac{A_{ol}}{1 + A_{ol}\beta} + \frac{A_{cm}}{1 + A_{ol}\beta}$$

If we note that CMRR $= A_{ol}/A_{cm}$, then A_{cm} can be expressed as

$$A_{cm} = A_{ol}/\text{CMRR}$$

Now substituting this for A_{cm} in the second term of the equation we obtain

$$A_{fb} = \frac{A_{ol}}{1 + A_{ol}\beta} + \frac{A_{ol}/\text{CMRR}}{1 + A_{ol}\beta} \tag{3-6}$$

where CMRR is expressed as a ratio rather than in dB.

EXAMPLE 3-2

Suppose we want a noninverting amplifier with $A_{fb} = 11$ where $R_1 = 10\ k\Omega$, $R_f = 100\ k\Omega$, $A_{ol} = 1000$, and CMRR $= 10{,}000$. What will be our actual gain?

$$A_{fb} = \frac{A_{ol}}{1 + A_{ol}\beta} + \frac{A_{ol}/\text{CMRR}}{1 + A_{ol}\beta}$$

$$\beta = \frac{R_1}{R_1 + R_f} = \frac{1}{11} = 0.091$$

Substituting into Equation 3-6 yields

$$A_{fb} = \frac{1 \times 10^3}{1 + (9.1 \times 10^{-2})(10^3)} + \frac{10^3/10^4}{1 + (9.1 \times 10^{-2})(10^3)}$$

$$= \frac{10^3}{92} + \frac{10^{-1}}{92} = 10.89 + 1.089 \times 10^{-3}$$

$$= 10.891$$

In this case CMRR was high enough that the error due to finite open loop gain was higher than the error due to common-mode gain.

EXAMPLE 3-3

To understand the importance of a high CMRR refer to Figure 3.4. This figure shows a differential input amplifier amplifying a bridge unbalance signal. Calculate V_{out} for common-mode error and from differential gain of $V_2 - V_1 = 1$ mV and $V_{cm} = 5$ V. Let $A_{ol} = 10{,}000$ and CMRR $= 20{,}000$, which is a CMRR $= 86$ dB (a high value of CMRR).

V_{out} from differential gain is

$$V_{out}(\text{diff}) = (V_2 - V_1)\frac{R_f}{R_1} = 1\text{ mV }(100) = 0.1\text{ V}$$

V_{out} from common-mode gain is

$$V_{outcm} = V_{cm}\frac{A_{ol}/\text{CMRR}}{1 + A_{ol}\beta}$$

$$= 5\text{ V}\left(\frac{10{,}000/20{,}000}{1 + 10{,}000\ (0.01)}\right)$$

$$= 5\text{ V }(0.00495) = 0.0247\text{ V}$$

The output error from common-mode gain for this set of inputs is 24.7%. An amplifier with a higher CMRR should be used to reduce the error. Note, however,

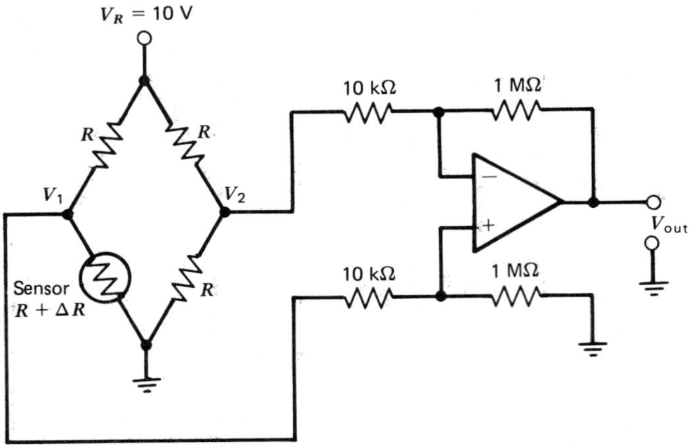

FIGURE 3.4 Bridge amplifier for Example 3-3.

that the output from the common-mode gain is low even though the common-mode input voltage is much larger than the differential input voltage.

Often, as in Example 3-2, error due to common-mode gain is low. Note that if the open loop gain is high enough the CMRR is high even though the common-mode gain is not particularly low. In a good amplifier, CMRR should be at least 10 times higher than the open loop gain. The exception to this is an amplifier that is to be used in the inverting mode.

The reason why common-mode error is virtually absent in an inverting amplifier is because both inputs to the amplifier are at zero volts, since the noninverting input is grounded. Since the output voltage due to A_{cm} is $A_{cm}V_2$ and $V_2 = 0$, V_{out} due to common-mode gain is approximately zero.

3-3 MEASURING CMRR

The technique of measuring CMRR is simply applying the same voltage to both inputs and measuring the output. A balanced differential input circuit is used as shown in Figure 3.5.

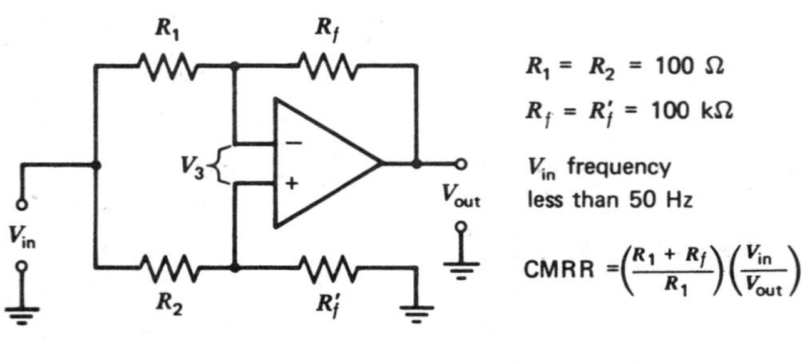

FIGURE 3.5 CMRR measuring circuit.

The gain of the CMRR measuring circuit must be high since the common-mode gain is low. Since both inputs have the same voltage applied to them, the output voltage will be low because of the common-mode gain of the amplifier. Since there is no differential input voltage, the gain of the amplifier of Figure 3.5 can be written as

$$A_{fb} = \frac{V_{out}}{V_{in}} = \frac{A_{ol}/CMRR}{1 + A_{ol}\beta}$$

where

$$\beta = \frac{R_1}{R_1 + R_f}$$

Substituting in the expression for β, we obtain

$$\frac{V_{out}}{V_{in}} = \frac{A_{ol}/CMRR}{1 + A_{ol}\left(\frac{R_1}{R_1 + R_f}\right)}$$

since

$$A_{ol}\beta >> 1$$

we can write

$$\frac{V_{out}}{V_{in}} \cong \frac{A_{ol}/CMRR}{A_{ol}\left(\frac{R_1}{R_1 + R_f}\right)} = \frac{1}{CMRR\left(\frac{R_1}{R_1 + R_f}\right)}$$

multiplying through by $R_1/(R_1 + R_f)$ and rearranging yields

$$CMRR = \frac{V_{in}}{V_{out}}\left(\frac{R_1 + R_f}{R_1}\right) \tag{3-7}$$

EXAMPLE 3-4

If in Figure 3.5 $V_{in} = 10\ V_{pp}$ and $V_{out} = 1.02\ V_{pp}$, calculate CMRR and CMRR (dB)

$$CMRR = \frac{V_{in}}{V_{out}}\left(\frac{R_1 + R_f}{R_1}\right) = \frac{10\ V}{1.02\ V}\left(\frac{100\ k\Omega + 100\ \Omega}{100\ \Omega}\right)$$

$$= 9.804\ (1001) = 9814$$

$$CMRR\ (dB) = 20 \log CMRR$$

$$= 20 \log 9814$$

$$= 79.8\ dB$$

3-4 TEMPERATURE-SENSITIVE PARAMETERS

The primary source of offset voltage change with temperature is the change of V_{BE} with temperature. The V_{BE} of a silicon transistor decreases approximately 2 m V/°C. This decrease is not exactly the same for each transistor, causing an offset voltage. Since both inputs drift at about the same rate with temperature, the offset is usually several microvolts per degree centigrade.

Another source of drift caused by temperature is due to the transistor betas (β). β dc increases with increasing temperature, and the β of every transistor does not increase at the same rate. This causes the bias current to be different for each transistor, resulting in increased input offset current. As seen in Chapter 2, this causes an offset voltage. The thermal leakage currents of the transistors are not the same; this adds to the input offset current.

FET input operational amplifiers also drift with temperature. The primary causes of $\Delta V_{os}/\Delta T$ for an FET amplifier is $\Delta V_{gs}/\Delta T$ and $\Delta gm/\Delta T$. The primary cause of $\Delta I_{os}/\Delta T$ is the change in gate leakage current with temperature. This is a very small current for FETs, but still there.

3-5 ERRORS CAUSED BY $\Delta V_{os}/\Delta T$ AND $\Delta I_{os}/\Delta T$

$\Delta V_{os}/\Delta T$, the change in offset voltage with temperature, and $\Delta I_{os}/\Delta T$, the change in input offset current with temperature, are given in the manufacturer's specifications of an operational amplifier. If $\Delta V_{os}/\Delta T$ and $\Delta I_{os}/\Delta T$ are not constant with temperature, a graph will be given.

For example, $\Delta V_{os}/\Delta T = \pm 20\ \mu V/°C$ and $\Delta i_{os}/\Delta T = \pm 0.5$ nA/°C in the Burr Brown 3500A operational amplifier.

To see the effect of temperature change, refer to Figure 3.6, an inverting amplifier that has an external compensation circuit. At one temperature we can completely compensate offset by adjusting P_1. We see that with $V_{in} = 0$, the offset voltage and compensating voltage are felt on the noninverting input so

$$V_{out} = \frac{R_1 + R_f}{R_1}(V_c - V_{os}) + \left(\frac{R_1 + R_f}{R_1}\right)\left(\frac{R_1 R_f}{R_1 + R_f}\right) I_{os}$$

which we adjust to zero with P_1. When the temperature changes, V_{out}, which is adjusted to zero with $V_{in} = 0$, varies with temperature. We can then write

$$\frac{\Delta V_{out}}{\Delta T} = \frac{R_1 + R_f}{R_1}\left|\frac{\Delta V_{os}}{\Delta T}\right| + R_f\left|\frac{\Delta I_{os}}{\Delta T}\right| \tag{3-8}$$

The absolute value signs indicate that we will consider the worst case,when both terms drift in the same direction. Often $\Delta V_{os}/\Delta T$ and $\Delta I_{os}/\Delta T$ drift with temperature in such a manner that they tend to cancel each other, but we must assume they both drift in the same direction.

Now let E be the error voltage because of $\Delta V_{os}/\Delta T$ and $\Delta I_{os}/\Delta T$.

$$E = \frac{\Delta V_{out}}{\Delta T}(\Delta T) = \left(\frac{R_1 + R_f}{R_1}\right)\left(\frac{\Delta V_{os}}{\Delta T}\right)\Delta T + R_f\left(\frac{\Delta I_{os}}{\Delta T}\right)\Delta T \tag{3-9}$$

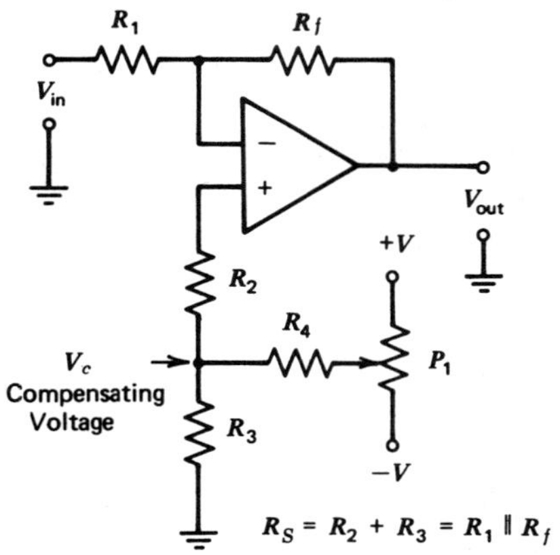

FIGURE 3.6 Offset compensated inverting amplifier.

EXAMPLE 3-5

If $R_1 = 10\ \text{k}\Omega$ and $R_f = 100\ \text{k}\Omega$ in Figure 3.6, with $\Delta V_{os}/\Delta T = \pm 10$ mV/°C and $\Delta I_{os}/\Delta T = \pm 1$ nA/°C and $\Delta T = 25$°C from room temperature ($T = 50$°C).

$$E = \frac{110\ \text{k}\Omega}{10\ \text{k}\Omega}(10\ \mu\text{V/°C})(25\text{°C}) + 100\ \text{k}\Omega\ (1\ \text{nA/°C})(25\text{°C})$$

So

$$E = 2.750\ \text{mV} + 2.5\ \text{mV} \simeq 5.25\ \text{mV}$$

Our output voltage with V_{in} applied at 50°C will be

$$V_{out} = -\frac{R_f}{R_1}V_{in} \pm E$$

We see the percentage error will vary with the input voltage, with the error larger in percentage as the input voltage decreases.

In our previous equation, if $V_{in} = 0.1$ V

$$V_{out} = -10(0.1\ \text{V}) \pm E = -1\ \text{V} \pm 5.25\ \text{mV}$$

so

$$E = 0.525\%$$

If $V_{in} = 10$ mV, $V_{out} = -0.1\ \text{V} \pm 5.25$ mV, an error of 5.25% results, which is a larger percentage error.

It is easier to estimate the error for a given input if it is referred to the input voltage. We define the error referred to the input as E_I. To do this, we merely find the amount of input voltage necessary on the inverting input to cause the error voltage on the output with $V_{in} = 0$. Since with $V_{in} = 0$, any output is from temperature drift alone

$$V_{out} = A_{fb}V_{in}(\text{drift}), \qquad E = A_{fb}E_I$$

where V_{in}(drift) is the apparent input from variations inside the op-amp from temperature change.
Hence

$$E_I = E\frac{1}{A_{fb}} = E\frac{R_1}{R_f}$$

for an inverting amplifier. Now we can put the error expression in a more useful form, since

$$V_{out} = -\frac{R_f}{R_1}(V_{in}) \pm E = -\frac{R_f}{R_1}V_{in} \pm \frac{R_f}{R_1}E_I$$

we see that

$$V_{out} = -\frac{R_f}{R_1}(V_{in} \pm E_I),$$

where

$$E_I = \frac{R_1}{R_f}E = \left(\frac{R_f + R_1}{R_f}\right)\left(\frac{\Delta V_{os}}{\Delta T}\right)\Delta T + R_1\left(\frac{\Delta I_{os}}{\Delta T}\right)\Delta T \tag{3-10}$$

Thus if $V_{in} = 1$ V, and $E_I = 10$ mV, we can see at a glance our error is $\pm 0.1\%$. That error will be evaluated at the operating temperature limits. Referring to Figure 3.7, a compensated noninverting amplifier, we see that

$$V_{out} = \left(\frac{R_1 + R_f}{R_1}\right)V_{in} \pm E$$

so

$$V_{out} = \left(\frac{R_1 + R_f}{R_1}\right)V_{in} \pm E_I$$

where

$$E_I = \left(\frac{\Delta V_{os}}{\Delta T}\right)\Delta T + R_f\left(\frac{\Delta I_{os}}{\Delta T}\right)\Delta T \tag{3-11}$$

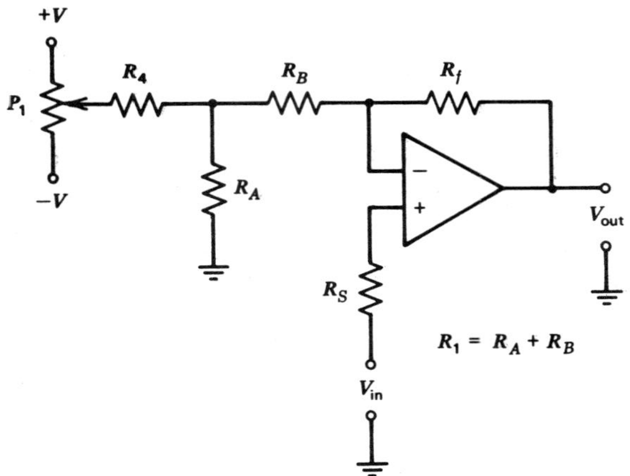

FIGURE 3.7 Offset compensated noninverting amplifier.

E_I is found for the noninverting amplifier in a manner similar to that used to find E_I for an inverting amplifier.

We have not, of course, included the effect of resistor termperature stability. If we choose very high stability resistors for critical applications, the amplifier drift should be the major error source.

3-6 CHOPPER STABILIZATION

One method used by manufacturers to reduce the temperature drift of operational amplifiers is chopper stabilization. Chopper stabilization consists of turning low frequency and dc signals into ac (chopping), amplifying the ac signal, changing the signal back to dc (demodulating), and applying the signal to the main amplifier. The chopper-modulator/demodulator stabilized amplifiers have offset voltage and drift reduction one to three orders of magnitude over a similar unstabilized main amplifier.

To visualize how chopper stabilization is accomplished, refer to Figure 3.8. First, everything within the dashed lines is contained within the amplifier module, with A being the inverting input, *B* the noninverting input, and C the output terminal. R_1 and R_f are the feedback resistors setting the closed loop gain of the amplifier.

C_A and R_A are a high-pass filter that allows high frequency signals to pass directly to the main amplifier and forces dc and very low frequency signals to pass through the chopper channel. The input low-pass filter in the chopper channel allows only the low frequencies and dc signals to enter the chopper channel. S_1 and S_2 are switches driven in synchronization by the chopper driver. S_1 and S_2 close and open together. Note that S_1 and S_2 could be semiconductor switches such as switching transistors driven between cutoff and saturation. S_1 causes a dc or low frequency voltage (e_1 of Figure 3.8) to be converted to an ac voltage (e_2 of Figure 3.8) that is modulated by the value of e_1. This is passed to the narrow band ac stabilization amplifier. Narrow band ac amplifiers can be built with extremely low offsets and

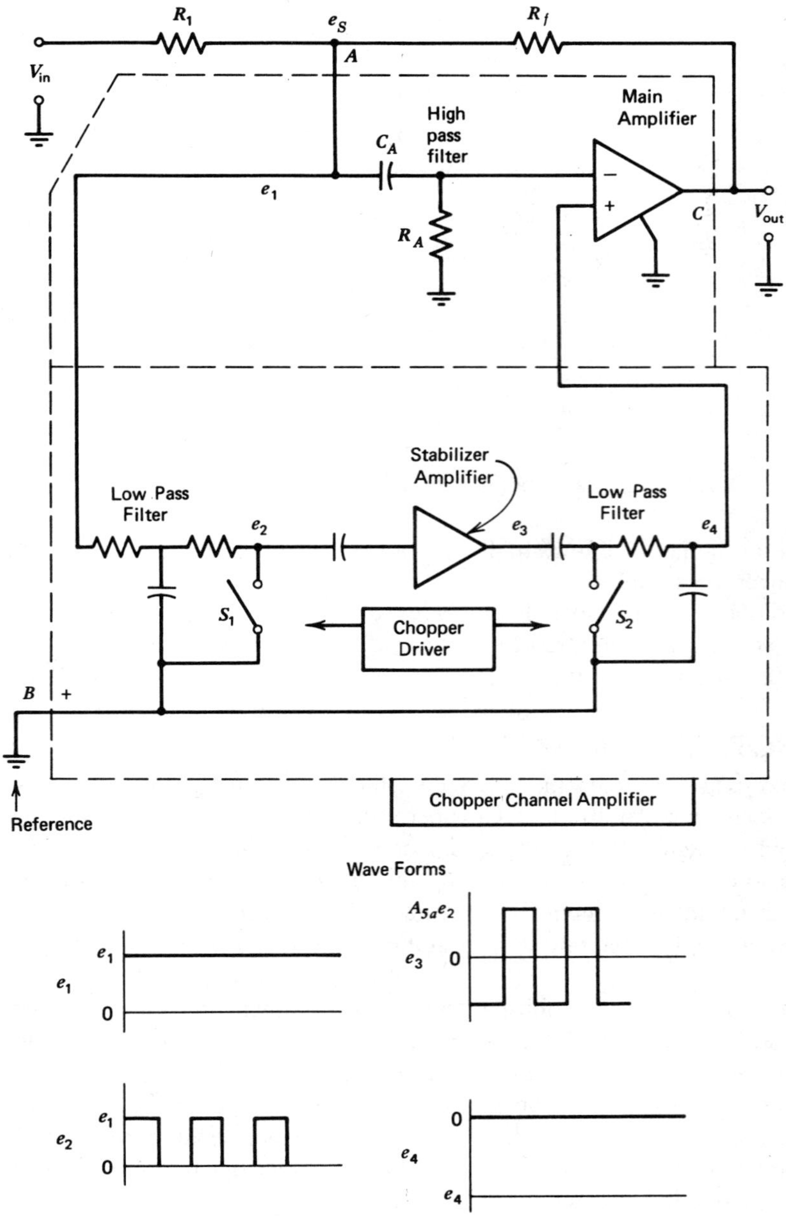

FIGURE 3.8 Chopper stabilization.

drift. The amplified output is ac coupled to S_2, which acts as a demodulator by clamping the output to zero volts when the input is zero volts, thereby converting the output to a dc pulse train. The low-pass filter on the output of the chopper amplifier filters the dc pulse train into a smooth dc that is applied to the main amplifier. Note that the voltage applied to the chopper channel is e_S at the summing point—which is usually low.

To understand how the chopper-stabilized operational amplifier reduces the effect of drift, suppose e_S is at some particular value and V_{os} starts to drift. The change in V_{os} causes the output voltage to change. The change in the output is coupled back to the summing point by R_1 and R_f, causing e_S to change. The chopper channel amplifies the change in e_S and applies it to the noninverting terminal of the main amplifier as a correction voltage that drives the output voltage (and therefore e_S) back to its original value. The same process acts on V_{os} of the main amplifier as well as changes in V_{os}.

The improvement in V_{os} is

$$V_{os(cs)} = V_{os(sa)} + \frac{V_{os(ma)}}{A_{(sa)}} \tag{3-12}$$

where

$V_{os(cs)}$ = offset voltage, chopper-stabilized amplified
$V_{os(sa)}$ = offset voltage of the stabilizing amplifier (very low)
$V_{os(ma)}$ = offset voltage of the main amplifier
$A_{(sa)}$ = voltage gain of the stabilizing amplifier

The improvement in offset voltage drift with temperature is by the same factor,

$$\frac{\Delta V_{os}}{\Delta T} \cong \frac{\Delta V_{os(sa)}}{\Delta T} + \frac{\Delta V_{os(ma)}/\Delta T}{A_{sa}}$$

Unfortunately, the addition of chopper stabilization limits the frequency response of the amplifier.

Chopper-stabilized amplifiers have input offset voltages as low as 20 μV, input offset temperature drift as low as 0.1 μV/°C, and drifts over time as low as 1 μV/month. Input bias currents are typically around 50 pA and input offset current drift about 1 pA/°C. Full power frequency response is typically from 15 kHz to 100 kHz. Some chopper-stabilized amplifiers must be used in the inverting configuration only since the noninverting terminal must be grounded. Other chopper-stabilized amplifiers do not have this restriction.

3-7 VARACTOR DIODE CARRIER OP-AMP

Varactor diode carrier op-amps are used to amplify millivolt level signals when extremely low bias currents are necessary. Examples of signals of this type are outputs from some pressure transducers, temperature-sensing transducers, and electrochemical cells. The low bias current comes from the fact that if the voltage across a diode is much less than the contact potential, less than 10 mV, very little diode current flows whether the bias is forward or reverse.

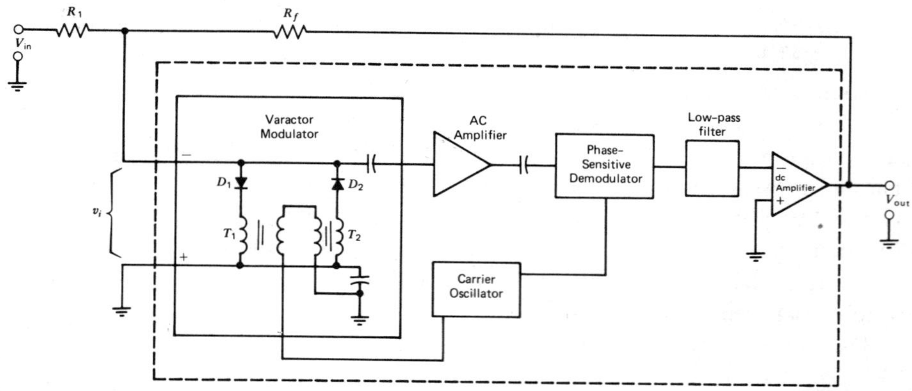

FIGURE 3.9 Varactor diode carrier op-amp block diagram.

Referring to Figure 3.9, the input voltage to the varactor carrier amplifier, v_i, is applied to two varactor diodes D_1 and D_2. T_1 and T_2 transformers couple the low-voltage high-frequency carrier signal ($f_c \cong 130$ kHz). The input imbalances the capacitance of the varactor diodes by increasing the forward bias on one and the reverse bias on the other. The varactor capacitance unbalance causes the phase and amplitude of the carrier to vary with v_i, thus the carrier is modulated.

The amplitude-modulated carrier is amplified by the ac amplifier, demodulated with a phase-sensitive demodulator, to sense polarity as well as amplitude; filtered and fed to the dc output amplifier. The entire unit is used as an op-amp. The gain is set by the feedback resistors in the usual fashion.

Varactor diode carrier amplifiers can have input bias currents as low as 0.01 pA, which is lower than many FET amplifiers, with very low noise. Input resistance will typically be 10^9 to 10^{11} ohms. This type of amplifier generally has a high CMRR.

3-8 CURRENT-TO-VOLTAGE CONVERTERS

Very low bias current amplifiers such as FET and MOSFET input op-amps and varactor diode carrier amplifiers are used to measure the output current of high output impedance sources such as electrochemical cells and some photodiodes. The short-circuit current is the quantity that one desires to measure from these high output impedance sources to maximize their frequency response. A high input resistance op-amp can be used to convert the short-circuit current into an output voltage as shown in Figure 3.10. The source is a Norton equivalent of the circuit where $I_s = V_s/R_s$. The op-amp inverting terminal presents a virtual ground to the source since $v_i \cong 0$. The current provided by the source that must flow through R_f is I_B which is much less than I_s. Thus the output voltage of the op-amp is

$$V_{out} = -I_s R_f$$

where

I_s = source output current V_s/R_s

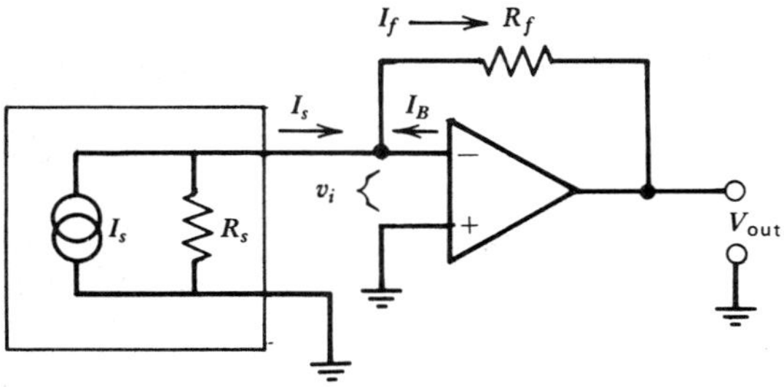

$V_i \cong 0$ so inverting input is a virtual ground or short circuit.

$$V_{out} = -I_s R_f = -I_f R_f \qquad \text{if } I_o >> I_B$$

FIGURE 3.10 Current-to-voltage convertor.

If the source current of the source in Figure 3.10 was 10 μA and $R_f = 1$ MΩ, the output voltage of the circuit would be

$$V_{out} = -\ 10 \text{ uA } (1 \text{ M}\Omega) = -10 \text{ V}$$

The virtual ground at the inverting amplifier has allowed us to convert an input current into an output voltage.

SUMMARY

1. Bias current and input offset current can be measured with a voltage follower circuit. Resistors large enough that RI_B is a readable voltage are used in the follower circuit. The voltage follower circuit used for bias current measurement allows the voltages due to I_B to be measured with normal measuring equipment instead of extremely high input resistance meters.

2. Common-mode gain (A_{cm}) is the voltage gain of an operational amplifier with the same voltage on each input. The common-mode gain of an operational amplifier is usually expressed in the specifications as common-mode rejection ratio (CMRR). CMRR $= A_{ol}/A_{cm}$.

3. Common-mode gain causes an error when using noninverting amplifiers. This error in the gain is found by the equation

$$\frac{A_{ol}/\text{CMRR}}{1 + A_{ol}\beta}$$

which will give the gain due to CMRR.

4. Inverting amplifiers have no error due to common-mode gain because one input is grounded and $A_{cm}(0 \text{ V}) = 0$.

5. CMRR can be measured with a high gain differential input op-amp circuit with both inputs connected to the same signal source.

6. Input offset voltage can be fully compensated at only one temperature. Therefore, when an operational amplifier is used at more than one temperature, the change in offset current will cause error in the output voltage.

7. The major source of error due to temperature is the change in I_{os} and V_{os} with temperature. The change in V_{os} is caused by mismatched $\Delta V_{BE}/\Delta T$ in bipolar transistors and

$\Delta V_{gs}/\Delta T$ and $\Delta_{GM}/\Delta T$ in FET transistors. The change in I_{os} with temperature is from mismatched $\Delta\beta/\Delta T$ in bipolar transistors and mismatched gate leakage change in FET transistors.

8. The maximum error due to temperature can be calculated for an inverting amplifier with the formula

$$E = \left(\frac{R_f + R_1}{R_1}\right)\left(\frac{\Delta V_{os}}{\Delta T}\right)\Delta T + R_f\left(\frac{\Delta I_{os}}{\Delta T}\right)\Delta T \tag{3-9}$$

where

E	= error in volts of the output voltage
$(\Delta V_{os}/\Delta T)$	= change in V_{os}/°C
$(\Delta I_{os}/\Delta T)$	= change in I_{os}/°C
ΔT	= change in temperature of interest

9. The error can be referred to the input, so that error can be expressed easily as a percentage of the input voltage by dividing the error expression by A_{fb}. The expression for error referred to the input (E_I) is:

$$E_I = \left(\frac{R_1 + R_f}{R_f}\right)\left(\frac{\Delta V_{os}}{\Delta T}\right)\Delta T + R_1\left(\frac{\Delta I_{os}}{\Delta T}\right)\Delta T \tag{3-10}$$

for an inverting amplifier, and

$$E_I = \left(\frac{\Delta V_{os}}{\Delta T}\right)\Delta T + R_f\left(\frac{\Delta I_{os}}{\Delta T}\right)\Delta T \tag{3-11}$$

for a noninverting amplifier.

10. Chopper-stabilized operational amplifiers will have offset voltages and offset voltage drift one to three orders of magnitude less than unstabilized op-amps. Chopper stabilization is a feedback process whereby any change in the op-amps offset is detected at the inverting input, changed to an ac voltage, amplified, changed back to dc, and applied to the non-inverting input so as to cancel the change in offset.

Varactor diode carrier amplifiers have very low bias currents and are used when very low voltages from high impedance sources must be amplified.

SELF-TEST QUESTIONS

3-1 State briefly the principle behind the bias current measuring circuit of Figure 3.1.

3-2 State how R_1 and R_2 are chosen for the bias current measurement circuit.

3-3 For an op-amp set up as in Figure 3.1 with $R_1 = R_2 = 1\ \text{M}\Omega$, we measure on the output:

□ S_1 and S_2 closed	+0.04 V
□ S_1 open	+0.1 V
□ S_2 open	−0.06 V

Calculate I_{B1}, I_{B2}, and I_{os}.

3-4 Write the definition of CMRR.

3-5 State briefly why common-mode gain is undesirable.

3-6 A noninverting amplifier has $R_1 = 20\ k\Omega$, $R_f = 1\ M\Omega$, $A_{ol} = 50{,}000$, and CMRR $= 100{,}000$. Calculate the actual closed loop gain.
3-7 Common-mode error is negligible in the inverting amplifier. State why.
3-8 For the circuit of Figure 3.5 $V_{out} = 0.5$ V and $V_{in} = 12$ V. Calculate CMRR as a ratio and in dB.
3-9 Name the two major factors that contribute to temperature drift in an op-amp.
3-10 An inverting amplifier set up with $R_1 = 15\ k\Omega$ and $R_f = 300\ k\Omega$ has a $|\Delta V_{os}/\Delta T| = 1$ mV/°C and $|\Delta I_{os}/\Delta T| = 0.8$ nA/°C. Find the error in the output if the temperature increases from 25 to 100°C.
3-11 Calculate the error referred to the input (E_I) for the amplifier of question 3-10.
3-12 State the major advantage of chopper-stabilized amplifiers.
3-13 Briefly describe the operation of the chopper channel of a chopper-stabilized amplifier in reducing drift.
3-14 Give the reason the stabilizer amplifier must have a high gain.

If you cannot answer certain questions, place a check next to them and review appropriate parts of the text to find the answers.

LABORATORY EXERCISE

Objective

After completing this laboratory, the student should be able to measure the bias current, input offset current, and CMRR of an amplifier. The student will also be able to calculate the maximum error in output voltage due to temperature drift and the actual drift from the experimental results on completing this experiment.

Equipment

1. Fairchild μA 741 operational amplifier or equivalent and manufacturer's specifications.
2. Assortment of 2% tolerance resistors.
3. Power supply, ± 15 V dc.
4. Voltmeter or oscilloscope capable of measuring millivolt level signals, dc and ac.
5. Signal generator.
6. Capacitor assortment.
7. Environmental oven or thermostat-controlled hot plate.
8. Pyrometer or other temperature-monitoring device.
9. Breadboard such as EL Instruments SK-10L mounted on vectorboard.

Procedure

1. Bias current and input offset current measurement.

Using the circuit shown in Figure 3.11, measure the bias current and the input offset current. Use the procedure outlined in the text. (The capacitors should be metalized mylar or better, but not electrolytic. Why?)

If the voltage readings are too small to read, R may have to be increased. If the voltages are greater than 10 V, then R should be reduced to about 1 MΩ.

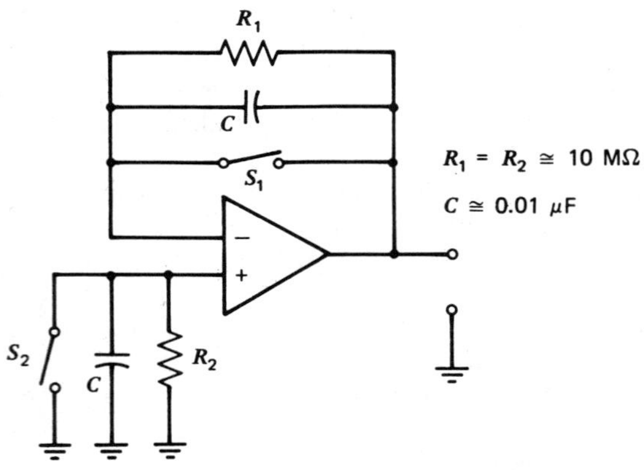

FIGURE 3.11 Circuit for I_B and I_{os} measurement.

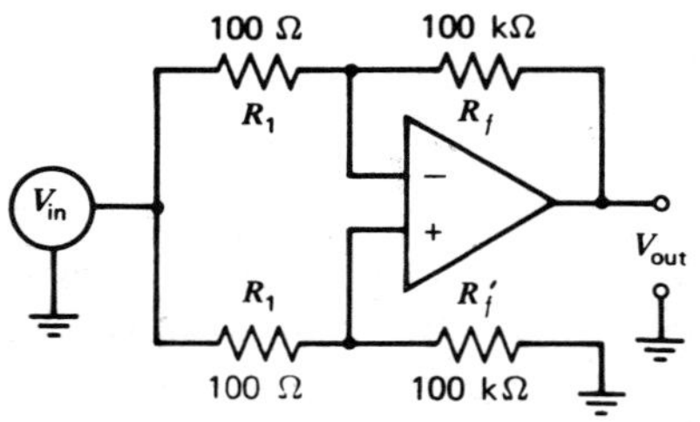

FIGURE 3.12 CMRR measurement circuit.

2. Measurement of CMRR and A_{cm}.

Construct the circuit shown in Figure 3.12, measure V_{in} and V_{out}, and do the calculation for CMRR. The output must not be clipped. V_{in} is an ac signal less than 5 Hz of about 10 V_p in magnitude. Recall:

$$\text{CMRR} = \left(\frac{V_{in}}{V_{out}}\right)\left(\frac{R_1 + R_f}{R_1}\right) \tag{3-7}$$

To measure the common-mode gain directly, set $R_1 = R_f$ in the circuit and use the relation $A_{cm} = V_{out}/V_{in}$.

Do your measured values of CMRR, I_B, and I_{os} meet those guaranteed by the manufacturers' specifications?

3. Measurement of error because of temperature change
   a. Build an inverter circuit of moderate gain ($A_{fb} \simeq 10$). If possible, construct the circuit so that the only component subjected to the temperature rise is the operational amplifier itself, so that the resistor drift does not contribute to the error voltage.
   b. Very carefully adjust the output voltage to zero, with the input grounded as shown in Figure 2.8 at room temperature.

c. Pick a temperature within the capability of your heating equipment (say 70°C) and slowly bring the temperature up, noting the output voltage (which will be the error voltage with zero volts input) every 5 or 10 degrees.

d. From the specifications of the amplifier, calculate the maximum error voltage, using Equation 3-9. Compare this to your measured values. Are the results different from those calculated? If so, why?

3. From the error voltage you measured, calculate the actual $\Delta V_{os}/\Delta T$. Recall that $(R_f \Delta I_{os})/\Delta T$ is part of the total $\Delta V_{os}/\Delta T$.

CHAPTER

4

FREQUENCY-RELATED CHARACTERISTICS

Even if we are going to use an amplifier with only dc signals, we must know under what circumstances it will be stable. If we are going to use the amplifier over a wide portion of its bandwidth, then of course we must know its behavior throughout the frequency range of interest. This chapter discusses the frequency-related characteristics of op-amps.

OBJECTIVES

After completing the study of this chapter and the self-test questions, the student should be able to:

1. List two factors that cause op-amps to be frequently dependent.
2. Draw the Bode plot of a three-stage dc coupled amplifier on semilog paper given the gain and cutoff frequency of each stage.
3. Use the gain-bandwidth product to find $f_{1_{fb}}$ at a given A_{fb} or A_{fb} at a given f_1.
4. State the factors that cause an op-amp to oscillate.
5. Calculate θ_{cl} given a complete Bode plot of A_{ol} or the information of objective 2.
6. Define slew rate (*S*) and calculate V_p(max) given operating frequency and *S* and minimum operating frequency at V_p given *S*.
7. Calculate the components necessary for lag compensation at a given A_{fb} given a Bode plot of an op-amp.
8. Define and state the operating principle of Miller-effect, input-stage, feedforward, and brute-force phase compensation.
9. Perform the laboratory for Chapter 4.

4-1 FREQUENCY RESPONSE

We know that no amplifier has the same voltage gain at all frequencies, although most amplifiers have a fairly constant voltage gain through some range of frequencies. The *frequency response* (change in voltage gain as frequency changes) of an operational amplifier has an important effect on the stability of practical op-amp circuits. Since the stages of almost all op-amps are dc coupled, they have no low frequency limit and we need only concern ourselves with the decrease in gain as frequency increases. Figure 4.1 shows the typical response of an amplifier and an op-amp. Frequency and gain are both plotted on a logarithmic scale to make the graphs more compact and readable. This is often done in op-amp specifications.

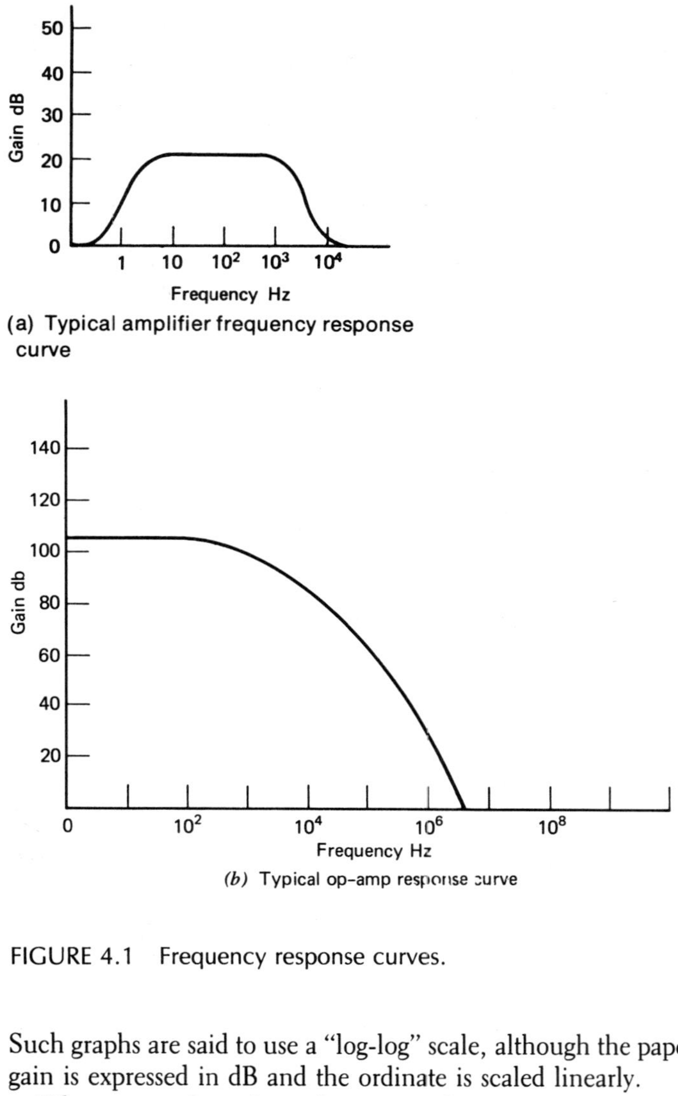

(a) Typical amplifier frequency response curve

(*b*) Typical op-amp response curve

FIGURE 4.1 Frequency response curves.

Such graphs are said to use a "log-log" scale, although the paper is semilog because gain is expressed in dB and the ordinate is scaled linearly.

What causes the gain to decrease as frequency increases in Figure 4.1*b*? If we refer to Figure 1.4, we notice that there are no capacitors in the circuit, but recall that in any circuit configuration—either integrated circuit or discrete component—there are metal leads separated from other metal leads by an insulator. This means that there will be some distributed stray capacitance due to wiring. Recall also that any semiconductor junction has some capacitance. As frequency increases, these stray capacitances shunt more and more of the ac signal to ground until at some frequency all the ac signal goes to ground through the stray capacitance and none reaches the load. For calculation these distributed stray capacitances can be lumped together as though they were a single capacitor, and each stage of the op-amp may be represented by an equivalent circuit consisting of a voltage source, a resistor, and the stray capacitance, as shown in Figure 4.2. Note that R_L could be the input resistance of the next stage or load resistance of the final stage.

As the frequency increases the reactance of the capacitor decreases, causing the parallel combination of R_L and C to become lower. There will be some frequency above which the voltage across the parallel combination of R_L and C will be less than $A_{oe}v_i$. The expression for the gain at any frequency (A) is

$$A = \frac{A_{ol}}{1 + j\dfrac{f}{f_1}}$$

where

A_{ol} = low frequency open loop gain
f = frequency of operation
f_1 = corner or 3 dB frequency, where A is 3 dB under A_{ol} or 0.707 A_{ol}

As long as $R_L >> R_{out}$, which is usually the case,

$$f_1 = \frac{1}{2\pi RC}$$

where

R = R_{out} of the amplifier
C = stray capacitance of the circuit plus junction capacitance

The frequency-dependent voltage gain equation is usually expressed in the rationalized form

$$|A| = \frac{A_{ol}}{\sqrt{1 + \left(\dfrac{f}{f_1}\right)^2}} \angle -\arctan\left(\frac{f}{f_1}\right) \tag{4-1}$$

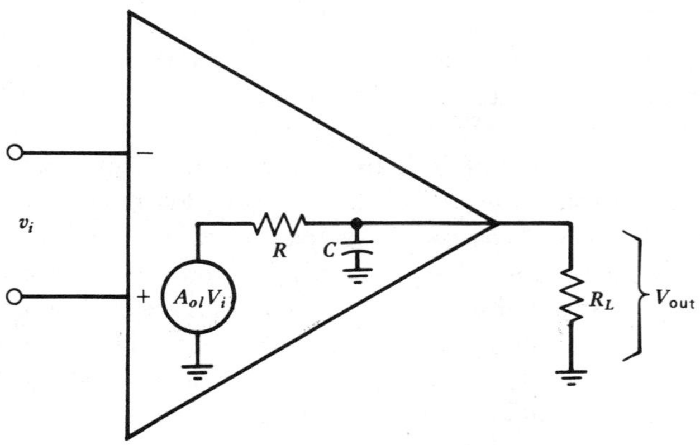

FIGURE 4.2 Equivalent circuit for frequency response.

We should note again that f is a variable that represents the frequency of interest and f_1 is a fixed frequency called the *corner frequency*. Recall that corner frequency, f_1, is defined as the frequency at which the amplifier voltage gain drops to 0.707 of the low frequency voltage gain, or −3 dB under A_{ol}. We can see that as frequency increases, the denominator of equation 4-1 increases so that the voltage gain decreases.

We also see from equation 4-1 that the phase of the output signal shifts with respect to the input with frequency and that the magnitude of the phase shift is arctan (f/f_1). The sign of the angle is negative since the output lags the input in phase. To see this, recall that a signal does not pass through an amplifier instantaneously, but is delayed by some small time due to the active device and other capacitances. Even if the delay time is constant for different frequencies, the delay time becomes a larger part of the signal period as frequency increases, and this therefore corresponds to a larger phase shift at high frequencies than at low frequencies.

Equation 4-1 is developed in Appendix D.

4-2 DECIBEL REVIEW

We noted in Section 4-1 that a logarithmic scale is used in drawing frequency response curves. A frequently used logarithmic scale used in dealing with gain is the decibel. The definition of voltage gain in decibels is:

$$\text{Voltage gain in decibels} = A\ (\text{dB}) = 20 \log_{10} A$$

where A = numerical voltage gain.

EXAMPLE 4-1

Find the voltage gain in dB of an amplifier with a numerical gain of 10.

$$A(\text{dB}) = 20 \log_{10} 10 = 20(1) = 20 \text{ dB}$$

TABLE 4.1

VOLTAGE GAIN AND EQUIVALENT DECIBEL (dB) GAIN

A	A(dB)	A	A(dB)
0.001	−60	10	20
0.01	−40	100	40
0.1	−20	1000	60
1	0	10,000	80

Henceforth we will use log for a base 10 logarithm and ln to represent a natural logarithm (base e).

Table 4.1 shows numerical voltage gains and their equivalent decibel value. Note that if the gain is less than one, the dB gain is negative.

Let us now express Equation 4-1 in decibels using the rules of logarithms

$$A(dB) = 20 \log A_{ol} - 20 \log \left[1 + \left(\frac{f}{f_1}\right)^2\right]^{1/2} \tag{4-2}$$

Now we will look at three different frequencies:

CASE 1
$f << f_1$

CASE 2
$f = f_1$

CASE 3
$f >> f_1$

CASE 1
$f << f_1$; f well below the corner frequency.

$$A(dB) = 20 \log A_{ol} - 20 \log \left[1 + \left(\frac{f}{f_1}\right)^2\right]^{1/2}$$

but $f << f_1$, so $\left(\frac{f}{f_1}\right)^2 \simeq 0$, therefore

$$\left[1 + \left(\frac{f}{f_1}\right)^2\right]^{1/2} \simeq 1 \qquad \text{and} \qquad 20 \log 1 = 0$$

Hence $A(dB) \simeq 20 \log A_{ol}$ when $f << f_1$.

CASE 2
$f = f_1$

$$\begin{aligned} A(dB) &= 20 \log A_{ol} - 20 \log \left[1 + \left(\frac{f}{f_1}\right)^2\right]^{1/2} \\ &= 20 \log A_{ol} - 20 \log \sqrt{2} \\ &= 20 \log A_{ol} - 20\,(0.15) \\ &= A_{ol}(dB) - 3\text{ dB} \end{aligned}$$

so when $f = f_1$, the gain has dropped 3 dB from the low frequency gain and the phase shift $\theta = -45°$ (since arctan 1 = 45°).

CASE 3
$f >> f_1$; f well above the corner frequency.

$$A(\text{dB}) = 20 \log A_{ol} - 20 \log \left[1 + \left(\frac{f}{f_1} \right)^2 \right]^{1/2}$$

When $f >> f_1$, $\left(\frac{f}{f_1}\right)^2 >> 1$, so $\left[1 + \left(\frac{f}{f_1} \right)^2 \right]^{1/2} \simeq \frac{f}{f_1}$ and we write

$$A(\text{dB}) = 20 \log A_{ol} - 20 \log \left(\frac{f}{f_1} \right)$$

which we see approaches zero as $f \rightarrow f_c$.

If we draw a curve of A(dB) versus frequency and phase shift θ versus frequency, we get a pair of curves as in Figure 4.3. We notice the phase angle is $-45°$ at f_1

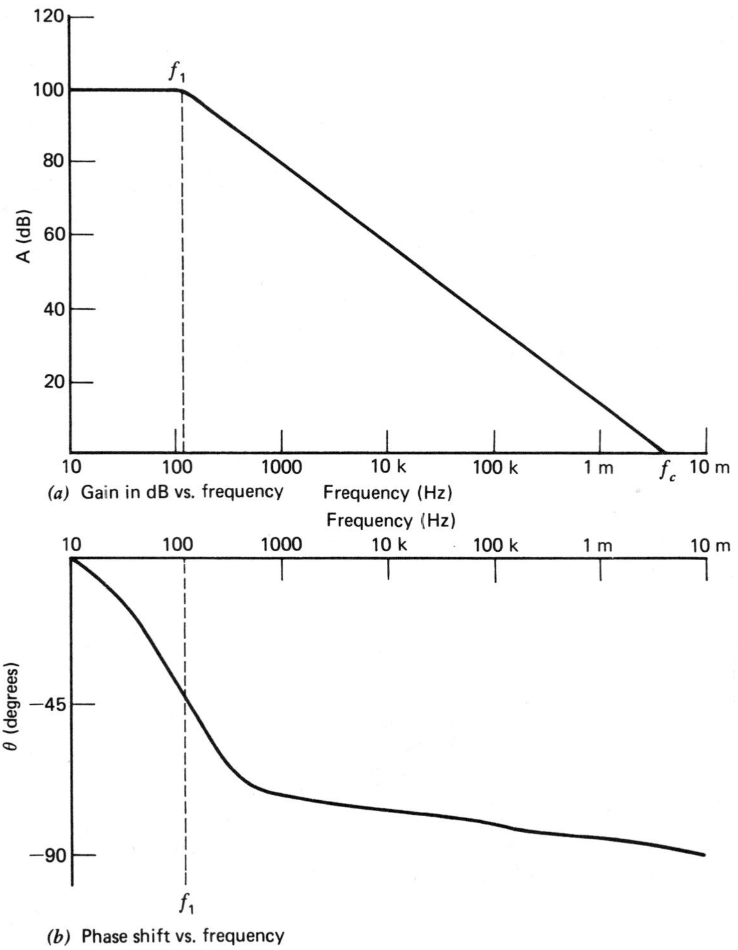

FIGURE 4.3 Frequency and phase response curves.

(the 3 dB point) and approaches 90° at the crossing frequency (f_c), which is the frequency at which the gain of the amplifier is 1, or 0 dB.

4-3 ROLL-OFF RATE

We now consider the rate at which the gain decreases with frequencies above f_1 in our equivalent circuit in Figure 4.4.

The decrease in gain with frequency is called *roll off*. Roll off is expressed as decibels per octave or decibels per decade. An octave is the doubling or halving of frequency. For example, if the frequency changes from 500 to 1000 Hz, it has increased one octave. A decade is a tenfold increase or decrease in frequency. If the frequency increases from 100 to 1000 Hz, it has increased one decade.

Let us see what happens to the gain of the circuit in Figure 4.4 as the frequency goes from f_a to f_b where f_a and $f_b > f_1$, and $f_b > f_a$. We can express the change in gain as:

$$\begin{aligned}\Delta A(\text{dB}) &= A(\text{dB}) \text{ at } f_b - A(\text{dB}) \text{ at } f_a \\ &= A_{ol}(\text{dB}) - 20 \log\left(\frac{f_b}{f_1}\right) - \left[A_{ol}(\text{dB}) - 20 \log\left(\frac{f_a}{f_1}\right)\right] \\ &= A_{ol}(\text{dB}) - 20 \log\left(\frac{f_b}{f_1}\right) - A_{ol}(\text{dB}) + 20 \log\left(\frac{f_a}{f_1}\right) \\ &= 20 \log\left(\frac{f_a}{f_1}\right) - 20 \log\left(\frac{f_b}{f_1}\right) = 20 \log\left(\frac{f_a}{f_b}\right)\end{aligned}$$

If f_b is 10 times f_a, we see

$$\Delta A(\text{dB}) = 20 \log (1/10) = 20 \log (0.1) = -20 \text{ dB}$$

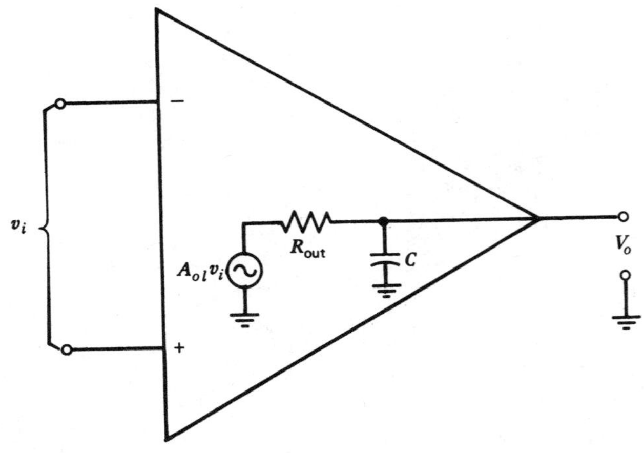

FIGURE 4.4 Equivalent circuit for frequency response and roll off.

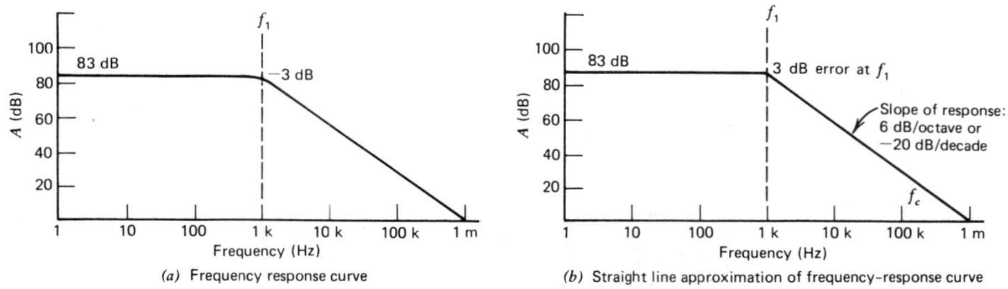

FIGURE 4.5 Frequency response curves.

So the roll-off rate of the equivalent circuit of Figure 4.4 is 20 db/decade.
If f_b is twice f_a, then

$$\Delta A(\text{dB}) = 20 \log (1/2) = -6 \text{ dB}$$

The equivalent circuit of Figure 4.4 has a roll off of −6 dB/octave, which is the same roll-off rate as 20 dB/decade.

This is also the natural roll-off rate of a simple *RC* series low-pass network. Figure 4.5 illustrates the frequency response curves of an amplifier with a 6 dB/octave or 20 dB/decade roll-off rate. Note in Figure 4.5 that at $f = 1$ kHz, A(dB) = 80 dB and at $f = 10$ kHz, A(dB) = 60 dB; thus, in one decade the drop in gain is −20 dB.

4-4 MULTISTAGE ROLL-OFF RATE

Most operational amplifiers are made up of two or more cascaded stages, each of which has a roll-off rate of 6 dB/octave. The roll off of the composite amplifier is more complicated than the preceding example might suggest. Let us use the straight-line approximation of the frequency response curve to analyze this situation. A plot of this sort, log gain versus log frequency, is called a *Bode plot*.

Consider the three-stage amplifier in Figure 4.6. The frequency response of each amplifier is plotted in Figure 4.7.

The gain of the overall amplifier is the product of the gain of each individual stage. The gain of each stage is:

$$A_{v_1} = \frac{A_{l_1}}{\left[1 + \left(\frac{f}{f_{1_1}}\right)^2\right]^{1/2}} \angle -\arctan\left(\frac{f}{f_{11}}\right)$$

$$A_{v_2} = \frac{A_{l_2}}{\left[1 + \left(\frac{f}{f_{1_2}}\right)^2\right]^{1/2}} \angle -\arctan\left(\frac{f}{f_{12}}\right)$$

$$A_{v_3} = \frac{A_{l_3}}{\left[1 + \left(\frac{f}{f_{1_3}}\right)^2\right]^{1/2}} \angle -\arctan\left(\frac{f}{f_{13}}\right)$$

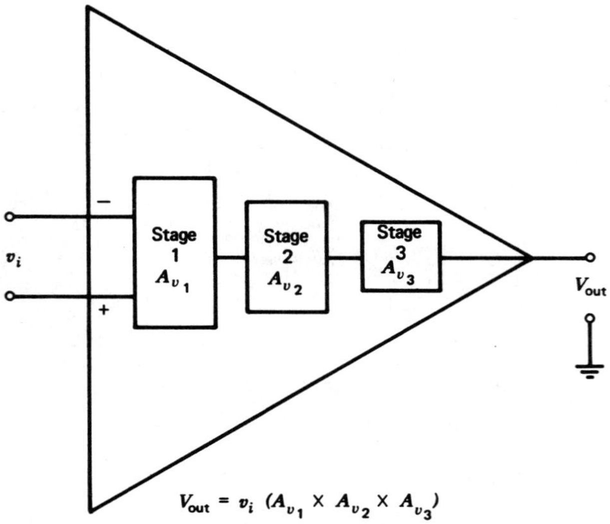

FIGURE 4.6 Three-stage operational amplifier.

where

f_{1_1} = f_1 of amplifier stage 1
f_{1_2} = f_1 of amplifier stage 2
f_{1_3} = f_1 of amplifier stage 3
A_{l_1} = low frequency gain of stage 1
A_{l_2} = low frequency gain of stage 2
A_{l_3} = low frequency gain of stage 3

The total gain, including the effect of frequency, is

$$A_T = \frac{A_{l_1}A_{l_2}A_{l_3}}{\left[1+\left(\frac{f}{f_{1_1}}\right)^2\right]^{1/2}\left[1+\left(\frac{f}{f_{1_2}}\right)^2\right]^{1/2}\left[1+\left(\frac{f}{f_{1_3}}\right)^2\right]^{1/2}} \angle -\arctan\left(\frac{f}{f_{1_1}}\right) - \arctan\left(\frac{f}{f_{1_2}}\right) - \arctan\left(\frac{f}{f_{1_3}}\right) \tag{4-3}$$

Equation 4-3 is unwieldy to handle, so in order to see the overall frequency response we will use Bode plots—making use of the fact that decibels are logarithmic functions so we can add decibel gains. This is equivalent to multiplying gain ratios. To find the amplifiers overall response, we draw the response to each stage on the same graph and add graphically the gains point by point.

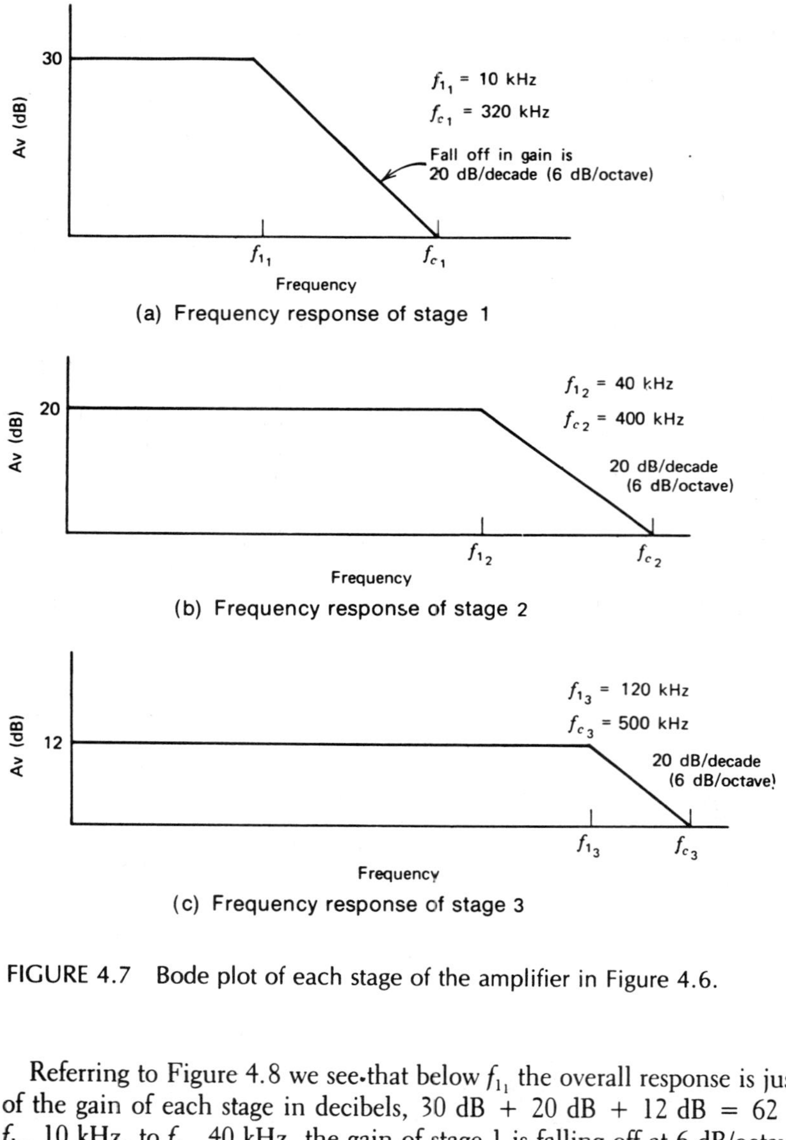

FIGURE 4.7 Bode plot of each stage of the amplifier in Figure 4.6.

Referring to Figure 4.8 we see that below f_{1_1} the overall response is just the sum of the gain of each stage in decibels, 30 dB + 20 dB + 12 dB = 62 dB. From f_{1_1}, 10 kHz, to f_{1_2}, 40 kHz, the gain of stage 1 is falling off at 6 dB/octave whereas the gain of stages 2 and 3 remain constant. The overall gain, which is equal to the point-by-point sum of the three-stage gains on the Bode plot, falls off at 6 dB/octave. From f_{1_2}, 40 kHz, to f_{1_3}, 120 kHz, both stage 1 and stage 2 gains are falling off at 6 dB/octave, whereas the stage 3 gain remains constant at 12 dB. Therefore, the overall response falls of at 12 dB/octave between f_{1_1} and f_{1_2}. From f_{1_3}, 120 kHz, to f_c of the overall amplifier, 390 kHz, all three stages have a roll-off rate of 6 dB/octave; therefore the roll-off rate of the three-stage amplifier is 18 dB/octave.

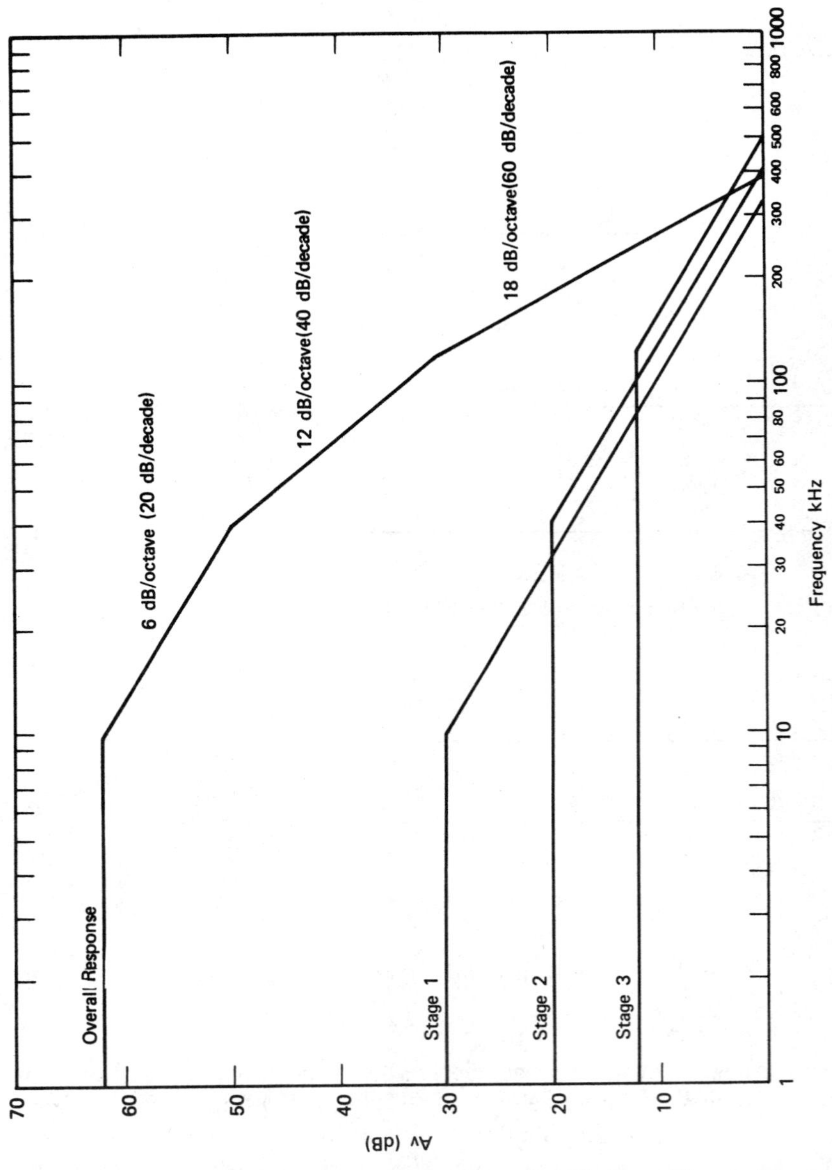

FIGURE 4.8 Overall frequency response of the cascaded amplifiers in Figure 4.7.

This method of analysis can be used for any multistage amplifier where the response of the individual stages is known.

4-4.1 Phase Response

We know that as a signal passes through an amplifier it is delayed in time and that this delay time corresponds to some lagging phase angle for any particular frequency. This phase angle is given as part of the gain equation in the form $-\arctan\left(\frac{f}{f_1}\right)$

We see from Figure 4.3 that for a single stage the lag is 90° at f_c. In a multistage amplifier each stage adds a further delay causing an increased phase lag. From the total gain equation (4.3), we see the total phase lag is expressed as

$$-\theta_T = -\arctan\frac{f}{f_{1_1}} - \arctan\frac{f}{f_{1_2}} - \arctan\frac{f}{f_{1_3}} \tag{4-4}$$

for our example three-stage amplifier.

We see from this expression that the maximum phase lag is −180° for two stages and −270° for three stages. The phase lag of a stage is less than 45° until the corner frequency where $f = f_1$ (arctan 1 = 45°), becoming larger as f increases above f_1, approaching the maximum (−90°) at f_c.

4-4.2 Frequency Response with Feedback

So far we have discussed only the frequency response of an operational amplifier with no feedback. Now we wish to look at the effect of feedback on the frequency response.

The addition of negative feedback, such as in an inverting or noninverting amplifier, increases the effective bandwidth of an operational amplifier.

To verify this, consider the expression for open loop gain in an amplifier with a 6 dB/octave roll off.

$$A = \frac{A_{ol}}{1 + j\frac{f}{f_1}} \tag{4-5}$$

where

A = open loop gain at f
A_{ol} = low frequency open loop gain
f_1 = the corner frequency

With the addition of feedback

$$A_{fb} = \frac{A_{ol}}{1 + A_{ol}\beta}$$

When we substitute equation 4-5 into this expression we find

$$A_{fb} = \left(\frac{A_{ol}}{1 + j\dfrac{f}{f_1}}\right) \Big/ \left(1 + \frac{\beta A_{ol}}{1 + j\dfrac{f}{f_1}}\right)$$

$$= \frac{A_{ol}}{1 + \beta A_{ol} + j\dfrac{f}{f_1}}$$

$$= \left(\frac{A_{ol}}{1 + A_{ol}\beta}\right) \Big/ \left(1 + j\frac{f}{f_1(1 + A_{ol}\beta)}\right) \qquad (4\text{-}6)$$

This equation is of the form

$$A_{fb} = \frac{A_{fb_l}}{1 + j\dfrac{f}{f_{1_{fb}}}} \qquad (4\text{-}7)$$

where

$$f_{1_{fb}} = f_1(1 + A_{ol}\beta)$$

A_{fb_l} = low frequency closed loop gain
$f_{1_{fb}}$ = closed loop corner frequency

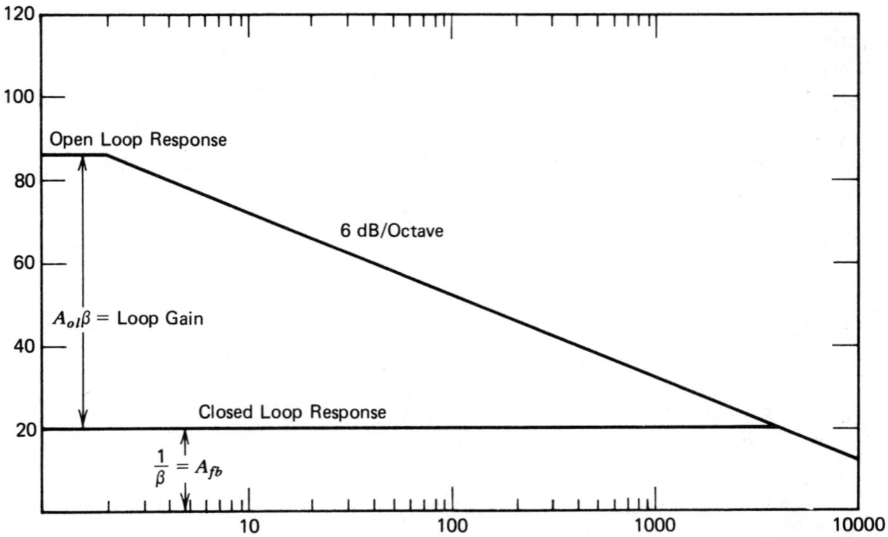

FIGURE 4.9 Effect of feedback on bandwidth.

Equation 4-7 shows that $f_{1_{fb}} = f_1\,(1 + A_{ol}\beta)$, or the corner frequency with feedback is equal to the corner frequency without feedback multiplied by the loop gain plus one. Since $(1 + A_{ol}\beta) > 1$ usually with feedback, $f_{1_{fb}} > f_1$, thus the effective bandwidth is increased by the use of feedback.

To see this, refer to Figure 4.9. Here we have an op-amp with an open loop gain of 86 dB and closed loop gain of 20 dB. The corner frequency of the op-amp with no feedback is 2 kHz. Since the closed-loop gain is limited to 20 dB, even at frequencies well above 2 kHz, the open loop gain is much larger than the closed loop gain, so the actual gain described by Equation 4-6 is almost the low frequency closed loop gain. When the frequency reaches 3.8 MHz, the closed loop gain and open loop gain are the same. The closed loop gain can be no larger than the open loop gain, so the amplifier with feedback now has the same roll-off rate as the amplifier would have without feedback. Note that the frequency response of the op-amp has not increased, but that by limiting the gain with feedback we do not run out of gain until a much higher frequency, thus increasing the effective bandwidth of the amplifier with feedback. In this case, we increased from $f_1 = 2$ kHz to $f_{1_{fb}} = 3.8$ MHz, a very large increase in bandwidth, but this bandwidth increase required a decrease in gain from 86 dB to 20 dB. Generally we must give up gain to increase bandwidth with the use of feedback.

Recall from chapter 2 that $A_{ol}\beta$ is called the loop gain. In Figure 4.9 we see the loop gain is the difference in dB of the open loop gain and ideal closed loop gain,

$$(A_{ol}\beta)(\text{dB}) = A_{ol}(\text{dB}) - (1/\beta)\,(\text{dB}) \tag{4-8}$$

where A_{ol} = open loop gain.

This is a convenient way to find the loop gain for use in calculation of $f_{1_{fb}}$. To verify this relationship note that

$$\frac{\text{Open loop gain}}{\text{Ideal closed loop gain}} = \frac{A_{ol}}{\frac{1}{\beta}} = A_{ol}\beta = \text{loop gain}$$

Therefore

$$A_{ol}\beta(\text{dB}) = \frac{A_{ol}(\text{dB})}{(1/\beta)(\text{dB})} = A_{ol}(\text{dB}) - (1/\beta)(\text{db})$$

using the rules of logarithms. Observe the loop gain increased as closed loop gain decreases.

EXAMPLE 4-2

To verify our graphic results from Figure 4.9, where $A_l = 86$ dB, $A_{fb} = 20$ dB, and $f_1 = 2$ kHz, we will calculate $f_{1_{fb}}$.

SOLUTION

$$A_{ol} = 86 \text{ dB} = 19{,}953$$
$$A_{ol}\beta = 166 \text{ dB} = 1995.3$$

so $f_{1_{fb}} = f_1 (1 + A_l\beta) = 2 \text{ kHz } 1996 = 3.999 \text{ MHz}$, which is in good agreement with the graph.

EXAMPLE 4-3

Suppose we wish to know the maximum closed loop gain we can have for a $f_{1_{fb}}$ of 300 kHz. Referring to Figure 4.9, we note the gain at which 300 kHz intersects the overall frequency response to the amplifier. We see this is 43 dB, or A_{fb} = 141, expressed as a ratio.

We also could have calculated this by using the relation

$$f_{1_{fb}} = f_1 (1 + A_{ol}\beta),$$

and solving for β, which gives us

$$\beta = \frac{f_{1fb} - f_1}{f_1 A_{ol}} = \frac{298 \text{ kHz}}{39.9 \text{ MHz}} = 0.00747$$
$$A_{fb} = \frac{1}{\beta} = \frac{1}{0.00747} = 134$$

which is close to the answer we obtained using the Bode plot.

4-4.3 Gain-Bandwidth Product

If the roll-off rate of an amplifier is 6 dB/octave, the gain-bandwidth product is a constant,

$$Af_1 = K$$

To verify this, let us multiply the ideal low frequency gain times the upper cutoff frequency of that amplifier with feedback. This yields the gain-bandwidth product

$$A_{fb} f_{1_{fb}} = \frac{A_{ol}}{1 + A_{ol}\beta} f_1(1 + A_{ol}\beta) = A_{ol} f_1 \tag{4-9}$$

where

A_{ol} = low frequency open loop gain

We found in the previous section that to increase bandwidth by using feedback, we must reduce gain. We now have a tool in Equation 4-9 to tell us how much gain we must give up for a certain desired bandwidth.

EXAMPLE 4-4

An op-amp has an open loop gain of 10,000 and an upper cutoff frequency of 400 Hz. Calculate the closed loop gain possible at an upper cutoff frequency of 150 kHz with feedback.

SOLUTION

$$A_{ol} f_1 = A_{fb} f_{1_{fb}} \tag{4-9}$$

So

$$A_{fb} = \frac{A_{ol} f}{f_{1_{fb}}} = \frac{(10^4)(400 \text{ Hz})}{(15 \times 10^4 \text{ Hz})}$$
$$= 26.6$$

Note that we could have obtained this same answer from a Bode plot of the op-amp frequency repsonse, but we would have had to convert gain from decibels.

It is important to understand that the linear relationship of the gain-bandwidth product is true only for a 6 dB/octave roll-off rate.

4-4.4 Oscillation

An op-amp is usually used with feedback. The advantages of feedback for an amplifier are: an increase in R_{in}, a decrease in R_{out}, reduced distortion, increased stability, and more precise gain settings, if the feedback is negative. Should the feedback become positive, the amplifier will become an oscillator and all the properties reversed. To cause oscillations, the feedback signal must be large enough so that the signal returned to the input is larger than the input signal and in phase

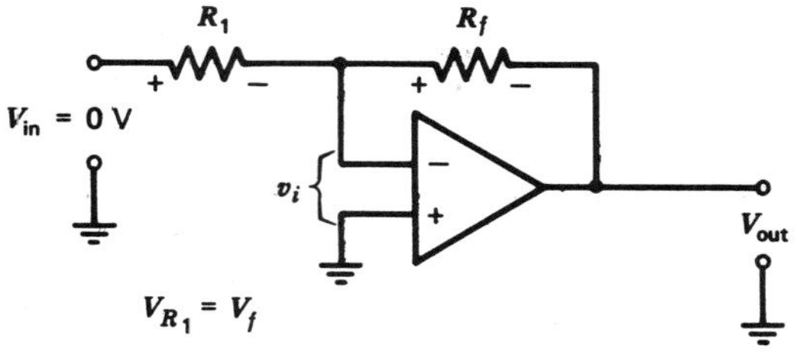

FIGURE 4.10 Inverting amplifier with feedback.

with the input signal (phase shift—360°). Under positive feedback conditions, noise at the input is enough to start oscillations.

Another way of stating that the voltage fed back must be larger than the input voltage is that the loop gain is greater than one. Referring to Figure 4.10, we see that any voltage fed back to the inverting terminal with no input voltage is of the same phase as the output and 180° out of phase with v_i (an offset voltage or noise) that caused the input with $V_{in} = 0$. For oscillations to occur, V_f must be larger than v_i,

$$V_f > v_i$$

Recall that $V_f = \beta V_o = A_{ol}\beta v_i$, so

$$\beta A_{ol} v_i > v_i$$

Therefore

$$\beta A_{ol} > 1$$

for oscillations to occur. A loop gain greater than one is not, by itself, enough to cause oscillations. To produce oscillations, the voltage fed back must also be in phase with the input (which was not the case in our example). This criteria can be expressed by saying that for oscillation to occur, the *loop gain* must be greater than one with a phase shift of 180°, that is,

$$A_{ol}\beta > 1 \angle 180° \qquad \text{at the loop gain crossing frequency, } f_{cl}$$

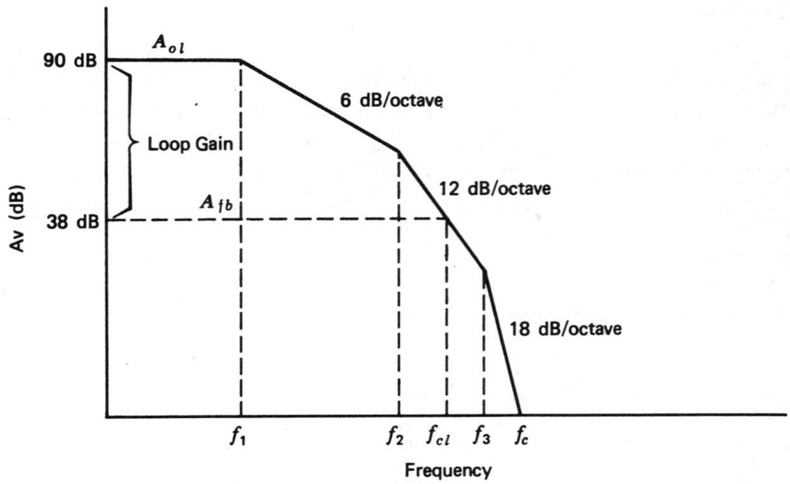

FIGURE 4.11 Loop gain crossing frequency.

The loop gain crossing frequency is that frequency at which the closed loop gain is equal to the open loop gain. For example, in Figure 4.11 the loop gain crossing frequency (f_{cl}) is the frequency at which the closed loop gain line (A_{fb} = 38 dB) intersects the open loop response curve. This point is marked f_{cl} on the graph. f_1, f_2, and f_3 are the upper cutoff frequencies of stages in the op-amp.

Another way of looking at the phase shift is as follows. An amplifier such as the one in Figure 4.10 will oscillate if the loop gain is larger than one and any output is fed back in phase with the input. Recall that it takes a finite time for a signal to propagate through an amplifier because of the various capacitances of the components. This delay time is fairly constant with frequency. If the delay is 1 μs, then the delay represents a half cycle or 180° of a 500 kHz signal. If the amplifier inversion provides 180° and the propagation delay provides 180° of phase shift, the total phase shift is 360° and the amplifier will oscillate at the frequency for which the propagation delay equals half a period. For a delay of 0.25 μs, that frequency will be 2 MHz. If the frequency response of the amplifier is such that it cannot amplify 2 MHz, then oscillations will not occur, so oscillation will occur only if the frequency at which the propagation delay equals one half a period is within the op-amp bandwidth.

We have previously expressed the loop gain in terms of the low frequency open loop gain, but now we must express it in terms of the frequency-dependent open loop gain, which we will call A. Therefore

$$
\begin{aligned}
A\beta &= \left[\frac{A_{ol}}{\sqrt{1 + \left(\frac{f}{f_1}\right)^2}} \angle -\arctan\left(\frac{f}{f_1}\right) \right] \beta \\
&= \frac{A_{ol}\beta}{\left[1 + \left(\frac{f}{f_1}\right)^2\right]^{1/2}} \angle -\arctan\left(\frac{f}{f_1}\right)
\end{aligned}
\tag{4-10}
$$

for an amplifier with a 6 dB/octave roll-off rate such as in Figure 4.4. Since β is a resistor ratio ($1/\beta = A_{fb}$), it introduces no significant phase shift, so the amplifier is the only source of phase shift. We can see from the expression of Equation 4-10 that the maximum phase shift that can occur in an amplifier with a 6 dB/octave roll-off rate is 90°. This means that if the roll-off rate is 6 dB/octave at f_{cl}, the amplifier will not oscillate no matter what the loop gain is. The loop gain must have a phase shift of 180° for oscillations to occur.

Not all amplifiers, however, have a roll-off rate of 6 dB/octave. We have seen one example (see Figures 4.6 and 4.8) where the roll-off rate is 6 dB/octave at some frequencies, 12 dB/octave at others, and 18 dB/octave at others. What about these cases?

Where the roll-off rate is 12 dB/octave, the loop gain would have two frequency-dependent terms (between f_{1_2} and f_{1_3} in Figure 4.8) and be written for the amplifier of Figures 4.6 and 4.8 as

$$A\beta = \frac{A_{l_1}A_{l_2}A_{l_3}\beta}{\left[1+\left(\frac{f}{f_{1_1}}\right)^2\right]^{1/2}\left[1+\left(\frac{f}{f_{1_2}}\right)^2\right]^{1/2}}$$

$$\angle -\arctan\left(\frac{f}{f_{1_1}}\right) - \arctan\left(\frac{f}{f_{1_2}}\right)$$

We see now that the maximum phase shift available between these two frequencies is 180°. The relative values of these cutoff frequencies, f_{1_1} and f_{1_2}, and f_{cl} determine if the phase shift reaches 180°. With a roll-off rate of 12 dB/octave at f_{cl} the amplifier may or may not oscillate, so this is a conditionally stable condition. Due to circuit conditions that may add to the phase shift (such as lead capacitance), it is best to avoid operating with a 12 dB/octave roll off at f_{cl}.

Referring once again to the amplifier in Figure 4.6 and its response curve in Figure 4.8, we see that above f_{1_3} the roll-off rate is 18 dB/octave. Between f_{1_3} and f_c the loop gain is

$$A\beta = \frac{A_{l_1}A_{l_2}A_{l_3}\beta}{\left[1+\left(\frac{f}{f_{1_1}}\right)^2\right]^{1/2}\left[1+\left(\frac{f}{f_{1_2}}\right)^2\right]^{1/2}\left[1+\left(\frac{f}{f_{1_3}}\right)^2\right]^{1/2}} \qquad (4\text{-}11)$$

$$\angle -\arctan\left(\frac{f}{f_{1_1}}\right) - \arctan\left(\frac{f}{f_{1_2}}\right) - \arctan\left(\frac{f}{f_{1_3}}\right)$$

The maximum possible phase shift in this frequency range is to 270°. The circuit will almost certainly oscillate if operated where the roll-off rate is 18 dB/octave at f_{cl}.

4-4.5 Stability Criteria

We need a reliable, easy-to-use method to determine if a given amplifier will be stable in operation. We are using stable operation to mean without oscillations in this context.

Since most op-amp applications use loop gains greater than one, we will develop our stability criteria in terms of the phase angle of the loop gain. We will consider three closed loop gain conditions for the amplifier whose response is shown in Figure 4.12:

1. f_{cl} on the 6 dB/octave slope.
2. f_{cl} on the 12 dB/octave slope.
3. f_{cl} on the 18 dB/octave slope.

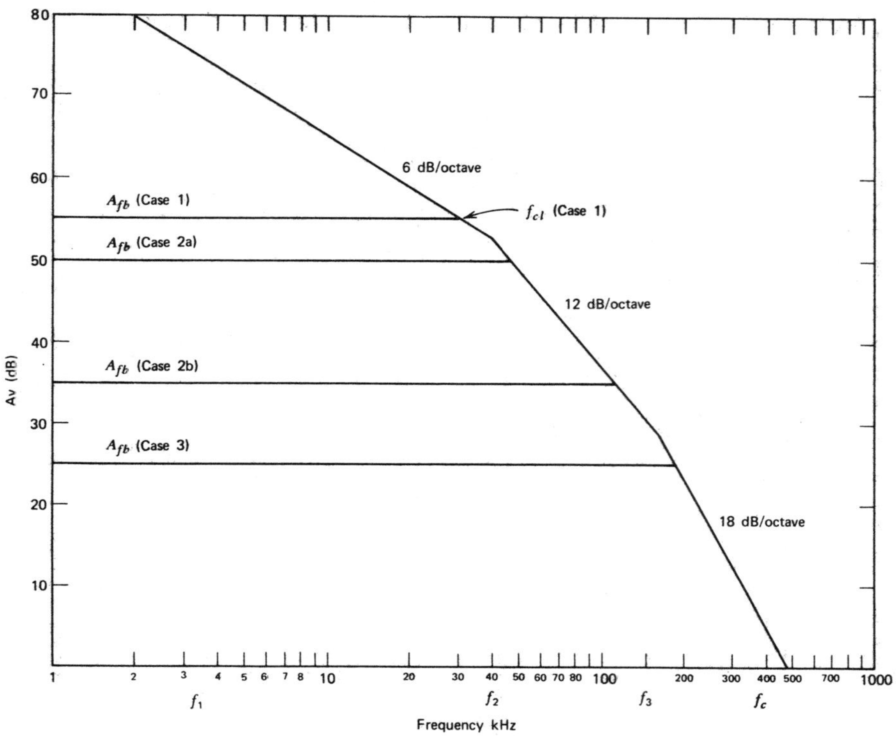

FIGURE 4.12 Frequency response of op-amp.

We will use the phase-angle information of the generalized loop gain equation, Equation 4-11, to calculate the phase angle of the loop gain at f_{cl}. We will call this quantity θ_{cl}. We will not calculate the magnitude of the loop gain since we can read it on the Bode plot, and it will be larger than one in all three cases.

EXAMPLE 4-5

In Figure 4.12, f_1 of the operational amplifier is 2 kHz. f_2 and f_3 are the break frequencies of some stages in the amplifier, but for our application we do not care which stages. Note $f_2 = 40$ kHz, $f_3 = 160$ kHz, and $f_c = 480$ kHz. Determine if the op-amp is stable by examining θ_{cl} at the following closed loop gains.

CASE 1

In case 1 we see that the closed loop gain is 55 dB, the loop gain is 24 dB, which is greater than 1, and $f_{cl} = f_{1_{fb}} = 31$ kHz. Note that f_{cl} is very close to f_2 where the roll-off rate becomes 12 dB/octave. Calculating the loop gain phase shift we see

$$\theta_{cl} = -\tan^{-1}\left(\frac{31\text{ kHz}}{2\text{ kHz}}\right) - \tan^{-1}\left(\frac{31\text{ kHz}}{40\text{ kHz}}\right) - \tan^{-1}\left(\frac{31\text{ kHz}}{160\text{ kHz}}\right)$$

$$\theta_{cl} = -86.3° - 37.8° - 11° = -135.1° < 180°$$

The phase margin for stability is 180° − 135.1° = 44.90°. A positive phase margin combined with $A_{ol}\beta > 1$ ensures stability. Thus the amplifier is stable. It is generally true that if the closed loop gain intersects the frequency response where the roll-off rate is 6 dB/octave, the amplifier is stable. Now we can use the Bode plot to find the closed loop gains at which the amplifier is stable, without the need for calculation. For example, the amplifier whose response is shown in Figure 4.12 will be stable at any closed loop gain between 53 dB and 79 dB.

CASE 2

We will look at two closed loop gains for Case 2:

a. f_{cl} just above f_2.
b. f_{cl} closer to f_3.

CASE 2a

A_{fb} = 50 dB, loop gain equals 29 dB, which is greater than one, and f_{cl} = 43.5 kHz. Calculating θ_{cl} we find

$$\theta_{cl} = -\tan^{-1}\left(\frac{43.5\text{ kHz}}{2\text{ kHz}}\right) - \tan^{-1}\left(\frac{43.5\text{ kHz}}{40\text{ kHz}}\right) - \tan^{-1}\left(\frac{43.5\text{ kHz}}{160\text{ kHz}}\right)$$
$$= -87.37° - 47.4° - 15.2° = -149.96°$$

This is a stable condition since $\theta_{cl} < 180°$ and the phase margin is positive.

CASE 2b

$$A_{fb} = 35\text{ dB},\ A_{ol}\beta = 44\text{ dB} >> 1,\text{ and } f_{cl} = 115\text{ kHz}$$
$$\theta_{cl} = -\tan^{-1}\left(\frac{115\text{ kHz}}{22\text{ kHz}}\right) - \tan^{-1}\left(\frac{115\text{ kHz}}{40\text{ kHz}}\right) - \tan^{-1}\left(\frac{115\text{ kHz}}{160\text{ kHz}}\right)$$
$$\theta_{cl} = -89° - 71.8° - 35.8° = -196.6°$$

The phase margin (180° − 196.6° = − 16.6°) is now negative. Since $\theta_{cl} > 180°$, the amplifier will oscillate if used at A_{fb} = 35 dB.

We see that when the closed loop gain intersects the frequency response of the operational amplifier where the operational amplifier gain is decreasing at a 12 dB/octave rate, the amplifier may be stable, or it may not. We must calculate θ_{cl} for each A_{fb} that we use in this region. It is best to avoid operating where the roll-off rate is 12 dB/octave.

CASE 3

$$A_{ol} = 25\text{ dB},\ A_{ol}\beta = 54\text{ dB} >> 1,\ f_{cl} = 190\text{ kHz}$$

Now

$$\theta_{cl} = -\tan^{-1}\left(\frac{190\text{ kHz}}{2\text{ kHz}}\right) - \tan^{-1}\left(\frac{190\text{ kHz}}{40\text{ kHz}}\right) - \tan^{-1}\left(\frac{190\text{ kHz}}{160\text{ kHz}}\right)$$
$$= -89.4° - 78.3° - 49.9° = -217°$$

The phase margin is now $180° - 217° = -37.6°$.
When f_{cl} occurs where the amplifier roll off is 18 dB/octave, $|\theta_{cl}| > 180°$ and will clearly oscillate. This condition is always to be avoided.

We may now summarize our stability requirements in terms where a line representing the closed loop gain intersects the frequency response curve of the op-amp. If the roll off at crossing is:

a. −6 dB/octave → stable.
b. −12 dB/octave → conditionally stable.
c. −18 dB/octave → unstable.

Question. Referring to Figure 4.8, what is the lowest value of A_{fb} at which we can be sure the amplifier will operate stably? Your answer should have been 53 dB because this is the lowest gain at which the roll off is −6 dB/octave.

4-4.6 Phase Margin

Often the stability criteria is expressed as phase margin, θ_{pm}. Phase margin is

$$\theta_{pm} = 180° + \theta_{cl} = 180° - |\theta_{cl}|$$

A positive phase margin is an indication of stability. For example, in Case 1, $\theta_{pm} = 180° - 135.1° = 44.9°$, which indicates stability. A phase margin of about 45° is desirable for maximum response to a pulse input without ringing or instability.

4-4.7 Slew Rate

The slew rate is defined as the maximum rate of change of output voltage per unit time.

$$S = \text{slew rate} = \left(\frac{\Delta V_{out}}{\Delta t}\right)_{max} \tag{4-12}$$

Slew rate is usually expressed in volts/microsecond (V/μs).

Because of its internal capacitances, an amplifier cannot respond instantly to a change in the input voltage. The internal capacitances must be charged, and there is a limit to the rate at which they will charge and therefore the rate at which the output voltage will change. Slew rate differs from the frequency limitations in that the cutoff frequency is the small signal gain limitation of the amplifier; the slew rate is a measure of the ability of an amplifier to handle large signals without distortion. This latter ability varies with frequency and output voltage. The slew rate limitation affects the gain only when a distortion is extreme. The type of distortion introduced on the output by exceeding the slew rate limit is shown in Figure 4.13.

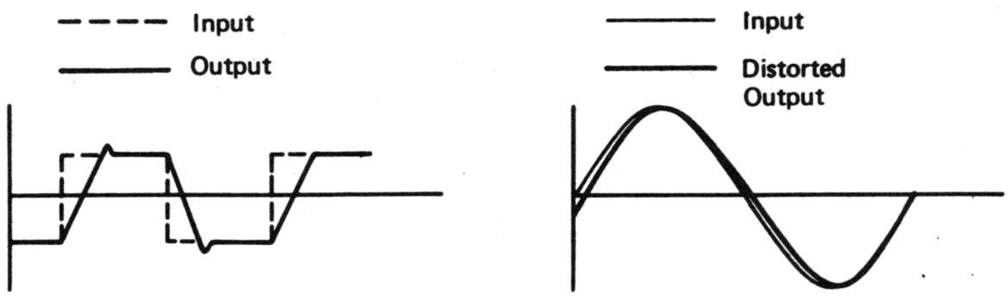

(a) Distortion of square wave (b) Distortion of sine wave

FIGURE 4.13 Effect of exceeding slew rate limit.

To use our amplifier's full bandwidth, we must keep our output voltage small enough to avoid exceeding the slew rate limit. To do this, we must know how to relate output voltage frequency, and slew rate.

For a sinusoidal signal the instantaneous voltage may be written

$$V = V_p \sin 2\pi ft$$

where V_p = peak voltage. Slew rate is defined as

$$S = \left(\frac{\Delta V}{\Delta t}\right)_{max}$$

so we differentiate the instantaneous voltage to obtain

$$\frac{dv}{dt} = 2\pi f V_p \cos 2\pi ft$$

We are interested in $(dv/dt)_{max}$, which occurs as the signal passes through zero for a sinusoidal waveform. At the zero crossing point, $\cos 2\pi ft = 1$, so

$$\left(\frac{dv}{dt}\right)_{max} = 2\pi f V_p = S \tag{4-13}$$

EXAMPLE 4-6

If the slew rate (S) of an amplifier is 20 V/μs, calculate the maximum undistorted peak output voltage we can obtain at 1 MHz (assuming this is within the bandwidth of the amplifier).

SOLUTION

$$S = 2\pi f V_p$$

solving for V_p,

$$V_p = \frac{S}{2\pi f} = \frac{20 \text{ V}/\mu\text{s}}{(6.28)(10^6 \text{ Hz})} = \frac{20 \text{ V}(0.159 \times 10^{-6} \text{ Hz})}{10^{-6} \text{ s}}$$
$$= 3.18 \text{ V}_p = 6.36 \text{ V}_{pp}$$

Example 4-6 shows that if we wish an undistorted output voltage at 1 MHz from the amplifier in this example, we must limit its output to 3.18 V_p or 6.36 V_{pp}.

EXAMPLE 4-7

An amplifier has a slew rate of 10 V/μs and a closed loop upper cutoff frequency (f_{1fb}) of 800 kHz at the closed loop gain we need. We must have an output of 5 V_p up to 250 kHz. Can we use this amplifier?

SOLUTION

$$S = 2\pi f V_p \tag{4-13}$$

We can substitute in V_p = 5 V_p and solve for the frequency at which we can get 5 V_p out with S = 10 V/μs.
Solving:

$$f = \frac{S}{2\pi V_p} = \frac{10 \text{ V}/\mu s}{(6.28)\ 5\text{V}} = \frac{0.318}{\mu s} = 318 \text{ kHz}$$

Thus we can use this amplifier up to 318 kHz at 5 V_p, so it is usable for our purposes.

Most complete op-amp specifications supply the slew rate and a Bode plot of a typical amplifier of that type.

4-5 LAG-PHASE COMPENSATION

To use an op-amp at low closed loop gains, where the roll-off rate is often 18 dB/octave, some phase compensation is necessary. Some amplifiers, such as the Fair-

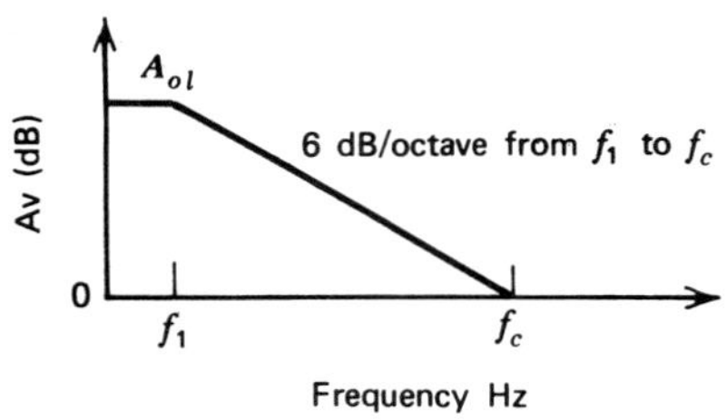

FIGURE 4.14 Fully phase-compensated op-amp response curve.

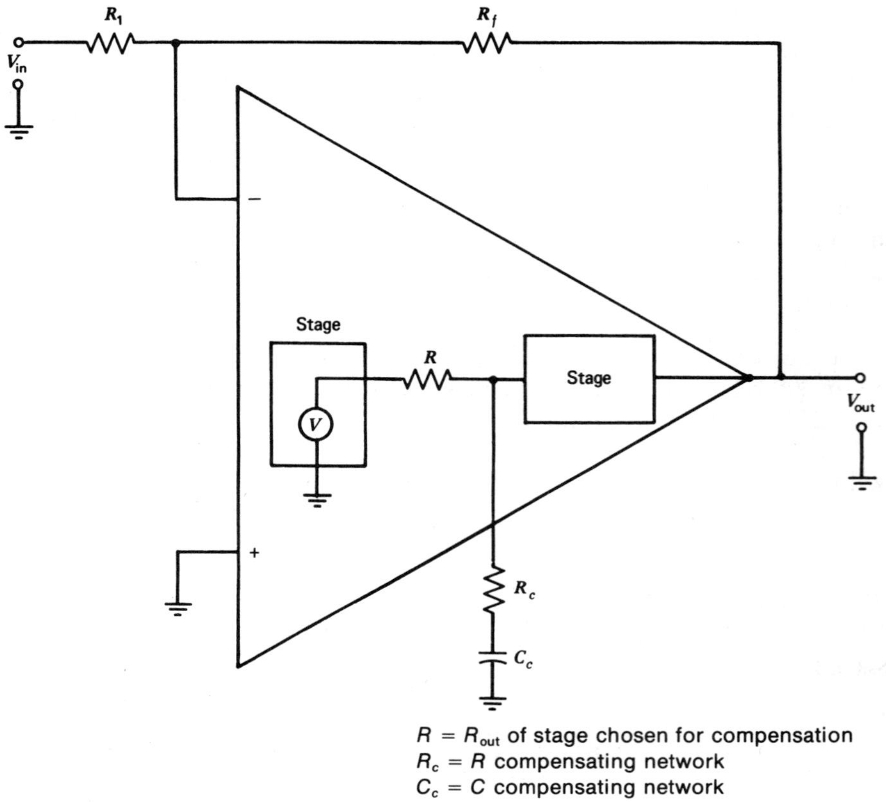

FIGURE 4.15 Lag-compensation network.

child μA 741, require no external phase compensation since they are compensated internally to have a 6 dB/octave roll-off rate from f_1 to f_c, as illustrated in Figure 4.14. Uncompensated amplifiers will have response curves similar in form to those we saw in Figures 4.9 and 4.12. These uncompensated amplifiers have one or more terminals, labeled roll-off, phase, or frequency compensation, to which external *RC* compensation networks are connected. Usually the manufacturer of an uncompensated amplifier will specify the values of R and C necessary for operation at a particular closed loop gain. It is instructive, however, to look at the philosophy and some methods of phase compensation.

The principle of phase compensation is illustrated in Figure 4.16. In this illustration a single *RC* network is used such as in Figure 4.15 for compensation. The compensating network is placed between the output of one stage of the amplifier and ground. The first breakpoint frequency of the compensation network (f_x) is lower than that of any uncompensated stage of the op-amp. The lower cutoff frequency of the amplifier stage is connected to the compensating network, and thus the compensated op-amp, is f_x, the lower 3 dB frequency of the compensating

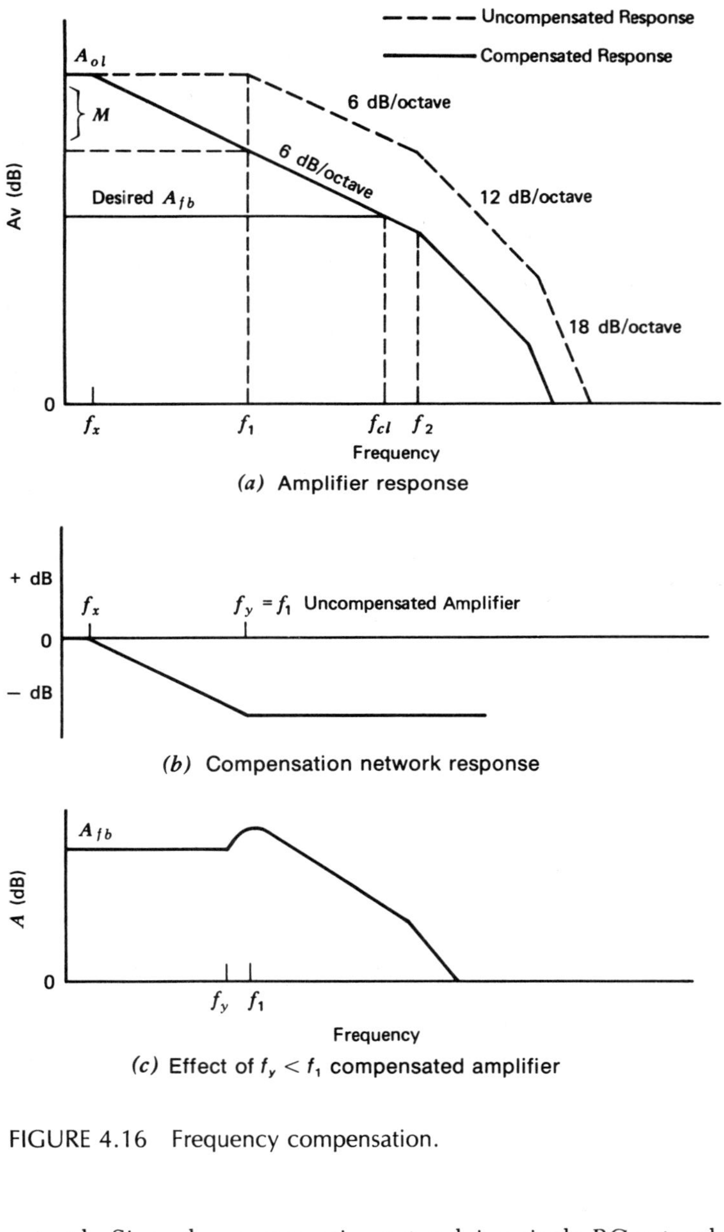

FIGURE 4.16 Frequency compensation.

network. Since the compensating network is a single *RC* network, it has a roll-off rate of 6 dB/octave, as shown in Figure 4.16*b*.

The job of the compensating network is to make sure the compensated amplifier has a 6 dB/octave roll-off rate where the closed loop response (A_{fb}) intersects the op-amp open loop response curve. This assures us that $\theta_{cl} < 180°$ and the amplifier operation is stable.

Let us look at the operation of the compensation network. At frequencies below

f_x, the compensated gain and uncompensated gain are the same since X_{c_c} is very high. At f_x the reactance of the compensating capacitor (X_{c_c}) starts falling off and the compensated amplifier response follows at 6 dB/octave. At f_1 the uncompensated amplifier gain starts falling off at 6 dB/octave, so the compensating network must stop rolling off. If it does not, the compensated amplifier will roll off at 12 dB/octave above f_1, since the roll-off rate will be the sum of the uncompensated amplifier and compensating network roll offs. To prevent this, we set the frequency at which X_{cc} becomes negligible or the second break frequency of the compensating network (f_y) equal to f_1 of the uncompensated amplifier (see Figure 4.16). Now the 6 dB/octave roll off due to the compensating network stops at f_1, and the roll off due to the uncompensated op-amp characteristics takes over from f_1 to f_2. The gain of the compensated amplifier at f_1 is much lower, because of the compensating network attenuation, than the gain of the uncompensated amplifier at f_1, so that at f_2 the open loop gain will be equal to or less than the closed loop gain we plan to use.

To achieve phase compensation, we must give up bandwidth. The closed loop bandwidth of the compensated amplifier is now

$$f_{1fb} = f_x\,(1 + A_{ol}\beta)$$

However, a wide bandwidth is useless in an amplifier that is oscillating.

We now outline a procedure to find values for R_c and C_c. Recall from our low pass networks study that

$$f_x = \frac{1}{2\pi C_c(R + R_c)}$$

and

$$f_y = \frac{1}{2\pi R_c C_c}$$

for the compensating network break frequencies. We must calculate $R = R_{out}$ of the stage the compensating network is connected to if it is not given. The frequency that we want f_x to be is found from the Bode plot in the following manner:

1. Draw a vertical line at f_2.
2. Draw a horizontal line at A_{fb} desired.
3. Draw a 6 dB/octave line from the point of intersection of lines 1 and 2 up to the A_{ol} line. Where this line and the A_{ol} line intersect is f_x; f_y is of course f_1.

If f_y is less than f_1, we will have a rise in our closed loop response as shown in Figure 4.16c. This is due to the roll off of the compensating network ceasing before the amplifier gain starts falling off. Therefore between f_y and f_1 the amplifier gain tends to increase. This condition is usually avoided.

To achieve a smooth 6 dB/octave roll off as we pass through f_1, the ratio of R_c to R must be correct. This ratio must provide an attenuation equal to the reduction in gain necessary to have a smooth roll-off curve at f_1 in the compensated op-amp. This attenuation is marked M on the ordinate in Figure 4.16*a* and is given by the formula

$$M(\text{dB}) = -20 \log \left(\frac{R + R_c}{R_c}\right) \tag{4-14}$$

From this we find that the value of R_c is:

$$R_c = \frac{R}{\left[\text{antilog}\left(\frac{M}{20}\right)\right] - 1} \tag{4-15}$$

The derivation of the equations for M and R_c is given in Appendix D.

After we have calculated R_c we can find C_c by using the low pass break frequency f_y, which must be equal to f_1. Since

$$f_y = \frac{1}{2\pi R_c C_c}$$

$$C_c = \frac{1}{2\pi R_c f_y}$$

EXAMPLE 4-8

This example demonstrates lag compensation. The op-amp specifications are the frequency response shown in Figure 4-17.

$A_{ol} = 60$ dB, $f_1 = 12$ kHz, $f_2 = 100$ kHz
R_{out} stage to be compensated $= 4$ kΩ
A_{fb} desired $= 23$ dB. Calculate R_c and C_c

SOLUTION

To compensate, first find f_x.

(a) Draw a vertical line to f_2 on the Bode plot.
(b) Draw a horizontal line at desired A_{fb}.
(c) Draw a 6 dB/octave line from the intersection of lines 1 and 2 to A_{ol}.
See $f_x = 1.2$ kHz from the Bode plot.

Next we find the attenuation required of the compensation network.

(a) Note the gain where the 6 dB/octave line, which will be the compensated response, crosses f_1. This is 40 dB.
(b) Find the attenuation M

$$M = A_{ol}(\text{dB}) - 40 \text{ dB} = 20 \text{ dB}$$

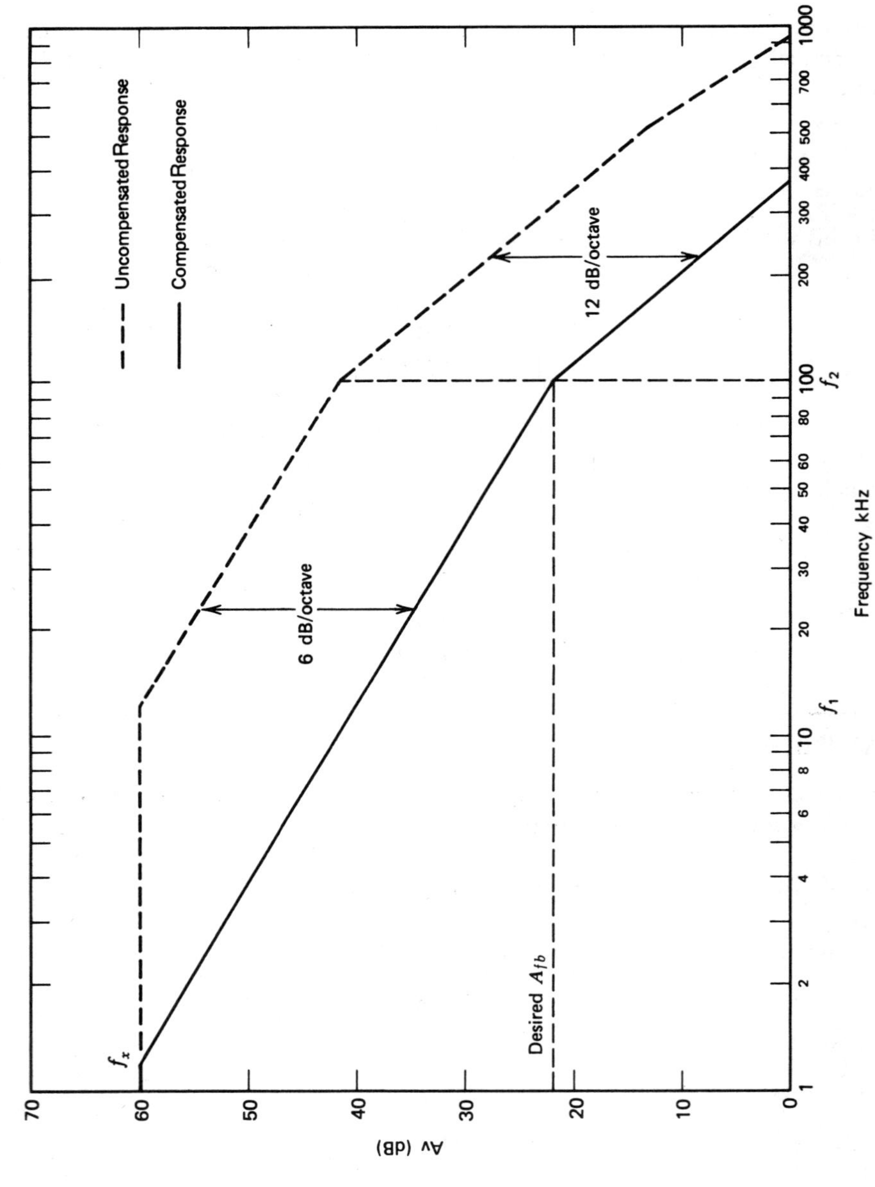

FIGURE 4.17 Op-amp frequency response curve for Example 4.8.

Calculate R_c:

$$R_c = \frac{R}{\left(\text{antilog}\,\frac{M}{20}\right) - 1} = \frac{4\text{ k}\Omega}{9} = 445\ \Omega$$

Calculate C_c:

$$C_c = \frac{1}{2\pi R_c f_y} = \frac{1}{6.28(445\ \Omega)(1.2 \times 10^4\text{ Hz})}$$
$$= 0.03\ \mu\text{F}$$

Check f_x:

$$f_x = \frac{1}{2\pi C_c(R + R_c)} = \frac{1}{2\pi 0.03\ \mu\text{F})(4.45\text{ k}\Omega)} = 1.195\text{ kHz}$$

Notice that f_{1fb} is

$$f_{1fb} = f_x\,(1 + A_{ol}\beta) = 1.2\text{ kHz}\,(1 + \text{antilog } 38\text{ dB})$$
$$= 1.2\text{ kHz}\,(1 + 79.5) = 1.2\text{ kHz}\,(80.5) = 96.7\text{ kHz}$$

which is in fair agreement with the Bode plot.

Example 4-8 shows only one method of phase compensation and the only one we will examine in detail. We will look briefly at some other methods in the next section.

4-6 OTHER METHODS OF PHASE COMPENSATION

4-6.1 Miller-Effect Compensation

The fully compensated monolithic integrated circuit operational amplifiers such as the Fairchild μA 741 use the Miller effect to help achieve compensation. The

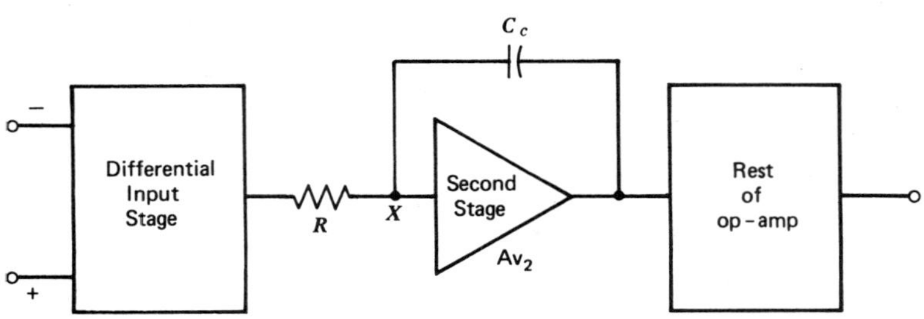

FIGURE 4.18 Miller-effect compensation.

response achieved is illustrated in Figure 4.14. The method used involves connecting the compensating capacitor from the output to the input of an intermediate stage of the op-amp, as shown in Figure 4.18. The Miller effect causes the compensating capacitor at the input of stage 2 to appear much larger than if C_c were connected to ground from that point. The compensating capacitor can be much smaller as a result. Keeping C_c small is very important when manufacturing an IC op-amp because large values are difficult to fabricate. The first break frequency for this compensation will be

$$f_x = \frac{1}{2\pi R\,(1 - A_{v_2})\,C_c} \tag{4-16}$$

The response of the second stage must be planned to have its cutoff frequency occur at the frequency at which X_c Miller becomes negligible. A resistor R_c may be used in series with C_c to set f_y of the compensating network.

4-6.2 Input-Stage Compensation

Frequently *RC* phase compensation is used on the first (differential) stage of the operational amplifier rather than on a later stage. The advantage of compensating the first stage is that the slew rate is optimized. This faster slew rate results because the low voltage swings at the first stage output allow compensation to be achieved without moving as much charge (resulting in low current) through C_c. Amplifiers using such compensation usually list the values of R_c and C_c needed for specific closed loop gains. The equations for f_x and f_y are given on Figure 4.19*b*.

Some amplifiers use two *RC* networks to achieve phase compensation. These will have either the formulas or R_c and C_c values necessary for compensation in the specifications.

4-6.3 Feedforward Compensation

Another type of compensation used is feedforward. This type of compensation is used to obtain wide bandwidths. Feedforward compensation is obtained by ac coupling of the higher frequency signals around the highest gain portion of the op-amp. Low frequency signals are fed through both the high gain and low gain, high frequency portions of the op-amp, allowing very good accuracy with feedback. By feeding the high frequency signals around the high gain portion of the op-amp, the frequency limitations and large phase shift of the high gain portion is bypassed. The high frequency portion will have a low phase shift for good high frequency stability. Feedforward compensation is illustrated in Figure 4.20.

4-6.4 Brute-Force Compensation

The next type of compensation we will discuss is brute-force compensation. It is used as a last resort to obtain stable operation at low frequencies with low closed loop gains when, for some reason, more elegant methods cannot be used. The

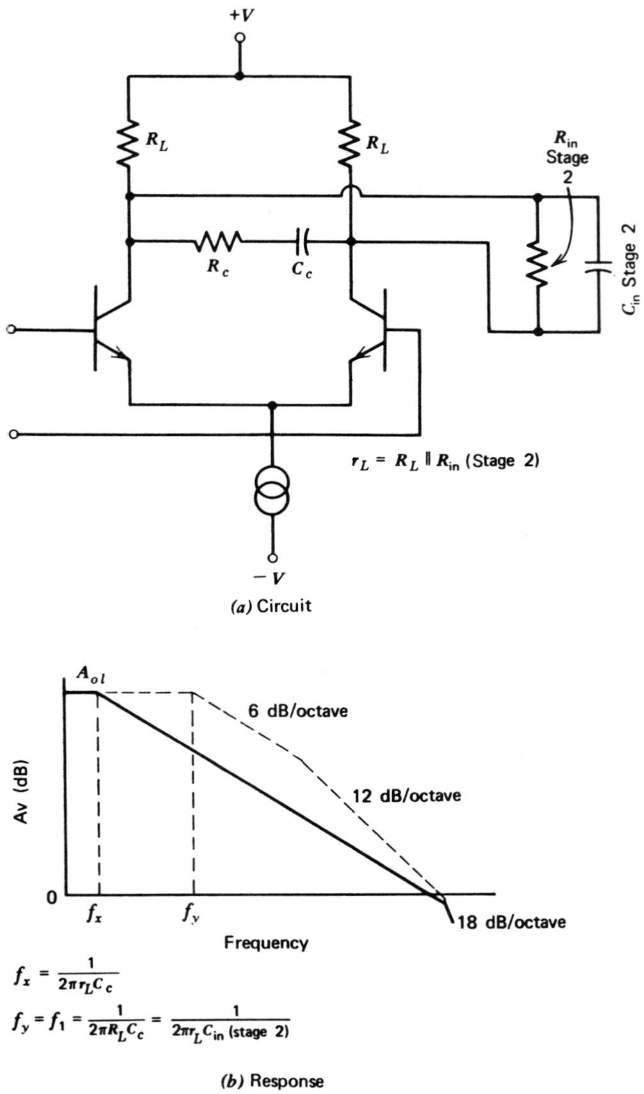

FIGURE 4.19 *RC* compensation applied to the differential input stage.

method is also used where the numerical solution for stability is very difficult, such as when using nonlinear feedback networks.

Referring to Figure 4.21, we see that an *RC* network directly across the input terminals of the op-amp is used. R_c is experimentally picked to stop the oscillations (R_c decreases the loop gain), and C_c is chosen to have a small reactance compared to R_c at the oscillating frequency. The resistor across the input will increase the amplifier noise.

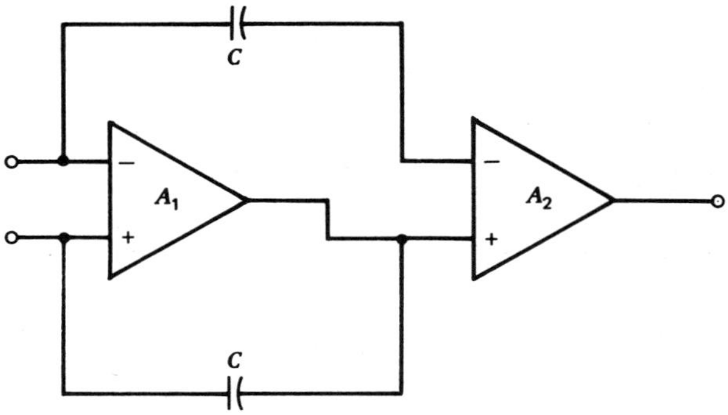

FIGURE 4.20 Feedforward compensation.

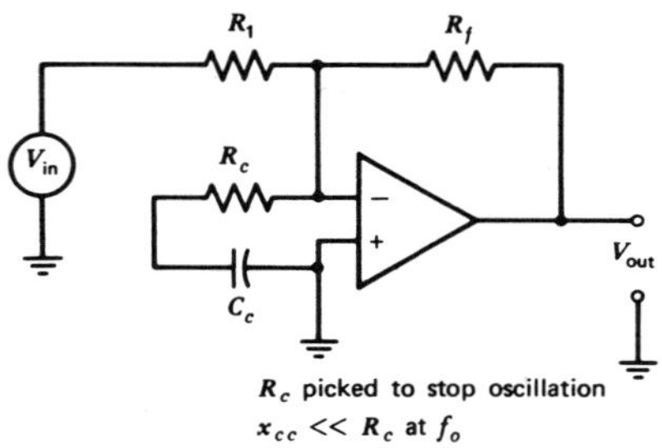

FIGURE 4.21 Brute-force compensating network.

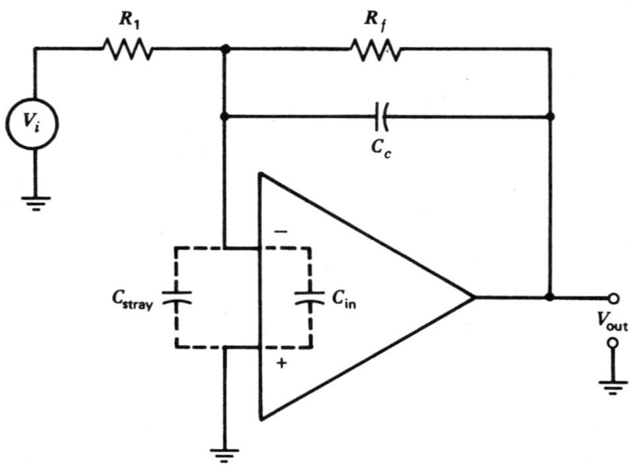

$$C_c R_f = R_1 (C_{stray} + C_{in})$$

$$C_c = \frac{R_1 (C_{stray} + C_{in})}{R_f}$$

FIGURE 4.22 Equalizing input capacitance.

4-6.5 Equalizing Input Capacitance

A method of equalizing the phase shift due to input and stray capacitance is shown in Figure 4.22. A capacitor C_c is added across R_f such that

$$C_c R_f = R_1 (C_{in} + C_{stray})$$

We recognize this as similar to compensating an attenuator for a wider frequency range. This is also called lead compensation.

4-7 *RC* SELECTION NOTE

Most manufacturers of op-amps (IC or module) include within the specifications of the amplifier a set of charts, graphs, or equations that easily enable one to find the components necessary to compensate the amplifier at any desired closed loop gain. The values of R_c and C_c given in the specifications are generally for a typical op-amp. It may be necessary to adjust these values to assure complete stability, especially with IC op-amps.

The Fairchild μA 791 is an uncompensated power op-amp capable of 1 A output current. The specifications are shown in Appendix C. Pins 4 and 9 are for frequency compensation in the metal can package, and pins 7 and 12 in the DIP package. The compensating capacitor is connected between the compensating pins. A voltage gain versus frequency graph on the last page of the specifications shows what size compensating capacitor should be used at a given closed loop gain. For example, if $A_{fb} = 10$ (20 dB), $C_c = 10$ pF from the graph. If an in-between A_{fb} is desired, the capacitor value of the next lower gain can be used or the graph can be interpolated. For example, if $A_{fb} = 5$ (13.9 dB), then C_c can be 100 pF or a value can be selected and tried between 10 pF and 100 pF. A reasonable value to try would be about 60 pF. Note that for all circuits shown in the specification sheet the manufacturer recommends that 0.0033 μF in series with 3.9 Ω be connected from the output to +V. This compensates for output reactance. The manufacturer also recommends a 10 μF solid tantalum capacitor be connected from +V and −V to ground for two supply operation to keep power supply noise and loading from causing instability.

The Texas Instruments TL 080 JFET input op-amp is the only op-amp of the TL 080 to TL 085 series that is not internally compensated. The specifications recommended the use of a 12 pF compensation capacitor for unity gain and above. The capacitor goes from the compensation pin (pin 8) to the combination offset null (N_1) and compensation pin (pin 1). Feedforward compensation can be used for wider bandwidth as is shown in Figure 5 of the specification sheet in Appendix C. For feedforward compensation, $C_1 = 500$ pF and $C_2 = 3$ pF from Figure 14 of the specification sheet. Note on $A_{fb} < 10$, C_2 will have to be increased somewhat.

When op-amps are used on printed circuit boards, or in other high density circuit layouts, it may be necessary not only to compensate each individual amplifier but to bypass the power supply leads to ground periodically with capacitors. The bypass capacitors should be connected as close to the amplifier +V and −V connections as possible and are normally between 0.001 and 0.1 μF. The only way to tell if bypass capacitors are necessary, what size, and how often (every amp, every other amp, or whatever) is through experimentation with the breadboarded circuit.

4-8 STABILIZING CAPACITIVE LOADS

When operational amplifers must drive relatively heavy capacitive loads, instability may result due to the $R_{out}\,C_{load}$ phase shift. To avoid this, a resistor is used to isolate the output capacitance from the amplifer. The resistor, R_c, is chosen to equal the open loop output resistance of the amplifier. C_c is still chosen so that

$$R_f C_c = R_L\, C_L$$

Two configurations of capacitive load decoupling are shown in Figure 4.23. R_c can be either inside or outside the feedback loop as shown in Figure 4.23*a* and *b*.

A high capacitance load that might require this compensation would be a long transmission line. The lower the output resistance of an operational amplifer, the more capacitance it can drive without instability.

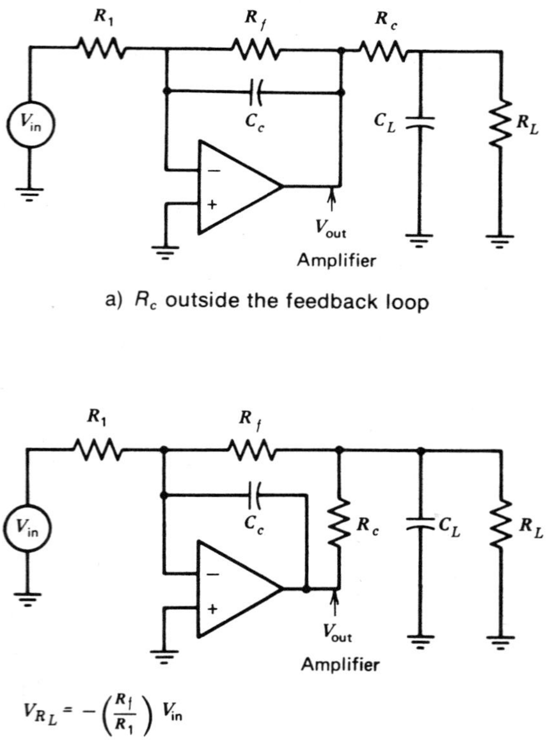

a) R_c outside the feedback loop

$$V_{R_L} = -\left(\frac{R_f}{R_1}\right) V_{in}$$

b) R_c inside the feedback loop

FIGURE 4.23 Capacitive load decoupling.

SUMMARY

1. An op-amp's voltage gain decreases with frequency. As frequency increases, a point will be reached where junction capacitance and stray capacitance of the amplifier are no longer negligible. These capacitances shunt a part of the signal to ground. As frequency increases, the part of the signal shunted to ground increases, causing voltage gain to decrease.
2. The frequency-dependent gain of a single stage is given as:

$$|A| = \frac{A_{ol}}{\sqrt{1 + \left(\frac{f}{f_1}\right)^2}} \angle -\arctan\left(\frac{f}{f_1}\right) \tag{4-1}$$

where

A	=	frequency dependent gain
A_{ol}	=	low frequency gain
f_1	=	corner (3 dB) frequency of amplifier
f	=	frequency of interest
$-\arctan(f/f_1)$	=	phase difference between input and output voltage. The minus sign indicates output lags input.

3. As long as amplifier frequency dependence is represented by a single *RC* network, its gain decreases at a rate of -6 dB/octave, which is the same as -20 dB/decade, at frequencies greater than f_1.
4. The frequency response of a multistage amplifier can be drawn easily (if the Bode plot of each stage is known) by graphically adding all the stage responses. To draw the Bode plot of a single stage, we need only know f_1 and the gain of that stage in decibels.
5. As long as an amplifier has a roll-off rate of -6 dB/octave, it has a constant gain-bandwidth product.
6. The addition of feedback to an amplifier causes its bandwidth to increase. If the amplifier roll-off rate is -6 dB/octave, the corner frequency with feedback ($f_{1_{fb}}$) is:

$$f_{1_{fb}} = f_1(1 + A_{ol}\beta) \tag{4-7}$$

where

f_1	=	corner frequency without feedback
$A_{ol}\beta$	=	loop gain

7. If the loop gain of an amplifier is greater than one, which always is the case with negative ($-180°$) feedback, and the loop gain phase shift is greater than 180°, an amplifier oscillates. Loop gain increases as closed loop gain decreases.
8. We can evaluate whether or not an amplifier will be stable at a given closed loop gain by the roll-off rate of the open loop gain curve at that closed loop gain.
   a. If the roll-off rate is -6 dB/octave, the amplifier will be stable.

b. If the roll-off rate is −12 dB/octave, the amplifier may or may not be stable. This condition is to be avoided.

c. If the roll-off rate is −18 dB/octave or greater, the amplifier will oscillate.

9. Phase compensation consists of adding various *RC* networks to the op-amp so that its roll-off rate is −6 dB/octave at the closed loop gain of interest. Op-amp manufacturers provide in the op-amp specifications the proper network and values of the components that must be used to keep an amplifier from oscillating.

10. Many operational amplifiers are fully compensated and require no external frequency compensation for normal applications. A similar amplifier that is externally compensated will normally provide a wider bandwidth at a given closed loop gain.

11. Slew rate (S) is the maximum change of output voltage per unit of time (normally given in volts/microsecond). The full bandwidth of an amplifier cannot be realized if its output voltage change per unit of time reaches the slew rate. For a given output voltage the maximum frequency at which the slew rate is not exceeded, or the maximum voltage output for a given frequency, can be found from:

$$S = 2\pi f V_p \tag{4-13}$$

where

S = slew rate
V_p = peak output voltage
F = frequency

SELF-TEST QUESTIONS

4-1 List two factors that cause op-amp gain to be frequency dependent.

4-2 If an amplifier has an $A_{ol} = 15{,}000$ and $f_1 = 1$ kHz, calculate A_{ol} at 10 kHz.

4-3 An amplifier has an $A_{ol} = 80$ dB. Calculate A_{ol} as a ratio.

4-4 A three-stage op-amp has the following characteristics for each stage.

Stage 1	$A_v = 20$ dB, $f_{1_1} = 1$ kHz
Stage 2	$A_v = 15$ dB, $f_{1_2} = 8$ kHz
Stage 3	$A_v = 10$ dB, $f_{1_3} = 15$ kHz

Draw the Bode plot of A_{ol} on four-cycle semilog graph paper.

4-5 An op-amp has an $A_{ol} = 65$ dB and $A_{fb} = 20$ dB. Calculate the loop gain at low frequency.

4-6 An amplifier has a gain-bandwidth product of 1.5 MHz.
(a) Calculate the gain at 1 MHz.
(b) Calculate the bandwidth at a voltage gain of 15.

4-7 State the conditions that cause an op-amp to oscillate.

4-8 State the relationship between the roll-off rate of A_{ol} at f_{cl} and amplifier stability.

4-9 An amplifier has an A_{ol} as drawn in Figure 4.12. Calculate θ_{cl} for $A_{fb} = 40$ dB.

4-10 An amplifier has an A_{ol} as shown in Figure 4.8. Calculate θ_{cl} for (a) $A_{fb} = 35$ dB, (b) $A_{fb} = 50$ dB.

4-11 An amplifier has a slew rate of 1 V/μs. Calculate the undistorted peak voltage V_p available at 40 kHz.
4-12 An amplifier has a slew rate of 0.8 V/μs. Calculate the maximum frequency at which V_{out} will be undistorted if the maximum output voltage is to be 3 V_p.
4-13 In Figure 4.15, $R = 10$ kΩ, $f_1 = 19$ kHz, $f_2 = 110$ kHz, and $A_{ol} = 70$ dB. Calculate R_c and C_c so the amplifier will be stable at $A_{fb} = 20$ dB. (*Hint.* Draw the Bode plot.)
4-14 State the advantage of applying compensation to the input stage of an op-amp.
4-15 List four methods of phase compensation and briefly describe each.

If you cannot answer certain questions, place a check next to them and review appropriate parts of the text to find the answers.

LABORATORY EXERCISE

OBJECTIVE

After completing this laboratory, the student should be able to verify the constancy of the gain-bandwidth product and the effect of exceeding the slew rate limit. The student should also be able to observe oscillation in an uncompensated amplifier and phase compensate an amplifier.

Equipment

1. Fairchild μA741 operational amplifier or equivalent and manufacturer's specifications.
2. Texas Instruments TL080 or similar operational amplifier and specifications.
3. Power supply, ±15 V dc adjustable.
4. General purpose oscilloscope capable of measuring 2 mV dc and ac.
5. Signal generator.
6. 10 kΩ potentiometers.
7. Breadboard such as EL Instruments SK-10 mounted on vectorboard.
8. Precision resistor assortment.
9. Metalized mylar capacitor assortment.

Procedure

1. Open loop gain measurement.
   (a) Due to the wide range of open loop gains possible for an IC op-amp, we will first measure the open loop gain to achieve laboratory precision. Connect the circuit of Figure 4.24 and measure the open loop gain as outlined in Chapter 2.
   (b) Use this amplifier through part 2 of this laboratory.
   Note. If instability occurs, it may be necessary to lower the 1 MΩ resistor or vary C.

$$A_{ol} = \frac{V_{out}}{V_s/100} = \frac{V_{out}}{v_i}$$

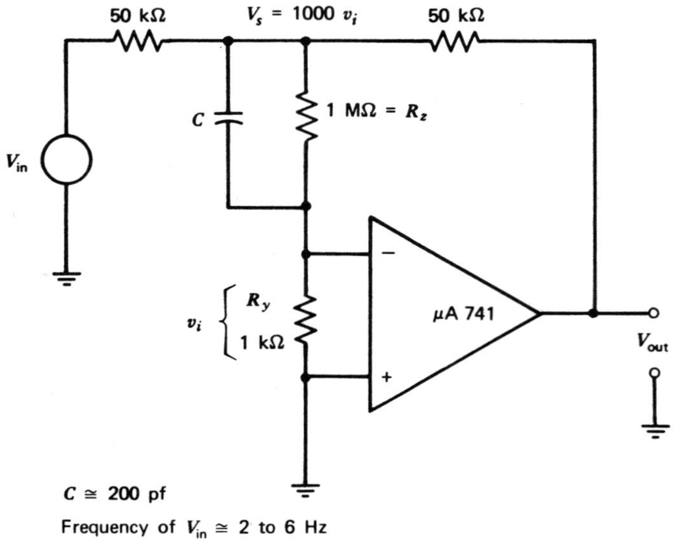

FIGURE 4.24 Open loop gain measuring circuit.

for $R_z = 1\ M\Omega$.
General equation:

$$A_{ol} = \frac{V_{out}}{V_s \Big/ \frac{R_z}{R_y}}$$

2. Gain-Bandwidth Product.
   (a) Connect a Fairchild μA 741 or equivalent operational amplifier as a noninverting amplifier (Figure 4.25) with $R_1 = 10\ k\Omega$. Pick R_f so that $A_{fb} = 200$ and null the offset.

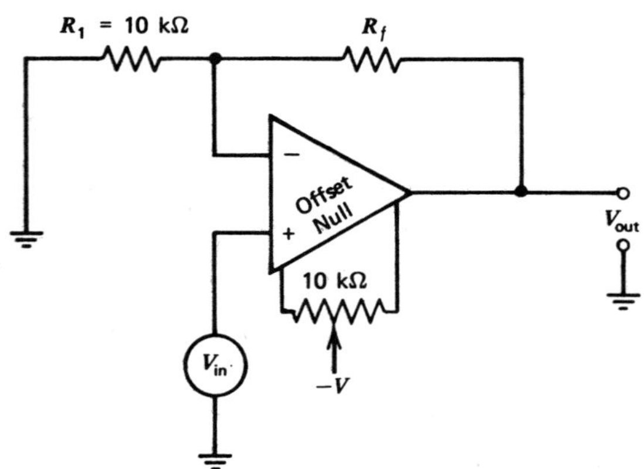

FIGURE 4.25 Noninverting amplifier for part 2 of Laboratory Exercise.

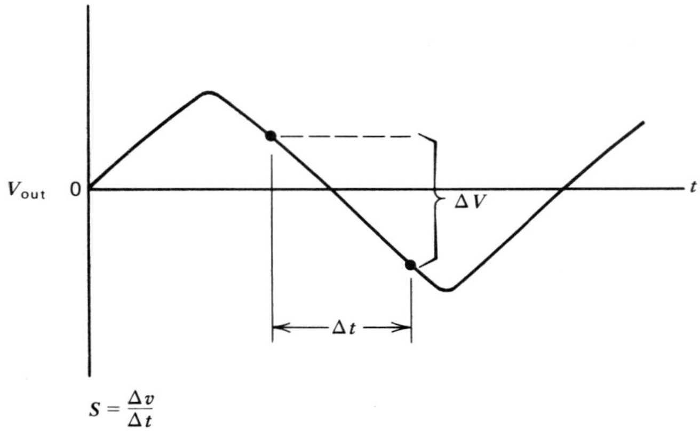

FIGURE 4.26 Slew rate measurement.

Note. Adjust the input voltage so that $V_{out} \leq 1\ V_p$ for all of part 2.
(b) Measure f_{1fb} and calculate f_1 using $f_{1fb} = f_1(1 + A_{ol}\beta)$. (This is easier than measuring f_1 directly.)
(c) Calculate the gain-bandwidth products $f_1 A_{ol}$ and $A_{fb} f_{1fb}$.
(d) Calculate the gain-bandwidth products and f_{1fb} for $A_{fb} \simeq 100$ and $A_{fb} \simeq 50$.
(e) Measure f_{1fb} with $A_{fb} = 100$ and $A_{fb} = 50$. Compare the measured and calculated values of f_{1fb} and the gain-bandwidth products.
Note. Calculate the gain-bandwidth product from the measured f_{1fb} at the chosen A_{fb}.

3. Slew Rate.
   (a) Set the circuit of part 2 up so that $A_{fb} \simeq 20$.
   (b) Measure the slew rate of your amplifier. To do this:
      (1) Put a sine wave into the amplifier.
      (2) Adjust the input voltage until the output voltage is near the maximum output of which the amplifier is capable.
      (3) Increase the frequency of the sine wave until the output looks triangular as in Figure 4.26.
      (4) Measure the slope of the linear portion of the output. This slope ($\Delta V_{out}/\Delta t$) is the slew rate.
   (c) Calculate the frequency at which a sine wave will begin to distort with an output voltage of 5 V_p (10 V_{pp}). Use the relation $S = 2\pi f V_p$.
   (d) Verify your answer to step (c) by measurement.
   (e) Calculate the peak voltage at which we can operate without distortion due to slew rate limiting at 50 kHz.
   (f) Verify your answer to step (e) by measurement.
   (g) What are the possible reasons for any disagreement between calculated and measured results?
4. Phase Compensation.
   (a) Connect a Texas Instruments TL080 or equivalent operational amplifier as an inverting amplifier with a closed loop gain of about 5. Do *not* compensate.
   (b) Observe the oscillations.
   (c) Insert $C_c = 12$ pF as recommended by the specification sheet and shown in Figure 4.27.

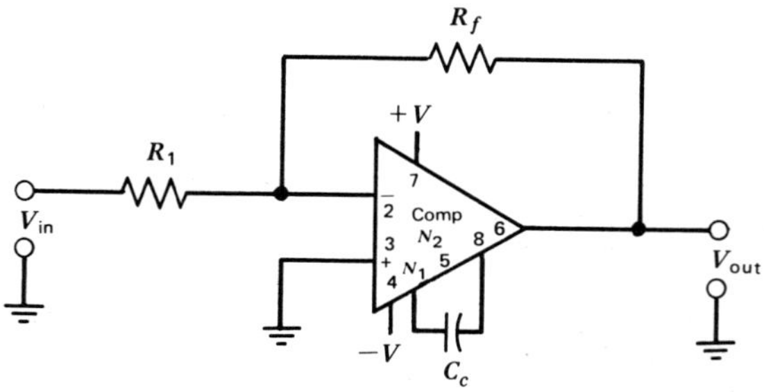

FIGURE 4.27 Texas Instruments TI TL080.

(d) Observe the stable operation.

Note. Due to variations in IC op-amp manufacture, some moderate adjustment to the size of C_c may be necessary to eliminate completely oscillations in some amplifiers.

CHAPTER

5

SUMMING CIRCUITS

Summing circuits include adders and subtracters. These can be used for solving algebraic equations and as proportional combiners for control circuits.

OBJECTIVES

After completing the study of this chapter and the self-test questions, the student should be able to:

1. Given the resistor values, calculate the output voltage of the inverting adder, scaling adder, averager, and adder-subtracter.
2. Given the desired response, calculate the resistors necessary to set up an inverting adder, scaling adder, averager, and adder-subtracter.
3. Draw the circuits of objective 2 from memory.
4. Solve a simultaneous equation of two unknowns using summing circuits.
5. Draw from memory a block diagram of a proportional control and state the principle of operation.
6. Perform the laboratory for Chapter 5.

5-1 INVERTING ADDER

The inverting adder does just as its name implies—algebraically sums two voltages and inverts the result. To verify the operation, refer to Figure 5.1. Note the input resistors have the same value as the feedback resistor.

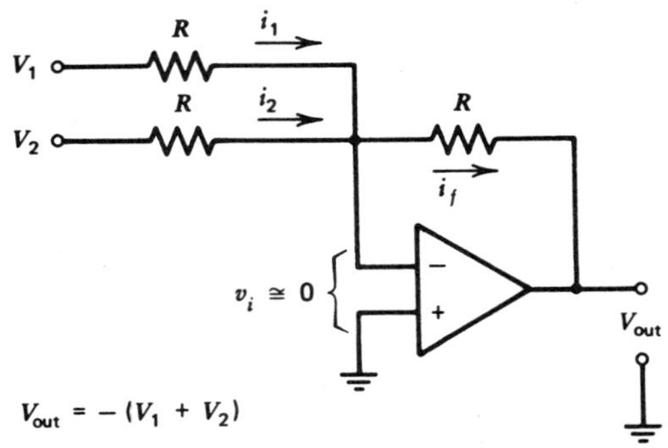

FIGURE 5.1 Inverting adder.

If R_{in} of the op-amp is high enough to make the bias current negligible compared to the feedback current (which is usually the case), then

$$i_1 + i_2 = i_f \tag{5-1}$$

by Kirchoff's current law. Furthermore, if the open loop gain is high enough that $v_i \simeq 0$, which is usually true of op-amps used in summing applications

$$i_1 = \frac{V_1}{R}, \quad i_2 = \frac{V_2}{R}, \quad \text{and} \quad i_f = -\frac{V_{out}}{R}$$

We can now write Equation 5-1 as

$$\frac{V_1}{R} + \frac{V_2}{R} = -\frac{V_{out}}{R}$$

Multiplying through by R we find

$$V_1 + V_2 = -V_{out}$$

therefore

$$V_{out} = -(V_1 + V_2)$$

This logic can be extended to any number of inputs, so that for n inputs we have

$$V_{out} = -(V_1 + V_2 + \cdots + V_n) \tag{5-2}$$

EXAMPLE 5-1

(a) If, in Figure 5.1, $V_1 = 3$ V and $V_2 = -4$ V, find V_{out}.
(b) In Figure 5.1, $V_1 = 3$ V and $V_2 = 2$ V, find V_{out}.

SOLUTION

(a) $V_{out} = -(V_1 + V_2) = -(3\text{ V} - 4\text{ V}) = 1$ V.
(b) $V_{out} = -(V_1 + V_2) = -(3\text{ V} + 2\text{ V}) = -5$ V.

All the summing circuits in this section work for ac or dc voltages. When summing ac voltages, we must calculate answers in the form $V = V_p \sin \omega t$, or, if all the ac voltages are *in phase*, we can use peak voltage, or rms voltage, whichever is convenient.

5-2 SCALING ADDER

If we wish to give different weights to various inputs, we will use a scaling adder. The scaling adder allows us to add, for example, $V_1 + 3\ V_2 + 4\ V_3$. From

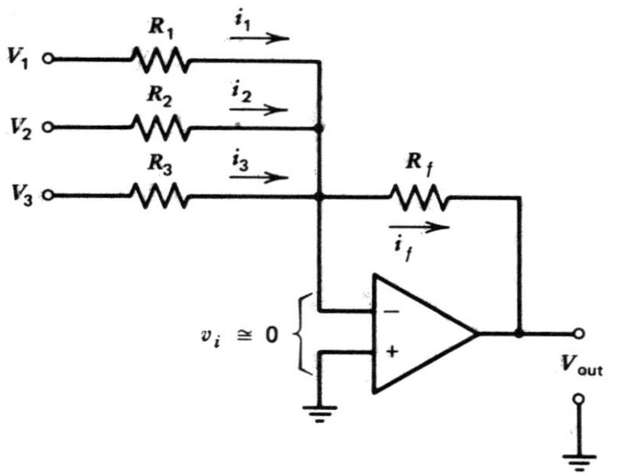

FIGURE 5.2 Scaling adder.

Kirchoff's current law we see that $i_1 + i_2 + i_3 = i_f$, assuming the amplifier bias current is negligible. Assuming a high open loop gain so that $v_i \simeq 0$, we see

$$i_1 = \frac{V_1}{R_1}, \qquad i_2 = \frac{V_2}{R_2}, \qquad i_3 = \frac{V_3}{R_3}$$

and

$$i_f = -\frac{V_{out}}{R_f}$$

therefore

$$\frac{V_1}{R_1} + \frac{V_2}{R_2} + \frac{V_3}{R_3} = -\frac{V_{out}}{R_f}$$

Solving this equation for V_{out} we find

$$V_{out} = -\left(V_1\frac{R_f}{R_1} + V_2\frac{R_f}{R_2} + V_3\frac{R_f}{R_3}\right) \tag{5-3}$$

EXAMPLE 5-2

Let $R_1 = 20\ k\Omega$, $R_2 = 50\ k\Omega$, $R_3 = 25\ k\Omega$, and $R_f = 100\ k\Omega$ in Figure 5.2. Let $V_1 = 1$ V, $V_2 = 2$ V, and $V_3 = -3$ V, and find V_{out}

$$V_{out} = -\left(V_1\frac{R_f}{R_1} + V_2\frac{R_f}{R_2} + V_3\frac{R_f}{R_3}\right)$$

$$= -\left[1\ V\left(\frac{100\ k\Omega}{20\ k\Omega}\right) + 2\ V\left(\frac{100\ k\Omega}{50\ k\Omega}\right) - 3\ V\left(\frac{100\ k\Omega}{25\ k\Omega}\right)\right]$$

$$= -[1\text{ V}(5) + 2\text{ V}(2) - 3\text{ V}(4)]$$

$$= -(5\text{ V} + 4\text{ V} - 12\text{ V}) = -(-3\text{ V}) = 3\text{ V}$$

EXAMPLE 5-3

Find R_1, R_2, and R_3 so that the circuit in Figure 5.2 will have a $V_{out} = -(6\ V_1 + 3\ V_2 + 4\ V_3)$. $R_f = 200\ \text{k}\Omega$.

SOLUTION

Consider the gain seen by each input. V_1 should have a gain of 6, V_2 of 3, and V_3 of 4. Solving,

$$6\ V_1 = V_1\frac{R_f}{R_1}$$

implies

$$R_1 = \frac{R_f}{6} = \frac{200\ \text{k}\Omega}{6} = 33.3\ \text{k}\Omega$$

similarly

$$R_2 = \frac{200\ \text{k}\Omega}{3} = 66.6\ \text{k}\Omega$$

and

$$R_3 = \frac{200\ \text{k}\Omega}{4} = 50\ \text{k}\Omega$$

We can extend our analysis of the scaling adder to show that for n inputs

$$V_{out} = -\left(V_1\frac{R_f}{R_1} + V_2\frac{R_f}{R_2} + \cdots + V_n\frac{R_f}{R_n}\right) \tag{5-3a}$$

AVERAGER

If in the circuit of Figure 5.2, we let $R_1 = R_2 = R_3 = \cdots = R_n$ for all inputs, and if $R_f = R_1/n$, where n is the number of inputs, we find

$$V_{out} = -\left(\frac{V_1 + V_2 + V_3 + \cdots + V_n}{n}\right) \tag{5-4}$$

In other words, the circuit will become an averager. We can show this by noting that

$$-\frac{V_{out}}{R_f} = \frac{V_1}{R_1} + \frac{V_2}{R_1} + \cdots + \frac{V_n}{R_1}$$

thus

$$V_{out} = -R_f\left(\frac{V_1 + V_2 + \cdots + V_n}{R_1}\right)$$

But $R_f = R_1/n$, therefore

$$V_{out} = -\frac{R_1}{n}\left(\frac{V_1 + V_2 + \cdots + V_n}{R_1}\right)$$

and

$$V_{out} = -\left(\frac{V_1 + V_2 + \cdots + V_n}{n}\right) \tag{5-4}$$

EXAMPLE 5-4

Set up the circuit in Figure 5.2 so that it averages the three inputs.

SOLUTION

Let $R_1 = R_2 = R_3 = 200\ \text{k}\Omega$, $R_f = \dfrac{R_1}{n}$ so $R_f = \dfrac{200\ \text{k}\Omega}{3} = 66.6\ \text{k}\Omega$.

Note that in the averager $R_f = R_1 \| R_2 \| \cdots \| R_n$. We can weight the average as long as this condition is met. By weighting the average we mean, for example,

$$V_0 = -\left(\frac{V_1 + 2\,V_2 + V_3 + \cdots + V_{in}}{\text{number of inputs}}\right)$$

The averager is just a special case of the inverting adder.

Note. The resistance of the feedback network is chosen so that the feedback current is much larger than the op-amp bias current but such that the amplifier can easily supply both the feedback current and any load current necessary. A wide range of possible feedback resistors can be used with most op-amps.

5-3 ADDER-SUBTRACTER

The adder-subtracter is illustrated in Figure 5.3. This circuit is a generalization of the differential input amplifier shown in Figure 1.11. The general solution for V_{out} is quite lengthy, so we will discuss the conditions that must be met for good operation.

The conditions are, in effect, that the sum of the gains on the inverting side of the amplifier must equal the sum of the gains on the noninverting side. In other words, the inverting and noninverting gains must be balanced.

Symbolically we can state this as:

$$\frac{R_f}{R_1} + \frac{R_f}{R_2} + \cdots + \frac{R_f}{R_m} = \frac{R_f'}{R_1'} + \frac{R_f'}{R_2'} + \cdots + \frac{R_f'}{R_n'}$$

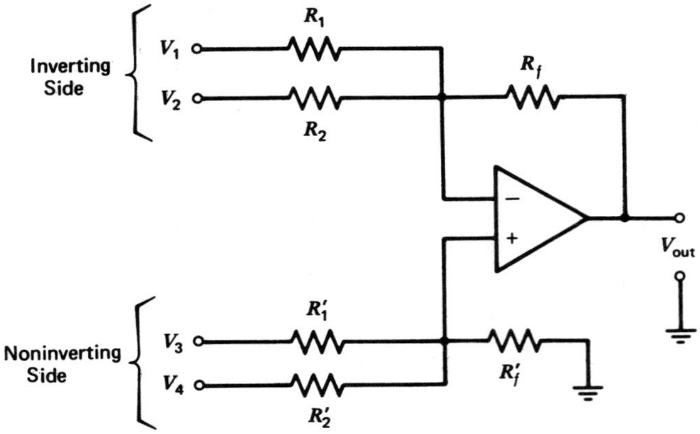

$$V_{out} = V_3\frac{R_f'}{R_1'} + V_4\frac{R_f'}{R_2'} - V_1\frac{R_f}{R_1} - V_2\frac{R_f}{R_2}$$

If $\quad \dfrac{R_f}{R_1} + \dfrac{R_f}{R_2} = \dfrac{R_f'}{R_1'} + \dfrac{R_f'}{R_2'}$

FIGURE 5.3 Adder-subtracter.

for m inverting inputs and n noninverting inputs, then

$$V_{out} = -\left(V_1\frac{R_f}{R_1} + V_2\frac{R_f}{R_2} + \cdots + V_m\frac{R_f}{R_m}\right) + \left[V_{(m+1)}\left(\frac{R_f'}{R_1'}\right) + V_{(m+2)}\left(\frac{R_f'}{R_2'}\right) + \cdots + V_{(m+n)}\left(\frac{R_f'}{R_n'}\right)\right] \tag{5-5}$$

An example may help clarify this.

EXAMPLE 5-5

In Figure 5.3 $V_1 = V_2 = 1$ V, $V_3 = V_4 = 2$ V, $R_f = 200$ kΩ, $R_f' = 100$ kΩ, $R_1 = 100$ kΩ, $R_2 = 25$ kΩ, $R_3 = 25$ kΩ, and $R_4 = 16.67$ kΩ. (a) Is balance maintained? (b) What is V_{out}?

SOLUTION

(a) Check balance:

$$\frac{R_f}{R_1} + \frac{R_f}{R_2} \stackrel{?}{=} \frac{R_f'}{R_1'} + \frac{R_f'}{R_2'}$$

$$\frac{200\text{ k}\Omega}{100\text{ k}\Omega} + \frac{200\text{ k}\Omega}{25\text{ k}\Omega} \stackrel{?}{=} \frac{100\text{ k}\Omega}{25\text{ k}\Omega} + \frac{100\text{ k}\Omega}{16.67\text{ k}\Omega}$$

$$2 + 8 = 10 = 4 + 6 = 10$$

Thus, balance is maintained.
(b) Now, using Equation 5-5,

$$V_{out} = V_3\frac{R'_f}{R'_1} + V_4\frac{R'_f}{R'_2} - V_1\frac{R_f}{R_1} - V_2\frac{R_f}{R_2}$$

$$= 2\text{ V}\left(\frac{100\text{ k}\Omega}{25\text{ k}\Omega}\right) + 2\text{ V}\left(\frac{100\text{ k}\Omega}{16.67\text{ k}\Omega}\right) - 1\text{ V}\left(\frac{200\text{ k}\Omega}{100\text{ k}\Omega}\right) - 1\text{ V}\left(\frac{200\text{ k}\Omega}{25\text{ k}\Omega}\right)$$

$$= 2\text{ V}(4) + 2\text{ V}(6) - 1\text{ V}(2) - 1\text{ V}(8)$$

therefore

$$V_{out} = 8\text{ V} + 12\text{ V} - 2\text{ V} - 8\text{ V} = 20\text{ V} - 10\text{ V} = 10\text{ V}$$

EXAMPLE 5-6

In Figure 5.3, $R_1 = R_2 = R'_1 = R'_2 = R_f = R'_f$. Write the expression for V_{out}.

SOLUTION

Using Equation 5-5,

$$V_{out} = V_3\frac{R'_f}{R'_1} + V_4\frac{R'_f}{R'_2} - V_1\frac{R_f}{R_1} - V_2\frac{R_f}{R_2}$$

Since all resistors are the same size, we substitute R for each resistor.

$$V_{out} = V_3\frac{R}{R} + V_4\frac{R}{R} - V_1\frac{R}{R} - V_2\frac{R}{R}$$
$$= (V_3 + V_4) - (V_1 + V_2)$$

The student should verify that balance is maintained.

Example 5-5 was planned to come out with balance maintained, but how do we maintain balance when we are building a circuit from scratch? We force a balance by adding zero at the proper gain to maintain balance. This is added to the side with the lowest gain. The method is shown in Example 5-7.

EXAMPLE 5-7

Set an adder-subtracter so that its output is

$$V_{out} = -4\text{ V}_1 - 2\text{ V}_2 + 10\text{ V}_3 + \text{V}_4$$

SOLUTION

It is convenient to choose $R_f = R'_f$, so let $R_f = R'_f = 100\ \text{k}\Omega$.

Now we can use Equation 5-5, in the form shown in Figure 5.3, since there are only four inputs, to find R_1, R_2, R'_1, and R'_2.

The coefficient of V_1 is $(R_f/R_1) = 4$, so

$$R_1 = \frac{R_f}{4} = \frac{100\ \text{k}\Omega}{4} = 25\ \text{k}\Omega$$

similarly

$$\frac{R_f}{R_2} = 2$$

so

$$R_2 = \frac{R_f}{2} = \frac{100\ \text{k}\Omega}{2} = 50\ \text{k}\Omega$$

$$R'_1 = \frac{R'_f}{10} = \frac{100\ \text{k}\Omega}{10} = 18\ \text{k}\Omega$$

and

$$R'_2 = \frac{R'_f}{1} = \frac{100\ \text{k}\Omega}{1} = 100\ \text{k}\Omega$$

Checking the balance we see that

$$\frac{R_f}{R_1} + \frac{R_f}{R_2} = 4 + 2 = 6$$

and

$$\frac{R'_f}{R'_1} + \frac{R'_f}{R'_2} = 10 + 1 = 11$$

The circuit is not balanced since the sum of the noninverting gains is five larger than the inverting gains. We must change our circuit so that its output is actually

$$V_{out} = -(4\ V_1 + 2\ V_2 + 5\ V_x) + (10\ V_3 + V_4)$$

but we will set $V_x = 0$ V so that the circuit has the output we desire. We now add a resistor R_x to the inverting input such that $R_f/R_x = 5$ as shown in Figure 5.4 to achieve our balance. The size of R_x is $R_f/5 = 20\ \text{k}\Omega$. Now

$$\frac{R_f}{R_1} + \frac{R_f}{R_2} + \frac{R_f}{R_x} = \frac{R'_f}{R'_1} + \frac{R'_f}{R'_2}$$

$$4 + 2 + 5 = 11 = 10 + 1 = 11$$

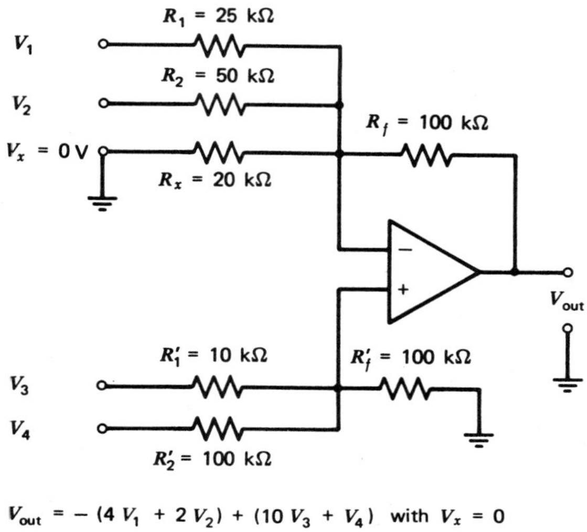

FIGURE 5.4 Balanced adder-subtracter.

so the circuit will operate accurately.

If the noninverting side has a smaller total gain, we add an R_x' from the noninverting input to ground of a size that will maintain a gain balance in a similar manner.

For stability the sum of the gains on each side should be one or greater, unless we know that the amplifier is stable for $A_{fb} < 1$.

5-4 DIRECT ADDER

If we wish to add two or more voltages directly so that $V_{out} = V_1 + V_2 + \cdots + V_n$, we can arrange a special case of the adder-subtracter to do so. Suppose we wish V_{out} to be $V_1 + V_2$. We set $R_f' = R_1' = R_2'$, and set $R_1 = R_f/n$ where n is equal to the number of inputs, in our case two. This is shown in Figure 5.5.

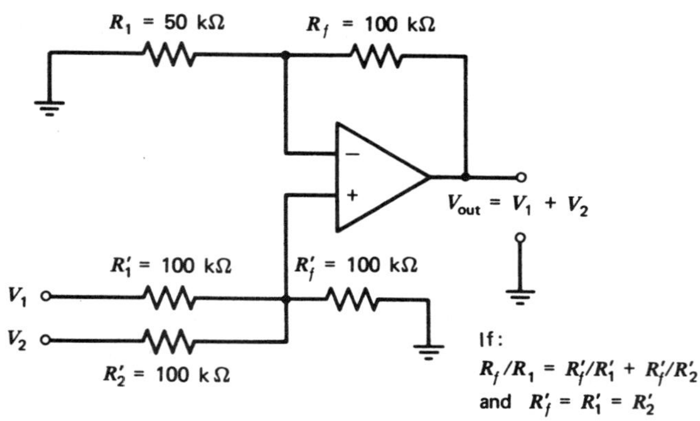

FIGURE 5.5 Two-input direct adder.

We can weight the inputs; for example, let $V_{out} = V_1 + 2\,V_2$, as long as

$$\frac{R_f}{R_1} = \frac{R_f'}{R_1'} + \frac{R_f'}{R_2'} + \cdots + \frac{R_f'}{R_n'}$$

for n inputs.

5-5 SIMULTANEOUS EQUATION SOLUTION

We mentioned earlier in this chapter that summing circuits can be used to solve simultaneous equations. We will demonstrate the method by doing an example, using adder-subtracters.

EXAMPLE 5-8

Solve the simultaneous equations for X and Y:
(a) $2X + 3Y = 40$.
(b) $2X + Y = 5$.

SOLUTION

First we must decide on a scale that can include all of our possible answers over the range of coefficients we will use. For instance, if the output voltage swing of the op-amp is ± 15 V, and X or Y will never be greater than 150, we can assign a scale of 0.1 V = 1. Thus the number X = 15 would correspond to 1.5 V on the X output. For this example we will use 0.1 V = 1. You may have to solve the equation algebraically for the limits that the variables may take. This does not limit the use of the circuit in that it is usually used to provide a continuous solution as part of a control network. If you reach the output limits, the scale of the coefficients can be changed until the answer fits the output limits.

Now let us solve one equation for X and the other for Y. Solving (a) for X we see

$$X = \frac{40 - 3Y}{2} = 20 - 1.5Y$$

Solving (b) for Y we find

$$Y = -2X + 5$$

We now set up one adder-subtracter so that its output is X,

$$X = 20\frac{R_f'}{R_1'} - \frac{R_f}{R_1}Y$$

The positive numbers will be put in on the noninverting side, and the negative numbers on the inverting side. Twenty is just a positive number, so we will put it in with a gain of one on the noninverting side. For the gain of one we just set $R'_f = R'_1$. For convenience we will set $R_f = R'_f$. Since

$$-1.5Y = \frac{R_f}{R_1} Y$$

we solve for R_1 to obtain

$$R_1 = \frac{R_f}{1.5}$$

We let $R'_f = R_f = 100\ \text{k}\Omega$, and obtain

$$R_f = R'_f = R'_1 = 100\ \text{k}\Omega$$

$$R_1 = \frac{100\ \text{k}\Omega}{1.5} = 66.7\ \text{k}\Omega$$

We now connect the first amplifier, as shown in Figure 5.6. Note that an R_x has been added to the noninverting input such that $R'_f/R'_x = 0.5$, so that the inverting and noninverting gains each equals 1.5.

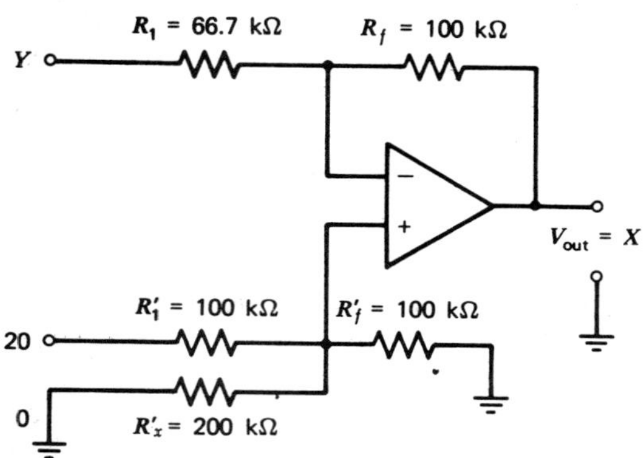

20 = 2 V for our scale of 1 V = 10
0 = 0 V in the balance input
Y = output from amplifier solving for Y (Fig. 6.7)

FIGURE 5.6 $X = 20 - 1.5Y$ connection.

Next we set up another amplifier to solve for Y. Since $Y = -2X + 5$,

$$Y = \frac{-R_f}{R_1}X + 5\frac{R_f'}{R_1'}$$

so

$$\frac{R_f}{R_1} = 2$$

and

$$\frac{R_f'}{R_1'} = 1$$

Letting $R_f = R_f' = 100\ \text{k}\Omega$, and solving for R_1 and R_1' we find

$$R_1 = \frac{100\ \text{k}\Omega}{2} = 50\ \text{k}\Omega$$

and

$$R_1' = \frac{100\ \text{k}\Omega}{1} = 100\ \text{k}\Omega$$

The inverting gain is 2, but the noninverting gain is only 1. To achieve balance we must add a zero input with a gain of one to the noninverting side, that is, $R_f'/R_x' = 1$.

By inspection we see that $R_x' = 100\ \text{k}\Omega$. Now we build the Y circuit as shown in Figure 5.7.

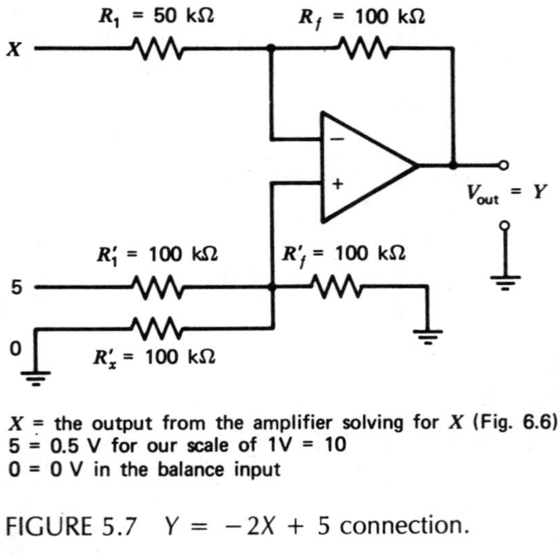

X = the output from the amplifier solving for *X* (Fig. 6.6)
5 = 0.5 V for our scale of 1V = 10
0 = 0 V in the balance input

FIGURE 5.7 $Y = -2X + 5$ connection.

To solve the simultaneous equation, we simply connect the output of the amplifier solving for X to the X input of the amplifier solving for Y and the output of the amplifier solving for Y to the Y input of the amplifier solving for X. The answer is read on the appropriate output. The complete solution circuit is shown in Figure 5.8.

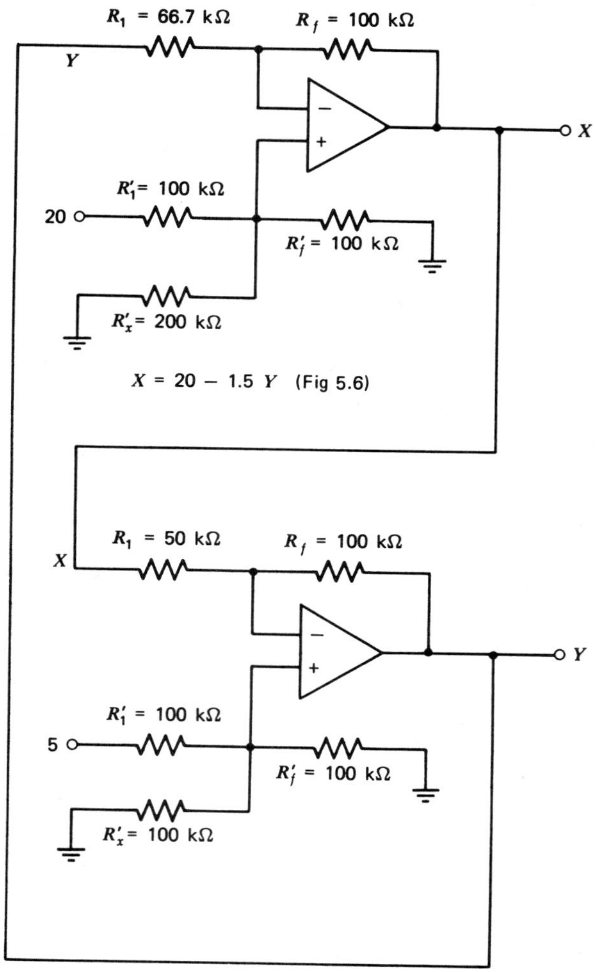

$Y = -2x + 5$ (Fig. 5.7)
$X = 6.25 = 0.625$ V
$Y = 17.5 = 1.75$ V

FIGURE 5.8 Circuit for solving simultaneous equations $2X + 3Y = 40$ and $2X + Y = 5$.

Note that it would require three adder-subtracters with three inputs each to solve three simultaneous equations containing three unknowns.

We can solve simultaneous equations using only inverters and inverting adders. This often takes more amplifiers than if we use adder-subtracters, but common-mode error and balance problems are eliminated. This method is shown in Figure 5.9, where the equations of Example 5-8 are solved. Note that some inputs are of opposite polarity to their sign in the equations. This gives them the correct polarity at the output of the inverting adder. For instance, to get X out of an inverting adder we must put $-X$ in, or to get 5 out we must put -5 in.

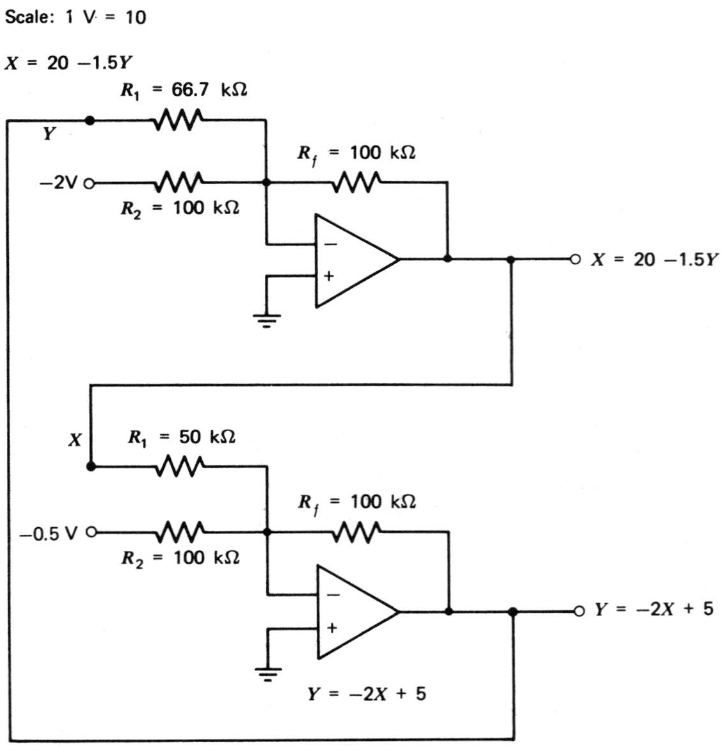

FIGURE 5.9 Simultaneous solution of $2X + 3Y = 40$ and $2X + Y = 5$ using inverting adders.

5-6 PROPORTIONAL CONTROL

Op-amp summing circuits are ideal for proportional control circuits. A *proportional control circuit* is one in which the output voltage of the control (which drives the controlled element) is proportional to the voltage difference between the system's set-point voltage (which tells the controlled element what to do) and the system's state voltage (which indicates what the controlled element is doing).

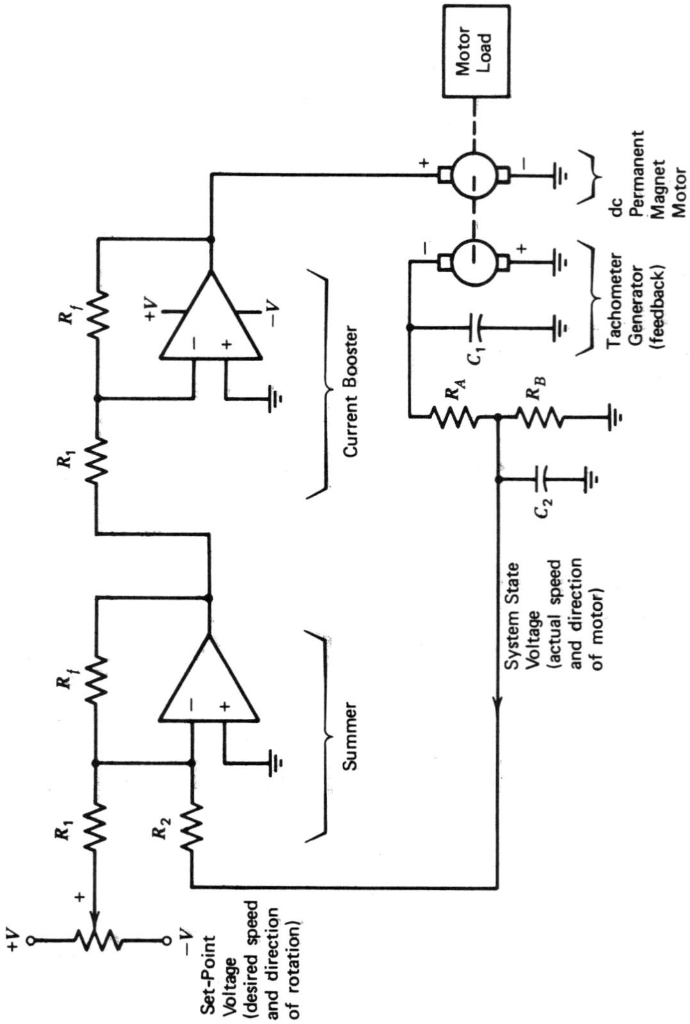

FIGURE 5.10 Proportional control example, dc motor speed control.

The concept of proportional control is illustrated with a dc motor speed control. A diagram of the control is shown in Figure 5.10. The parts of the system are the following:

1. A *set-point voltage* tells the motor which direction and how fast to turn. The *magnitude* of the set-point voltage tells the motor how *fast* to turn, and the *polarity* of the set-point voltage tells the motor which *direction* to turn.
2. A *summing circuit* compares the set-point voltage to the system state voltage and develops an output voltage proportional to the difference between the set-point and state voltages. This voltage is called the *error voltage*.
3. A *current booster*, often called a servoamplifier, provides the motor driving voltage with sufficient current to operate the motor. For our system, the motor must reverse direction so the current booster must supply both a positive and negative voltage. The exact configuration of this circuit is unimportant as long as it provides the necessary voltage and current to operate the motor. It can be an op-amp with a current booster such as those described in Chapter 9.
4. The *motor* for our example is a permanent magnet (pm) dc reversible motor. A shunt motor could be used just as easily.
5. A *signal* proportional to the motor speed is needed for the control. Frequently a small permanent magnet motor called a *tachometer generator* is used for this purpose. The tach generator shaft is connected to the motor shaft, either directly or through gearing, so that the motor drives the tach generator. If the tach generator provides dc output, it will have a commutator; if it is an ac type, it will have slip rings to tap off the armature voltage. In our example we are using a dc tach generator; an ac tach generator could be used if its output is rectified.

The voltage out of the tach generator will be proportional to the speed of its armature that is connected to the motor shaft. The polarity of the voltage from the tach generator will reverse if the direction of rotation reverses. The tach generator will be connected so that its output is opposite in polarity to the set-point voltage. Capacitors C_1 and C_2 filter the brush noise out of the tach generator output whereas R_A and R_B reduce the magnitude of the tach generator output voltage if necessary. The output of the tach generator is fed to the other input of the summer.

The circuit operates as follows:

1. The set-point voltage is set to the value that provides the desired motor speed in the desired direction.
2. The set-point voltage and state voltage will be different, and an error voltage results at the summer input. The summer amplifies the error voltage (the amount of amplification is set by the ratios of R_f/R_1 and R_f/R_2) and applies it to the input of the current amplifier.
3. The current amplifier amplifies the voltage by the desired amount and provides a voltage of the proper polarity and current to start the motor turning in the desired direction.
4. The motor causes the tach generator to turn and supply a state voltage opposite in phase to set-point voltage. This voltage will be proportional to the speed of the motor.
5. The state voltage is fed back to the input of the summer through R_2 and cancels a portion of the set-point voltage, since the state and set-point voltages are out of phase. When the state voltage and set-point voltage difference provides an error voltage of the proper magnitude, the system is in equilibrium and the motor runs at the desired speed. The state voltage must be set so that it will not cancel *all* of the set-point voltage or the error voltage would go to zero and the motor would stop. This would cause the sequence of 1 through

5 to repeat, and the motor would stop and start, stop and start in a very jerky manner that is not desired. The resistor ratios in the summer and the tachometer output are set so the state voltage cancels just enough of the set-point voltage to provide smooth operation.

The advantage of a proportional control in a motor speed control is that it keeps the motor speed constant under varying motor loads. Assume the motor is running at a constant speed and the load on the motor increases. The increased load causes the speed of the motor to start to decrease, which causes the tach generator output to decrease. The state voltage decreases—it is the tach generator output—while the set-point voltage remains constant, causing the error voltage increase. The increase in error voltage is amplified and applied to the motor, causing its speed to increase until a new equilibrium is reached in which the motor speed is nearly the same speed as it was before the increased load occurred. If the motor load is decreased, the process is reversed to keep the motor speed from increasing.

This example is one of very many methods of applying summing circuits to proportional speed control. A similar type of circuit is used to provide proportional control to position valves, antennas, and other remotely positioned equipment. Proportional control is used to control processes throughout industry. The concept of using a summing circuit to compare the set-point and state signals and provide an error voltage proportional to the difference is central to all proportional control systems.

5-7 SIGNAL MIXER

The adders shown in this chapter can be used to mix signals linearly. One application is in audio-mixing. Instruments picked up at several microphones can be mixed with voice, combined, and routed to a single power amplifier by a circuit such as the one shown in Figure 5.11*a*. Switching to cut out amplifier channels and ac input coupling can be used as shown in Figure 5.11*b*. Each input has independent gain adjust.

EXAMPLE 5-9

Calculate the components in Figure 5.11*a* so that each channel can be adjusted in gain from 1 to 50 and the minimum overall gain is 10 with all inputs adjusted to minimum gain.

The overall gain of 10 can be obtained by setting the inverting summer gain to 10. If $R_f' = 500\ \text{k}\Omega$ then

$$R_2 = R_3 = R_4 = \frac{R_f'}{10} = \frac{500\ \text{k}\Omega}{10} = 50\ \text{k}\Omega$$

The channel gain is set at 1 by letting $R_1 + P = R_f$. If $R_f = 500\ \text{k}\Omega$, then $R_1 + P = 500\ \text{k}\Omega$.

$$R_1 = \frac{R_f}{A_{fb}(\text{max})} = \frac{500\ \text{k}\Omega}{50} = 10\ \text{k}\Omega$$

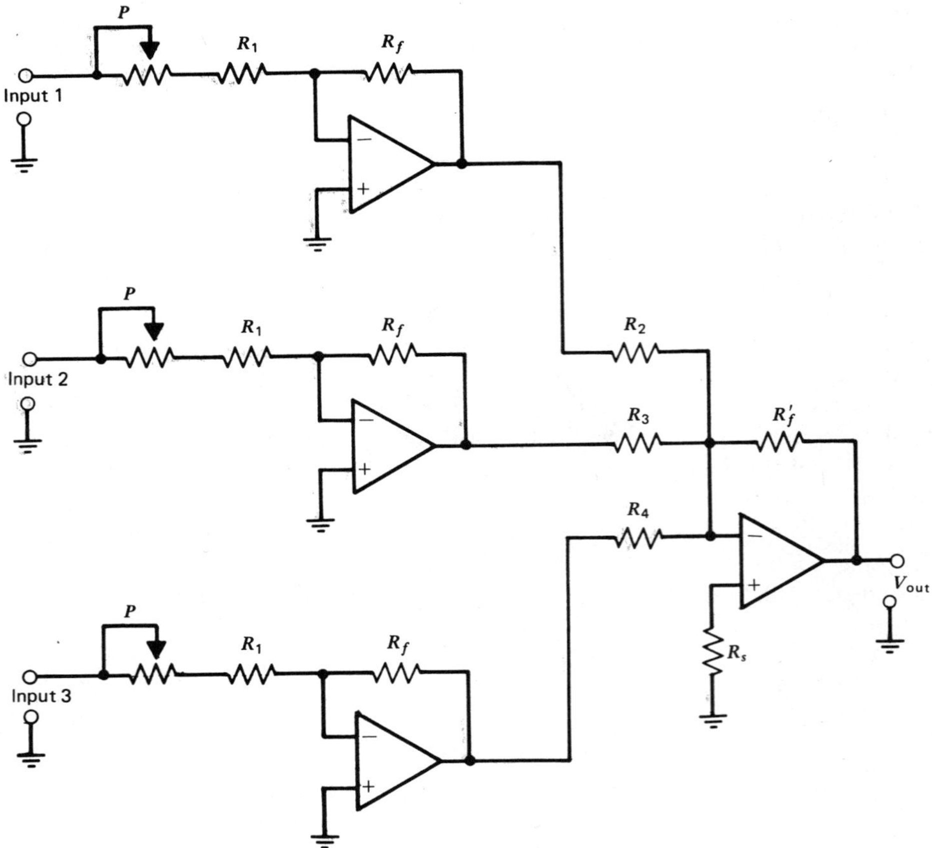

(a) Amplifiers and summer

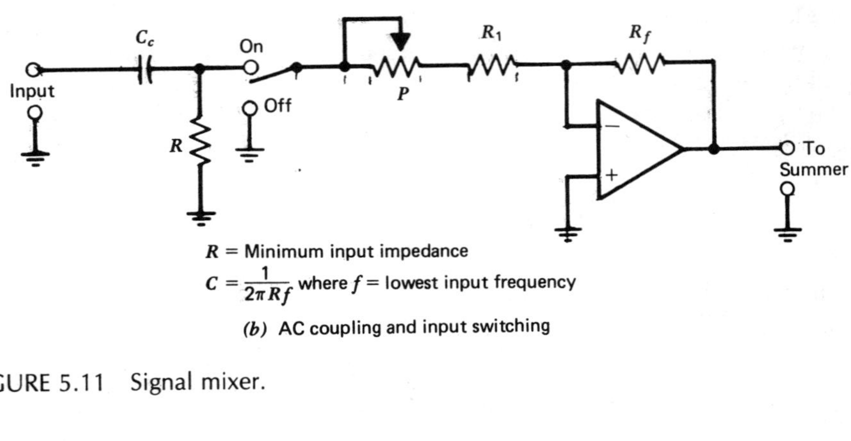

R = Minimum input impedance

$C = \frac{1}{2\pi Rf}$ where f = lowest input frequency

(b) AC coupling and input switching

FIGURE 5.11 Signal mixer.

Thus

$$P = R_f - R_1 = 500\ \text{k}\Omega - 10\ \text{k}\Omega = 490\ \text{k}\Omega$$

Since a 500 kΩ potentiometer is a standard value and 490 kΩ is not, we would probably settle for a minimum channel gain of $R_f/(P + R_1) = 500\ \text{k}\Omega/510\ \text{k}\Omega = 0.98$

R_s for the summer provides compensation for bias current induced offset voltage and is

$$R_s = R_f' \| R_2 \| R_3 \| R_4$$
$$= 500\ \text{k}\Omega \| (50\ \text{k}\Omega/3) = 16.1\ \text{k}\Omega$$

SUMMARY

1. The inverting adder sums its input voltages and inverts the result.
2. The inverting scaling adder, a special case of the inverting adder, assigns weights to the various inputs.
3. The averager is also a special case of the inverting adder. Its output voltage is the numerical average of its inputs.
4. The adder-subtracter can add and subtract simultaneously, but the sum of the inverting gains must equal the sum of the noninverting gains. We can easily force the sum of the inverting gains to equal the sum of the noninverting gains by adding one resistor.
5. The direct adder is a special case of the adder-subtracter in which only the noninverting inputs of the circuit are used.
6. Summing circuits can be used to solve algebraic equations or for providing proportional control.

SELF-TEST QUESTIONS

5-1 The inverting adder of Figure 5.1 has $R = 100\ \text{k}\Omega$, $V_1 = 4$ V, and $V_2 = -1$ V. Calculate V_{out}.

5-2 The scaling adder of Figure 5.2 has $R_f = 1\ \text{M}\Omega$, $R_1 = 100\ \text{k}\Omega$, $R_2 = 50\ \text{k}\Omega$, and $R_3 = 200\ \text{k}\Omega$. Calculate V_{out} if $V_1 = 0.1$ V, $V_2 = 0.3$ V, and $V_3 = -0.4$ V.

5-3 Calculate R_1, R_2, and R_3 for the scaling adder of Figure 5.2 so that $V_{out} = -(4\ V_1 + 3\ V_2 + 6\ V_3)$ if $R_f = 1\ \text{M}\Omega$.

5-4 The adder-subtracter of Figure 5.3 has $R_f = 200\ \text{k}\Omega$, $R_f' = 200\ \text{k}\Omega$, $R_1 = 10\ \text{k}\Omega$, $R_2 = 25\ \text{k}\Omega$, $R_1' = 20\ \text{k}\Omega$, and $R_2' = 40\ \text{k}\Omega$. Calculate the R_x necessary to balance the circuit and V_{out} if $V_1 = 0.1$ V, $V_2 = -0.3$ V, $V_3 = 0.2$ V, and $V_4 = 0.1$V.

5-5 The adder-subtracter of Figure 5.3 has $R_f = R_f' = 1\ \text{M}\Omega$. Calculate R_1, R_2, R_1', R_2', and R_x so that $V_{out} = 3\ V_3 + 4\ V_4 - (2\ V_1 + V_2)$.

5-6 Calculate R_1, R_1', and R_2' of the direct adder on Figure 5.5 so that $V_{out} = 2\ V_1 + 3\ V_2$.

5-7 Set up an adder-subtracter circuit to solve the equations: $X + Y = 8$; $3X - 2Y = 4$. Scale the inputs so the outputs fit between ±15 V.

5-8 State the advantage of using only inverting amplifiers for analog computations. State the disadvantage.

5-9 State the purpose of the summing amplifier of Figure 5.10.

5-10 State the function of the tachometer generator of Figure 5.10.

5-11 For the circuit of Figure 5.11*a*, calculate R_1, P, R_2, R_3, and R_4 if the gain for each channel of amplification is to be variable from 2 to 20 and the summer gain is one. Let $R_f = R_f' = 200\ k\Omega$.

If you cannot answer certain questions, place a check next to them and review appropriate parts of the text to find the answers.

LABORATORY EXERCISE

Objective

After completing this laboratory, the student should be able to calculate the components for and set up inverting adder, scaling adder, adder-subtracter, and direct adder circuits to perform summations of the student's choosing. The student should also be able to solve simultaneously two equations with two unknowns using adder-subtracters.

Equipment

1. Two Fairchild μA741 operational amplifiers or equivalent.
2. 2% tolerance resistor assortment.
3. Power supply, ±15 V dc.
4. Voltmeter or oscilloscope capable of measuring as low as 2 mV dc.
5. Potentiometer assortment including 10 kΩ.
6. Breadboard such as the EL Instruments SK-10 mounted on vectorboard.

For this laboratory ac voltages of moderate frequency can be used if they are all of the same phase, however dc voltages are more convenient. The easiest way to obtain a desired value of dc voltage as an input is to use a potentiometer and adjust it as shown in Figure 5.12.

Be sure to null the offset carefully with all circuit inputs grounded throughout this laboratory.

Procedure

1. Inverting adder (scaling).
   (a) Set up an inverting adder (Figure 5.13) such that $V_{out} = -(3\ V_1 + 2\ V_2 + 5\ V_3)$.

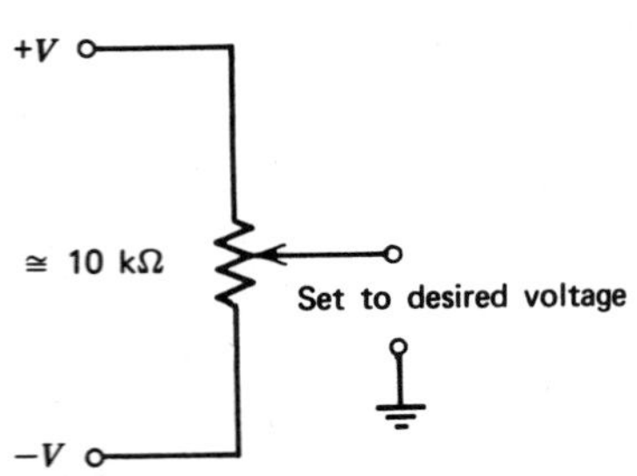

FIGURE 5.12 Use potentiometer to obtain dc voltage.

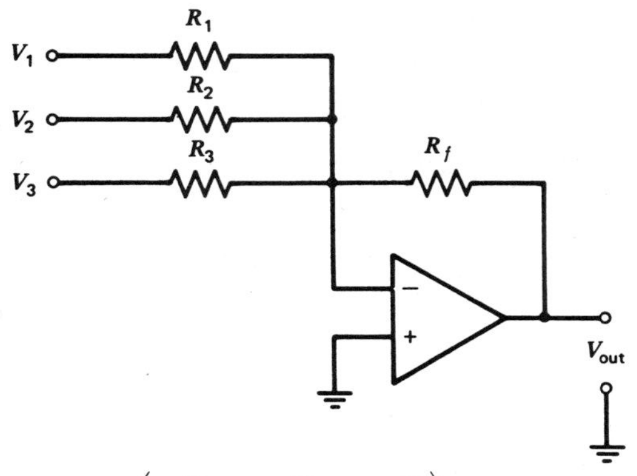

$$V_{out} = -\left(V_1\frac{R_f}{R_1} + V_2\frac{R_f}{R_2} + V_3\frac{R_f}{R_3}\right)$$

FIGURE 5.13 Inverting adder.

Use $R_f \simeq 100\ \text{k}\Omega$, calculate R_1, R_2, and R_3.

(b) Calculate and measure V_{out} for $V_1 = 3$ V, $V_2 = 2$ V, $V_3 = -1$ V.

2. Averager.

(a) Set $R_1 = R_2 = R_3$ in the circuit in Figure 5.13. Calculate R_f so that the circuit will average the inputs.

(b) With $V_1 = V_2 = V_3$, calculate and measure the output voltage.

3. Adder-subtracter.

(a) Let $R_f' = R_f \simeq 100\ \text{k}\Omega$ in the circuit of Figure 5.14. Calculate R_1, R_2, R_1', and R_2' so that

$$V_{out} = -(2\ V_1 + 3\ V_2) + (4\ V_3 + 2\ V_4)$$

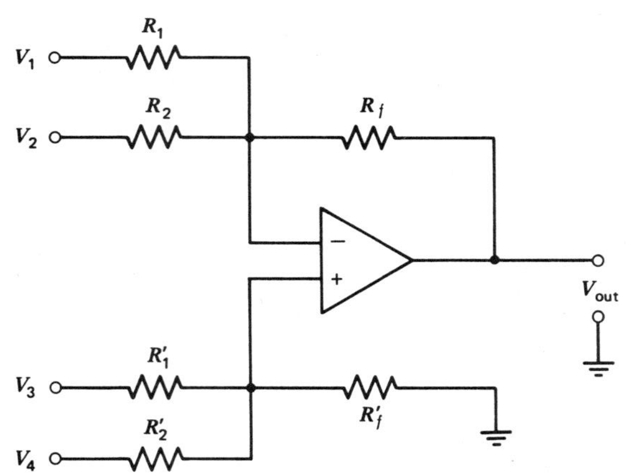

$$V_{out} = -\left(V_1\frac{R_f}{R_1} + V_2\frac{R_f}{R_2}\right) + \left(V_3\frac{R_f'}{R_1'} + V_4\frac{R_f'}{R_2'}\right)$$

FIGURE 5.14 Adder-substracter.

Don't forget to balance the circuit.

(b) Construct the circuit. Set $V_1 = V_2 = V_3 = V_4 = 2$ V. Calculate and measure the output voltage.

4. Simultaneous equation solution.

(a) Choose two simultaneous equations with two unknowns that you can solve. If you can't think of any you want to solve use

$$2X + 3Y = 12$$

and

$$3X - 2Y = 6$$

Pick a suitable scale for your amplifier supply voltage and answer.

(b) Construct a solution circuit using adder-subtracter circuits. Don't forget to balance the amplifiers.

(c) Check your measured and calculated answers.

Note. Each adder-subtracter must be nulled with its output open circuited and all inputs grounded.

The low closed loop gain of the adder-subtracters connected to solve the equation may cause oscillation in some amplifiers. A 0.01 μF capacitor connected across R_f of each amplifier will usually stop the oscillation. If the capacitor does not stop the oscillation, it will be necessary to raise the total closed loop gain of the amplifiers. This can be accomplished by connecting a resistor (≅ 10 kΩ) from *both* the inverting and noninverting input terminals to ground.

Discussion

For each part discuss the possible cause of any discrepancies between the measured and calculated results.

CHAPTER

6

INTEGRATORS AND DIFFERENTIATORS

Two of the most important analog computing circuits are the integrator and differentiator. The integrator is useful in control networks anywhere a differential equation must be solved or the integral of a voltage is needed. Integrators are also useful for generating sawtooth and triangle waveforms. A differentiator is useful whenever an output porportional to a rate of change is needed.

OBJECTIVES

After completing the study of this chapter and the self-test questions, the student should be able to:

1. Draw from memory a compensated integrator and differentiator.
2. Given the desired response, calculate the components for a compensated integrator and differentiator.
3. Given the components, find the output voltage of a compensated integrator and differentiator.
4. Calculate the components for an integrator used as a low-pass filter and a differentiator used as a high-pass filter and as a band-pass filter.
5. State the operation of a three-mode integrator in each mode.
6. Define and state the purpose of bounding.
7. Calculate the frequency given the components and the components given the frequency of a double integrator used as an oscillator.
8. Draw from memory the frequency response of a compensated integrator and differentiator.
9. List and draw from memory three different types of integrators and differentiators (i.e., augmenting integrator, summing integrator).
10. Perform the laboratory for Chapter 6.

6-1 INTEGRATOR

We can think of integration as determining the area under a curve. Since an op-amp integrator operates on voltage over a period of time, we can think of it as providing a sum of voltage over time as illustrated in Figure 6.1.

An operational amplifier integrator has a circuit configuration as shown in Figure 6.2. To determine the integrating property of this circuit, let us first recall some

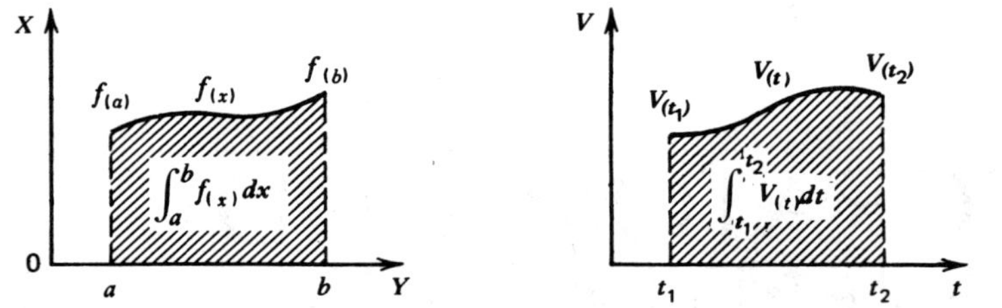

(a) Integral as area under a curve (b) Integral of voltage over time

FIGURE 6.1 The integral as an area.

relations resulting from the definition of capacitance. Capacitance, C, is defined as

$$C = \frac{Q}{V}$$

where

Q = electric charge
V = voltage

Thus we see that charge is

$$Q = CV$$

and a change in charge/unit time, that is, the capacitor current, is

$$i_c = \frac{dQ}{dt} = C\frac{dV}{dt} \tag{6-1}$$

If the op-amp is close to ideal with $I_B \simeq 0$, A_{ol} very high so that $v_i \simeq 0$, then

$$i_R = i_c$$

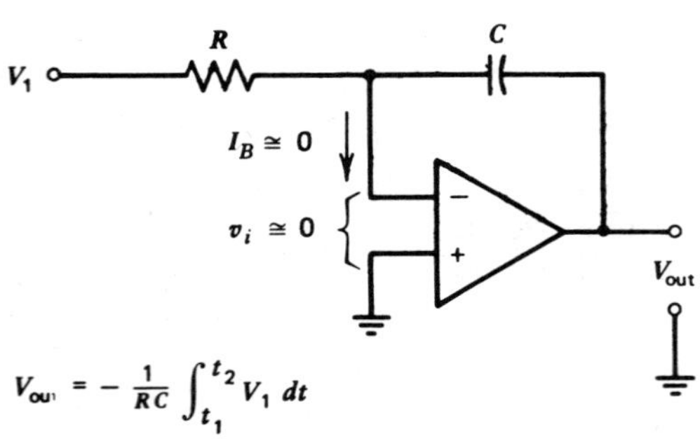

FIGURE 6.2 Op-amp integrator.

From Equation 6-1 we find

$$i_c = \frac{dQ}{dt} = C\frac{dV_c}{dt} = i_R$$

Since $v_i \simeq 0$, and $V_c = -V_{out}$,
we write

$$i_c = -C\frac{dV_{out}}{dt} = \frac{V_1}{R} = i_R$$

Solving now for dV_{out} we find

$$dV_{out} = -\frac{1}{RC}V_1\,dt$$

Integrating to obtain V_{out}, we find

$$V_{out} = -\frac{1}{RC}\int V_1\,dt \tag{6-2}$$

The limits of the integral in Equation 6.2 are the times, t_1 to t_2, that we look at the signal. To calculate the integral of a voltage waveform, we must first express that voltage as a function of time.

We will now look at the integral of some common waveforms.

EXAMPLE 6-1

(a) What will the output waveform of an integrator look like if the input is a step waveform such as that in Figure 6.3*a*?

(b) If $R_1 = 1\ \text{M}\Omega$, $C = 0.1\ \mu\text{F}$, and $V_{in} = 1$ V, what value will V_{out} be 3 ms after t_0?

SOLUTION

(a) Writing the step input as a function of time, we see

$$V_1 = V \qquad t \geq t_0$$
$$V_1 = 0 \qquad t < t_0$$

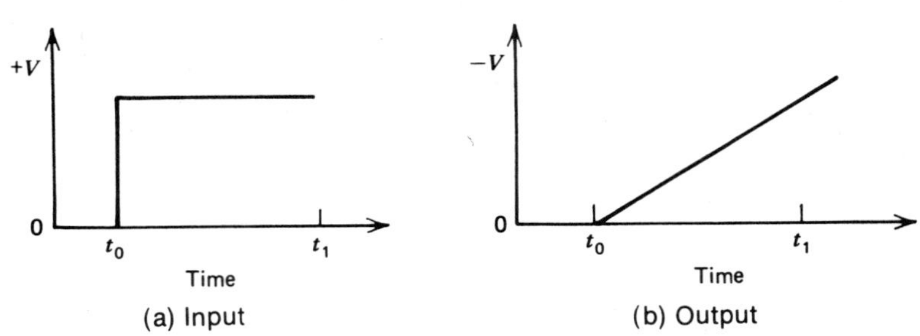

FIGURE 6.3 Step waveform response.

Integrating the first part,

$$V_{out} = -\frac{1}{RC}\int V\,dt = -\frac{1}{RC}(Vt)$$

This is a linear ramp voltage of opposite polarity to the input.

(b) To answer this, we simply evaluate our V_{out} equation from $t_0 = 0$ to $t_1 = 3$ ms.

$$\begin{aligned} V_{out} &= -\frac{1}{RC}Vt\,\Bigg|_{t=0}^{t=3\text{ ms}} = -\frac{1}{(1\text{ M}\Omega)(0.1\ \mu\text{F})}(1\text{ V})\,\Bigg|_{t=0}^{t=3\text{ ms}} \\ &= -10(1\text{ V})(3\text{ ms}) - [-10(1\text{ V})(0)] \\ &= -30\text{ mV} \end{aligned}$$

Note the integral results in the expression

$$V_{out} = -\frac{1}{RC}(t_1 - t_0) = -\frac{1}{RC}\Delta t$$

which can be used when V_{in} is a square wave or a pulse.

Of course, integration ceases when, after a time, the output voltage reaches the maximum output voltage the amplifier can supply.

EXAMPLE 6-2

An integrator has $R = 10\text{ k}\Omega$ and $C = 0.1\ \mu\text{F}$. V_{in} is a 1 kHz square wave of 5 V amplitude (10 V_{pp}). What is the output?

SOLUTION

Since this is a repetitive waveform, we can characterize the output by looking at one full period. First we write the input as a function of time (see Figure 6.4*a*).

$$V_{in} = 5\text{ V} \qquad t_1 < t \leq t_2$$
$$V_{in} = -5\text{ V} \qquad t_2 < t \leq t_3$$

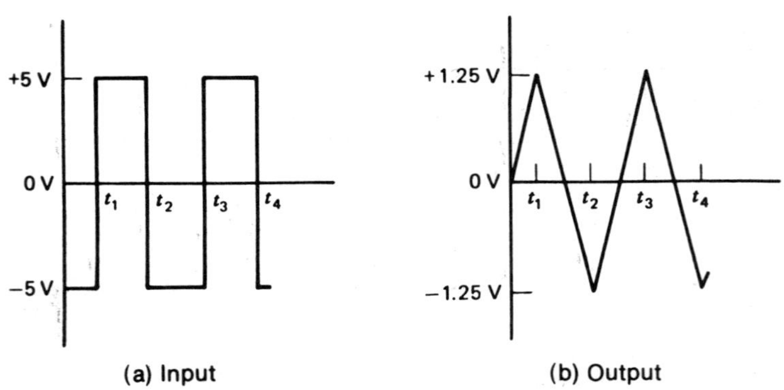

FIGURE 6.4 Integrator response to square wave.

We can integrate this one half period at a time. To characterize the output, we need the shape of the waveform and the voltages at the end of each half period.

Substituting into Equation 6-2 and integrating, we see

$$V_{out} = -\frac{1}{RC}\int V\,dt = -\frac{V}{RC}t$$

which is a linear ramp over each half period. We would expect this from our first example since V_{in} is a step function over each half period.

The voltage at the end of the first half period, t_1 to t_2, is

$$\begin{aligned} V_{out} &= -\frac{V}{RC}t \Bigg|_{t\,=\,0}^{t\,=\,0.5\text{ ms}} = -\frac{5\text{ V}(0.5\text{ ms})}{(10\text{ k}\Omega)(0.1\ \mu\text{F})} \\ &= \frac{2.5(10^{-3})\text{V}}{1 \times 10^{-3}} = -2.5\text{ V} \end{aligned}$$

The voltage at the end of the second half period, t_2 to t_3, is

$$\begin{aligned} V_{out} &= -\frac{(-5\text{ V})}{RC}t \Bigg|_{t\,=\,0.5\text{ ms}}^{t\,=\,1\text{ ms}} \\ &= -\frac{(-5\text{ V})}{RC}(t_3 - t_2) \\ &= -\frac{(-5\text{ V})(0.5\text{ ms})}{(10\text{ k}\Omega)(0.1\ \mu\text{F})} \\ &= 2.5\text{ V} \end{aligned}$$

The output is illustrated in Figure 6.4*b*. The output will be 2.5 V_{pp} since the output for any input must start from where it is at the time (that is, when V_{in} goes to -5 V, the output starts positive from -1.25 V).

If the square wave in Example 6-2 had been 5 V_{pp}, 2.5 V_p, the output would have been 1.25 V_{p-p}, 0.625 V_p centered on zero volts.

EXAMPLE 6-3

An integrator has a ramp voltage input such as in Figure 6.4*b*. What will be the shape of the output?

SOLUTION

Expressing V_{in} as a funciton of time between t_1 and t_2 we get

$$V_{in} = -\frac{V}{RC}t = -Kt \qquad t_1 \leqslant t \leqslant t_2$$

where $K = -\dfrac{V}{RC}$

FIGURE 6.5 Response to linear ramp.

for the waveform in Figure 6.4*b*.

Using Equation 6-2, we see

$$V_{out} = -\frac{1}{RC}\int_{t_1}^{t_2} -Ktdt = \frac{K}{2RC}\,t^2\Big|_{t_1}^{t_2}$$

The output is an exponential function of time, as shown in Figure 6.5.

6-2 PRACTICAL INTEGRATOR CONSIDERATIONS

So far we have considered the operational amplifier in our integrator as ideal. Actually a practical op-amp has some offset voltage and requires some bias current. The offset voltage is integrated as a step function causing a linearly increasing ramp on the output with a polarity that depends on the polarity of V_{os} and a value that depends on the size of V_{os}. The bias current flows through the feedback capacitor causing a ramp output. Over a period of time these two effects cause the capacitor to charge to the maximum output the amplifier can supply. This slow charging of the feedback capacitor (by the integral of V_{os} and I_B) sets a limit on the amount of time we can integrate accurately. In addition, V_{os} adds to the capacitor voltage, which is also V_{out}, and causes an error by that amount. Thus the expression for V_{out} of Equation 6-2 is

$$V_{out} = -\frac{1}{RC}\int V_1 dt + \frac{1}{RC}\int V_{os} dt + \frac{1}{C}\int I_B dt + V_{os} \qquad (6\text{-}3)$$

The last three terms of Equation 6-3 are error terms. The first term is the useful output, the other three terms add to or subtract from the useful output causing error.

The offset voltage term is reduced by (1) using an op-amp of inherently low V_{os}, (2) periodically resetting the integrator (discharging the capacitor to some preselected value), or (3) shunting C with a resistor R_d as shown in Figure 6.6. All three corrective measures may be used to reduce the error terms.

Shunting the feedback capacitor with R_d limits the offset voltage contribution to $(R_d/R)\ V_{os}$ instead of $V_{os}\ (A_{ol})$ at low frequencies, where the capacitor is virtually

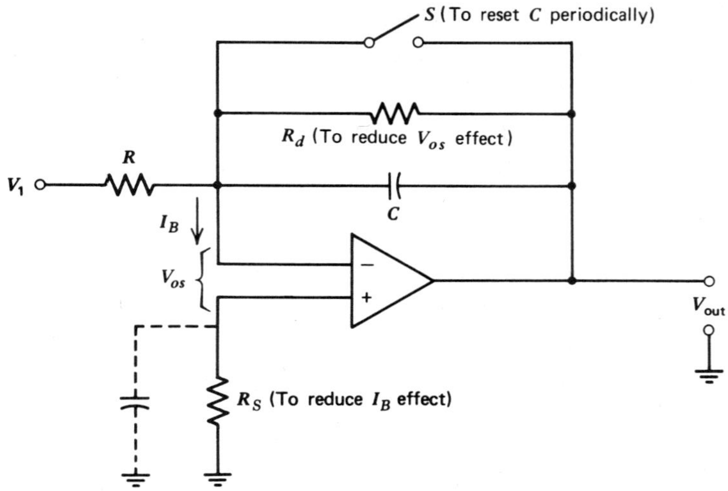

FIGURE 6.6 Integrator dc error correction.

an open circuit. Unfortunately, this also limits the lowest frequency at which we can integrate. For example:

$$f(\text{operation}) = \frac{3}{2\pi R_d C}$$

will allow us an accuracy of about 5%. As operating frequency increases above $f = 1/2\pi R_d C$, so will accuracy.

The bias current error term of Equation 6-3 can be reduced by using an FET input op-amp and by adding a resistor R_s from the noninverting input to ground equal to $R \| R_d$ if R_d is used.[1] This compensates for I_B in the same way I_B is compensated in an amplifier (see Chapter 2). This is illustrated in Figure 6.6. With R_s, the third term of Equation 6-3 now becomes

$$\frac{1}{C} \int I_{os} dt$$

Sometimes R_s is shunted by a capacitor C_s such that $(R_{in}\text{ amp})(C_{in}\text{ amp}) \simeq R_s C_s$ for frequency compensation.

The use of a chopper-stabilized amplifier also helps eliminate the error due to V_{os}, I_B, and I_{os}.

When an open loop integrator (no R_d) is nulled for offset prior to use, C must be shunted with a resistor, which is removed after nulling.

The capacitor used in long-term integrators must be of very high shunt resistance. Polystyrene and teflon dielectrics offer good long-term stability. For shorter term integration of ac signals, around 1 kHz or greater, good quality mylar capacitors often are satisfactory.

[1] Let $R_s = R$, if R_d is not used.

6-3 INTEGRATOR FREQUENCY RESPONSE

The frequency response of an integrator, using a fully frequency-compensated op-amp, is shown in Figure 6.7. We see that for an open loop integrator (Figure 6.2) the integrating range is from its first break frequency to somewhat beyond the integrator crossing frequency. The first break frequency of the integrator is $1/(2\pi RA_{ol}C)$, with the Miller effect causing A_{ol} to appear in the equation. Accuracy improves to about 5% at about three times this frequency and holds to less than 1% from 10 times this frequency through the crossing frequency for a very good op-amp. The crossing frequency of the integrator is $1/(2\pi RC)$.

When R_d is added for low frequency stability, we see, in Figure 6.7, that the first break frequency is raised. This is due to the Miller effect reduction caused by the low frequency gain reduction from A_{ol} to R_d/R. The first break frequency of the compensated integrator is

$$f_x = \frac{1}{2\pi R\left(\frac{R_d}{R}\right)C} = \frac{1}{2\pi R_d C} \tag{6-4}$$

The integrating range is reduced to between $1/(2\pi R_d C)$ and a little above $1/(2\pi RC)$.

EXAMPLE 6-4

An integrator is to be used at about 20 kHz. The accuracy should be within 2%, and the response desired is $V_{out} = -5000 \int V_{in}dt$. Find R, C, and R_d.

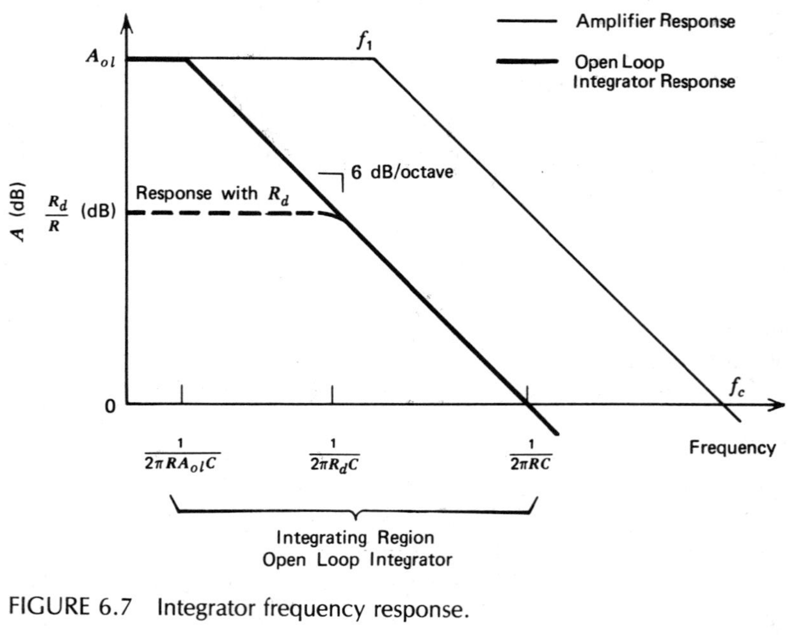

FIGURE 6.7 Integrator frequency response.

SOLUTION

Pick a reasonable, obtainable value of C and calculate R. Let $C = 0.1\ \mu F$.

$$V_{out} = -\frac{1}{RC}\int V_{in}dt$$

$$-5000 = -\frac{1}{RC}$$

therefore

$$R = \frac{1}{5000\ C} = \frac{1}{5000(0.1\ \mu F)} = 2\ k\Omega$$

We may have to try several values of C to get reasonable values for both R and C, but $R = 2\ k\Omega$ and $C = 0.1\ \mu F$ are reasonable.

Since the desired accuracy is 2%, if the first break frequency of the compensated integrator is less than one tenth the integrating frequency, we will have more than the desired accuracy. Set the lowest integrating frequency to 2 kHz and calculate R_d.

$$R_d = \frac{1}{2\pi f_1 C}$$

where f_1 = first break frequency

$$R_d = \frac{1}{2\pi f_1 C} = \frac{1}{6.28(2\ kHz)(0.1\ \mu F)} = 796\ \Omega$$

This is a ridiculously low value for R_d as our low frequency gain is less than 1. This is not too unusual a result. We now pick R_d for a reasonable low frequency gain and check the first break frequency of the integrator to assure that it is below 2 kHz. Let $R_d/R = 2000$, then $R_d = 4\ M\Omega$, and check f_1.

$$f_1 = \frac{1}{2\pi R_d C} = 0.39\ Hz < 2\ kHz$$

If we let $R = 2\ k\Omega$, $C = 0.1\ \mu F$, and $R_d = 4\ M\Omega$, the integrator will have the desired accuracy at 20 kHz.

Note that the frequency response of a compensated integrator (with R_d, Figure 6.6) is that of a low-pass filter having a 6 dB/octave roll-off, with the advantage of gain. This circuit can be used whenever this active filter requirement is met. Set R and R_d for the desired A_{fb} and choose C so that

$$C = \frac{1}{2\pi R_d f_1} \tag{6-5}$$

EXAMPLE 6-5

We wish to use an integrator as a low-pass filter with an $f_1 = 3$ kHz and $A_{fb} = 20$.

SOLUTION

Let $R_1 = 10\ k\Omega$ and $R_d = R_1(20) = 10\ k\Omega(20) = 200\ k\Omega$.

$$C = \frac{1}{2\pi R_d F_1} = \frac{1}{2\pi(200\ k\Omega)(3\ kHz)} = 265\ pF$$

The values to put in the circuit are therefore $R_1 = 10\ k\Omega$, $R_d = 200\ k\Omega$, and $C = 265$ pF.

6-4 BOUNDING

Bounding means limiting the maximum output voltage of an op-amp to some voltage less than it is capable of putting out, even though the input exceeds its maximum allowed voltage. Ideally the bound circuit has no effect on the output until the assigned bounding voltage is reached, whereupon the bound circuit prevents the output voltage from going any higher. Bounding is used to prevent the op-amp from *saturating,* that is, ever reaching the maximum output voltage that the op-amp is capable of delivering. This is desirable because when an amplifier saturates, it recovers to its initial condition (once the excessive input voltage is removed) much more slowly than an amplifier that is held out of saturation by a bound. Without bounding, chopper-stabilized amplifiers can require several seconds to recover from saturation.

Bounding is used on many types of op-amp circuits *other* than integrators, but it is particularly useful on integrators since the stored change on the feedback capacitor tends to hold the op-amp in saturation once it occurs.

One type of bounding circuit is shown in Figure 6.8*a*. The back-to-back zener diodes of Figure 6.8*a* comprise the simplest bound circuit. The output voltage is limited to

$$V_{out}(max) \simeq V_Z + 0.7\ V = V_{Bound}$$

with this circuit. Once V_{out} exceeds the bound voltage, the zener diode begins to conduct. Since the zener has a very low impedance during conduction, the gain of the amplifier is very low to any further increase in input voltage. If V_{out} is below the bound voltage, the nonconducting zener impedance is very high and the amplifier gain is set by the impedance of the feedback element, C.

The zener diode bound of Figure 6.8*a* may have too much leakage for precision applications. The diode bound shown in Figure 6.8*b* uses small signal diodes that can be chosen for low leakage. The diodes are reverse biased except when the bound voltage, the zener voltage plus the diode forward voltage drops, is exceeded. The

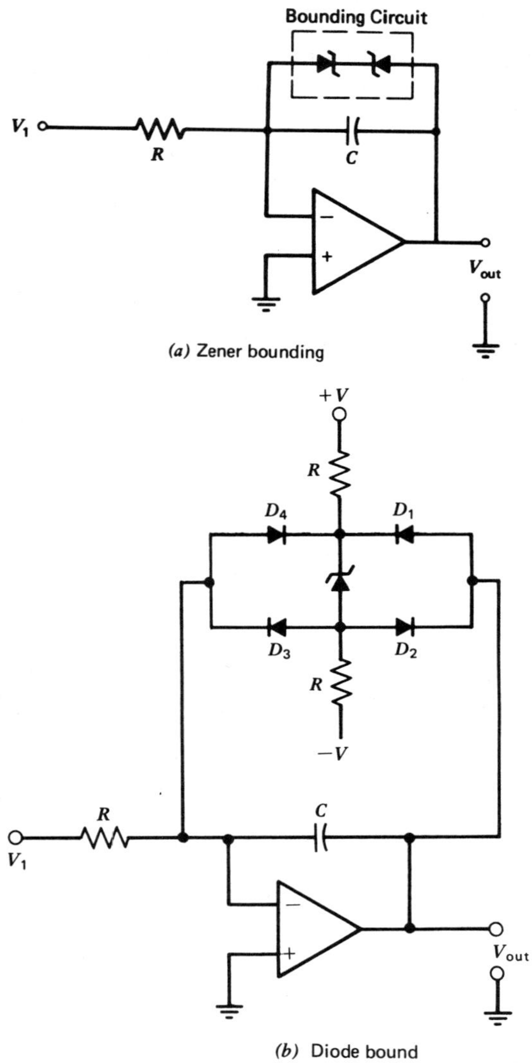

FIGURE 6.8 Bounding circuits.

diodes then conduct, switching the zener into the circuit and reducing the gain. When V_{out} exceeds the positive bound, D_1 and D_3 conduct; and when V_{out} exceeds the negative bound, D_2 and D_4 conduct. The resistors are chosen as follows:

$$R = \frac{+V - V_Z/2}{I_{zener}}$$

The zener current must be less than the output current of the op-amp.

6-5 SUMMING INTEGRATOR

Integrators are not limited to a single input. A summing integrator with n inputs is shown in Figure 6.9. Referring to Figure 6.9 we see that

$$i_c = i_{R_1} + i_{R_2} + \cdots + i_{R_n}$$

so

$$-C\frac{dV_{out}}{dt} = \frac{V_1}{R_1} + \frac{V_2}{R_2} + \cdots + \frac{V_n}{R_n}$$

We let $R_1 = R_2 = \cdots = R_n$ and find

$$-C\frac{dV_{out}}{dt} = \frac{V_1 + V_2 + \cdots + V_n}{R_1}$$

thus

$$dV_{out} = -\frac{(V_1 + V_2 + \cdots + V_n)}{CR_1}dt$$

We now integrate to obtain

$$V_{out} = -\frac{1}{R_1C}\int_{t1}^{t2}(V_1 + V_2 + \cdots + V_n)dt \tag{6-6}$$

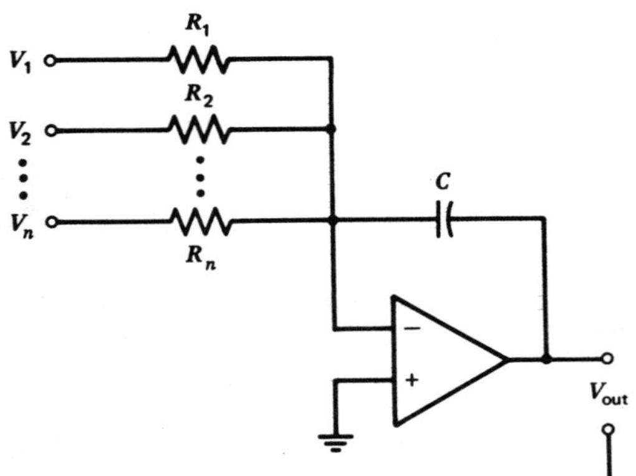

$V_{out} = -\frac{1}{R_1C}\int (V_1 + V_2 + \ldots + V_n)\,dt$

If $R_1 = R_2 = \ldots = R_n$

FIGURE 6.9 Summing integrator.

6-6 AUGMENTING INTEGRATOR

If a resistor is added in series with the feedback capacitor, as shown in Figure 6.10, an output signal proportional to the input and the time integral of the input is

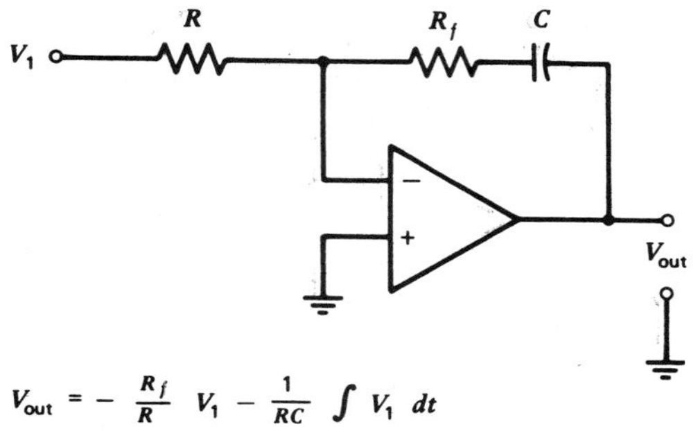

FIGURE 6.10 Augmenting integrator.

obtained. An integrator and amplifier essentially have been combined into one circuit. The output voltage is now given as

$$V_{out} = -\frac{R_f}{R} V_1 - \frac{1}{RC} \int V_1 dt \tag{6-7}$$

The augmenting integrator can also have more than one input.

6-7 DIFFERENTIAL INTEGRATOR

The differential integrator takes the time integral of the difference between two signals. The circuit, shown in Figure 6.11, has a response of

$$V_{out} = \frac{1}{RC} \int (V_2 - V_1) dt \tag{6-8}$$

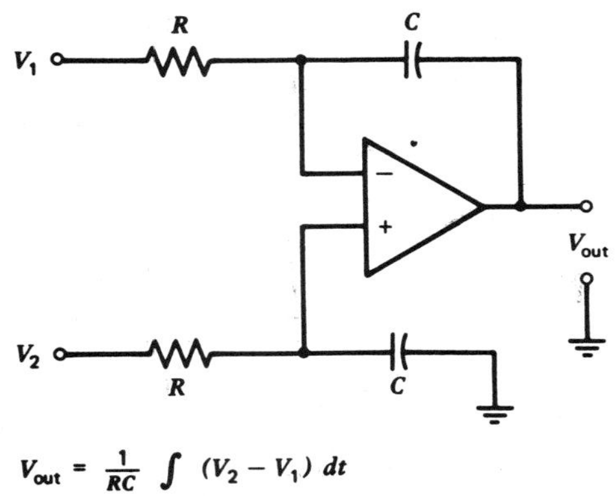

FIGURE 6.11 Differential integrator.

6-8 THREE-MODE INTEGRATOR

Any integrator used for long-term integration must be periodically reset to some desired initial condition (which may be zero). In addition, the output may be periodically held constant for a while to allow sequential monitoring of several outputs or to assure that no change in the output occurs during the time needed to read the output. The three-mode integrator, shown in Figure 6-12, has provisions for integrating, holding the output, and resetting the integrator periodically.

The modes are:

- □ *Run*—the integrator is integrating.
- □ *Hold*—no output change for specified period.
- □ *Set* (or *reset*)—the integrator is returned to its initial condition.

In the *run* mode the integrator integrates normally. Any of the integrators mentioned in this section may be used as a three-mode integrator. The longer the integrator runs, the more error accumulates from bias current, offset voltage, and leakage through the capacitor. The maximum amount of time the integrator is allowed to run is determined by the allowable error for a particular application. During the run mode the output voltage of Figure 6.12 is

$$V_{out} = -\frac{1}{RC}\int V_1 dt + V_{IC} \tag{6-9}$$

where V_{IC} is the voltage the integrator is reset to the reset mode. The reset voltage

$$V_{IC} = -\frac{R_B}{R_A}V_B \tag{6-10}$$

This value is zero if $V_B = 0$.

The maximum integrator run time can be found as follows. Since

$$C = \frac{It}{V}$$

$$t = \frac{CV}{I}$$

so

$$t_{run}(\text{max}) = \frac{CV_{error}}{I_B}$$

where

I_B = op-amp bias current
V_{error} = maximum error voltage

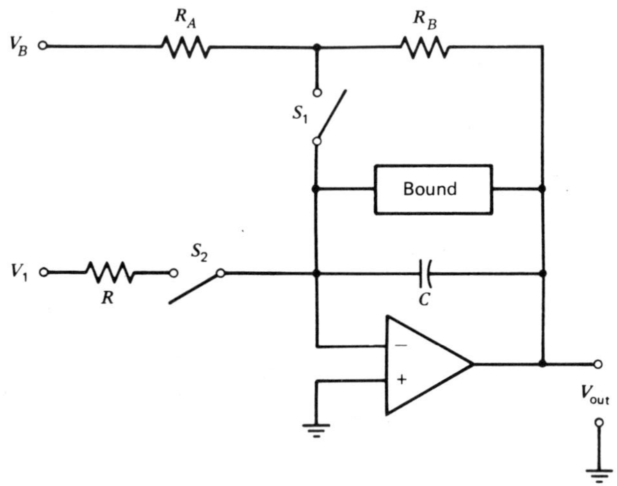

S_1	S_2	Condition
Open	Closed	Run $V_{out} = -\frac{1}{RC}\int V_1\,dt + V_{ic}$ $V_{ic} = V_{\text{initial condition}} = -\frac{R_B}{R_A} V_B$
Open	Open	Hold $V_{hold} = V_{out}$ when S_2 opened
Closed	Open	Reset $V_{out} = -V_B \frac{R_B}{R_A} = V_{ic}$ After $7R_bC$

FIGURE 6.12 Three-mode integrator.

In the *hold* mode the input resistor is disconnected from the integrator as shown in Figure 6.12. Since the input resistance of the op-amp will be high, the capacitor voltage is held at the last voltage it had on it. This voltage cannot be held indefinitely since the amplifier input resistance is not infinite, nor is the internal shunt resistance of the capacitor. The hold voltage will decrease exponentially with a time constant $\tau \simeq CR_{in(op\text{-}amp)}$, which is usually quite a long time.

During the *reset* mode the capacitor is forced to charge or discharge to a voltage set by the feedback network R_A and R_B. The capacitor voltage resets to

$$V_{IC} = -\frac{R_B}{R_A} V_B$$

It is desirable usually to reset quickly, so R_A and R_B are as low as the amplifier can drive. A time of 7 R_BC should be allowed for resetting the integrator. R_B should not be so low that the capacitor discharge current will damage the switch or relay contacts used to select the mode.

The switch or relay contacts used to select the modes could be transistor or FET switches.

6-9 DOUBLE INTEGRATOR

The use of a low-pass T network as the input element of an integrator along with a T feedback element allows double time integrals to be taken. The circuit is shown in Figure 6.13. The double integrator has a response given by

$$V_{out} = -\frac{4}{(RC)^2} \iint V_1 dt \tag{6-11}$$

This circuit can be used to solve differential equations of the form

$$\frac{d^2x}{dt^2} + ax = f(t)$$

If the output of the double integrator is connected to the input, the circuit becomes a phase-shift oscillator with an oscillation frequency of

$$f = \frac{1}{2\pi R(C/2)} \tag{6-12}$$

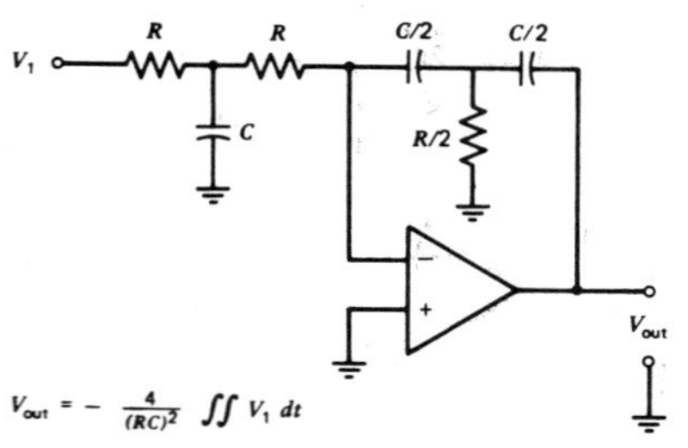

FIGURE 6.13 Double integrator.

EXAMPLE 6-6

The double integrator is to be used as a 1 kHz oscillator, C is picked at 0.01 μF. Calculate *R*.

SOLUTION

$$\text{Since } f = \frac{1}{2\pi R(C/2)}$$

$$R = \frac{1}{2\pi f(C/2)} = \frac{1}{6.28(1\text{ kHz})(0.005\ \mu\text{F})} = 3.18\text{ k}\Omega$$

The components necessary are:

$$C = 0.01\ \mu\text{F},\ C/2 = 0.005\ \mu\text{F},\ R = 3.18\text{ k}\Omega,\ R/2 = 1.59\text{ k}\Omega$$

6-10 DIFFERENTIATOR

The differentiator, Figure 6.14, produces an output proportional to the rate of change of the input. To differentiate, we must pass only the changing portion of the input voltage to the amplifier and the gain of the differentiating circuit must increase as the rate of change of the input increases. The use of a capacitor C as the input element to the op-amp accomplishes this. To obtain the response equation, we recall that the capacitor current

$$i_c = C\frac{dV_c}{dt}$$

The capacitor voltage, is, of course, the input voltage V_1. Assuming an ideal op-amp, the current through the feedback resistor is equal to the capacitor current, so we may write

$$i_R = -i_c$$

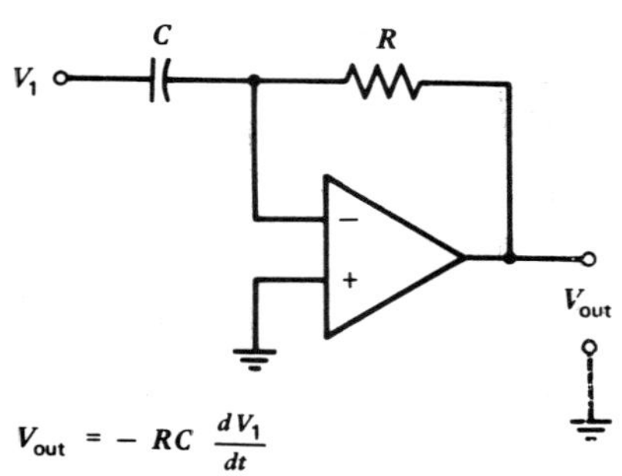

FIGURE 6.14 Differentiator.

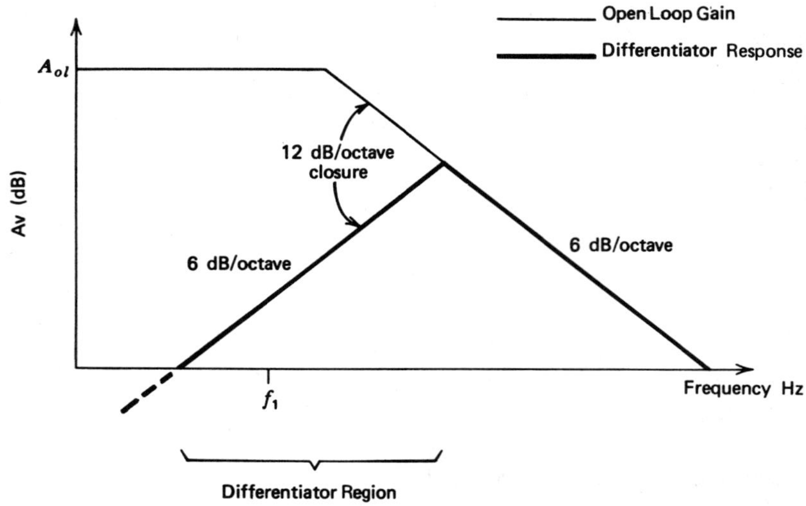

FIGURE 6.15 Uncompensated differentiator response.

But

$$V_{out} = Ri_R = -i_c R$$

therefore

$$V_{out} = -RC\,\frac{dV_1}{dt} \tag{6-13}$$

Since X_c decreases as frequency increases, the differentiator circuit will exhibit a high gain to high frequency components of the input signal, even though that component is above the useful differentiating frequency of the circuit. Noise from resistors and semiconductors that are present in the circuit as well as in the input are therefore amplified. The circuit also tends to be unstable where the differentiator response (through which gain rises at 6 dB/octave) approaches the 6 dB/octave roll off of a compensated amplifier (see Figure 6.15). If the operational amplifier open loop response is rolling off at 12 dB/octave over some part of its frequency range, oscillation may well occur.

6-11 DIFFERENTIATOR STABILIZATION

To avoid the undesired characteristics of the differentiator, the stabilization measures shown in Figure 6.16 are taken.

The capacitor C_c is set to start a 6 dB/octave roll off above the desired maximum differentiating frequency, reducing the presence of high frequency noise. This roll off starts at

$$f_2 = \frac{1}{2\pi RC_c} \tag{6-14}$$

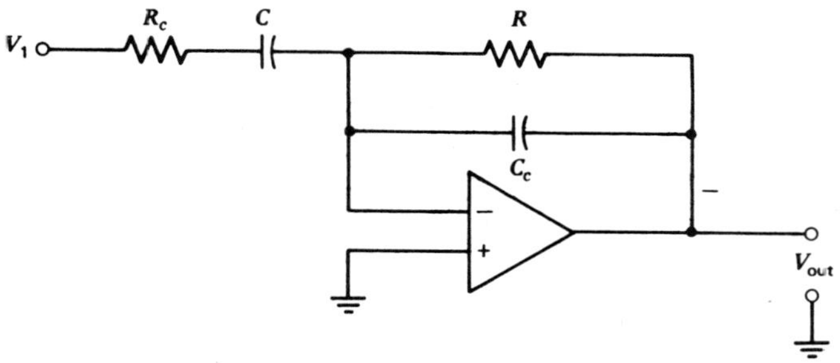

FIGURE 6.16 Compensated differentiator.

The resistor R_c sets a limit to high frequency gain, provides phase margin for stability, and reduces the current demands on the input voltage. The effect of the compensation on the differentiator response is shown in Figure 6.17. We see that R_c causes the gain to level out and differentiation to cease at

$$f_1 = \frac{1}{2\pi R_c C} \tag{6-15}$$

Note that the circuit in Figure 6.16 looks like a compensated ac coupled integrator and in Figure 6.17 we see that from f_2 (where $f = 1/(2\pi RC_c)$) to f_c the circuit can be used as an integrator. From f to $f_1 = 1/(2\pi R_cC)$ the circuit will act as a stable differentiator. The frequency $f_1 = 1/(2\pi R_cC)$ should be set as low as possible to give the desired range and accuracy. The expected error in relation to $f_1 = 1/(2\pi R_cC)$ is given in Table 6.1.

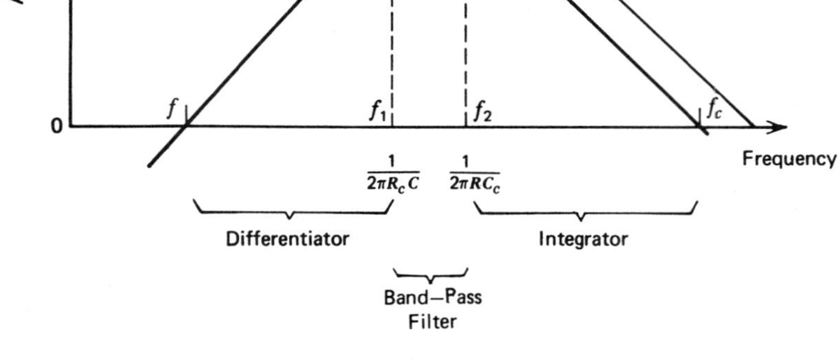

FIGURE 6.17 Compensated differentiator response.

TABLE 6.1

DIFFERENTIATOR ERROR TABLE

$f =$	$0.01\ f_1$	$0.1\ f_1$	$0.33\ f_1$	f_1
Error =	Negligible	1%	5%	50%

Often R_c and C_c are chosen so that $R_cC = RC_c$, so $f_1 = f_2$.

As with the integrator, high quality polystyrene or teflon capacitors (mylar for higher frequencies) and low noise metal film resistors are necessary for optimum performance. Bounding circuitry is useful where recovery time is of any importance.

6-12 COMPENSATED DIFFERENTIATOR AS A FILTER

Notice that the response (Figure 6.17) of the compensated differentiator is that of an active band-pass filter with 6 dB/octave slopes. The circuit can be used as a band-pass filter with the separation between $f_1 = 1/(2\pi R_cC)$ and $f_2 = 1/(2\pi RC_c)$ setting the bandwidth for constant output. Over this range the gain will be R/R_c approximately.

EXAMPLE 6-7

We wish to use a differentiator as a band-pass filter with $f_1 = 1$ kHz, $f_2 = 5$ kHz, and $A_{fb} = 30$. Find R, C, C_c, R_c.

SOLUTION

Since $A_{fb} = R/R_c$, we must pick R or R_c. Let $R = 30\ \text{k}\Omega$ and

$$R_c = \frac{R}{A_{fb}} = \frac{30\ \text{k}\Omega}{30} = 1\ \text{k}\Omega$$

Since

$$f_1 = \frac{1}{2\pi R_cC}$$

$$C = \frac{1}{2\pi R_cf} = \frac{1}{2\pi(1\ \text{k}\Omega)(1\ \text{kHz})} = 0.159\ \mu\text{F}$$

and

$$f_2 = \frac{1}{2\pi RC_c}$$

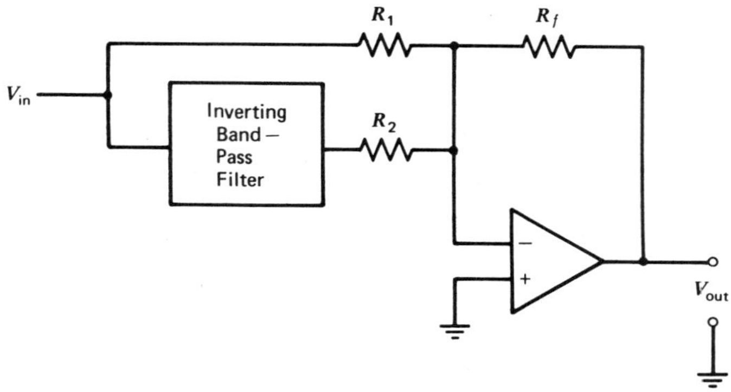

FIGURE 6.18 Band reject filter.

so

$$C_c = \frac{1}{2\pi Rf} = \frac{1}{2\pi(30\ \text{k}\Omega)(5\ \text{kHz})} = 0.0011\ \mu\text{F}$$

Therefore $R = 30\ \text{k}\Omega$, $R_c = 1\ \text{k}\Omega$, $C = 0.159\ \mu\text{F}$, and $C_c = 0.0011\ \mu\text{F}$.

The compensated differentiator can be used as a band-reject (band-stop) filter by putting it in one input of an adder shown in Figure 6.18. The signal will pass through the band-pass filter and cancel the signal through R_1 at frequencies in the pass band of the band-pass filter. The band-pass filter must invert, as the compensated differentiator does, and the adder must be set so that

$$\frac{R_f}{R_1} = \text{band-pass filter gain} + \frac{R_f}{R_2} \tag{6-16}$$

for good rejection.

If $f_2 = 1/(2\pi RC_c)$ can be set much higher than $f_1 = 1/(2\pi R_c C)$, the compensated differentiator can be used as a high-pass filter with a gain of R/R_c.

6-13 SOME DIFFERENTIATOR WAVEFORMS

It is instructive to look at the differentiator output waveforms for some common input signals. We will do this by looking at some examples.

EXAMPLE 6-8

The differentiator of Figure 6.16 has $R = 0.1\ \text{M}\Omega$, $C = 0.1\ \mu\text{F}$, and R_c and C_c set at the appropriate value to stabilize the circuit. The input is a 3 V peak, 60 Hz sine wave, $V = 3Vp \sin 2\pi(60)t$. What is the output voltage and waveform?

SOLUTION

$$
\begin{aligned}
V_{out} &= -RC\frac{dV_{in}}{dt} \\
&= -RC\frac{d(3V \sin 2\pi(60)t)}{dt}
\end{aligned}
\qquad (6\text{-}17)
$$

so

$$V_{out} = -RC\ 3[2\pi(60)] \cos 2\pi(60)t$$

Thus the output is a cosine wave, which we expect since $d \sin u = \cos u\ du$. The size of the output is then

$$V_{out} = (0.01)(3\text{ V})\ 120\pi \cos 2\pi ft = -\ 11.31\ V_p \cos 2\pi ft$$

EXAMPLE 6-9

For the differentiator of Figure 6.16, $R = 10\text{ k}\Omega$, $C = 0.1\ \mu\text{F}$, R_c and C_c are of the appropriate size. The input is a triangular waveform shown in Figure 6.19*a*. What is the output?

SOLUTION

First express the waveform as a function of time over the period of interest. Since this is a repetitive waveform that is symmetric about t_1, we need only solve one half period. The output for the next half period will look the same, but the polarity will be reversed. We see that the voltage rises linearly to 2 V during 0.5 ms, so we can write

$$V_{out} = \frac{2\text{ V}}{0.5\text{ ms}}\, t = (4 \times 10^3\text{ V/s})t$$

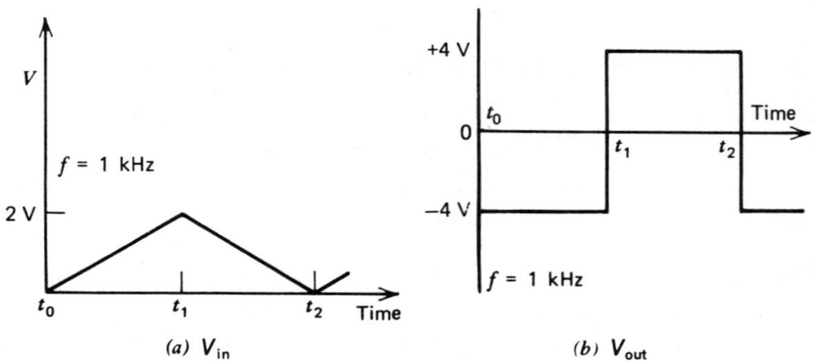

FIGURE 6.19 Differentiator response to triangular input.

where t is time in seconds. Since the differentiator reacts only to changes in voltage, we can neglect the dc component of the input signal. We can now solve for the output using Equation 6.12.

$$
\begin{aligned}
V_{out} &= -RC\frac{d(4 \times 10^3)t}{dt} \\
&= -RC(4 \times 10^3 \text{ V/s}) \\
&= -(10 \text{ k}\Omega)(0.1 \text{ }\mu\text{F})(4 \times 10^3 \text{ V/s}) \\
&= -(0.001 \text{ s})(4 \times 10^3 \text{ V/s}) = -4 \text{ V}
\end{aligned}
$$

The output is then a square wave of 4 V*p*, 8 V*pp*, amplitude as shown in Figure 6.19*b*, with the same frequency as the input. From this example we can generalize that any linear ramp causes the differentiator to have a constant output, proportional to the slope of the ramp, throughout the duration of the ramp.

EXAMPLE 6-10

The differentiator of Example 6-9 has a square wave input of 5 V amplitude, 5 kHz prf, and rise and fall times of 1 μs. Draw the output voltage.

SOLUTION

The input, Figure 6.20*a*, must be broken into parts to differentiate. The constant 5 V and 0 V amplitude portions of the input result in no differentiator output since the derivative of a constant is zero. The rise and fall times of the pulse may be approximated as linear ramps. Since $t_r = t_f$, the output voltage is equal during t_r and t_f, but of opposite polarity, and occurs during t_r and t_f.

To find V_{out} during the rise and fall times we must express them as a function of time, as follows

$$t_r = -t_f = (5 \text{ V}/1 \text{ }\mu\text{s})t = (5 \times 10^6 \text{ V/s})t$$

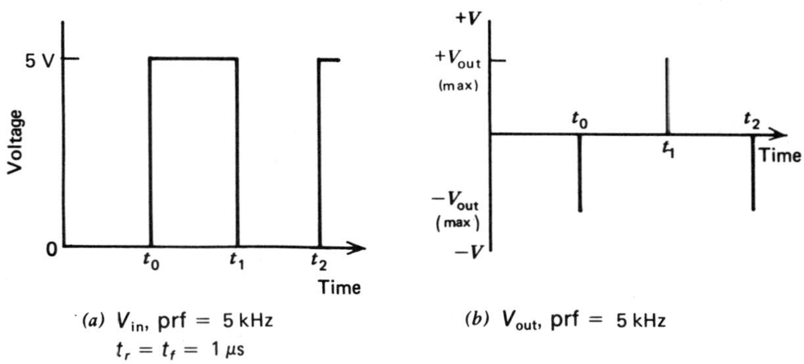

FIGURE 6.20 Differentiator output with square wave input.

Now, from Equation 6-12,

$$V_{out} = -RC\frac{dV_{in}}{dt} = -RC(5 \times 10^6 \text{ V/s})$$
$$= (0.001)(-5 \times 10^6 \text{ V/s}) = -5 \times 10^3 \text{ V}$$

during t_r, and $V_{out} = +5 \times 10^3$ V during t_f. It would be very unusual to have an op-amp that would put out 5 kV. The output then will be two 1 μs wide pulses of opposite polarity with an amplitude equal to the op-amps' maximum output voltage, or the bound voltage if a bound is used. As noted in Section 6-4, a bound is desirable in many differentiator circuits to speed recovery times.

If the op-amp used in the differentiator has too low a slew rate to respond as fast as the input voltage change, we may not see an output that reaches the maximum output voltage, if the input lasts only for a short time. For instance, if our square wave had a 1 ns rise time, we might see no output change at all.

EXAMPLE 6-11

What values of R_c and C_c should be used in Example 6-5, where $R = 10$ kΩ and $C = 0.1$ μF, if the maximum gain allowed is 1000 and the maximum frequency of differentiation is 10 kHz at a 1% error?

SOLUTION

From Table 6.1, we see that for a 1% error the maximum operating frequency should be $f_{max} = (0.1)\ 1/(2\pi R_c C)$. Since we are interested only in differentiating, we set

$$1/2\pi R_c C = 1/2\pi RC_c = 10(10 \text{ kHz}) = 100 \text{ kHz}$$

R_c is chosen for the maximum allowable gain, which is approximately R/R_c, so

$$R_c = R/1000 = 10\ \Omega$$

Now we find C_c from the relation

$$R_c C = RC_c$$

to find

$$C_c = \frac{R_c C}{R} = \frac{(10\ \Omega)(0.1\ \mu\text{F})}{10\ \text{k}\Omega} = 0.001\ \mu\text{F}$$

6-14 SUMMING DIFFERENTIATOR

Like the other computing circuits we have studied, the differentiator is not limited to a single input. Referring to Figure 6.21 we see that

$$-i_R = i_{i_1} + i_{c_2} + \cdots + i_{c_n}$$

for an n input differentiator. Since $V_{out} = i_R R$ and $i_c = C dV_c/dt$, we can write the output voltage equation as

$$-V_{out} = R\left(C_1\frac{dV_1}{dt} + C_2\frac{dV_2}{dt} + \cdots + C_n\frac{dV_n}{dt}\right)$$

This circuit must, of course, be stabilized.

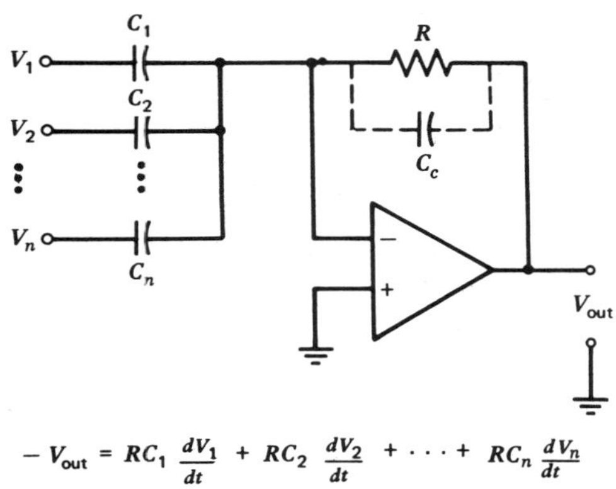

FIGURE 6.21 Summing differentiator.

6-15 AUGMENTING DIFFERENTIATOR

The augmenting differentiator combines a rate and direct response in one amplifier. Essentially the circuit is an *amplifier* and *differentiator* combined. Like the differentiator it can have more than one input. The circuit response is

$$V_{out} = -\frac{R}{R_1}V_1 - RC\frac{dV_1}{dt} \tag{6-18}$$

The circuit is shown in Figure 6.22.

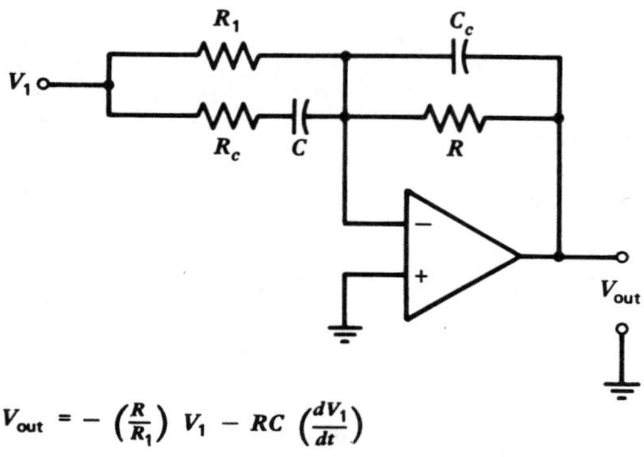

FIGURE 6.22 Augmenting differentiator.

6-16 DIFFERENCE DIFFERENTIATOR

Like the integrator, the differentiator can be used in a differential input mode. This circuit is shown in Figure 6.23. The response is

$$V_{out} = RC\,\frac{d(V_2 - V_1)}{dt} \tag{6-19}$$

Care must be taken to maintain good matching in the components to minimize error.

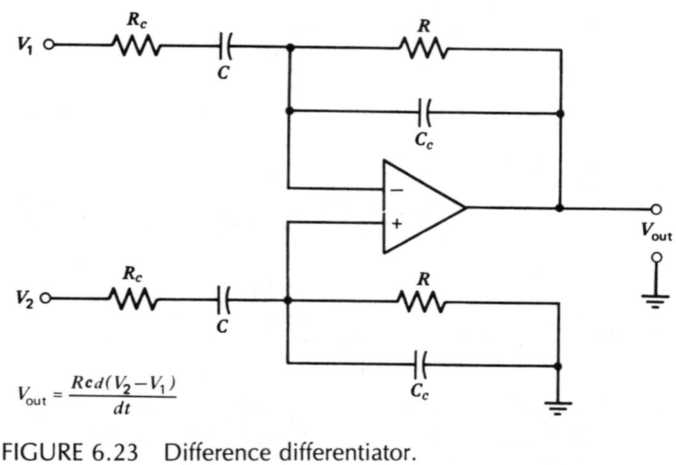

FIGURE 6.23 Difference differentiator.

6-17 SOLVING A DIFFERENTIAL EQUATION

One use of differentiators and integrators is for solving equations that include rate changes of the variables. For example, suppose we wish to solve for the current in the *RCL* circuit of Figure 6.24. To do this, we will express the condition of the circuit in terms of time derivatives of some variable.

We use Kirchhoff's voltage law to express the voltage drops of the circuit

$$V_{in} = V_L + V_c + V_R \tag{6-20}$$

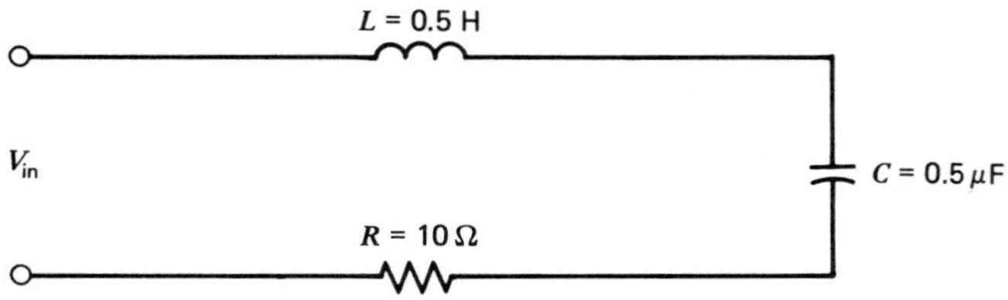

FIGURE 6.24 *RCL* circuit.

The resistor voltage is expressed as just *iR*, which can be expressed as

$$V_R = \frac{dQ}{dt} R \tag{6-21}$$

The instantaneous voltage across the inductor is

$$V_L = -L \frac{di}{dt} = -L \frac{d^2Q}{dt^2} \tag{6-22}$$

Capacitance is defined as

$$C = Q/V_c$$

so the voltage across the capacitor is

$$V_c = \frac{Q}{C} \tag{6-23}$$

We now substitute Equations 6-1, 6-2, and 6-3 back into Equation 6-20 to obtain

$$V_{in} = -L \frac{d^2Q}{dt^2} + R \frac{dQ}{dt} + \frac{Q}{C}$$

We now have an equation arranged in decreasing orders of derivatives of a single variable. We can rearrange this equation to obtain a solution for Q.

$$Q = LC \frac{d^2Q}{dt^2} - RC \frac{dQ}{dt} + CV_{in} \tag{6-24}$$

Now we set up a summer to solve for Q with each input being one of the terms of Equation 6-24. First we substitute into Equation 6-24 the values from the circuit of Figure 6-24, obtaining

$$Q = (0.5 \text{ H})(0.5 \ \mu\text{F}) \frac{d^2Q}{dt} - 10 \ \Omega(0.5\mu\text{F}) \frac{dQ}{dt} + (0.5 \ \mu\text{F})V_{in}$$

$$= \left[0.25 \frac{d^2Q}{dt^2} - 5 \frac{dQ}{dt} + 0.5 \ V_{in}\right] (10^{-6}) \text{ coulombs}$$

To ease the gain requirements on the circuit, we will solve for an answer in microcoulombs (μc). The answer for Q will be differentiated to obtain current in

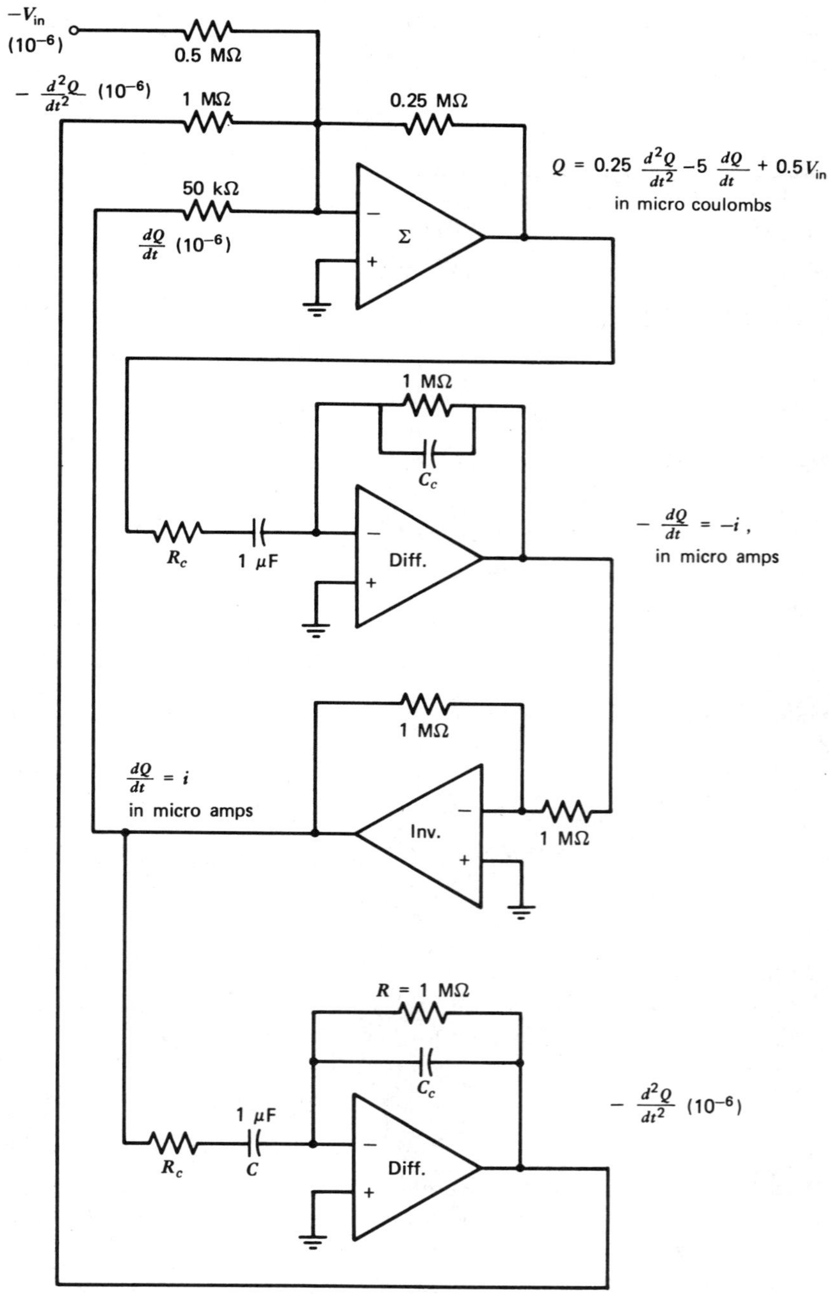

FIGURE 6.25 Solution of the differential equation $V_{in} = -L\frac{d^2Q}{dt^2} + R\frac{dQ}{dt} + \frac{Q}{C}$.

microamps and to supply one of the inputs to the adder. We will use inverting adders to avoid the balance problem of an adder-subtracter. Recall that if we want a particular sign out of an inverting adder we must put the opposite sign in, for example, to obtain $V_{out} = -X + Y$ we must put $+X$ and $-Y$ into the inputs.

The solution of the circuit of Figure 6.24 is shown in Figure 6.25. The circuit must be scaled so as to reach the answer, as in solving simultaneous equations. With the circuit shown in Figure 6.25, we can now examine the current over time for a variety of input voltage conditions.

The equation for the circuit of Figure 6.24 can be written in terms of integrals and solved by using integrators. The integrator solution circuit will generally be more stable than the circuit of Figure 6.25. The solution is as follows:

$$V_c = \frac{Q}{C}$$

but

$$i_c = \frac{dQ}{dt}$$

so

$$dQ = i\,dt$$

Integrating

$$Q = \int i\,dt$$

Now we can write

$$V_{in} = V_R + V_L + V_c$$

$$= iR - L\frac{di}{dt} + \frac{1}{C}\int i\,dt$$

Integrating both sides with respect to time, we obtain

$$\int V_{in}\,dt = R\int i\,dt - L\,i + \frac{1}{C}\iint i\,dt$$

since the integral of di/dt is just i. We now solve for i and get

$$L\,i = R\int i\,dt + \frac{1}{C}\iint i\,dt - \int V_{in}\,dt$$

$$i = \frac{R}{L}\int i\,dt + \frac{1}{LC}\iint i\,dt - \frac{1}{L}\int V_{in}\,dt$$

Substituting values

$$i = \frac{10\ \Omega}{0.5\ \text{H}} \int i\,dt + \frac{1}{0.5\ \mu\text{F}\ 0.5\ \text{H}} \iint i\,dt - \frac{1}{0.5\ \text{H}} \int V_{in}\,dt$$

$$= 20 \int i\,dt + 4 \times 10^6 \iint i\,dt - 2 \int V_{in}\,dt$$

The solution is shown in Figure 6.26. The 10^6 factor of the second term must be kept in. Since this is such a large value, it is split between an amplifier with a gain of 200 and an integrator with a coefficient of 1000.

If V_{in} is an ac voltage, compensating resistors R_d should be used on the integrators. If V_{in} is a step voltage, the integrator capacitors must be periodically reset as for a three-mode integrator. If the integrators are FET input op-amps, the integrating times can be longer before reset is necessary.

SUMMARY

1. An integrator output is proportional to the time average of its input voltage. An integrator must have a −6 dB/octave decrease in gain over the frequency range of which it is used as an integrator.
2. The output voltage of an integrator is found by the equation

$$V_{out} = -\frac{1}{RC} \int_{t_1}^{t_2} V_{in}\,dt \tag{6-2}$$

3. If an integrator is used to integrate ac voltages, a compensating resistor (R_d) is used across C to make it less sensitive to offset voltage drift and bias current charging the capacitator. For good accuracy the break frequency of R_dC is

$$f = \frac{1}{2\pi R_d C}$$

and is set to about one tenth of the *lowest* frequency to be integrated.
4. If an integrator is used to integrate signals near dc, the capacitor of the integrator must be discharged periodically (reset) to keep the capacitor voltage due to bias current from causing excessive error.
5. If R and R_d are chosen for a desired voltage gain, and C is chosen so that $1/(2\pi R_d C)$ is a desired break frequency, an integrator can be used as a low-pass RC filter with gain.
6. The output voltage of a differentiator is proportional to the rate of change of the input voltage. The equation for the output voltage of a differentiator is

$$V_{out} = -RC\,\frac{dV_{in}}{dt} \tag{6-13}$$

7. A differentiator must have a 6 dB/octave rise in gain over the frequency range at which it is used as a differentiator. A capacitor input provides this response.
8. A differentiator must be compensated to avoid unwanted high frequency noise on its output. A capacitor C_c is added across R to accomplish this. To obtain good accuracy, C_c is picked so that $1/(2\pi RC_c)$ is about ten times the highest frequency to be integrated.

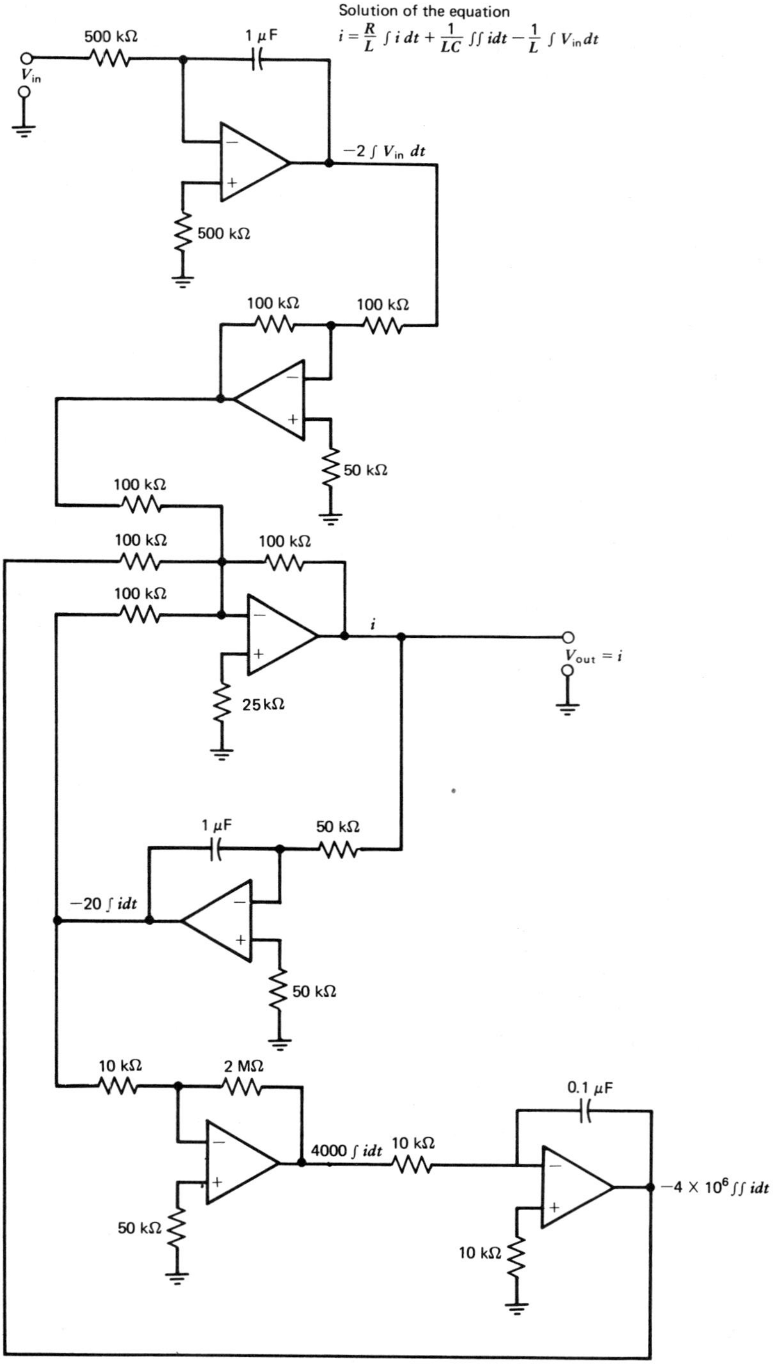

FIGURE 6.26 Solution of the equation $i = \frac{R}{L}\int i\,dt + \frac{1}{LC}\iint i\,dt - \frac{1}{L}\int u\,dt.$

9. Since a differentiator has a capacitive input, a resistor R_c is used in series with C to avoid loading V_{in} excessively; R_c is chosen so that $1/(2\pi R_c C) = 1/(2\pi RC_c)$.
10. A differentiator can be used as a band-pass filter if the ratio of R/R_c is chosen for midband gain, C is chosen so that $\frac{1}{2}\pi R_c C$ is the lower 3 dB frequency, and C_c is chosen so that $1/(2\pi RC_c)$ is the upper 3 dB frequency of the filter.

SELF-TEST QUESTIONS

6-1 State the function of an integrating circuit.
6-2 Find R of the integrator of Figure 6.2 so that $V_{out} = -10 \int V_{in}\, dt$ if $C = 1\ \mu F$.
6-3 For the integrator of question 6-2, calculate V_{out} after 2 ms if (a) $V_{in} = 3t$; (b) $V_{in} = 2t^2$; (c) $V_{in} = 2e^t$.
6-4 List three considerations that limit practical integrating time.
6-5 State the corrections used for each of the factors listed in question 6-4.
6-6 An integrator has $R = 20\ k\Omega$. Calculate R_d and C if the integrator is to be used as a low-pass filter with a gain of 10 and f_1 of 10 kHz.
6-7 State the purpose of bounding.
6-8 For the augmenting integrator of Figure 6.10, $R = 100\ k\Omega$. Calculate R_f and C so that $V_{out} = -5\ V_1 - 10 \int V_1\, dt$.
6-9 State the function of each of the modes of a three-mode integrator.
6-10 Calculate R so that the double integrator set up as an oscillator oscillates at 20 kHZ if $C = 0.1\ \mu F$.
6-11 Calculate C so that the differentiator output will be $V_{out} = -0.001\ dV_i/dt$ if $R = 100\ k\Omega$.
6-12 If $V_{in} = 4\ t$, what will V_{out} of question 6-11 be at (a) 2 ms; (b) 40 s.
6-13 State the purpose of compensating a differentiator.
6-14 If the differentiator of Figure 6.16 has $C = 0.1\ \mu F$, calculate R_c, R, and C_c so that the integrator will have a maximum differentiation frequency of 20 kHz and $A_v(max) = 1000$.
6-15 Calculate R_c, R, and C_c so that the differentiator of Figure 6.16 will act as a band-pass filter with a passband between 500 Hz and 2 kHz and a passband gain of 20. Let $C = 0.1\ \mu F$.
6-16 We wish to operate a differentiator at 1 kHz with 1% accuracy. Find the minimum upper differentiation frequency (f_1).

If you cannot answer certain questions, place a check next to them and review appropriate parts of the text to find the answers.

LABORATORY EXERCISE

Objective

After completing this laboratory, the student should be able to set up an integrator and differentiator to obtain a desired integral or derivative. The student will verify that differentiation and integration are inverse operations using an integrator and differentiator. The student will also connect a double integrator as an oscillator.

The student has the option of solving a differential equation using adders, inverters, and differentiators.

Equipment

1. Two Fairchild μA741 operational amplifiers or equivalent and manufacturer's specifications.
2. 2% resistor assortment.
3. Power supply, ±15 V dc.
4. Mylar capacitor assortment.
5. Signal generator that will produce sine, square, and triangular waves.
6. Dual trace oscilloscope.
7. 10 kΩ potentiometer.
8. Breadboard with IC sockets such as EL Instruments SK-10.

Procedure

1. Integrator.
   (a) Connect the integrator shown in Figure 6.27 for a response of

$$V_{out} = -2000 \int V_1 \, dt$$

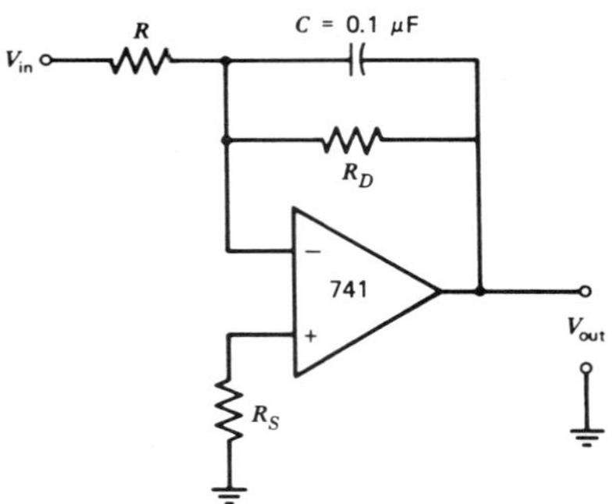

FIGURE 6.27 Integrator.

Compensate the integrator so that the lowest frequency of integration is about 30 Hz. Null the offset.

   (b) Calculate, and verify by measurement, the output using the following inputs:
      (1) Sine wave of 1 kHz, 2 V_p.
      (2) Square wave of 1 kHz, 5 V_{pp} ac.
   (c) Choose R and C so the integrator will act as an active low-pass filter with a gain of 5 and f_1 = 500 Hz. Verify this action by measuring the low frequency gain and f_{co}, the cut-off frequency, using a sine wave input.
2. Differentiator.
   (a) Connect the circuit shown in Figure 6.29 for a response of

$$V_{out} = -0.1 \times 10^{-2} \frac{dV_{in}}{dt}$$

Compensate the differentiator for a maximum differentiating frequency of 1 kHz.
(b) Calculate the output for a 1 kHz, 2.5 V_p triangular waveform. Verify by measuring the output.
(c) Set the compensated differentiator up as a band-pass filter, by properly choosing R_c and C_c, for a response as shown in Figure 6.28. Verify the filter action by measurement. Use a sine wave input and plot your results.

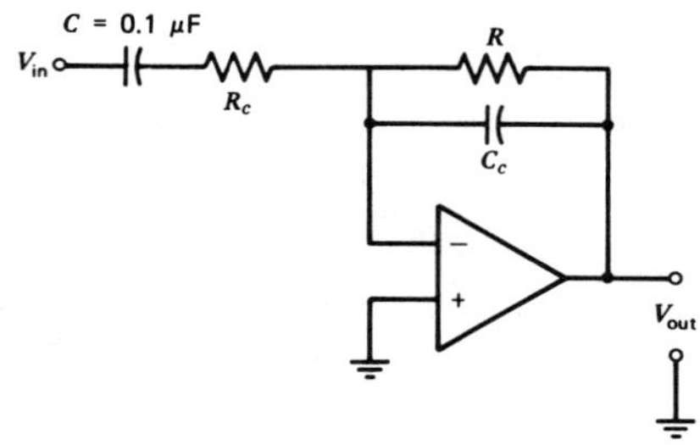

FIGURE 6.28 Differentiator.

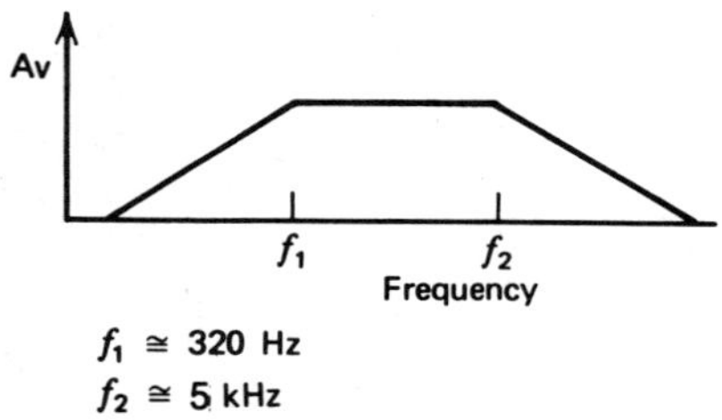

FIGURE 6.29 Bandpass filter response.

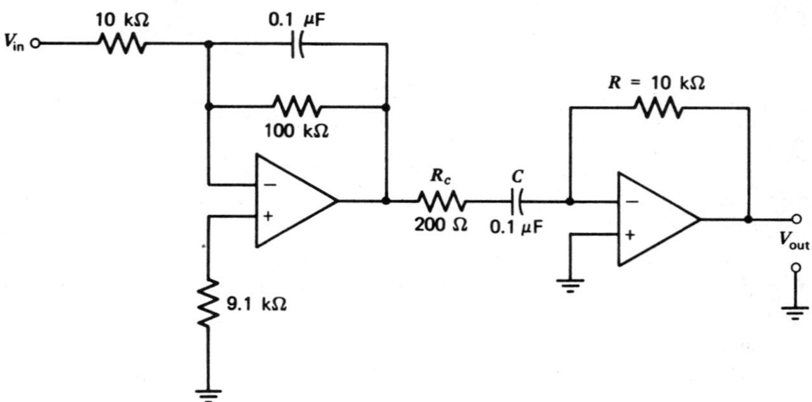

FIGURE 6.30 Inverse operation proof.

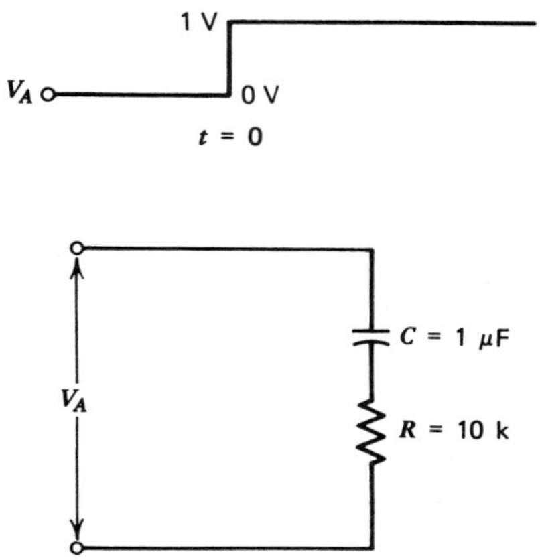

FIGURE 6.31 Circuit for solution.

3. Inverse operations.

If differentiation and integration are inverse operations, we should be able to put a square wave into an integrator then differentiate the integrator output and get our original square wave back. Connect the circuit shown in Figure 6.30 using a 5 V_p 1 kHz square wave input. What is the output of the integrator? The differentiator? Record the waveforms.

4. Double integrator oscillator.

Using R = 10 kΩ and C = 0.1 μF, connect the double integrator of Figure 6.13 as an oscillator. Calculate and measure the oscillation frequency.

This circuit is very sensitive to the size of C. If the circuit does not oscillate at first, increase C by about 20%. (Precision capacitors eliminate this problem.)

5. Optional section.

This involves solving a differential equation. Connect a circuit to solve for the current of the circuit in Figure 6.31 as a function of time. Use $V_{in} = 1\ V_{dc}$.

Note. The output measuring device (oscilloscope) must be triggered when the input voltage is applied.

Discussion

For each part of the laboratory exercise, discuss and account for any discrepancies between theory and your measurements.

CHAPTER

7

LOGARITHMIC CIRCUITS

Logarithm and antilogarithm circuits are used to perform analog multiplication and division, provide signal compression, and take logarithms and exponentials. Although we will not go into detail concerning temperature-compensated precision circuits, we will look at the basic forms of these circuits.

OBJECTIVES

After completing the study of this chapter and the self-test questions, the student should be able to:

1. Draw from memory a diode and transistor logarithm and antilog circuit.
2. Given the component and component specifications, calculate the output voltage of a logarithm and antilog circuit.
3. Draw the block diagram of and state the principle of a divide and multiply circuit using logarithm and antilog circuits.
4. Given the components, draw the V_{in} versus V_{out} graph of a function synthesizer.
5. State the use and operating principle of a signal compression circuit.
6. Perform the laboratory for Chapter 7.

7-1 THE LOGARITHM CIRCUIT

To produce a logarithmic amplifier response, one must have a device with a logarithmic characteristic to place within the feedback loop. The semiconductor *pn* junction is a device that provides this characteristic. Recall from semiconductor theory that the current through a semiconductor diode is

$$I_D = I_s(e^{qV_D/kT} - 1) \cong I_s e^{qV_D/kT} \tag{7-1}$$

where

I_s = leakage current under low reverse bias (due to thermal electron-hole pair generation)
q = electron charge (1.6×10^{-19} coulombs)
V_D = diode voltage
k = Boltzmann's constant (1.38×10^{-23} joule/°K)
T = absolute temperature in degrees Kelvin

Similarly, the collector current in a common-base transistor is

$$I_C = I_{ES}(e^{qV_{BE}/kT} - 1) \tag{7-2}$$

$$I_C \cong I_{ES}e^{qV_{BE}/kT}$$

where

V_{BE} = emitter-base voltage
I_{ES} = Emitter-base diode current under small reverse bias when the collector-base leads are shorted. $I_{ES} \cong I_s$.

The current relations for both the diode current and transistor collector current just presented are of the same form, so the argument used for one applies to the other. Both the diode and transistor can be used to provide a logarithmic response. A diode, D, is connected as shown in Figure 7.1 to provide a logarithmic amplifier.

To see how the use of the diode in the feedback loop makes the response logarithmic, we solve Equation 7-1 for V_D, which is V_{out}. Solving

$$I_D = I_s e^{qV_D/kT}$$

we obtain

$$\ln I_D = \ln I_s + qV_D/kT$$

$$\ln I_D - \ln I_s = qV_D/kT$$

thus

$$V_{out} = V_D = kT/q\ (\ln I_D - \ln I_s)$$

since

$$I_D = I_{R_1} = V_1/R_1$$

$$V_{out} = kT/q\ (\ln V_1/R_1 - \ln I_s)$$

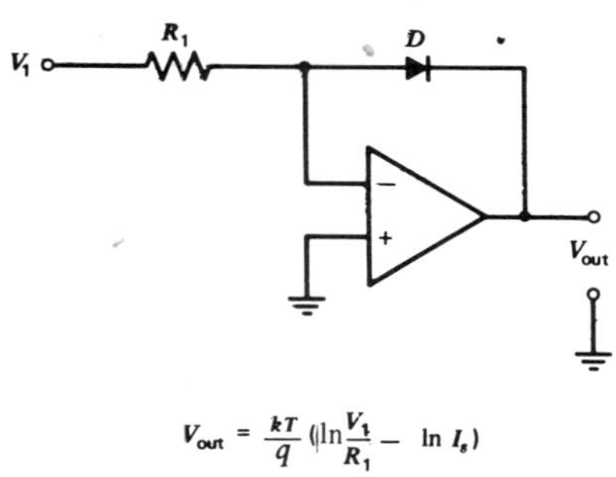

FIGURE 7.1 Logarithmic amplifier.

kT/q is usually about 26 mV at 25°C. It is instructive to look at the output voltage waveform. Plotting I_D versus V_{out}, $V_{out} = V_D$, on a linear scale, we obtain the characteristic logarithmic *VI* plot of a diode (Figure 7-2*a*). If we plot V_{out} versus log *I* (on semilog paper) (Figure 7-2*b*), we obtain a straight line with a slope of 26 mV. Note that V_{out} goes to a maximum of around 0.6 V. If a larger output is necessary, we must amplify the output. The logarithmic amplifier will have a logarithmic response over about three decades of input current, depending on the diode. The logarithmic response of the diode usually ceases at about 1 mA.

The ln I_s term will be a constant, small error term, which must be known for the diode if it is appreciable.

The output of a logarithmic amplifier is in one direction only, which is determined by the direction of the diode. For instance, the circuit of Figure 7.1 has a negative output voltage in response to a positive input voltage. If the diode is reversed, the output will be positive with a logarithmic response to a negative input voltage.

A common-base-connected transistor can also be used as the logarithmic element in the feedback, as shown Figure 7.3, to obtain a greater input range. If we note that $I_C = -I_{R_1}$ we can start with Equation 7-2 and solve for V_{BE} to obtain

$$V_{out} = V_{BE} = kT/q\,(\ln V_1/R_1 - \ln I_{ES}) \tag{7-3}$$

In Figure 7.3 the output voltage is negative going in response to a positive input. A *pnp* transistor would result in a positive output in response to a negative input.

The logarithm circuits of this section work but are not temperature compensated or corrected to cancel ln I_s. For precise operation over a wide temperature range this would be done; it requires more extensive circuitry. A usable, wide range logarithmic amplifier requires an op-amp with an inherently low offset voltage and bias current.

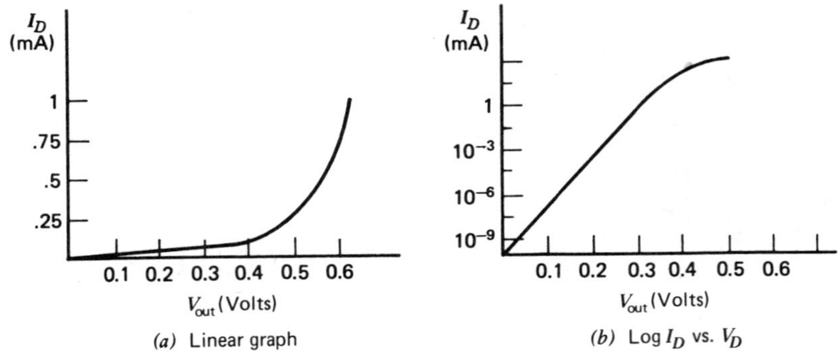

FIGURE 7.2 Logarithmic amplifier output.

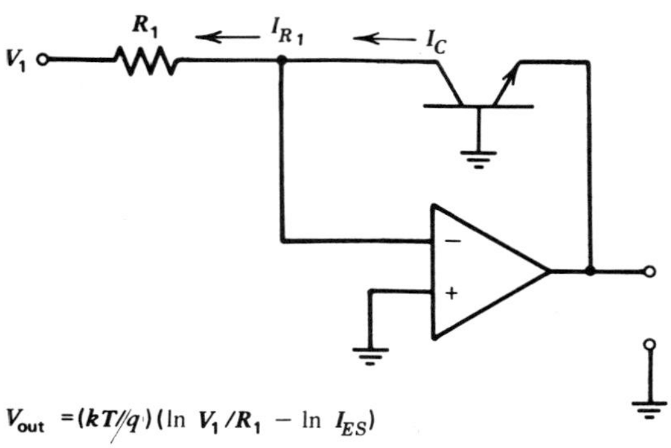

(a) Basic logarithmic amplifier

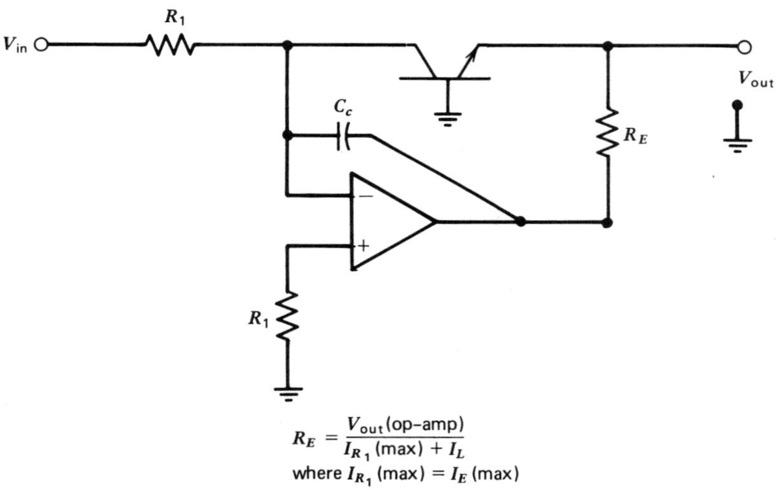

(b) Improved basic logarithmic amplifier.

FIGURE 7.3 Transistor logarithmic amplifier.

EXAMPLE 7-1

Set up a logarithmic amplifier of the type shown in Figure 7.3*a*, and calculate V_{out} with +2 V dc input.

SOLUTION

First we must pick R_1 so that the V_{BE} of the transistor (V_{BE} versus I_E curve) is still logarithmic at the maximum input voltage we desire. Assume this is at $I_E = I_C = 0.1$ mA. Since

$$I_C = I_{R_1}$$

and

$$I_{R_1} = \frac{V_{in}}{R_1}$$

$$R_1 = \frac{V_{in}(\max)}{I_E}$$

If we let $V_{in}(\max) = 10$ V,

$$R_1 = \frac{10 \text{ V}}{0.1 \text{ mA}} = 100 \text{ k}\Omega$$

Assume the measured $I_{ES} = 40$ nA. The value of kT/q is about 26 mV for a transistor at room temperature.

Find V_{out} if $V_{in} = +2$ V dc.

$$\begin{aligned} V_{out} &= kT/q\,(\ln V_{in}/R_1 - \ln I_s) \\ &= 0.026 \text{ V } [\ln(2 \times 10^{-5}) - \ln(4 \times 10^{-8})] \\ &= 0.026 \text{ V } [\ln(2 \times 10^{-5}) - \ln(4 \times 10^{-8})] \\ &= 0.026 \text{ V } \ln(2 \times 10^{-5}/4 \times 10^{-8}) \\ &= 0.026 \text{ V } \ln 5(10^2) \\ &= 0.026 \text{ V } [\ln 5 + 2(2.303)] \\ &= 0.026 \text{ V } (1.61 + 4.606) = 0.1616 \text{ V} \end{aligned}$$

The procedure and answer to Example 7-1 would be the same whether a diode or transistor is used provided $I_s = I_{ES}$.

The logarithm amplifier of Figure 7.3*a* can be improved by the addition of two components as shown in Figure 7.3*b*. The resistor R_1 connected to the noninverting input helps to compensate for bias current. R_E provides a large load resistor for the op-amp since r_e, the ac emitter resistance of the transistor (26 mV/I_E), can be very low at moderate emitter currents. At 1 mA, $r_e = 26\ \Omega$ and at 0.1 mA, $r_e = 260\ \Omega$. R_E is chosen so that the load current and the maximum emitter current can be supplied.

$$R_E = \frac{V_{out}(\text{op-amp})}{I_E(\max) + I_{load}}$$

If $V_{out}(\max)$ of the op-amp in Example 7-1 is 14 V and $I_{load} = 1$ mA, then $R_E = 12.7$ kΩ. C_c is used for additional stability when needed. Normally $C_c = 100$ pF.

FET input op-amps, because of their low bias currents, are preferable for logarithmic amplifiers.

7-2 A MORE SOPHISTICATED LOG CIRCUIT

One approach to a high accuracy logarithmic amplifier is shown in Figure 7.4. The circuit uses the fact that

$$V_{EB} = kT/q\,(\ln I_C - \ln I_s) \tag{7-4}$$

where

$$I_s = \text{emitter-base saturation current } I_{EBO}$$

The difference in the emitter-base voltage of the matched differential transistor pair Q_1 and Q_2 is

$$\begin{aligned} V_{BE_1} - V_{BE_2} &= kT/q\,(\ln I_{C_1} - \ln I_s) - kT/q\,(\ln I_{C_2} - \ln I_s) \\ &= kT/q \ln I_{C_1} - kT/q \ln I_{C_2} \\ &= kT/q \ln I_{C_1}/I_{C_2} \end{aligned}$$

Since $V_{BE_1} - V_{BE_2}$ is V_x, and $I_{C_1} = V_1/R_1$, we see that

$$\begin{aligned} V_X &= kT/q \ln (V_1/R_1 I_{C_2}), \\ &= kT/q\,[\ln V_1 - \ln (1/R_1 I_{C_2})] \end{aligned}$$

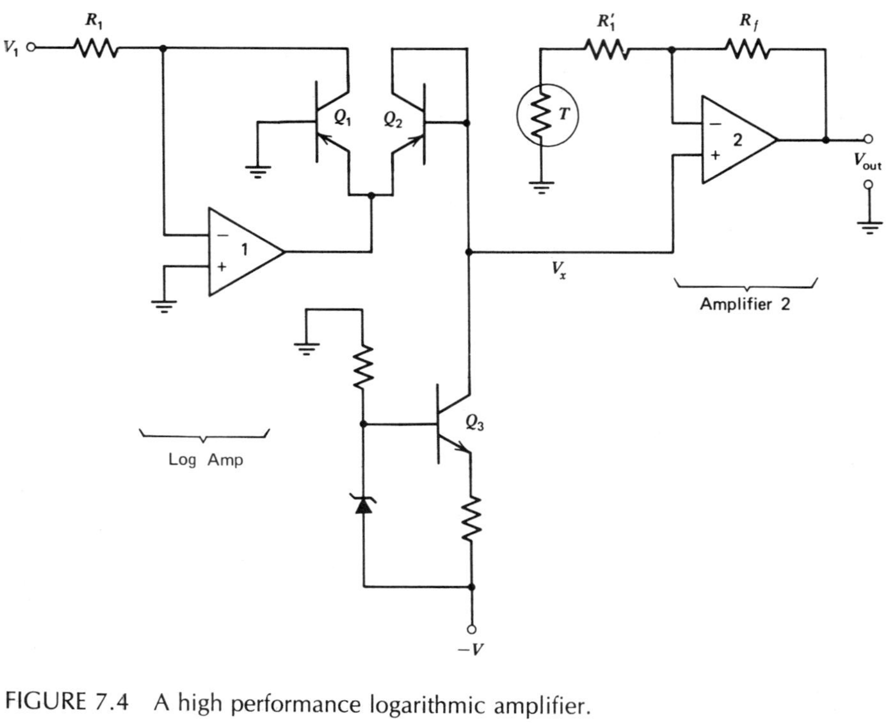

FIGURE 7.4 A high performance logarithmic amplifier.

I_{C_2} is the output of the constant current source I_{C_3}. Ideally I_{C_2} is set such that $R_1/I_{C_2} = 1$, otherwise the offset of amplifier 2 is set so that it is equal to (kT/q) ln $(1/RI_{C_2})$. If the gain of amplifier 2 is set so that it equals q/kT, then the output of the circuit in Figure 7.4 is $V_{out} = \ln V_1$. The thermistor in the feedback loop of op-amp 2 is used to compensate for the temperature variations of the kT/q term. If the thermistor is not used, then the circuit is less temperature constant. This circuit is accurate over 5 decades input current range.

The transistors, Q_1 and Q_2, can be replaced with matched diodes if the constant current source is adjusted so that its output current is equal to the reverse current of the diode.

7-3 ANTILOG AMPLIFIER

To convert from logarithms (i.e., to obtain antilogs), we must take the exponential of the logarithm, since $e^{\ln X} = X$. When we take the exponential of a logarithm we obtain the antilog. By using a logarithmic device as the input element of an amplifier, as shown in Figure 7.5, we obtain an exponential response, thus an antilog amplifier. Referring to Equation 7-2 we see

$$I_C = I_{ES} e^{qV_{BE}/kT}$$

Now

$$V_{out} = R_f I_f = -R_f I_C$$

thus

$$V_{out} = R_f I_{ES} e^{qV_{BE}/kT} = -R_f I_{ES} e^{qV_1/kT}$$

This is the same as saying

$$V_{out} = -R_f I_{ES} \text{ antilog } (V_1 q/kT) \tag{7-5}$$

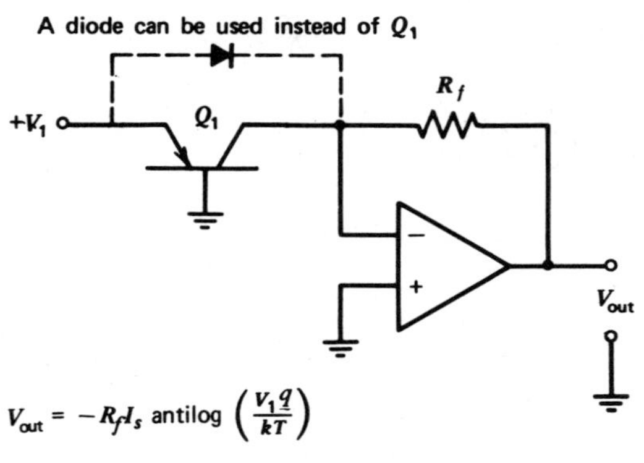

FIGURE 7.5 Antilog amplifier.

where V_1 is a logarithm.

A diode can be used as an input element instead of a transistor, in which case

$$V_{out} = -R_f I_s \text{ antilog } (V_1 q/kT) \quad (7\text{-}5a)$$

The transistor will perform better.

If the input is negative instead of positive, the input transistor must be an *npn*, or the diode, if used, must be used with the cathode toward V_1.

EXAMPLE 7-2

Connect an antilog circuit such as that in Figure 7.5. Set $I_f = 0.1$ ma when $V_{out} = 10$ V. I_{ES} of the transistor is 40 nA. Find the value of R.

SOLUTION

Since the summing-point voltage is approximately 0 V and $I_{R_f} = I_C$,

$$R = \frac{10 \text{ V}}{0.1 \text{ mA}} = 100 \text{ k}\Omega$$

EXAMPLE 7-3

Find V_{out} of the circuit in Example 7-2 if

$$V_{in} = 0.1616 \text{ V and } \alpha \cong 1.$$

SOLUTION

$$\begin{aligned} V_{out} &= -R_f I_{ES} \text{ antilog } (V_{in} q/kT) \\ &= -\ 100 \text{ k}\Omega(4 \times 10^{-8} \text{ A}) \text{ antilog } (0.1616/0.026) \\ &= -0.004 \text{ V antilog } 6.216 \\ &= -0.004 \text{ V } e^{6.216} \\ &= -0.004 \text{ V } (500) = -2 \text{ V} \end{aligned}$$

It may be instructive to look back at Example 7-1 at this time.

7-4 MULTIPLYING CIRCUIT

With logarithms we can use the relation

$$\ln (a \times b) = \ln a + \ln b$$

to construct a multiplying circuit.

The multiplying circuit is constructed as shown in Figure 7.6. The logarithm of V_1 and the logarithm of V_2 are summed to obtain $\ln V_1 + \ln V_2$. The antilog of the sum is then taken to find V_1V_2. The circuit is not limited to two inputs, but a separate logarithmic amplifier is necessary for each input.

The output of each logarithmic amplifier is

$$V_{out_1} = kT/q \ln V_1/R_1 - kT/q \ln I_{s_1}$$

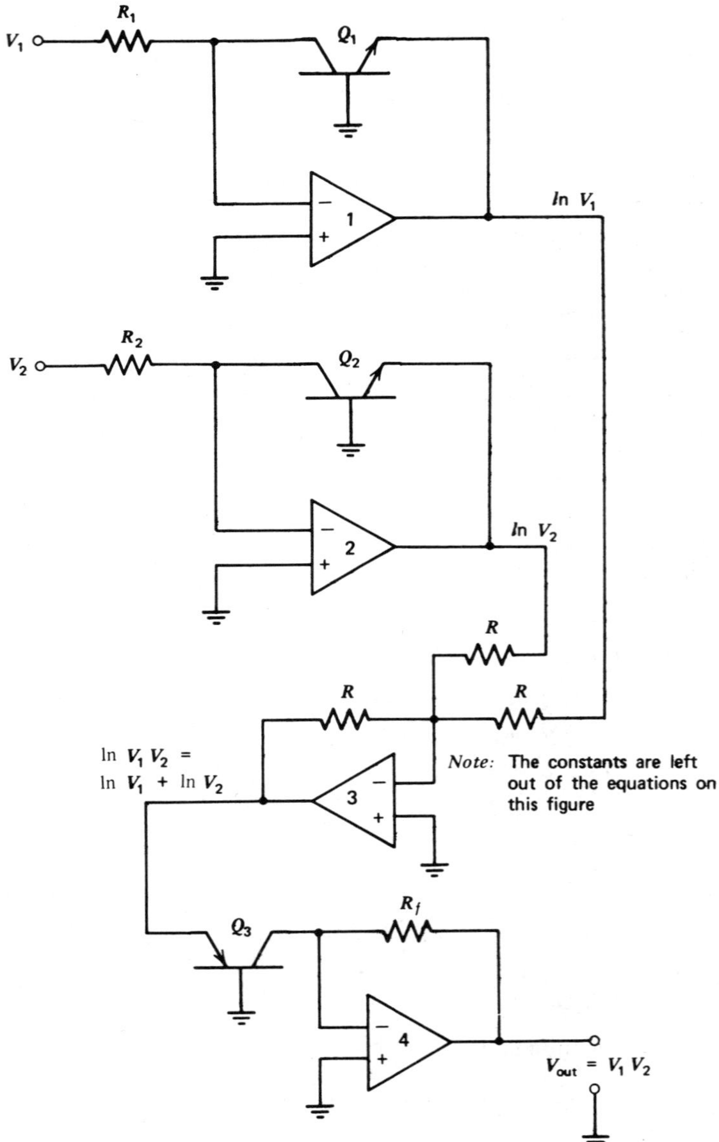

FIGURE 7.6 Multiplying circuit.

and

$$V_{out_2} = kT/q \ln V_2/R_2 - kT/q \ln I_{s_2}$$

The output of the adder is then

$$V_{out_3} = kT/q\,(\ln V_1/R_1 + \ln V_2/R_2 - \ln I_{s_1} - \ln I_{s_2})$$

The output of the circuit is

$$\begin{aligned} V_{out_4} &= R_f I_{s_4} \text{ antilog } (\ln V_1/R_1 + \ln V_2/R_2 - \ln I_{s_1} - \ln I_{s_2}) \\ &= R_f I_{s_4} \text{ antilog } [\ln (V_1V_2/R_1R_2I_{s_1}I_{s_2})] \\ &= (R_f I_{s_4}/R_2 I_{s_1} I_{s_2})\, V_1V_2 \end{aligned}$$

If

$$R_f I_{s_4} = R_1R_2I_{s_1}I_{s_2}$$

then

$$V_{out_4} = V_1V_2 \tag{7-6}$$

The I_s values must be measured and are very close to I_{EBO} measured at a low reverse emitter-base voltage. The circuit can be built using diodes instead of transistors in the log and antilog amplifiers. All the amplifiers in the circuit are assumed to be fully phase-compensated.

Commercial multipliers are available that are stable over a wide range of operating conditions and temperatures. These circuits are complex and thoroughly engineered circuits with simple block diagrams that resemble Figure 7.6. Some multipliers can operate with two inputs of only one polarity and are called *two-quadrant multipliers*, whereas others operate with two inputs of either polarity and are called *four-quadrant multipliers*.

Multipliers are used in modulator circuitry, demodulating circuitry, phase detection, analog computing circuitry for industrial control, nonlinear waveform generation, and linearization of nonlinear transducer outputs in data acquisition to name but a few applications. An excellent source of detailed information on the construction and use of nonlinear circuits is the *Nonlinear Circuits Handbook, Designing With Analog Function Modules and IC's* by the engineering staff of Analog Devices, Inc.

7-5 DIVIDER CIRCUIT

Since

$$\ln a/b = \ln a - \ln b$$

we can use the principle behind the multiplier to build a dividing circuit. The only change is the use of an adder-subtracter in place of the inverting adder in the multiplying circuit. The circuit diagram is shown in Figure 7.7.

The outputs of the logarithmic amplifiers are

$$V_{out_1} = kT/q\,(\ln V_1/R_1 - \ln I_{s_1})$$

and

$$V_{out_2} = kT/q\,(\ln V_2/R_2 - \ln I_{s_2})$$

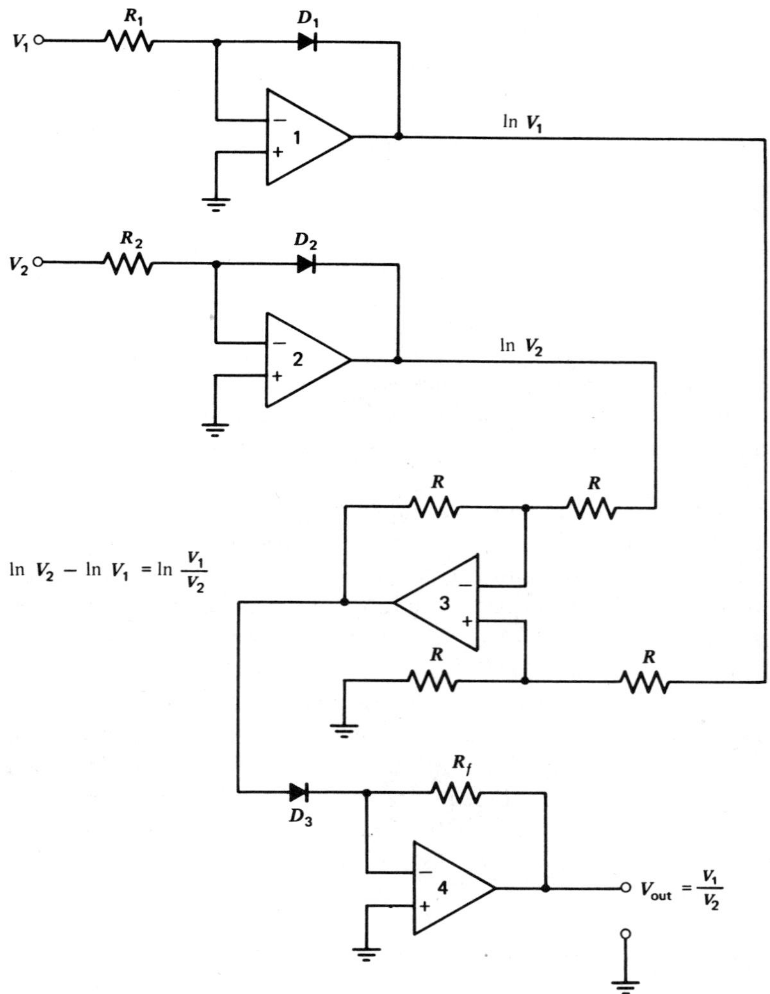

FIGURE 7.7 Divider circuit. (*Note:* Constants are left out of equations in this figure.)

To find V_1/V_2, the output of log amp 1 is fed into the positive input of the adder-subtracter, and the output of log amp 2 into the negative input. The output of the adder-subtracter is

$$V_{out_3} = kT/q\,(\ln V_1/R_1 - \ln I_{s_1} - \ln V_2/R_2 + \ln I_{s_2})$$

Usually

$$I_{s_1} \cong I_{s_2}$$

and

$$R_1 = R_2$$

so

$$\begin{aligned} V_{out_3} &= kT/q\,[\ln (V_1/R_1) - \ln (V_2/R_2)] = kT/q \ln (V_1R_2/V_2R_1) \\ &= kT/q \ln V_1/V_2 \end{aligned}$$

The output of the antilog amplifier is

$$\begin{aligned} V_{out_4} &= RI_{s_3} \text{ antilog } (\ln V_1/V_2) && \text{(7-7)} \\ &= RI_{s_3} V_1/V_2 && \text{(7-7a)} \end{aligned}$$

If I_{s_3} is such that $R_f I_{s_3}$ can be made to equal one, then $V_{out_4} = V_1/V_2$.

As in the case of multipliers, high quality divider circuit modules are commercially available. The list of applications for them is virtually the same as for multipliers.

7-6 LOG-RATIO CIRCUIT

The output of the adder-subtracter of Figure 7.7 is a log ratio with

$$V_{out} = (R_f/R_1)(kT/q) \ln V_1/V_2$$

where

R_f = both feedback resistors of the adder-subtracter
R_1 = both input resistors of the adder-subtracter

The connection of the two log amps and the adder-subtracter as shown in Figure 7.7 is known as a *log-ratio circuit.*

7-7 FUNCTION SYNTHESIZER

Any nonlinear output may be approximated by a series of linear outputs of varying slope. In this manner a wide variety of nonlinear functions can be realized with operational amplifiers. A simple synthesis circuit illustrating the principle is shown in Figure 7.8*a*.

The slope of the output versus the input voltage plot is just

$$S_1 = V_{out}/V_1 = -R_f/R_1$$

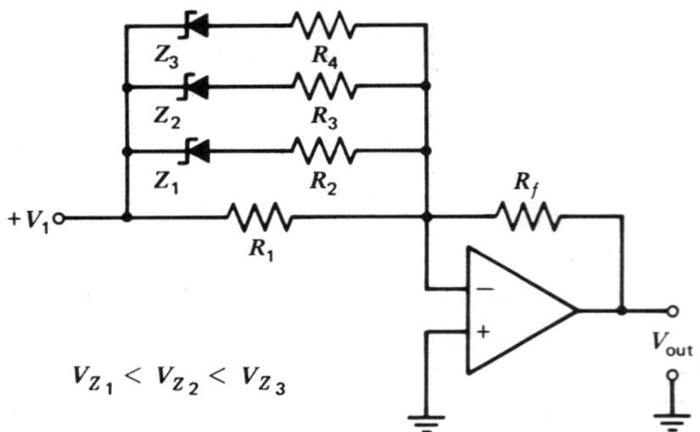

(*a*) Synthesis circuit

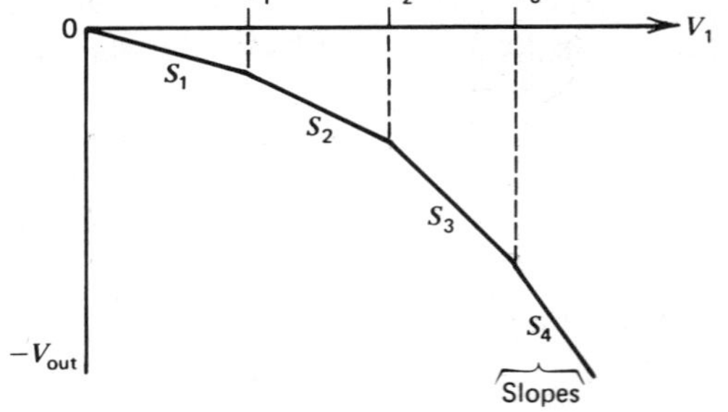

$$S_1 = -\frac{R_f}{R_1}, \quad S_2 = -\left(\frac{R_f}{R_1} + \frac{R_f}{R_2}\right), \quad S_3 = -\left(\frac{R_f}{R_1} + \frac{R_f}{R_2} + \frac{R_f}{R_3}\right)$$

$$S_4 = -\left(\frac{R_f}{R_1} + \frac{R_f}{R_2} + \frac{R_f}{R_3} + \frac{R_f}{R_4}\right)$$

(*b*) Out put vs. input

FIGURE 7.8 Simple function synthesizer.

until the first zener diode, Z_1, breaks down, as shown in Figure 7.8*b*. While the input voltage V_1 is between V_{Z_1} and V_{Z_2}, the output voltage is

$$V_{out} = -V_1R_f/R_1 - V_1R_f/R_2$$

making the slope of the transfer curve

$$S_2 = V_{out}/V_1 = -(R_f/R_1 + R_f/R_2)$$

Similarly, when $V_{Z_2} \leqslant V_1 < V_{Z_3}$,

$$S_3 = -(R_f/R_1 + R_f/R_2 + R_f/R_3)$$

and when $V_1 > V_{Z_3}$

$$S_4 = -(R_f/R_1 + R_f/R_2 + R_f/R_3 + R_f/R_4)$$

The shorter each line segment is, the better the nonlinear approximation is and the more complex the circuit becomes. The circuit will respond to negative input voltage if the zener diodes are reversed.

The circuit of Figure 7.8 is not a practical circuit because of the noise generated by zener diodes operated in reverse breakdown and the sharp corners produced by the sharp zener breakdown voltages. The breakdown voltages available from zener diodes are too limited for this circuit to have great flexibility.

A more versatile form of this circuit is shown in Figure 7.9*a*. The use of the forward conduction characteristic of the diodes to switch slopes results in rounded corners, which is often an advantage. The circuit can be set up to respond to both negative and positive input circuits—also an advantage. Essentially, the zener diode is replaced with a voltage divider and a diode that is reverse biased until the input voltage exceeds the voltage set by the voltage divider. Note that a negative voltage is used to reverse bias the diode that is to conduct with a positive going input signal, and vice versa.

The function synthesizing networks of Figures 7.8 and 7.9 are in parallel with R_1. If they are placed in parallel with R_f, the function synthesizing circuits will have a gain that decreases as V_{in} increases, so the output slope will decrease with increasing V_{in}. This is shown in Figure 7.10. If enough segments are included in a function synthesizing circuit for which gain decreases as V_{in} increases, a very low distortion sine wave can be generated from a triangular input.

Synthesis circuits can provide very stable nonlinear functions of wide dynamic range. They are used to produce nonlinear functions such as $\log_{10}$, ln, antilog, square roots, and powers. There are many variations and types of synthesis circuits. Interested students should consult *Operational Amplifiers, Design and Applications* by Tobey, Graeme, and Huelsman of the Burr Brown Research Corporation and the previously mentioned *Nonlinear Circuits Handbook* from Analog Devices, Inc. for further details.

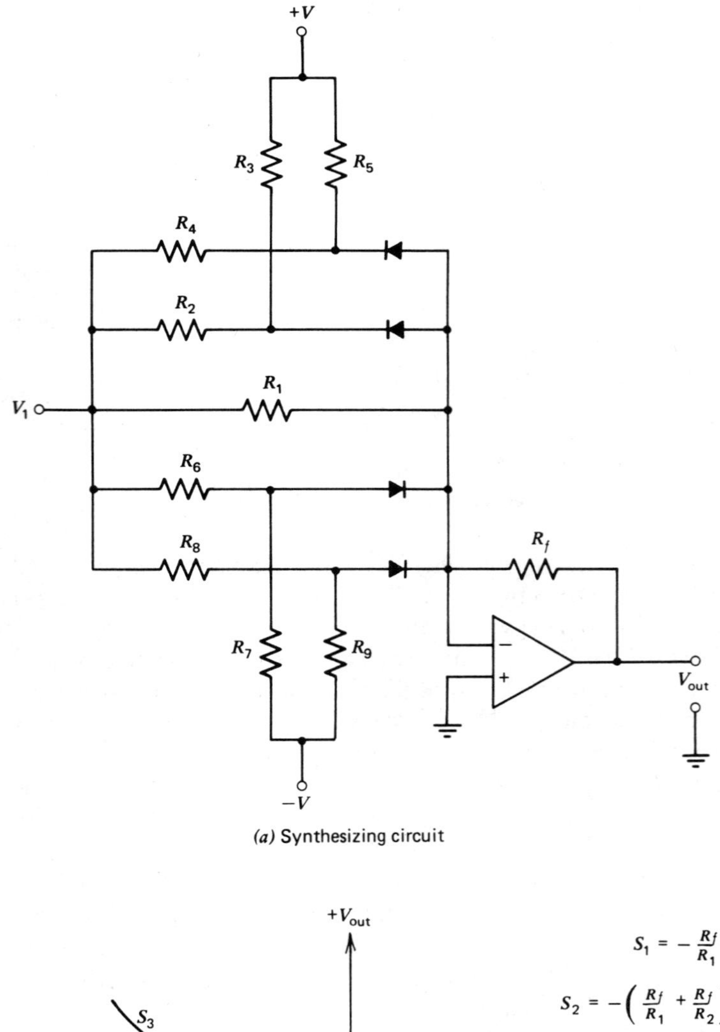

(a) Synthesizing circuit

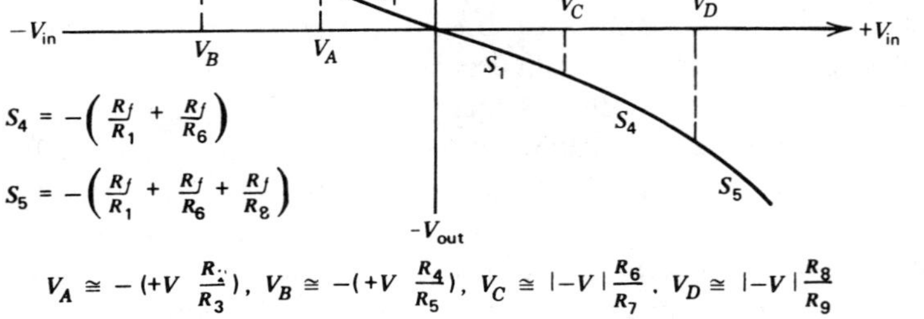

(b) Output vs. input

FIGURE 7.9 A more flexible synthesizing circuit.

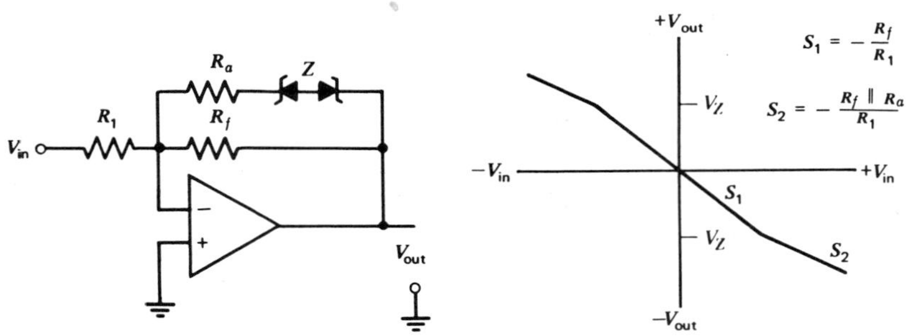

(*a*) Synthesizing circuit (*b*) Outcome vs. input

FIGURE 7.10 Function synthesizer, decreasing slope.

7-8 SIGNAL COMPRESSION

Sometimes in a system a signal has too wide a dynamic range to be handled conveniently. If the signal is scaled down linearly, the lower voltage information may become obscured by noise or difficult to distinguish. If the signal is scaled down with a logarithmic response (compressed), the higher voltage information is reduced more in amplitude than the lower voltage information, as shown in Figure 7.11*c*.

A simple compression circuit is shown in Figure 7.12. Note that the circuit is bilateral. Since one diode is reverse biased when the other is forward biased, this is possible. Even though the circuit is basically a logarithmic amplifier, it is not a true log amp since the logarithm function is discontinuous at zero. R_f provides a linear zero crossing and provides for less attenuation of very small signal levels.

The circuit of Figure 7.12 will act as an expansion circuit if D_1 and D_2 are put in parallel with R_1. An expansion circuit is used to convert compressed signals back to their original form or to expand small signals of similar amplitude to enable one to distinguish between them.

SUMMARY

1. A logarithmic amplifier is obtained by using a feedback network with a logarithmic response. A semiconductor diode or the emitter-base junction of a transistor can be used for natural logarithms. The output voltage of a logarithmic amplifier is proportional to the logarithm of its input voltage.
2. An antilog amplifier must have an exponential response to input voltage amplitude. A logarithmic input network to an amplifier provides the proper response. This input network can be a diode or transistor emitter-base junction for natural logarithm antilog.
3. A combination of logarithm amplifiers, antilog amplifiers, and summers can be used to construct circuits that will multiply, divide, and take various ratios.
4. Function synthesizers are circuits that can provide many nonlinear input versus output responses. They are built by using input or output networks that provide the response desired. The networks are made of smaller networks, each of which provides the desired output over a narrow range of the input voltage.

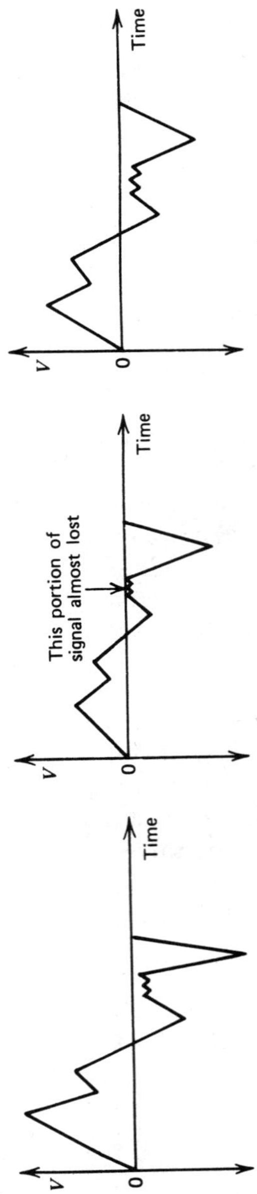

(a) Input

(b) Linear compression

(c) Logarithmic compression

FIGURE 7.11 Signal compression waveforms.

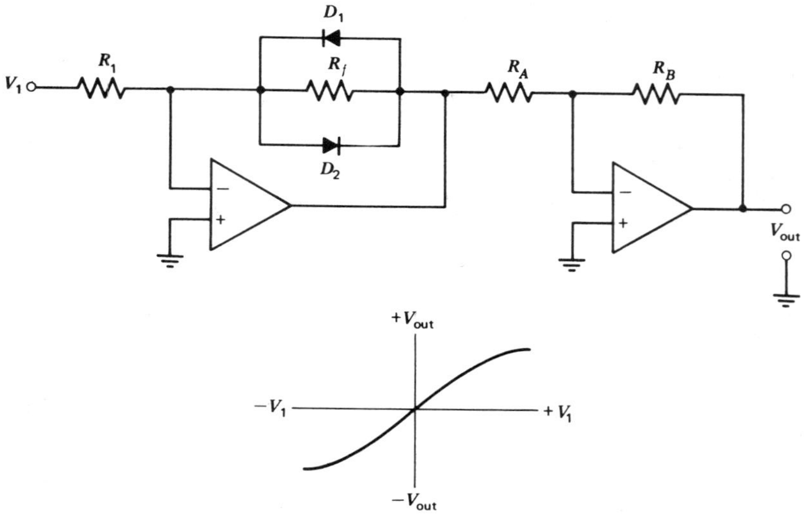

FIGURE 7.12 Compression circuit.

5. Function synthesizers are used to provide precision logarithmic, antilog, power, root, and other nonlinear amplifier responses.
6. Signal compression allows a signal to be processed with circuitry that has a smaller dynamic range than would otherwise be possible. A special logarithmic amplifier provides the signal compression.

SELF-TEST QUESTIONS

7-1 Name the components used to provide the logarithmic response of log amplifiers.
7-2 Without compensating, circuitry log amplifiers are very temperature sensitive. Give the two major reasons for this condition.
7-3 State the difference between an antilog and log amplifier.
7-4 State the operating principle behind the function synthesizing circuit of Figure 7.9.
7-5 State the use of a signal compressor.
7-6 A logarithm circuit such as the one in Figure 7.1 has an $R_1 = 200\ k\Omega$ and the diode has an $I_s = 15$ nA. If $V_{in} = 5$ V, calculate V_{out}.
7-7 An antilog circuit such as the one in Figure 7.5 has an $R_f = 100\ k\Omega$ and the diode has an $I_s = 4$ nA. If $V_{in} = 0.1$ V, calculate V_{out}.
7-8 Draw the block diagram of a circuit that will raise a number to a power (i.e., X^n). Do not worry about the coefficients.
7-9 State what must be done to the function synthesizing circuit of Figure 7.9 to obtain an output in which gain decreases as V_{in} increases.
7-10 How is the signal from a signal compressor such as the one shown in Figure 7.12 uncompressed?
7-11 State the function of the resistor R_E in Figure 7.3*b*.

If you cannot answer certain questions, place a check next to them and review appropriate parts of the text to find the answers.

LABORATORY EXERCISE

Objective

After completing this laboratory, the student should be able to compute the components for and build a logarithmic amplifier, an antilogarithm amplifier, and a simple synthesizing amplifier, and build a signal compressor.

Equipment

1. Two Fairchild μA741 operational amplifiers or equivalent.
2. 2% resistor assortment.
3. Power supply, ±15 V dc.
4. Audio signal generator.
5. Two zener diodes, $V_Z \cong 2$ V and $V_Z \cong 4$ V.
6. Two *npn* transistor, 2N3710 or equivalent (a matched pair such as a 2N2461 is best).
7. Two diodes, 1N914 or equivalent.
8. Oscilloscope.
9. Two 10 kΩ potentiometers, carbon or metal film, *not wire wound.*
10. Breadboard with IC sockets such as EL Instruments SK-10.
11. Transistor curve tracer (optional, but convenient).

Procedure

1. Logarithmic amplifier.
   (a) Measure I_{ES} of one *npn* transistor at a low reverse emitter-base voltage, $V_{BE} \cong 1$ V.
   (b) Connect the circuit shown in Figure 7.13, picking R_1 such that $I_{R1} = 0.1$ mA with $V_1 = 10$ V. *Note.* A 20 MΩ resistor across the transistor emitter to collector terminals will ease offset null problems but introduce some error. If oscillations occur, a small capacitor (≅0.001 μF) across the transistor, emitter to collector, should stop them.
   (c) Calculate and measure the output voltage for $V_{in} = +3$ V and $V_{in} = +6$ V.
   (d) Do not tear the circuit down.

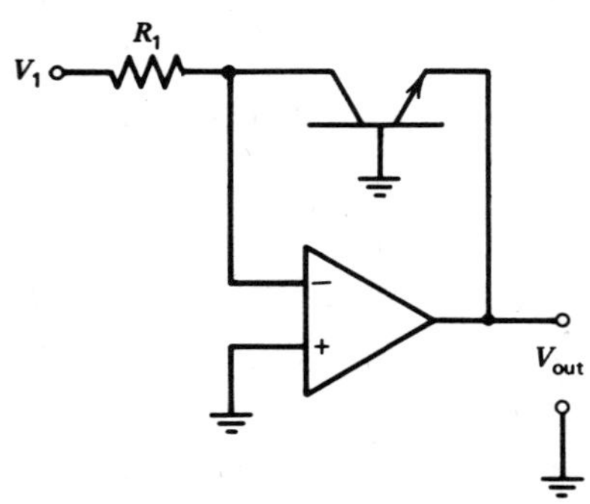

FIGURE 7.13 Log amp.

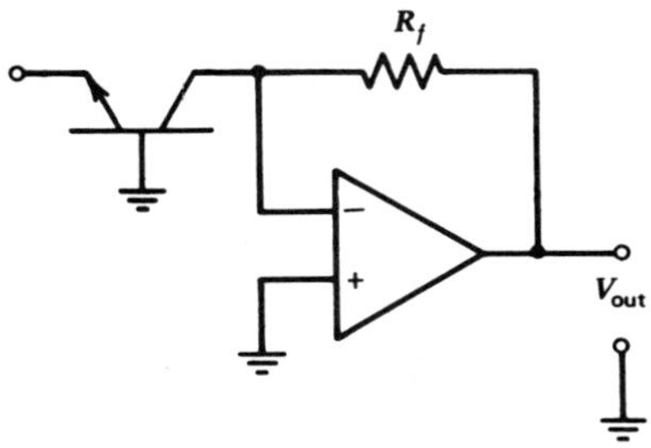

FIGURE 7.14 Antilog amp.

2. Antilog amplifier.
   (a) Measure I_{ES} of the transistor. It should be near the value of the transistor used in part 1.
   (b) Connect the antilog circuit shown in Figure 7.14. Set R_f so that I_f = 0.1 mA at V_{out} = 10 V.
   (c) Connect the output of the logarithmic amplifier constructed in part 1 to the input of the antilog amplifier.
   (d) Calculate and measure V_{out} with the logarithmic amplifier input at +3 V and +6 V. Record your measurements.
3. Function synthesizer.
   (a) Connect the circuit shown in Figure 7.15.

$$V_{Z_1} \cong 2 \text{ V} \qquad V_{Z_2} \cong 4 \text{ V}$$

$$R_1 = R_2 = R_3 = R_f = 100 \text{ k}\Omega$$

   (b) Plot V_{out} versus $+V_{in}$ from V_{in} = 0 to V_{in} = +8 V. Using the information in Figure 7.8, compare your results to those you would expect from calculations.

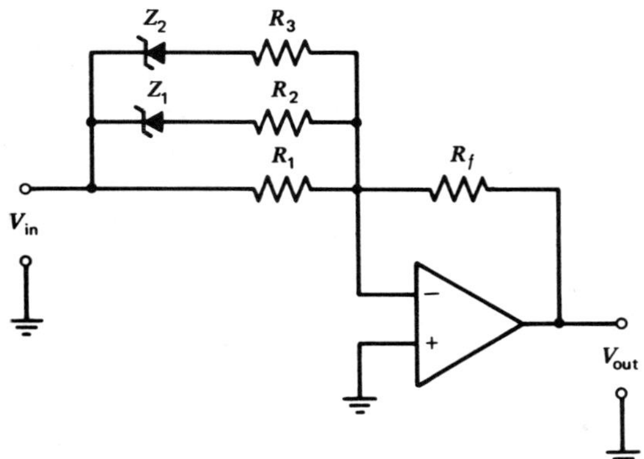

FIGURE 7.15 Function synthesizer.

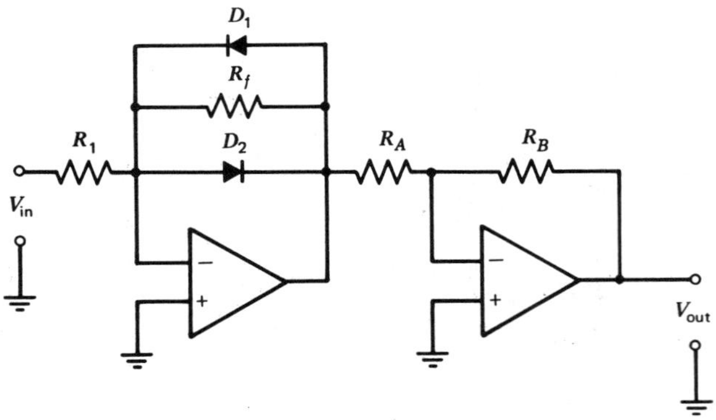

FIGURE 7.16 Compression circuit.

4. Signal compression.
   (a) Buld the signal compressor shown in Figure 7.16. Let $R_1 = R_f = 100\ \text{k}\Omega$, $R_A = 20\ \text{k}\Omega$, $R_B = 200\ \text{k}\Omega$, and D_1 and D_2 be 1N914 or equivalent.
   (b) Plot V_{out} versus V_{in}.
   (c) Observe the signal compression with a sine wave input signal and any other waveform of interest to you. This will be observed as a change in voltage gain as the amplitude of V_{in} changes.

Discussion

Discuss the probable causes of any deviation of your measurements from the theoretical values for each part of the laboratory.

CHAPTER

8

ACTIVE FILTERS

Active filters are circuits using resistors, capacitors, and amplifiers, usually op-amps, to allow only selected frequencies to pass from the filter's input to its output. These frequency-selective circuits are used to boost or attenuate certain frequencies in audio circuits, electronic music generators, seismic instruments, communications circuits, and in research to study the frequency components of such diverse signals as brain waves and mechanical vibrations. Active filters are used in almost every area of electronics, so they are worth our attention.

OBJECTIVES

After completing the study of this chapter and the self-test questions, the student should be able to:

1. State the major characteristics of Bessel, Butterworth, and Chebyshev filters.
2. State the advantages and disadvantages of active filters.
3. Calculate the components for each active filter type given in this chapter, including cascaded filters.
4. Identify from memory the active filter types discussed in this chapter.
5. Perform the laboratory for Chapter 8.

8-1 BASIC FILTER DEFINITIONS

A filter, whether active or passive (containing no amplifiers), allows a certain portion of the frequency spectrum to pass through to its output. The filter is classified by which portion of the frequency spectrum it passes.

Low-pass filters allow frequencies from dc up to some selected cutoff frequency, f_c, to pass through and attenuate all frequencies above f_c, as shown in Figure 8.1*a*. The range of frequencies from zero to f_c is called the *passband*. The range of frequencies above f_b is called the *stop band*. The range of frequencies from f_c to f_b is called the *transition region*. The rate at which attenuation changes in the transition region is an important filter characteristic. The frequency at which the output voltage of the filter drops to a value of 0.707 of its value in the passband (or has dropped 3 dB) is the cutoff frequency, f_c. The frequency at which the output voltage is 3 dB above the stop band value is f_b.

A high-pass filter attenuates all frequencies up to f_c and passes all frequencies above f_c up to the frequency limit of the high-pass filter. A high-pass frequency characteristic is shown in Figure 8.1*b*.

A band-pass filter, as illustrated in Figure 8.1*c*, passes all frequencies between a lower cutoff frequency, f_1, and an upper cutoff frequency, f_2. All frequencies

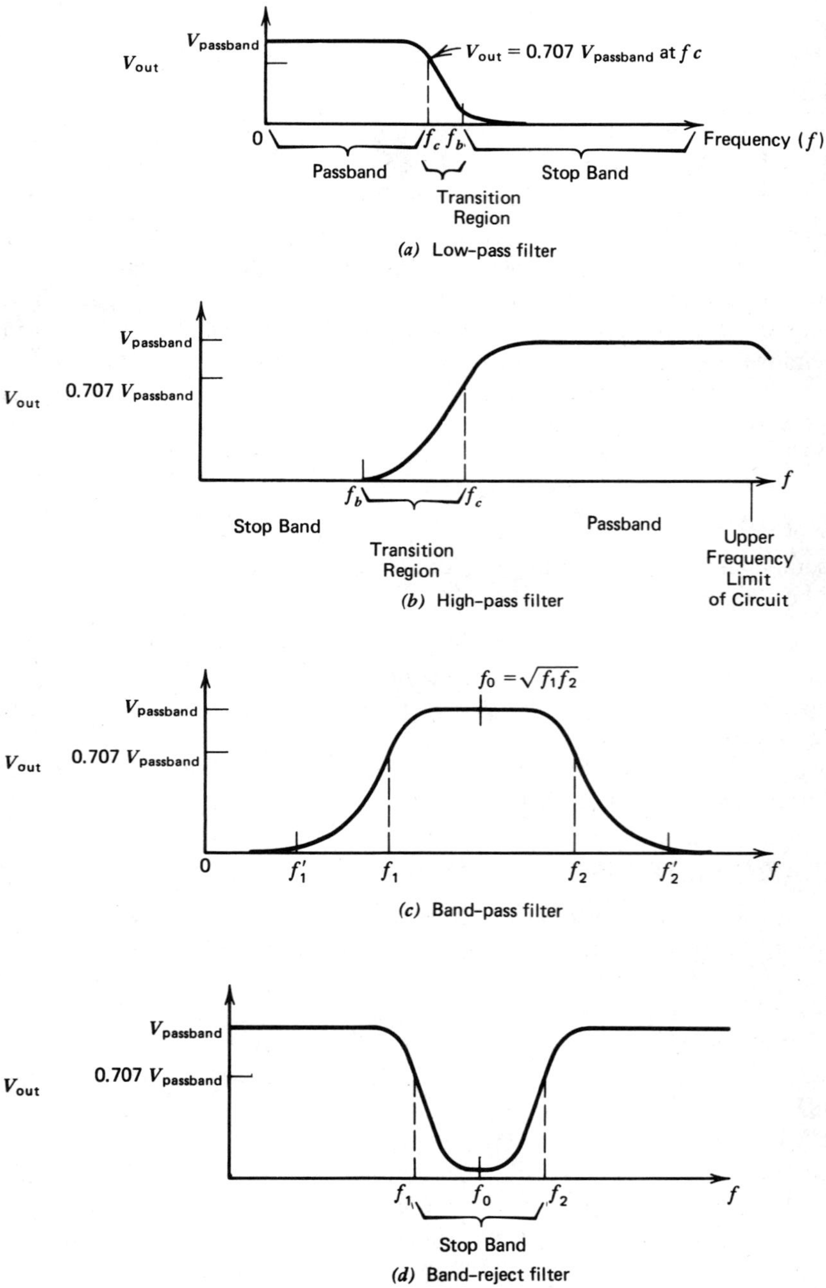

FIGURE 8.1 Filter characteristics.

below f_1 and above f_2 are attenuated. The frequency ranges from f_1' to f_1, and f_2 to f_2' are the transition regions. The center frequency (f_o) is considered to be the geometric mean of f_1 and f_2 and is found from the following equation:

$$f_o = \sqrt{f_1 f_2} \tag{8-1}$$

A band-reject filter attenuates all frequencies between f_1 and f_2 and passes all other frequencies as shown in Figure 8.1*d*. A band-reject filter with a narrow band of attenuated frequencies is called a *notch filter*. Band-reject filters are useful for eliminating undesired frequencies, such as 60 Hz, from audio systems.

8-1.1 Advantages of Active Filters

Passive filters are constructed from inductors, capacitors, and resistors. In the frequency range in which active filters are useful, most passive filters require large, heavy, costly inductors and attenuate frequencies in the passband (even though the stop-band frequencies are attenuated more). The inductors used in passive filters have winding resistance, core losses, and interwinding capacitance that cause them to behave far from ideally.

The advantages of active filters over passive filters are:

1. They use resistors and capacitors that behave more ideally than do inductors.
2. They are relatively inexpensive.
3. They can provide gain in the passband and seldom have any severe loss (as do passive filters).
4. The use of op-amps in active filters provides isolation from input to output. This allows active filters to be easily cascaded to obtain higher performance.
5. Active filters are relatively easy to tune.
6. Very low frequency filters can be constructed using modest value components.
7. Active filters are small and light.

Active filters do have some disadvantages. They require a power supply and are limited in maximum frequency to the highest operating frequency of the op-amp. This limits most op-amp active filters to a few megahertz at most. With discrete amplifiers this frequency can be exceeded. As manufacturers improve the frequency response of op-amps, the upper frequency limit of active filters will be extended.

8-1.2 Limitations of Chapter 8

Entire books have been written on active filters. Some very useful ones are listed at the end of this chapter. You will not be an active-filter expert on completing this chapter, but you will understand some basic active-filter types and be able to construct some active filters that work well and be in a position to dig deeper. The equations used in this chapter are presented without development, for to do so would require a considerable amount of calculus. They are developed in many of the references at the end of this chapter.

8-2 POLES AND TRANSITION REGION RESPONSE

Any discussion of active filters is accompanied by reference to poles. For example, in this chapter we will discuss chiefly two-pole filters. The word pole is a reference to a graph of the mathematics used to derive the equations used for calculating active-filter responses. For all practical purposes, a *pole* refers to the rate of attenuation change in the transition region caused by each *RC* network used for determining frequency response in the filter. Recall from Chapter 4 that each *RC* network in a multistage amplifier contributes 6 dB/octave to the amplifier's roll off. The same is true of active filters. Each pole (*RC* network of the filter) contributes about 6 dB/octave to the active filter's rate of attenuation change in its transition region. For example, a two-pole, low-pass filter has a 12 dB/octave increase in attenuation between f_c and f_b, and a five-pole, high-pass filter has a 30 dB/octave decrease in attenuation between f_b and f_c. This is illustrated for one type of low-pass filter, a Butterworth, in Figure 8.2*a*.

The order of a filter simply indicates its number of poles. For example, a second-order low-pass filter is a two-pole, low-pass filter and has a rate of attenuation change in the transition region of 12 dB/octave. A sixth-order filter has six poles and a rate of attenuation change in the transition region of 36 dB/octave.

Low-order filters can be cascaded to make higher-order filters. Three second-order filters can be cascaded to make a sixth-order filter. We will discuss cascading in some detail later in this chapter.

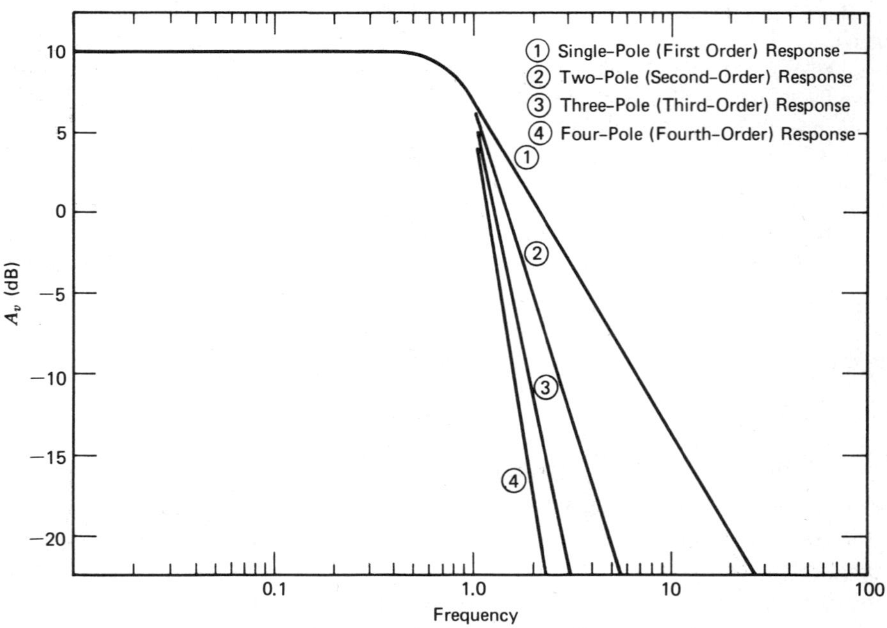

(*a*) Butterworth low-pass filter response curves, f_c = 1kHz

FIGURE 8.2a Butterworth filter response curve, f_c = 1 kHz.

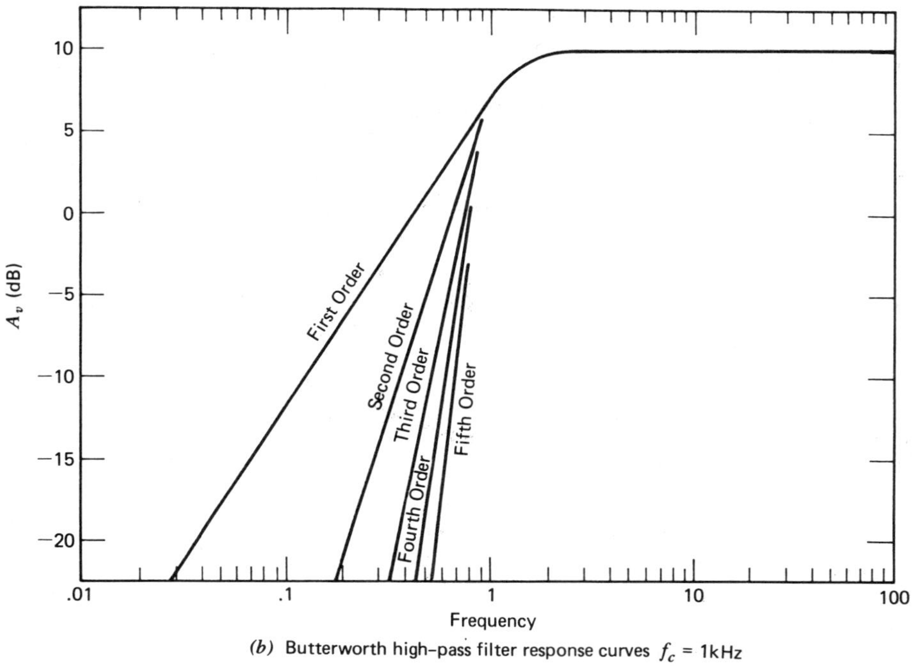

(b) Butterworth high-pass filter response curves f_c = 1kHz

FIGURE 8.2b Butterworth high-pass filter response (f_c = 1 kHz).

8-3 TYPES OF FILTER RESPONSES

8-3.1 Butterworth

A Butterworth filter response is very flat in the passband. Its response is called maximally flat. The transition region attenuation change of a Butterworth filter is 6 dB/octave per pole. Thus an eighth-order Butterworth will have an attenuation change of 48 dB/octave in the transition region.

The phase response of a Butterworth filter is not linear; in other words, the time required for a signal to propagate through the filter is not linear with frequency. Therefore, a step response or pulse applied to a Butterworth filter input will cause overshoot on the output. A Butterworth filter is used when all frequencies in the passband must have the same gain. Figure 8.2*a* shows a low-pass Butterworth response and Figure 8.2*b* shows a high-pass Butterworth response.

8-3.2 Chebyshev

A Chebyshev filter will have ripples in the passband but not in the stop band. The higher the filter order, the more ripples will appear in the passband. The ripple amplitude can be fixed in the filter by design and is usually fixed at 0.5 dB, 1 dB, 2 dB, or 3 dB. As more ripple is allowed, more attenuation is obtained in the

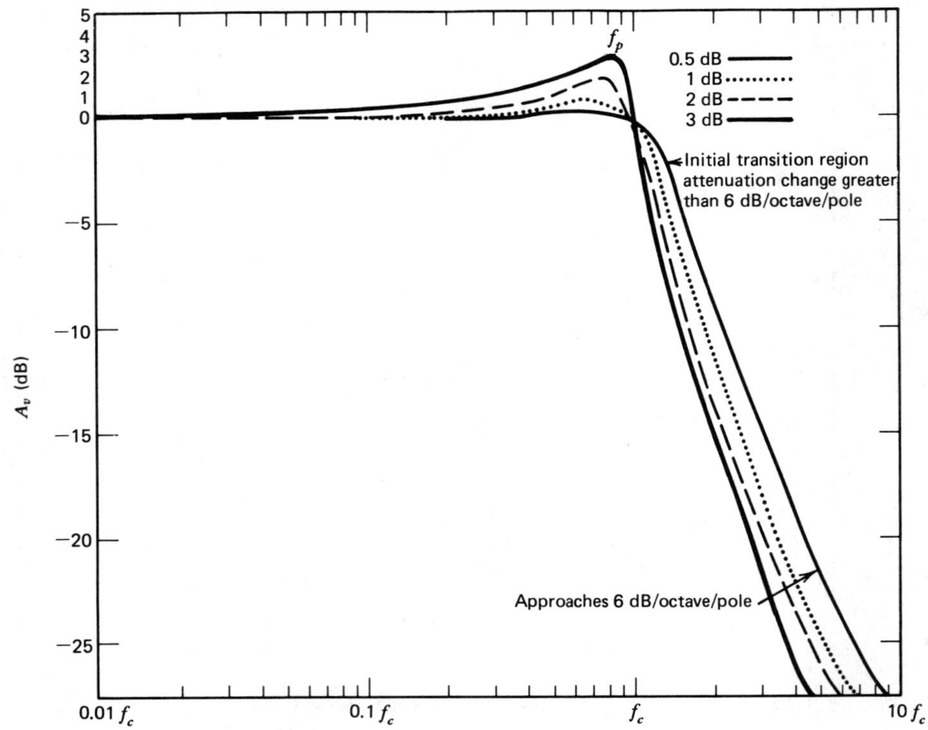

(a) Second order Chebyshev low-pass response curves

transition region. This is illustrated in Figure 8.3*a* for a second-order Chebyshev low-pass filter.

The Chebyshev filter has a transition region attenuation change (TRA) of more than 6 dB/octave/pole. The Chebyshev filter is very useful when the attenuation in the transition region must be very rapid. Ripples in the passband are the price one pays for the transition region response. The transition region attenuation change of a Chebyshev filter is:

$$\text{TRA} = 20 \log \varepsilon + 6(n - 1) + 20n \log \omega/\omega_c$$

where

n = filter order
ε = a constant between 1 and 0 that sets the ripple of the filter
0.5 dB ripple, $\varepsilon = 0.3493$
3 dB ripple, $\varepsilon = 0.9976$

This is greater than the Butterworth transition region response by a factor of 20 log ε + 6(n − 1). Chebyshev filters can have fewer poles than Butterworth filters

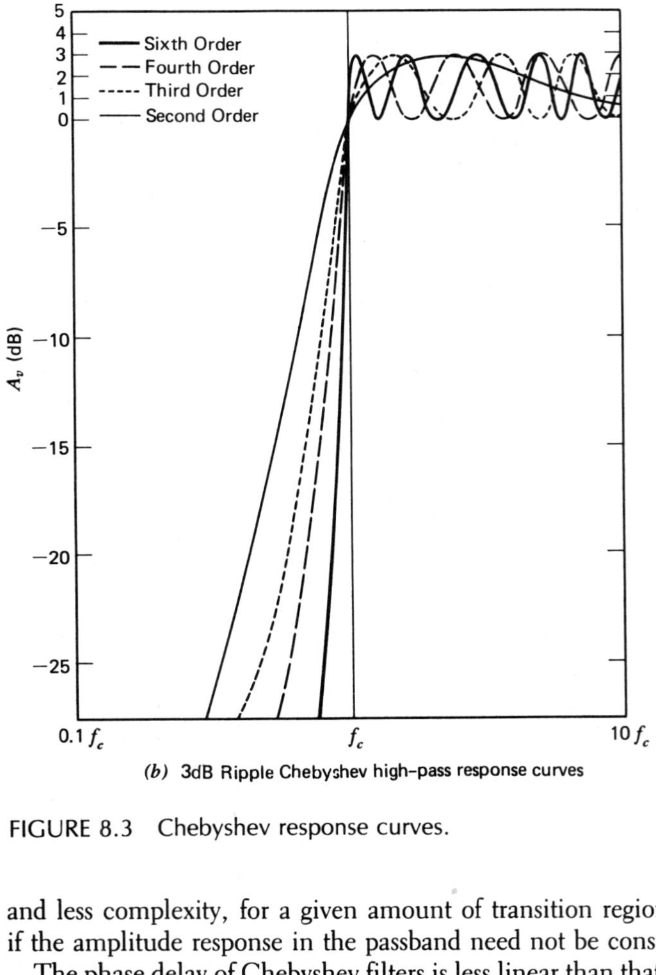

(b) 3dB Ripple Chebyshev high-pass response curves

FIGURE 8.3 Chebyshev response curves.

and less complexity, for a given amount of transition region attenuation change, if the amplitude response in the passband need not be constant.

The phase delay of Chebyshev filters is less linear than that of Butterworth filters. The higher the order of a Chebyshev filter and the more ripple it has, the less linear its phase delay. Thus the more overshoot it will have with step and pulse inputs.

A high-pass Chebyshev response is shown in Figure 8.3*b* for several filter orders. Note that the peak response is not at the cutoff frequency and varies with the ripple as shown in Figure 8.3*a*. The peak frequency (f_p) is related to the cutoff frequency (f_c) by:

$$f_p = f_c\sqrt{1 - \frac{\alpha^2}{2}} \qquad \text{for a low-pass filter}$$

$$f_p = f_c \Big/ \sqrt{1 - \frac{\alpha^2}{2}} \qquad \text{for a high-pass filter}$$

where α = damping factor.

The damping factor will be discussed in greater detail later in this chapter.

8-3.3 Bessel

Bessel filters are referred to as linear phase or linear time-delay filters. The phase delay of a signal from the input to the output increases linearly with frequency. Bessel filters, therefore, have almost no overshoot with a step-response input. This characteristic makes them the best choice for filtering rectangular waveforms without altering the shape of the waveform.

Bessel filters have less than 6 dB/octave/pole transition region attenuation change. The cutoff frequency of a Bessel filter is defined to be the frequency at which the phase delay of the filter is one half the maximum phase delay.

$$\theta(f_c) = \theta\text{max}/2 = \frac{(n\pi/2)}{2} \text{ radians}$$

where

θ = phase delay
n = filter order

The 3 dB frequency of a Bessel filter is not the defined f_c. This is illustrated in Figure 8.4, the response of a Bessel filter.

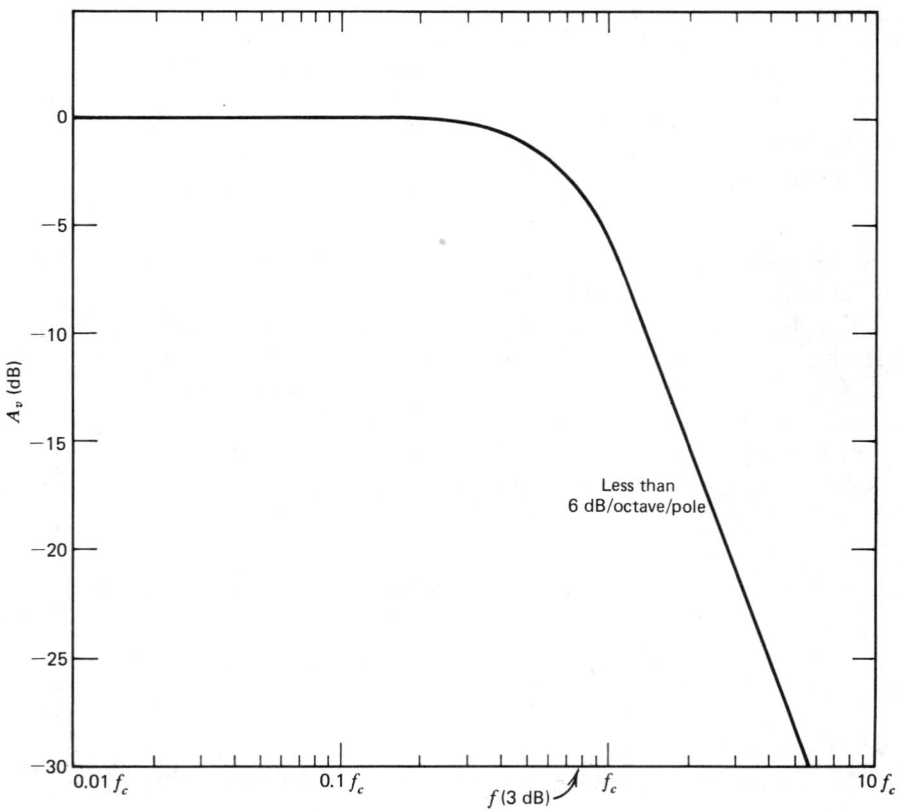

FIGURE 8.4 Second-order Bessel low-pass response.

8-3.4 Other Responses

The Bessel, Butterworth, and Chebyshev are the only responses covered in this chapter. Other possible responses include the *inverse Chebyshev*, which has a flat passband but ripples in the stop band; the *elliptic*, which has ripples in the stop and passbands but very large transition region attenuation changes; and the *parabolic*, which has a very good pulse response. If you are interested in these responses, refer to the books listed at the end of this chapter.

8-3.5 Some Definitions

Damping factor, α

The damping factor sets the shape of the transition region and the overshoot of the pass-band response near the transition region. Thus the damping factor sets the shape of the filter's response and the type of filter. A second-order Butterworth filter will have a damping factor (α) of 1.414; a 3 dB ripple second-order Chebyshev will have $\alpha = 0.766$.

A Bessel filter, a Butterworth filter, and a Chebyshev filter might all have the same schematic, differing only in component values. The damping factor sets the filter response. Several low-pass responses with different damping factors are shown in Figure 8.5.

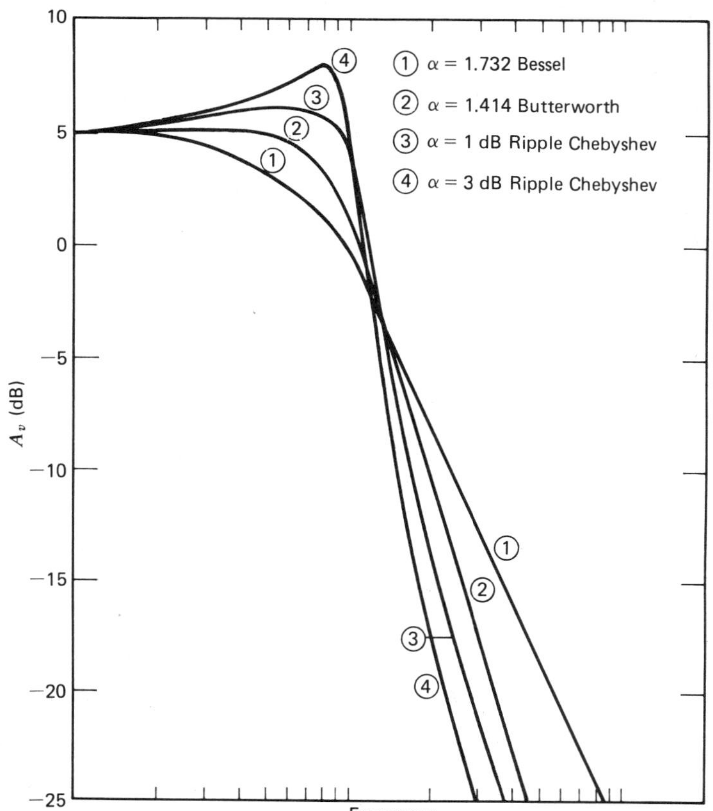

FIGURE 8.5 Low-pass responses with damping factor varied.

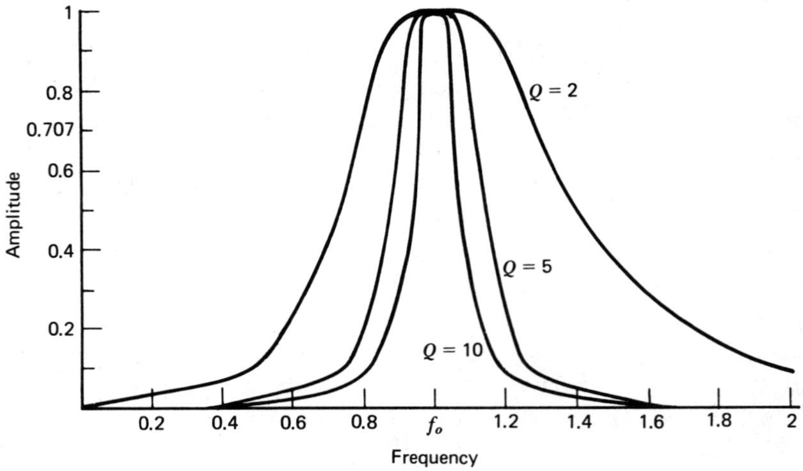

FIGURE 8.6 Band-pass filter responses.

Quality factor, Q

The Q is the relationship between the passband center frequency and the 3 dB frequencies in a band-pass circuit as shown in Figure 8.6.

$$Q = \frac{f_0}{f_2 - f_1} = \frac{\sqrt{f_1 f_2}}{f_2 - f_1} \qquad (8\text{-}2)$$

where

$f_0 = \sqrt{f_1 f_2}$ = center frequency
f_1 = lower 3 dB frequency
f_2 = upper 3 dB frequency

For active filters

$$Q = 1/\alpha$$

Passband gain, A_p

The gain of the active filter in its passband is the passband gain.

$$A_p = \frac{V_{out}}{V_{in}} \qquad (8\text{-}3)$$

Sensitivity, S

Sensitivity is the amount one filter parameter changes as another parameter is varied. For example:

$$S_{R_1}^{\omega_0} = -0.5$$

where

$\omega_0 \ = 2\pi f_0$
$R_1 \ = R_1$ in the active filter circuit

indicates ω_0 will decrease 0.5% if R_1 is increased 1%.

Sensitivities of the filters f_c, or ω_0 for band-pass filters, and α, Q for band-pass filters, are commonly calculated. The procedure is tedious and time-consuming, but it must be done if a filter is to be used over a wide range of ambient temperatures or if loose component tolerances are to be used. The equations for S vary with each filter type and are presented in several of the references listed at the end of this chapter. We will not present the many sensitivity equations in this chapter.

8-3.6 Components

For optimal performance and high quality, close tolerance components must be used in active filters. The resistors and capacitors used must have low temperature coefficients and little drift with age.

For second-order filters ±5% capacitor tolerances and ±2% resistor tolerances are recommended. For higher order filters ±1% resistors and ±2% capacitors are best. Even with component tolerances this close, some trimming may be required to set the damping factor and frequency to the desired value in an active filter.

The best choice of resistors for active filters are metal film resistors. Metal film resistors have low noise, very good frequency response, and fairly low temperature coefficients of resistance (TC). The temperature coefficients of ±100 ppm/°C (parts per million) and ±50 ppm/°C are readily available and TC of ±10 ppm/°C can be purchased.

Wire-wound resistors are excellent for low-frequency filters since they are available in high precision, have low noise, and low temperature drift (TC about ±10 ppm/°C). For even moderate frequencies, wire-wound resistors must be noninductively wound.

Carbon film resistors are available in high precision, are fairly low in noise, have good frequency response, and fair temperature coefficients. Carbon composition resistors are best avoided except for experimentation, since they are noisy and have poor temperature stability.

The best capacitors for active filters are polystyrene, NPO ceramic, and mica. These are all quite expensive and large in size for a given capacitance value, but they have low dissipation factors and low temperature coefficients. Mica capacitors are only available to about 0.01 μF and polystyrene to about 10 μF. For noncritical uses, such as laboratories in school, metallized mylar or polycarbonate capacitors may be used. The physically small disk ceramic capacitors should be avoided for active filter use since their capacitance changes up to several percent with voltage, temperature, time, and frequency.

Metal or carbon film resistors and mylar capacitors will suffice for the laboratory at the end of this chapter since temperature stability is not a factor.

8-4 SOME TYPES OF ACTIVE FILTERS

In this section we will discuss the characteristics of the active filters to be calculated in the next section.

8-4.1 Sallen and Key (VCVS)

VCVS stands for voltage-controlled voltage source. The op-amp in these circuits is used as a VCVS. The Sallen and Key low-pass and high-pass second-order active filter circuits, shown in Figure 8.7, are popular, inexpensive, and easy to adjust. In each circuit each RC provides 6 dB/octave of transition region attenuation change. Since there are two RC circuits, R_1C_1 and R_2C_2, the circuits shown are second order. In the low-pass circuit, R_1C_1 and R_2C_2 are integrators. In the high-pass circuit, R_1C_1 and R_2C_2 are differentiators. R_A and R_B set the damping factor. The feedback from the amplifier through C_1 in the low-pass filter and R_1 in the high-pass filter provides the response shape near the passband edge. If $R_1 = R_2$ and $C_1 = C_2$, the components of this filter are easy to calculate.

The Sallen and Key filters must have a fixed gain as R_A and R_B set the damping factor, therefore the filter type.

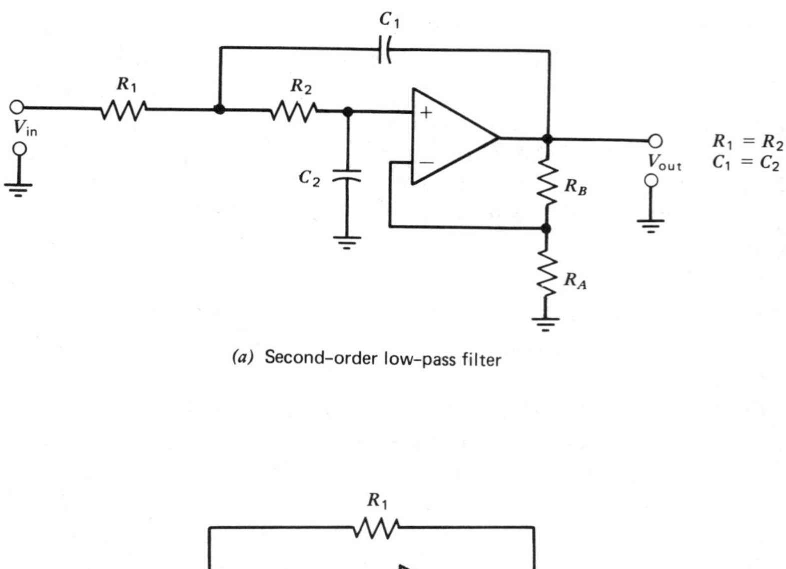

(a) Second-order low-pass filter

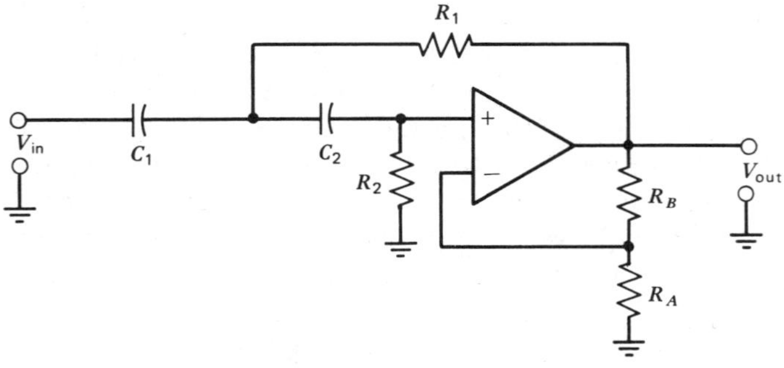

(b) Second-order high-pass filter

FIGURE 8.7 Sallen and Key active filters.

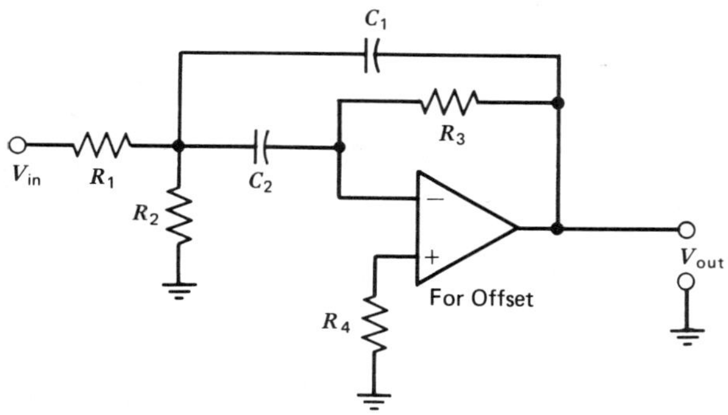

FIGURE 8.8 Multiple-feedback band-pass filters.

8-4.2 Multiple Feedback

The multiple-feedback active filter is a simple, well-behaved band-pass filter for low to moderate Qs, up to about 10. This circuit is shown in Figure 8.8. Note that feedback is through C_1 and R_3 (thus multiple feedback for a name). R_1 and C_1 provide the low-pass response, and R_3 and C_2 provide the high-pass response. The feedback provides the peaking (Q) near f_0. R_2 can be omitted, but the component calculation procedure changes. R_2 in the circuit raises R_{in} and provides a controllable passband gain. We will calculate the circuit both ways in the next section. Multiple feedback filters can be constructed as high-pass and low-pass active filters, too.

8-4.3 State Variable

The state-variable active filter circuit shown in Figure 8.9*a* is called a unity gain state variable, since its passband gain is one. The state variable provides simultaneously a second-order high-pass, low-pass, and band-pass response. The band-pass response varies with the high-pass and low-pass response. For example, if the high-pass and low-pass responses are a Butterworth type, the band-pass will be first order. The high-pass and low-pass responses will be the same, both Butterworth or both 3 dB ripple Chebyshev, since the same components set all three responses. The band-pass filter cannot be optimized simultaneously with the high-pass and low-pass response.

The state-variable filter is very stable, has low Q and α sensitivities, and little interaction between adjustments of frequency and Q. As a band-pass filter, stable Qs up to 100 can be obtained. The state variable, sometimes called the "universal active filter," is used in many commercial active filters.

The state variable is complicated, requiring three op-amps for the unity gain version and four op-amps for the version with independent gain and α (Q for band-pass) adjustment. The latter version is shown in Figure 8.9*b*.

The theory of operation of the state-variable filter can be explained in two ways. The first is illustrated in Figure 8.10. The state variable can be considered as a circuit solving the second-order differential equation of an active filter using integrators much like the example in Chapter 6-17.

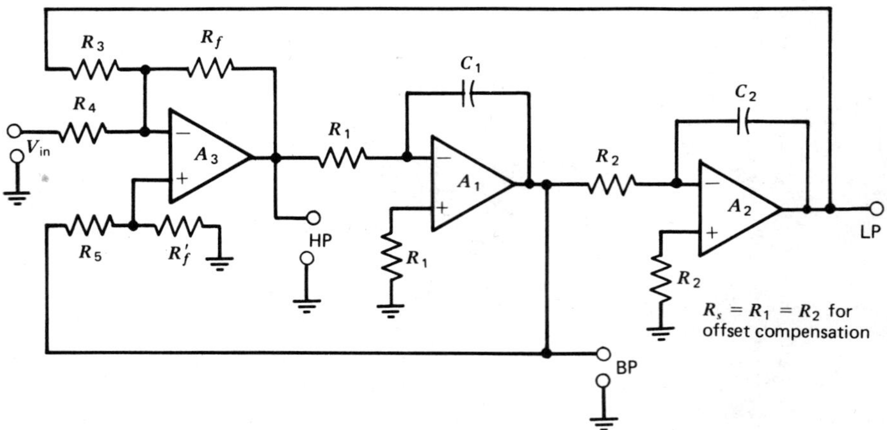

(a) Unity gain state-variable second-order filter

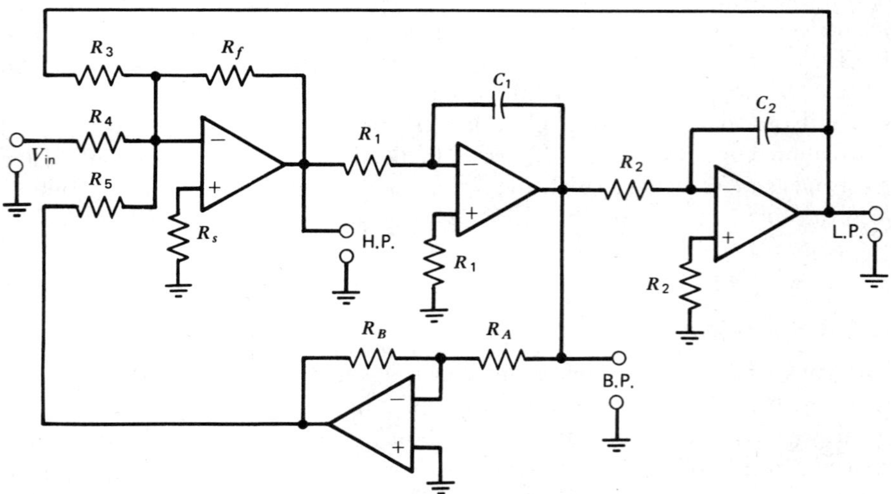

(b) State-variable, variable gain

FIGURE 8.9 State variable active filter.

The basic state-variable filter consists of a summer and two integrators. The integrators provide the coefficients of the solution and are fed back to the summer, the band-pass coefficient being multiplied by α. A more intuitive feel for the filter can be obtained by considering each response separately.

Two cascaded integrators will provide a second-order low-pass response. If the output of the first integrator is fed back with a variable feedback factor and summed with the input, the response near the cutoff frequency can be adjusted. The low-pass response is taken from the second integrator.

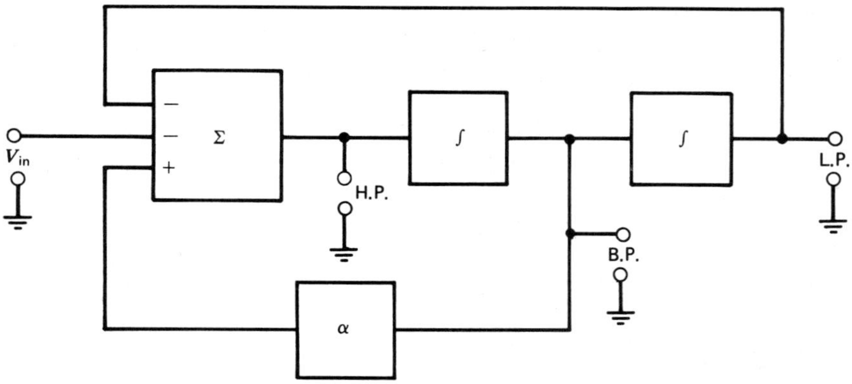

FIGURE 8.10 State-variable block diagram.

The high-pass response is obtained by summing, out of phase, the low-pass response with the input. The two cancel at all frequencies up to f_c. Above f_c the low-pass response disappears, allowing V_{in} to pass through the summer as a high-pass output.

The band-pass response can be considered as the integral of the sum of the high-pass and low-pass outputs. The high-pass output provides a decrease in attenuation as frequency increases up to f_c, being integrated to provide an increase in attenuation above f_c. Since f_c is the same for both integrators, an output will be obtained only where the responses overlap, as shown in Figure 8.11. If $\alpha = 1/Q$ is low, then $Q = 1/\alpha$ will be high from the peaking.

In the unity gain state-variable circuit, Figure 8.9*a*, the f_c of the integrators sets the filter f_c, and R_5 and R_f' set α (or Q for the band-pass). Normally, $R_1 = R_2$ and $C_1 = C_2$. In the state-variable filter with gain, Figure 8.9*b*, R_A and R_B of the inverting amplifier (A_4) set α. The output of the inverting amplifier is now summed directly with V_{in} and the low-pass output. The passband gain is set with R_4 and R_f.

A notch filter can be constructed from the state-variable active filters by merely summing the high-pass and low-pass outputs that are out of phase. They will cancel

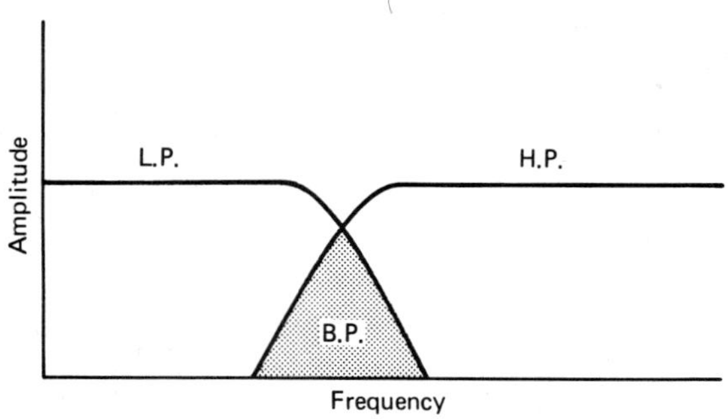

FIGURE 8.11 State-variable band-pass explanation.

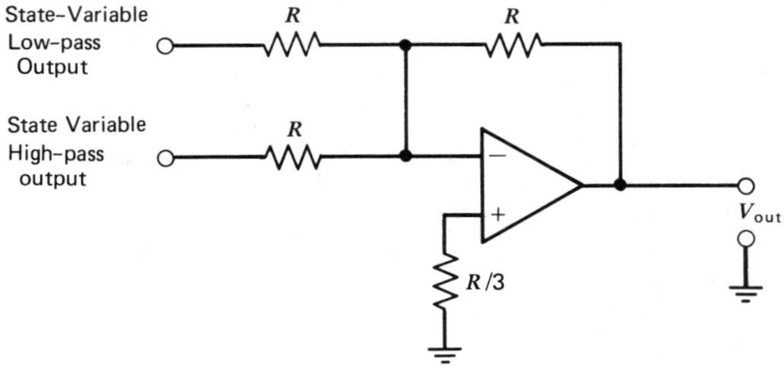

FIGURE 8.12 Summing amplifier for constructing a state-variable notch filter.

each other only where their responses overlap. If the filter is set up as a band-pass filter, the notch filter response will be the opposite of the band-pass and very controllable. A summer suitable for addition to the state variable to obtain a notch response is shown in Figure 8.12.

8-4.4 Biquadratic (Biquad)

The biquadratic (biquad) filter is a very stable, easily cascadable, active filter capable of Qs greater than 100 in the band-pass realization. One characteristic of a biquad is that its bandwidth stays constant as its frequency is varied, thus its Q increases with frequency in adjustable filters. The biquad bandpass filter, shown in Figure 8.13, consists of a summing integrator feeding an inverting amplifier, feeding a

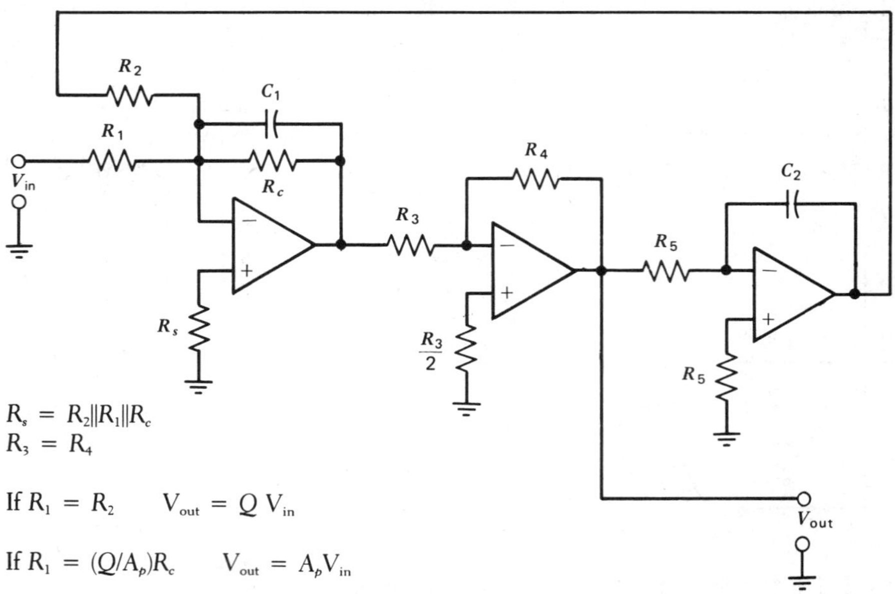

$R_s = R_2 \| R_1 \| R_c$
$R_3 = R_4$

If $R_1 = R_2 \qquad V_{out} = Q\,V_{in}$

If $R_1 = (Q/A_p)R_c \qquad V_{out} = A_p V_{in}$

FIGURE 8.13 Biquad band-pass filter.

second integrator. If $R_1 = R_2$, the passband gain of the circuit is R_c/R_1. The center frequency may be adjusted with R_2. R_c sets the circuit Q.

The circuit works as follows. The summing integrator subtracts a low-pass signal from V_{in}, 180° out of phase, until the transition region is reached. As the second-order transition frequency band near f_0 is reached, the integrator's decreasing output no longer cancels the input, so an output occurs. Above f_0 the roll off of the two series integrators attenuates the output signal, thus providing a band-pass response.

8-5 COMPONENT CALCULATION PROCEDURES AND EXAMPLES

In this section we will go through a component calculation procedure for each type of second-order active filter discussed in Section 8-4. An example calculation will follow each procedure.

8-5.1 Sallen and Key Equal Component Low-Pass Filter

Since this is the equal component filter, $R_1 = R_2$ and $C_1 = C_2$. Begin by choosing a filter type and f_c. The procedure is as follows:

1. Look up the $f_{(3\,dB)}/f_c$ ratio from Table 8-1 for the selected filter type. Calculate f_c if $f_{(3\,dB)}/f_c \neq 1$.

$$f_c = f_{(3\,dB)}/\text{ratio}$$

2. Select C and calculate R from

$$f_c = \frac{1}{2\pi RC} \tag{8-4}$$

where

$$R = R_1 = R_2$$
$$C = C_1 = C_2$$

Repeat this step if necessary until a reasonable value of R is found.

3. Select the damping factor from Table 8-1 for the type of filter chosen.
4. Choose an appropriate value of R_A. Often it is convenient to let $R_A = R$. Calculate R_B from

$$R_B = (2 - \alpha)R_A \tag{8-5}$$

5. Calculate passband gain from

$$A_p = (R_B/R_A) + 1$$

TABLE 8-1

SECOND-ORDER FILTER DAMPING FACTORS AND $f_{(3\ dB)}/f_c$

Filter Type	α	$f_{(3\ dB)}/f_c$
Butterworth	1.414	1.00
Bessel	1.732	0.785
Chebyshev		
0.5 dB ripple	1.578	1.390
1 dB ripple	1.059	1.218
2 dB ripple	0.886	1.074
3 dB ripple	0.766	1.000

Low pass $f_c = f_{(3\ dB)}/\text{ratio}$
High pass $f_c = f_{(3\ dB)}\ (\text{ratio})$

where

$f_{(3\ dB)}$ = desired 3 dB cutoff frequency
f_c = frequency used in calculations
ratio = $f_{(3\ dB)}/f_c$ from table

EXAMPLE 8-1

Calculate the components for a Sallen and Key equal component second-order low-pass filter. The filter is to be a Butterworth with $f_{(3\ dB)} = 2$ kHz.

SOLUTION

From Table 8-1

$$f_{(3\ dB)}/f_c = 1$$

so

$$f_{(3\ dB)} = f_c$$

Choose

$$C = 0.1\ \mu F = C_1 = C_2$$

Let

$$R_1 = R_2 = R$$

From

$$f_c = \frac{1}{2\pi RC}$$

$$R = \frac{1}{2\pi f_c C} = \frac{1}{2\pi(2\ \text{kHz})(0.1\ \mu F)} = 796\ \Omega$$

This value of R is a little lower than desired for a 741 op-amp, so we choose $C = 0.047\ \mu\text{F}$ and recalculate R.

$$R = \frac{1}{2\pi f_c C} = \frac{1}{2\pi(2\ \text{kHz})(0.047\ \mu\text{F})} = 1.69\ \text{k}\Omega$$

Choose a 1.69 kΩ ± 2% resistor.
R_A is arbitrarily chosen at 10 kΩ, so

$$R_B = R_A(2 - \alpha) = 10\ \text{k}\Omega(2 - 1.414) = 5.86\ \text{k}\Omega$$

Choose a 5.9 kΩ ± 2% resistor.
The passband gain is fixed by the choice of filter type.

$$A_p = (R_B/R_A) + 1 = (5.9\ \text{k}\Omega/10\ \text{k}\Omega) + 1 = 1.59$$

Construct the filter in Figure 8.14*a* using the components calculated. The result will be a second-order Butterworth low-pass filter.

Note. Once again, some adjustment of R_1 and R_2 and R_B may be necessary to set the filter f_c and α at the precise values desired because of component tolerances.

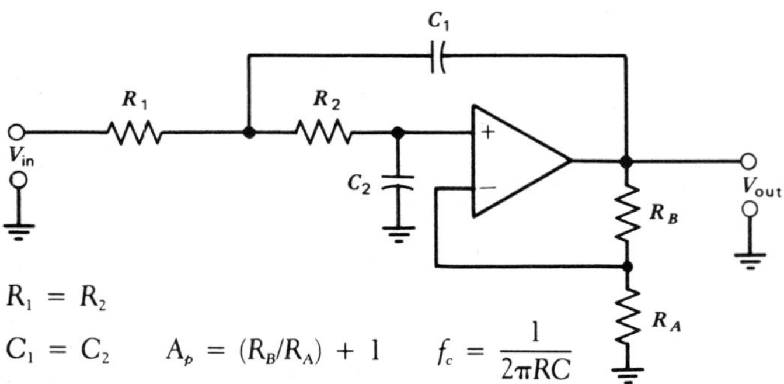

$R_1 = R_2$

$C_1 = C_2 \qquad A_p = (R_B/R_A) + 1 \qquad f_c = \dfrac{1}{2\pi RC}$

(*a*) Sallen and Key equal component second-order low-pass active filter.

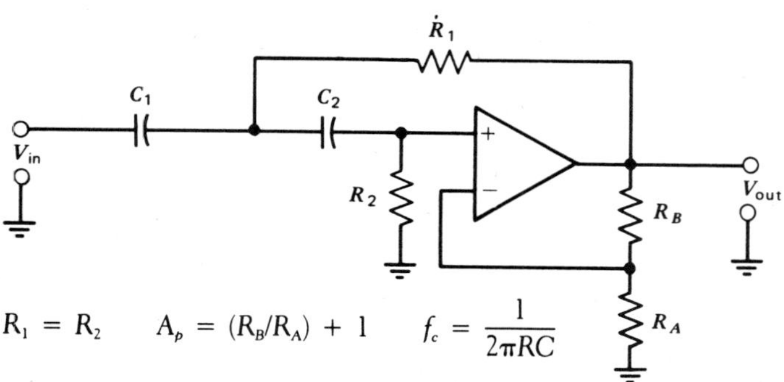

$R_1 = R_2 \qquad A_p = (R_B/R_A) + 1 \qquad f_c = \dfrac{1}{2\pi RC}$

$C_1 = C_2$

(*b*) Sallen and Key equal component second-order high-pass filter.

FIGURE 8.14 Sallen and Key active filters.

8-5.2 Sallen and Key Equal Component Second-Order High-Pass Filter

The procedure for the circuit shown in Figure 8.14*b* is as follows:

1. Select filter type and $f_{(3\text{ dB})}$. Look up $f_{(3\text{ dB})}/f_c$ for filter type from Table 8-1. Calculate f_c if the ratio is not one from

$$f_c = f_{(3\text{ dB})}\,(\text{ratio}) \tag{8-6}$$

2. Letting $C = C_1 = C_2$ and $R = R_1 = R_2$, select C and calculate R from

$$f_c = \frac{1}{2\pi RC}$$

3. Look up α from Table 8-1 for your filter type. Select R_A and calculate R_B from

$$R_B = (2 - \alpha)R_A$$

4. Calculate the passband gain from $A_p = (R_B/R_A) + 1$.

Note. Both the high-pass and low-pass Sallen and Key filters are adjusted as follows:

- ☐ Adjust f_c by varying C_1 and C_2 or R_1 and R_2 together.
- ☐ Adjust α by varying R_B.

EXAMPLE 8-2

Calculate the components for a second-order Sallen and Key high-pass filter. The filter is to be a 1 dB ripple Chebyshev with $f_{(3\text{ dB})} = 3$ kHz.

SOLUTION

From Table 8-1

$$f_{(3\text{ dB})}/f_c = 1.218$$
$$\alpha = 1.059$$
$$f_c = f_{(3\text{ dB})}(1.218) = 3\text{ kHz }(1.218) = 3.654\text{ kHz}$$

Let $C = C_1 = C_2$ and $R = R_1 = R_2$. If C is chosen to be 0.022 μF

$$R = \frac{1}{2\pi f_c C} = \frac{1}{2\pi(3.654\text{ kHz})(0.022\ \mu\text{F})}$$
$$R = 1.979\text{ k}\Omega$$

Use 1.96 kΩ ± 2%.
Let

$$R_A = 10\text{ k}\Omega$$
$$R_B = (2 - \alpha)R_A = (2 - 1.059)\,10\text{ k}\Omega = 9.41\text{ k}\Omega$$

Use 9.53 kΩ ± 2%.

$$A_p = (R_B/R_A) + 1 = (9.53\ \text{k}\Omega/10\ \text{k}\Omega) + 1$$
$$A_p = 1.953$$

8-5.3 Multiple-Feedback Band-Pass Filter

The multiple-feedback circuit shown in Figure 8.15 may or may not have R_2 in it. Calculation procedures are given for the circuit with and without R_2. R_2 allows the passband gain to be set at a desired value.

CALCULATION PROCEDURE WITHOUT R_2

1. Select f_1 and f_2 and an op-amp with $A_{ol} > 2\,Q^2$ at the desired f_1 and f_2.
2. From the desired f_1 and f_2, calculate f_0 and Q from

$$f_0 = \sqrt{f_1 f_2}$$
$$Q = \frac{f_0}{f_2 - f_1}$$

If Q is greater than 15, choose a state-variable or biquad bandpass filter. If Q is less than 15, proceed.

3. Select $C_1 = C_2 = C$ and calculate

$$R_1 = \frac{1}{4\pi f_0 QC} \tag{8-7}$$
$$R_3 = \frac{2\,Q}{2\pi f_0 C} \tag{8-8}$$

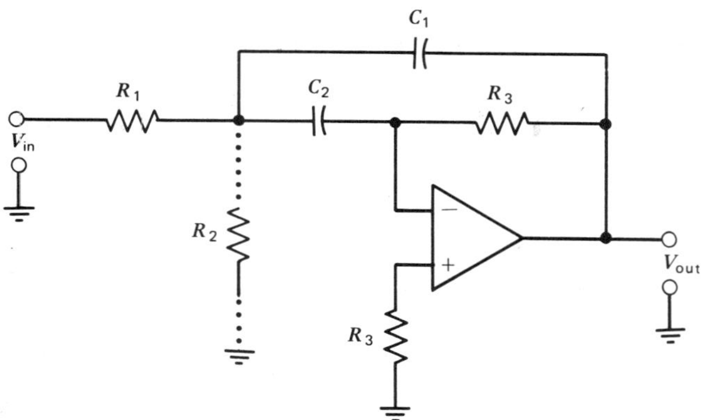

FIGURE 8.15 Multiple-feedback band-pass filter.

4. Calculate

$$A_p = 2\,Q^2 \tag{8-9}$$

CALCULATION PROCEDURE WITH R_2

1. Select f_1 and f_2. Once again the op-amp A_{ol} should be greater than $2\,Q^2$ at f_1 and f_2.
2. Calculate f_0 and Q from

$$f_0 = \sqrt{f_1 f_2}$$

$$Q = \frac{f_0}{f_2 - f_1}$$

3. If $Q < 15$, select the desired passband gain $A_p < 2\,Q^2$.
4. Choose $C_1 = C_2 = C$ and find

$$R_1 = \frac{Q}{2\pi f_0 C A_p} \tag{8-10}$$

$$R_2 = \frac{Q}{2\pi f_0 C(2\,Q^2 - A_p)} \tag{8-11}$$

$$R_3 = \frac{2\,Q}{2\pi f_0 C} \tag{8-12}$$

5. Check A_p from

$$A_p = R_3/2\,R_1 \tag{8-13}$$

For the multiple feedback circuit, adjustments are as follows:

- ☐ Vary f_0 by adjusting C_1 and C_2 or R_1 and R_2 simultaneously.
- ☐ Adjust Q by varying the R_3/R_1 ratio holding the R_3R_1 product constant.
- ☐ Adjust gain in the circuit with R_2.

EXAMPLE 8-3

Calculate the components of a multiple-feedback band-pass circuit without R_2. This will provide the maximum passband gain. Let $f_1 = 4.5$ kHz and $f_2 = 5.5$ kHz.

SOLUTION

$$f_0 = \sqrt{f_1 f_2} = \sqrt{(4.5\text{ kHz})(5.5\text{ kHz})} = 4.975\text{ kHz}$$

$$Q = \frac{f_0}{f_2 - f_1} = \frac{4.975\text{ kHz}}{1\text{ kHz}} = 4.975$$

Select $C = C_1 = C_2 = 0.001\ \mu$F.

From Equations 8-7 and 8-8

$$R_1 = \frac{1}{4\pi f_0 QC} = \frac{1}{4\pi(4.975\ \text{kHz})(4.975)(0.001\ \mu\text{F})}$$

$$= 3.215\ \text{k}\Omega$$

Use 3.32 kΩ ± 2%.

$$R_3 = \frac{2\,Q}{2\pi f_0 C} = \frac{2(4.975)}{2\pi(4.975\ \text{kHz})(0.001\ \mu\text{F})}$$

$$= 318\ \text{k}\Omega$$

Use 316 kΩ ± 2%.

From Equation 8-9,

$$A_p = 2\,Q^2 = 2(4.975)^2 = 49.5$$

Almost every op-amp will have $A_{ol} >> 50$ at 5 kHz, so a wide choice of amplifiers can be used.

EXAMPLE 8-4

Calculate the components for a multiple-feedback band-pass filter with $f_1 = 760$ Hz, $f_2 = 890$ Hz, and $A_p = 10$. With a specified passband gain, the circuit with R_2 must be used.

SOLUTION

$$f_0 = \sqrt{f_1 f_2} = \sqrt{760\ \text{Hz}\ 890\ \text{Hz}} = 822.4\ \text{Hz}$$

$$Q = \frac{f_0}{f_2 - f_1} = \frac{822.4\ \text{Hz}}{890\ \text{Hz} - 760\ \text{Hz}} = 6.326$$

Choose $C_1 = C_2 = C = 0.0047\ \mu\text{F}$.

From Equations 8-10, 8-11, and 8-12, calculate R_1, R_2 and R_3.

$$R_1 = \frac{Q}{2\pi f_0 C A_p} = \frac{6.326}{2\pi(822.4\ \text{Hz})(0.0047\ \mu\text{F})10}$$

$$= 26\ \text{k}\Omega$$

Use 26.1 kΩ ± 2%.

$$R_2 = \frac{Q}{2\pi f_0 C(2\,Q^2 - A_p)} = \frac{6.326}{2\pi(822.4\ \text{Hz})(0.0047\ \mu\text{F})(80 - 10)}$$

$$= 3.84\ \text{k}\Omega$$

Use 3.83 kΩ ± 2%.

$$R_3 = \frac{2\,Q}{2\pi f_0 C} = \frac{2(6.326)}{2\pi(822.4\text{ Hz})(0.0047\ \mu\text{F})}$$
$$= 521\text{ k}\Omega$$

Use 511 kΩ ± 2%.

Check A_p from Equation 8-13.

$$A_p = R_3/2\,R_1 = \frac{511\text{ k}\Omega}{2(26.1\text{ k}\Omega)} = 9.79$$

If $A_p = 9.79$ is close enough to the desired 10, then the circuit is finished. Otherwise, select a slightly higher value for R_3 and lower value for R_1.

8-5.4 State-Variable Filters

The calculation procedure for unity gain state-variable circuits of Figure 8.16 is as follows.

STATE-VARIABLE UNITY GAIN HIGH-PASS AND LOW-PASS CALCULATION PROCEDURE

1. Select $f_{(3\,\text{dB})}$ and filter type.
2. Find $f_{(3\,\text{dB})}/f_c$ from Table 8-1. Calculate f_c if $f_{(3\,\text{dB})}/f_c \neq 1$.
3. Let $R_1 = R_2 = R_3 = R_4 = R_f = R_f' = R$. Select $C = C_1 = C_2$ and calculate

$$R = \frac{1}{2\pi f_c C}$$

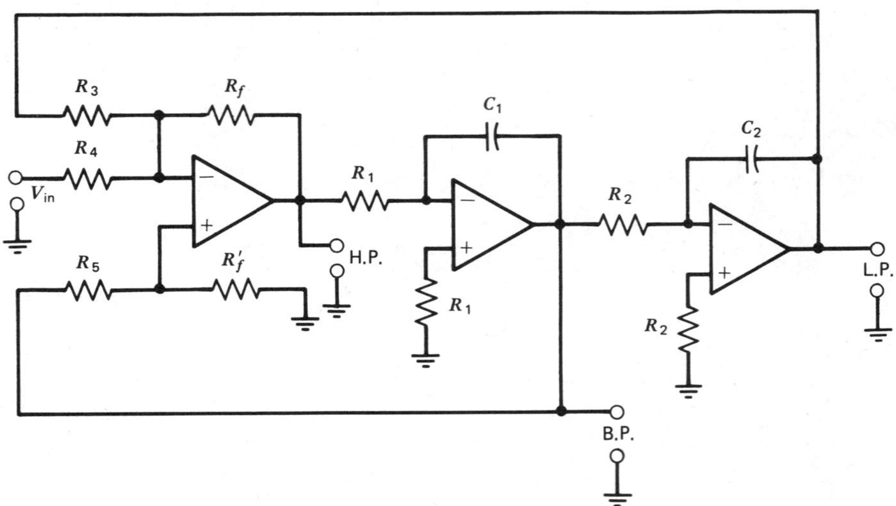

FIGURE 8.16 Unity gain second-order state-variable active filter.

4. Select α from Table 8-1 for filter type and calculate R_5 from

$$R_5 = R_f' [(3/\alpha) - 1] \tag{8-14}$$

To adjust filter parameters:

- □ Adjust f_c by varying R_1 and R_2 or C_1 and C_2 simultaneously.
- □ Adjust α by varying R_f'.

STATE VARIABLE UNITY GAIN BAND-PASS CALCULATION PROCEDURE

1. Select f_1 and f_2. $A_p = Q$ for the unity gain state-variable filter.
2. Calculate f_0 and Q. Q can be as high as 100.
3. Select $C = C_1 = C_2$ and calculate $R = R_1 = R_2 = R_3 = R_4 = R_f = R_f'$ from

$$R = \frac{1}{2\pi f_0 C}$$

4. Calculate R_5 from

$$R_5 = R_f' (3\,Q - 1) \tag{8-15}$$

To vary filter parameters:

- □ Adjust f_0 by varying R_1 and R_2 or C_1 and C_2 simultaneously.
- □ Adjust Q with R_f'.

EXAMPLE 8-5

Calculate the components for a second-order unity gain state-variable low-pass Chebyshev filter. Let $f_{(3\,\text{dB})} = 12$ kHz and the ripple amplitude = 2 dB. The pass-band gain of the circuit is fixed at one.

SOLUTON

From Table 8-1

$$f_{(3\,\text{dB})}/f_c = 1.074$$

$$\alpha = 0.886$$

$$f_c = f_{(3\,\text{dB})}/1.074 = 12\text{ kHz}/1.074 = 11.174\text{ kHz}$$

Let $C_1 = C_2 = C = 0.001\ \mu\text{F}$ and $R = R_1 = R_2 = R_3 = R_4 = R_f = R_f'$.

$$R = \frac{1}{2\pi f_c C} = \frac{1}{2\pi(11.174\text{ kHz})(0.001\ \mu\text{F})}$$

$$= 14.2\text{ k}\Omega$$

Use 14 kΩ ± 2%.

$$R_5 = R_f' (3/\alpha - 1) = 14\text{ k}\Omega\,(3/0.886 - 1)$$

$$= 33.4\text{ k}\Omega$$

Use 33.2 kΩ ± 2%.

If appreciable offset occurs in the first stage, use a balancing resistor (R_x) as shown in the adder-subtracter circuit of Chapter 5, Section 3.

Note. $Q = 1/\alpha = 1.13$. The band-pass output will have a Q of 1.13, which is very low. This is why the state-variable filter should normally be used as either a high pass and/or low pass, or band pass, but not both simultaneously.

EXAMPLE 8-6

Calculate the components for a unity gain state-variable band-pass filter with f_1 = 940 Hz and f_2 = 1 kHz.

SOLUTION

$$f_0 = \sqrt{f_1 f_2} = \sqrt{940\text{ Hz } 1\text{ kHz}} = 969.5\text{ Hz}$$

$$Q = \frac{f_0}{f_2 - f_1} = \frac{969.5\text{ Hz}}{1\text{ kHz} - 940\text{ Hz}} = 16.15$$

Let $C_1 = C_2 = C = 0.033\ \mu\text{F}$ and let $R = R_1 = R_2 = R_3 = R_4 = R_f = R_f'$.

$$R = \frac{1}{2\pi f_0 C} = \frac{1}{2\pi(969.5\text{ Hz})(0.033\ \mu\text{F})}$$

$$= 4.97\text{ k}\Omega$$

Use 5.11 kΩ ± 2%.

$$R_5 = (3\,Q - 1)\,R_f' = [3(16.15) - 1]\ 5.11\text{ k}\Omega$$

$$= 243\text{ k}\Omega$$

Use 237 kΩ ± 2%.

Balance the adder-subtracter if necessary.

$$A_p = Q = 16.15$$

CALCULATION PROCEDURE FOR STATE-VARIABLE, VARIABLE GAIN, ACTIVE FILTERS

As in the unity gain state-variable active filter, the calculation procedure for the high-pass and low-pass circuits of the variable gain state-variable active filter (Figure 8.17) are identical. The band-pass circuit calculation procedure is different.

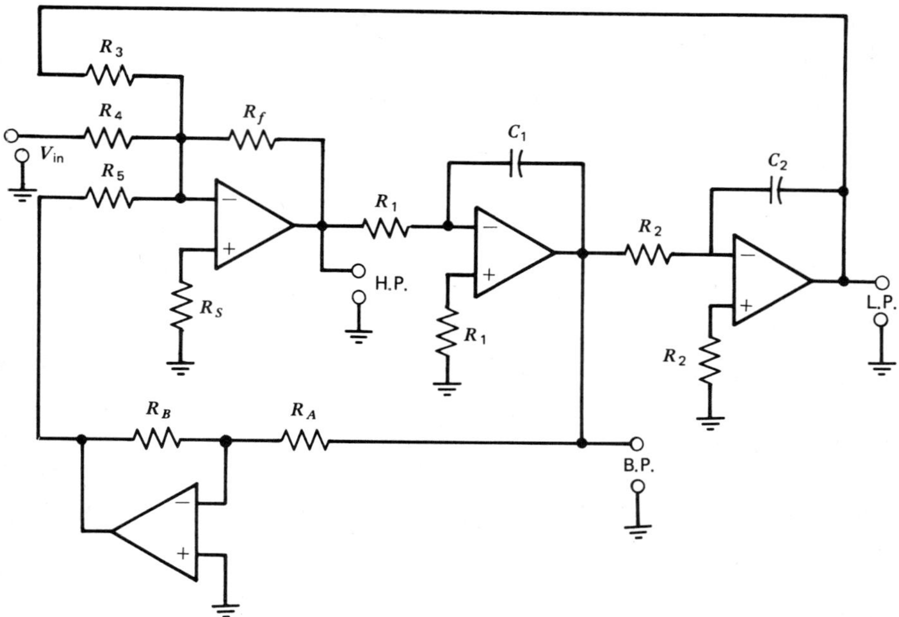

FIGURE 8.17 State-variable, variable gain second-order active filter circuit.

STATE VARIABLE, VARIABLE GAIN, HIGH-PASS AND LOW-PASS CALCULATION PROCEDURES

1. Look up $f_{(3\,dB)}/f_c$ and α for filter type from Table 8-1. Calculate f_c if $f_{(3\,dB)}/f_c \neq 1$.
2. Select $C = C_1 = C_2$ and $R = R_1 = R_2 = R_3 = R_5 = R_f = R_A$; calculate R from

$$f_c = \frac{1}{2\pi RC}$$

3. Calculate R_4 for A_p. Since

$$A_p = \frac{R_f}{R_4}$$

$$R_4 = R_f/A_p$$

4. Calculate R_B where

$$R_B = \alpha R_A \tag{8-16}$$

Adjustments:

- □ Adjust α with R_A or R_B.
- □ Adjust A_p with R_4.
- □ Adjust f_c by varying C_1 and C_2 or R_1 and R_2 simultaneously.

STATE VARIABLE, VARIABLE GAIN, BAND-PASS CALCULATION PROCEDURE

1. Select f_1 and f_2, and A_p. Calculate f_0 and Q where

$$f_0 = \sqrt{f_1 f_2}$$

$$Q = \frac{f_0}{f_2 - f_1} \leqslant 150$$

2. Calculate G, the proportionality constant between A_p and Q. Since

$$A_p = GQ \tag{8-17}$$

$$G = A_p/Q$$

3. Select $C = C_1 = C_2$. Let $R = R_1 = R_2 = R_3 = R_5 = R_f = R_A$ and calculate R from

$$R = \frac{1}{2\pi f_0 C}$$

4. Calculate R_4 from

$$R_4 = R_f/G \tag{8-18}$$

5. Calculate R_B

$$R_B = R_A/Q \tag{8-19}$$

Note. If $R_B < V_{out}/I_{out}$ (max) of the op-amp, select a larger R_A and repeat the calculation for R_B.

EXAMPLE 8-7

Calculate the components for a second-order variable gain state-variable high-pass Butterworth filter with $f_c = 1.5$ kHz and $A_p = 5$.

SOLUTION

From Table 8-1

$$f_{(3\,\text{dB})}/f_C = 1$$

$$\alpha = 1.414$$

Choose $C = C_1 = C_2 = 0.0068\ \mu\text{F}$. Let $R = R_1 = R_2 = R_3 = R_5 = R_f = R_A$.

$$R = \frac{1}{2\pi f_c C} = \frac{1}{2\pi(1.5\text{ kHz})(0.0068\ \mu\text{F})}$$

$$= 15.6\text{ k}\Omega$$

use 15.4 kΩ ± 2%.

$$R_B = \alpha R_A = 1.414\ (15.4\ \text{k}\Omega)$$
$$= 21.8\ \text{k}\Omega$$

Use 21.5 kΩ ± 2%.

$$R_4 = R_f/A_p = 15.4\ \text{k}\Omega/5 = 3.08\ \text{k}\Omega$$

Use $R_4 = 3.01$ kΩ ± 2%.

$$R_s = R_3 \parallel R_f \parallel R_4 \parallel R_5 = 1.934\ \text{k}\Omega$$

Use $R_s = 1.96$ kΩ ± 2%.

EXAMPLE 8-8

Calculate the components of a variable gain state-variable band-pass filter with $f_1 = 2.2$ kHz, $f_2 = 2.3$ kHz, and $A_p = 10$.

SOLUTION

$$f_0 = \sqrt{f_1 f_2} = \sqrt{2.2\ \text{kHz}\ 2.3\ \text{kHz}}$$
$$= 2.249\ \text{kHz}$$
$$Q = f_0/(f_2 - f_1)$$
$$= 2.249\ \text{kHz}/(2.3\ \text{kHz} - 2.4\ \text{kHz}) = 22.5$$
$$G = A_p/Q = 10/22.5 = 0.444$$

Let $C = C_1 = C_2 = 0.0033\ \mu\text{F}$ and $R = R_1 = R_2 = R_3 = R_5 = R_f = R_A$

$$R = \frac{1}{2\pi f_0 C} = \frac{1}{2\pi(2.249\ \text{kHz})(0.0033\ \mu\text{F})}$$
$$= 21.44\ \text{k}\Omega$$

Use 21.5 kΩ ± 2%.

$$R_B = R_A/Q = 21.5\ \text{k}\Omega/22.5 = 956\ \Omega$$

Use 953 Ω ± 2%.

Note. A 741 op-amp can drive 953 Ω, although this is as low a value as we would care to use for R_B.

$$R_4 = R_f/G = 21.5\ \text{k}\Omega/0.444$$
$$= 48.4\ \text{k}\Omega$$

Use 48.7 kΩ ± 2%.

$$R_s = R_5 \parallel R_3 \parallel R_4 \parallel R_f = 2.9\ \text{k}\Omega$$

Use 2.87 kΩ ± 2%.

8-5.5 Biquad Band-Pass Filter

The biquad band-pass calculation procedure for the circuit of Figure 8.18 is as follows:

1. Choose f_1, f_2, and A_p. Calculate f_0 and Q where

$$f_0 = \sqrt{f_1 f_2}$$
$$Q = f_0/(f_2 - f_1)$$

2. Calculate $G = Q/A_p$. (8-20)
3. Calculate R_1 and R_c as follows

$$R_1 = \frac{G}{2\pi f_0 C} \quad (8\text{-}21)$$

$$R_c = \frac{Q}{2\pi f_0 C} \quad (8\text{-}22)$$

4. Let $R = R_2 = R_3 = R_4 = R_5$. Calculate R from

$$R = \frac{1}{2\pi f_0 C}$$

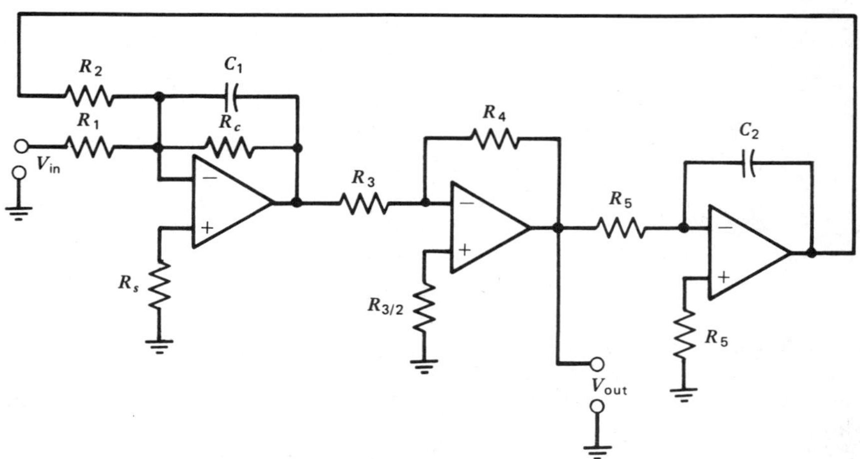

FIGURE 8.18 Biquad band-pass filter.

Adjustments:

- ☐ Adjust f_0 with R_2.
- ☐ Adjust Q with R_c.
- ☐ Adjust A_p with R_1.

EXAMPLE 8-9

Calculate the components of a biquad band-pass filter such that $f_1 = 97$ Hz, $f_2 = 102$ Hz, and $A_p = 10$.

SOLUTION

$$f_0 = \sqrt{f_1 f_2} = \sqrt{102\text{ Hz } 97\text{ Hz}} = 99.47\text{ Hz}$$
$$Q = f_0/(f_2 - f_1) = (99.47\text{ Hz})/5\text{ Hz} = 19.9$$
$$G = Q/A_p = 19.9/10 = 1.99$$

Let $C = C_1 = C_2 = 0.047\ \mu\text{F}$. Then $R = R_2 = R_3 = R_4 = R_5$.

$$R = \frac{1}{2\pi f_0 C} = \frac{1}{2\pi(99.47\text{ Hz})(0.047\ \mu\text{F})} = 34\text{ k}\Omega$$

Use 34.8 kΩ ± 2%.

$$R_1 = \frac{G}{2\pi f_0 C} = \frac{1.99}{2\pi(99.47\text{ Hz})(0.047\ \mu\text{F})} = 67.7\text{ k}\Omega$$

Use 68.1 kΩ ± 2%.

$$R_c = \frac{Q}{2\pi f_0 C} = \frac{19.9}{2\pi(99.47\text{ Hz})(0.047\ \mu\text{F})} = 677\text{ k}\Omega$$

Use 681 kΩ ± 2%. For offset balance, let op-amp noninverting input resistor

$$R_3/2 = 34.8\text{ k}\Omega/2 = 17.4\text{ k}\Omega$$

Use 17.4 kΩ ± 2%.

$$R_s = R_c \parallel R_1 \parallel R_2 = 21.9\text{ k}\Omega$$

Use 21.5 kΩ ± 2%.

8-6 CASCADING FILTERS

When filters with more than two poles are required, they can be obtained simply by cascading two or more lower order filters. This section explains how to cascade active filters to obtain higher order filters.

8-6.1 Cascading for Higher Order Responses

Second- and first-order active filters can be cascaded to construct all higher order filters. As shown in Figure 8.19, a second-order filter and a first-order filter can be cascaded to form a third-order filter and two second-order filters can be cascaded to form a fourth-order filter. This process can be continued for any order filter desired. Odd-order filters are usually constructed using a first-order first stage and second-order stages for the rest. For example, a seventh-order filter is constructed from a first-order stage and three second-order stages. Even-order filters are constructed from $n/2$ second-order stages, where n is the order desired. The stages in a cascaded active filter are not identical since cascading causes bandwidth shrinkage.

To understand bandwidth shrinkage, refer to Figure 8.20. If two first-order filters with gains A_1 and A_2 and equal bandwidths are cascaded, the total bandwidth will be less than the bandwidth of each stage. Since $A_T = A_1A_2$ at f_1, $A_1 = 0.707\ A_1$ midband and $A_2 = 0.707\ A_2$ midband, and at f_2, $A_1 = 0.707\ A_1$ midband and $A_2 = 0.707\ A_2$ midband. Therefore, at f_1, $A_T = 0.5\ A_1A_2 = 0.5\ A_T$ midband and

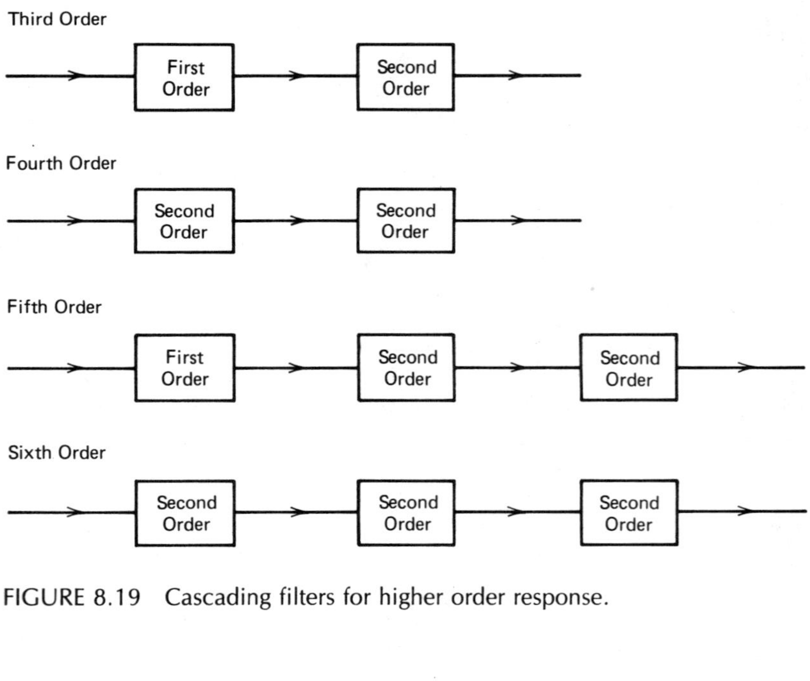

FIGURE 8.19 Cascading filters for higher order response.

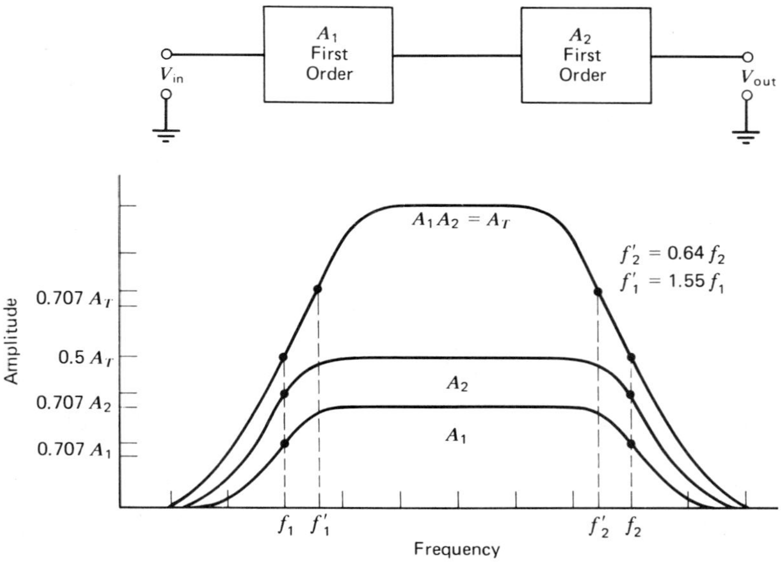

FIGURE 8.20 Bandwidth shrinkage.

at f_2, $A_T = 0.5\ A_T$ midband. The new $f_1' = 0.707\ A_T$ is at a higher frequency than f_1, and the new $f_2' = 0.707\ A_T$ is at a lower frequency than f_2 of the individual amplifiers. For n first-order stages, the cascaded upper cutoff frequency f_2' and the cascaded lower cutoff frequency f_1' are related to f_2 and f_1 by

$$f_2' = f_2\sqrt{2^{1/n} - 1} \tag{8-23}$$

and

$$f_1' = f_1/\sqrt{2^{1/n} - 1} \tag{8-24}$$

These equations are not useful for most cascaded active filters since second-order stages are usually used and Equations 8-23 and 8-24 work only for first-order stages. The same bandwidth shrinkage effect occurs regardless of the filter order.

A filter with ripple in the passband will, on cascading, have even more ripple in the passband. Since $A_T = A_1A_2 = A_1(\text{dB}) + A_2(\text{dB})$, a 3 dB ripple in a second-order filter will become a 6 dB ripple if two 3 dB ripple filters are cascaded.

For optimal results in cascading filters, it has been found that varying the α of each stage, and f_c in non-Butterworth filters, is necessary. So for a sixth-order filter, the three cascaded second-order stages will look alike but have some component values that are different in each stage.

Table 8.2 gives the α and f_c factor used for cascading filters from third-order to eighth-order. A couple of examples will follow to clarify the use of the table.

TABLE 8-2

CASCADED FILTER CHARTS

Butterworth

	First Stage		Second Stage		Third Stage		Fourth Stage	
Order	α	f_c factor	α	f_c factor	α	f_c factor	α	f_c factor
2	1.414	1						
3	1.00	1	1.00	1				
4	1.848	1	0.765	1				
5	1.00	1	1.618	1	0.618	1		
6	1.932	1	1.414	1	0.518	1		
7	1.00	1	1.802	1	1.247	1	0.445	1
8	1.962	1	1.663	1	1.111	1	0.390	1

Chebyshev, 0.5 dB Ripple

	First Stage		Second Stage		Third Stage		Fourth Stage	
Order	α	f_c factor	α	f_c factor	α	f_c factor	α	f_c factor
3	1	0.626	0.586	1.069				
4	1.418	0.597	0.340	1.031				
5	1	0.362	0.849	0.690	0.220	1.018		
6	1.463	0.396	0.552	0.768	0.154	1.011		
7	1	0.256	0.916	0.504	0.388	0.823	0.113	1.008
8	1.478	0.296	0.621	0.599	0.288	0.861	0.087	1.006

Chebyshev, 1 dB Ripple

	First Stage		Second Stage		Third Stage		Fourth Stage	
Order	α	f_c factor	α	f_c factor	α	f_c factor	α	f_c factor
3	1	0.494	0.496	0.997				
4	1.275	0.529	0.281	0.993				
5	1	0.289	0.715	0.655	0.180	0.994		
6	1.314	0.353	0.455	0.747	0.125	0.995		
7	1	0.205	0.771	0.480	0.317	0.803	0.092	0.996
8	1.328	0.265	0.511	0.584	0.234	0.851	0.702	0.997

TABLE 8-2

CASCADED FILTER CHARTS CONT.

Chebyshev, 2 dB Ripple

	First Stage		Second Stage		Third Stage		Fourth Stage	
Order	α	f_c factor	α	f_c factor	α	f_c factor	α	f_c factor
3	1	0.369	0.392	0.941				
4	1.076	0.471	0.218	0.964				
5	1	0.218	0.563	0.627	0.138	0.976		
6	1.109	0.316	0.352	0.730	0.096	0.983		
7	1	0.155	0.607	0.461	0.243	0.797	0.070	0.987
8	1.206	0.238	0.395	0.572	0.179	0.842	0.054	0.990

Chebyshev, 3 dB Ripple

	First Stage		Second Stage		Third Stage		Fourth Stage	
Order	α	f_c factor	α	f_c factor	α	f_c factor	α	f_c factor
3	1	0.299	0.326	0.916				
4	0.929	0.443	0.179	0.950				
5	1	0.178	0.468	0.614	0.113	0.967		
6	0.958	0.298	0.289	0.722	0.078	0.977		
7	1	0.126	0.504	0.452	0.199	0.792	0.057	0.983
8	0.967	0.224	0.325	0.566	0.147	0.839	0.044	0.987

Bessel

	First Stage		Second Stage		Third Stage		Fourth Stage	
Order	α	f_c factor	α	f_c factor	α	f_c factor	α	f_c factor
3	1	2.322	1.447	2.483				
4	1.916	2.067	1.241	1.624				
5	1	3.647	1.775	2.874	1.091	2.711		
6	1.959	2.872	1.636	3.867	0.977	3.722		
7	1	4.972	1.878	3.562	1.513	5.004	0.888	4.709
8	1.976	3.701	1.787	4.389	1.407	0.637	0.816	5.680

Note. On all odd-order filters the first stage is a first-order filter stage. All other stages are second-order filters.

For low-pass filters
$f_c = f_{(3\,dB)}\,(f_c \text{ factor})$

For high-pass filters
$f_c = f_{(3\,dB)}/(f_c \text{ factor})$

where

$f_c = f_c$ used in calculations for stage
$f_{(3\,dB)} = f_{(3\,dB)}$ desired

Cascaded band-pass filters normally use only even-order filters. If band-pass filters have a passband of more than about 50% of f_0, they can be constructed with more control by using cascaded high-pass and low-pass filters. A fourth-order band-pass filter constructed in this fashion would require two second-order high-pass filters and two second-order low-pass filters. This is illustrated in Figure 8.21. Note the low-pass filter sets f_2 and the high-pass filter sets f_1. The passband consists of the overlap area.

The first-order stages used in cascading are shown in Figure 8.22. Each filter shown in Figure 8.22 has an $\alpha = 1$. Note that the first-order bandpass filter is a compensated differentiator.

The f_c ratio of Table 8.2 is the reciprocal of the $f_{(3\,dB)}/f_c$ of Table 8.1

EXAMPLE 8-10

Construct a very stable fifth-order low-pass Butterworth filter with $A_p = 10$ and $f_{(3\,dB)} = 750$ Hz.

State-variable second-order stages will be used for good stability. The block diagram and circuit we will use is shown in Figure 8.23.

SOLUTION

From Table 8-2 we find the α and f_c ratio for each stage.

- □ *Stage 1* $\alpha = 1$, f_c ratio $= 1$
- □ *Stage 2* $\alpha = 1.618$, f_c ratio $= 1$
- □ *Stage 3* $\alpha = 0.618$, f_c ratio $= 1$

Thus, $f_{(3\,dB)} = f_c$ for all three stages.

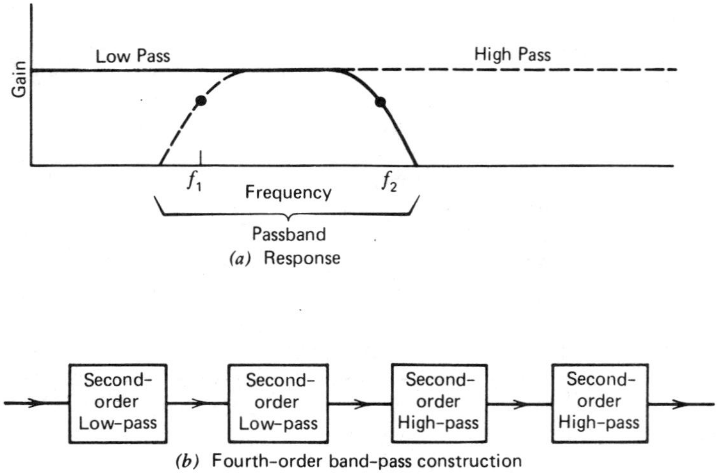

FIGURE 8.21 Wide bandwidth band-pass filters using high-pass and low-pass filters.

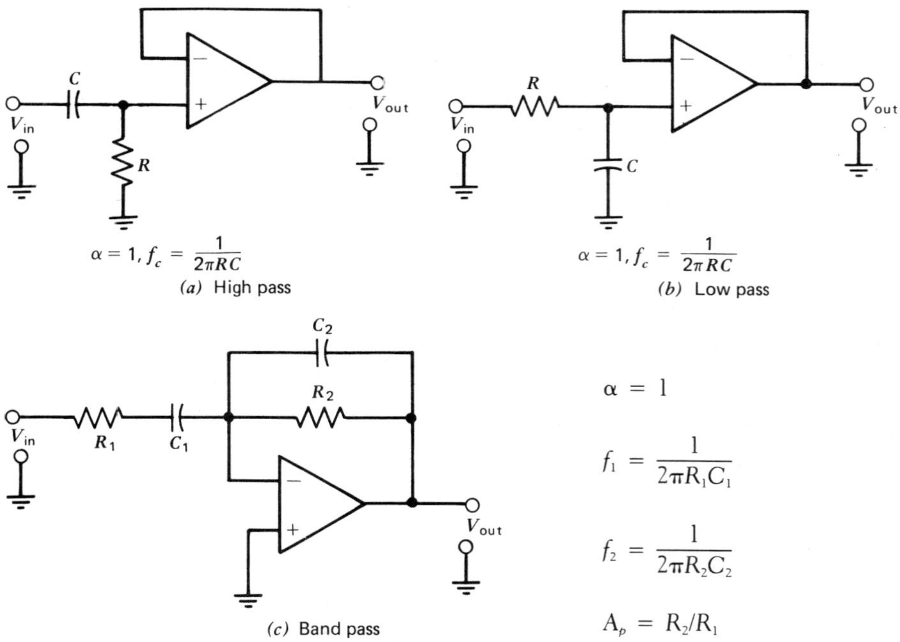

FIGURE 8.22 First-order stages.

The first-order stage has $A_p = 1$. We divide the gain evenly between the second-order stages. $A_2 = A_3 = \sqrt{A_T}$ thus

$$A_{p_2} = \sqrt{10} = 3.16$$

$$A_{p_3} = \sqrt{10} = 3.16$$

Since A_p is specified, the variable-gain state-variable will be used.

STAGE 1
Choose $C = 0.01\ \mu F$.

$$R = \frac{1}{2\pi f_c C} = \frac{1}{2\pi(750\text{ Hz})(0.01\ \mu\text{F})} = 21.2\text{ k}\Omega$$

Use 21.5 kΩ ± 2%. Stage 1 is complete.

STAGE 2
Let $C_1 = C_2 = C = 0.01\ \mu F$. Since $R = R_1 = R_2 = R_3 = R_5 = R_f = 1/2\pi f_c C$, and $C = 0.01\ \mu F$, $R = 21.5\text{ k}\Omega$ as in the first stage.

$$R_4 = R_f/A_p = 21.5\text{ k}\Omega/3.16 = 6.715\text{ k}\Omega$$

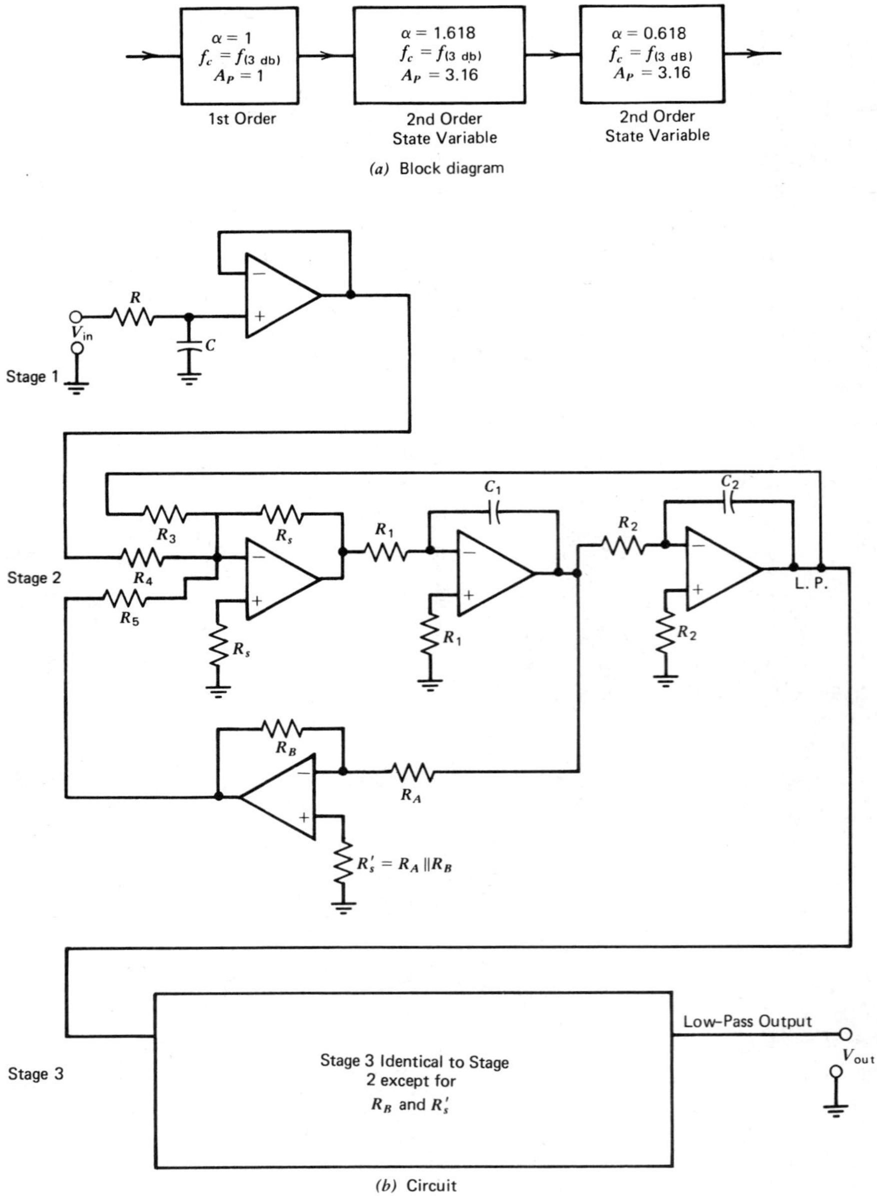

FIGURE 8.23 Fifth-order low-pass Butterworth, Example 8-10.

Use $R_4 = 6.81\ \text{k}\Omega \pm 2\%$. Let $R_A = R$.

$$R_B = \alpha R_A = 1.618\ (21.5\ \text{k}\Omega) = 34.8\ \text{k}\Omega$$

Use $R_B = 34.8\ \text{k}\Omega \pm 2\%$.

$$R_s = R_3 \parallel R_f \parallel R_4 \parallel R_5 = 3.49\ \text{k}\Omega$$

Use $3.48\ \text{k}\Omega \pm 2\%$. Since offset is amplified in cascaded stages, we will use an R_s' to control offset in the inverting amplifier.

$$R_s' = R_A \parallel R_B = 13.28\ \text{k}\Omega$$

Use $13.3\ \text{k}\Omega \pm 2\%$. Stage 2 is complete.

STAGE 3
Stage 3 will be identical to Stage 2 except for R_B and R_s' since only α changes.

$$\begin{aligned} R_B &= \alpha R_A = 21.5\ \text{k}\Omega\ (0.618) \\ &= 13.3\ \text{k}\Omega \end{aligned}$$

Use $13.3\ \text{k}\Omega \pm 2\%$.

$$R_s' = R_A \parallel R_B = 8.21\ \text{k}\Omega$$

Use $8.25\ \text{k}\Omega \pm 2\%$. Stage 3 is complete. Adjustments for α are made with R_B of Stage 3.

EXAMPLE 8-11

Construct a sixth-order Chebyshev high-pass filter with a 3 dB ripple. Let $f_{(3\ \text{dB})} = 1$ kHz. To illustrate the use of another circuit, we will use three Sallen and Key second-order high-pass stages as shown in Figure 8.24.

Each stage will look alike, but the components will vary since $f_c \neq f_{(3\ \text{dB})}$ and α varies for each stage.

SOLUTION

From Table 8-2:

STAGE 1
$\alpha = 0.958$, f_c factor $= 0.298$

STAGE 2
$\alpha = 0.289$, f_c factor $= 0.722$

STAGE 3
$\alpha = 0.078$, f_c factor $= 0.977$.

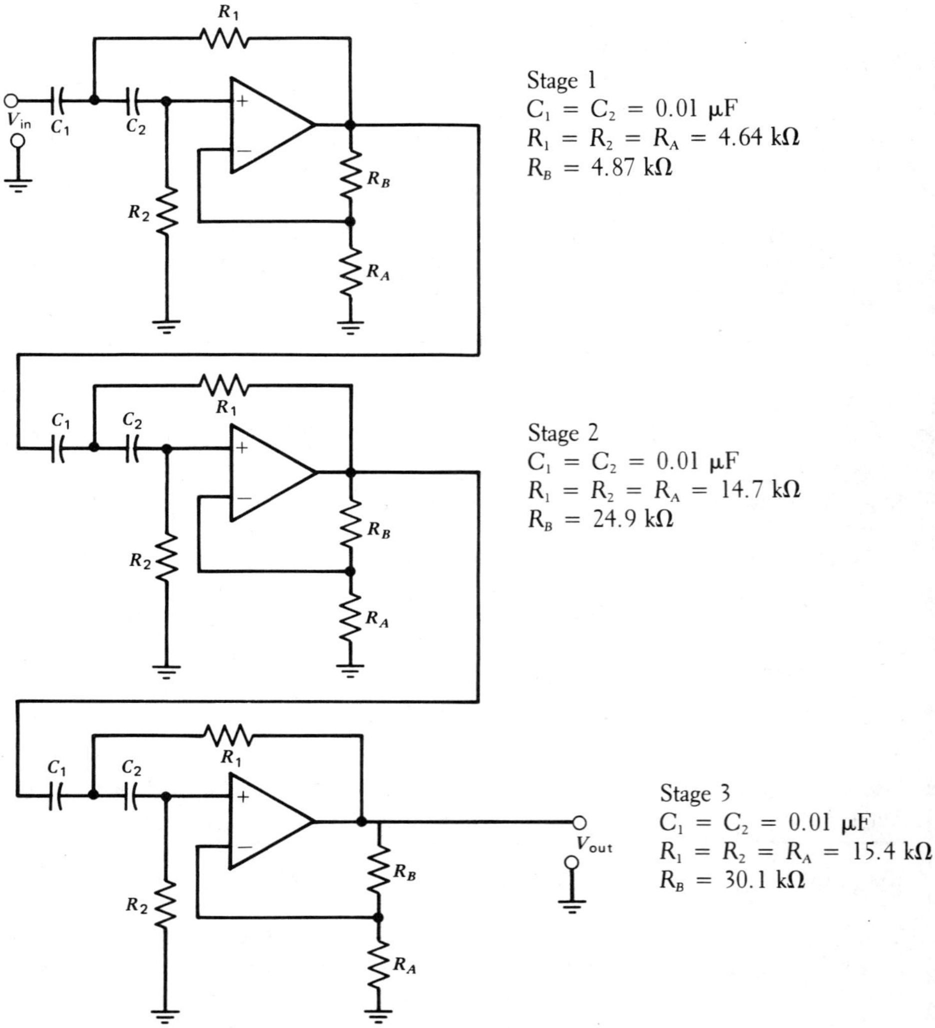

FIGURE 8.24 Sallen and Key sixth-order-cascaded filter, Example 8.11.

STAGE 1

$$f_c = f_{(3\ \text{dB})}/f_c \text{ factor} = 1\ \text{kHz}/0.298 = 3.356\ \text{kHz}$$

Let $R = R_1 = R_2 = R_A$. Choose $C_1' = C_2 = 0.01\mu\text{F}$.

$$R = \frac{1}{2\pi f_c C} = \frac{1}{2\pi(3.356\ \text{kHz})(0.01\mu\text{F})}$$

$$= 4.74\ \text{k}\Omega$$

Use 4.64 kΩ ± 2%.

$$R_B = (2 - \alpha)R_A = (2 - 0.958)(4.64\ \text{k}\Omega)$$
$$= 4.83\ \text{k}\Omega$$

Use 4.87 kΩ ± 2%.

$$A_{p1} = (1 + R_B/R_A) = (1 + 4.83\ \text{k}\Omega/4.64\ \text{k}\Omega) = 2.049$$

STAGE 2

$$f_c = f_{(3\ \text{dB})}/f_c\ \text{factor} = 1\ \text{kHz}/0.722$$
$$= 1.385\ \text{kHz}$$

Let $C = C_1 = C_2$, $R = R_1 = R_2 = R_A$.

$$R = \frac{1}{2\pi f_c C} = \frac{1}{2\pi(1.385\ \text{kHz})(0.01\ \mu\text{F})}$$
$$= 14.49\ \text{k}\Omega$$

Use 14.7 kΩ ± 2%.

$$R_B = (2 - \alpha)R_A = (2 - 0.289)\ 14.7\ \text{k}\Omega$$
$$= 25.15\ \text{k}\Omega$$

Use 24.9 kΩ ± 2 %.

$$A_{p2} = 1 + R_B/R_A = 1 + 24.9\ \text{k}\Omega/14.7\ \text{k}\Omega = 2.69$$

STAGE 3

$$f_c = f_{(3\ \text{dB})}/f_c\ \text{factor} = 1\ \text{kHz}/0.977$$
$$= 1.024\ \text{kHz}$$

Let $C = C_1 = C_2 = 0.01\ \mu\text{F}$, $R = R_1 = R_2 = R_A$.

$$R = \frac{1}{2\pi f_c C} = \frac{1}{2\pi(1.024\ \text{kHz})(0.01\ \mu\text{F})}$$
$$= 15.54\ \text{k}\Omega$$

Use 15.4 kΩ ± 2 %.

$$R_B = (2 - \alpha)R_A = (2 - 0.078)\ 15.4\ \text{k}\Omega$$
$$= 29.6\ \text{k}\Omega$$

Use 30.1 kΩ ± 2 %.

$$A_{p_3} = 1 + R_B/R_A = 1 + 30.1\ \text{k}\Omega/15.4\ \text{k}\Omega = 2.954$$

$$A_T = A_{p_1}A_{p_2}A_{p_3} = (2.049)(2.69)(2.954)$$

$$= 16.28$$

This gain will be at the peak of the 3 dB ripples.

Adjustments (probably needed):

- □ Adjust ripple with R_B of Stage 3.
- □ Adjust $f_{(3\ \text{dB})}$ with R_1 and R_2 of Stage 1.

8-7 GYRATORS

Gyrators are *RC* networks with feedback from an amplifier connected to simulate an inductor. Gyrators are sometimes called synthetic inductors.

A gyrator circuit enables one to construct a large value of inductance in a small, light, inexpensive package. Such an active inductor can simplify some filter realizations by allowing the use of classic filters using inductors. The inductors in these filters are replaced by gyrators. The gyrator's function is to make the capacitor look like an inductor at its input.

The major disadvantages of gyrators are that few have frequency responses above a few kilohertz, and circuits that do not need one terminal grounded are quite complex.

Refer to Figure 8.25 for a discussion of how the gyrator works. Recall that X_L increases as frequency increases. Thus the voltage across the inductor rises as frequency rises. The voltage follower in the gyrator circuit reproduces the voltage across *R*. As frequency increases, X_c decreases, so V_{R_1} increases. Thus, V_{out} of the follower increases. The follower output voltage is fed back to the input through R_2, thus the terminal voltage increases as frequency increases, as it would for an in-

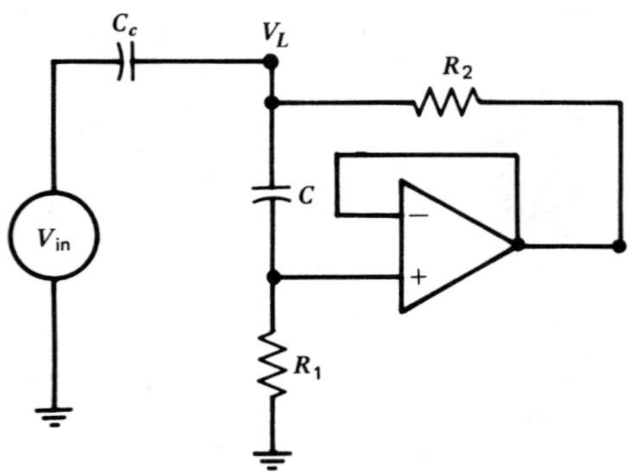

$$Q = \tfrac{1}{2}\sqrt{\frac{R_1}{R_2}}$$

$$L = R_1R_2C$$
$$R_1 >> R_2$$

FIGURE 8.25 Gyrator (Berndt and Dutta-Roy circuit).

ductor. R_2 should be as small as possible, the smaller, the better. Performance is improved if the op-amp is followed by a current amplifier in the feedback loop that drives R_2.

8-7.1 Calculation of Gyrator Components

Select R_2 = minimum resistance the op-amp can drive. Let $R_1 >> R_2$ (up to 200 R_2), and $R_1 < 0.1\ R_{in}$ op-amp. Select L from

$$L = R_1 R_2 C \tag{8-25}$$

Calculate C

$$C = L/R_1 R_2$$

$$\text{Gyrator } Q = \frac{1}{2}\sqrt{\frac{R_1}{R_2}} \tag{8-26}$$

EXAMPLE 8-12

Construct a series tuned LCR circuit with $f_0 = 300$ Hz using a gyrator. See Figure 8.26.

SOLUTION

Let $C_r = 0.1\ \mu\text{F}$

From $f_0 = \dfrac{1}{2\pi\sqrt{LC}}$

$$\begin{aligned} L &= \frac{1}{4\pi^2 f_0^2 C_r} \\ &= \frac{1}{4\pi^2 (300\ \text{Hz})^2\ (0.1\ \mu\text{F})} \\ &= 2.8\ \text{H} \end{aligned}$$

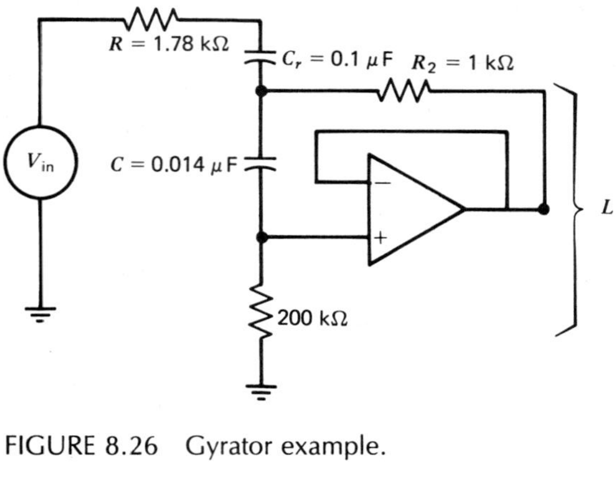

FIGURE 8.26 Gyrator example.

Let the inductor $Q = 10$

$$R_2 = 1\ \text{k}\Omega \text{ for a 741 op-amp}$$

From $Q = \frac{1}{2}\sqrt{\frac{R_1}{R_2}}$

$$R_1 = 2Q^2R_2 = 2(100)\ 1\ \text{k}\Omega = 200\ \text{k}\Omega$$

From $L = R_1R_2C$

$$C = \frac{L}{R_1R_2} = \frac{2.8\ \text{Hz}}{(1\ \text{k}\Omega)(200\ \text{k}\Omega)} = 0.014\ \mu\text{F}$$

If precision is required, parallel two capacitors such as a 0.012 μF and a 0.002 μF.

If circuit $Q = 3$, then

$$\begin{aligned} R &= X_c/\text{circuit } Q \\ &= 2\pi f_0 L/3 = 5.28\ \text{k}\Omega/3 \\ &= 1.76\ \text{k}\Omega \end{aligned}$$

Use 1.78 kΩ ± 2%.

If R is too small compared to R_2, the circuit Q will not be realized.

SUMMARY

1. Active filters are used in almost every phase of electronics and are worth study.
2. Although active filters have many advantages over passive filters, they do have limitations, too, particularly in maximum frequency of operation. (It is hoped this will become less of a limitation as op-amps are improved.)
3. The procedures for calculating active filters, as we have seen, are not too complex, even when the circuit looks complex, as does a state variable. To predict completely the behavior of an active filter, one should also calculate the sensitivities; the S equations are found in many of the references at the end of this chapter.
4. The Sallen and Key and multiple-feedback active filters are simple, reliable circuits, but not as stable as the more complex state-variable and biquad active filters. The second-order filters may be cascaded (with first-order filters for odd orders) to obtain higher order filters. We have seen that the procedure for cascading is tedious, but not hard. Now you should be able to construct high performance active filters.

ACTIVE FILTER REFERENCE LIST

Allen, Phillip E., *Modern Design Techniques and Applications of Active Filters*, Department of Electrical Engineering and Computer Science, Santa Barbara, Calif., 1974.

Berlin, Howard M., *Design of Active Filters with Experiments*, Howard W. Sams & Co., Indianapolis, Ind., 1977.

Hilburn, J. L., and Johnson, D.E., *Manual of Active Filter Design*, McGraw-Hill, New York, N.Y., 1973.

Johnson, David E., *Introduction to Filter Theory*, Prentice-Hall, Englewood Cliffs, N.J., 1978.

Johnson, David E., and Hilburn, John L., *Rapid Practical Design of Active Filters*, John Wiley & Sons, New York, N.Y., 1975.

Johnson, D. E., Johnson, J. R., and Moore, H. P., *A Handbook of Active Filters*, Prentice-Hall, Englewood Cliffs, N.J., 1980.

Lancaster, Don, *Active Filter Cookbook*, Howard W. Sams & Co., Indianapolis, Ind., 1975.

Toby, Gene E., Graeme, Jerald G., and Huelsman, Lawrence P., *Operational Amplifiers, Design and Application*, McGraw-Hill, New York, N.Y., 1971.

Wong, Yu Jen, and Ott, William E., *Function Circuits, Design and Application*, McGraw-Hill, New York, N.Y., 1976.

SELF-TEST QUESTIONS

8-1 State four advantages of active filters over passive filters.

8-2 State two disadvantages of active filters.

8-3 Draw from memory a low-pass, high-pass, and band-pass filter response curve. Label the passband, stop-band, and transition region.

8-4 State the relationship between a pole and the transition region attenuation change of an active filter.

8-5 List an advantage of each of the following filter responses: Butterworth, Chebyshev, and Bessel.

8-6 State the relationship between the damping factor (α) of a filter and its response near f_c.

8-7 Calculate the components for a second-order Sallen and Key equal component Butterworth, low-pass filter where $f_{(3\text{ dB})} = 5$ kHz. Let $C = 0.015\ \mu F$.

8-8 Calculate the components for a second-order Sallen and Key high-pass, Butterworth filter. Let $f_{(3\text{ dB})} = 7$ kHz and $C = 0.012\ \mu F$.

8-9 Calculate the components for a second-order, multiple-feedback, band-pass filter if $f_1 = 1$ kHz, $f_2 = 1.2$ kHz, and $A_p = 5$. Let $C = 0.0033\ \mu F$.

8-10 Calculate the components for a second-order, unity gain, state-variable high-pass filter. The filter is to be a 1 dB ripple Chebyshev with $f_{(3\text{ dB})} = 8$ kHz, and $C = 0.001\ \mu F$.

8-11 Calculate the components for a unity gain, state-variable, band-pass filter where $f_1 = 1.1$ kHz and $f_2 = 1.15$ kHz. Let $C = 0.022\ \mu F$.

8-12 Calculate the components for a second-order, state-variable, variable gain low-pass filter. Let the filter be a 2 dB ripple Chebyshev with $f_{(3\text{ dB})} = 2.5$ kHz, $C = 0.033\ \mu F$, and $A_p = 8$.

8-13 Calculate the components for a state-variable, variable gain, band-pass filter where $f_1 = 500$ Hz, $f_2 = 525$ Hz, $A_p = 10$, and $C = 0.01\ \mu F$.

8-14 Calculate the components for a biquad band-pass filter where $f_1 = 740$ Hz, $f_2 = 760$ Hz, $A_p = 10$, and $C = 0.0047\ \mu F$.

8-15 Calculate the components for a fourth-order, 2 dB ripple Chebyshev, low-pass filter. Use two second-order Sallen and Key stages. Let $f_{(3\text{ dB})} = 3$ kHz and $C = 0.0033\ \mu F$.

8-16 List the α and f_c ratio for each stage of an eighth-order 1 dB ripple Chebyshev, high-pass filter.
8-17 State the function of a gyrator.
8-18 State the advantages of metal film resistors.

If you cannot answer certain questions, place a check next to them and review appropriate parts of the text to find the answers.

LABORATORY EXERCISE

Objective

After completing this laboratory, the student should be able to calculate the components, breadboard, and check the frequency response of a Sallen and Key, multiple-feedback, state-variable, and biquad active filter.

Equipment

1. 5 Fairchild μA741 operational amplifiers or equivalent.
2. ±2% tolerance resistor assortment.
3. Power supply, ±15 V.
4. Oscilloscope with external horizontal input.
5. Audio sweep generator (an audio signal generator can be used if a sweep generator is not available).
6. Metallized mylar capacitor assortment, preferably ±5% tolerance.
7. Breadboard such as EL Instruments SK-10.

Throughout this laboratory use the component calculation procedures in Section 8-5.

Procedure

1. Test equipment connection.

The sweep generator and oscilloscope can be connected to display the frequency response of an active filter. The connection for this is shown in Figure 8.27. The frequency analog of the sweep generator is a ramp voltage that is used to drive the horizontal amplifier of the oscilloscope. The horizontal sweeps will represent frequency and the vertical display will represent amplitude as shown in Figure 8.28. The variable gain adjustment of the horizontal amplifier will need adjustment to make the horizontal trace size fit the CRT display.
Note. If the sweep generator does not have retrace blanking, a lighter frequency pattern will occur on the CRT during retrace. This can be minimized by using a slow sweep rate, but not entirely eliminated. Even with retrace blanking, excessive sweep rates will cause an unstable oscilloscope display.

Precise measurement of $f_{(3\ dB)}$, f_1, f_2, and gain is best accomplished by putting the sweep generator on continuous, the oscilloscope in the normal voltage versus time mode, and manually adjusting the frequency.

Connect the test equipment as shown in Figure 8.27. Adjust the oscilloscope horizontal gain for a full CRT width signal from the sweep generator's input.

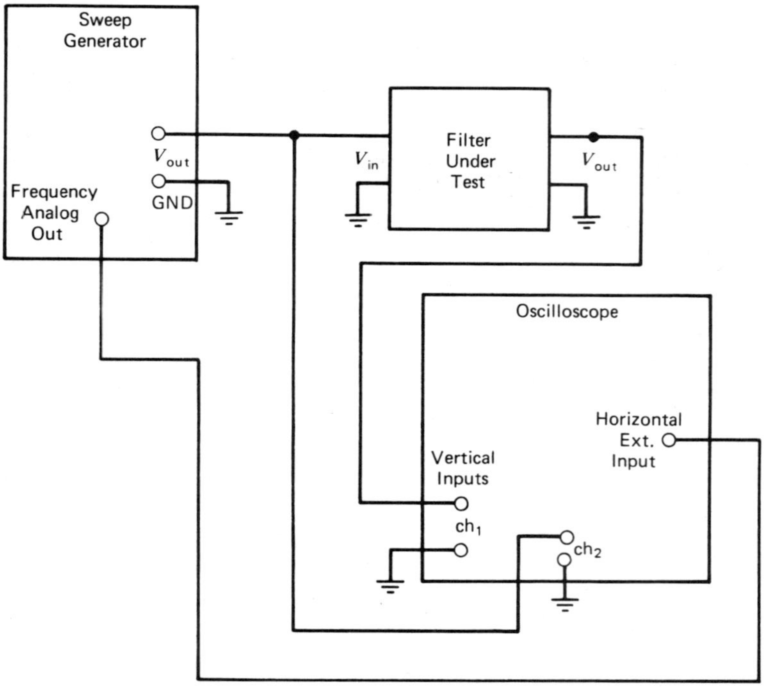

FIGURE 8.27 Oscilloscope setup for frequency response display.

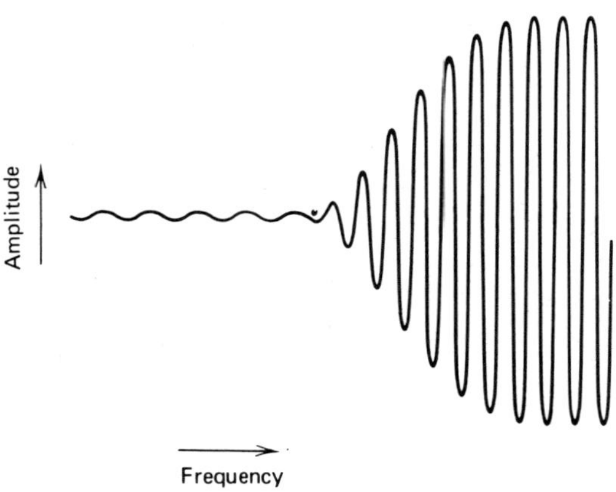

FIGURE 8.28 Oscilloscope display of a high-pass filter in the equipment setup of Figure 8.27.

2. Sallen and Key filter.
   (a) Calculate the components for a Sallen and Key, second-order Butterworth active low-pass filter. Set $f_{(3\text{ dB})} = 1$ kHz. The circuit is shown in Figure 8.29. Look up α in Table 8-1.
   (b) Breadboard the circuit.
   (c) Check the frequency response $f_{(3\text{ dB})}$, transition region attenuation change from 300 to 600 Hz (it should be 12 dB), and the passband gain. Compare your results with your calculations. For ±2% resistors and ±5% capacitors, your results should be within 10%.

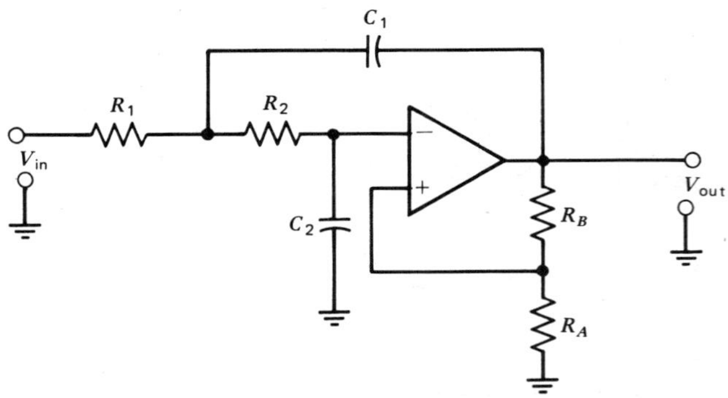

$R_1 = R_2 = R$
$C_1 = C_2 = C$

$$f_c = \frac{1}{2\pi RC}$$

$$R_B = (2 - \alpha)R_A$$

$$A_p = R_B/R_A + 1$$

FIGURE 8.29 Sallen and Key low-pass filter.

3. Multiple feedback filter.
   (a) Calculate the components for a multiple-feedback filter (Figure 8.30) where $f_1 = 900$ Hz and $f_2 = 1100$ Hz. Let $A_p = 6$.
   (b) Breadboard the circuit.
   (c) Check f_0, f_1, f_2, Q, and A_p.

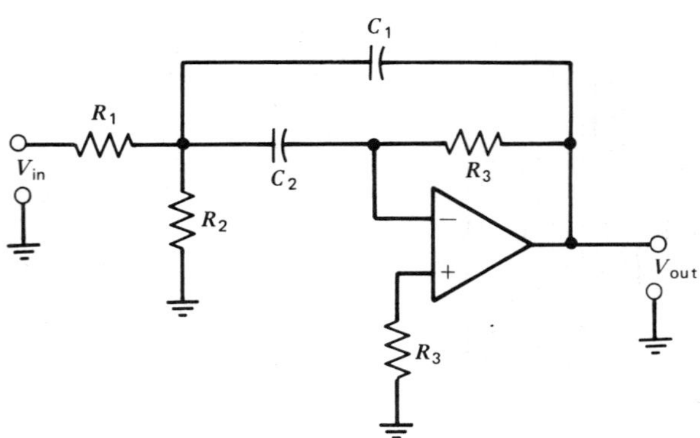

$$f_o = \sqrt{f_1 f_2}$$

$$Q = f_o/(f_2 - f_1)$$

$$R_1 = \frac{Q}{2\pi f_o C A_p}$$

$$R_2 = \frac{Q}{2\pi f_o C(2Q^2 - A_p)}$$

$$R_3 = \frac{2Q}{2\pi f_o C}$$

$$C_1 = C_2$$

$$A_p = R_3/2\,R_1.$$

FIGURE 8.30 Multiple-feedback filter.

4. Unity gain state variable.
   (a) Calculate the components for a second-order, unity gain, state-variable Chebyshev high-pass filter with 2 dB ripple. The circuit is shown in Figure 8.31. Recall that the $f_{(3\ \mathrm{dB})}/f_c$ ratio and α must be found in Table 8-1.
   (b) Connect the circuit.
   (c) Check $f_{(3\ \mathrm{dB})}$, the ripple amplitude in dB and A_p at the ripple peak.

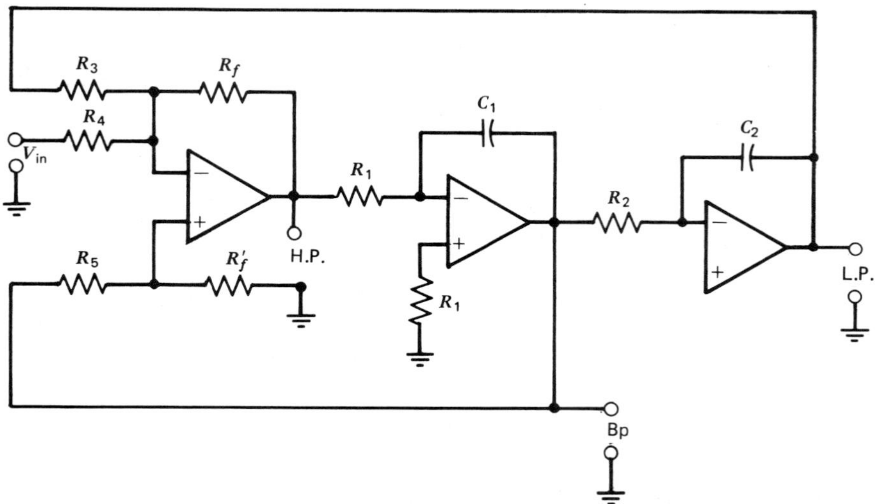

$R_1 = R_2 = R_3 = R_4 = R_f = R_f' = R$ $\qquad f_c = \dfrac{1}{2\pi RC}$

$C_1 = C_2 = C$

$R_5 = R_f'(3/\alpha - 1) = R_f'(3Q - 1)$

FIGURE 8.31 Unity gain state-variable filter.

5. State variable, variable gain.
   (a) Connect a variable gain, state-variable circuit as a band-pass filter with $f_1 = 900$ Hz, $f_2 = 920$ Hz, and $A_p = 20$ (see Figure 8.32*a*).
   (b) Connect the circuit.
   (c) Check f_0, f_1, f_2, Q, and A_p.
   (d) Connect the summer of Figure 8.32*b* to the high-pass and low-pass outputs.
   (e) Check f_0, f_1, f_2, Q, and $1/A_p$ of the notch response.
6. Biquad.
   (a) Calculate the components of a biquad band-pass filter for $f_1 = 800$ Hz, $f_2 = 820$ Hz, and $A_p = 10$ (Figure 8.33).
   (b) Connect the circuit.
   (c) Check f_0, f_1, f_2, Q, and A_p.
7. Optional section, cascading.
   (a) Calculate the components for a third-order cascaded, low-pass Butterworth active filter. Use a first-order stage from Section 8-7.1 and a variable gain, state-variable second-order stage. Set $f_{(3\ \mathrm{dB})} = 2$ kHz and $A_p = 5$. Do not forget to look up the f_c ratio and α from Table 8-2.
   (b) Connect the circuit.
   (c) Check $f_{(3\ \mathrm{dB})}$, A_p and the rate of attenuation change from 2.5 to 5 kHz.

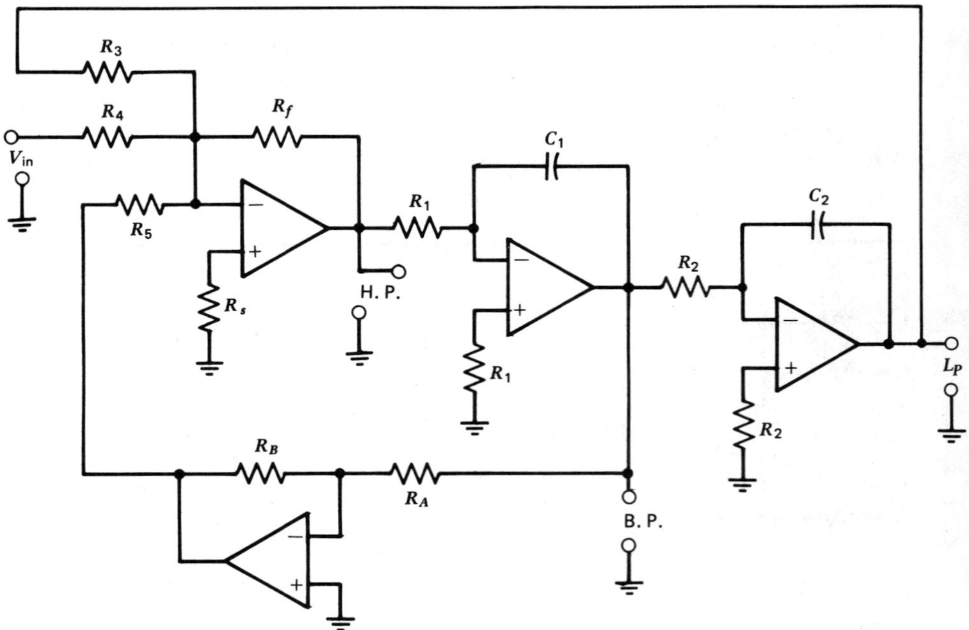

(a) Circuit diagram of the filter

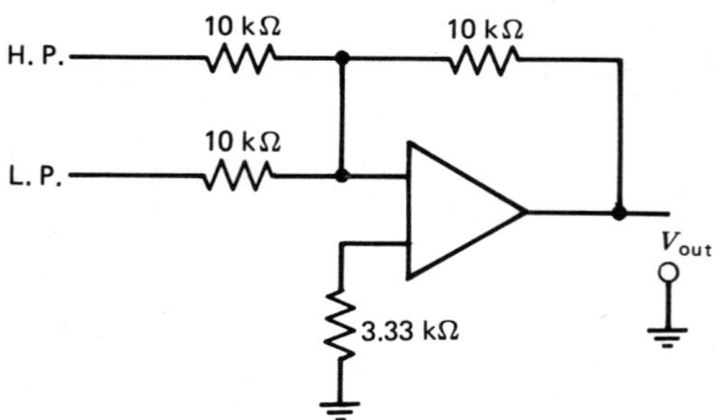

(b) Summer for a notch filter

$$f_o = \sqrt{f_1 f_2} \qquad Q = f_o/(f_2 - f_1)$$

$$G = A_p/Q$$

$$C_1 = C_2 = C \qquad R_1 = R_2 = R_3 = R_5 = R_f = R_A = R$$

$$f_o = \frac{1}{2\pi RC}$$

FIGURE 8.32 State-variable, variable gain band-pass filter.

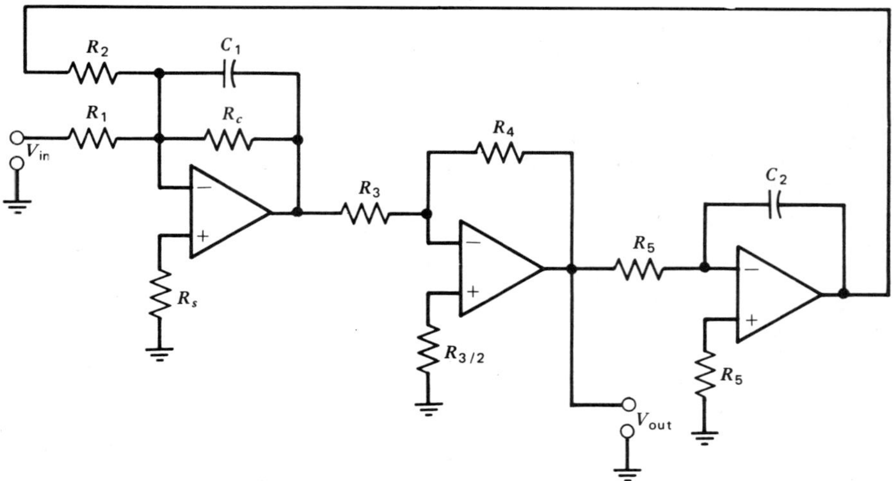

$f_o = \sqrt{f_1 f_2}$ $\quad Q = f_o/(f_2 - f_1)$

$G = Q/A_p$ $\quad C_1 = C_2 = C$

$R_2 = R_3 = R_4 = R_5 = R$

$$R = \frac{1}{2\pi f_o C}$$

$$R_1 = \frac{G}{2\pi f_o C}$$

$$R_c = \frac{Q}{2\pi f_o C}$$

FIGURE 8.33 Biquad band-pass filter.

(d) Calculate the components for a fourth-order 1 dB ripple high-pass Chebyshev filter. Use Sallen and Key second-order stages. Let $f_{(3\ \mathrm{dB})} = 4$ kHz.

(e) Connect the circuit of step (d).

(f) Check $f_{(3\ \mathrm{dB})}$, ripple amplitude in dB, A_p, and transition region attenuation change from 1.5 to 3 kHz.

CHAPTER

9

CIRCUIT SELECTION

The various applications to which operational amplifier modules have been applied can fill several books. In this chapter we will look at just a few circuits from the following types:

- Peak detector.
- Peak-to-peak detector.
- Zero-crossing detector.
- Precision rectifier.
- Current source.
- Voltage sources.
- Staircase generator.
- Wien-bridge oscillator.
- Square-wave generator.
- Triangle-wave generator.
- Linear ramp generator.

OBJECTIVES

After completing the study of this chapter and the self-test questions, the student should be able to:

1. Briefly state the operating principle of each of the circuits in the list just presented.
2. Perform the laboratory for Chapter 9.

9-1 PEAK DETECTOR

A peak detector's output is the highest voltage that the input sees in a given time period until the circuit is reset. A relatively simple high performance positive peak detector is shown in Figure 9.1*a*, and its output in Figure 9.1*b*.

The two amplifiers (in Figure 9.1*a*) within a single feedback loop operate as a unity gain follower. Amplifier A_1 is a unity gain follower that charges C_1 to the peak voltage. D_1 prevents the discharge of C_1. D_2 provides a feedback path to A_1 when its output is less than the peak voltage, preventing its saturation. Amplifier A_2 is a unity gain follower that acts as a buffer between C_1 and the output of the detector. The high input impedance of A_2 prevents loading of C_1 by the output circuit. Where long storage times are involved, A_2 should be an FET input amplifier. The resistor R_f provides some isolation between V_{out} and the summing point voltage of A_1 when V_1 is lower than V_{c_1}. The capacitor C_1 of Figure 9.1*a* should be chosen so that $I_{max}/C_1 \leq S_1$ where S_1 is the slew rate of A_1. If D_1 and D_2 are

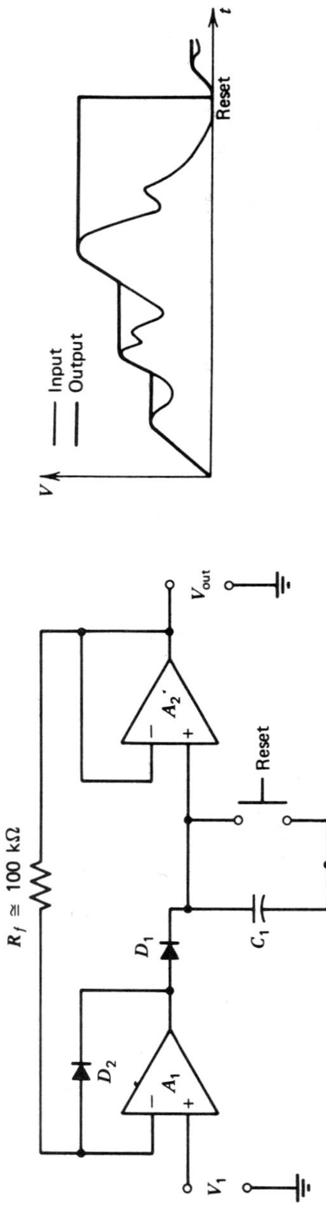

(a) Circuit

(b) Input and output waveforms

FIGURE 9.1 Peak detector.

reversed, the circuit of Figure 9.1*a* becomes a negative peak detector. The reset button can be replaced with a transistor switch.

9-2 PEAK-TO-PEAK DETECTOR

A precision peak-to-peak detector can be built as shown in Figure 9.2. A positive peak detector and a negative peak detector are the inputs to an adder-subtracter circuit that takes the difference in the positive and negative peaks.

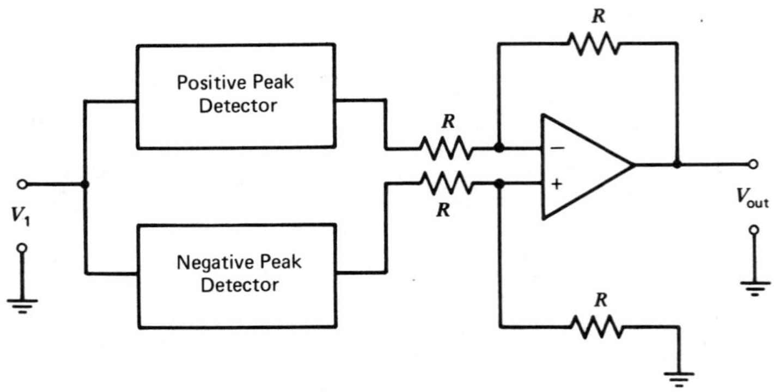

FIGURE 9.2 Peak-to-peak detector.

9-3 ZERO-CROSSING DETECTOR

In some electronic systems it is important to detect the instant a signal crosses zero. Many phase-detecting systems make use of zero-crossing detectors to obtain relative phase information. A zero-crossing detector is shown in Figure 9.3*a*, and V_{out} versus V_{in} in Figure 9.3*b*.

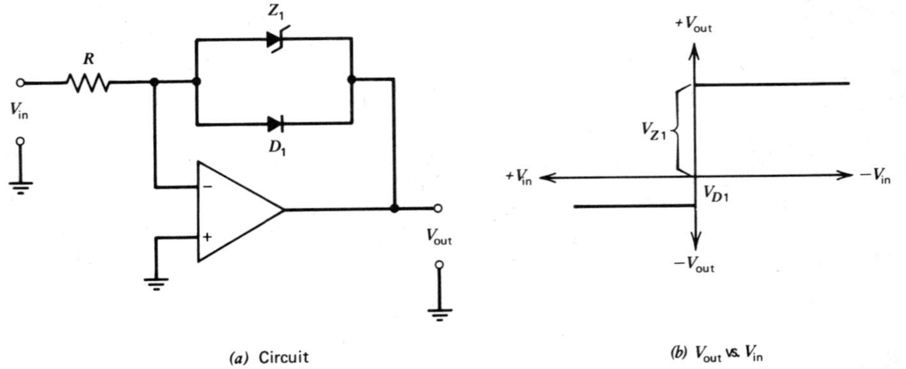

FIGURE 9.3 Zero-crossing detector.

When V_{in} is greater than zero, the output voltage is clamped to the forward voltage drop of D_1. As the input voltage goes to zero, D_1 loses its forward bias. The zener diode Z_1 has not yet started conducting, so as the input crosses zero volts the amplifier gain is approximately equal to the amplifier's open loop gain, thus giving good zero-crossing accuracy. As the input goes negative from zero volts, the amplifier output drives to the breakdown voltage of Z_1. As the input goes from negative to positive, the output voltage goes from $+V_{Z_1}$ to $-V_{D_1}$. The diode D_1 is chosen for a low forward voltage drop. For instance, D_1 might be an ion-implanted silicon diode. If the voltage drop of D_1 is not critical, the forward bias characteristics of Z_1 can be used and D_1 eliminated. The input resistor R is chosen to ensure that the input voltage source is not loaded. A resistor equal to R may be placed from the noninverting input to ground to help eliminate offset due to bias current. Diodes D_1 and Z_1 may be turned around to obtain a negative output of V_{Z_1}.

9-4 PHASE DETECTOR

A circuit to detect the phase difference between a reference input signal and a test signal may be constructed from two zero-crossing detectors as shown in Figure 9.4. The zero-crossing detector outputs of Figure 9.4 can be obtained by differentiating the output of the circuit of Figure 9.3. The phase difference of the test and reference signals is found by measuring the time difference $(t_2 - t_1)$ between their zero-crossing points. The gate generator might be a flip-flop that is set by the pulse at t_1 and reset by the pulse at t_2. The pulse from the gate generator is $t_2 - t_1$ seconds long. The gate generator then enables a sampling gate that allows a high frequency pulse train from the pulse generator to pass to the counter as long as the gate generator output is present. The counter then counts the pulses that passed through the sampling gate during the time between the pulses occurring at t_1 and t_2. The

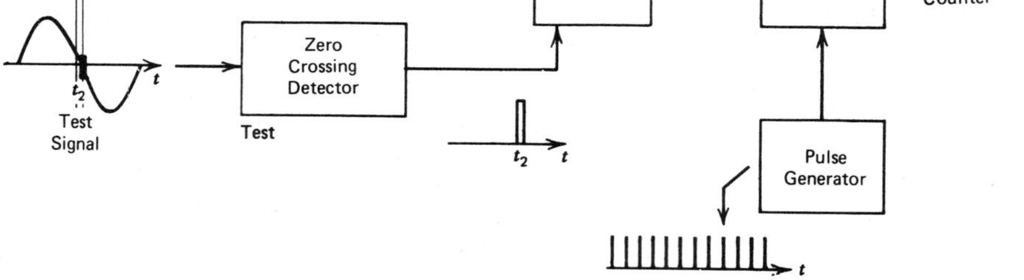

FIGURE 9.4 Phase detector.

number of pulses counted during the time $(t_2 - t_1)$ is proportional to the phase difference of the signal. Once we know the proportionality constant (K), which is determined by the pulse generator, the phase difference ($\Delta\phi$) is just

$$\Delta\phi = \omega_{REf}(t_2 - t_1) = K \text{ (number of pulses)} \tag{9-1}$$

9-5 SAMPLE AND HOLD

A modification of the peak detector of Figure 9.1 will result in a sample and hold circuit, shown in Figure 9.5*a*. The job of the sample and hold circuit is to charge a capacitor quickly to the circuit input voltage when given a command to do so and hold that input voltage on the output for an extended length of time. Sample and hold circuits are used extensively in data acquisition systems, in digital industrial control, and digital communications. The sample and hold circuit holds its output steady until its analog output voltage can be changed to a digital number by an analog-to-digital converter (analog-to-digital converters are discussed in Chapter 11).

There are many variations of sample and hold circuits providing different levels of speed and accuracy. The circuit of Figure 9.5*a* provides moderate accuracy but good speed. A_1 is a buffer for the input and provides capacitor charging current through the junction field effect transistor (JFET) switch when Q_1 is turned on. FET input amplifier A_2 acts as an output buffer so that the capacitor need not provide output current. The FET input amplifier is essential if the capacitor is to hold its charge for any length of time. R_1 isolates the capacitor from the input of A_2 if the power is turned off with the capacitor charged. Q_3 and Q_2 are transistor switches that apply $-V$ to the gate of the JFET switch when they are off and that allow the gate to see the JFET drain voltage through R_2 (R_2 will be between 1 and 10 MΩ) when they are on. When Q_2 and Q_3 are on, the collector voltage of Q_3 is nearly $+V$ and diode D is off. Q_3 allows the circuit to operate from TTL (transistor-transistor logic) output voltages.

When the control voltage is at 5 V, Q_2 and Q_3 are on and the JFET gate is tied to its drain through R_2. The JFET is on and A_1 charges C to V_{in} through the Q_1. C will charge more quickly if V_{in} is positive than negative, since if V_{in} is negative the capacitor can charge at a maximum current of I_{dss}, drain current with the gate connected to the source, of Q_1. Q_1 should be a JFET of the type that the source and drain can be reversed for optimum bipolar performance. The control voltage should be high long enough for the capacitor to charge fully to the maximum input voltage. The sample time should be at least $10\,[R_{out(A_1)} + r_{ds\,on\,(Q_1)}]C$.

When the control voltage is zero Q_2 and Q_3 turn off, the gate voltage of Q_1 goes to $-V$ so it turns off, and the hold time begins. The output will remain about at the last input voltage level until the next sample period. The capacitor will slowly lose voltage by discharge through the JFET, its own leakage resistance, and from the bias current of A_2. The voltage lost from the capacitor at the end of the hold period will be

$$\Delta V_c \text{ (hold)} = \frac{I_c\, t_{hold}}{C}$$

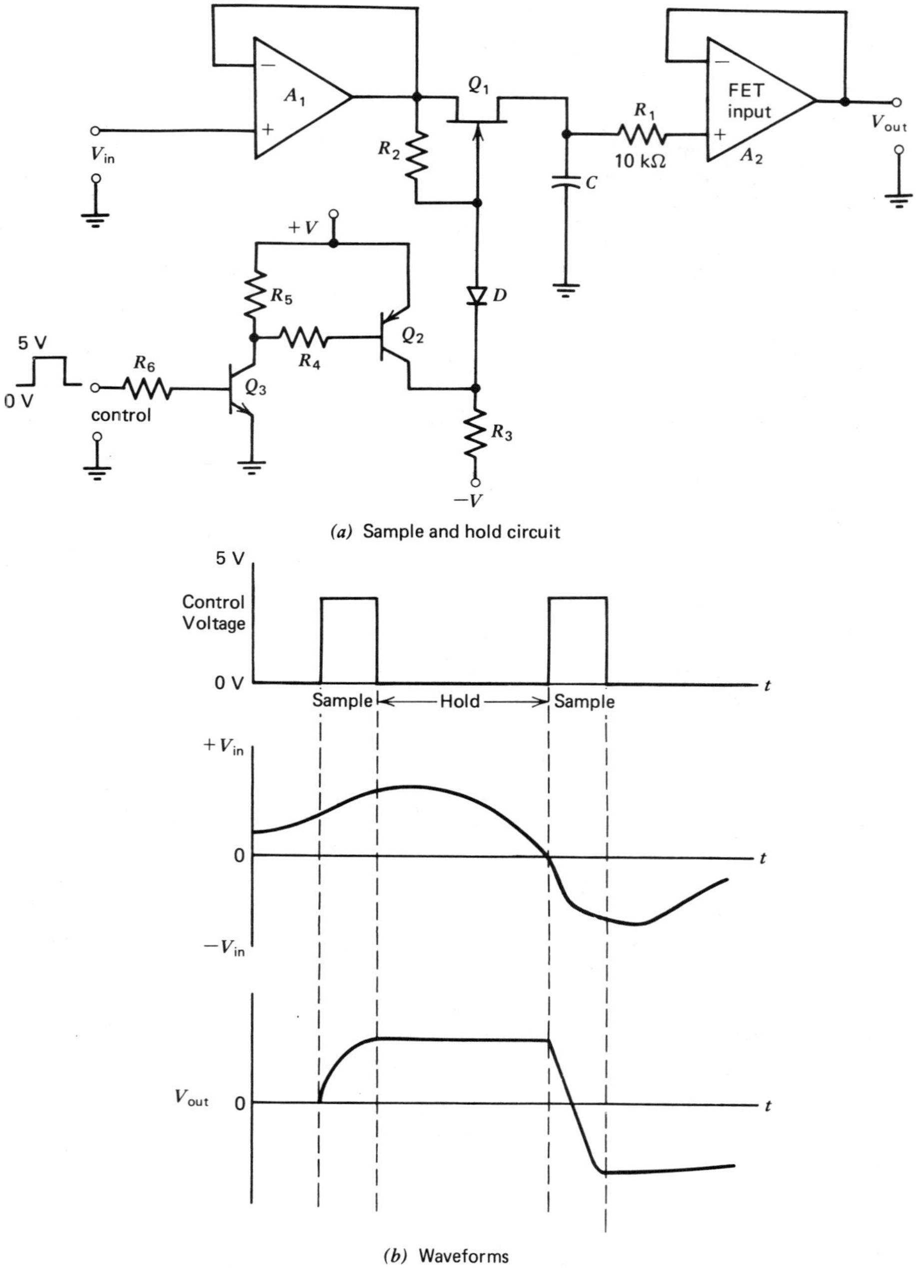

FIGURE 9.5 Sample and hold.

where

I_c = leakage current of JFET + A_2 bias current.

The capacitor should be a high quality teflon, polyethylene, or polycarbonate dielectric in larger values, or glass or silvered mica dielectric in smaller values.

A_1 and A_2 could be dependent voltage followers by putting A_1 and A_2 in the same feedback loop during the sample period to reduce offset errors. If A_1 and A_2 are independent voltage followers, the circuit can operate at higher frequencies.

EXAMPLE 9-1

Calculate the components of the circuit of Figure 9.5. $\pm V = \pm 15$ V, A_1 and A_2 are TI TL 081. Q_2 and Q_3 have the following specifications:

$$BV_{CEO} = 40 \text{ V}, \quad \beta_{min} = 40, \quad V_{CE(sat)} = 0.5 \text{ V}, \quad V_{BE(sat)} = 0.7 \text{ V}$$

The JFET specifications are:

$$V_{GS}(\text{max}) = 30 \text{ V}, \quad I_{dss} = 20 \text{ mA}, \quad I_{gss} = 50 \text{ pA}, \quad I_d(\text{off}) = 40 \text{ pA}$$

The TL 081 will provide at best 20 mA output current. $r_{ds}(\text{on})$ is 100 Ω. The maximum hold time is to be 10 ms with a 0.1% error.

$$C = \frac{I_c\, t_{hold}}{\Delta V_{(hold)}}$$

$I_c = I_{BA_2} + I_{D(off)Q_1} = 200 \text{ pA} + 40 \text{ pA}$
$\Delta V_{(hold)} = 0.1\%\ V_{in}(\text{max}) = 0.1\%\ (10 \text{ V}) = 0.01 \text{ V}$

$$C = \frac{240 \text{ pA } 10 \text{ ms}}{0.01 \text{ V}} = 240 \text{ pF}$$

Since other surprise leakage paths may exist, we will triple this value and use a silvered mica capacitor, readily available in this size. Let C = 750 pF. The sample time must be the greater of

$$t_{sample} = \frac{2CV_{in}}{I_{out\,A_1}}$$

$$t_{sample} = 10\,C\,(R_{out(A_1)} + r_{ds(on)Q_1})$$

$$t_{sample} = 2V_{in}/S_{(A_1)}$$

Since V_{out} may have to go from $+V_{in}$ to $-V_{in}$.

$$t_{sample} \geq \frac{750 \text{ pF } (20 \text{ V})}{20 \text{ mA}} = 7.5\ \mu\text{s}$$

or

$$t_{sample} \geq 10\ (750\ \text{pF})\ 100\ \Omega = 0.75\ \mu\text{s}$$

or

$$t_{sample} \geq 20\ \text{V}/(13\ \text{V}/\mu\text{s}) = 1.54\ \mu\text{s}$$

so

$$t_{sample} \text{ must be greater than } 7.5\mu\text{s}.$$

The rest of the calculations are straightforward switching transistor calculations. Let $R_2 = 1\ \text{M}\Omega$, $I_{C_{Q2}} = I_{C_{Q3}} = 1\ \text{mA}$

$$R_3 = \frac{2(+V) - V_{CE(sat)Q_2}}{I_{C_{Q2}}} = \frac{29.5\ \text{V}}{1\ \text{mA}} = 29.5\ \text{k}\Omega$$

$$R_4 = \frac{+V - V_{BE(sat)Q_2} - V_{CE(sat)Q_3}}{I_{C_{Q2}}/\beta \min Q_2} = 560\ \text{k}\Omega$$

$$R_5 = \frac{+V - V_{CE(sat)Q_3}}{I_{C_{Q3}}} = \frac{14.5\ \text{V}}{1\ \text{mA}} = 14.5\ \text{k}\Omega$$

$$R_6 = \frac{V_{control} - V_{BE(sat)Q_3}}{I_{C_{Q3}}/\beta \min Q_3} = \frac{5\ \text{V} - 0.7\ \text{V}}{1\ \text{mA}/40} = 172\ \text{k}\Omega$$

Sample and hold circuits are available in both hybrid and monolithic IC packages. Monolithic IC sample and hold circuits must have an external capacitor.

9-6 PRECISION RECTIFIER

Precision rectifiers, also called absolute value circuits, provide an output equal to the absolute value of the input without the loss of voltage required to forward bias a rectifier diode. Thus a precision rectifier such as the one shown in Figure 9.6*a* can be used to rectify ac signals in the millivolt range. The forward voltage drop of the diodes is reduced by A_{ol} since they are in the feedback loop of amplifier A_1.

Precision rectifiers are used in ac voltmeters, power measurement, and signal monitoring. Precision rectifiers are also used in demodulating amplitude modulated waveforms when the carrier frequency is relatively low.

The circuit works as follows. Let $R_1 = R_2 = R_3 = R_4 = R_5$. When V_{in} goes positive, the output of A_1 swings negative. D_1 is forward biased and D_2 is reversed biased. The voltage at point a is then $-V_{in}$. Since the noninverting terminal of A_2 is connected through R_3 to the inverting terminal of A_1, it is at virtual ground, so A_2 is just an inverting amplifier with $A_{fb} = -1$ (recall that $R_4 = R_5$). V_{out} is then equal to V_{in}.

If V_{in} swings negative, the output of A_1 goes positive, causing D_1 to be reverse biased and D_2 to be forward biased. A_2 now becomes a noninverting amplifier since

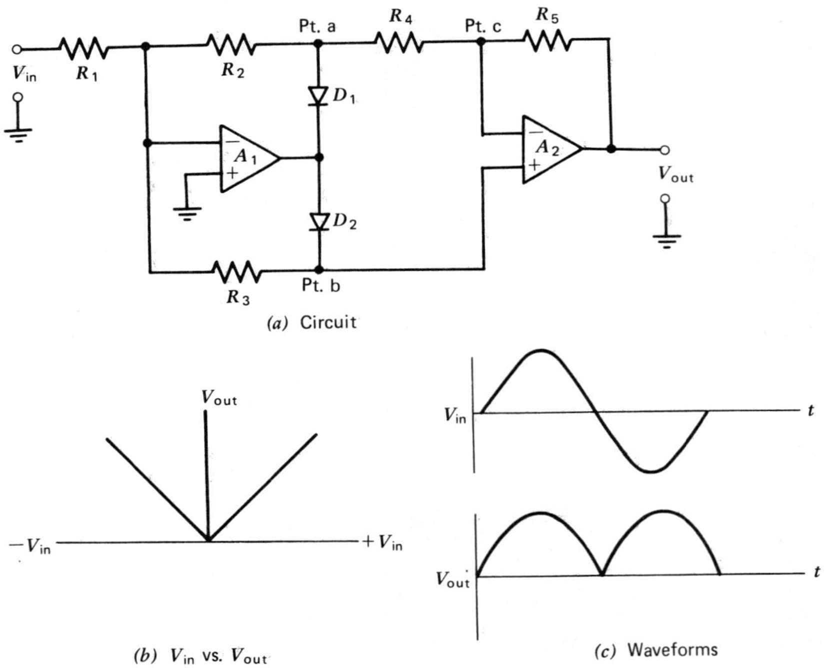

FIGURE 9.6 Precision rectifier (absolute value) circuit.

its inverting terminal is connected to the virtual ground at the inverting terminal of A_1 through R_2 and R_4. The gain of A_2 is now

$$A_{fb_2} = 1 + \frac{R_5}{R_2 + R_4}$$

Since all the resistors are equal in value, $A_{fb_2} = 3/2 = 1.5$. The gain of inverting amplifier A_1 is no longer one since not all of the current through R_1 flows through R_3. The voltage at point c is equal to the voltage at point b since V_i of A_2 is almost zero. With $R_2 = R_3 = R_4$, $R_1 + R_4 = 2\ R_3$, so one third of the input current flows through R_2 and R_4 and two thirds of the input current from R_1 flows through R_3. The voltage at point b is then 2/3 V_{in}. Since point c and point b are at the same voltage, one could say that

$$A_{fb_1} = \frac{(R_2 + R_4) \parallel R_3}{R_1} = 2/3 \text{ or } 0.66$$

since all the resistors are equal in value. The overall gain of the circuit when V_{in} is negative is then

$$A_{fb_T} = A_{fb_1} A_{fb_2} = (2/3)\,(3/2) = 1$$

If all the resistors are equal in value, the circuit gain is one. If $R_1 = R_2 = R_4 = R$ and $R_3 = R_5 = 2\ R$, the circuit gain is two.

9-7 OUTPUT CURRENT BOOSTERS

Applications occur where more output current is needed than the best operational amplifier for that particular job can supply. If the current booster is used inside the feedback loop, the amplifier performance is substantially unaffected.

The simplest way to increase current, if only a single polarity of current is needed, is to drive an emitter follower within the feedback loop as shown in Figure 9.7*a*.

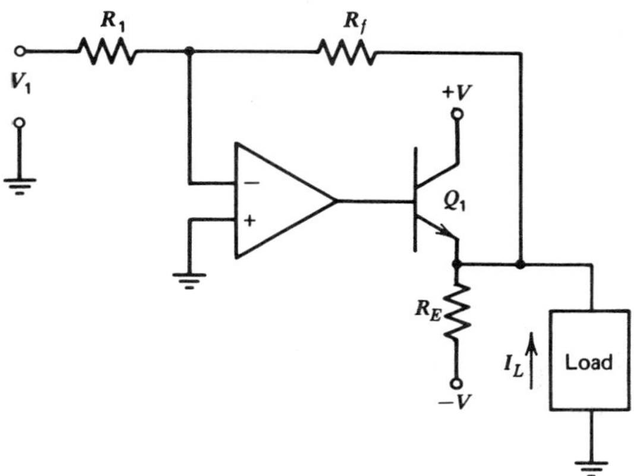

(a) Unipolar current booster

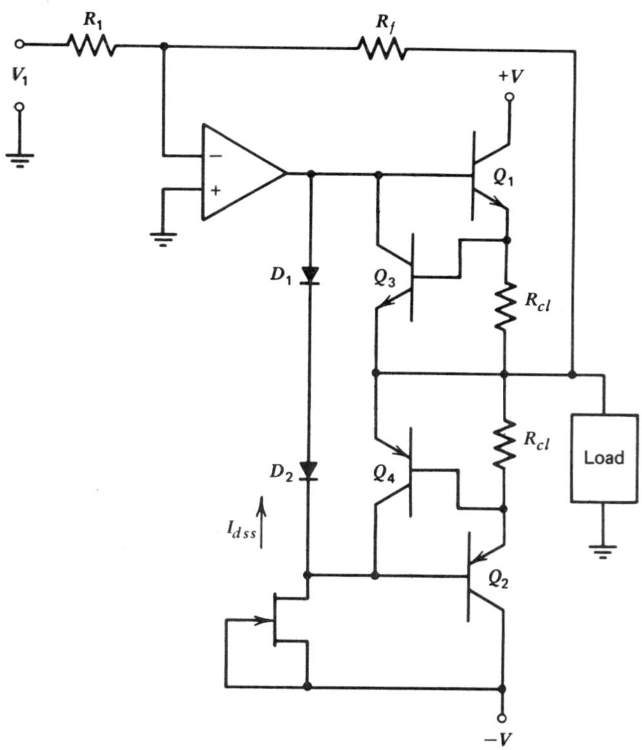

(b) Bipolar current booster

FIGURE 9.7 Current boosters.

The V_{BE} error of the emitter follower is reduced by the open loop gain of the amplifier so that the output offset is V_{BE}/A_{ol}. If the current required is such that the op-amp cannot supply sufficient drive for Q_1, then Q_1 of Figure 9.7*a* can be replaced with a Darlington pair. The output offset is doubled with a Darlington pair, however. A *pnp* transistor would be used to reverse the output current polarity.

A bipolar current booster is shown in Figure 9.7*b*. This is just a complementary output stage with current limiting. The resistors R_{cl} are chosen so that V_R is sufficient to turn on Q_3 or Q_4 when the load current is at the maximum allowed value. R_{cl} is picked then such that

$$R_{cl} = V_{BE}/I_{max} \tag{9-2}$$

The FET constant current source is chosen so that $I_{dss} \geq I_{BQ_2}$. For increased output current, Q_1 and Q_2 can be replaced by Darlington pairs, but four diodes instead of two are necessary to prevent crossover distortion. As in the unipolar booster, the output offset by the transistor V_{BE} is reduced by the open loop gain of the op-amp, so that it is $V_{BE_{Q1}}/A_{ol}$. Often amplifiers that must work into high capacitive loads must use a booster. One example is a line-driving amplifier that must work into a rather long transmission line.

9-8 CURRENT SOURCES

Operational amplifiers are convenient current sources. A floating current source is shown in Figure 9.8*a*. Since $I_{R_1} = I_f$ and $v_i \cong 0$, the current through the load will be V_1/R_1. A booster circuit can be used on the op-amp if the current desired is more than the amplifier can supply. Note that this current source is voltage controllable so it can be remotely programmed or fixed as desired. R_2 and C may not be necessary unless the load is inductive.

Figure 9.8*b* shows an emitter follower type unipolar current source. This circuit controls the current by causing the resistor voltage to equal the difference between the supply voltage and the input voltage. Feedback will force v_i to approximately zero volts so that the voltage at the inverting input equals the input voltage, which equals the voltage on the transistor emitter. Since $I_L = (|V_{supply}| - V_1)/R$, the current source is easily programmed and is as stable as V_1. If V_1 is supplied by a temperature-compensated zener diode regulator referenced to the supply voltage, this can be very stable. A big advantage of the current sources of Figure 9.8*b* is that the transistor supplies a current boost so that the load current can be greater than the amplifier output current. The transistors can be replaced with Darlington pairs to further boost the current capability of the source. These sources can be used with a grounded load if V_1 is sufficiently above ground to allow the voltage range necessary for the application.

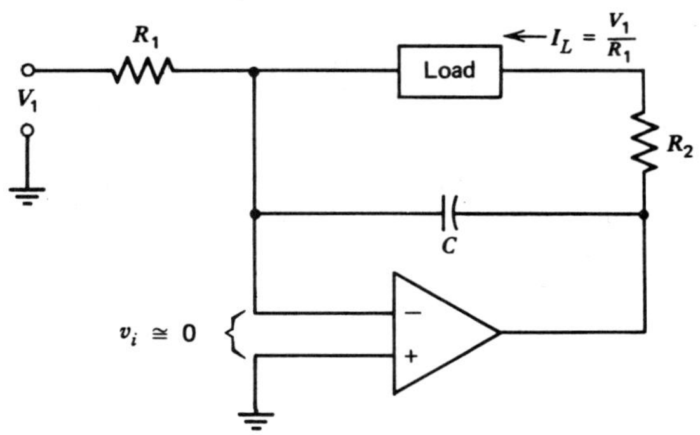

(a) Floating current source

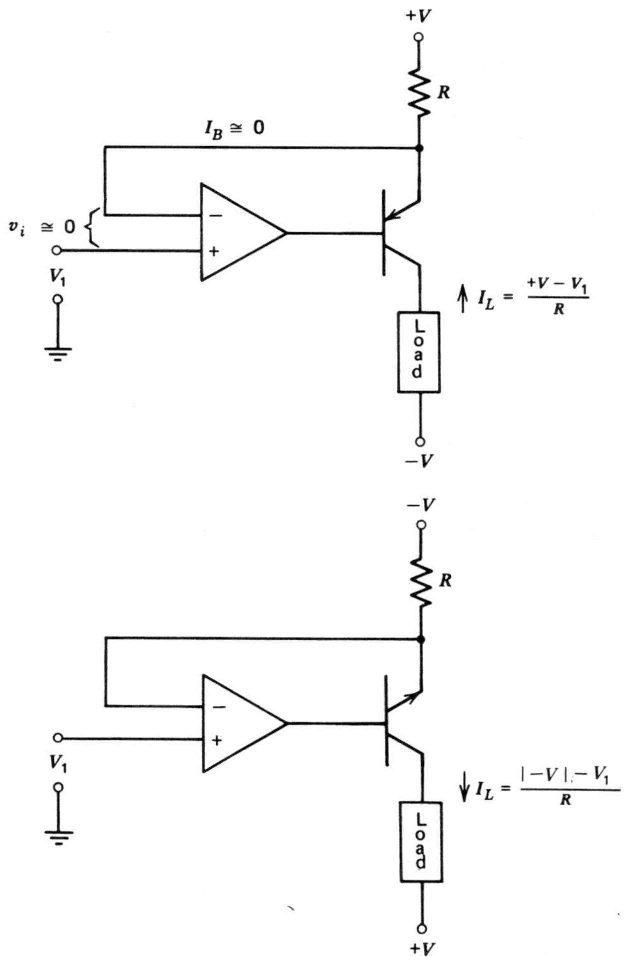

(b) Unipolar current sources

FIGURE 9.8 Current sources.

9-9 VOLTAGE SUPPLIES

Operational amplifiers may be used as low output impedance reference voltage sources. In addition, a load voltage regulation due to the amplifier feedback is supplied, and

$$\% \text{ regulation} = \frac{R_o}{A_{ol}\beta R_L}(100) \tag{9-3}$$

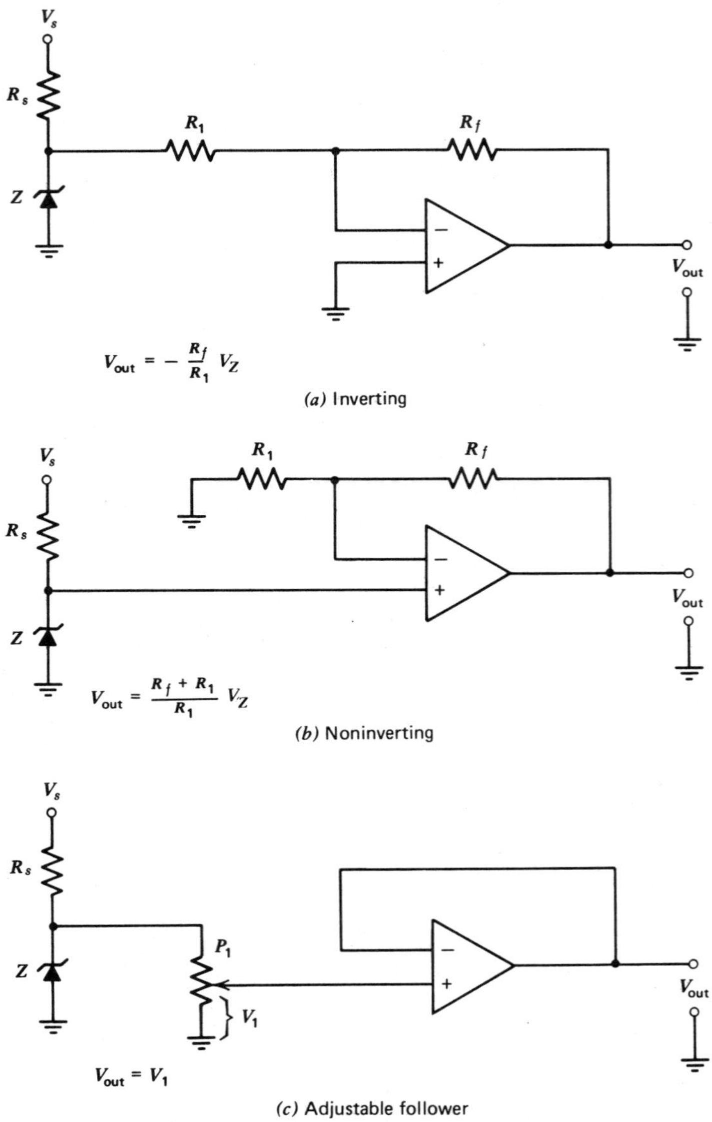

FIGURE 9.9 Zener reference sources.

where

R_o = op-amp open loop output resistance
β = feedback factor (1 for a follower)
R_L = lowest load resistance

This regulation assumes a stable input voltage. The reference voltage sources, shown in Figure 9.9, are essentially amplifiers with a regulated input voltage. Since the zener current is not dependent on the load current, the performance is better than the zener reference alone. Other voltage sources than zener diodes may, of course, be used. Current boosters may be used to increase the load capacity of the reference sources shown in Figure 9.9. The zener voltage will remain more stable if R_s is replaced by a constant current source set for optimum zener bias current.

9-10 STAIRCASE GENERATOR

A simple staircase generator can be built as shown in Figure 9.10. This generator must be driven by a current pulse train. The circuit including Q_1 and Q_2 is one way to do this. Q_1 and Q_2 are, respectively, a constant current source and a reset transistor. The circuit consists of a current source (A_1), peak detector and FET input op-amp buffer (A_2), and a reset amplifier (A_3). The current source rapidly charges C linearly to each voltage step when Q_1 is on. When Q_1 turns off, the capacitor holds the voltage step until the next step arrives. When the last voltage step has been reached, the output of A_3 swings positive, discharging C to reset the circuit. When the output of A_3 swings positive, Q_3 turns on, holding the output of A_3 positive until the capacitor discharges to zero volts. When C discharges to zero volts, the output of A_3 swings negative, turning both Q_2 and Q_3 off. The voltage on the noninverting terminal then rises to the reset voltage again and the process repeats. The diode is to keep the emitter-base breakdown of Q_3 from being exceeded when the output of A_3 is negative. The waveforms are shown in Figure 9.10*b*. A_1 must be a high slew rate op-amp such as a JFET input TI TL 081.

EXAMPLE 9-2

Construct a staircase generator using the circuit in Figure 9.10*a*. Use a Texas Instruments TL 082 dual FET input op-amp for A_1 and A_3 and a TL 081 FET input op-amp for A_2. The specifications for the TI TL 081 and TL 082 are in Appendix C. Q_1 is a *pnp* silicon transistor with $\beta_{min} = 200$, Q_2 an FET with $V_{gs\,off} = 3$ V, $I_{dss} = 20$ mA, and Q_3 a switching transistor with $V_{CE}(sat) = 0.2$ V, $V_{BE}(sat) = 0.7$ V, and $\beta_{min} = 30$. Let the staircase generator output consist of 10 one-volt steps lasting 10 ms with a rise time of 20 μS. The leakage currents of the transistor Q_1 and the FET Q_2 are respectively 0.1 nA and 50 pA. The bias current of A_2 is 200 pA.

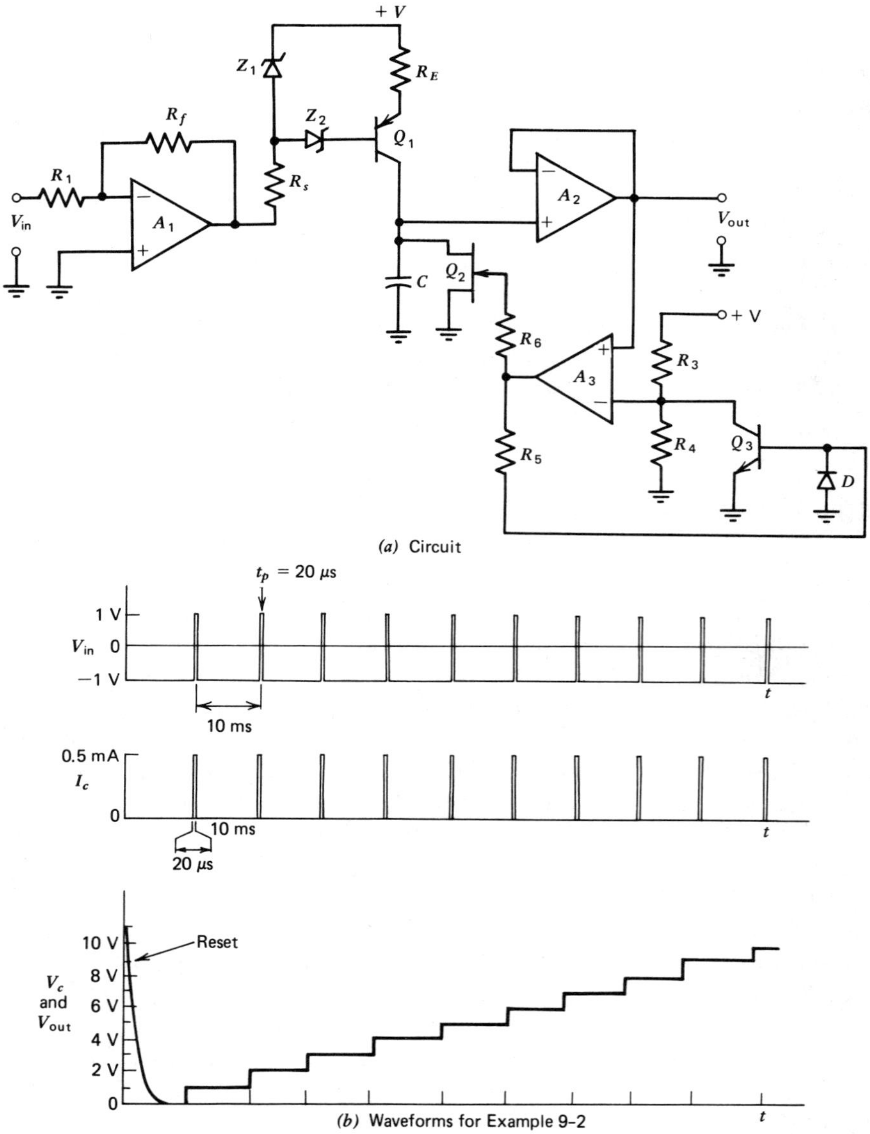

FIGURE 9.10 Staircase generator.

SOLUTION

If the maximum droop in the step for 10 ms from the discharge of C is 10 mV, the C is found for the maximum change in the step voltage allowed. Since

$$C = \frac{It}{V}$$

if

$$I = \text{total leakage current}$$

$$t = \text{step duration}$$

$$V = \text{maximum change in step voltage}$$

$$C = \frac{350\ \text{pA}\ 10\ \text{ms}}{10\ \text{mV}} = 350\ \text{pF}$$

Since we are very conservative, we will use a 0.01 μF capacitor.

The capacitor must be charged to the step voltage within 20 μs. Since

$$C = \frac{It}{V}$$

$$I = \frac{CV}{t}$$

where

I = constant current source current
V = step voltage
t = charge time for step

$$I = \frac{0.01\ \mu\text{F}\ 1\ \text{V}}{20\ \mu\text{s}} = 0.5\ \text{mA}$$

$V_{Z_1} - V_{Z_2}$ is the bias voltage for Q_1. Without Z_2, Q_1 will not completely turn off. If V_{Z_1} is chosen to be 10 V and $V_{Z_2} = 4.7$ V, then

$$R_E = \frac{V_{Z_1} - (V_{BE} + V_{Z_2})}{I_E} = \frac{4.6\ \text{V}}{0.5\ \text{mA}} = 9.2\ \text{k}\Omega$$

Let $I_Z = I_{R_s} = 1$ mA. This value is much, much greater than $I_{B_{Q_1}}$, so if $\pm V = \pm 15$ V, then $\pm V_{out}(\text{max}) \cong \pm 12$ V and

$$R_s = \frac{+V_{out}(\text{max}) + |-V_{out}(\text{max})| - V_{Z_1}}{I_Z}$$

$$R_s = \frac{24\ \text{V} - 10\ \text{V}}{1\ \text{mA}} = 14\ \text{k}\Omega$$

If $+V_{in} = 1$ V and $-V_{in} = -1$ V, then for a maximum output swing of A_1 the closed loop gain must be at least 15. If R_f is chosen to be 470 kΩ, then for an inverting amplifier

$$R_1 = \frac{R_f}{A_{fb}}$$

$$R_1 = \frac{470\ \text{k}\Omega}{15} = 31.3\ \text{k}\Omega$$

R_3 and R_4 are chosen so that the output of A_3 does not swing positive until the buffer output voltage is greater than the tenth-step voltage. Thus V_{R_4} with Q_3 off is set at 10.5 V. The current through R_3 and R_4 must be much greater than the bias current. Letting $I_{R_3} = I_{R_4} = 1$ mA and solving the voltage divider

$$R_3 = \frac{+V - 10.5\ \text{V}}{1\ \text{mA}} = 4.5\ \text{k}\Omega$$

$$R_4 = \frac{10.5\ \text{V}}{1\ \text{mA}} = 10.5\ \text{k}\Omega$$

When Q_3 is on $I_{C_{Q_3}} = +V/R_3 = 3.33$ mA. R_5 must supply the base current for Q_3 as well as the reset voltage for Q_2 when the output of A_3 is positive.

$$R_5 = \frac{+V_{out}(\max) - V_{BE}(\text{sat})_{Q3}}{I_{C_{Q3}}/\beta_{\min_{Q^3}}}$$

$$R_5 = \frac{12\ \text{V} - 0.7\ \text{V}}{3.33\ \text{mA}/30} \cong 100\ \text{k}\Omega$$

R_6 simply isolates the gate of Q_2 from the output voltage of A_3 when it is positive. Any value A_3 can drive is satisfactory. Let $R_6 = 100$ kΩ.

9-11 SIGNAL SOURCES

9-11.1 Wien-Bridge Oscillator

The Wien-bridge can be used to construct a very good oscillator in the audio range. The Wien-bridge is well covered in many texts, so we will only consider its use as a frequency determining device for an oscillator. The basic components of a Wien-bridge oscillator are shown in Figure 9.11*a*. Recall that the balance frequency of a Wien-bridge (which is the oscillation frequency of the oscillator) is

$$f_d = 1/(2\pi\sqrt{R_1R_2C_1C_2}) \tag{9-4}$$

usually $R_1 = R_2$ and $C_1 = C_2$ so that

$$f_d = 1/(2\pi\ R_1C_1) \tag{9-4a}$$

As long as the feedback factor of the amplifier is ⅓, the circuit will oscillate at any preset amplitude. If $\beta < 1/3$, oscillation amplitude will diverge with time until severe distortion occurs. If $\beta > 1/3$, the oscillation will converge to zero with time. Some form of automatic amplitude control is required. This usually works by varying the negative feedback gain (β) to stabilize the oscillations and reduce distortion. These stabilization schemes can be quite complex, involving as many as four additional operational amplifiers.

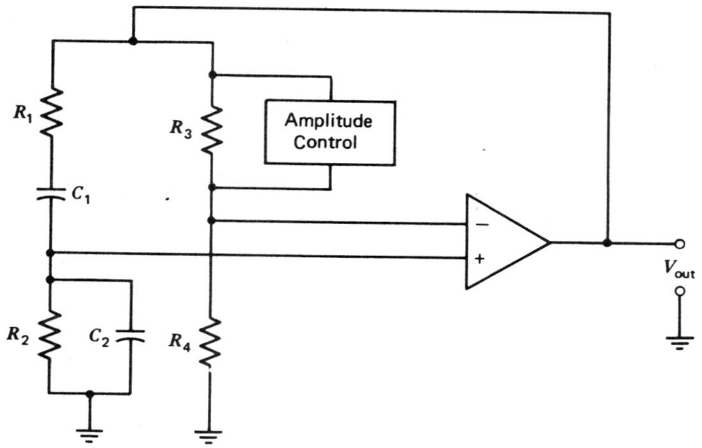

(a) Basic Wien–bridge oscillator

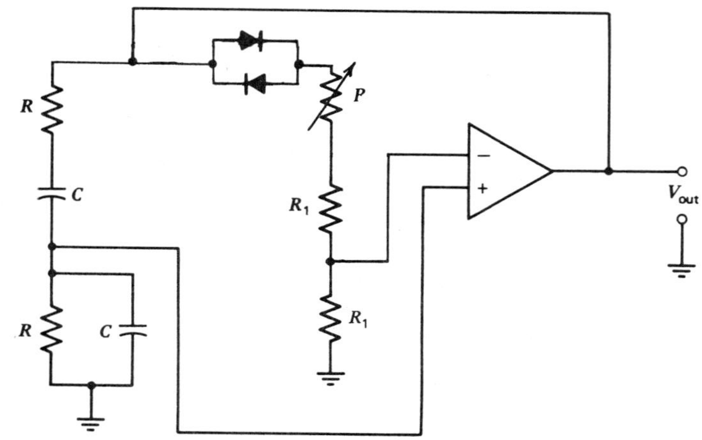

(b) One amplifier Wien–bridge oscillator

FIGURE 9.11 Wien-bridge oscillator.

One single amplifier Wien-bridge oscillator is shown in Figure 9.11*b*. The positive feedback loop has equal resistors and capacitors, whereas the negative feedback loop has a gain of 2. This provides a β slightly greater than ⅓ to assure oscillation.

The back-to-back diodes provide the automatic gain control. As V_{in} increases the diode, ac resistance decreases from the fact that

$$r_D = \frac{26 \text{ mV}}{I_D}$$

where

r_D = ac diode resistance
I_D = instantaneous diode current
26 mV = diode junction noise voltage at room temperature

Thus, as V_{out} increases, R_f total decreases, stabilizing the output amplitude and preventing clipping. $P \cong 2R_1$ and is used for amplitude adjust. The feedback resistor sizes are chosen to limit distortion and may be experimentally selected for the least distortion. The maximum output frequency is limited by the amplifier slew rate. Distortion as low as about 0.05% can be obtained with this circuit. A buffer amplifier on the output of this circuit is almost a necessity. There are many other types of op-amp sine wave generators.

EXAMPLE 9-3

Set the Wein-bridge oscillator of Figure 9.11*b* to oscillate at 10 kHz. Select C = 0.001 μF.

SOLUTION

$$f_0 = \frac{1}{2\pi RC}$$

$$R = \frac{1}{2\pi f_0 C}$$

$$= \frac{1}{2\pi(10\text{ kHz})(0.001\ \mu\text{F})}$$

$$= 15.9\text{ k}\Omega$$

for $\beta = 1/3$ $\quad A_{fb} = 3$
Let

$$R_1 = 25\text{ k}\Omega$$

Then

$$R_1 + \frac{1}{2}P = 75\text{ k}\Omega$$

Therefore

$$R_1 = 25\text{ k}\Omega$$
$$P = 75\text{ k}\Omega$$

9-11.2 Square-Wave Generator

There are many types of op-amp nonlinear signal generators. We list only one representative, simple circuit from each class in this chapter. The circuit shown in Figure 9.12 is a simple square-wave generator. The back-to-back zeners control the output voltage amplitude limiting it to $+V_Z$ and $-V_Z$. These need not be symmetrical, but usually are. R_f and C provide the circuit timing with the amplifier acting as a comparator. The feedback is provided to the noninverting input by the R_3 and R_4 voltage divider. The feedback factor is

$$\beta = R_4/(R_3 + R_4) \qquad \text{where } R_1 >> R_3 \text{ or } R_4 \tag{9-5}$$

The input resistors R_1 are to assure a high input impedance to op-amps with input protection. To see the operation, assume the output has just switched from negative to positive. The capacitor, which has a negative charge on it, starts to charge positive. When the capacitor voltage reaches the voltage on the noninverting input, which will be $+V_{out}\beta = +V_Z\beta$, the circuit switches negative and the process repeats on the negative cycle. The period for this square-wave generator is

$$\tau = 2\, R_f C \ln\left(\frac{1 + \beta}{1 - \beta}\right) \tag{9-6}$$

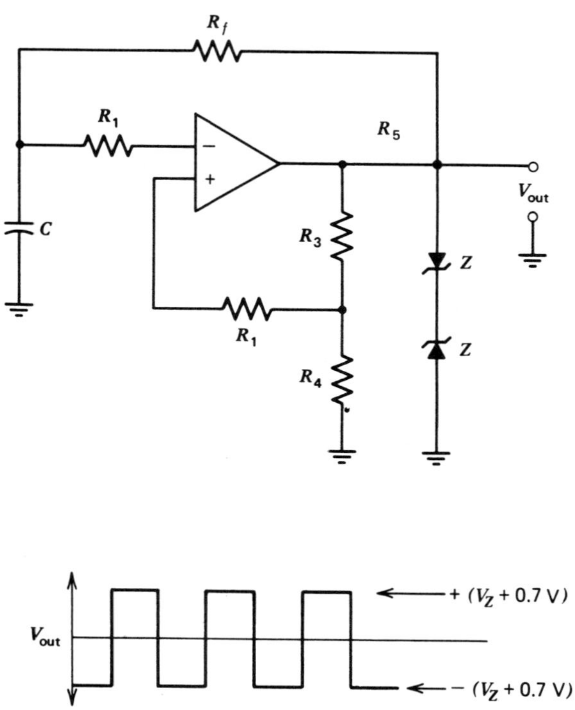

FIGURE 9.12 Square-wave generator.

for a symmetrical square wave. If one arranges for $\beta = 0.473$, then

$$\tau = 2\,R_fC \text{ and } P_{rf} = 1/\tau = \frac{1}{2\,R_fC} \qquad (9\text{-}6a)$$

The maximum frequency this circuit can operate at is limited by the operational amplifier's slew rate. The frequency stability is dependent primarily on the zener diode and capacitor stability. R_5 prevents excessive current through the zener diodes.

EXAMPLE 9-4

Calculate the components for the square-wave generator shown in Figure 9.12 so that f_{osc} (oscillating frequency) = 1 kHz with $V_{out} = 8$ V peak.

SOLUTION

The zeners are chosen to be $V_{out} - 0.7$ V or about 7.3 V.

R_1 is chosen to be greater than $R_3 + R_4$. Let $R_1 = 100$ kΩ, and we will arrange for $R_3 + R_4 << 100$ kΩ.

If $V_{R_4} = 0.473\ V_{out}$ then

$$\tau = 2R_fC$$

so we will arrange this. Let $I_{R_4} = I_{R_3} = 1$ mA

$$R_4 = \frac{V_{R4}}{I_{R4}} = \frac{0.473\ V_{out}}{I_{R4}}$$

$$= \frac{0.473\ (8\ \text{V})}{1\ \text{mA}} = \frac{3.784\ \text{V}}{1\ \text{mA}} = 3.78\ \text{k}\Omega$$

$$R_3 = \frac{V_{out} - V_{R4}}{I_{R3}} = \frac{8 - 3.784\ \text{V}}{1\ \text{mA}} = 4.22\ \text{k}\Omega$$

Let $C = 0.01\ \mu$F.

$$R_f = \frac{\tau}{2C} = \frac{1\ \text{ms}}{2(0.0/\mu\text{F})} = 50\ \text{k}\Omega$$

Lastly, if $I_L = 3$ mA, V_{out}(max op-amp) = 14 V, and $I_Z = 5$ mA.

$$R_5 = \frac{V_{out}(\text{max}) = V_{out}}{I_Z + I_L + I_{R_3} + I_{R_f}}$$

$$= \frac{6\ \text{V}}{9.24\ \text{mA}} = 649\ \Omega$$

9-11.3 Triangle-Wave Generator

The student may recall from Chapter 6 that by integrating a square wave one obtains a triangular waveform. This is the easiest way to obtain a good triangular waveform. The circuit shown in Figure 9.13 uses this approach. Amplifier A_2 of Figure 9.13 is just an integrator with the noninverting input connected to a potentiometer R_s to provide a symmetry adjustment. The integrator components R_1 and C provide the timing, so that adjusting R_1 changes the frequency of operation. Since the voltage on R_1 is $\pm V_Z$ and the summing point voltage of A_2 remains at V_s (due to feedback) the current through R_1 is constant; therefore the charging current for C

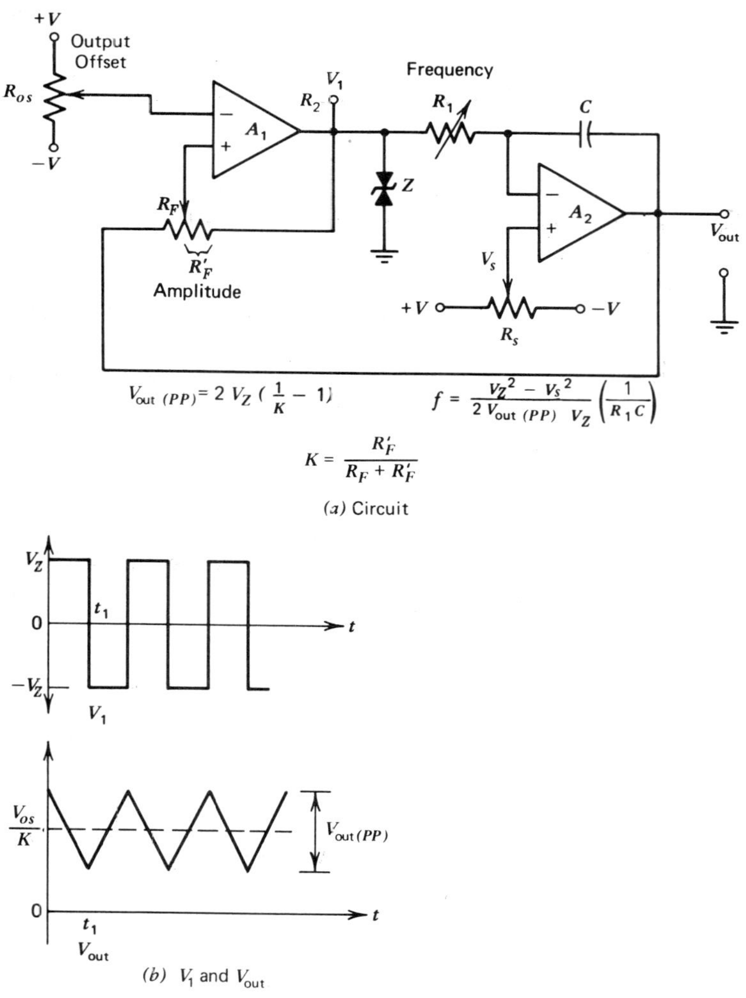

FIGURE 9.13 Traingle-wave generator.

is constant. The constant charging current of C means the output waveform is linear. The output of A_2 is fed back to the noninverting input of A_1, which is a comparator, through potentiometer R_f. R_f adjusts the feedback factor of the loop and, therefore, the amplitude of the triangular wave. The potentiometer R_{os} on the inverting input of A_1 adjusts the offset of the output waveform by acting as the reference voltage for the comparator. The symmetry of the output waveform is adjusted with R_s.

The pertinent equations are listed in Figure 9.13. We will not derive these, but we will look at one cycle of operation to obtain an intuitive understanding of the circuit operation. Assume the output of A_1, which is labeled V_1 on the drawing, is positive at $+V_Z$. The integrator capacitor C will begin to charge linearly through R_1. The voltage feedback to the noninverting input of the comparator A_1 from the output of A_2 is

$$(V_Z - V_{out})(1 - K) + V_{out}$$

When this voltage, because of the negative going output voltage from the charge on C, is equal to V_{os}, the output of A_1 will change polarity from $+V_Z$ to $-V_Z$. Now C will charge in the opposite direction causing the output ramp to be positive going. When the voltage fed back to the noninverting input of A_1 is equal to V_{os}, the output of A_1 will go from $-V_Z$ back to $+V_Z$. This will cause the output ramp to change in direction from positive to negative. Note that a square-wave output is available from A_1 as well as a triangular waveform from A_2. In addition, almost every parameter of the triangle wave is adjustable. The maximum output frequency is limited by the slew rate of A_2, or the current driving capability of A_1, whichever results in the lowest frequency.

EXAMPLE 9-5

Set the circuit of Figure 9.13 up for an output of 10 V_{pp} centered on zero volts at a frequency of 2 kHz.

SOLUTION

The equation for frequency given in Figure 9.13 is difficult to solve for component values. If we recall that C is to be charged by a constant current of V_Z/R_1 from

$$C = \frac{It}{V}$$

where

$t = \tau/2$
$V = V_{out\ pp}$
$I = V_Z/R_1$

we can choose C at 0.01 μF and calculate

$$I = \frac{CV}{t} = \frac{0.01\ \mu F\ 10\ V}{0.25\ mS} = 0.4\ mA$$

If $V_Z = 10$ V, then

$$R_1 = 10\ V/0.4\ mA = 25\ k\Omega$$

Since the voltage is to be symmetrical about zero, let R_s and R_{os} both be 100 kΩ to adjust output offset and symmetry. The inverting input of A_1 would be grounded through a 10 kΩ resistor since no output offset is desired.

The output of amplifier A_1 must change polarity when the output of $A_2 = 5$ V. When V_{out} of $A_1 = +10$ V and V_{out} of $A_2 = -5$ V, the noninverting input of A_1 must be zero volts. Thus, at the switching point of A_1, $V_{R_F} = 5$ V and $V_{R_F'} = 10$ V. Let $I_{R_F} = I_{R_F'} = 0.1$ mA. Then

$$R_F = \frac{5\ V}{0.1\ mA} = 50\ k\Omega$$

$$R_F' = \frac{10\ V}{0.1\ mA} = 100\ k\Omega$$

If $I_Z = 4$ mA and $V_{out}(max)A_1 = 14$ V, then R_2 must be

$$R_2 = \frac{V_{out}(max) - V_Z}{I_Z + I_{R1} + I_{RF}}$$

$$= \frac{14\ V - 10\ V}{4\ mA + 0.4\ mA + 0.1\ mA} = 888\ \Omega$$

9-11.4 Linear Ramp Generator

A circuit very similar to the triangular waveform generator can be used to generate a linear ramp. This circuit is shown in Figure 9.14. This circuit varies from the triangle generator in that the current for the positive going output ramp comes from $-V$ through R_1 and the current for the negative going ramp comes from the output of A_1 through R_2 and D_1. Since $R_2 << R_1$, the negative going ramp occurs much more quickly. Otherwise the circuit operation is the same. No symmetry control is needed, since the output is supposed to be asymmetrical. If the discharge of C through R_2 is not fast enough (because of A_1 output current limitations), the positive transition of A_1 after the ramp can be used to drive a transistor reset. The peak-to-peak output voltage is

$$V_{pp} = 2\ V_Z((1/K) - 1) = \frac{\left(\frac{|-V|}{R_1}\right)T}{C} \tag{9-7}$$

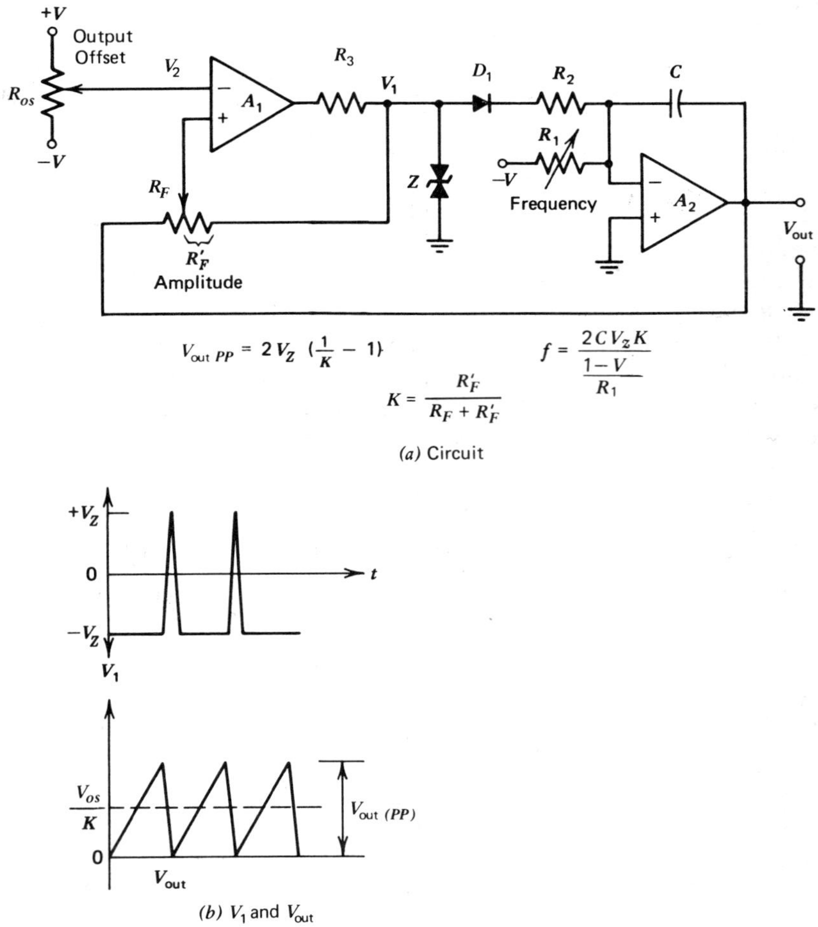

FIGURE 9.14 Linear ramp generator.

EXAMPLE 9-6

Construct a sawtooth generator with a peak output voltage of 10 V at 2 mA, a period of 10 ms, and a reset time of 0.1 ms. The circuit to be used is shown in Figure 9.14. Let $\pm V = \pm 15$ V and $V_Z = 10$ V.

SOLUTION

Once again, the component calculations are performed most easily from common-sense application of simple equations than from the use of the more complex derived equations. If the maximum current available from A_1 is 20 mA, and 4 mA is used to bias the zener with another 0.1 mA to drive R_F, then the most current that can be used to discharge C during the reset time is 15.9 mA. If I_{R_2} is chosen as 12 mA, then the value of R_2 is

$$R_2 = \frac{V_Z - V_D}{12\text{ mA}} = \frac{9.3\text{ V}}{12\text{ mA}} = 775\ \Omega$$

and

$$C = \frac{It}{V} = \frac{12\text{ mA}(0.1\text{ ms})}{10\text{ V}}$$
$$= 0.12\ \mu\text{F}$$

R_1 is chosen to charge C to 10 V in 10 ms.

$$I_{R1} = \frac{CV}{t}$$
$$= \frac{0.12\ \mu\text{F}\ (10\text{ V})}{10\text{ ms}} = 0.12\text{ mA}$$
$$R_1 = |-V|/I_{R_1} = 15\text{ V}/0.12\text{ mA}$$
$$= 125\text{ k}\Omega$$

The output polarity of A_1 must switch from negative to positive when V_{out} of A_2 = 10 V and from positive to negative when V_{out} = 0 V. To solve for V_2, R_F, and R'_F, we must use the equation

$$V_{out\ pp} = 2\ V_Z[(1/K) - 1]$$
$$\frac{1}{K} = \frac{V_{out\ pp}}{2\ V_Z} + 1$$
$$\frac{1}{K} = \frac{10\text{ V}}{20\text{ V}} + 1 = 1.5$$
$$K = 0.667$$

Since

$$K = \frac{R'_F}{R_F + R'_F}$$

$$R_f = \frac{R'_F - KR'_F}{K}$$

Let

$$R'_F = 100\text{ k}\Omega$$
$$R_F = \frac{100\text{ k}\Omega - 0.667\ (100\text{ k}\Omega)}{0.667} = 49.9\text{ k}\Omega$$

The voltage on the inverting input (V_2) must be found. It can be supplied by a voltage divider. When $V_{out} = 0$ V and $V_1 = +10$ V, the output of op-amp A_1 must switch negative. Since

$$\frac{V_2}{V_{out} + V_1} = \frac{R_F}{R_F + R_F'}$$

$$V_2 = \frac{R_F(V_{out} + V_1)}{R_F + R_F'}$$

$$= \frac{50\ \text{k}\Omega\ (10\ \text{V})}{150\ \text{k}\Omega} = 3.33\ \text{V}$$

Finally, if the maximum output voltage of A_1 is 14 V

$$R_3 = \frac{V_{out}(\text{max})\ A_1 - V_Z}{I_{R2} + I_{out} + I_{R_F'}}$$

$$= \frac{14\ \text{V} - 10\ \text{V}}{12\ \text{mA} + 2\ \text{mA} + 0.133\ \text{mA}} = 283\ \Omega$$

Note that if V_1 is an adjustable voltage, the current through R_1 will vary with V_1. If I_{R_1} varies, the time required for C to charge varies and the circuit becomes a voltage controlled oscillator (VCO).

9-11.5 Voltage-Controlled Oscillator (VCO)

We have seen in the last section that by varying V_1 in Figure 9.14 the frequency of the linear ramp generator can be controlled by the input voltage.

The circuit of Figure 9.15 operates by varying the input voltage to an integrator. The higher the input voltage, the faster the capacitor charges because of increased charging current and the higher the frequency. A_1 is an amplifier that works as a voltage follower when FETs Q_1 and Q_2 are off and as an inverting amplifier with a gain of one when Q_1 and Q_2 are on. A_1 alternately supplies $+V_{in}$ and $-V_{in}$ to the integrator, causing its capacitor to charge negative with $+V_{in}$ and positive with $-V_{in}$. A_3 is a comparator that switches Q_1 and Q_2 on when the integrator output V_B reaches

$$V_B = -V_Z\left(\frac{R_4}{R_4 + R_5}\right)$$

and switches Q_1 and Q_2 off when the output of A_2 reaches

$$V_B = V_Z\left(\frac{R_4}{R_4 + R_5}\right)$$

The circuit operates as follows. With Q_1 and Q_2 off, which means the output voltage is at $-V_Z$, V_{in} is applied to the integrator, causing the capacitor to charge

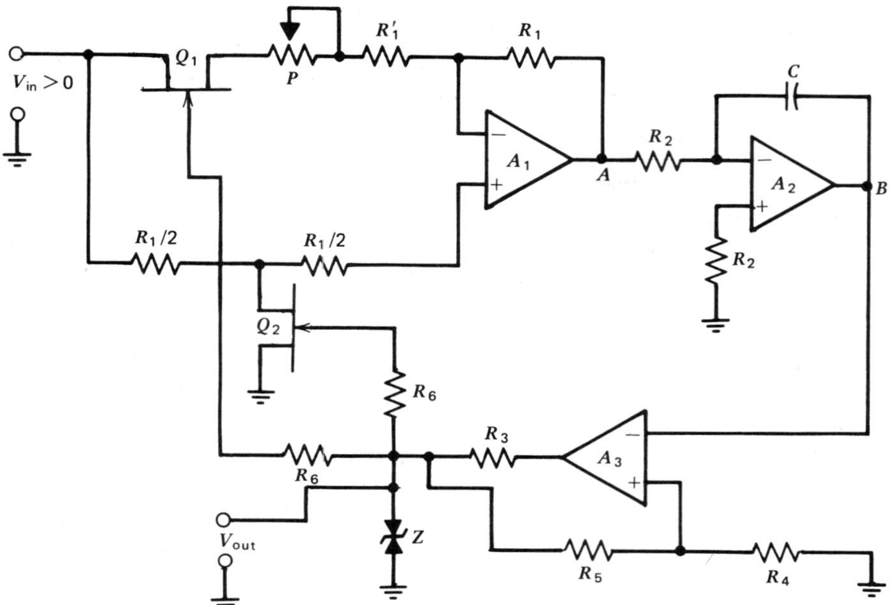

(a) Circuit

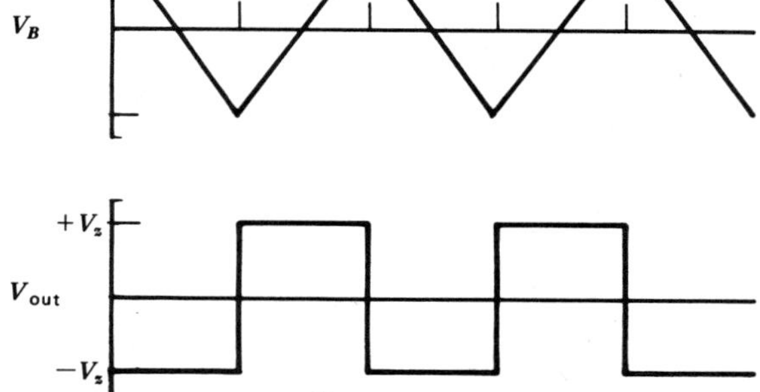

(b) V_A, V_B, and V_{out}

FIGURE 9.15 Voltage-controlled square-wave generator.

linearly negative. The integrator output continues going negative until it equals the negative voltage across R_4. The comparator (A_3) output then switches to $+V_Z$. Q_1 and Q_2 turn on, converting A_1 into an inverting amplifier with a gain of one. The integrator input is now $-V_{in}$ and its output starts going positive. The integrator output goes positive until the integrator output reaches the positive voltage across R_4 at which time the comparator (A_3) output switches negative, causing Q_1 and Q_2 to turn off. The process then repeats.

For good low frequency performance A_2 should be a FET input op-amp. For high slew rate switching, A_1 and A_2 are best FET input op-amps. V_{in} must be less than V_Z.

EXAMPLE 9-7

Calculate the components for the VCO of Figure 9.15 so that the frequency is 10 kHz with $V_{in} = 5$ V. The op-amps are to be TL 081 JFET input shown in Appendix C. Let $V_Z = 10$ V and $\pm V = \pm 15$ V. The JFETs, Q_1 and Q_2, have $V_{gs\ off} = -3$ V and are selected for $R_{ds\ Q2} = 200\ \Omega$. The output current of the VCO is to be 10 mA.

SOLUTION

The switching voltage for the comparator is selected and set by R_5 and R_4. This voltage is the peak output voltage of the integrator. We will arbitrarily set this voltage at 5 V. R_4 and R_5 will now be equal. Let $I_{R4} = I_{R_5} = 0.1$ mA.

$$R_4 = R_5 = \frac{5\ \text{V}}{0.1\ \text{mA}} = 50\ \text{k}\Omega$$

R_6 for driving the JFETs is selected to be 100 kΩ.

The total current drawn from A_3 is now known, so R_3 can be calculated. If $I_Z = 3$ mA

$$R_3 = \frac{V_{out}(\text{max})\ A_3 - V_Z}{2I_{R6} + I_{out} + I_Z + I_{R4}}$$

$$= \frac{3.5\ \text{V}}{13.3\ \text{mA}} = 263\ \Omega$$

The resistors for A_1 are almost arbitrarily chosen. Let $R_2 = 50$ kΩ, $R_1' = 49$ kΩ, $P = 25$ kΩ for adjustment of $+V_A = -V_A$. For bias current induced offset voltage cancellation, the resistors going to the noninverting input of A_1 are $R_1/2 = 25$ kΩ.

The output current available from the TL 081 is about 20 mA. If $I_{R_2} = 1$ mA at the maximum frequency of 10 kHz, we can find C from

$$C = \frac{It}{V}$$

where

$I = I_{R1}(\max)$
$t = \tau/2$
$V = V_{cpp}$

$$C = \frac{1\ \text{mA}\ 0.05\ \text{ms}}{10\ \text{V}} = 0.005\ \mu\text{F}$$

$$R_1 = \frac{5\ \text{V}}{1\ \text{mA}} = 5\ \text{k}\Omega$$

Larger charging currents would have resulted in a very small capacitor.
For curiosity, let us see what the frequency will be when $V_{in} = 1$ V. From

$$C = \frac{It}{V}$$

$$t = \frac{CV}{I}$$

where

$t = \tau/2$
$I = V_{in}/R_1$
$V = V_{cpp}$

so

$$t = \frac{0.005\ \mu\text{F}\ (10\ \text{V})}{1\ \text{V}/5\ \text{k}\Omega} = 0.25\ \text{ms}$$

$$f = \frac{1}{2t} = \frac{1}{0.5\ \text{ms}} = 2000\ \text{Hz}$$

SUMMARY

1. Operational amplifiers are useful in many applications in all areas of electronics.
2. Peak detectors are circuits that detect and remember their peak input voltage. The high input impedance of operational amplifiers, especially FET input op-amps, makes them ideal for this application.
3. Zero-crossing detectors are circuits designed to have a large change in output voltage at the instant an input signal crosses zero volts. The high voltage gain of op-amps makes them useful in this application. Zero-crossing detectors are useful in constructing many types of test equipment such as phase detectors.
4. Discrete transistors can be used to boost the output current of operational amplifiers. As long as the transistors are within the feedback loop, negligible error results from their use.

5. The high gain and ease with which feedback is applied to operational amplifiers make them ideal for use as precision current sources and voltage supplies.
6. Operational amplifiers can be used to simplify the circuitry and improve the performance of many waveform generators. Square-wave generators made of op-amps are essentially self-resetting comparators. Op-amp triangle and ramp generators are essentially integrators with a square-wave input supplied by a comparator used as a square-wave generator. If the input voltage of an op-amp triangle-wave or linear ramp generator is varied, a voltage controlled oscillator (VCO) results.

SELF-TEST QUESTIONS

9-1 State the operating principle of a peak detector.
9-2 Why should A_2 of Figure 9.1 be a FET input op-amp?
9-3 State the function of the adder-subtracter in Figure 9.2.
9-4 Explain why current boosters cause very little error in the A_{fb} of an op-amp.
9-5 What properties of the op-amp improve the stability of the reference voltage sources in Figure 9.9?
9-6 State the reason an automatic amplitude control is necessary with a Wien-bridge oscillator.
9-7 Name the essential function performed by the op-amp in the square-wave generator of Figure 9.12.
9-8 Name the functions performed by amplifiers A_1 and A_2, respectively, in the triangle-wave generator of Figure 9.13.
9-9 Why must A_1 of Figure 9.14 have a high slew rate?
9-10 If V_1 is varied, the frequency of the linear ramp circuit of Figure 9.14 varies. Why?
9-11 State the function of each amplifier of Figure 9.15.

If you cannot answer certain questions, place a check next to them and review appropriate parts of the text to find the answers.

LABORATORY EXERCISE

Objective

After completing this laboratory, the student should be able to construct and operate three of the circuits listed in the laboratory, and report their operation.

Equipment

The equipment needed for this laboratory will vary due to the nature of the laboratory assignment. Most of the equipment listed here will definitely be needed.

1. Three Texas Instruments TL081, or equivalent, operational amplifiers.
2. $\pm 2\%$ resistor assortment.
3. Power supply, ± 15 V dc.
4. 10 kΩ and several 50 kΩ potentiometers.
5. Two JFET *n* channel transistors.
6. Two 4.7 V zeners.
7. Oscilloscope.
8. Breadboard with IC sockets such as EL Instruments SK-10.
9. Audio signal generator.

Procedure.

1. Peak detector.
 (a) Construct a peak detector as shown in Figure 9.16.
 (b) Insert an audio signal and slowly adjust the input voltage while monitoring the input and output voltage on an oscilloscope.
 (c) Set the input voltage to zero volts and observe how long the output holds the peak voltage.

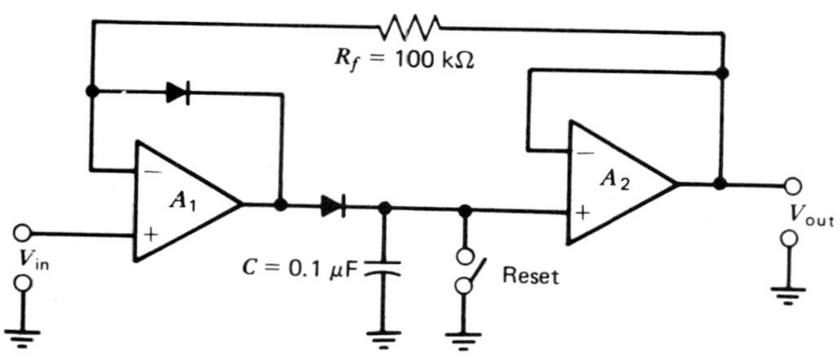

FIGURE 9.16 Peak detector.

2. Wien-bridge oscillator.
 (a) Calculate the components for the Wien-bridge oscillator shown in Figure 9.17. Let $C = 0.1\ \mu F$ and $f_0 = 2$ kHz.
 (b) Adjust P until oscillation begins. Measure the frequency of oscillation and compare the measured and calculated value.
 (c) Observe the waveform at point A. Compare it to the waveform on the op-amp output.

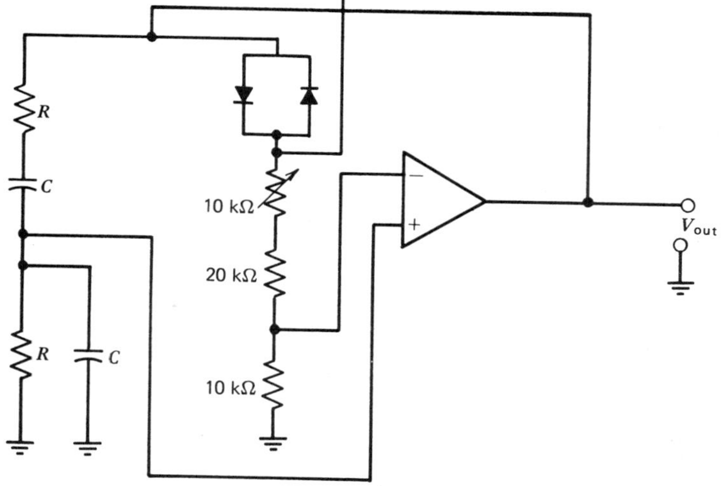

FIGURE 9.17 Wien-bridge oscillator.

3. Square-wave generator.
   (a) Construct the square-wave generator shown in Figure 9.18. Set the frequency of oscillation at 1 kHz.
   (b) Observe the output and compare the actual frequency of oscillation with the calculated value.
   (c) Connect an integrator to the output of the square-wave generator as shown in Figure 9.19. Calculate R_1 and C so the triangle wave will be 5 V_p, 10 V_{pp}.

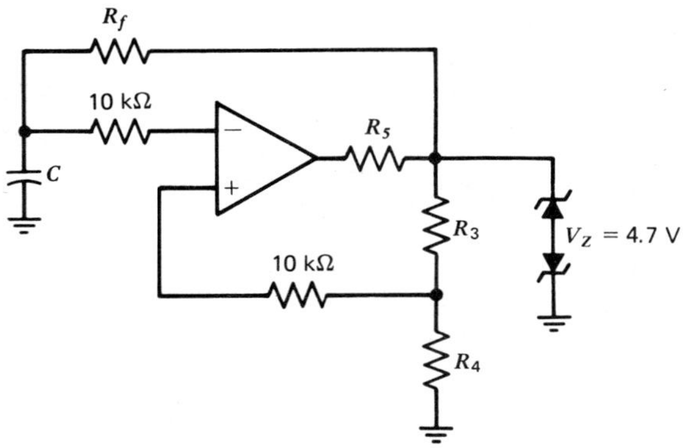

Let $\dfrac{R_4}{R_4 + R_5} = 0.473$

$$f = \frac{1}{2R_fC}$$

FIGURE 9.18 Square-wave generator.

4. Optional section.
   (a) Calculate the components for and construct the VCO of Figure 9.15 for an oscillation frequency of 1 kHz with $V_{in} = 5$ V.
   (b) Plot the oscillator frequency versus the input voltage as V_{in} varies in 0.5 V steps from 0.5 V to 5.5 V.

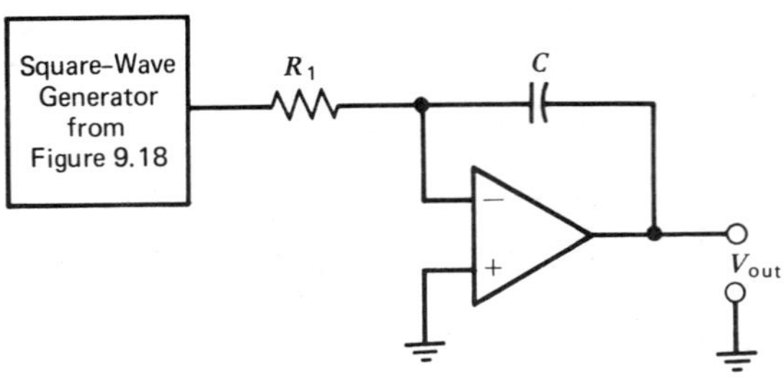

$$C = \frac{I_{R1}t}{V_{pp}} \qquad R_1 = \frac{V_Z + 0.7\text{ V}}{I_{R1}}$$

FIGURE 9.19 Triangle-wave generator.

CHAPTER

10

VOLTAGE REGULATOR INTEGRATED CIRCUITS

Operational amplifiers may be thought of as general purpose ICs with a multitude of uses in a large variety of analog circuits. In applications where a specific set of op-amp characteristics must be emphasized, special purpose operational amplifiers have been developed if the market seemed sufficient. One could fill a book discussing all the special purpose integrated circuits now available. We will discuss linear and switching IC power supply regulators in this chapter. If a person becomes an enthusiastic browser of linear IC catalogs, that person will become familiar with the available special purpose ICs.

OBJECTIVES

After completing the study of this chapter and the self-test questions, the student should be able to:

1. Calculate the component values for and state the operating principle of an op-amp linear voltage regulator.
2. State the operating principle and calculate the auxiliary component values of the included IC voltage regulators.
3. State the operating principle and calculate the component values for the IC switching regulators included in this chapter.
4. Perform the laboratory for Chapter 10.

10-1 VOLTAGE REGULATORS

10-1.1 Types of Regulators

A voltage source that remains constant with variations of input voltage and load current is essential for almost all modern electronic circuitry. Power supply variations can cause circuit output variations indistinguishable from output variations caused by the proper circuit input, and excessive voltage can destroy the entire circuit. Well-regulated dc power supplies are then essential to the proper performance of both linear and digital circuits.

In recent years, integrated circuit voltage regulators that require only a few external components have become widely available. These IC regulators offer excellent performance at a very reasonable price. As a result, each printed circuit board in a system can have its own voltage regulator or regulated power supply, the printed circuit board being supplied only unregulated dc or ac voltage. The system reliability is enhanced since the failure of a single voltage regulator may not totally disable the entire system. The problem of long wires distributing regulated dc picking up noise and transient voltages is reduced by locating the voltage regulators near the circuitry using the regulated voltage.

Two major types of voltage regulators are available in IC form, linear and switching. Linear regulators use a *series-pass* or *shunt* circuit element to perform the regulation. Referring to Figure 10.1*a*, a block diagram of a series-pass regulator, the series-pass element (usually a transistor) will drop more or less voltage as needed to hold the output voltage constant. A shunt regulator, shown in block diagram

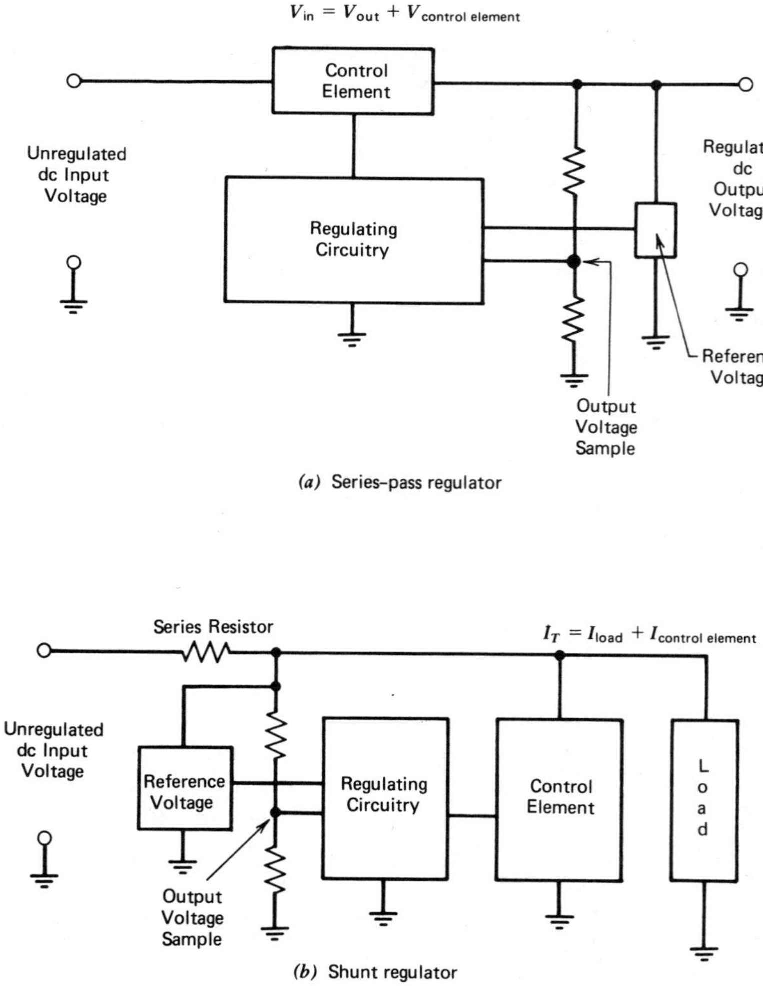

FIGURE 10.1 Linear voltage regulator types.

form in Figure 10.1*b*, uses a control element, usually a transistor, in parallel with the load. The power supply current is held constant. If the load voltage drops, the shunt element conducts less, forcing more current through the load. The increased load current causes the load voltage to increase to its original value. The constant current drain of the shunt regulator causes it to be very inefficient with low current loads, so it is seldom used today. The details of shunt regulators will not be covered in this chapter.

The second major type of voltage regulator is the *switching regulator*. The regulating element (a transistor or power MOSFET) is operated either full on or full off and dissipates appreciable power only during switching. A switching regulator block diagram is shown in Figure 10.2*a*. If the switching times are short compared

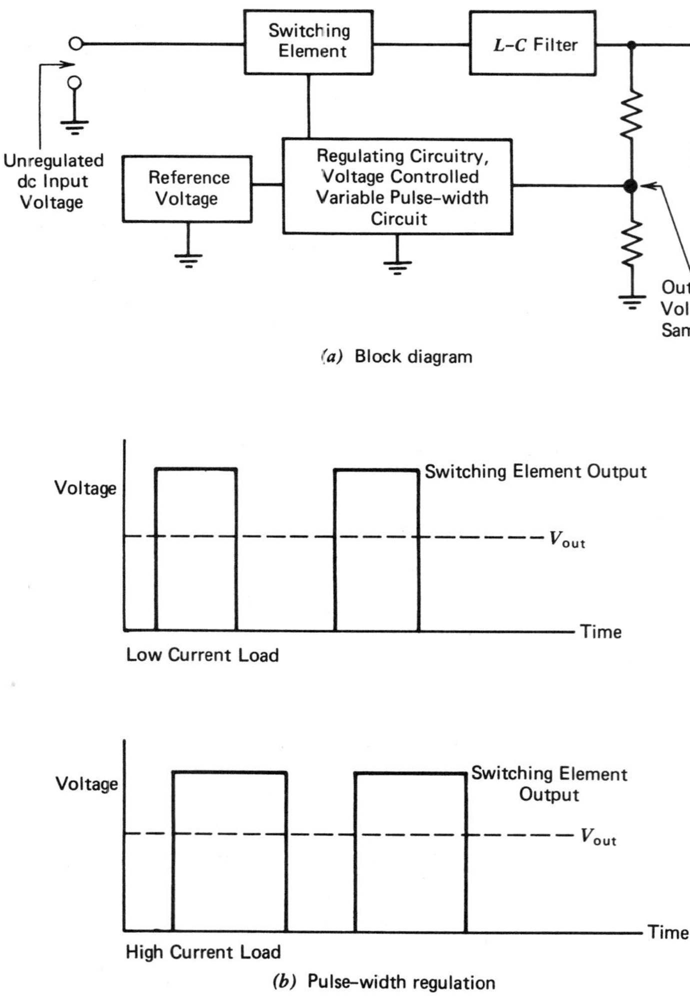

FIGURE 10.2 Switching power regulator.

to the on and off times of the transistor, very little input power is used by the control element. Thus switching power supplies are much more efficient than linear power supplies. The efficiency of switching power supplies is often greater than 90%, whereas series-pass power supplies seldom have efficiencies greater than 70 to 80%, and often much less because of the power consumed by the series-pass regulating transistor. The regulation in switching power supplies is accomplished by varying the on and off times of the control element. The longer the control element is on, the higher the output voltage. This is called *pulse-width modulation* and is illustrated in Figure 10.2*b*. Inductors and capacitors are used to filter the output voltage. If the frequency is high enough, small inductors and capacitors can be used. Most switching regulators operate between 5 and 100 kHz, so the filter elements are quite small and lightweight. Switching regulators are in general smaller, lighter, more complicated, and more expensive than linear regulators, but linear regulators offer somewhat better regulation and faster transient response.

There are many variations of switching regulators. We will cover only a few basic types in this chapter.

10-1.2 Series-Pass Linear Voltage Regulators

The basic series-pass regulator is shown in Figure 10.3*a*. It is a closed loop feedback circuit. Q_1 is the series-pass regulating transistor. Q_2 is the current amplifier for Q_1. Since most series-pass power transistors have fairly low β's, in the order of 20 to 30, Q_1 and Q_2 are connected as a Darlington pair. The output of Q_1 to the load is taken from the emitter, so Q_1 and Q_2 provide current gain, but no voltage gain. R_1 and R_2 form a voltage divider from which a sample voltage (V_S) is developed. The sample voltage is compared to the reference voltage (V_R), developed across zener diode Z_1, by the operational amplifier. The input voltage to the operational amplifier is the difference between the reference voltage and the sample voltage,

$$V_{in} = V_R - V_S = V_Z - V_{out}\left(\frac{R_2}{R_1 + R_2}\right) \tag{10-1}$$

The op-amp input, $V_R - V_S$, is amplified by the amplifier and fed to the series-pass transistor to correct for any change in output voltage, whether from V_{in} or the load changing.

The operation of the series-pass regulator is as follows: If V_{out} decreases, either from V_{in} dropping or increased load current demand, V_S decreases and V_R remains constant. $V_R - V_S$ increases, with the voltage at the inverting terminal of the op-amp going negative with respect to V_R. The output voltage of the op-amp then is driven positive, causing the emitter voltages of Q_1 and Q_2 to go positive, until $V_S \cong V_R$ again. This will occur when V_{out} is at the voltage it was before the change in load or input. The op-amp output causes Q_1 and Q_2 to turn on harder, increasing the load current to make up for V_{out} dropping. Thus the collector to emitter voltage (V_{CE}) of Q_1 drops to adjust for the decrease in output voltage. The opposite occurs if V_{in} increases or I_L decreases.

Since V_{CE} of Q_1 is equal to $V_{in} - V_{out}$, and all of the load current flows through Q_1, the efficiency of the circuit is directly related to the magnitude of V_{CE} of Q_1.

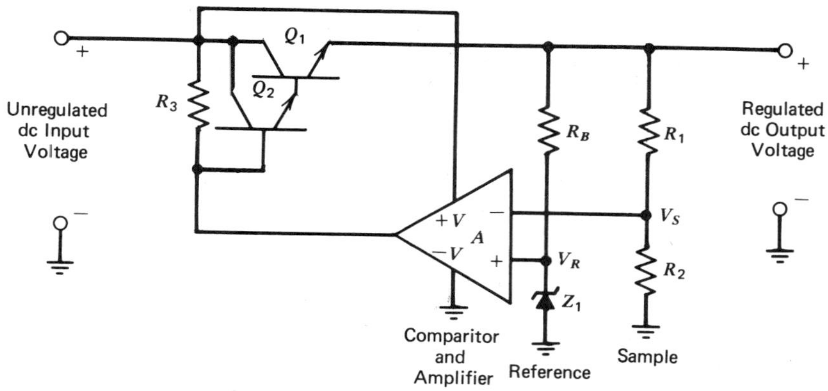

(a) Basic regulator using an op-amp

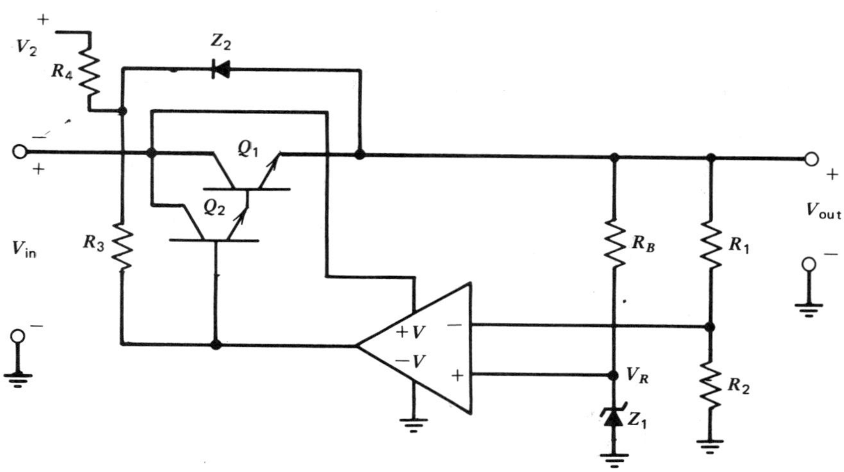

(b) Basic regulator improved with preregulation

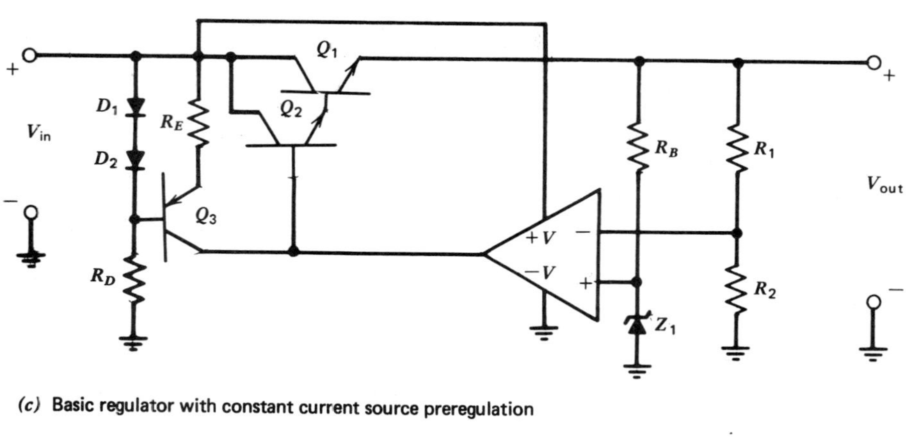

(c) Basic regulator with constant current source preregulation

FIGURE 10.3 Series-pass regulator.

If a series-pass regulator is providing 15 V regulated from an unregulated input of 30 V, the maximum efficiency $(P_{out}/P_{in})(100) = 50\%$. For good efficiency $V_{in} - V_{out}$ should be kept as small as possible, but there is a limit to how small $V_{in} - V_{out}$ can be. $V_{in} - V_{out}$ must be greater than $V_{BEQ_1} + V_{BEQ_2} + V_{CEQ_1}$ (min) for linear operation. A $V_{CE} \cong 2$ V is enough to provide linear action for most power transistors. Thus, $V_{in} - V_{out} \geqslant 3.5$ V for safe, linear operation of Q_1 and Q_2. The unregulated input voltage will normally have some ripple, so V_{in}(min) including ripple must be about 3.5 V greater than V_{out}.

Many types of op-amps can be used in this type of voltage regulator including a 741 or equivalent. It is powered from a single supply by using the unregulated output of the supply as $+V$ and ground for $-V$. The op-amp output must be able to reach the regulated output voltage plus the emitter-base voltage drop of Q_1 and Q_2. If the op-amp has a ground connection, a voltage divider or zener diode regulator will be used from the regulated output to set the ground terminal voltage of the op-amp at a voltage about $V_{out}/2$. V_{out} is limited to the op-amp maximum supply voltage. If R_1 in Figure 10.3*a* is a potentiometer, the basic supply can be adjusted from $V_{in} - 3.5$ V to a voltage just above V_{Z_1}. As drawn, the supply output cannot be adjusted to less than V_{Z_1} since V_R would always be greater than V_S.

If the voltage gain of the Darlington pair, Q_1 and Q_2, is assumed to be one, without R_1 and R_2 connected the open loop gain of the amplifier consisting of the op-amp A, Q_1 and Q_2, is A_{ol} of the op-amp. If V_{Z_1} is the amplifier input and R_1 and R_2 the feedback resistors, the output voltage is $A_{fb} \times V_{Z_1}$. In equation form

$$V_{out} = V_{Z1}\left(\frac{A_{ol}}{1 + A_{ol}\beta}\right)$$

but

$$\beta = \frac{R_2}{R_1 + R_2} \quad \text{and} \quad A_{ol} >> \frac{R_1 + R_2}{R_2}, \text{ so} \tag{10-2}$$

$$V_{out} \cong V_{Z1}\left(\frac{R_2 + R_1}{R_2}\right)$$

Thus when V_{in} is applied V_{out} will drive until $V_S = V_R$, as though the entire regulation circuit were a noninverting op-amp amplifier.

$\Delta V_{out}/\Delta V_{in}$ and $\Delta V_{out}/\Delta I_{out}$

The equation for the change in regulator output voltage for a change in input voltage is

$$\frac{\Delta V_{out}}{\Delta V_{in}} = \frac{R_L}{R_C}\left(\frac{R_2 + R_1}{A\, R_2}\right) \tag{10-3}$$

where

R_L	= Load resistance
R_C	= The ac collector resistance of the series-pass transistor
A	= op-amp voltage gain
R_1 and R_2	= the resistors in the voltage divider that provide V_S

If a Darlington pair is used as the series-pass element, R_c of Q_1 is increased to

$$R_{CQ1}\text{eff} = R_{CQ1}\beta_{Q2}$$

where β_{Q_2} is the common emitter-current gain of the Darlington driver.

The equation for the change in output voltage for a change in output current is

$$\frac{\Delta V_{out}}{\Delta I_{out}} = \frac{R_L}{A}\left(\frac{R_1 + R_2}{R_2}\right) \qquad (10\text{-}4)$$

Equations 10-3 and 10-4 are derived in Appendix F.

EXAMPLE 10-1

Before going further, an example of the circuit of Figure 10.3*a* is in order.

Construct a power supply using a 741 op-amp that has $V_{out} = 10$ V, $I_{out} = 1$ A, $V_{in}(\text{min}) = 14$ V, and $V_{in}(\text{max}) = 20$ V. Let $\beta_{Q_1} = 20$, $R_{cQ_1} = 120\ \Omega$, $V_{BEQ_1} = 0.85$ V, $\beta_{Q_2} = 50$, $R_{cQ_2} = 1\ \text{k}\Omega$, $V_{BEQ_2} = 0.7$ V.

SOLUTION

The zener diode, Z_1, is chosen to be a 5.6 V zener because the temperature drift of a zener is lower in the 5 to 8 V range. From the specification sheet, the zener has minimum temperature drift at 10 mA, so that is the bias current we will use in calculating R_B. The current through R_1 and R_2 must be much greater than the maximum bias current of the 741 op-amp so $I_{R_1} >> 0.5\ \mu\text{A}$. The values of R_1 and R_2 must be small enough so that they do not develop a significant amount of noise voltage. If $I_{R_1} = 0.1$ mA, this condition is met.

R_3 must provide adequate base current for Q_2 at the minimum input voltage.

We are now ready to calculate.

$$I_{BQ1} = I_{out}/\beta_{Q1} = 1\text{ A}/20 = 50\text{ mA}$$

$$I_{BQ2} = I_{BQ1}/\beta_{Q2} = 50\text{ mA}/50 = 1\text{ mA}$$

This value can easily be driven by a 741 op-amp.

$$R_3 = \frac{V_{in}(\text{min}) - (V_{out} + V_{BEQ1} + V_{BEQ2})}{I_{BQ2} + I_{out}(\text{op-amp})}$$

I_{out} of the op-amp is selected to be a value less than its maximum output current (about 10 mA for a 741) and above to I_{BQ_2} so that V_{out} of the op-amp is a voltage source to Q_2. Select I_{out}(op-amp) = 3 mA, which is greater than I_{BQ_2}.

$$R_3 = \frac{14\ V - (10\ V + 0.85\ V + 0.7\ V)}{3\ mA}$$

$$R_3 = 816\ \Omega$$

Note. $V_{in}(min) - V_{out} > 3.5$ V needed for operation. Also V_{in}(max) must be less than V_{max} of the op-amp. $20\ V < |+V| + |-V| = 36$ V.

We now check to make sure the 741 can supply the maximum current needed for R_3.

$$I_{max} = \frac{V_{in}(max) - (V_{out} + V_{BEQ1} + V_{BEQ2})}{R_3} - I_{BQ2}$$

$$= \frac{8.45\ V}{816\ \Omega} - 1\ mA = 10.3\ mA - 1\ mA = 9.3\ mA$$

This value of I_{max} is okay. Otherwise, additional current amplification would be needed to drive Q_2.

R_B provides bias current to the zener and is found from

$$R_B = \frac{V_{out} - V_Z}{I_B} = \frac{10\ V - 5.6\ V}{10\ mA} = 440\ \Omega$$

R_1 and R_2 provide the sampling voltage to the inverting input. Let $I_{R_1} \cong I_{R_2} = 0.1$ mA because the bias current of the 741 is much less than the 0.1 mA chosen.

$$R_2 = V_Z/I_{R_2} = 5.6\ V/0.1\ mA = 56\ k\Omega$$

and

$$R_1 = \frac{V_{out} - V_{Z1}}{I_{R1}} = \frac{4.4\ V}{0.1\ mA} = 44\ k\Omega$$

We now check the regulation.

$$\Delta V_{in} = V_{in}(max) - V_{in}(min) = 20\ V - 14\ V = 6\ V$$

from Equation 10-3

$$\Delta V_{out} = \frac{R_L \Delta V_{in}}{R_c(eff)}\left(\frac{R_2 + R_1}{A\ R_2}\right) = \frac{\Delta V_{in}[V_{out}/I_{out}(max)](R_2 + R_1)}{R_{CQ1}\beta_{Q2} A\ R_2}$$

where $V_{out}/I_{out}(max) = R_L(min)$ and A = 20,000, the minimum open loop gain of a μA 741.

$$\Delta V_{out} = \frac{10\ \Omega\ 100\ k\Omega\ (6\ V)}{120\ \Omega(50)(20{,}000)\ 56\ k\Omega}$$

$$= 0.89\ \mu V \qquad \text{from } \Delta V_{in}$$

ΔV_{out} from ΔI_L is found from Equation 10-4.

$$\Delta V_{out} = \Delta I_L \frac{R_L}{A}\left(\frac{R_1 + R_2}{R_2}\right) = 0.89\ mV$$

for a load current change of 1 A.

The excellent performance of the example circuit is due to the very high open loop gain of the op-amp. The actual performance of the circuit will vary with offset voltage drift and reference voltage drift as temperature varies.

The circuit of Figure 10.3*b* offers improved performance compared to the circuit of Figure 10.3*a*. The input voltage change causes the base current of Q_2 to vary slightly because R_3 is connected to the unregulated input. This was ignored in our previous regulation calculations. Connecting R_3 to a voltage source with no ripple will reduce the ripple on the output. A zener diode regulator powered by an auxiliary secondary winding with rectifiers and a capacitive filter (V_2) provides a stabilized voltage source. The additional voltage is necessary because the voltage supplying R_3 must be greater than $V_{out} + V_{BEQ1} + V_{BEQ2}$. R_4 is found as follows:

$$R_4 = \frac{V_2 + V_{in} - (V_{Z2} + V_{out})}{I_{R3}(max) + I_B Z_2}$$

where

$I_{BZ2} \cong 0.5\ I_{R3}(max)$.

A constant current source supply for I_{BQ_2} can replace the auxiliary supply V_2 as shown in Figure 10.3*c*. The constant current source causes the base current of Q_2 to remain constant with input ripple. D_1 and D_2 provide a constant voltage to the base of Q_3. A low voltage zener would be acceptable in place of D_1 and D_2. R_D sets the current through the diodes at a level much greater than the base current of Q_3. Setting

$$R_D = \frac{V_{in}(min) - (V_{D1} + V_{D2})}{20\ I_{BQ3}}$$

will work. The emitter resistor of Q_3 is found by the following equation:

$$R_E = \frac{(V_{D1} + V_{D2}) - V_{BEQ3}}{I_{out}/\beta_{Q1}\beta_{Q2}}$$

where β_{Q1} and β_{Q2} are the minimum specification sheet values. At all times V_{CQ3} must be greater than $V_{out} + V_{BEQ1} + V_{BEQ2}$, and V_{CEQ3} must be greater than about 1.5 V to keep Q_3 from saturating. The constant current source and the V_2 circuit are called preregulators.

10-1.3 Constant Current Limiting

If the output of a power supply is shorted, the excessive current can destroy the power supply regulator because of the excessive heat generated in the series-pass transistor. Two types of current limiting circuits are used to prevent this, *constant current limiters* and *foldback current limiters*. The constant current limiter is shown in Figure 10.4*a*. As I_{out} exceeds the maximum current of the power supply, the voltage drop across R_{cl} causes V_{BE} of Q_{cl} to increase. Q_{cl} then turns on, forcing the

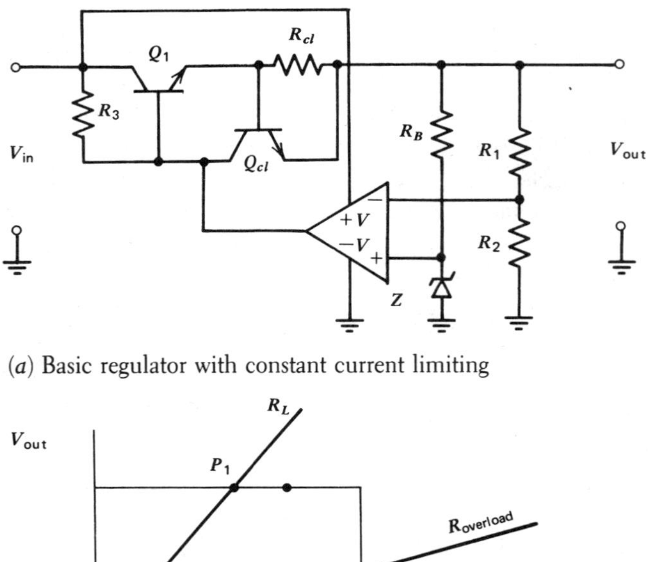

(*a*) Basic regulator with constant current limiting

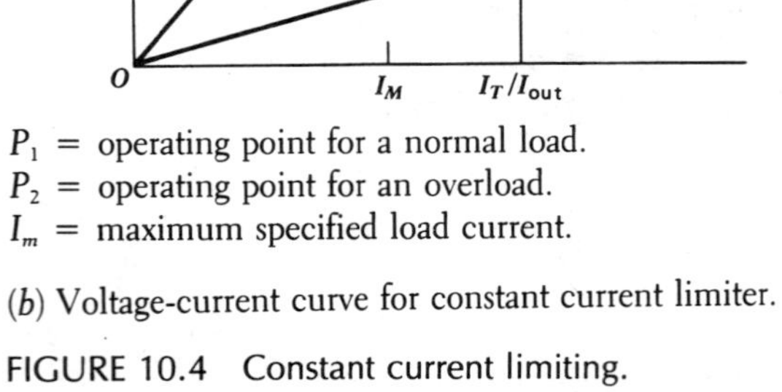

P_1 = operating point for a normal load.
P_2 = operating point for an overload.
I_m = maximum specified load current.

(*b*) Voltage-current curve for constant current limiter.

FIGURE 10.4 Constant current limiting.

voltage across R_3 to rise, thus V_{BQ1} and V_{out} drop. The collector current of Q_{cl} through R_3 decreases the base current of emitter-follower Q_1, causing the V_{CE} of Q_1 to increase. If the output is shorted, $I_{out}R_{cl}$ causes Q_{cl} to saturate, dropping V_{out} to zero. R_{cl} is selected as follows:

$$R_{cl} = \frac{V_{BE\ off\ (max)\ Q_{cl}}}{I_m} \tag{10-5}$$

where

I_m = maximum specified load current
V_{off} (max) $_{Q_{cl}}$ = maximum V_{BE} at which Q_{cl} is off. Usually 0.3 to 0.4 V.

The output voltage will not begin to drop appreciably until I_m, where $V_{R_{cl}}$ is around 0.5 V. The series-pass transistor must be able to dissipate a power of

$$P_{Q1} = V_{in}(\max)\left(\frac{V_{BE_{on}\ Q_{cl}}}{V_{BE_{off}\ Q_{cl}}}\right)I_m \tag{10-6}$$

which for $V_{BE_{off}} = 0.35$ V and $V_{BE_{on}} = 0.6$ V is

$$P_{Q1} = V_{in}(\max)\left(\frac{0.6\text{ V}}{0.35\text{ V}}\right)I_m = 1.7\ V_{in}(\max)I_m$$

The maximum collector current of Q_{cl} is $V_{in}(\max)/R_3$, and it must be able to handle a power dissipation of

$$\begin{aligned} P_{Q_{cl}} &= [0.5\ V_{in}(\max)][0.5\ V_{in}(\max)/R_3] \\ &= (0.25)[V_{in}(\max)]^2/R_3 \end{aligned}$$

To reduce the power dissipation in Q_1, foldback current limiting is used.

10-1.4 Foldback Current Limiting

Foldback current limiting reduces the maximum power dissipation of Q_1 by reducing the short circuit current of the power supply to a value less than I_m, as shown in Figure 10.5*a*. As I_{out} rises above I_m, the current in Q_1 is reduced to a value less than I_m. Note that current limiting action does not start until $I_{out} > I_m$ by about 40% in most power supplies so that maximum specified load current will not start current limiting action, but a genuine overload where $I_{out} > I_T$, the specified current limit, will.

The operation of the foldback current limit circuit Q_{cl}, R_{cl}, R_A, and R_B, shown in Figure 10.5*b*, is as follows. V_{BE} of Q_{cl} is equal to $V_{R_{cl}} + V_{R_A}$. R_A is less than R_B so $V_{R_A} < V_{R_B}$. In normal operation, $V_{R_A} + V_{R_{cl}}$ is less than the voltage required to turn on Q_{cl}. As I_{out} increases to the limiting current I_T, $V_{R_{cl}}$ becomes large enough to forward bias Q_{cl}, turning Q_{cl} on. As the output voltage drops, V_{R_A} begins to rise

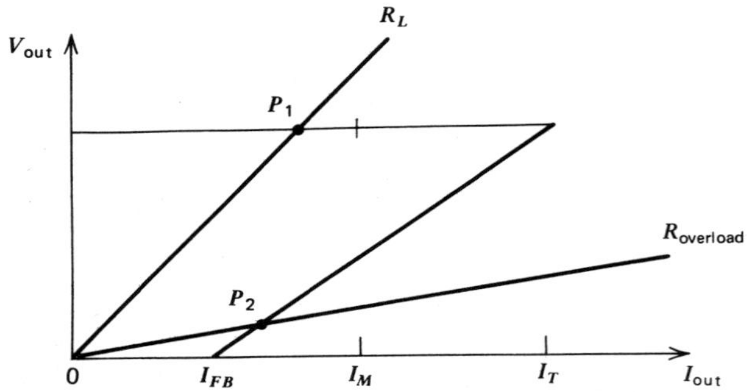

I_{FB} = Foldback current.
I_M = Maximum specified load current.
I_T = Limiting current.

(a) Foldback V_{out} versus I_{out} characteristic

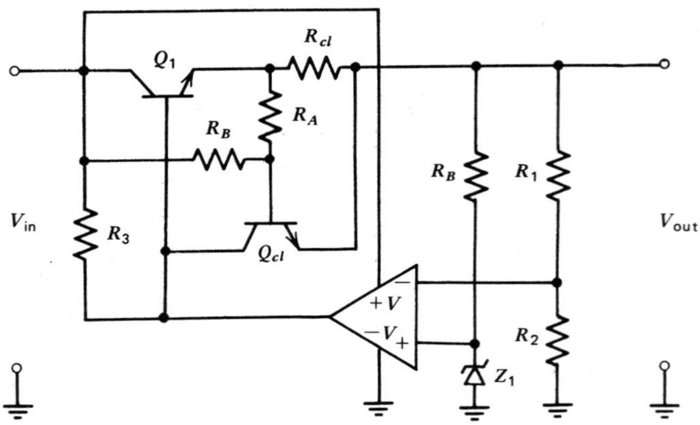

(b) Foldback limiting circuit

FIGURE 10.5 Foldback current limiting.

because of the increase in the current through R_B. V_{R_A} is of the same polarity as $V_{R_{cl}}$ and aids in holding Q_{cl} on. Since

$$V_{BEQ_{cl}} = V_{R_A} + V_{R_{cl}}$$

less current is needed through R_{cl} to hold Q_{cl} on as V_{R_A} increases. I_{out} is then decreased as V_{R_A} increases.

The foldback current is usually set at about 20% of I_m. The reason it is not set at less is because a nonlinear load whose resistance is less at turn on than at full-load current can cause the power supply to latch in a low voltage state (P_2 in Figure 10.5*a*). An example of such a load is an incandescent light bulb, which may draw 10 times as much current cold as when hot. Another example is a circuit with many power supply bypass capacitors at IC power supply connections. Until charged, these capacitors will draw a great deal of current, even though only for a short time. The low effective resistance of the capacitors can latch the power supply in a low voltage state because the supply immediately begins current limiting. Setting $I_{FB} = 20\%, I_m$ prevents most latch-up situations with normal loads.

To calculate R_{cl}, R_A, and R_B, first calculate R_{cl} to initiate current limiting at I_m, as shown in Equation 10-5. R_A and R_B are chosen for a short circuit condition. If $V_{out} = 0$, then

$$V_{BEQcl} = V_{RA} + V_{Rcl} = 0.65 \text{ V}$$

for Q_{cl} to be fully on. With $R_L = 0$, we want

$$V_{Rcl} = 0.2\, I_m R_{cl}$$

thus

$$V_{RA} = V_{BEQcl} - 0.2\, I_m R_{cl}$$

I_{RA} must be set to provide the V_{BE} for Q_{cl} not provided by R_{cl} in the short circuit condition.

Set $I_{RA} \cong 20\, I_{BQcl}$, where

$$I_{BQcl} = \left[\frac{V_{in}(\text{max})}{R_3}\right] \Big/ \beta_{\text{min } Qcl}$$

now

$$R_A = \frac{V_{RA}}{I_{RA}} = \frac{V_{BEQcl} - 0.2\, I_m R_{cl}}{I_{RA}} \qquad (10\text{-}7)$$

and since $I_{RB} \cong I_{RA}$

$$R_B = \frac{V_{in}(\text{min}) - V_{BE \text{ on } Qcl}}{I_{RB}} \qquad (10\text{-}8)$$

The short circuit power dissipation of Q_1 is now reduced to

$$P_{Q1} = 0.2\, I_m V_{in}(\text{max}) \qquad (10\text{-}9)$$

If Q_1 is replaced with a Darlington pair, the current limit transistor collector is connected to the base of Q_2.

EXAMPLE 10-2

Calculate R_{cl}, R_A, and R_B for foldback limiting for the power supply of Example 10-1 (shown in Figure 10.6).

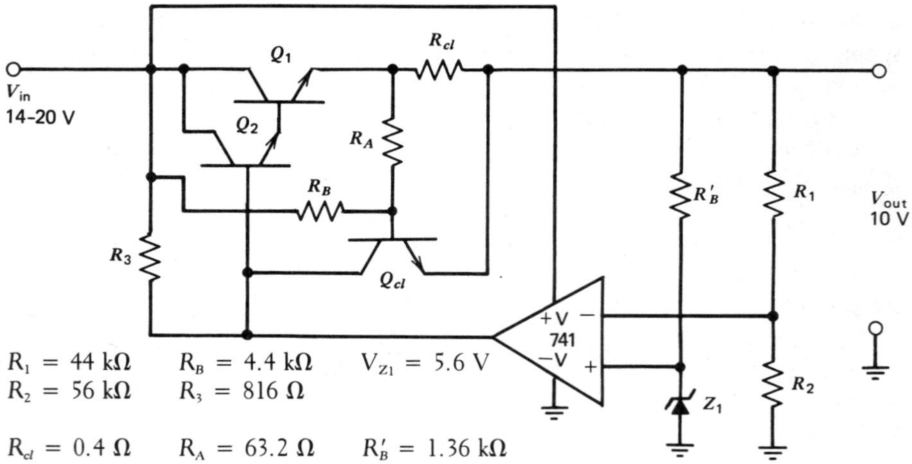

FIGURE 10.6 Power supply of Example 10-1 with foldback current limiting.

SOLUTION

Let $\beta_{\min\ Q_{cl}} = 50$.

$$V_{BE\ \text{off}} = 0.4\ V$$
$$V_{BE\ \text{on}} = 0.7\ V$$
$$I_{FB} = 0.2\ I_m$$

The $V_{BE\ \text{on}}$ and $V_{BE\ \text{off}}$ of Q_{cl} should be measured for the transistor type used.

From Example 10-1,

$$V_{in} = 14 \text{ to } 20\ V$$
$$V_{out} = 10\ V$$
$$I_m = 1\ A$$
$$R_3 = 816\ \Omega$$

First we calculate R_{cl}

$$R_{cl} = \frac{V_{BE\ \text{off}\ Q_{cl}}}{I_m} = \frac{0.4\ V}{1\ A} = 0.4\ \Omega$$

Note. If this were a constant current limit circuit, we would be finished. Continuing:

$$I_{BQcl} = \left[\frac{V_{in}(\max)}{R_3}\right] \Big/ \beta_{Qcl}$$
$$= \left(\frac{20\ V}{816\ \Omega}\right) \Big/ 50 = \frac{24.5\ mA}{50} = 0.49\ mA$$

Let $I_{R_A} = 20\ I_{BQ_{cl}} = 20(0.49\ \text{mA}) = 9.8\ \text{mA}$
From Equation 10-7

$$R_A = \frac{V_{BEQ_{cl}} - 0.2\ I_m R_{cl}}{I_{RA}}$$

$$= \frac{0.7\ \text{V} - 0.2(1\ \text{A})(0.4\ \Omega)}{9.8\ \text{mA}}$$

$$= \frac{0.62\ \text{V}}{9.8\ \text{mA}} = 63.2\ \Omega$$

From Equation 10-8

$$R_B = \frac{V_{in}(\text{min}) - V_{BEQ_{cl}}}{I_{RB}}$$

$$= \frac{14\ \text{V} - 0.7\ \text{V}}{9.8\ \text{mA}} = 1.36\ \text{k}\Omega$$

We now check to see that V_{R_A} is negligible at I_m with V_{in} at its maximum.

$$V_{RA} \text{ at } V_{in}(\text{max}) = [V_{in}(\text{max}) - V_{out}]\left(\frac{R_A}{R_A + R_B}\right) - V_{Rcl}$$

$$= 10\ \text{V}\ \frac{63.2\ \Omega}{1.36\ \text{k}\Omega + 63.2\ \Omega} - 1\ \text{A}\ (0.4\Omega)$$

$$= 0.44\ \text{V} - 0.4\ \text{V} = 0.04\ \text{V}$$

V_{R_A} is negligible at $V_{in}(\text{max})$.

Often R_A is a potentiometer, so I_{FB} can be adjusted to the desired value.

10-2 IC LINEAR VOLTAGE REGULATORS

All of the IC linear voltage regulators we will consider operate in the same way as the op-amp regulator of the preceding section. The circuit details vary from one manufacturer to another and from one IC to another, but all must have the same elements. These common elements are:

1. A reference voltage that is stable with variations in input voltage, temperature, and time. The voltage reference stability is the key to the regulator's performance.
2. A comparator to compare a sample of the output voltage with the reference voltage. The comparator is usually a differential amplifier with a large voltage gain.
3. A series-pass transistor or Darlington pair.

The voltage divider used to sample the output voltage is supplied by the IC user except in fixed output voltage regulators.

New, more versatile voltage regulators capable of higher voltages (2 to 50 V) and current (more than 10 A in some hybrid units and to 5 A in some monolithic units) are being introduced at a rapid rate. Most adjustable IC linear regulators can provide around 100 mA at up to about 50 V if $V_{in} - V_{out}$ is kept at its minimum value, so the regulator power dissipation is less than the 500 to 800 mW typically specified. We will consider three linear IC voltage regulators whose performance is typical: the National Semiconductor LM105 and LM309 and the Fairchild μA723.

10-2.1 LM105/LM205/LM305

The National Semiconductor LM105 is sold in three temperature ranges; each range has a different device number. The LM105 operates from −55 to 125°C, the LM205 from −25 to 85°C, and the LM305 from 0 to 70°C. The LM105 and LM205 specifications are identical over their respective temperature ranges, but some specifications of the LM305 are reduced. We will refer to the devices as the LM105, identifying LM305 specification variances when applicable.

The LM105 is a variable output voltage series-pass linear regulator. V_{out} is adjustable from 4.5 to 40 V (LM305, 4.5 to 30 V). The maximum input voltage is 50 V (LM305, 40 V), and the minimum input voltage is 8.5 V. The minimum differential input voltage ($V_{in} - V_{out}$) for proper device operation is about 3 V, although the device will operate with ($V_{in} - V_{out}$) as low as 1.8 V. The maximum differential input voltage is 30 V. The maximum device output current must be reduced as ($V_{in} - V_{out}$) is increased to hold the device power dissipation to within 800 mW. The LM105 is normally supplied in a low profile TO-5 metal can package, although a flat package is also available.

The LM105 provides 0.1% load regulation (I_L < 12 mA), 1% temperature regulation maximum, and 0.06%/V input line regulation. Output currents up to 10 A can be obtained with external pass transistors. External pass transistors improve load regulation by a factor of β of the external transistor. The specifications are reproduced in Appendix C.

CIRCUIT OPERATION

The LM105 internal circuitry is shown in Figure 10.7*a*, and the basic regulator connection in Figure 10.7*b*. Much of the internal circuitry is devoted to providing a stable reference voltage.

Starting at the output, Q_{15} is the series-pass output transistor. Q_{14} is a current amplifier for Q_{15}. Q_{14} and Q_{15} are in a Darlington configuration with no external current booster transistor because R_{10} is shorted out as shown in Figure 10.7*b*. Q_{16} is the current limiting transistor. It can be connected for constant current or foldback limiting.

Q_{12} is a multicollector constant current source. One collector provides a high impedance collector load to driving transistor Q_5 through the diode connected dc level shifting transistor Q_6. The high effective R_L provides a very high voltage gain. Q_5 is driven by the differential amplifier Q_2 and Q_3. The base of Q_2 is connected to the reference voltage and the base of Q_3 to the sample of the power supply output voltage.

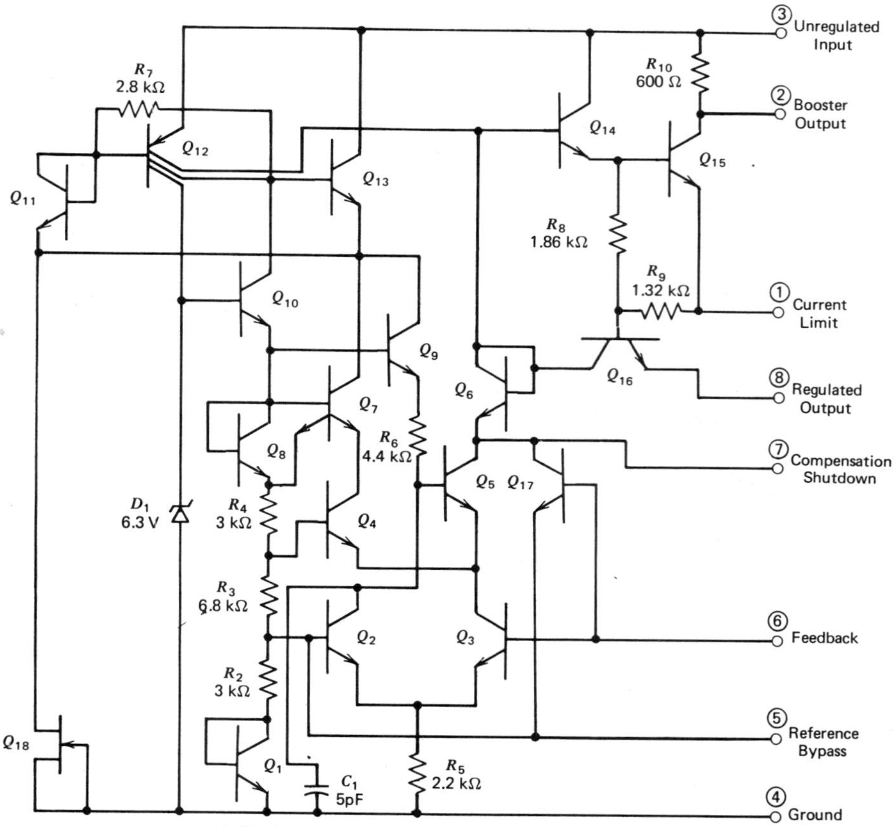

(a) Circuit reprinted courtesy of National Semiconductor

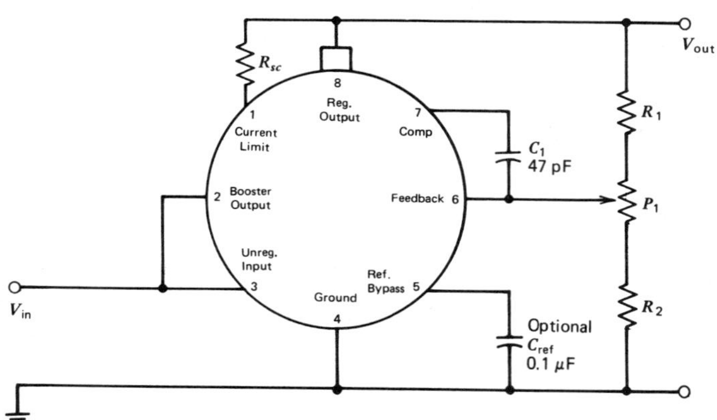

(b) Basic regulator

FIGURE 10.7 National Semiconductor LM 105/205/305.

The bottom collector of Q_{12} provides a constant current to reference zener D_1. The middle collector of Q_{12} provides a constant current to the reference voltage divider string Q_{10}, Q_9, Q_1, R_2, R_3, and R_4.

Q_{13}, Q_{11}, and constant current generator Q_{18}, an FET operating at I_{dss}, provide temperature compensation for Q_{12}.

The base of Q_{10} is at 6.3 V from reference zener D_1. Q_8 provides temperature compensation for dual emitter transistor Q_7.

Q_7 provides bias current for Q_4, which provides a temperature stabilized dc emitter voltage for gain transistor Q_5. Diode connected Q_1 provides temperature compensation for Q_2 so the voltage across R_5 is constant even if the V_{BE} of Q_2 and Q_3 vary with temperature. R_2, R_3, and R_4 simply provide the proper voltages at the base of Q_2 and Q_4. Q_{17} will turn on and shut the circuit down if the feedback pin exceeds the reference bypass pin by 0.7 V. This would occur if an external series-pass transistor short circuited.

Resistors R_8 and R_9 shunt current away from the base of Q_{15} during over current conditions. The current limiting resistor, R_{sc}, goes between pins 1 and 8. When I_{out} exceeds I_m, Q_{16} turns on limiting current. The current-limiting voltage varies between 0.23 V at 125°C and 0.48 V at −55°C, but is about 0.3 V at room temperature.

BASIC LM105 REGULATOR

The basic LM105 regulator circuit is shown in Figure 10.7*b*. It works just like the op-amp circuit of the first section. R_1 and R_2 provide a sample of the regulated output voltage to the inverting input of the differential amplifier. The differential amplifier drives the regulating transistors, Q_{14} and Q_{15}, until the base voltage of Q_3 is equal to the reference voltage on the base of Q_2. V_{out} is then equal to V_{ref} $(R_1 + R_2)/R_2$. The reference voltage will vary from regulator to regulator between 1.63 and 1.81 V, with the typical value being 1.7 V. A potentiometer in the sampling divider is then useful for adjusting V_{out} to the exact value desired. The parallel value of R_1 and R_2 should be 2 kΩ for good differential amplifier stability. A graph of R_1 and R_2 versus V_{out} is given in the specifications.

A convenient way to solve for a voltage divider with a given parallel resistance is found by simultaneously solving the voltage divider equation and the parallel resistance equation. The results follow:

Let

$$\eta = \frac{V_D}{V_A} \tag{10-10}$$

where

V_D = desired divider voltage
V_A = voltage applied to the divider

Then

$$R_1 = \frac{R_p}{\eta} \tag{10-11}$$

$$R_2 = \frac{R_p}{1 - \eta} \tag{10-12}$$

where

R_p = desired parallel resistance.

R_{sc}, the current limiting resistor, is found for constant current limiting by $R_{sc} = 0.3\ \text{V}/I_m$.

C_1 prevents oscillation; its recommended value is 47 pF. C_{ref} eliminates noise from the reference voltage and, therefore, the regulated output. Pins 2 and 3 are shorted together to provide a Darlington output transistor pair.

EXAMPLE 10-3

Set the LM105 up as a 10 V, 10 mA constant current limited supply. $V_{in}(\text{min}) = 15$ V, $V_{in}(\text{max}) = 20$ V. The circuit is shown in Figure 10.7*b*.

SOLUTION

First calculate R_1 and R_2. From the specification sheet, $R_1 \parallel R_2 = 2\ \text{k}\Omega$ and V_{fb} (pin 6) = 1.7 V typically. From Equations 10-10, 10-11, and 10-12

$$\eta = \frac{V_D}{V_A} = \frac{1.7\ \text{V}}{10\ \text{V}} = 0.17$$

$$R_1 = \frac{R_p}{\eta} = \frac{2\ \text{k}\Omega}{0.17} = 11.76\ \text{k}\Omega$$

$$R_2 = \frac{R_p}{1 - \eta} = \frac{2\ \text{k}\Omega}{0.83} = 2.41\ \text{k}\Omega$$

The feedback voltage varies from 1.63 to 1.81 V, giving a $\Delta V_{fb} = 0.18$ V. To adjust V_{out} to 10 V, a potentiometer (P_1) is required.

$$P_1 = (R_1 + R_2)\Delta V_{fb}/V_{out}$$

$$P_1 = 14.17\ \text{k}\Omega(0.18\ \text{V})/10\ \text{V} = 255\ \Omega$$

Use a standard value 500Ω potentiometer.

This size potentiometer will allow an adjustment in voltage (ΔV) of

$$\Delta V = V_A\left(\frac{P}{R_1 + R_2}\right) = 10\ \text{V}\left(\frac{500\ \Omega}{14.17\ \text{k}\Omega}\right) = 0.36\ \text{V}$$

which is more than the 0.11 V needed to adjust V_{fb} from 1.7 to 1.81 V, so one half of P_1 may be subtracted from R_1 and R_2 to get the final values.

$$R_1 = 11.76\ \text{k}\Omega - 250\ \Omega = 11.51\ \text{k}\Omega$$

$$R_2 = 2.41\ \text{k}\Omega - 250\ \Omega = 2.16\ \text{k}\Omega$$

$$P_1 = 500\ \Omega$$

The current through R_1, P, and R_2 is

$$I_{R1} = V_A/(R_1 + P + R_2) = 10\ \text{V}/14.17\ \text{k}\Omega = 0.7\ \text{mA}.$$

Since I_{out} is only 10 mA, this amount must be added to I_{out} for the calculation of R_{sc}.

$$R_{sc} = \frac{0.3\ \text{V}}{I_{out} + I_{R1}} = \frac{0.3\ \text{V}}{10.7\ \text{mA}} = 28\ \Omega$$

The maximum power dissipation of the LM105 with a short-circuited output is

$$P_D = V_{in}(\max)(I_{out} + I_{R1} + I_{SB})$$

where

I_{SB} = standby current drain of the IC, 3 mA max
P_D = 20 V (10 mA + 0.7 mA + 3 mA) = 274 mW

CURRENT BOOSTING THE LM105

Figure 10.8 illustrates a LM105, 28 V, 2 A power supply with foldback current limiting. The use of the *pnp* (Q_B) transistor to drive the *npn* series-pass transistor (Q_A) is called *complementary drive.* The advantage of complementary drive over the Darlington configuration is that a lower V_{CE} is required across Q_A for efficient operation and $[V_{in}(\min) - V_{out}]$ can be a lower voltage.

The foldback current-limiting circuit works as follows. When V_{out} is at the regulated value, the current through R_6 is fairly high. R_6 is much larger than R_4 and $I_{R_6} \cong I_{R4}$. The voltage across R_4, $V_{R_4} = I_{R_6}R_4$, opposes the voltage developed across R_{sc}, as shown in Figure 10-8*b*. Thus, $I_{out}R_{sc}$ must be greater than V_{R_4} by the V_{BE} of Q_{16} before current limiting begins. Therefore

$$I_m R_{sc} = V_{R4} + V_{BE\ \text{on}\ (Q16)}$$

at the current limit. When the output voltage drops under an overload condition, the current through R_6 decreases. Thus the voltage drop across R_4 drops and the

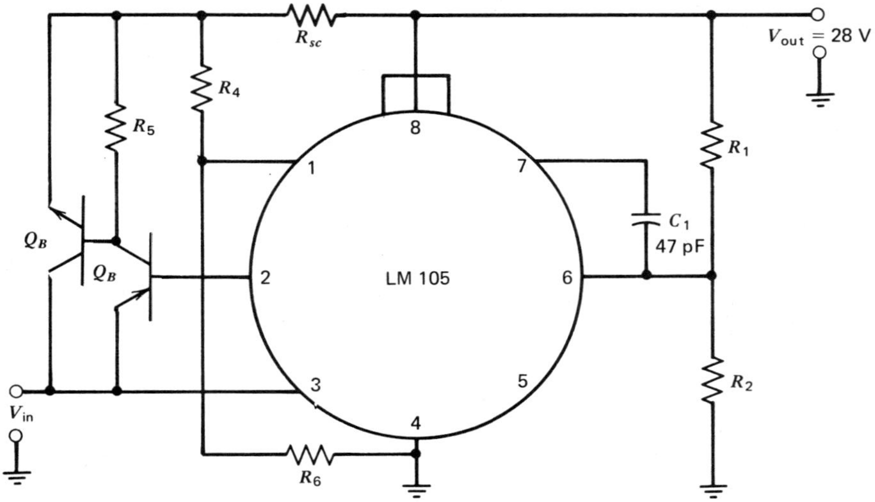

(a) Connections for 28V, 2A supply

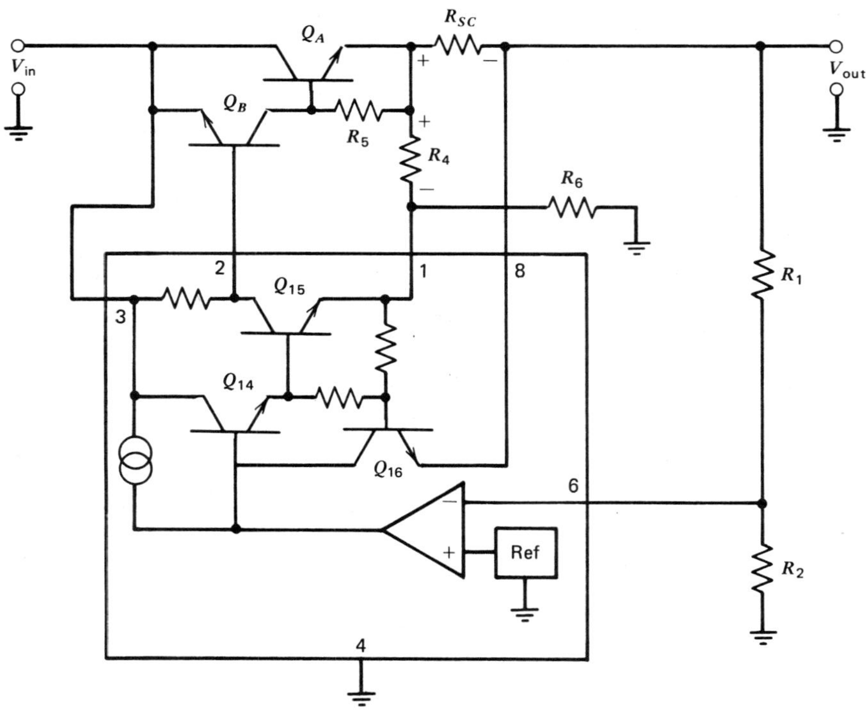

(b) Foldback current limiting circuit

FIGURE 10.8 LM 105 with current boost and foldback current limiting.

current through R_{sc} required to hold $V_{R_{sc}} = V_{BEQ16}$ drops. When $V_{out} = 0$ V, if $I_{R_6} >> I_{BQ16}$, which we will arrange, then

$$V_{R4} = I_{R_6}R_4 = (V_{BEQ16}/R_6)R_4$$

Since we wish the foldback current to equal about 20% of I_m when $V_{out} = 0$ let

$$V_{R_{sc}} = 0.2\, I_m R_{sc} + (V_{BEQ16}/R_6)R_4$$

If $V_{R_{sc}} \cong 5\, V_{BEQ16}$ and $V_{R_4} \cong 4V_{BEQ16}$ at V_{out} and I_m specified, the calculations will provide foldback current limiting beginning at I_m and folding back to $0.2\, I_m$ at $V_{out} = 0$. Let

$$R_{sc} = \frac{5\, V_{BEQ16}}{I_m} \tag{10-13}$$

$$R_6 = \frac{V_{out} - 4\, V_{BEQ16}}{I_{R6}} \tag{10-14}$$

where $I_{R_6} >> I_{BQ16} = 20$ mA for the LM105. Since $I_{R_6} \cong I_{R_4}$

$$R_4 = \frac{4\, V_{BEQ16}}{I_{R4}} \tag{10-15}$$

The selection of the transistors Q_A and Q_B is best shown by example.

EXAMPLE 10-4

Calculate the components for a 28 V, 2 A regulator with foldback current limiting as shown in Figure 10.8*a*. Let $V_{in} = 35$ V.

SOLUTION

A series-pass *npn* transistor is selected with the following specifications:

$$I_c(\text{max}) = 3\text{ A}$$
$$V_{CEO}(\text{sus}) = 45\text{ V}$$
$$\beta = 20 \text{ at } I_C = 2\text{ A}$$
$$V_{BE} = 0.8\text{ V at } I_C = 2\text{ A}$$
$$P_D = 30\text{ W}$$

This power dissipation rating is greater than the power dissipation under a short-circuit condition, which is

$$P_{D\,(V_{out}=0)} = 0.2\, I_m\, V_{in} = 0.4\text{ A } 35\text{ V} = 14\text{ W}$$
$$P_{D\,(V_{out}=28\text{V})} \doteq (V_{in} = V_{out})\,(I_m) = 7\text{ V } 2\text{ A} = 14\text{ W}$$

Now Q_B must be selected. $I_{C_{Q_B}}$ must be greater than $I_{B_{Q_A}}$.

$$I_{B_{Q_A}} = I_m/B_{Q_A} = 2\ \text{A}/20 = 0.1\ \text{A}$$

Q_B is selected with the following specifications:

$$\begin{aligned} I_C(\text{max}) &= 0.3\ \text{A} \\ V_{CEO}(\text{sus}) &= 45\ \text{V} \\ \beta &= 35 \quad \text{at} \quad I_C = 0.15\ \text{A} \\ V_{BE} &= 0.7\ \text{V} \quad \text{at} \quad I_C = 0.15\ \text{A} \\ P_D &= 3\ \text{W} \end{aligned}$$

The current needed to drive Q_B is $I_{C_{Q_B}}/B_{Q_B}$. This is 4.2 mA, which the LM 105 can supply easily.

$$R_5 = V_{BE_{QA}}/(I_{C_{QB}} - I_{B_{QA}}) = 0.8\ \text{V}/0.05\ \text{A} = 16\ \Omega$$

Note. $I_{C_{Q_B}}$ was selected at 0.15 A. Any value in excess of $I_{B_{Q_A}} = I_m/\beta_{Q_A}$ that allows enough margin for error in the V_{BE} of Q_A is acceptable. If 0.12 A > $[V_{BE_{Q_A}}(\text{max})/R_5 + I_m/B_{Q_A}]$, then 0.12 A is acceptable for $I_{C_{Q_B}}$.

We now select R_1 and R_2 for the typical $V_{fb} = 1.7$ V:

$$\eta = \frac{V_D}{V_A} = \frac{1.7\ \text{V}}{28\ \text{V}} = 0.0607$$

$$R_1 \parallel R_2 = R_p = 2\ \text{k}\Omega$$

$$R_1 = \frac{R_p}{\eta} = \frac{2\ \text{k}\Omega}{0.0607} = 32.9\ \text{k}\Omega$$

$$R_2 = \frac{R_p}{1 - \eta} = \frac{2\ \text{k}\Omega}{1 - 0.0607} = 2.13\ \text{k}\Omega$$

The foldback current limiting resistors are selected using Equations 10-13, 10-14, and 10-15. $V_{BE_{Q16}}$ is typically 0.3 V at room temperature. I_{R_6} is set at 20 mA.

$$R_{sc} = \frac{5\ V_{BEQ16}}{I_m} = \frac{5(0.3\ \text{V})}{2\ \text{A}} = 0.75\ \Omega$$

$$R_6 = \frac{V_{out} - 4\ V_{BEQ16}}{I_{R_6}} = \frac{28\ \text{V} - 1.2\ \text{V}}{20\ \text{mA}} = 1.34\ \text{k}\Omega$$

$$R_4 = \frac{4\ V_{BEQ16}}{I_{R_4}} = \frac{1.2\ \text{V}}{20\ \text{mA}} = 60\ \Omega$$

C_1 is selected at the manufacturer's recommended value of 47 pF.

10-2.2 LM309

The National Semiconductor LM309 is a fixed 5 V regulator intended for supplying logic circuits. The LM309 is sold in two packages, a TO-5 case and a TO-3 case. The LM309 in the TO-5 case can supply up to 200 mA and dissipate up to 2 W with the proper heat sink. The TO-3 case can provide up to 2 A and dissipate up to 10 W with the proper heat sink. The LM309 has constant current limiting that automatically reduces I_m as $V_{in} - V_{out}$ increases to prevent excessive regulator

FIGURE 10.9 LM 309.

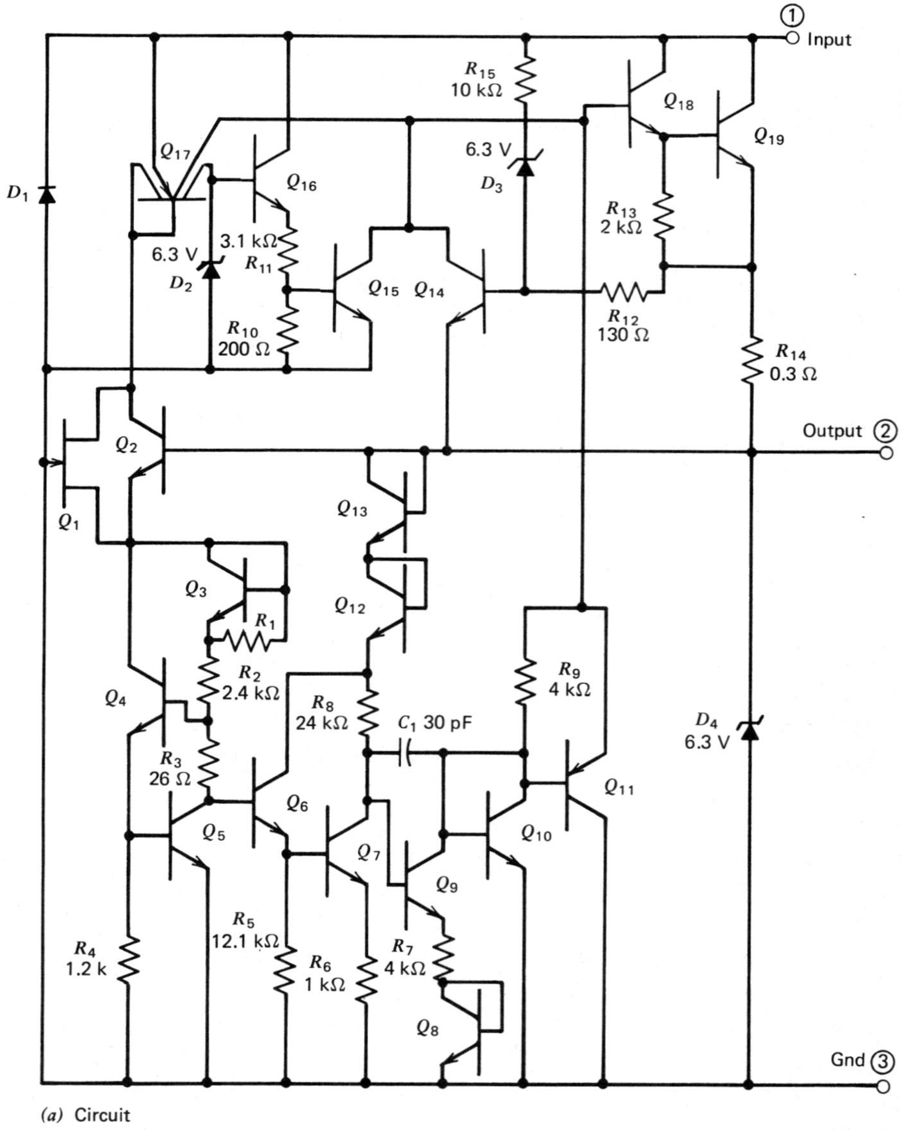

(*a*) Circuit

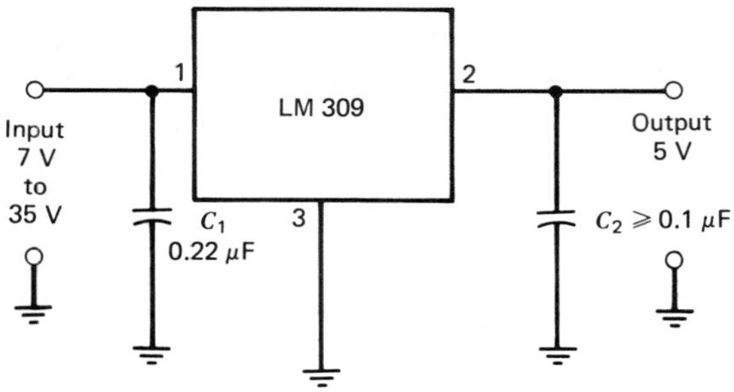

(*b*) Typical connection

C_1 required only if regulator is at a distance from supply filter.

C_2 is not needed for stability; it improves transient response.

power dissipation. The regulator also features thermal overload protection so that the power dissipation limits of the package cannot be exceeded.

The LM309 operates over a range of junction temperatures from 0 to 125°C (a similar LM109 operates from −55 to 150°C). The output voltage of the LM309 is between 4.8 and 5.2 V, with 5.05 V being typical. The output voltage will vary no more than 50 mV as V_{in} goes from 7 to 25 V, and no more than 50 mV (100 mV for TO-3 case) as the load current varies from 5 mA to 0.5 A (5 mA to 1.5 A for TO-3 case).

The LM309 is designed for easy use, as can be seen from the typical application shown in Figure 10.9*b*. The IC can be used as an adjustable regulator and with current-boosting transistors also. Complete LM309 specifications are contained in Appendix C.

CIRCUIT OPERATION

The circuitry of the LM309 is shown in Figure 10.9*a*. D_1 provides protection for the device (and the load) if the input voltage is reversed. Q_{18} and Q_{19} are the output Darlington pair. Q_{17} is a constant current source that is temperature-compensated with one of its collectors diode connected. Q_1 and Q_2 provide base current for Q_{17}.

Current limiting is provided by R_{14} and Q_{14}. Current limiting works conventionally if $(V_{in} - V_{out}) < 6.3$ V. If $(V_{in} - V_{out}) > 6.3$ V, D_3 conducts, providing current to R_{12}. The voltage drop across R_{12} is of the same polarity as the voltage across R_{14}. Thus, as $V_{in} - V_{out}$ increases, the voltage required across R_{14} to initiate current limiting decreases. I_m then decreases as $V_{in} - V_{out}$ increases above 6.3 V, limiting device power dissipation.

D_2 provides a constant voltage to Q_{16}, which provides a constant voltage for R_{11} and R_{10}. The variation of the V_{BE} of Q_{16} with temperature provides compensation for any temperature drift of D_2. R_{10} provides a constant, temperature-stabilized 0.3 V to the base of Q_{15}. As temperature increases, V_{BE} of Q_{15} decreases. At T_{max}, $V_{R_{10}} = V_{BE\,on}$ of Q_{15}. As Q_{15} turns on, current limiting begins and the power dissipation of the LM309 is limited to a safe value.

D_4 provides overvoltage protection for the LM309 load. If V_{out} rises above 6.3 V, D_4 conducts and fuses into a short circuit. The LM309 is destroyed in the process, but a complex digital system will be saved from overvoltage destruction.

The LM309 uses a band-gap reference voltage. This is a reference that uses a voltage equal to the band-gap voltage of silicon (1.205 eV) as a reference. Actually, the emitter-base voltages of transistors provide the reference voltage, but the band-gap voltage set from the emitter-base voltages provides a temperature stable operating point. Refer to Figure 10.10*c* for a simple band-gap reference illustration. The current through Q_A is much greater than the current through Q_B, so that V_{BE} of Q_A is greater than the V_{BE} of Q_B. Q_B is an amplifier with a large current gain; its emitter current is then

$$I_{E_{QB}} = V_{BE_{QA}} - V_{BE_{QB}}/R_2 = \Delta V_{BE}/R_2$$

The voltage across R_3 is

$$V_{R_3} = I_{E_{QB}} R_3 = (\Delta\, V_{BE}/R_2)R_3$$

The collector voltage of Q_C is then

$$V_{CE_{QC}} = V_{BE_{QC}} + V_{R_3} = 1.205 \text{ V}$$

Referring back to Figure 10.9*a*, as the temperature increases, the negative temperature coefficient of the V_{BE} of Q_3 is compensated by the positive temperature coefficient of V_{R_2} ($V_{BE_{QA}}$ is to be constant compared to Q_B in Figure 10.10*c*). In the LM309 circuit, Q_5 and Q_6 provide a fixed, temperature-compensated voltage to the base of Q_6 and Q_7. Q_6 and Q_7 jointly act as amplifier Q_B of Figure 10.10*c*. The ΔV_{BE} of the band-gap regulator appears across R_8. Q_9 acts as a voltage comparator and Q_C of the band-gap regulator.

Any change of V_{out} is coupled to the top of R_8 by diode connected transistors Q_{12} and Q_{13}. The output voltage change causes the base voltage of Q_9 to vary, causing the complementary drive pair Q_{10} and Q_{11} to vary the base voltage of Darlington pair Q_{18} and Q_{19}, thus the output voltage. Q_8 is for temperature compensation of Q_{10}, and C_1 provides frequency compensation.

LM309 CONNECTIONS

The typical application of the LM309 shown in Figure 10.9*b* indicates the simplicity of use for which the IC was designed. C_1 is required only if the regulator is an appreciable distance from the unregulated supply, and C_2 is needed only for good transient response, that is, fast return to original output after a sudden input voltage change.

The LM309 can be used as a variable supply, as shown in Figure 10.10*a*, or fixed supply with $V_{out} > 5$ V if R_2 is fixed. The voltage from V_{out} to ground (V_{R_1}) will be kept at 5 V by the voltage regulator. Since V_{R_1} varies if the load current or the input voltage varies, the output voltage is regulated even though it is greater than 5 V. The quiescent current of the LM309 is typically 5.2 mA, but it can be as high as 10 mA. The voltage divider R_1 and R_2 must provide this current. R_1 and R_2 may be chosen as follows.

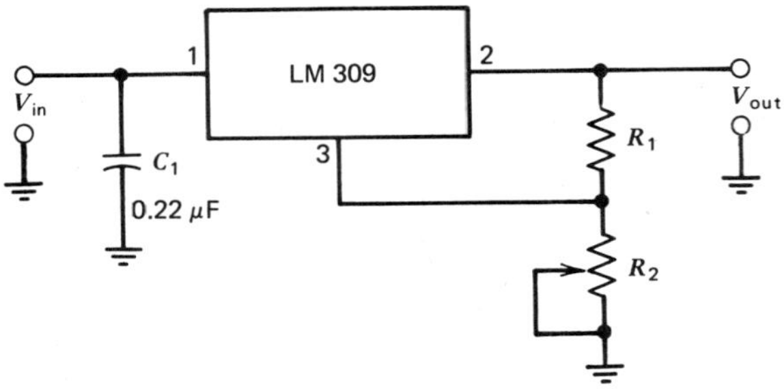

(a) Adjustable output regulator

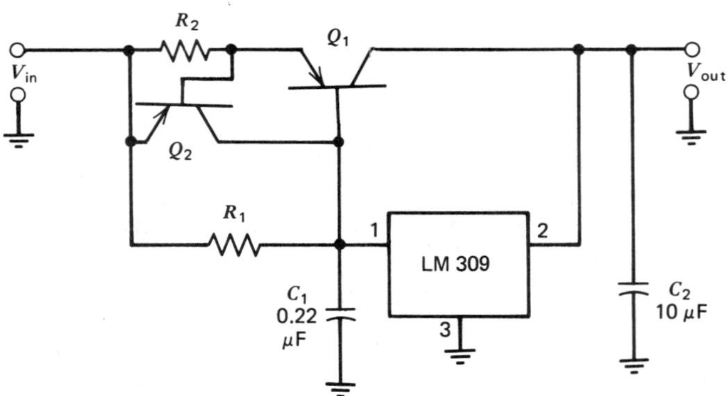

(b) High current 5 V regulator with constant current limiting

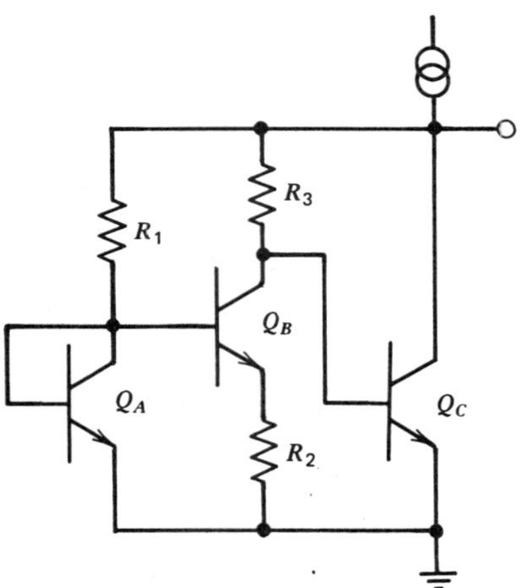

$$V_{Ref} = V_{BE\,Q_C} + \frac{R_3}{R_2}\,\Delta V_{BE} = 1.205\text{ V}$$

$$\Delta V_{BE} = V_{BE\,Q_A} - V_{BE\,Q_B} = V_{R3} = \left(\frac{\Delta V_{BE}}{R_2}\right) R_3$$

$$I_{Q_A} \gg I_{Q_B}$$

(c) Simple band-gap voltage reference

FIGURE 10.10 LM 309 applications.

EXAMPLE 10-5

Set the LM309 up as a fixed-voltage regulator with an output voltage of 10 V at 1 A. V_{in} = 15 V.

The 1 A = I_{out} specifications require the use of an LM309 in the TO-3 case.

SOLUTION

The voltage across R_1 and R_2 is the same, but the currents are different. I_{R_1} and I_{R_2} must be greater than 10 mA for sufficient quiescent current to be supplied for the LM309. Using Ohm's law and letting I_{R_2} = 20 mA

$$R_2 = \frac{V_{out} - 5\text{ V}}{I_{R_2}} = \frac{5\text{ V}}{20\text{ mA}} = 250\ \Omega$$

$$R_1 = \frac{5\text{ V}}{I_{R_2} - 5.2\text{ mA}} = 338\ \Omega$$

From the specification sheet, the maximum quiescent current change is 0.8 mA and the quiescent current could be as high as 10 mA. Thus the maximum that the quiescent current will vary from device to device will normally be 5.6 mA. Nortonizing R_1 and R_2, this can cause an output voltage change of $\Delta I_Q(R_1 \parallel R_2)$ = 0.8 V. If this is unacceptable, a potentiometer will have to be used in the R_1, R_2 divider for adjusting V_{out} to 10 V. The potentiometer should be about 0.8 V/ΔI_Q or 150 Ω.

The operating power dissipated in the LM309 is V_{in} I_m, under maximum load circuit conditions

$$P_D = I_m(V_{out} - V_{in}) = 1\text{ A }(5\text{ V}) = 5\text{ W}$$

The package must have a heat sink (see the specification sheet).

A high current 5 V regulator can be constructed as shown in Figure 10.10*b*. The current limiting provided by Q_2 is for the protection of Q_1 since the LM309 protects itself very well. An example will show how to select the components for a high current 5 V regulator.

EXAMPLE 10-6

Calculate the components for a 5 V, 10 A constant current limited power supply using the LM309 as shown in Figure 10.10*b*. V_{in} = 15 V.

SOLUTION

First, Q_1 must be selected. Q_1 must be able to dissipate under shorted output conditions

$$P_D = V_{in}\, I_m = (15\text{ V})(10\text{ A}) = 150\text{ W}$$

For safety, we select a 200 W transistor. The maximum I_{out} of the LM309 with $(V_{in} - V_{out}) = 10$ V is shown in the specifications for the TO-3 package to be about 1 A. The power transistor must have a β of 10 at $I_C = 10$ A.

A transistor (Q_1) with the following specifications is selected.

$$V_{CE}\ (\text{sus}) = 20\ \text{V}, \quad I_C(\text{max}) = 15\ \text{A}$$

$$V_{BE}\ (\text{at } I_C = 10\ \text{A}) = 0.9\ \text{V}, \quad P_D = 200\ \text{W}$$

$$\beta = 15 \text{ at } I_C = 10\ \text{A}$$

$$I_B = I_m/B_{Q_1} = 10\ \text{A}/15 = 0.67\ \text{A}$$

$$R_1 = V_{BE_{Q1}}/(I_{out\ (309)} - I_{B_{Q1}})$$

I_{out} of the LM309 is chosen to be in excess of $I_{B_{Q1}}$, enough to cover component tolerances and V_{BE} variations of Q_1. If these are 20%, the I_{out} of the LM309 is 1.2 $I_{B_{Q2}}$ and $I_{R_1} = 0.2\ I_{B_{Q2}}$, thus

$$R_1 = V_{BE_{Q1}}/0.2\ I_{B_{Q2}} = 0.9\ \text{V}/134\ \text{mA} = 6.7\ \Omega$$

R_2 is chosen as in previous constant current limit examples. Let $V_{BE\,on}$ of $Q_2 =$ 0.3 V, then

$$R_2 = V_{BE\ on Q_2}/I_m = 0.3\ \text{V}/10\ \text{A} = 0.03\ \Omega$$

Q_2 must be chosen for $I_C > I_B$ of Q_1 and dissipate a maximum power of

$$P_{D_{Q2}} = V_{in}\ I_{B_{Q1}} = 15\ \text{V}(0.67\ \text{A}) = 10\ \text{W}$$

The output capacitor is chosen from the definition of capacitance

$$C = \frac{Q}{V} = \frac{It}{V}$$

where

V = maximum voltage change allowed
I = maximum load current drawn
t = t switching for logic load, approximately $t_r + t_f$

For most TTL loads, 10 μF is sufficient.

10-2.3 Fairchild Semiconductors μA723

The Fairchild μA723 is a very versatile IC voltage regulator. The equivalent circuit and simplified equivalent circuit are shown in Figure 10.11*a* and *b*. The μA723

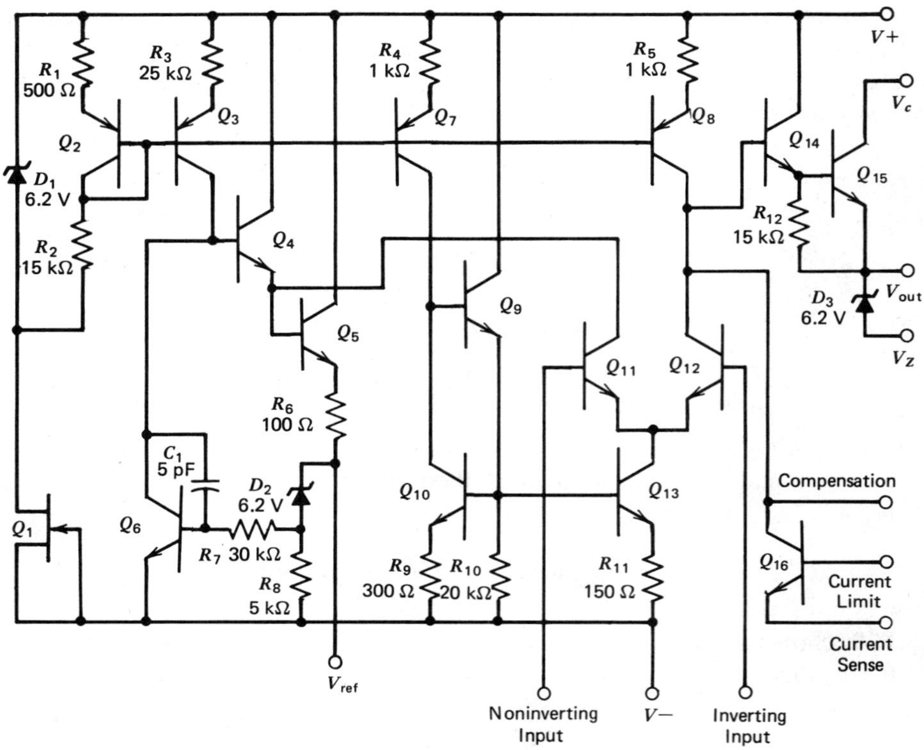

(a) Equivalent circuit

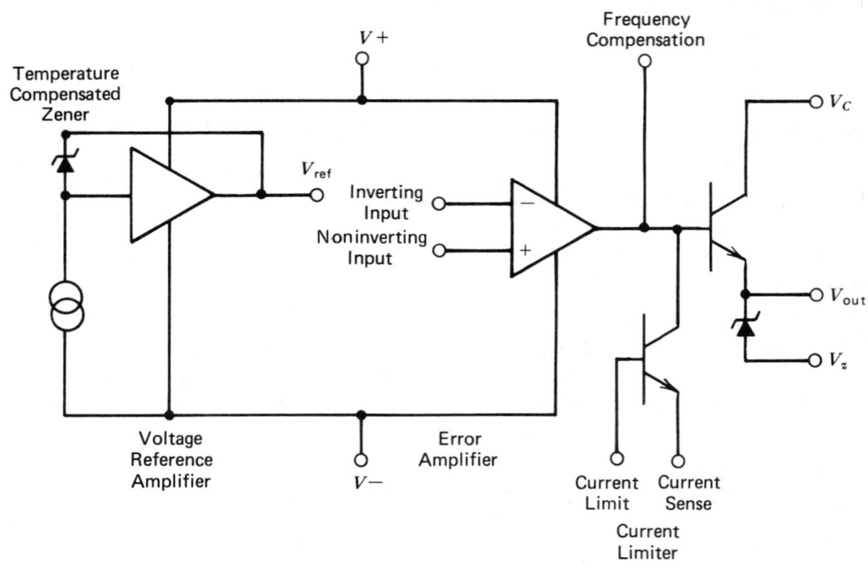

(b) Simplified equivalent circuit

FIGURE 10.11 Fairchild μA 723.

is supplied in a military temperature range from −55 to 125°C and a commercial temperature range from 0 to 70°C. The commercial device, called the μA723C, is the one we will use for our illustrations. The μA723 is supplied in two packages: a metal can that can dissipate 800 mW and a plastic 14 pin DIP that can dissipate 1 W.

The μA723C can operate with an input voltage from 9.5 to 40 V and provide an output voltage from 2 to 37 V. The μA723C can provide an output current of 150 mA for $V_{in} - V_{out} = 3$ V, but I_{out} drops to 10 mA for $V_{in} - V_{out} = 38$ V. The device features worst-case load regulation of 0.6% V_{out} and worst-case line regulation of 0.5% V_{out}. Typical regulation will be considerably better. The quiescent current drain of the μA723C is 3.5 mA maximum, typically about 2.3 mA, and relatively constant with applied voltage changes.

The μA723 is very versatile with many applications shown in the specifications in Appendix C. We will show only four in this section. Note in the specifications that resistor values and equations are listed for various applications and output voltages. Also, note that the μA723, if powered by an ungrounded zener diode regulator, can control an external pass transistor to regulate voltages up to 250 V.

CIRCUIT OPERATION

Referring to Figure 10.11*b*, we see the μA723 consists of a temperature compensated 6.2 V zener biased with a constant current source. A buffer amplifier provides the zener voltage as a reference voltage output that provides up to 15 mA. An error amplifier is provided to compare the reference voltage and a sample of the regulated output voltage. The error amplifier drives a series-pass Darlington output pair. A current limit transistor is also provided.

The zener by V_{out} shown in Figure 10.11*b* is used when the μA723 is connected as a negative voltage regulator. The zener has a 6.2 V breakdown voltage and reduces the power dissipation of the series-pass transistor. The zener can provide up to 25 mA.

Refer to Figure 10.11*a* for the circuit details. Q_{14} and Q_{15} are the Darlington output transistors when V+ and V_c are connected for normal voltage regulator use or when the regulator drives an external *npn* series-pass transistor. When the μA723 is used to drive a *pnp* or a complementary *pnp-npn* output, V_c goes to the base of the *pnp* and a drive resistor is used between V_c and V+.

Q_{16}, the current limit transistor, can be used for either constant or foldback current limiting. The current limit sense voltage varies with temperature from 0.45 V at −50°C to 0.8 V at 150°C, but is 0.65 V at room temperature. A graph of the current limit sense voltage versus temperature is provided in the specifications.

The error amplifier consists of Q_{11}, Q_{12}, Q_8, Q_7, Q_{10}, Q_9, Q_{13}, and resistors R_{10} and R_{11}. Q_{11} and Q_{12} are the actual differential amplifier. Q_8 is a constant current source for Q_{12} providing a large collector load impedance, thus a high gain, for Q_{12}. Q_{12} provides the base drive for Q_{14}. Q_{13} is a temperature-compensated constant current source for the emitter current of Q_{11} and Q_{12}. Q_7 provides constant current to the Q_{13} temperature-compensation transistors Q_9 and Q_{10}. As the V_{BE} of Q_{13} drops with increasing temperature, so does the V_{BE} of Q_{10} so that $V_{R_{11}}$, thus I_C of Q_{13}, stays constant. Q_9 holds the base current of Q_{10} constant as the V_{BE} of Q_{10} changes with temperature.

Q_1 provides constant current to D_1 so that D_1 provides a constant voltage to R_1, R_2, and diode connected transistor Q_2. Q_2 provides a constant base voltage to constant current source transistors Q_3, Q_7, and Q_8. Q_2 compensates for the V_{BE} variation with temperature of Q_3, Q_7, and Q_8.

Q_3 provides a constant current to the reference zener, D_2, and the reference amplifier Q_4, Q_5, and Q_6. Most of the current from Q_3 flows through Q_6. Q_4 and Q_5 provide a low impedance at the reference voltage output. The reference voltage, $V_{ref} = V_{R_8} + V_{D_2}$. Any change in V_{R_8}, either from temperature drift of the V_{BE} of Q_6 or D_2 or reference-voltage loading, is sensed by Q_6, which provides an opposite polarity correction voltage at the base of Q_4. The reference voltage is specified to be between 6.8 and 7.5 V, with 7.15 V as the typical value. Thus the regulated output voltage sample divider will need a potentiometer to set V_{out} precisely.

10-2.4 μA723C Applications

Four applications of the μA723C are presented here: the basic low-voltage regulator (V_{out} between 2 and 7 V), the basic high-voltage regulator (V_{out} between 7 and 37 V), a positive voltage regulator with an external current boost transistor and fold-back current limiting, and a negative voltage regulator with constant current limiting. These applications are presented as examples. Many more applications are presented in the μA723 specification sheet.

EXAMPLE 10-7

Construct a 5 V, 100 mA power supply using a μA723C as shown in Figure 10.12*a*. $V_{in} = 10$ V.

SOLUTION

The reference voltage, V_{ref}, must be reduced to 5 V for the noninverting input of the error amplifier by means of a voltage divider. For maximum temperature stability, a resistor $R_3 = R_1 \parallel R_2$ must be provided for the inverting input of the error amplifier. R_3 may be left out if parts must be kept to a minimum. The divider resistance, $R_1 \parallel R_2 = R_3$, must be less than 10 kΩ for stability, but greater than 490 Ω to avoid loading the reference voltage. The values in the specification sheet resistor chart show 966 Ω ≤ $R_1 \parallel R_2$ ≤ 3.52 kΩ with I_D at about 1 mA. We will follow Fairchild's lead and set the voltage divider current (I_D) at about 1 mA. Since the error amplifier draws very little current, we will neglect its input bias current.

$$R_1 = \frac{V_{ref} - V_{out}}{I_D} = \frac{2.15 \text{ V}}{1 \text{ mA}} = 2.15 \text{ k}\Omega$$

$$R_2 = \frac{V_{out}}{I_D} = \frac{5 \text{ V}}{1 \text{ mA}} = 5 \text{ k}\Omega$$

$$R_3 = R_1 \parallel R_2 = 2.15 \text{ k}\Omega \parallel 5 \text{ k}\Omega = 1.5 \text{ k}\Omega$$

$$R_{sc} = \frac{V_{\text{current sense}}}{I_m} = \frac{0.65 \text{ V}}{100 \text{ mA}} = 6.5\ \Omega$$

$$P_D(\text{max}) = V_{in}\,(I_m + I_{\text{standby}}) = 10 \text{ V}\,(103 \text{ mA}) = 1.03 \text{ W}$$

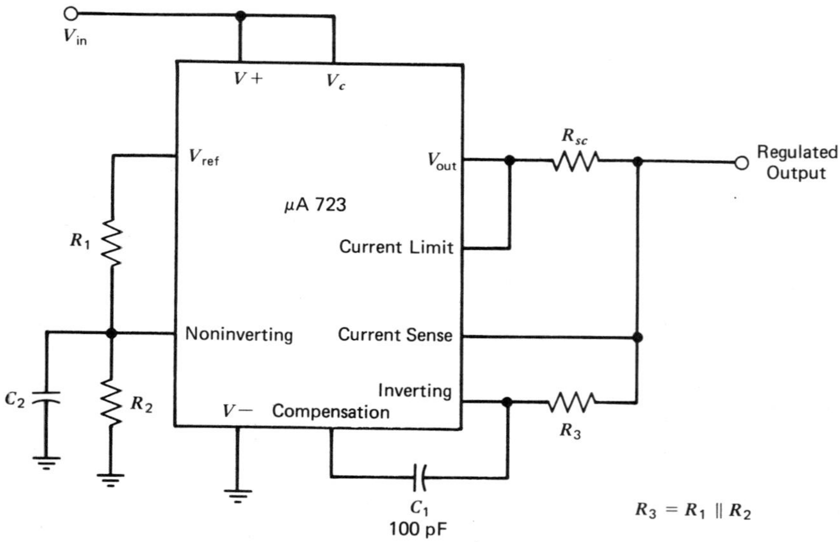

(a) Basic low-voltage regulator (V_{out} = 2 to 7 V)

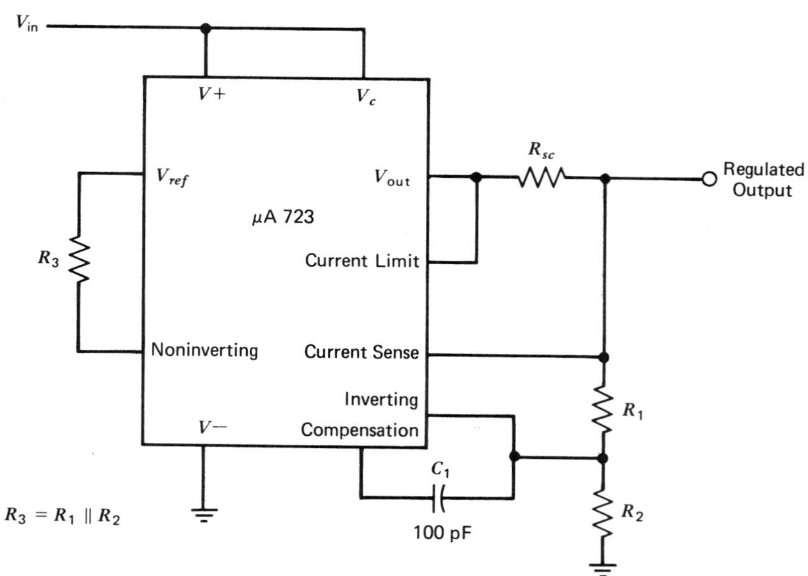

(b) Basic high-voltage regulator (V_{out} = 7 to 37 V)

FIGURE 10.12 Basic μA 723 regulators.

A heat sink is advisable for the μA723C package since short-circuit power dissipation is at the limit the DIP can stand. Normal dissipation is a safe 500 mW if the case temperature is held to 70°C.

From the specifications, $C_2 = 5\ \mu F$ to reduce reference noise voltage, and $C_1 = 100$ pF for frequency stability. *Note.* C_2 should be a solid tantalum capacitor for optimum performance since solid tantalum has a lower high-frequency series resistance than other electrolytic capacitors.

EXAMPLE 10-8

Construct a 15 V, 50 mA power supply using a μA723C as shown in Figure 10.12*b*. $V_{in} = 20$ V.

SOLUTION

From the specification sheet, we find that since $V_{in} - V_{out} = 5$ V, the μA723C can easily supply 50 mA if P_D(max) is not exceeded. The short circuit power dissipation is

$$P_D(\text{max}) = V_{in}\ (I_m + I_{\text{standby}}) = 15\text{ V }(53\text{ mA})$$
$$= 795\text{ mW}$$

This value is safe indefinitely for the DIP package if the case temperature is held at or below 50°C.

Since $V_{out} > V_{ref}$, the voltage fed to the inverting input of the μA723C must be reduced with a voltage divider. The reference voltage can be applied directly to the noninverting input, although using $R_3 = R_1 \parallel R_2$ will reduce temperature drift. We will use the typical reference voltage ($V_{ref} = 7.15$ V) and $I_D = 1$ mA as in the preceding example.

$$R_1 = \frac{V_{out} - V_{ref}}{I_D} = \frac{15\text{ V} - 7.15\text{ V}}{1\text{ mA}} = 7.85\text{ k}\Omega$$

$$R_2 = \frac{V_{ref}}{I_D} = \frac{7.15\text{ V}}{1\text{ mA}} = 7.15\text{ k}\Omega$$

$$R_3 = R_1 \parallel R_2 = 7.85\text{ k}\Omega \parallel 7.15\text{ k}\Omega = 3.74\text{ k}\Omega$$

$$R_{sc} = \frac{V_{\text{sense}}}{I_m + I_D} = \frac{0.65\text{ V}}{51\text{ mA}} = 12.74\ \Omega$$

The resistors chosen will be the nearest standard value to the calculated value from the resistors available in the tolerance used. ±5% is the loosest tolerance recommended.

C_1 is chosen at 100 pF from the specification sheet.

EXAMPLE 10-9

Construct a 28 V, 0.5 A dc power supply, as shown in Figure 10.13*a*, using a μA723C. The power supply is to have foldback current limiting. V_{in}(max) = 38 V, V_{in}(min) = 34 V.

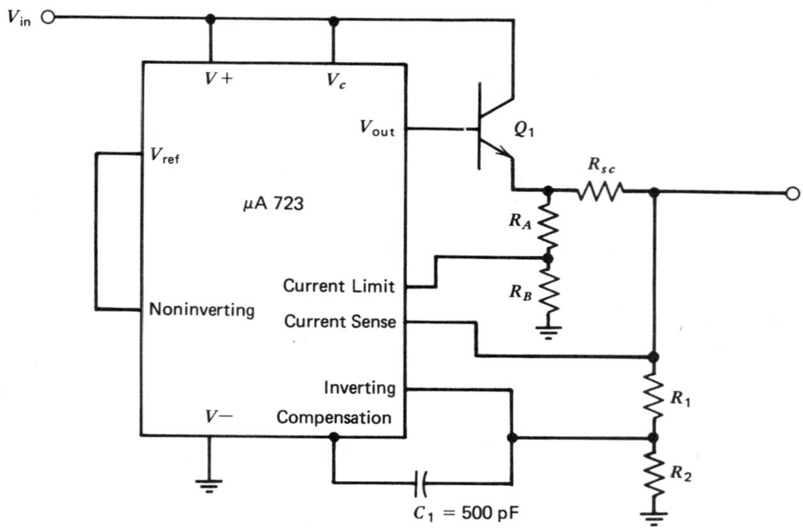

Note: For $V_{out} < 7$ V connect inverting input (Inv) to regulated output, R_1 to V_{ref}, junction of R_1, R_2 to noninverting input (Noninv), R_2 to ground, and compensation (Comp) to ground. See specifications.

(a) Positive voltage regulator with foldback current limiting and external current boost transistor

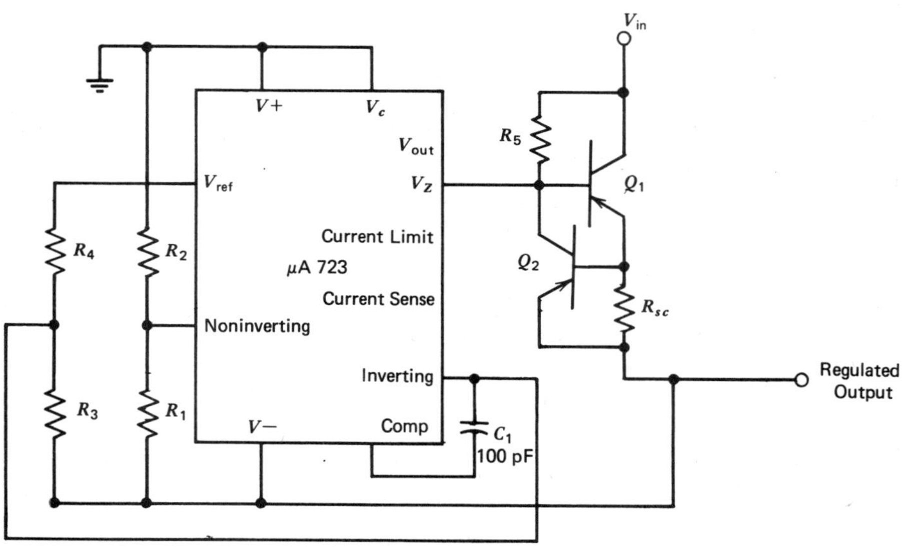

(b) Negative voltage regulator with external current boost transistor and constant current limiting.

FIGURE 10.13 μA 723 applications.

SOLUTION

The first job is to select a series-pass power transistor.

$$P_{D\,(V_{out}=28\,V)} = (V_{in}(\max) - V_{out})\,I_m$$

$$P_{D\,(V_{out}=28\,V)} = 10\text{ V }(0.5\text{ A}) = 5\text{ W}$$

$$P_{D\,(V_{out}=0)} = V_{in}(\max)\,I_{FB} = V_{in}(\max)\,0.2\,I_m$$

$$P_{D\,(V_{out}=0)} = 38\text{ V }(0.1\text{ A}) = 3.8\text{ W}$$

A 10 W transistor with a proper heat sink will work nicely.

Now β_{min} of the pass transistor must be chosen. The μA723C can supply 40 mA with $V_{in} - V_{out} = 10$ V.

$$\beta_{min} = I_m/I_{723} = 0.5\text{ A}/40\text{ mA} = 12.5$$

A transistor with $\beta_{min} = 20$ at $I_C = 0.5$ A is chosen. $V_{CE}(\text{sus}) > V_{in}(\max)$ so $V_{CE}(\text{sus})$ is chosen to be 50 V.

Now R_1 and R_2 can be chosen. We will use $V_{ref} = 7.15$ V and $I_D = 1$ mA. Since $V_{out} > 7$ V, V_{ref} is connected to the noninverting input of the error amplifier and the V_{out} sample is set to V_{ref} with R_1 and R_2. The output sample is applied to the inverting terminal of the error amplifier. R_3 of Figure 10.12*b* could have been used.

$$R_1 = \frac{V_{out} - V_{ref}}{I_D} = \frac{20.85\text{ V}}{1\text{ mA}} = 20.85\text{ k}\Omega$$

$$R_2 = \frac{V_{ref}}{I_D} = \frac{7.15\text{ V}}{1\text{ mA}} = 7.15\text{ k}\Omega$$

Since $R_1 \parallel R_2 < 10\text{ k}\Omega$, the voltage divider is satisfactory.

The foldback current limiting resistors, R_A and R_B, are found as in Example 10-4. The current for R_A and R_B need only be about 1 mA for the μA723C, however. From Equations 10-13, 10-14, and 10-15, let

$$R_{sc} = \frac{5\,V_{sense}}{I_m} = \frac{5(0.65)}{0.5\text{ A}} = \frac{3.25\text{ V}}{0.5\text{ A}} = 6.5\ \Omega$$

$$R_B = \frac{V_{out} - 4\,V_{sense}}{I_{R_B}} = \frac{28\text{ V} - 2.6\text{ V}}{1\text{ mA}} = \frac{25.4\text{ V}}{1\text{ mA}} = 25.4\text{ k}\Omega$$

$$R_A = \frac{4\,V_{sense}}{I_{R_B}} = \frac{2.6\text{ V}}{1\text{ mA}} = 2.6\text{ k}\Omega$$

C_1 is chosen as 500 pF from the specification sheet.

This regulator could have been set up for $V_{out} < 7$ V by changing the R_1, R_2, and reference connections to those shown in Figure 10.12*a*.

EXAMPLE 10-10

Construct a −15 V, 250 mA negative voltage regulator as shown in Figure 10.13*b*. The regulator is to be constant current limited. $V_{in} = -24$ V.

SOLUTION

Once again the series-pass transistor is chosen first. The zener output can provide up to 25 mA. I_Z is also I_{out} of the μA723C. V− is the regulated output voltage (−15 V) and V+ is ground, so $V_{in} - V_{out}$ for the μA723C is −15 V. With $V_{in} - V_{out} = 15$ V, the μA723C can supply the entire 25 mA the zener (D_3) can carry. Note $I_{R_5} = I_{BQ_1} + I_{out}(723)$, so the μA723 must carry almost the entire base current of Q_1 when the load current is minimum. Thus, I_B of Q_1 is limited to a value less than 25 mA. β_{min} of Q_1 is therefore

$$\beta_{min} = \frac{I_{out}}{I_{723}} = \frac{0.25 \text{ A}}{25 \text{ mA}} = 10$$

With constant current limiting, $P_D(\text{max})$ of Q_1 is

$$P_D(\text{max}) = V_{in}I_m = 24 \text{ V } (0.25 \text{ A}) = 6 \text{ W}$$

Thus, Q_1 is chosen with the following specifications:

$$P_D = 10 \text{ W}, \quad \beta_{min} = 30 \quad \text{at} \quad I_c = 0.25 \text{ A}$$
$$V_{CE}(\text{sus}) = 35 \text{ V}, \quad V_{BE} = 0.65 \text{ V} \quad \text{at} \quad I_c = 0.25 \text{ A}$$

Next, Q_2 is chosen with the following specifications:

$$P_D \geqslant I_{BQ_1}(V_{in}) = 24 \text{ V } (0.25 \text{ A}/30) = 0.2 \text{ W}$$

$$P_D = 0.5 \text{ W}, \quad \beta_{min} = 60 \quad \text{at} \quad I_c = 10 \text{ mA}$$
$$V_{BE\ on} = 0.4 \text{ V}, \quad V_{CE}(\text{sus}) = 40 \text{ V}$$

Now, R_{sc} is calculated from

$$R_{sc} = \frac{V_{BEQ2}}{I_m} = \frac{0.4 \text{ V}}{0.25 \text{ A}} = 1.6\ \Omega$$

R_5 must carry $I_B(\text{max})$ of Q_1. The voltage across R_5 is $V_{in} - (V_{out} - V_{BEQ_1})$. Therefore, R_5 can be found from Ohm's law as follows using absolute values:

$$R_5 = \frac{V_{in} - (V_{out} - V_{BEQ_1})}{I_m/\beta_{min\,Q1}}$$

$$= \frac{24 \text{ V} - (15 \text{ V} - 0.65 \text{ V})}{0.25 \text{ A}/30} = \frac{9.65 \text{ V}}{8.33 \text{ mA}} = 1.16 \text{ k}\Omega$$

The maximum differential voltage of the error amplifier and the maximum voltage between V− and the error amplifier inputs is 8 V. With V− at −15 V and V+ at ground if V_{out} goes to zero under short-circuit conditions with V_{ref} tied directly to an error amplifier input, one of these values could be exceeded. Thus both the reference voltage and the sample voltage are reduced with voltage dividers. One end of both voltage dividers is tied to the regulated output voltage to avoid a large differential voltage between the error amplifier inputs under any output voltage condition. Normally the reference voltage is just halved, which means $R_3 = R_4$ in Figure 10.13*b*.

Note that the reference voltage is applied to the inverting input and the output sample is applied to the noninverting input of the error amplifier. The polarity of the output is now negative and the direction the output must go to correct for an output change is now opposite that of a positive regulator, thus the signals to the error amplifier inputs must be reversed from the positive regulator connections.

Calculations of R_3 and R_4 proceed as follows:

$$R_3 = R_4 \quad \text{and} \quad I_{R3} \cong I_{R4} = 1 \text{ mA}$$

$$R_3 + R_4 = \frac{V_{ref}}{1 \text{ mA}} = \frac{7.15 \text{ V}}{1 \text{ mA}} = 7.15 \text{ k}\Omega$$

$$R_3 = R_4 = \frac{R_3 + R_4}{2} = \frac{7.15 \text{ k}\Omega}{2} = 3.575 \text{ k}\Omega$$

Choose any standard value near the calculated value such as 3.33 kΩ for R_3 and R_4. The voltage applied to the inverting input is $-|V_{out} - 0.5\ V_{ref}| = -11.425$ V so $V_{R_2} = 11.425$ V and V_{R_1} and $= 0.5\ V_{ref}$. Let $I_D = 1$ mA

$$R_1 = \frac{0.5\ V_{ref}}{I_D} = \frac{3.575 \text{ V}}{1 \text{ mA}} = 3.575 \text{ k}\Omega$$

$$R_2 = \frac{V_{R2}}{I_D} = \frac{11.425 \text{ V}}{1 \text{ mA}} = 11.425 \text{ k}\Omega$$

The μA723C can also be used as a switching regulator. This will be discussed in the next section.

VOLTAGE REGULATOR CHOICES

Many linear voltage regulators are available from nearly every major semiconductor manufacturer and many hybrid circuit manufacturers. Those we looked at are merely a small sample to understand how they work. Regulators are available for ±V supplies such as ±15 V for linear circuits, in many fixed positive and negative voltages in many current ranges, and in many adjustable ranges. A thorough perusal of a few linear IC catalogs will help familiarize the curious and interested student with the currently available linear voltage regulators.

10-3 SWITCHING REGULATORS

The basic concept of a switching voltage regulator is to use the semiconductor control device as a switch, as illustrated in Figure 10.2. Since the voltage drop across a transistor switch (V_{CE}(sat)) is low, the power lost in the switch is low and the efficiency of the supply is high. This is contrasted with a linear voltage regulator whose series-pass transistor must dissipate (V_{in} − V_{out}) I_{out} at all times. If the switching regulator is operated at a high frequency, normal frequencies are 5 to 100 kHz, the filter inductor and capacitors can be smaller. Since the efficiency of a switching regulator does not vary much as V_{in} changes, switching regulators can tolerate more input ripple than linear supplies, so the rectifier filter capacitor can be smaller.

Two basic types of regulation are used with switching regulators: constant pulse-width variable frequency and constant frequency variable pulse width. The constant pulse-width regulation scheme uses a constant on time for the switch and varies the frequency for voltage regulation, a higher frequency for heavy loads and a lower frequency for light loads. The varying frequency causes the harmonics produced by the switching element to be hard to predict, often producing unwanted interference. The constant frequency variable pulse-width regulation method, called *pulse-width modulation* (PWM), is preferred since switching harmonics can more easily be predicted. In PWM the switching element is on longer for heavy loads than for light loads.

The most commonly used switch in switching regulators is the bipolar transistor. The MOSFET power transistors (VMOS and DMOS) are becoming more popular as switching regulator switches. The MOS switches, in general, switch faster than bipolar transistors, have no storage time, and so have lower switching losses. The high input impedance of MOS power switches allows the use of lower power driving circuitry. At present, MOS power switches are available that can switch about 9 A at 500 V to 28 A at 50 V. Higher power MOSFET switches are becoming available at a rapid rate. Bipolar power transistor switches can handle much more power than the MOSFET switches at the present time. Power transistors capable of switching 100 A at 120 V to 3.6 A at 800 V can be purchased. Bipolar transistors have a lower on resistance at any given power level than MOSFET power switches, so often the MOSFET power switches have more advantages in switching power supplies at higher frequencies where bipolar transistor switching losses become quite significant. The on resistance of a power MOSFET has a positive temperature coefficient as opposed to a bipolar transistor that has a negative temperature coefficient. MOSFET's are therefore less prone to thermal run away.

10-3.1 Calculation of Switching Dissipation

At high frequencies, the switching time of a transistor can become a significant portion of the period, and power dissipation during switching can become a significant portion of the total transistor power dissipation. A simple method of calculating the average switching power dissipation and the average dissipation of the transistor is shown in Figure 10.14. The maximum power dissipation occurs when I_c and V_{CE} are maximum during turn on. With a resistive load this is when I_c = $0.5\ V_{cc}/R_L$ and $V_{CE} = 0.5\ V_{cc}$. The power dissipation versus time during switching looks like a triangle. Averaging the area in watt-seconds under the triangle over a

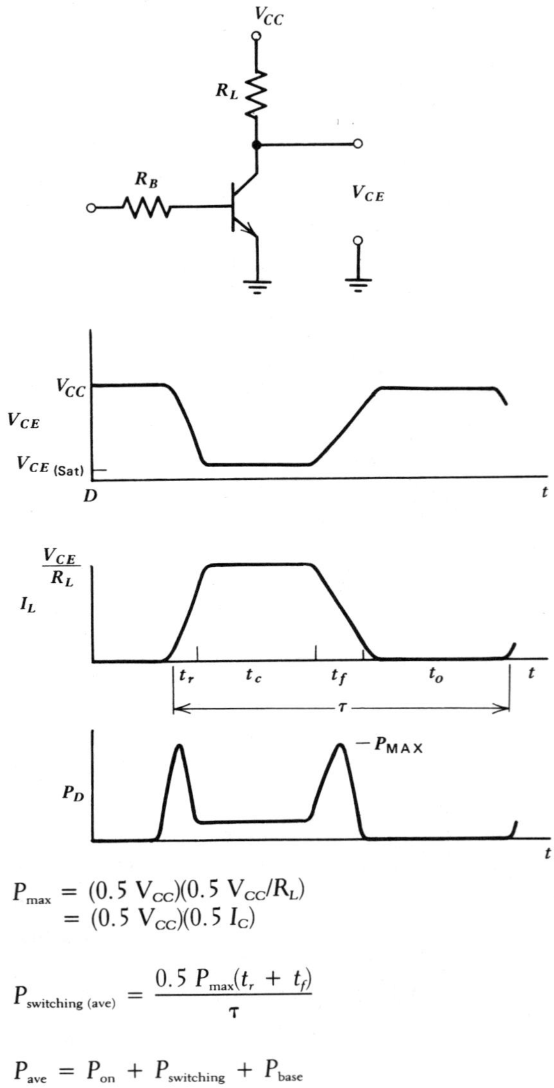

$$P_{max} = (0.5\ V_{CC})(0.5\ V_{CC}/R_L)$$
$$= (0.5\ V_{CC})(0.5\ I_C)$$

$$P_{switching\ (ave)} = \frac{0.5\ P_{max}(t_r + t_f)}{\tau}$$

$$P_{ave} = P_{on} + P_{switching} + P_{base}$$

$$P_{ave} = \frac{V_{CE(sat)} I_L t_c}{\tau} + \frac{0.5\ P_{max}(t_r + t_f)}{\tau} + \frac{V_{BE(sat)} I_B t_c}{\tau}$$

FIGURE 10.14 Transistor switching dissipation.

full cycle gives the average switching dissipation. An example will illustrate the calculation.

EXAMPLE 10-11

Calculate the average power dissipation of the transistor in Figure 10.14 if t_r = 1 μs, t_f = 2 μs, t_c = 5 μs, and t_0 = 6 μs (τ = 14 μs). I_c(sat) = V_{cc}/R_L = 1 A, V_{cc} = 10 V, I_B = 0.1 A, V_{BE}(sat) = 0.8 V, V_{CE}(sat) = 0.5 V.

SOLUTION

$$P_{max} = (0.5\ V_{cc}/R_c)\ 0.5\ V_{cc} = 0.5\ \text{A}\ (5\ \text{V}) = 2.5\ \text{W}$$

$$P_{ave\ t_r\ \text{and}\ t_f} = 0.5P_{max}(t_r + t_f)\tau = 1.25\ \text{W}\ (3\ \mu\text{s})/14\ \mu\text{s} = 0.26\ \text{W}$$

$$P_{on\ (ave)} = V_{CE}(\text{sat})I_c(\text{sat})t_c/\tau = 0.5\ \text{V}\ (1\ \text{A})(5\ \mu\text{s})/14\ \mu\text{s} = 0.17\ \text{W}$$

$$P_{base\ (ave)} = V_{BE}(\text{sat})I_B t_c/\tau = 0.8\ \text{V}\ (0.1\ \text{A})5\ \mu\text{s}/14\ \mu\text{s} = 0.028\ \text{W}$$

$$P_{ave} = P_{(t_r + t_f)\ ave} + P_{on\ (ave)} + P_{base\ (ave)} = 0.458\ \text{W}$$

10-3.2 Basic Switching Supply Configurations

This section looks at three basic switching supply configurations: step-down, step-up, and voltage inverter.

STEP-DOWN SWITCHING SUPPLY

The step-down inverter is shown in Figure 10.15*a*. It provides an output voltage at any value less than the input. The output voltage depends on V_{in} and the ratio

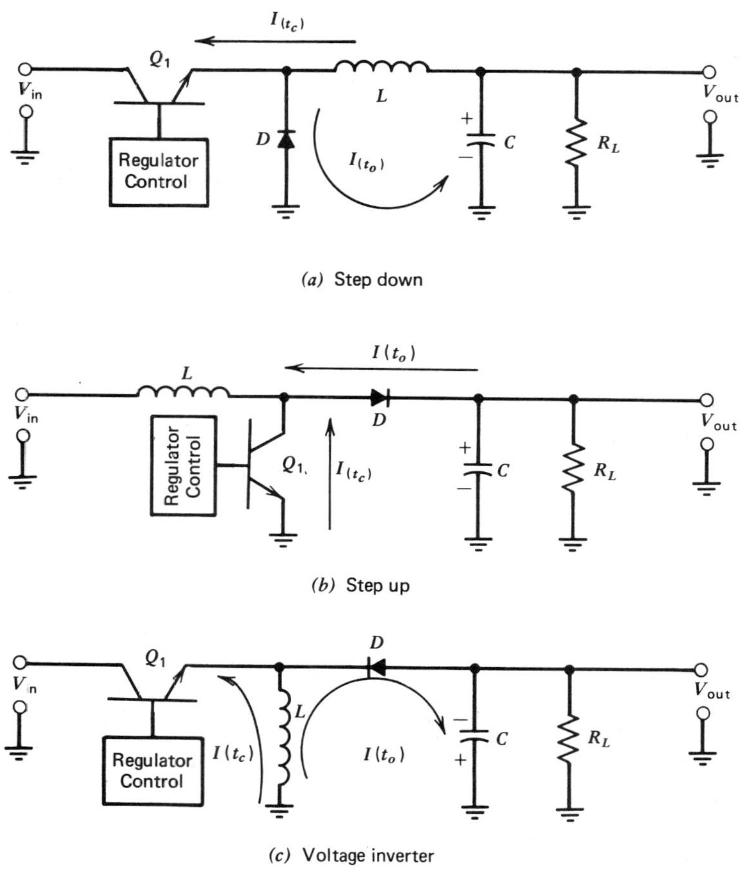

FIGURE 10.15 Basic switching regulator configurations.

of the transistor on time (t_c) and the switching period (τ).

$$V_{out} = V_{in}\left(\frac{t_c}{\tau}\right) \tag{10-16}$$

where

t_c = transistor on time
τ = $t_c + t_o$ = period
t_o = transistor off time

The operation of the step-down invertor is as follows. When Q_1 is on the inductor (L), current flows through Q_1 and R_L. I_L rises at a rate of V_L/L. The voltage across L is $V_{in} - V_{CE}(\text{sat})_{Q1} - V_{out}$. During t_c the inductor charges to a peak current (I_p) of

$$I_p = \frac{V_{in} - V_{CE}(\text{sat})_{Q1} - V_{out}}{L}$$

C provides load current until $I_L > I_{out}$. When Q_1 is turned off, the collapsing magnetic field of L causes its polarity to reverse and L becomes the source for the load. L discharges through D, which is now forward biased because of the polarity reversal of L, R_L, and C at a rate of V_L/L. The voltage across L is now $V_{out} + V_D$. L supplies current to the capacitor and load until $I_L < I_{out}$, after which C provides the load current. Q_1 will be turned on again just before $I_L = 0$.

The maximum output current of the supply is half I_p, the peak inductor current. The average current through L during t_c must be the same as during t_o. The t_c/t_o ratio is adjusted to provide the load current and voltage desired.

The inductor can be calculated by recalling that it must provide the maximum output current during the transistor off time t_o. From the well-known inductor relationship $V_L = L(\Delta I/\Delta t)$, we can find L as follows:

$$L = V_L\left(\frac{\Delta t}{\Delta I}\right) = \frac{(V_{out} + V_D)t_o}{I_p} \tag{10-17}$$

The filter capacitor can be found by recalling that:

$$C = \frac{Q}{V} = \frac{I_t}{V} = \frac{I\Delta t}{\Delta V}$$

Since C must supply I_{out} for about ½ t_c + ½ t_o, and the maximum current supplied is $I_{out} = 0.5\ I_p$, we can find C by substituting our values to get

$$C = \frac{I_p/2\ (t_c/2 + t_o/2)}{V_{ripple}} = \frac{I_p\ (t_c + t_o)}{4\ V_{ripple}} \tag{10-18}$$

where V_{ripple} is the peak-to-peak ripple.

The average output current must equal the average inductor charging current during t_c. Inductor current during a time t is $V_L(t/L)$. Equating the charge and discharge current

$$\frac{2(V_{in} - V_{CE}(sat) - V_{out})t_c}{L} = \frac{2(V_{out} + V_D)t_o}{L}$$

we get

$$\frac{t_c}{t_o} = \frac{V_{out} + V_D}{V_{in} - V_{CE}(sat) - V_{out}} \tag{10-19}$$

EFFICIENCY

If switching losses are neglected (allowed if $t_c + t_o >> t_r + t_f$), the efficiency can be found as follows.

$$V_{out} = V_{in}(t_c/\tau)$$
$$P_{out} = V_{in}(t_c/\tau)I_{out} = V_{out}I_{out}$$
$$P_{in} = P_{out} + P_{switch} + P_D$$
$$P_{in} = V_{in}(t_c/\tau)I_{out} + I_pV_{CE}(sat)(t_c/\tau) + V_D(t_o/\tau)I_p$$

where

$$(t_c/\tau)I_p = I_{in\ (ave)} \doteq I_{out\ (ave)} = I_p(t_o/\tau)$$
$$P_{in} = V_{out}I_{out} + V_{CE}(sat)I_{out} + V_DI_{out}$$

thus

$$\text{Eff} = \frac{P_{out}}{P_{in}} = \frac{V_{out}I_o}{V_{out}I_{out} + V_{CE}(sat)I_{out} + V_DI_{out}}$$

$$= \frac{V_{out}}{V_{out} + V_{CE}(sat) + V_D} \tag{10-20}$$

Other expressions for efficiency exist, depending on the factors taken into account. Note that if $V_{CE}(sat) = 1$ V and $V_D = 1$ V for $V_{out} = 10$ V and $V_{in} = 20$ V

$$\text{Eff} = \frac{10\text{ V}}{10\text{ V} + 2\text{ V}} = 83.3\%$$

for a step-down switching power supply. A series-pass power supply would have an efficiency of only 50% under the same conditions.

STEP-UP SWITCHING SUPPLY

The basic step-up switching power supply is shown in Figure 10.15*b*. Its output voltage is

$$V_{out} = V_{in}\left(\frac{t_c + t_o}{t_o}\right) \tag{10-21}$$

When Q_1 is on, L charges to a peak current $I_p = (V_L/L)t_c$. D_1 is off since V_{out}

$> V_{CE}(\text{sat})_{Q_1}$.

$$I_p = \left(\frac{V_L}{L}\right)t_c = \left(\frac{V_{in} - V_{CE}(\text{sat})}{L}\right)t_c \tag{10-22}$$

When Q_1 is turned off, the collapsing magnetic field of L causes its polarity to reverse, forward biasing D. The voltage across L adds to V_{in}, causing V_{out} to be $V_{in} + V_L - V_D$. I_L now discharges at the rate of $(V_{out} + V_D - V_{in})/L = V_L/L$ until Q_1 is turned on again. Since the current is stepped down by the ratio that the voltage is stepped up

$$I_o(\max) = I_p/2\left(\frac{t_o}{t_o + t_c}\right)$$

Relating I_p to $I_{out}(\max)$

$$I_p = 2I_{out}(\max)\left(\frac{t_o + t_c}{t_o}\right)$$

The inductor can be found by realizing it must provide output current during t_o and recalling $V_L = L(\Delta I/\Delta t)$

$$L = \left(\frac{V_{out} + V_D - V_{in}}{I_p}\right)t_o \tag{10-23}$$

The capacitor must supply load current during t_c so from $C = It/V$

$$C = \frac{I_{out}t_c}{V_{ripple}}$$

The ratio of t_c to t_o is found by equating the average inductor current. As in the step-down supply, all terms cancel except the voltage across the inductor with Q_1 on and Q_1 off.

$$\frac{t_c}{t_o} = \frac{V_{out} + V_D - V_{in}}{V_{in} - V_{CE}(\text{sat})} \tag{10-24}$$

The efficiency of a step-up inverter, which can be greater than 90%, can be found to be (see Appendix F)

$$\text{Eff} = \frac{V_{in}}{V_{in} + V_{CE}(\text{sat}) + V_D\left(\frac{t_o}{t_c + t_o}\right)}$$

VOLTAGE INVERTER SWITCHING SUPPLY

The voltage inverter is illustrated in Figure 10.15c. It provides an output voltage of opposite polarity to the input.

When Q_1 is on, the inductor charges toward V_{in} at a rate V_L/L, D is reverse biased since its anode is negative. When Q_1 is switched off, the collapsing magnetic field of the inductor causes its polarity to reverse. The reverse in polarity forward biases D_1, since its cathode is now negative with respect to its anode. L now discharges at a rate of V_L/L until Q_1 is switched on again.

The peak current that L charges to during the time Q_1 is on is

$$I_p = \left(\frac{V_{in} - V_{CE}(sat)_{Q1}}{L}\right)t_c \tag{10-25}$$

The average input current is

$$I_{in\ (ave)} = I_p/2\left(\frac{t_c}{t_c + t_o}\right) \tag{10-26}$$

and the average output current is

$$I_{out\ (ave)} = I_p/2\left(\frac{t_o}{t_c + t_o}\right) \tag{10-27}$$

So the peak input current is related to I_{out} by

$$I_p = 2\, I_{out}\left(\frac{t_c + t_o}{t_o}\right) \tag{10-28}$$

If diode and transistor losses are neglected

$$V_{out} = V_{in}\left(\frac{t_c}{t_o}\right) \tag{10-29}$$

As in the step-up and step-down inverters, L must supply current during t_o, so from $V_L = L(\Delta I/\Delta t)$

$$L = \left(\frac{|V_{out}| + V_D}{I_p}\right)t_o \tag{10-30}$$

Note that L is always

$$L = \left(\frac{V_{L(Q1\ off)}}{I_p}\right)t_o$$

The output capacitor can be calculated as previously done by realizing C must supply the load current while Q_1 is on.

$$C = \frac{It}{V} = \frac{I_{out}t_c}{V_{ripple}}$$

From the voltage relationship between t_c, t_o, V_{in}, and V_{out}

$$\frac{t_c}{t_o} = \frac{|V_{out}| + V_D}{V_{in} - V_{CE}(sat)} \tag{10-31}$$

Efficiency As before,

$$P_o = V_o I_{out}$$

and

$$P_{in} = V_o I_{out} + V_D I_{out} + V_{CE}(\text{sat}) I_{in\ (ave)}$$

but

$$I_{in\ (ave)} \cong I_{out}\left(\frac{t_c}{T}\right)$$

$$P_{in} = V_o I_{out} + V_D I_{out} + V_{CE}(\text{sat})\left(\frac{t_c}{T}\right)$$

thus

$$\text{Eff} = \frac{P_{out}}{P_{in}} = \frac{V_D I_{out}}{V_o I_{out} + V_D I_{out} + V_{CE}(\text{sat}) I_{out}\left(\frac{t_c}{T}\right)}$$

$$\text{Eff} = \frac{V_o}{V_o + V_D + V_{CE}(\text{sat})\left(\frac{t_c}{T}\right)} \tag{10-32}$$

Efficiencies can be greater than 90%.

Switching power supplies cannot be operated with no load or $V_{out} = V_{in}$, since no inductor current flows.

10-3.3 Fairchild μA78S40 Switching Power Supply Control

The Fairchild μA78S40 universal switching regulator subsystem is a versatile pulse-width modulated switching regulator capable of line and load voltage regulation of 80 dB (0.01%). It has an input voltage range from 2.5 to 40 V and can control output voltages limited by the external components used. As a step-down or step-up regulator with no external switching transistors, it can supply output voltages from 1 to 40 V. The μA78S40 draws a maximum quiescent current of 2.5 mA at V_{in} = 5 V and 3.5 mA at V_{in} = 40 V, with typical values being 1.8 mA and 2.3 mA respectively. The regulator is supplied in a 16 pin DIP that can dissipate 1.5 W in the plastic version and 1 W in the hermetically sealed version. The device is supplied in a commercial version with an operating temperature range of 0 to 70°C and a military version with an operating temperature range of −55 to 125°C. The μA78S40 can supply up to 1.5 A without external transistors.

The equivalent circuit of the μA78S40 is shown in Figure 10.16. It consists of an oscillator, two transistors for output switches, Q_1 and a driver Q_2, an AND gate, and an RS flip-flop to drive the output transistors, a reference source, a diode, and two op-amps.

The oscillator can be set between 100 Hz and 100 kHz. The timing capacitor (C_T) sets the output transistors off time (t_o). The on time is set at eight times the

FIGURE 10.16 Fairchild μA 78540 universal switching regulator subsystem. *(Courtesy of Fairchild Camera and Instrument Corporation)*

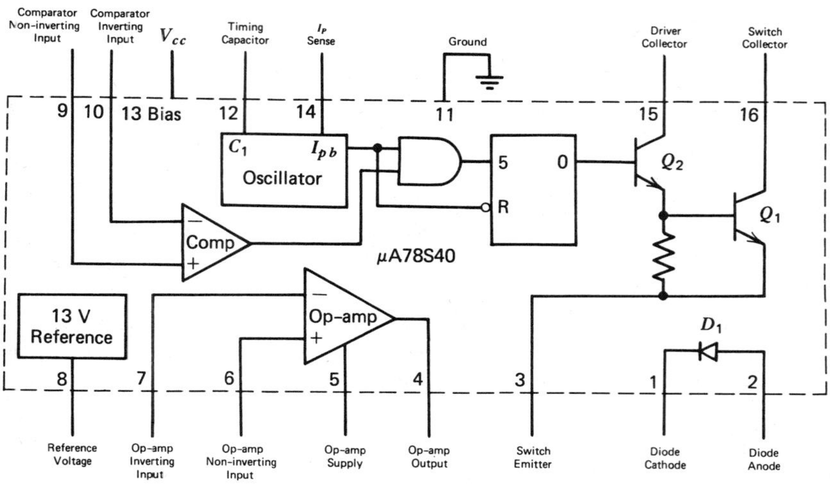

(a) Block diagram

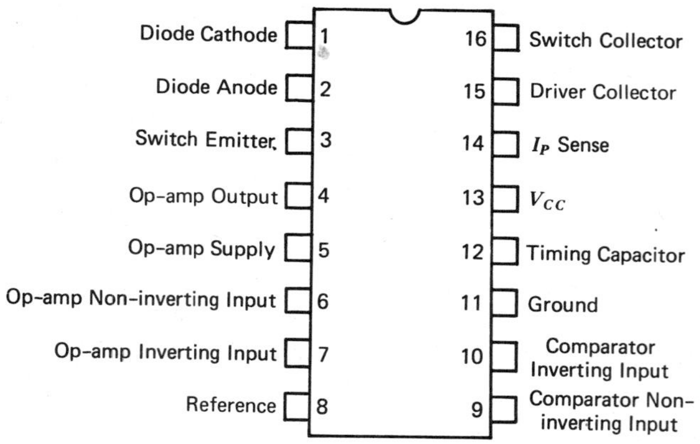

(b) Connection diagram 16-pin DIP (top view)

The μA78S40 is a Monolithic Regulator Subsystem consisting of all the active building blocks necessary for switching regulator systems. The device consists of a temperature-compensated voltage reference, a duty-cycle controllable oscillator with an active current limit circuit, an error amplifier, high-current, high-voltage output switch, a power diode, and an uncommitted operational amplifier. The device can drive external *npn* or *pnp* transistors when currents in excess of 1.5 A or voltages in excess of 40 V are required. The device can be used for step-down, step-up or inverting switching regulators as well as for series-pass regulators. It features wide supply voltage range, low standby power dissipation, high efficiency, and low drift. It is useful for any stand-alone, low part-count switching system and works extremely well in battery operated systems.

- Step-up, step-down, or inverting switching regulators
- Output adjustable from 1.3 to 40 V
- Output currents to 1.5 A without external transistors
- Operation from 2.5 to 40 V input
- Low standby current drain

- 80 dB line and load regulation
- High gain, high current, independent op-amp

(*c*) General description

---

off time internally. The oscillator has a temperature-compensated current sensor with a 0.33 V threshold (V_{cs} = 0.33 V) that turns off the output transistor whenever the peak current is sensed. I_p is used to vary the duty cycle and thus t_c in the regulator. The lower limit of t_c or t_o is 10 μs.

The output transistors can block 40 V and supply up to 1.5 A. Q_2 is the driver for Q_1. Q_1 and Q_2 can be connected as a Darlington pair, or Q_2 can be used with an external resistor to provide increased base drive to Q_1 as is necessary in the step-up supply. V_{CE}(sat) of Q_1 is 1.1 V typically, and 1.3 V max at I_C = 1 A. The β of Q_1 is 70 with I_C = 1 A and V_{CE} = 5 V, so for V_{CE} = 1 V at I_C = 1 A, a value of β_{Q_1} = 20 will be used for calculations.

The AND gate output is connected to the flip-flop input to turn Q_1 off when $V_{out} > V_{ref}$. It is part of the regulating circuitry.

The diode will block 40 V and drop 1.5 V in the forward direction with I_F = 1 A. The typical V_D = 1.25 V at 1 A.

The reference voltage is temperature compensated and equal to 1.31 V maximum, 1.24 V typically, and 1.18 V minimum.

A high gain-error amplifier is supplied as a comparator for regulation. An auxiliary op-amp for use with the voltage inverter configuration or for use as a driver for an auxiliary series-pass regulator is supplied. The auxiliary amplifier can supply up to 100 mA with its output positive and up to 35 mA with its output supplying current to +V through a resistor.

The μA78S40 specifications, in Appendix C, have design equations and typical applications.

OPERATION

We will discuss the operation of the μA78S40 in the step-down circuit shown in Figure 10.17*a*. C_T is chosen for the desired t_o, and R_{sc} is chosen so that $V_{R_{sc}}$ = 0.33 V at the desired peak inductor current I_p. Any time that $V_{out} > V_{ref}$ the comparator output sends a low to one input of the AND gate turning Q_1 off. After C_T times out, the oscillator turns Q_1 on again until either I_p is reached or $V_{out} > V_{ref}$. Regulation occurs as follows: If V_{out} decreases, the inductor's average discharge current increases, thus t_c is increased since it takes longer to charge the inductor to I_p. This causes V_{out} to rise to its original value. If V_{out} increases, then the inductor discharges less during t_o, so that I_p is reached sooner when Q_1 is on. This reduces t_c and, therefore, the output voltage. In addition, if $V_{out} > V_{ref}$, the comparator turns off Q_1, reducing t_c and, therefore, V_{out}. If t_o or t_c is less than 10 μs, switching losses become severe.

FIGURE 10.17 μA 78540 switching supply applications.

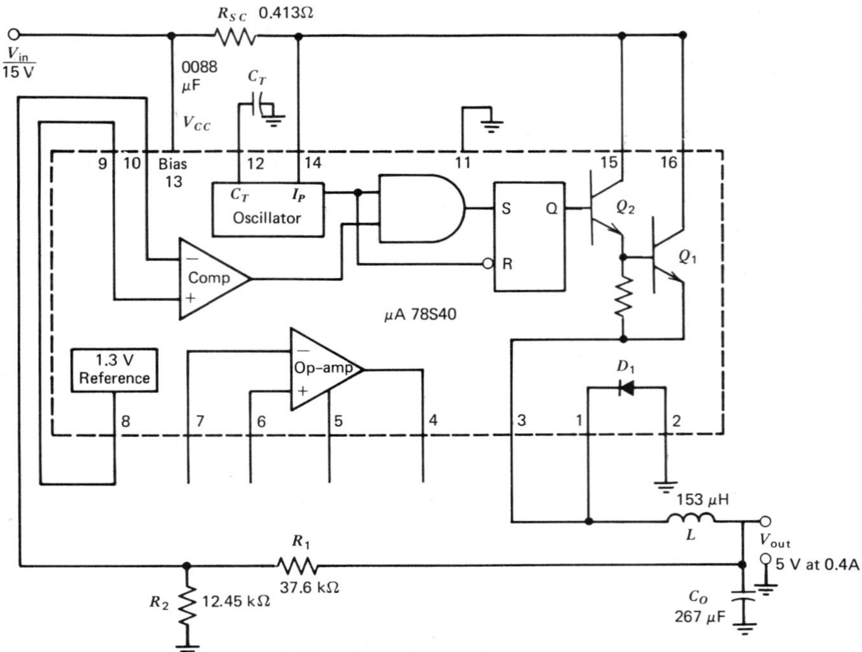

(a) Step-down voltage regulator

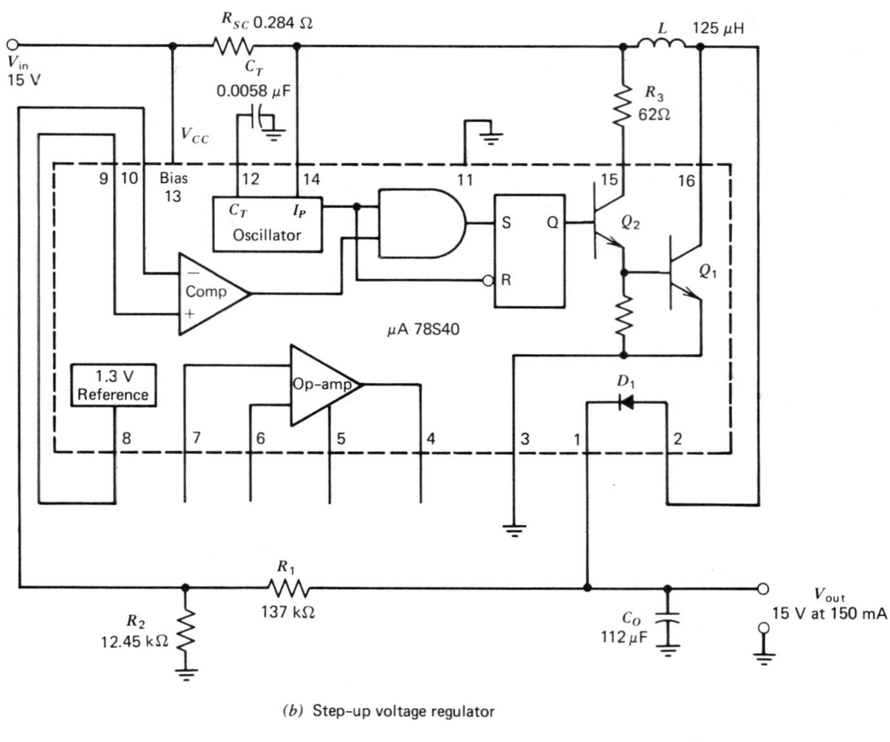

(b) Step-up voltage regulator

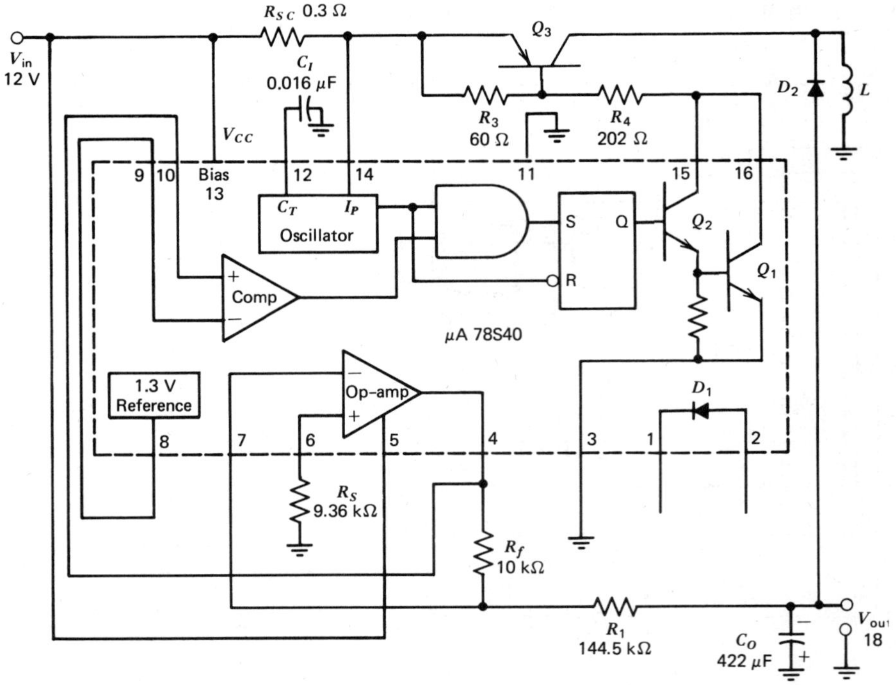

(c) Inverting voltage regulator

μA78S40 APPLICATIONS

We will show the calculation procedure for the μA78S40 applications by example. The calculations for any type of switching regulator IC are very similar in these configurations.

EXAMPLE 10-12: STEP DOWN

Set up a μA78S40 switching regulator as a 5 V, 400 mA step-down switching power supply. $V_{in} = 15$ V. Limit the output ripple to 25 mV_{pp}. Set $f_{osc} = 30$ kHz. See Figure 10.17*a*.

SOLUTION

We will use typical values from the specification sheet so

$$V_{CE}(\text{sat})_{Q_1} = 1.1 \text{ V}$$

$$V_D = 1.25 \text{ V}$$

$$V_{ref} = 1.245 \text{ V}$$

Since $I_{out}(\max) = I_p/2$

$$I_p = 2\,I_o(\max) = 800 \text{ mA} \quad \text{or} \quad 0.8 \text{ A}$$

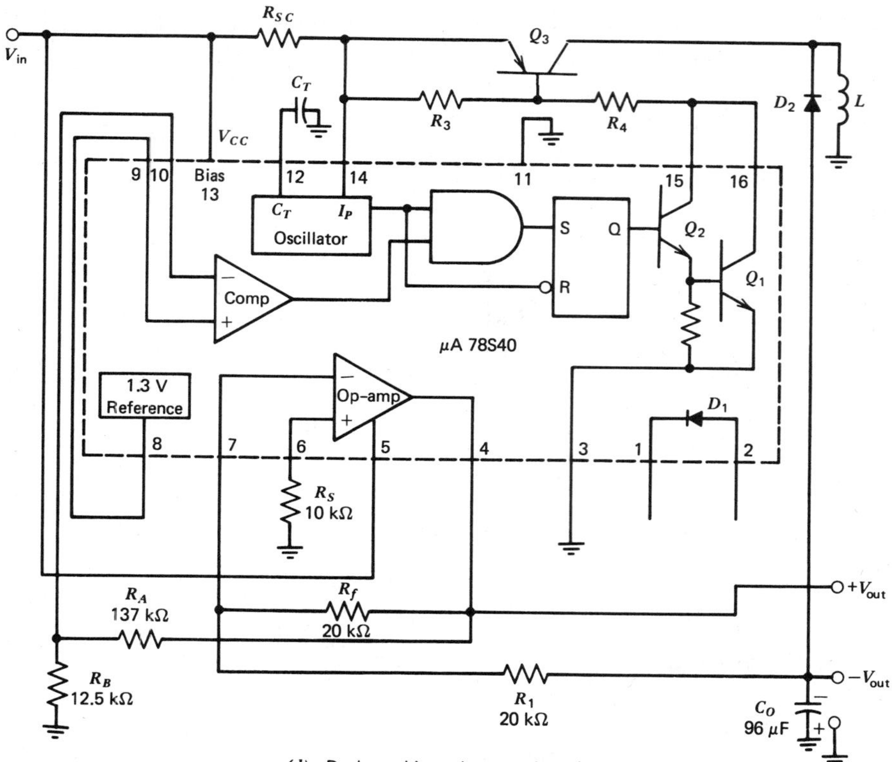

(d) Dual–tracking voltage regulator (voltage inverter and inverting op–amp)

Next, we need t_c and t_o

$$t_c + t_o = 1/f_{osc} = 1/30\text{ kHz} = 33.33\ \mu\text{s}$$

from Equation 10-19,

$$\frac{t_c}{t_o} = \frac{V_{out} + V_D}{V_{in} - V_{CE}(\text{sat}) - V_{out}} = \frac{5\text{ V} + 1.25\text{ V}}{15\text{ V} - 1.1\text{ V} - 5\text{ V}} = 0.702$$

$$t_c = 0.702\ t_o$$

Substituting into $t_c + t_o = 33.33\ \mu\text{s}$

$$0.702\ t_o + t_o = 33.33\ \mu\text{s}$$

$$t_o = \frac{33.3\ \mu\text{s}}{1.702} = 19.58\ \mu\text{s}$$

$$t_c = \tau - t_o = 33.33\ \mu\text{s} - 19.58\ \mu\text{s} = 13.74\ \mu\text{s}$$

Note. If the load current is to vary widely, we would have to select a longer τ. If the load current is relatively constant ($\Delta I_{out} < 40\%$), we can continue since t_c = 10 μs minimum will be adequate.

Calculate L from Equation 10-17

$$L = \left(\frac{V_{out} + V_D}{I_p}\right)t_o$$
$$= \left(\frac{5\text{ V} + 1.25\text{ V}}{0.8\text{ A}}\right)19.58\ \mu\text{s}$$
$$= 153\ \mu\text{H}$$

From Equation 10-18 we find C_o

$$C_o = \frac{I_{out}(t_c + t_o)}{4\,V_{ripple}} = \frac{0.8\text{ A }(33.33\ \mu\text{s})}{4\,(25\text{ mV})} = 267\ \mu\text{F}$$

R_{sc} is found from

$$R_{sc} = \frac{V_{cs}}{I_p} = \frac{0.33\text{ V}}{0.8\text{ A}} = 0.413\ \Omega$$

From the specifications

$$C_T = 45 \times 10^{-5}\, t_{off}\,(\mu\text{s})$$
$$= 45 \times 10^{-5}\,(19.58 \times 10^{-6}) = 0.088\ \mu\text{F}$$

A tight tolerance capacitor should be used.

Finally, if $I_{divider}\ (I_D) = 0.1$ mA

$$R_1 = \frac{V_{out} - V_{ref}}{I_D} = \frac{5\text{ V} - 1.245\text{ V}}{0.1\text{ mA}} = 37.55\text{ k}\Omega$$

$$R_2 = \frac{V_{ref}}{I_D} = \frac{1.245\text{ V}}{0.1\text{ mA}} = 12.45\text{ k}\Omega$$

$$P_D \cong \frac{I_D V_{CE}(\text{sat})\, t_c}{\tau} + \frac{I_{out} V_D t_o}{\tau} + V_{in}\, I_{quiescent}$$
$$= \frac{0.4\text{ A }(1.1\text{ V})\ 13.74\ \mu\text{s}}{33.33\ \mu\text{s}} + \frac{0.4\text{ A }1.25\text{ V }29.58\ \mu\text{s}}{33.33\ \mu\text{s}} + 15\text{ V }(2.5\text{ mA})$$
$$= 0.18\text{ W} + 0.29\text{ W} + 37.5\text{ mW} = 0.513\text{ W}$$

EXAMPLE 10-13: STEP UP

Connect a μA78S40 as a step-up inverter such as the one shown in Figure 10.17*b*. V_{out} = 15 V, I_{out} = 150 mA, V_{in} = 5 V, f_{osc} = 20 kHz (τ = 50 μs), and V_{ripple} = 50 mV_{pp}. Use typical values from specification sheet.

SOLUTION

From Equation 10-14

$$\frac{t_c}{t_o} = \frac{V_{out} + V_D - V_{in}}{V_{in} - V_{CE}(sat)} = \frac{15\text{ V} + 1.25\text{ V} - 5\text{ V}}{5\text{ V} - 1.1\text{ V}} = 2.88$$

$$t_c = 2.88\, t_o$$

so

$$t_o + 2.88\, t_o = 50\ \mu\text{s}$$

$$= \frac{50\ \mu\text{s}}{3.88} = 12.88\ \mu\text{s}$$

$$t_c = \tau - t_o = 50\ \mu\text{s} - 12.88\ \mu\text{s} = 37.12\ \mu\text{s}$$

For a step-up inverter,

$$I_p = 2\, I_{out}(\text{max})\left(\frac{t_c + t_o}{t_o}\right)$$

$$= 2\,(0.15\text{ A})\,\frac{50\ \mu\text{s}}{12.88\ \mu\text{s}} = 1.16\text{ A}$$

Thus

$$R_{sc} = \frac{V_{cs}}{I_p} = \frac{0.33\text{ V}}{1.16\text{ A}} = 0.284\ \Omega$$

From Equation 10-23,

$$L = \left(\frac{V_{out} + V_D - V_{in}}{I_p}\right)t_o = \left(\frac{15\text{ V} + 1.25\text{ V} - 5\text{ V}}{1.16\text{ A}}\right)12.88\ \mu\text{s} = 125\ \mu\text{H}$$

Since R_3 must supply Q_1 base current,

$$R_3 = \frac{V_{in} - V_{BE_{Q1}} + V_{CE_{Q2}}}{I_p/\beta_{Q2}} = \frac{5\text{ V} - 1.4\text{ V}}{1.16\text{ A}/20} = 62\ \Omega$$

From $C = It/V$,

$$C_o = \frac{I_{out} t_c}{V_{ripple}} = \frac{0.15\text{ A}\,(37.12\ \mu\text{s})}{50\text{ mV}} = 112\ \mu\text{F}$$

From the specification sheet,

$$C_T = 45 \times 10^{-5}\,(t_{off}) = 45 \times 10^{-5}\,(12.88\ \mu\text{s})$$

$$= 0.0058\ \mu\text{F}$$

Finally, with $I_D = 0.1$ mA,

$$R_1 = \frac{V_{out} - V_{ref}}{I_D} = \frac{15\text{ V} - 1.245\text{ V}}{0.1\text{ mA}} = 137\text{ k}\Omega$$

$$R_2 = \frac{1.245\text{ V}}{0.1\text{ mA}} = 12.45\text{ k}\Omega$$

Checking P_D,

$$P_D = I_{in\ (ave)} V_{CE}(\text{sat}) + I_{out} V_D + V_{in} I_{quiescent}$$

$$= (I_p/2)\, V_{CE}(\text{sat})\left(\frac{t_c}{\tau}\right) + (I_p/2)\, V_D\left(\frac{t_o}{\tau}\right) + V_{in} I_{quiescent}$$

$$= \frac{0.58\text{ A }(1.1\text{ V})\ 37.12\ \mu\text{s}}{50\ \mu\text{s}} + \frac{0.58\text{ A }(1.25\text{ V})\ 12.88\ \mu\text{s}}{50\ \mu\text{s}} + 5\text{ V }2.5\text{ mA}$$

$$= 0.473\text{ W} + 0.187\text{ W} + 12.5\text{ mW} = 0.68\text{ W}$$

EXAMPLE 10-14: VOLTAGE INVERTER

Construct an inverting power supply using a μA78S40 as shown in Figure 10.17*c*. Let $V_{in} = 12$ V, $V_{out} = -18$ V, $I_{out} = 200$ mA, $f_{osc} = 10$ kHz ($\tau = 100\ \mu$s), and $V_{ripple} = 30\ \text{mV}_{pp}$. Use typical values from the specification sheet.

SOLUTION

An auxillary *pnp* transistor and diode are needed so that the pins on the μA78S40 will not go negative. The auxiliary op-amp will be used to connect $-V_{out}$ to $+V_{ref}$. $R_s = R_1 \parallel R_2$ prevents offset voltage from bias current. The transistor and diode will be selected after I_p is found. Assume $V_{CE}(\text{sat}) = 1$ V and $V_D = 1$ V.

From Equation 10-31,

$$\frac{t_c}{t_o} = \frac{|V_{out}| + V_D}{V_{in} - V_{CE}(\text{sat})} = \frac{18\text{ V} + 1\text{ V}}{12\text{ V} - 1\text{ V}} = 1.73$$

$$t_c = 1.73\, t_o$$

solving for t_o from $t_c + t_o = \tau$

$$1.73\, t_o + t_o = 100\ \mu\text{s}$$

$$t_o = 100\ \mu\text{s}/2.73 = 36.6\ \mu\text{s}$$

$$t_c = 63.4\ \mu\text{s}$$

From Equation 10-28,

$$I_p = 2\, I_{out}\left(\frac{t_o + t_c}{t_o}\right) = 1.1\text{ A}$$

We now select the transistor to have the following specifications. $V_{BE}(\text{sat}) = 0.8$ V, $BV_{CEO} = 40$ V, $I_C(\text{max}) = 2$ A, $V_{CE}(\text{sat}) = 1$ V at $I_C = 1$ A, and $\beta_{min} = 30$ at

$I_C = 1$ A. The diode is selected to have a breakdown voltage of 50 V and forward voltage drop $V_F = 1$ V at 1 A.

We now calculate L from Equation 10-30.

$$L = \left(\frac{|V_{out}| + V_D}{I_p}\right)t_o = \left(\frac{18\text{ V} + 1\text{ V}}{1.1\text{ A}}\right)36.6\ \mu\text{s} = 632\ \mu\text{H}$$

C_o is calculated as before

$$C_o = \frac{I_{out}t_c}{V_{ripple}} = \frac{0.2\text{ A }(63.4\ \mu\text{s})}{30\text{ mV}} = 422\ \mu\text{F}$$

From the specifications

$$C_T = 45 \times 10^{-5}\,(t_o) = 45 \times 10^{-5}\,(36.6\ \mu\text{s}) = 0.016\ \mu\text{F}$$

$$R_{sc} = \frac{V_{cs}}{I_p} = \frac{0.33\text{ V}}{1.1\text{ A}} = 0.3\ \Omega$$

To calculate R_3 and R_4, we find $I_{B_{Q3}}$

$$I_{B_{Q3}} = I_p/\beta_{min} = 1.1\text{ A}/30 = 36.6\text{ mA}$$

If we allow Q_1 to supply 50 mA

$$R_4 = \frac{V_{in} - V_{CE}(\text{sat})_{Q1} - V_{BE_{Q3}}}{I_{C_{Q1}}} = \frac{10.1\text{ V}}{50\text{ mA}} = 202\ \Omega$$

$$R_3 = \frac{V_{BE_{Q3}}}{50\text{ mA} - I_{B_{Q3}}} = \frac{0.8\text{ V}}{13.4\text{ mA}} = 60\ \Omega$$

For the inverting amplifier

$$\frac{V_{out}}{V_{in}} = \frac{R_f}{R_1} = \frac{V_{ref}}{V_{out\ supply}}$$

The choice of the resistor R_f is not critical so let $R_f = 10\text{ k}\Omega$

$$R_1 = R_f\left(\frac{V_{out\ supply}}{V_{ref}}\right) = 10\text{ k}\Omega\left(\frac{18\text{ V}}{1.245\text{ V}}\right) = 144.5\text{ k}\Omega$$

$$R_s = R_1 \parallel R_f = 9.36\text{ k}\Omega$$

Since I_{Q_1} peak $= 50$ mA, the chip is well below $P_D(\text{max})$. The switching transistor should be able to dissipate $P_{max} = \frac{1}{2} V_{in}\ \frac{1}{2} I_p = 3.3$ W.

EXAMPLE 10-15: DUAL-TRACKING REGULATOR

Construct a dual-tracking ±15 V regulator with a μA78S40. *Note.* Tracking means that the positive and negative voltages always change together, so if the +15 V drops to +14.99 V, the −15 V goes to −14.99 V. Let $I_{out} = 100$ mA for both supplies. $V_{in} = 22$ V. Use typical specification sheet values and the same *pnp* transistor and diode that were used in Example 10-14. Let $f_{osc} = 15$ kHz ($\tau = 66.66\ \mu s$). The circuit is shown in Figure 10.17*d*.

SOLUTION

pnp transistor, Q_3:

$$P_D = 5\text{ W},\quad BV_{CEO} = 40\text{ V},\quad I_{out}(\max) = 2\text{ A},\quad V_{CE}(\text{sat}) = 1\text{ V},\quad V_{BE}(\text{sat}) = 0.8\text{ V},$$

and

$$\beta_{min} = 30 \text{ at } I_C = 1\text{ A}$$

Diode:

$$\text{Reverse breakdown voltage} = 40\text{ V},\quad V_F = 1\text{ V at } 1\text{ A}$$

The auxiliary op-amp of the μA78S40 will be used as the positive supply since its output cannot go negative. The switching regulator will be used in the voltage invertor configuration.

From Equation 10-31,

$$\frac{t_c}{t_o} = \frac{|V_{out}| + V_D}{V_{in} - V_{CE}(\text{sat})} = \frac{15\text{ V} + 1\text{ V}}{22\text{ V} - 1\text{ V}} = 0.76$$

From $t_c + t_o = \tau$

$$0.76\, t_o + t_o = \tau = 66.66\ \mu s$$

$$t_o = \frac{66.66\ \mu s}{1.76} = 37.88\ \mu s$$

$$t_c = \tau - t_o = 66.66\ \mu s - 37.88\ \mu s = 28.78\ \mu s$$

Since

$$I_p = 2\, I_{out}\left(\frac{t_c + t_o}{t_o}\right) = 200\text{ mA}\left(\frac{66.66\ \mu s}{37.88\ \mu s}\right) = 0.352\text{ A}$$

$$L = \frac{(|V_{out}| + V_D)t_o}{I_p} = \frac{16\text{ V }(37.88\ \mu s)}{0.352\text{ A}} = 1.72\text{ mH}$$

If $V_{ripple} = 30\text{ mV}_{pp}$

$$C_o = \frac{I_{out}\, T_c}{V_{ripple}} = \frac{100\text{ mA } 28.78\ \mu s}{30\text{ mV}} = 96\ \mu F$$

The voltage divider for the error amplifier will be supplied from the +15 V supply, which is the output of the auxiliary op-amp connected as an inverting amplifier with $A_v = 1$. Let $I_D = 0.1$ mA

$$R_1 = R_F = 20\text{ k}\Omega \text{ chosen arbitrarily}$$

$$R_A = \frac{V_{out} - V_{ref}}{I_D} = \frac{15\text{ V} - 1.245\text{ V}}{0.1\text{ mA}} = 137.5\text{ k}\Omega$$

$$R_B = \frac{V_{ref}}{I_D} = \frac{1.245\text{ V}}{0.1\text{ mA}} = 12.45\text{ k}\Omega$$

$$R_{sc} = \frac{V_{cs}}{I_p} = \frac{0.33\text{ V}}{0.352\text{ A}} = 0.938\ \Omega$$

Now we solve for R_4 and R_3.

$$I_{B_{Q3}}(\text{max}) = \frac{I_p}{\beta_{min}} = \frac{0.352\text{ A}}{30} = 11.8\text{ mA}$$

Let $I_{C_{Q1}} = 15$ mA.

$$R_3 = \frac{V_{BE}(\text{sat})_{Q3}}{I_{C_{Q1}} - I_{B_{Q3}}} = \frac{0.8\text{ V}}{3.2\text{ mA}} = 250\ \Omega$$

$$R_4 = \frac{V_{in} - V_{CE_{Q1}} - V_{BE_{Q3}}}{I_{C_{Q1}}} = \frac{22\text{ V} - 1.1\text{ V} - 0.8\text{ V}}{15\text{ mA}}$$

$$= 1.34\text{ k}\Omega$$

Finally,

$$C_T = 45 \times 10^{-5}\,(t_o)$$
$$= 45 \times 10^{-5}\,(37.88\ \mu\text{s}) = 0.017\ \mu\text{F}$$

10-3.4 Free-Running Switching Regulator

The circuit shown in Figure 10.18*a* is a self-oscillating switching regulator. It requires only a stable reference voltage, a high gain differential amplifier, and two voltage dividers to work. Linear IC regulators, such as the μA723, used as switching regulators, work on the same principle. The μA723, in Figure 10.18*b*, provides the stable reference voltage, the high gain op-amp comparator, and current to drive the switch.

The circuit of Figure 10.18*a* works as follows. When Q_1 is on, the voltage at point 3 is approximately V_{in}. Thus the voltage at point 1 is slightly above V_{ref} ($R_3 >> R_4$). The current in L rises at a rate $(V_{in} - V_{CE_{Q1}} - V_{out})/L$, providing current to C and the load. The voltage at point 2 rises as the output voltage rises from L, charging C. When the voltage at point 2 is greater than the voltage at point 1, the output of the op-amp goes low, turning Q_1 off. The voltage at point 3 now becomes $-V_D$ and the voltage at point 1 becomes slightly less than V_{ref} (once again $R_3 >> R_4$), so Q_1 is held off. As the inductor discharges at the rate of $(V_{out} - V_D)/L$, the voltage at point 2 drops. When the voltage at point 2 is less than the voltage at point 1, the op-amp output switches positive, turning Q_1 on again and the process repeats. The circuit operates with a small square-wave ripple at point 1; some output ripple is necessary. The output ripple can be reduced by making C larger. The capacitor and load current set the time required for the output voltage to decrease or increase to the voltage at point 1, and thus the oscillating

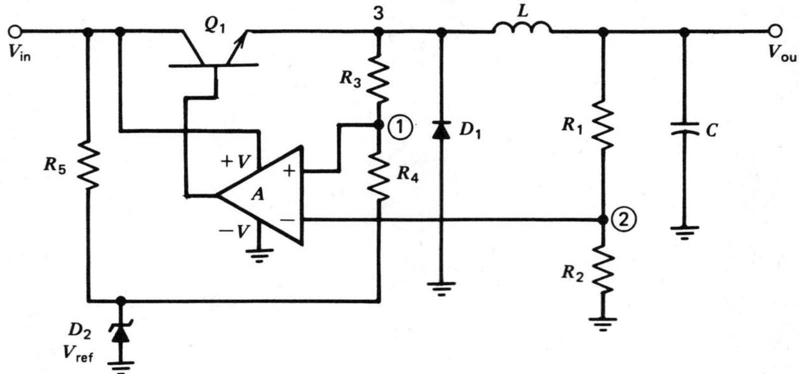

(a) **Op-amp free-running step-down switching regulator**

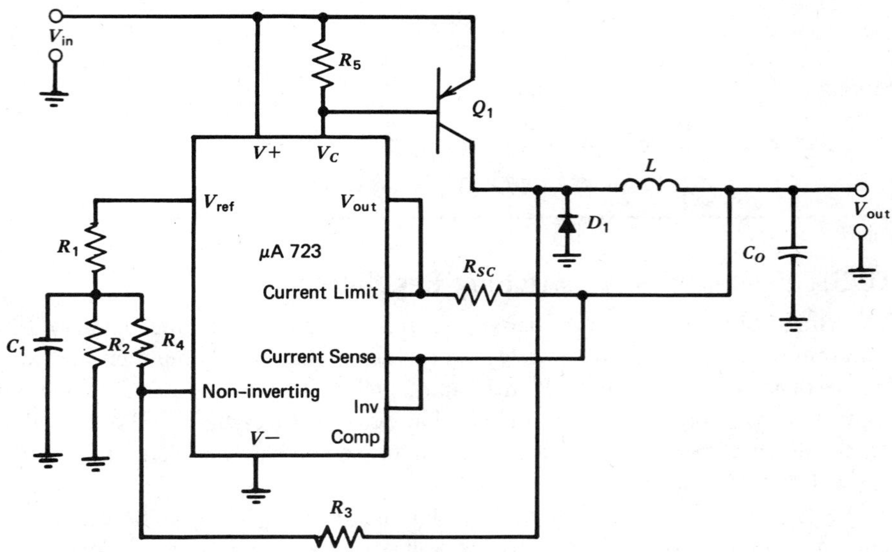

(b) **μA 723 as free-running step-down switching regulator.**

FIGURE 10.18 Free-running switching regulators.

frequency. The frequency changes with load current, and this can be a distinct disadvantage.

The center voltage of the square wave is V_R. From our previous discussion we know V_{out} will attempt to be

$$V_{out} = V_{ref}\left(\frac{R_1 + R_2}{R_2}\right)$$

The ripple at the output will then be the amplified ripple at point 1. Thus:

$$V_{ripple}\,(\text{out}) = V_{ripple}\,(\text{pt. 1})\left(\frac{R_1 + R_2}{R_2}\right) \tag{10-33}$$

The ripple at point 1 is

$$V_{ripple}\,(\text{pt. 1}) = \frac{(V_{in} - V_{CE_{Q1}} - V_{ref})R_4}{R_3 + R_4} + \frac{(V_{ref} + V_D)R_4}{R_3 + R_4}$$

therefore, after combining terms

$$V_{ripple}\,(\text{pt. 1}) = \frac{R_4}{R_3 + R_4}(V_{in} - V_{CE_{Q1}}) \tag{10-34}$$

If R_1 and R_2 are used ($V_{ref} \neq V_{out}$), then the ripple at point 2 is

$$V_{ripple}\,(\text{pt. 2}) = (V_{in} - V_{CE_{Q1}})\left(\frac{R_4}{R_3 + R_4}\right)\left(\frac{R_2}{R_1 + R_2}\right)\left(\frac{R_1 + R_2}{R_2}\right)$$

$$V_{ripple}\,(\text{pt. 2}) = V_{ripple}\,(\text{pt. 1})$$

The last condition should occur at the frequency of oscillation.

To calculate the components for the free-running switching regulator, find t_c and t_o as in the previous example for the type voltage regulator desired (step-down for Figures 10.18*a* and *b*). Calculate L as shown before. Calculate the V_{ripple} allowed at point 1 from the output ripple allowed from Equation 10-33. If R_1 and R_2 are not in the circuit ($V_{ref} = V_{out}$ as set up in Figure 10.18*b*), then V_{ripple} at point 1 is equal to the allowed output ripple. The capacitor is chosen as in the previous examples, and R_3 and R_4 are chosen from:

$$\frac{R_4}{R_3} \cong \frac{V_{ripple}\,(\text{pt. 1})}{V_{in} - V_{CE_{Q1}}} \tag{10-35}$$

if $R_3 >> R_4$. An example follows.

EXAMPLE 10-16

Calculate the components for a step-down voltage regulator using a Fairchild μA723 as the output transistor driver, regulated voltage reference, and amplifier, as shown in Figure 10.18*b*. Let $V_{out} = 5$ V and $V_{in} = 12$ V, $I_{out} = 400$ mA. $V_{ripple} = 30\ \text{mV}_{pp}$.

SOLUTION

First, find t_c and t_o. The diode and transistor Q_1 will be selected to have "on" voltage drops of about 1 V. From Equation 10-19

$$\frac{t_c}{t_o} = \frac{V_{out} + V_D}{V_{in} - [V_{out} + V_{CE}(\text{sat})]} = \frac{6\text{ V}}{6\text{ V}} = 1$$

Therefore

$$t_c = t_o$$

Select the oscillating frequency at full load. $f_{osc} = 10$ kHz, thus $t_c = t_o = 50\ \mu$s. Since for a step-down

$$I_p = 2\ I_{out}(\max) = 2\ (400\text{ mA}) = 0.8\text{ A}$$

$$L = \frac{(V_{out} + V_D)\ t_o}{I_p} = \frac{6\text{ V }50\ \mu\text{s}}{0.8\text{ A}} = 375\ \mu\text{H}$$

$$C_o = \frac{I_p\ \tau}{4\ V_{ripple}} = 667\ \mu\text{F}$$

The reference voltage of the μA723 is greater than V_{out} so V_{ref} is voltage divided down to V_{out} by R_1 and R_2. Let $I_D = 1$ mA

$$R_1 = \frac{V_{ref} - V_{out}}{I_D} = \frac{7.15\text{ V} - 5\text{ V}}{1\text{ mA}} = 2.15\text{ k}\Omega$$

$$R_2 = \frac{V_{out}}{I_D} = \frac{5\text{ V}}{1\text{ mA}} = 5\text{ k}\Omega$$

Select $R_3 = 500$ kΩ, from Equation 10-35

$$R_4 = R_3\left(\frac{V_{ripple}}{V_{in} - V_{CE_{Q1}}}\right) = 500\text{ k}\Omega\left(\frac{30\text{ mV}}{11\text{ V}}\right) = 1.36\text{ k}\Omega$$

The transistor switch, Q_1, must have $\beta \geqslant I_p/I_{out}$ (723) $= 0.8$ A/150 mA $= 5.3$. This condition is easily satisfied. Select Q_1 to have $\beta = 20$ at $V_{CE} = 1$ V and $I_C = 1$ A. I_B of Q_1 at I_p is now 0.8 A/20 = 40 mA. Let $I_{R_5} = 0.25\ I_{B_{Q1}}$ where $V_{BE_{Q1}} = 0.8$ V.

$$R_5 = \frac{V_{BE_{Q1}}}{0.25\ I_{B_{Q1}}} = \frac{0.8\text{ V}}{10\text{ mA}} = 800\ \Omega$$

As the load decreases, the frequency of the free-running PWM switching regulator will increase. The varying frequency output is often not acceptable because of the unpredictable harmonics the switching frequency variation causes.

10-4 MOTOROLA MC3420 SWITCHMODE REGULATOR

The MC3420 is a versatile switching regulator whose operation is similar to that of most IC switching power supply regulators.

10-4.1 More Inverter Configurations

The Motorola MC3420/3520 can be used to control any of the circuits in Section 10-3.1 as well as push-pull, half-bridge, and bridge-switching power supply configurations. The last three configurations for switching power supplies are shown in Figure 10.19. The configurations can control more power than the single-switch configurations since one of the transistors is on, supplying load current, most of the time. All three configurations provide input-to-output voltage isolation via the power transformer and can step up or step down the output voltage by transformer action.

PUSH-PULL

The push-pull switching power supply configuration, shown in Figure 10.19*a*, is a simple circuit capable of delivering high power. The output is a series of positive and negative square pulses that are rectified by the diodes in the secondary. A secondary with no center tap, with a full wave bridge, could just as easily have been used on the output. When Q_1 is on, current flows through the side of the transformer primary connected to the collector of Q_1. The autotransformer action of T_1 causes the collector voltage of Q_2 to be 2 V_{cc} when Q_1 is on and vice-versa. When Q_2 is turned on and Q_1 off, Q_2 delivers current to the other side of the T_1 primary and to the load by transformer action.

If Q_2 is turned on, just as the base of Q_1 goes to ground, or preferably negative, a current flows through the collector of Q_1, with V_{CE} of Q_1 at 2 V_{cc}, until the storage time of Q_1 is over. The same thing occurs if Q_2 is turned off just as Q_1 is turned on. The power dissipation during the storage time of the transistors (P_D = 2 V_{cc} I_C) is very high and can lead to reliability problems or the use of oversized transistors for Q_1 and Q_2. Most push-pull inverter controls provide for a *dead time* equal to the turn off time of the transistors between the turn off voltage of the "on" transistor and the application of base drive to the "off" transistor. This dead time greatly improves the reliability of the push-pull inverter.

Another problem with the push-pull inverter is that if the on times of the transistors are not equal, or if the saturation voltages of the two transistors are not fairly close, a small dc current flows in the T_1 primary each half-cycle. This can cause T_1 to saturate, which will cause the output transistors to burn up. A push-pull inverter must have close symmetry for t_c, V_{CE}(sat), and I_C of both Q_1 and Q_2.

HALF BRIDGE

The half bridge is shown in Figure 10.19*b*. The half bridge uses two capacitors of equal size to store energy. The half-bridge output transformer has an alternate polarity voltage applied each half-cycle, so it does not have the symmetry problem that the push-pull circuit has. The transistors Q_1 and Q_2 connect the T_1 primary alternately to ground and V_{in}. The voltage on the transformer is 0.5 V_{in} when Q_1 is on and (V_{in} − 0.5 V_{in}) = 0.5 V_{in} when Q_2 is on, since the capacitors each are charged to 0.5 V_{in}. The power out of the half bridge is 0.5 V_{in} (I_C), thus I_C must be high for high power from a given V_{in}. Q_1 and Q_2 must block only V_{in} when they are off.

Dead time between the turn off of one transistor and the turn on of the other is necessary; otherwise, there is no limit to the current that can flow through the transistor that is turned off during its storage time.

Diodes D_1 provide a limit to the voltage induced on the T_1 primary when Q_1 and Q_2 are both off (as during the dead time or with PWM). The collapsing magnetic field can cause the primary of T_1 to deliver a very high negative or positive

FIGURE 10.19 Other switching power supply configurations.

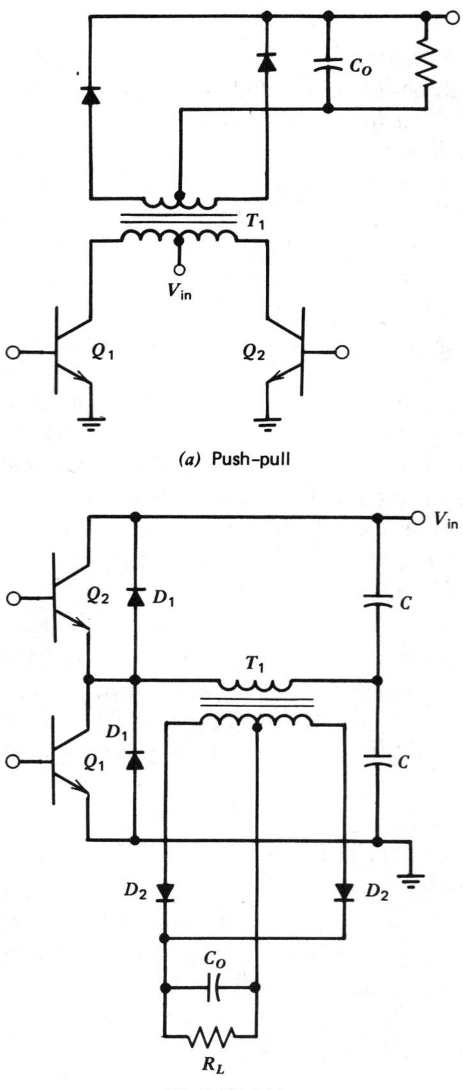

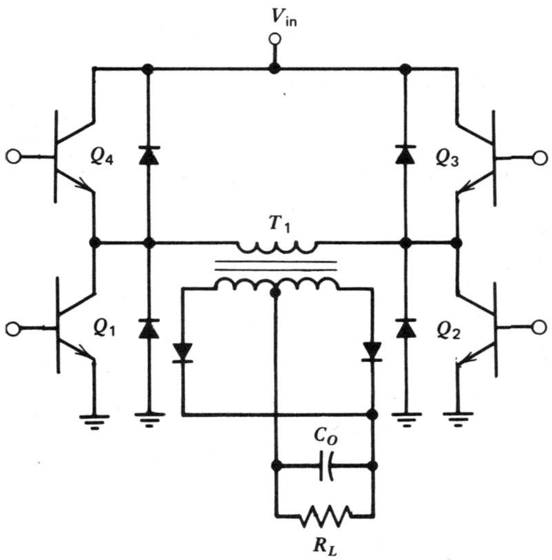

(*c*) Bridge

voltage to Q_1 and Q_2. Diodes D_1 limit the voltage from the primary of T_1 to V_{in} or ground when Q_1 and Q_2 are both off. These diodes ar called *free-wheeling diodes.*

BRIDGE
The bridge configuration shown in Figure 10.19*c* can deliver the same power as the half bridge with one-half the collector current, or as the push-pull with one-half the voltage rating for each transistor. The transistors are turned on in pairs, Q_1 and Q_3 provide one-half cycle of the output and Q_2 and Q_4 provide the other. If PWM is used, the free-wheeling diodes are necessary. Dead time is necessary for the bridge inverter for the same reason as in the half bridge. The bridge inverter requires two more transistors than the half-bridge or push-pull inverters and is more complex to control.

10-4.2 MC3420 Operation

The Motorola MC3420 is available in a commercial version that operates between 0 and 70°C and in a military version (MC3520) that operates between −55 and 125°C. The device is available in a ceramic or plastic 16 pin DIP. It can operate with a supply voltage between 10 and 30 V and draws 16 mA quiescent current typically.

The MC 3420 is a subsystem consisting of the following parts (see Figure 10.20*a*):

A temperature-compensated reference voltage that is used internally and available for external use. The reference voltage is typically 7.8 V and can supply 0.4 mA.

An oscillator that provides a triangular ramp voltage at the "ramp out" pin 8, and a pulse string to gate 1, for output pulse steering, and pin 3, F/F out. The

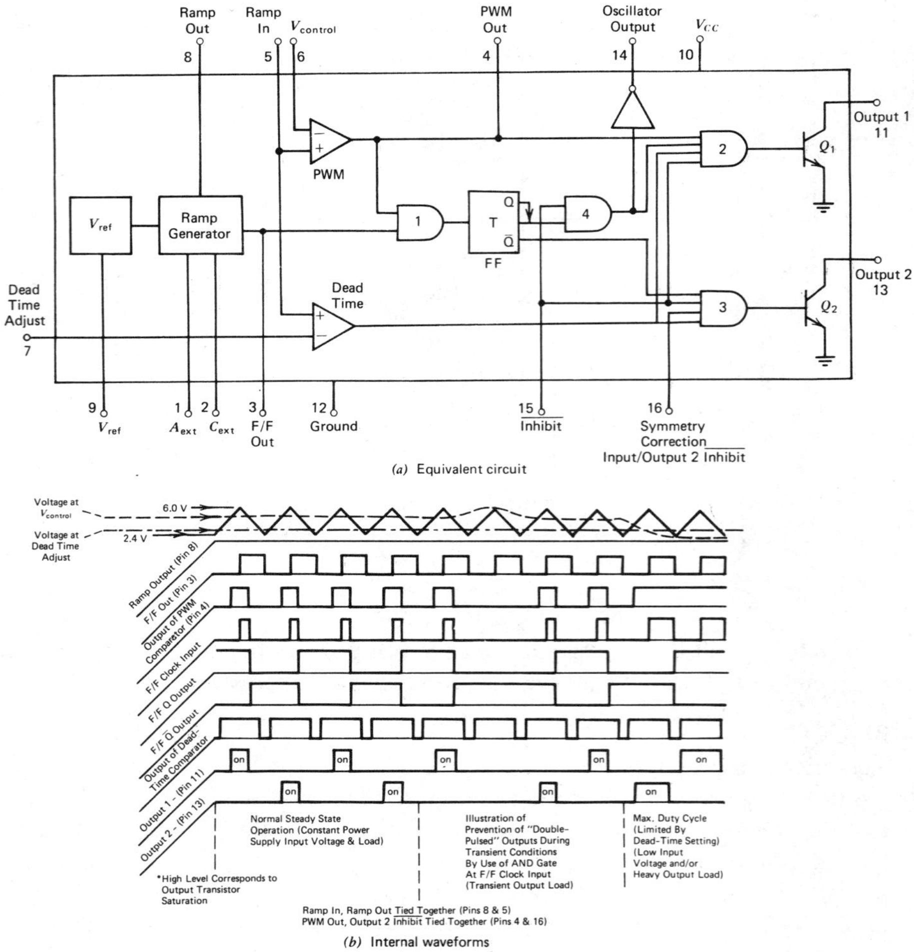

FIGURE 10.20 MC 3420/3520 switchmode regulator control circuit.

frequency of the oscillator can be set by the user from 2 to 100 kHz with an external resistor and capacitor.

A pulse-width modulation (PWM) comparator that compares the control voltage on pin 6 to the ramp voltage on pin 5 from pin 8. The user can provide a ramp if desired. Whenever the ramp voltage is above the control voltage, the output of the PWM comparator is positive and can turn on an output transistor via gates 2 and 3. If the output of the ramp is below the control voltage, the PWM comparator output is zero, disabling gates 2 and 3.

A dead-time comparator provides an adjustable dead time for Q_1 and Q_2. When $V_{ramp} < V_{pin\ 7}$, the dead-time comparator is zero, disabling gates 2 and 3. The control voltage should be greater than the pin 7 voltage in normal operation.

Gate 1 and the flip-flop (FF) steer the pulse-width modulation signal to Q_1 and Q_2 by alternately enabling gates 2 and 3. The flip-flop toggles on the trailing edge of the F/F and the PWM signals, enabling the gate that was disabled the previous half-cycle.

Gate 4 disables gate 2 if the inhibit not pin goes low. Gate 3 is disabled directly by the inhibit not (pin 15). Normally, pin 15 is at V_{in}. This pin can be used to shut the control off under various overload conditions.

Gates 2 and 3 control Q_1 and Q_2 respectively. To turn on a transistor, the PWM comparator output, inhibit not output, Q or $\overline{Q}$ of the FF, and the dead-time comparator must all be high. The PWM output, pin 4, is tied to pin 16 (symmetry correction) for normal operation.

Q_1 and Q_2 can provide 50 mA maximum collector current and can withstand a maximum V_{CE} of 40 V.

MC3420 OPERATION WITH A PUSH-PULL SWITCHING POWER SUPPLY

Refer to the waveforms of Figure 10.20*b* for this discussion. The MC3420 is shown connected in Figure 10.21 for regulation of a push-pull circuit. The circuit of Figure 10.21 will be used to describe the MC3420 operation.

A_1 compares a sample of the output to the reference to provide a control voltage. It is connected so that as V_{out} increases, $V_{control}$ increases. A_2 compares the voltage across R_{cs} to a reference voltage. As the voltage across R_{cs} increases above the reference, the control voltage increases to a voltage greater than the ramp voltage, turning the MC3420 off. Diodes D_3 act as an OR gate allowing the higher output of A_1 or A_2 to provide the control voltage. *Note.* The output of A_2 could have been used to ground the inhibit not pin when the output current reached I_m. The gain of A_1 is kept moderate (around 100), and the gain of A_2 is as high as desired for fast action to an overcurrent.

R_1 and R_2 provide a sample of the output voltage; R_3 and R_4 provide a reference to A_1. R_9 and R_{10} provide a reference to A_2; and R_{11} and R_{12} provide a reference to the dead-time comparator. Voltage dividers R_9, R_{10}, R_{11}, R_{12}, R_3, and R_4 are supplied by the reference voltage at pin 9 and will be set to draw 0.1 mA per divider. R_5 and R_6 set the voltage gain of A_1, and R_7 and R_8 set the voltage gain of A_2. R_s and R_s' prevent bias current generated offset voltage.

Q_3 and Q_4 are used to drive the output transistors Q_5 and Q_6. Since Q_1 and Q_2 provide only grounds, *pnp* transistors are used for Q_3 and Q_4 to reduce the driving circuit complexity. C_5 provides faster turn on and turn off times for Q_5 and Q_6. R_{15} provides a path to ground for the leakage current of Q_5 and Q_6 to increase their ability to block transient voltages. Diodes D_1 are also the free-wheeling diodes to reduce voltage transients caused by the collapsing magnetic field of T_1 when Q_5 and Q_6 are both off, as well as the output rectifiers.

R_T and C_T set the oscillator frequency.

REGULATION

Normal operation occurs as follows. When V_{out} is at its specified value, and Q of the FF is high, Q_1 is on while the ramp voltage is above the control voltage and turns off when the ramp voltage drops below the control voltage. At this time, the FF changes state, causing Q to go low and $\overline{Q}$ to go high, enabling one input of gate 3. The ramp will continue to drop until it is less than the dead-time adjust voltage. While the ramp is below the dead-time adjust voltage, both gates 2 and

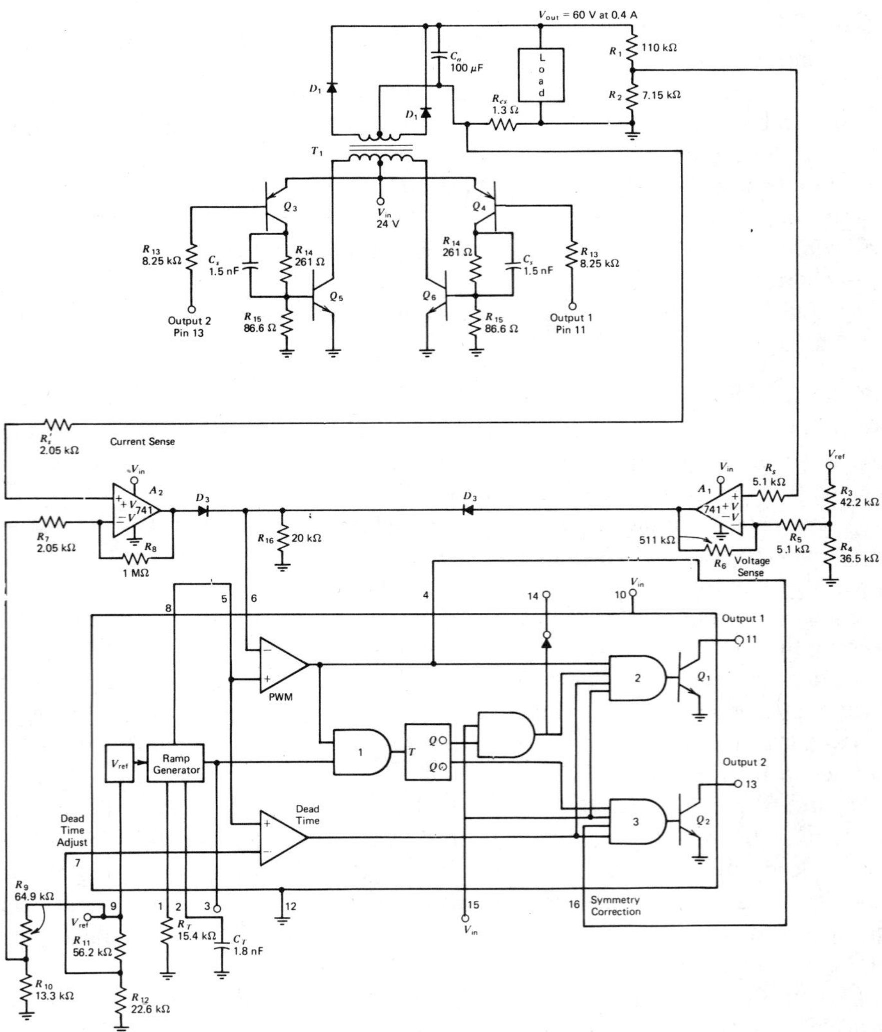

FIGURE 10.21 MC 3420 push-pull switching voltage regulator.

3 are disabled so neither Q_1 nor Q_2 can turn on. As the ramp voltage rises on the next half cycle, first the dead-time comparator goes positive and then, as V_{ramp} rises above $V_{control}$, the PWM output goes high. Q_1 cannot turn on now since Q is low, but Q_2 can since $\overline{Q}$, the inhibit, the dead-time comparator output, and the PWM comparator output are all high. When the ramp voltage drops below the control voltage, again the PWM comparator output goes low, turning off Q_2 and toggling the FF so that Q is now high and $\overline{Q}$ low. The next half cycle, Q_1, can turn on again as a result of the FF state change.

Regulation occurs as follows: If V_{out} drops, the control voltage drops. Therefore, the ramp voltage will be higher than the control voltage for a greater length of time and the PWM output will be high longer during each half cycle. The output pulses will increase in length and more average voltage will be delivered to the load, correcting the voltage drop. If the output voltage increases, the control voltage will increase. Now the ramp voltage will be above the control voltage for less time each half cycle. The PWM output will put out shorter pulses, decreasing the average voltage.

If the current rises above I_m, the voltage across R_{cs} will cause the output of A_2 to rise above the maximum ramp voltage so the PWM comparator never goes high. If the gain of A_2 is moderate, the maximum current is regulated in a constant current fashion. If the gain of A_2 is high and a latch is provided on its output, the control shuts off until it is reset.

EXAMPLE 10-17

Calculate the components for the push-pull switching regulator of Figure 10.21. Use an MC3420 switchmode regulation circuit to reduce the parts count. Set $V_{out} = 60$ V at 400 mA. $V_{in} = 24$ V. Assume T_1 is ideal and set the frequency at 20 kHz ($\tau = 50$ μs). $V_{ripple} = 60$ mV_{pp}.

SOLUTION

The transformer turns ratio must be (assuming $V_{D_1} = 1$ V and $V_{CE}(\text{sat})_{Q5} = 1$ V):

$$\frac{N_s}{N_p} = \frac{V_{out} + V_{D1}}{V_{in} - V_{CE}(\text{sat})_{Q5}} = \frac{61\text{ V}}{23\text{ V}} = 2.65$$

The primary current is

$$I_p = I_s\left(\frac{Ns}{Np}\right) = 400\text{ mA }(2.65) = 1.1\text{ A}$$

Choose a transformer that can handle currents larger than these by at least 30%.

Choose the duty cycle at maximum I_{out}. Maximum I_{out} is chosen to occur with the transistors on 80% of the time. The control voltage necessary for an 80% duty cycle is found in the specification sheet in Appendix C to be 3 V for 40% duty cycle each half cycle. If $t_s + t_f$, the turn off time for Q_5 and Q_6 is 1 μs, then the reference voltage for the dead-time comparator (V_{DT}) can be found with the following equation from the specification sheet.

$$\text{Dead time} = \frac{1}{f_o}\left(\frac{V_{DT} - 2}{4}\right)$$

Which, when solved for V_{DT}, is

$$V_{DT} = 4\,f_o\,(\text{dead time}) + 2$$

Letting the dead time be 3 μs > ($t_s + t_f$),

$$V_{DT} = 4(20\text{ kHz})(3\text{ μs}) + 2$$
$$= 2.24\text{ V}$$

The maximum ramp voltage is 6 V, so the control voltage should be 6 V from the current limit amplifier at I_m. Let $I_m = 1.2\ I_{out}(\text{max})$ so that full load current can be drawn. Summarizing the voltages necessary, $V_{DT} = 2.24$ V, $V_{control} = 3$ V, V_{out} of A_2 at $I_m = 6$ V.

Transistor specifications:

Q_5 and Q_6

$$V_{CE}(\text{sus}) > 2\ V_{in} = 60\ \text{V}$$
$$V_{CE}(\text{sat}) = 1\ \text{V at } I_C = 1\ \text{A}$$
$$V_{BE}(\text{sat}) = 0.85\ \text{V at } I_C = 1\ \text{A}$$
$$\beta_{min} = 15 \text{ at } I_C = 1\ \text{A}, \qquad V_{CE} = 1\ \text{V}$$
$$P_D > P_{max} = 30\ \text{W}$$

Q_3 and Q_4

$$V_{CE}(\text{sus}) > V_{in} = 40\ \text{V}$$
$$V_{CE}(\text{sat}) = 0.8\ \text{V at } I_C = 0.1\ \text{A}$$
$$V_{BE}(\text{sat}) = 0.75\ \text{V at } I_C = 0.1\ \text{A}$$
$$\beta_{min} = 30 \text{ at } I_C = 0.1\ \text{A} \quad \text{and} \quad V_{CE} = 0.8\ \text{V}$$
$$P_D = 1.2\ \text{W}$$

Diodes

$$D_{1\ \text{blocking voltage}} > 2\ V_{out} = 200\ \text{V}$$

$$V_F = 1\ \text{V at } I_F = 1\ \text{A}$$

D_2 Use same diodes as D_1

D_3 Small signal switching diodes such as 1N914.

$$V_F = 0.6\ \text{V}, \qquad V_R = 75\ \text{V}$$

The op-amps are μA741 or equivalent.

Resistor calculations:
From the specification sheet $R_{16} < 50$ kΩ, so R_{16} is selected at 20 kΩ, a value the op-amps can easily supply current to. Since the reference voltage is 7.8 V at 0.4 mA, the three reference voltage dividers are selected to draw 0.1 mA each to avoid loading the reference.

Calculating R_3 and R_4 for $V_{control} = 3$ V. The D_3 diode drop must be added into the voltage calculations, so let $V'_{control} = V_{control} + V_{D_3} = 3.6$ V.

$$R_3 = \frac{V_{ref} - V'_{control}}{I_D} = \frac{7.8\text{ V} - 3.6\text{ V}}{0.1\text{ mA}} = 42\text{ k}\Omega$$

$$R_4 = \frac{V'_{control}}{I_D} = \frac{3.6\text{ V}}{0.1\text{ mA}} = 36\text{ k}\Omega$$

Resistors R_1 and R_2 provide 3.6 V when $V_{out} = 60$ V. Let $I_D = 0.5$ mA for R_1 and R_2

$$R_1 = \frac{V_{out} - V'_{control}}{I_D} = \frac{60\text{ V} - 3.6\text{ V}}{0.5\text{ mA}} = 112.8\text{ k}\Omega$$

$$R_2 = \frac{V'_{control}}{I_D} = \frac{3.6\text{ V}}{0.5\text{ mA}} = 7.2\text{ k}\Omega$$

The reference voltage for the dead-time comparator (V_{DT}) is 2.24 V, so

$$R_{11} = \frac{V_{ref} - V_{DT}}{I_D} = \frac{7.8\text{ V} - 2.24\text{ V}}{0.1\text{ mA}} = 55.6\text{ k}\Omega$$

$$R_{12} = \frac{V_{DT}}{I_D} = \frac{2.24\text{ V}}{0.1\text{ mA}} = 22.4\text{ k}\Omega$$

Resistors R_9 and R_{10} provide the current-sense reference voltage. Since it is desirable to keep R_{cs} small, the gain of A_2 will be set high so that when $V_{R_{cs}}$ is greater than the current-sense reference, V_{out} will rapidly rise to 6.6 V. If $V_{R_{cs}} = 1.3$ V at $I_m = 1.2\,I_{o(max)}$, $R_{cs} = 1.3\ \Omega$, then the current-sense reference voltage = 1.3 V.

$$R_9 = \frac{V_{ref} - 1.3\text{ V}}{I_D} = \frac{7.8\text{ V} - 1.3\text{ V}}{0.1\text{ mA}} = 65\text{ k}\Omega$$

$$R_{10} = \frac{1.3\text{ V}}{I_o} = \frac{1.3\text{ V}}{0.1\text{ mA}} = 13\text{ k}\Omega$$

The gain of A_2 is set at 100 (a value used in the example circuits shown in the specification sheet). Let $R_6 = 500\text{ k}\Omega$

$$\frac{R_6}{R_5} = 100 = A_{fb(A2)}$$

Therefore

$$R_5 = \frac{R_6}{100} = \frac{500\text{ k}\Omega}{100} = 5\text{ k}\,\Omega$$

$$R_s = R_5 \| R_6 = 4.96\text{ k}\Omega \simeq 5\text{ k}\,\Omega$$

The gain of A_2 is set at 500, so it will override A_1 quickly if $I_{out} > I_m$. Let R_8 = 1 MΩ

$$R_7 = \frac{1\ \text{M}\Omega}{500} = 2\ \text{k}\Omega$$

$$R_s' = R_8 \| R_7 \simeq 2\ \text{k}\Omega$$

The resistors in the drive circuit must now be calculated. Let $I_{R_{15}} = 10$ mA.

$$R_{14} = \frac{V_{in} - V_{CE_{Q4}} - V_{BE_{Q6}}}{(I_p/\beta_{Q_6}) + I_{R_{15}}} = \frac{24\ \text{V} - 0.85\ \text{V} - 0.8\ \text{V}}{(1.1\ \text{A}/15) + 10\ \text{mA}}$$

$$= \frac{22.35\ \text{V}}{73.3\ \text{mA} + 10\ \text{mA}} = 268\ \Omega$$

$$R_{15} = \frac{V_{BE_{Q6}}}{I_{R_{15}}} = \frac{0.85\ \text{V}}{10\ \text{mA}} = 85\ \Omega$$

$$R_{13} = \frac{V_{in} - V_{BE_{Q4}} - V_{CE_{Q1}}}{(I_{CQ_4}/\beta_{Q_4})} = \frac{22.75\ \text{V}}{(83.3\ \text{mA}/30}$$

$$= 8.2\text{k}\ \Omega$$

Now the capacitors. C_o must provide I_{out} during the time the transistors are off. The maximum duty cycle is 80% at I_{out}(max). The minimum duty cycle depends on the minimum load current, but the current supplied by the capacitor decreases as the duty cycle decreases, tending to hold the ripple constant. To assure the ripple specification, assume that C_o must supply I_{out} 30% of τ or 15 μs out of each cycle.

$$C_o = \frac{I_{out}t_o}{V_{ripple}} = \frac{0.4\ \text{A}\ (15\ \mu\text{s})}{60\ \text{mV}} = 100\ \mu\text{F}$$

C_s is the speed up capacitor and can be approximated by

$$C_s = \frac{I_{BQ6}t_r}{V_{cs}} = \frac{73\ \text{mA}\ (0.5\ \mu\text{s})}{24\ \text{V}} = 0.0015\ \mu\text{F}$$

where

t_r = 0.5 μs
I_{BQ6} = $I_p/\beta\ \text{min}_{Q6}$ = 73 mA
V_{cs} ≅ V_{in} = 24 V

From the specifications, 5 kΩ < R_{ext} <20 kΩ. Choose C_{ext} = 0.0018 μF

$$f_o = \frac{0.55}{R_{ext}\,C_{ext}}$$

$$R_{ext} = \frac{0.55}{f_o\,C_{ext}} = \frac{0.55}{1.8\text{ nF }(20\text{ kHz})} = 15.3\text{ k}\Omega$$

SUMMARY

1. In this chapter we studied the basic op-amp linear voltage regulator and three linear IC regulators: the LM105, the LM309, and the μA723.
2. Linear voltage regulators offer superb line and load regulation at very reasonable cost.
3. Linear regulators are not very efficient because the regulating transistor must dissipate considerable power in the process of regulation.
4. Switching voltage regulators, by virtue of the low voltage drop of a transistor switch, are much more efficient than linear voltage regulators. Switching regulators are, however, more complex than linear regulators.
5. Switching voltage regulator IC's are available that greatly simplify the construction of switching power supplies. We studied the μA78S40 and the MC3420.
6. New voltage regulators, both switching and linear, with increased capabilities are becoming available at a rapid rate. A thorough review of various manufacturers' linear IC catalogues will enable the reader to select the best voltage regulator IC for a given application.

SELF-TEST QUESTIONS

10–1 Name the two basic types of voltage regulators and two advantages and one disadvantage of each.

10–2 State the function of Z_2 in Figure 10.3*b* and Q_3 in Figure 10.3*c*.

10–3 Assume Q_1, Q_2, and Z_1 in Figure 10.3a have the following specifications:

- □ Q_1: $V_{BE} = 0.8$ V at $I_C = 2$ A, $\beta_{min} = 20$ at $I_C = 2$ A.
- □ Q_2: $V_{BE} = 0.7$ V at $I_C = 0.1$ A, $\beta_{min} = 40$ at $I_C = 0.1$ A.
- □ Z_1: $V_Z = 6.2$ V at $I_B = 4$ mA.

Calculate R_1, R_2, R_B, and R_3 if A is a 741, $V_{in} = 12$ V, $V_{out} = 9$ V at 2 A.

10–4 Calculate the efficiency of the regulator in question 10–3 at $I_o = 2$ A.

10–5 Calculate R_{cl} for the circuit in Figure 10.4*a* if R_1, R_2, R_B, and R_3 are as calculated in question 10–3. In other words, add constant current limiting to the circuit of question 10–3. Let $I_m = 2.1$ A, $V_{BE(Q_{cl})} = 0.4$ V.

10–6 Add foldback current limiting to the circuit of question 10–3 as shown in Figure 10.5*b*. Let $I_m = 2.1$ A, $I_{FB} = 0.2\,I_m$, V_{BE} of $Q_{cl} = 0.4$ V. Let $I_{RA} = 1$ mA. Use R_{cl} from question 10–5.

10–7 Calculate the components for the circuit shown in Figure 10.8, a current-boosted LM105 circuit with foldback current limiting. Let $V_{in} = 30$ V, $V_{out} = 20$ V, $I_m = 1.5$ A, $I_{FB} = 0.2\,I_m$. Use the transistor specifications given in Example 10–4 for Q_A and Q_B.

10–8 Why must the circuit of Figure 10.10*b* be current-limited by Q_2 when the LM309 has a current-limit circuit built in?

10–9 Calculate the components for the LM723 circuit of Figure 10.12*b* if V_{in} = 24 V and V_{out} = 18 V at 20 mA.
10–10 Calculate the components for the circuit in Figure 10.13*a* if V_{in} = 24 V, V_{out} = 15 V, I_m = 300 mA, I_{FB} = 0.2 I_m, and Q_1 has the following specifications.

$$Q_1\text{: } \beta_{min} = 30 \text{ at } I = 0.3 \text{ A, } V_{BE} = 0.65 \text{ V at } I = 0.3 \text{ A.}$$

10–11 State the purpose of the different locations for R_1 and R_2 in Figures 10.12*a* and *b*.
10–12 State the reason the Inv and Noninv connections are reversed in Figure 10.13*b* as compared to Figure 10.13*a*.
10–13 A transistor such as shown in Figure 10.14 has V_{cc} = 20 V, R_c = 200 Ω, I_B = 10 mA, V_{CE}(sat) = 0.6 V, $V_{BE}(sat)$ = 0.8 V, t_c = 20 μs, t_o = 40 μs, t_r = 1 μs, and t_f = 2 μs. Calculate the average power dissipation.
10–14 Draw from memory and state the basic operating principle of the step-down, step-up, and voltage inverter switching supplies of Figure 10.15.
10–15 Calculate the components for a step-down switching supply using the μA78S40 as shown in Figure 10.17*a*. Let V_{in} = 12 V, V_{out} = 5 V at 500 mA. Use typical specification sheet values. Let f_{osc} = 10 kHz and V_{ripple} = 40 mV_{pp}.
10–16 Calculate the components for a step-up switching supply using the μA78S40 as shown in Figure 10.17*b*. Let V_{in} = 12V, V_{out} = 24 V at 200 mA, and f_{osc} = 20 kHz. Use typical specification sheet values. Let V_{ripple} = 20 mV_{pp}.
10–17 Calculate the components for a voltage inverting switching supply using the μA78S40 as shown in Figure 10.17*c*. Let V_{in} = 5 V, V_{out} = −5 V at 1 A and f_{osc} = 20 kHz. Let Q_3 be a *pnp* transistor with V_{CE}(sat) = 1 V at I_C = 5 A, V_{BE} (sat) = 1.2 V at I_C = 5 A, β_{min} = 10 at I_C = 5 A, V_{CE}(sus) = 20 V. D_2 blocks 20 V and has V_F = 1 V at 5 A. Let V_{ripple} = 40 mV_{pp}.
10–18 State the major disadvantage of the free-running PWM circuits in Figure 10.18.
10–19 What is dead time and the purpose of dead time?
10–20 State the major advantage the bridge inverter (Figure 10.19*c*) has over the push-pull (Figure 10.19*a*) and the half bridge (Figure 10.19*b*).
10–21 Can the MC3420 switchmode regulator be used to control a step-down inverter? If so, draw a schematic of its use as such. State one precaution that must be taken when it is used as a step-up or step-down inverter.

If you cannot answer certain questions, place a check next to them and review appropriate parts of the text to find the answers.

LABORATORY EXERCISE

Objective

After completing this laboratory, the student should be able to calculate the components for and construct an op-amp linear regulated power supply with constant and foldback current limiting. The student should also be able to calculate the components for and construct a step-up, step-down, and voltage inverting power supply.

Note. This laboratory is written so that no specialized IC regulators are required. The instructor will provide any specialized IC regulators desired for power supply construction for the laboratory.

EQUIPMENT

1. Fairchild μA741 operational amplifier or equivalent.
2. ±2% tolerance resistor assortment.
3. Electrolytic capacitor assortment.
4. High Q inductors of 350 μH, 570 μH, and 450 μH.
5. Pulse generator.
6. Multimeter, Simpson 260 or equivalent.
7. Oscilloscope.
8. Breadboard with IC sockets such as EL Instruments SK-10.
9. 110 V:12 V CT 2 A power transformer.
10. Three 2 A, 200 V diodes.
11. *npn* power transistor such as RCA 40310 mounted on a heat sink.
12. Two *npn* transistors such as RCA 40314.
13. *npn* power-switching transistor such as 2N5190.
14. *pnp* power switching transistor such as 2N3740.
15. Low power switching transistor such as the 2N2369A.
16. One 6.3 V, 1 W zener diode.
17. One each 10Ω–20 W, 20Ω–10 W, 40Ω–5 W, 80Ω–15 W, and 16Ω–2 W resistors.
18. One variac, 250 VA.

Procedure

1. Linear-regulated power supply.

(a) Calculate the components for the circuit shown in Figure 10.22*a*. Set $V_{out} = 10$ V at 1 A. Use the specifications for your Q_1, Q_2, and Q_{cl}. Calculate the filter capacitor C_f for $V_{ripple} = 1\ V_{pp}$ from

$$C_f = \frac{I_{out}t}{V_{ripple}} = \frac{1\text{ A }(8.33\text{ ms})}{1\text{ V}}$$

Set I_m at 1.1 I_o(max).

(b) Construct the circuit.

(c) Vary the load from $I_{out} = 0$ to $I_{out} = 1$ A (i.e., insert the 10 Ω −20 W power resistor) and measure ΔV^{out}.

(d) Calculate the load regulation

$$\%\text{ regulation} = \frac{\Delta V_{out}}{V_{loaded}}(100)$$

(e) Vary the input voltage with the variac from 120 V_{rms} to 100 V_{rms}. Read V_{out} at each setting.

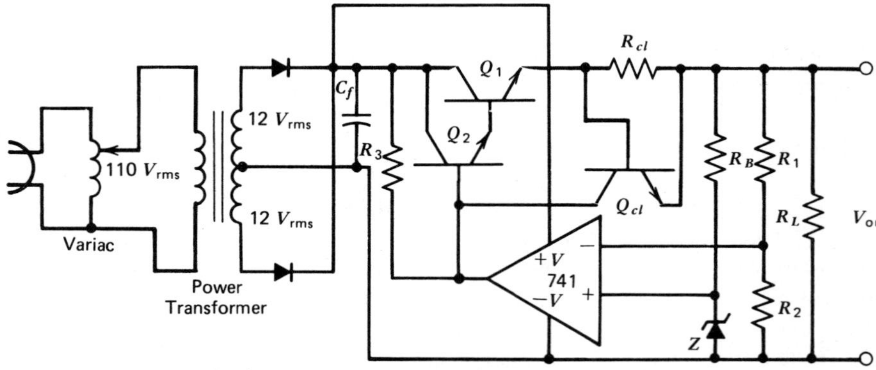

(a) Constant current-limited series-pass power supply

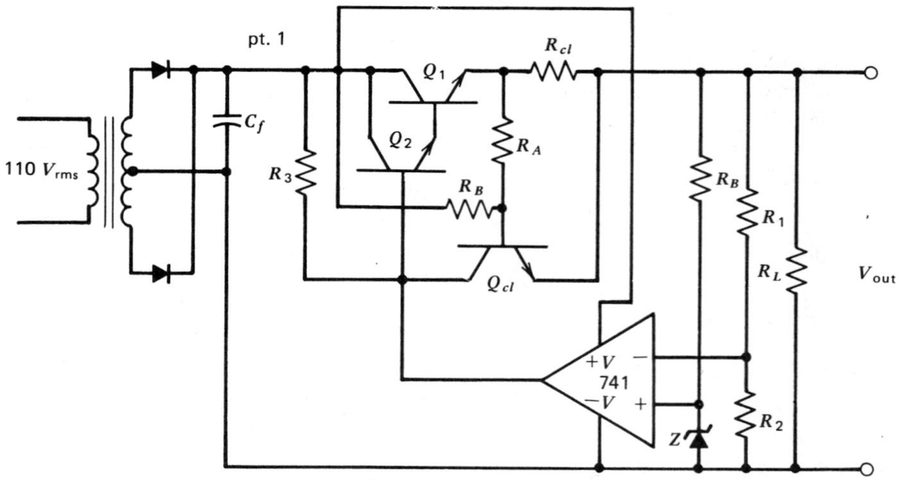

(b) Foldback current—limited series—pass power supply

FIGURE 10.22 Op-amp linear-regulated power supply.

(f) Calculate the line regulation

$$\% \text{ regulation/volt} = \left(\frac{\Delta V_{out}}{\Delta V_{in}(V_{out})}\right)(100)$$

(g) Put an ammeter in series with R_c. Set the ammeter on a range greater than 1 A. Short the load and quickly note the reading, then remove the short circuit.

(h) Modify the circuit, as shown in Figure 10.20*b*, for foldback current limiting. Set $I_{FB} = 0.2\ I_m$.

(i) Short circuit the output briefly and measure I_{FB}.

(j) Measure V_{in} and I_{in} at point 1 of the circuit in Figure 10.22*b* with the 10 Ω − 20 W resistor connected to the output. Calculate the circuit efficiency.

2. Step-down switching supply.

(a) Calculate the components for a step-down switching supply as shown in Figure 10.23. The pulse generator will provide the driving voltage for t_c. Record calculated t_c and t_o.

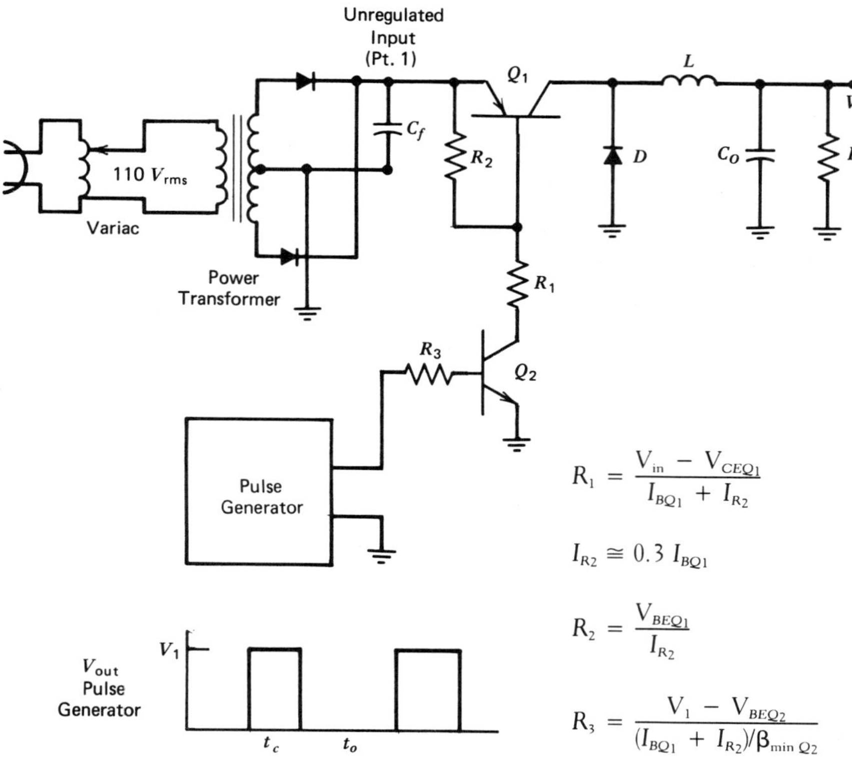

$$R_1 = \frac{V_{in} - V_{CEQ1}}{I_{BQ1} + I_{R2}}$$

$$I_{R2} \cong 0.3\ I_{BQ1}$$

$$R_2 = \frac{V_{BEQ1}}{I_{R2}}$$

$$R_3 = \frac{V_1 - V_{BEQ2}}{(I_{BQ1} + I_{R2})/\beta_{\min Q2}}$$

FIGURE 10.23 Step-down supply.

Note. Use the next larger size L you have over the calculated value.

Let $V_{in} = 16$ V, $V_{out} = 10$ V at 0.5 A, $V_{ripple} = 50$ mV, and $f_{osc} = 10$ kHz.

(b) Set the pulse generator so the output is high during t_c and low during t_o.

(c) Measure the output voltage with $R_L = 20\ \Omega - 120$ W in the circuit.

Note. Your measured value will be less than your calculated value by an amount $I_o R_{DC}$ of the inductor. Do not operate the supply with no load.

(d) Vary t_c slightly (±10%) and note the change in output voltage.

(e) Measure the input voltage and average input current (use a dc ammeter) at point 1. Measure V_{out} and I_{out}. Calculate the power supply efficiency.

(f) Measure the output ripple.

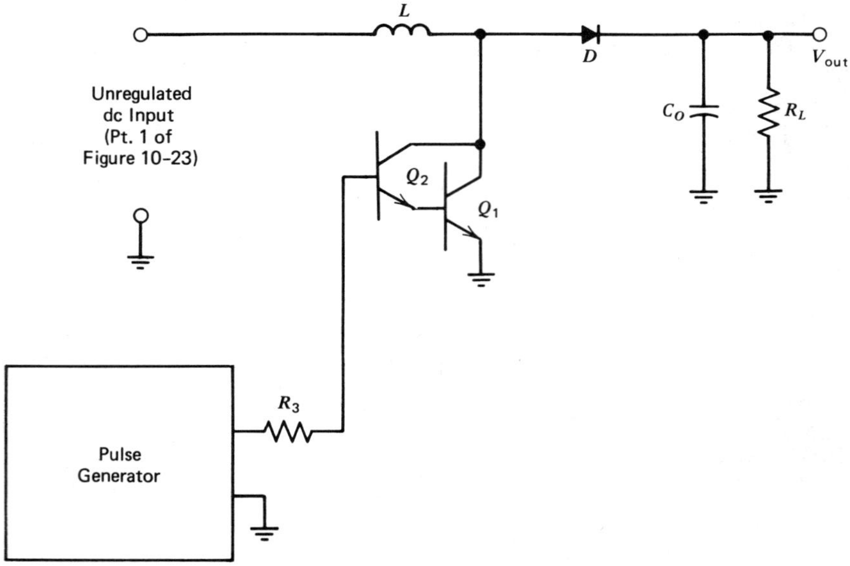

FIGURE 10.24 Step-up supply.

3. Step-up switching supply.
   (a) Calculate t_c, t_o, and the components for the circuit shown in Figure 10.24. Let f_{osc} = 10 kHz, V_{in} = 16 V, V_{out} = 24 V at 0.3 A, and V_{ripple} = 50 mV_{pp}.
   (b Connect the circuit with the pulse generator set high for t_c and 0 V for t_o. Do not operate the circuit with no load.
   (c) Measure the output voltage with R_L = 80 Ω − 15 W.
   (d) Repeat steps (d) and (e) of part 2.

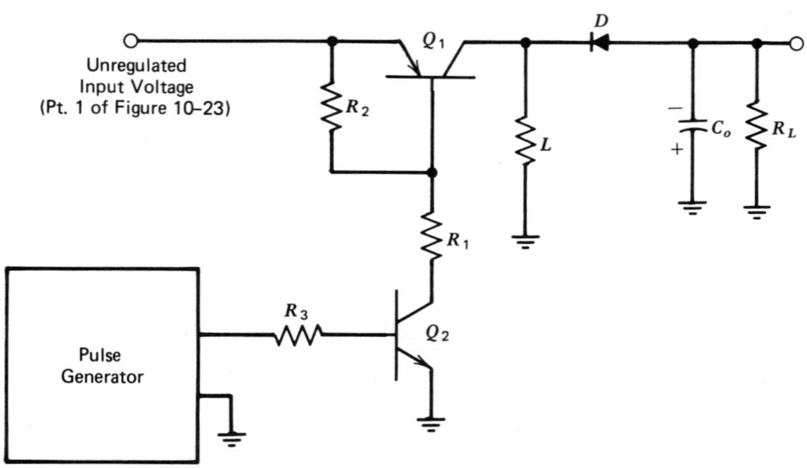

FIGURE 10.25 Voltage-inverting supply.

4. Voltage inverter.
   (a) Calculate t_c, t_o, and the components for the circuit shown in Figure 10.25.
   Note. Only L and C_o must be calculated; R_2, R_1, and R_3 from part 2 are the same if Q_1 and Q_2 are the same.
   Set $V_{out} = 5$ V at 0.3 A, $V_{in} = 16$ V, $f_{osc} = 10$ kHz, $V_{ripple} = 50$ mV_{pp}.
   (b) Connect the circuit. Set the pulse generator high during t_c and at 0 V during t_o. Do not operate the circuit with no load.
   (c) Measure V_{out} with $R_2 = 16\ \Omega - 2$ W.
   (d) Repeat parts (d) throught (f) of part 2.
5. Optional section: Free running PWM oscillator.
   (a) Calculate the components R_A, R_B, R_6, R_4, R_3, R_5, and C_o for the circuit of Figure 10.26. Set $V_{out} = 10$ V at 0.5 A, and $V_{in} = 16$ V. Let $V_{ripple} = 50$ mV. See Example 10-15 for a solved problem. Recall

$$V_{out} = V_{ref}\left(\frac{R_4 + R_5}{R_5}\right)$$

$$V_{ref} = V_{out}\left(\frac{R_5}{R_4 + R_5}\right)$$

$$R_B \text{ very large} \cong 500\ \text{k}\Omega$$

$$\frac{R_A}{R_B} = \frac{V_{ripple}}{V_{in} - V_{CEQ1}}$$

$$\text{Set } C_o = \frac{I_{RLtc}}{V_{ripple}}$$

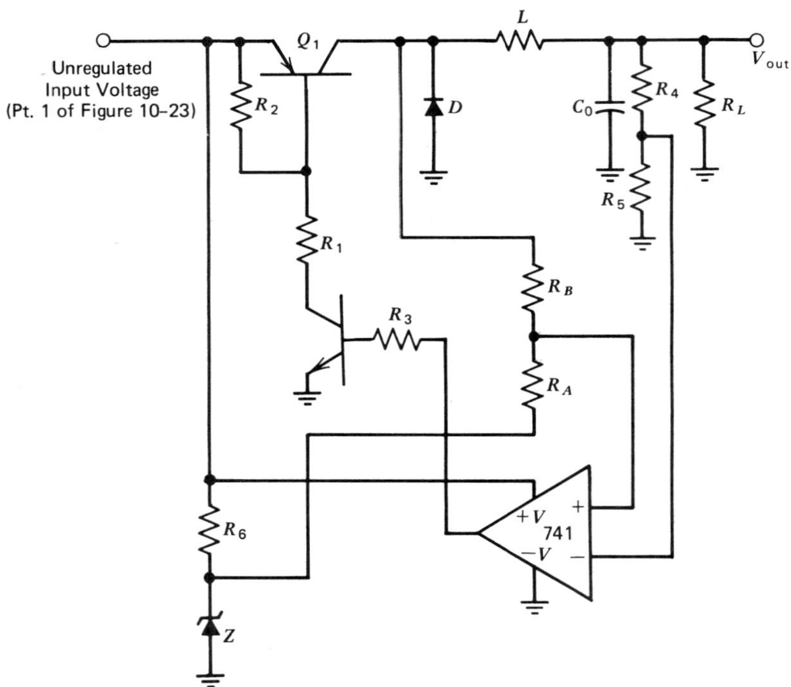

FIGURE 10.26 Free-running step-down inverter.

L should remain the same value as in part 2.

(b) Connect the circuit with $R_L = 20\ \Omega - 20$ W, and measure V_{out}.

(c) Change R_L to 40 Ω and measure V_{out}.

(d) Calculate the load regulation.

(e) Observe the frequency variation at the 741 output as R_L is changed from 20 Ω to 40 Ω.

CHAPTER

11

SOME SPECIAL PURPOSE ICS

As the market for a specific electronic function grows large enough to justify the development effort and cost, the circuit function will generally be put in an integrated circuit if possible. Very often the cost of the circuit function drops when this is accomplished, and in many instances the performance improves. In this chapter we will look at a few special purpose integrated circuits that fill a specific market need. Four of the ICs are special purpose op-amps. They are comparators, instrumentation amplifiers, isolation amplifiers, and current-differencing amplifiers. The first two are available in monolithic or hybrid form; the third in hybrid only; the last in monolithic only. We will also look at an extremely useful timing circuit, the 555 timer.

OBJECTIVES

After completing the study of this chapter and the self-test questions, the student should be able to:

1. State the function of the integrated circuits in this chapter.
2. State the major uses of the ICs in this unit.
3. Calculate the components for a comparator Schmitt trigger and comparator level detector.
4. Calculate the components for an inverting and noninverting current-differencing amplifier.
5. Calculate the components for a 555 timer astable and monostable circuit.
6. Calculate the components for an instrumentation bridge amplifier.
7. Perform the laboratory for Chapter 11.

11-1 COMPARATORS

Comparators are special purpose op-amps designed to compare one voltage level with another and change in output voltage when one input voltage is larger than the other. Any op-amp can be used as a comparator, but specially designed comparators ease the application problems.

A comparator should have low offset voltage, low offset voltage drift, stable oscillation-free operation, and low bias current. In many applications single-supply operation, which many comparators offer, is an advantage.

Comparators are an integral part of automatic testing, analog-to-digital conversion, power supply regulation (as error amplifiers), and logic-level shifting. The list of uses for comparators goes on and on.

An IC comparator, the LM311, is shown in Appendix C. It is suitable for most of the applications mentioned in this section. The specification sheet also has many applications in it. The device features single supply operation from 5 V (TTL compatible) to 36 V, or dual ± 15 V operation. It has an open collector output that can switch up to 50 mA at 40 V so it can operate relays. It can be strobed, which means it can be turned off when no voltage comparison is desired and turned on when two voltages are to be compared. It has typical ratings of V_{os} = 2 mV, I_{os} = 10 nA, I_B = 100 nA, and A_v = 200,000. The LM311 is stable under virtually any operating condition.

11-1.1 Comparator Operation

Figure 11.1*a* shows a single-supply comparator circuit. One input is connected to a reference voltage, the other to the input. Since V_{in} is connected to the inverting

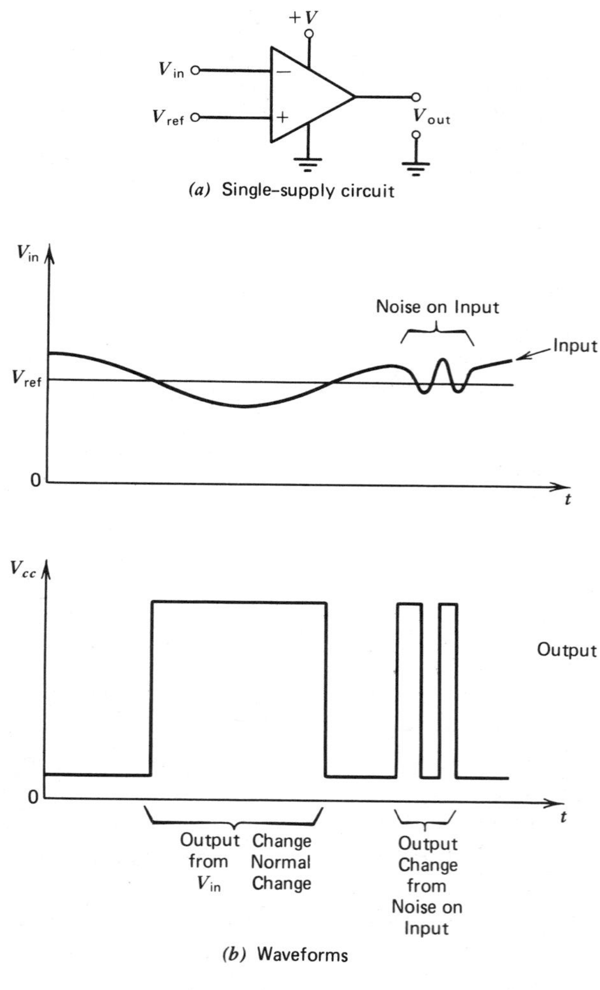

FIGURE 11.1 Comparator.

terminal, the output will be low when $V_{in} > V_{ref}$ and high when $V_{in} < V_{ref}$. If one wanted V_{out} high when $V_{in} > V_{ref}$, then one need only reverse the connections of the inverting and noninverting terminals of the comparator. As shown in Figure 11.1*b* , as V_{in} goes above V_{ref}, the output changes state or voltage level immediately. If the output voltage change is 5 V and the comparator gain is 10,000, the voltage difference needed between V_{in} and V_{ref} to cause an output voltage change is

$$V_{in} - V_{ref} = \frac{\Delta V_{out}}{A} = \frac{5\text{ V}}{10{,}000} = 0.5\text{ mV}$$

If the reference voltage is, for example 2.5 V, this represents a possible error of

$$\frac{2(V_{in} - V_{ref})(100)}{V_{ref}} = \frac{1\text{ mV }(100)}{2.5\text{ V}} = 0.04\%$$

The $2(V_{in} - V_{ref})$ is because the 0.5 mV could be above or below V_{ref}. The result is a very good precision in comparing two voltage levels. This precision has some drawback as shown in Figure 11.1*b*. If V_{in} is varying slowly and is near V_{ref}, noise on V_{in} can cause meaningless output voltage changes. Immunity to noise, but less precision, results if hysteresis is used in the comparator. *Hysteresis* is the use of a higher reference level when V_{in} goes from low to high than when V_{in} goes from high to low. The high reference level is called the *upper trip point* (UTP), and the lower reference level is called the *lower trip point* (LTP).

11-1.2 Schmitt Trigger and Window Comparator

The Schmitt trigger is a comparator with hysteresis. A two-supply Schmitt trigger is shown in Figure 11.2*a*; the input and output waveforms are shown in Figure 11.2*b*. When $V_{in} <$ UTP, V_{out} is high. The UTP is provided by voltage divider resistors R_1 and R_2. The upper trip point is

$$\text{UTP} = \left(\frac{R_2}{R_1 + R_2}\right)(+V_{sat}) \tag{11-1}$$

where

$+V_{sat} = +V_{out}$ (max) of the comparator, usually about 1 V less than + V

When V_{in} rises above the UTP, the output voltage goes negative to $-V_{sat}$, the maximum negative output voltage of the comparator. The negative output voltage causes the noninverting input of the comparator to drop to the lower trip point, which is

$$\text{LTP} = \left(\frac{R_2}{R_1 + R_2}\right)(-V_{sat}) \tag{11-2}$$

The comparator will not change state again until $V_{in} <$ LTP as shown in Figure 11.2*b*. Note that a small amount of noise on the input will not cause a state change

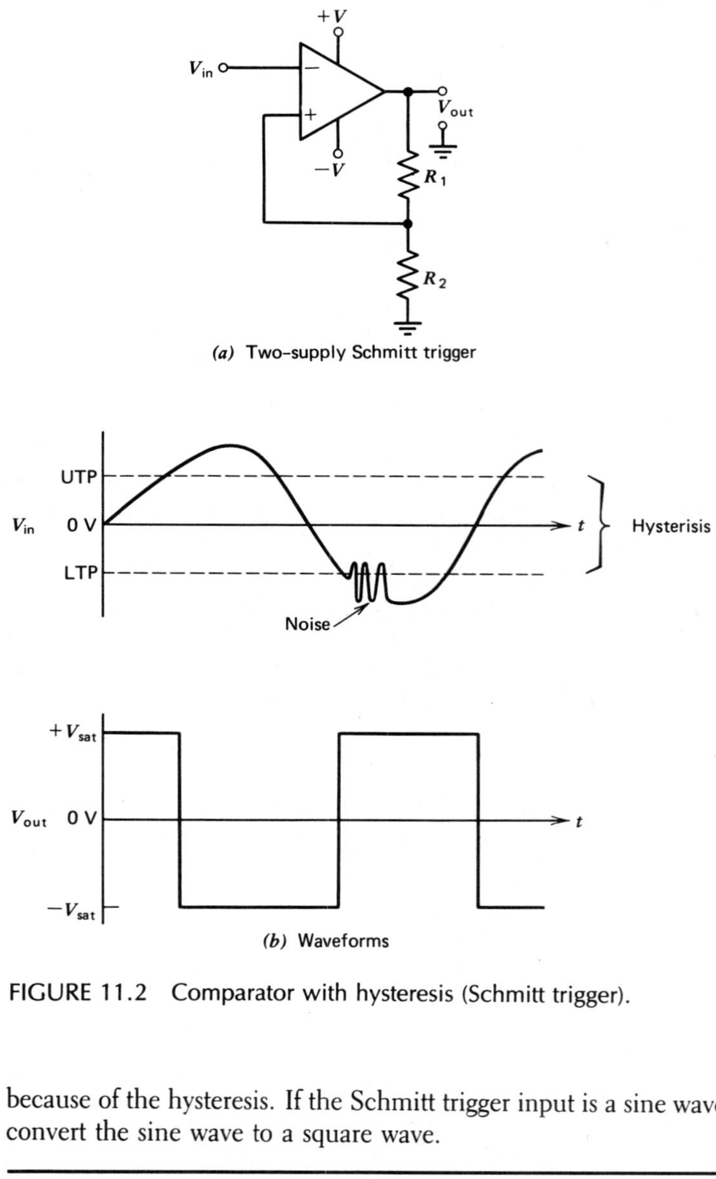

FIGURE 11.2 Comparator with hysteresis (Schmitt trigger).

because of the hysteresis. If the Schmitt trigger input is a sine wave, the circuit will convert the sine wave to a square wave.

EXAMPLE 11-1

Calculate the components for the Schmitt trigger circuit shown in Figure 11.2. Set the hysteresis to 2 V. $+V = 15$ V, $-V = -15$ V, $+V_{sat} = 14$ V, and $-V_{sat} = -14$ V.

SOLUTION

Since the hysteresis is 2 V and $|\text{UTP}| = |\text{LTP}|$ for this circuit, UTP $= 1$ V and

LTP = −1 V. Let $I_{R_1} = I_{R_2} = 0.1$ mA. The bias current is neglected since $I_{R_1} >> I_B$

$$R_1 = \frac{|V_{sat}| - UTP}{I_{R1}} = \frac{14\text{ V} - 1\text{ V}}{0.1\text{ mA}} = 130\text{ k}\Omega$$

$$R_2 = \frac{UTP}{I_{R2}} = \frac{1\text{ V}}{0.1\text{ mA}} = 10\text{ k}\Omega$$

Figure 11.3 shows a two-supply op-amp used as a Schmitt trigger where the UTP and LTP are of the same polarity. The output voltage change turns Q_1 on and off to set the upper and lower trip point. The zener diode Z_1 provides a unipolar output voltage. If $V_{Z1} = 5$ V, the output is TTL compatible. Operation is as follows: As V_{in} goes positive from a voltage less than the UTP, Q_1 is off since V_{out} is negative. R_3 is in the circuit so the

$$UTP = V_{ref}\left(\frac{R_2 + R_3}{R_1 + R_2 + R_3}\right) \qquad (11\text{-}3)$$

D_1 holds the negative emitter-base voltage of Q_1 to −0.7 V. The output of the circuit is −0.7 V because the zener diode is forward biased. When $V_{in} >$ UTP, the output of the comparator goes to $+V_{sat}$ and the circuit output to V_Z. The switching transistor turns on shorting R_3 out, thus reducing the voltage on the inverting terminal to the LTP, which is

$$LTP = V_{ref}\left(\frac{R_2}{R_1 + R_2}\right) + V_{CE}\text{ (sat) } Q_1 \qquad (11\text{-}4)$$

The output voltage does not change again until $V_{in} <$ LTP. If D_1 is left out, this circuit can be used for a single-supply comparator as easily as a dual-supply comparator.

EXAMPLE 11-2

Calculate the components for a Schmitt trigger as shown in Figure 11.3 using a μA741 as a comparator. Let ±V = ±15 V, UTP = 7 V, LTP = 4 V, V_{out} TTL compatible, and $V_{ref} = +15$ V. The comparator must supply 3 mA to the load.

SOLUTION

The gain of the μA741 could be limited to any value desired, but we will use the op-amp open loop to provide minimum trip-point uncertainty. Let $I_{divider}$ for R_1, R_2, and R_3 = 1 mA, and the specifications for Q_1 be

$$BV_{CEO} = 30\text{ V},\quad V_{CE\,(sat)} = 0.1\text{ V at } I_c = 1\text{ mA},\quad V_{BE} = 0.7\text{ V at } I_c = 1\text{ mA},$$
$$\beta_{min} = 50 \text{ at } I_c = 1\text{ mA}.$$

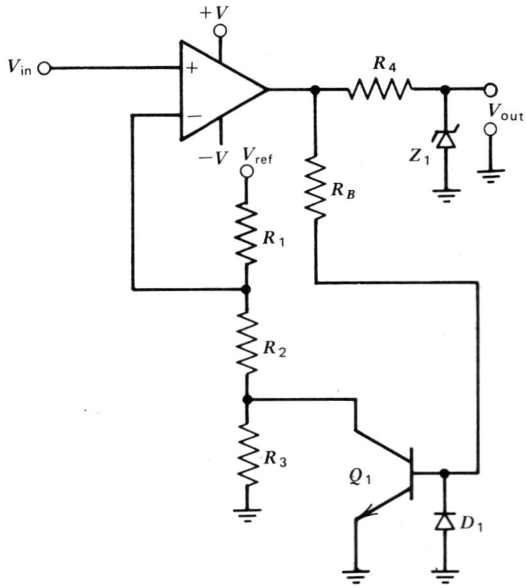

(a) Circuit

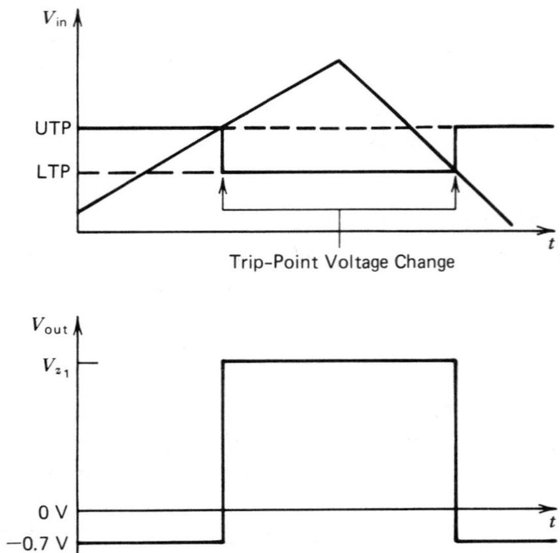

$$\mathrm{UTP} = V_{\mathrm{ref}}\left(\frac{R_2 + R_3}{R_1 + R_2 + R_3}\right)$$

$$\mathrm{LTP} = V_{\mathrm{ref}}\left(\frac{R_2}{R_1 + R_2}\right) + V_{CE(\mathrm{sat})}$$

(b) Waveforms

FIGURE 11.3 Schmitt trigger with more precise trip points and unipolar output.

Choose $V_{Z_1} = 4.7$ V at $I_Z = 2$ mA.
Since $I_D = 1$ mA

$$R_2 + R_3 = \frac{\text{UTP}}{I_D} = \frac{7\text{ V}}{1\text{ mA}} = 7\text{ k}\Omega$$

$$R_1 = \frac{V_{\text{ref}} - \text{UTP}}{I_D} = \frac{8\text{ V}}{1\text{ mA}} = 8\text{ k}\Omega$$

To find R_2, recall from Equation 11-4

$$\text{LTP} = V_{\text{ref}}\left(\frac{R_2}{R_1 + R_2}\right) + V_{C\text{ (sat)}}\, a_1$$

Solving for R_2 as follows, we find

$$\text{LTP} - V_{CE\text{ (sat)}} = V_{\text{ref}}\left(\frac{R_2}{R_1 + R_2}\right)$$

$$[\text{LTP} - V_{CE\text{ (sat)}}]\,(R_1 + R_2) = V_{\text{ref}}\, R_2$$

$$[\text{LTP} - V_{CE\text{ (sat)}}]\, R_1 + [\text{LTP} - V_{CE\text{ (sat)}}]\, R_2 = V_{\text{ref}}\, R_2$$

$$[\text{LTP} - V_{CE\text{ (sat)}}]\, R_1 = V_{\text{ref}}\, R_2 - [\text{LTP} - V_{CE\text{ (sat)}}]\, R_2$$

$$[\text{LTP} - V_{CE\text{ (sat)}}]\, R_1 = R_2\, [V_{\text{ref}} - \text{LTP} + V_{CE\text{ (sat)}}]$$

$$R_2 = \frac{(\text{LTP} - V_{CE\text{ (sat)}})\, R_1}{V_{\text{ref}} - \text{LTP} + V_{CE\text{ (sat)}}}$$

$$= \frac{(4\text{ V} - 0.1\text{ V})\, 8\text{ k}\Omega}{15\text{ V} - 4\text{ V} + 0.1\text{ V}} = 2.8\text{ k}\Omega$$

so

$$R_3 = 7\text{ k}\Omega - R_2 = 7\text{ k}\Omega - 2.8\text{ k}\Omega = 4.2\text{ k}\Omega$$

R_B is calculated from

$$R_B = \frac{+V_{\text{sat}} - V_{BE}}{I_D/\beta_{\text{min}}} = \frac{14\text{ V} - 0.7\text{ V}}{1\text{ mA}/50} = 665\text{ k}\Omega$$

and R_4 is

$$R_4 = \frac{V_{\text{sat}} - V_Z}{I_Z + I_L}$$

$$= \frac{14\text{ V} - 4.7\text{ V}}{2\text{ mA} + 3\text{ mA}}$$

$$= 1.86\text{ k}\Omega$$

A window comparator can be constructed as shown in Figure 11.4. A window comparator provides one output if the input is between the UTP and LTP, and another if the input is greater than the UTP or lower than the LTP. The window comparator can detect upper and lower test limits at the same time. For example, if a device output for a given test is supposed to be between 4.8 and 5.2 V, the UTP of the window comparator is set at 5.2 V and the LTP at 4.8 V. If the output is above 5.2 V, the output of A_1 is high; if the output is below 4.8 V, the output of A_2 is high. If the device under test (D.U.T.) output is between 4.8 and 5.2 V, the output of both A_1 and A_2 are low. Diodes *D* are an OR gate for the outputs of A_1 and A_2, so that if either amplifier output is high, the circuit output is high.

EXAMPLE 11-3

Set up a window comparator as shown in Figure 11.4 using the LM311 comparator. The UTP = 6 V and LTP = 5 V. V_{ref} = +V = 12 V, and no strobe is needed. R_L = 2 kΩ.

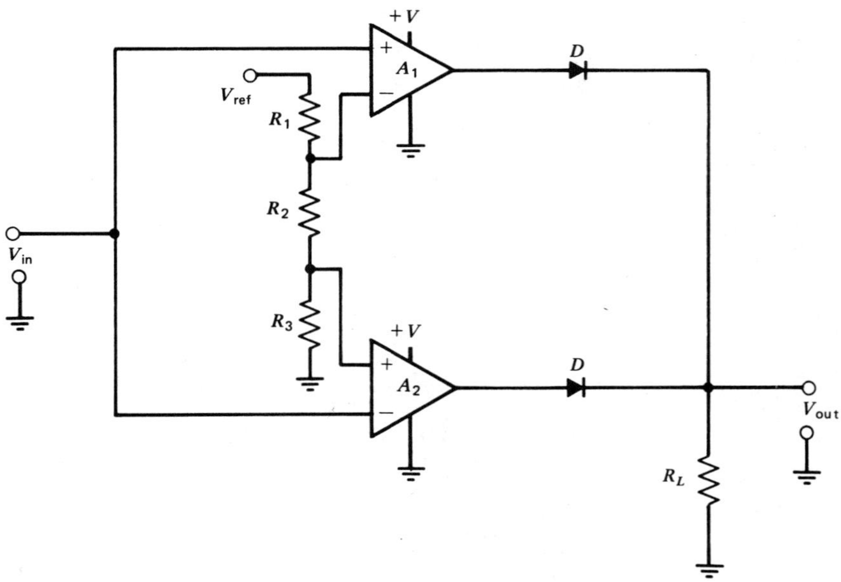

V_{in} > UTP, V_{in} < LTP Output High

UTP > V_{in} > LTP Output Low

$$\text{UTP} = V_{ref}\left(\frac{R_2 + R_3}{R_1 + R_2 + R_3}\right)$$

$$\text{LTP} = V_{ref}\left(\frac{R_3}{R_1 + R_2 + R_3}\right)$$

Note. Ground can be −V for a two-supply comparator if UTP and LTP are not both positive.

FIGURE 11.4 Window comparator.

SOLUTION

Since no strobe is needed, the strobe input of the LM311 is connected to pin 5. Since $R_L = 2\ k\Omega$, the LM311 can provide more than enough output current. All that is necessary is to calculate R_1, R_2, and R_3 and connect the circuit. Let $I_D = 0.5$ mA.

$$R_1 = \frac{V_{ref} - UTP}{I_D} = \frac{12\ V - 6\ V}{0.5\ mA} = 12\ k\Omega$$

$$R_2 = \frac{UTP - LTP}{I_D} = \frac{6\ V - 5\ V}{0.5\ mA} = 2\ k\Omega$$

$$R_3 = \frac{LTP}{I_D} = \frac{5\ V}{0.5\ mA} = 10\ k\Omega$$

11-1.3 Automatic Testing

At the heart of virtually all automatic test equipment, production or otherwise, is a precision comparator. With the test conditions set up, the comparator compares the voltage at a terminal of the unit under test with a reference voltage. If the voltage on the unit under test is above or below the reference voltage, depending on the desired response, the unit passes or fails the test.

For example, suppose that the 1 output of a digital NAND gate must be a minimum of 2.4 V with V_{cc} at its minimum value (4.75 V), and 0.8 V (O) on each input. We want a no-go, or negative, test result if the gate output is less than 2.4 V and a go output if the gate output is above 2.4 V at I_{out}(max) = 400 μA for a 7400 series gate.

We apply the test conditions to the unit under test and use a comparator to compare the output of the unit under test to a precisely adjusted 2.4 V reference as shown in Figure 11.5*a*. If a positive output of the comparator indicates a go and

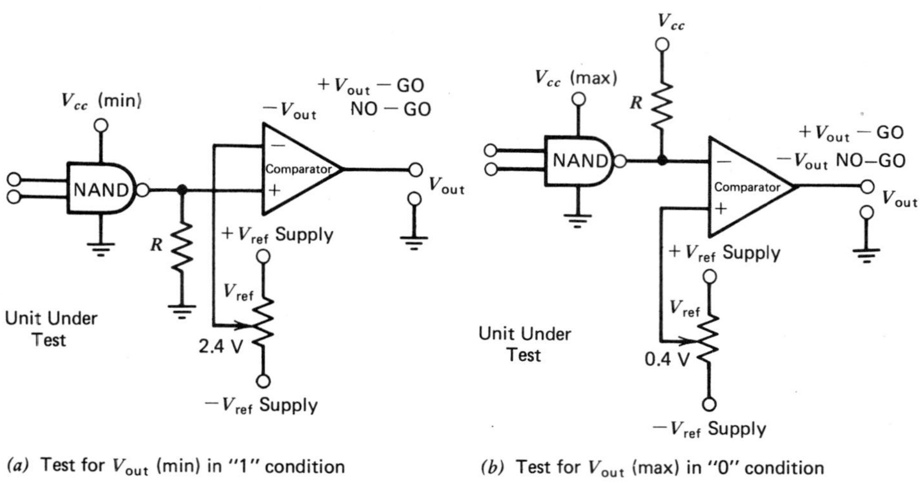

FIGURE 11.5 Comparator use in GO—NO GO testing.

a negative output a no-go condition, we must arrange the comparator inputs to give the proper output polarity for the test results. Since we are testing for a minimum, if the gate output is less than 2.4 V we want the op-amp output to be negative. To meet this condition, we connect the reference to the inverting input and the test point to the noninverting input. Now if the output voltage of the device under test is less than the reference voltage, we will have a negative output, which is what we desire for a no-go. The output, positive for go and negative for no-go, will be routed to indicator lights, entered into a computer, or otherwise recorded. The load resistor R is chosen as

$$R = \frac{V_{out}(\text{min})}{I_{out}(\text{max})} = \frac{2.4\ \text{V}}{400\ \mu\text{A}} = 6\ \text{k}\Omega$$

for this test.

The test conditions can be set up with programmable power supplies and current sources with relays or analog switches connecting the test conditions to the pins of the unit under test. The comparator can be switched to various pins of the unit under test and to the appropriate reference voltage with relays or analog switches. The reference could be a precision programmable power supply or an arrangement of precision resistors that is switched to provide various voltage dividers. The comparator amplifier must be very stable, with a low offset voltage. Chopper-stabilized amplifiers with internal offset null are preferred for low-speed applications, but other types may be necessary for high-speed testing.

The comparator is often used with open loop so a high degree of "trip-point" accuracy is achieved with moderate open loop gains.

As another example look at Figure 11.5*b*. Here we are testing the same NAND gate of Figure 11.5*a* for its maximum zero voltage that is specified at 0.4 V with $I_{sink} = 16$ mA and $V_{cc}(\text{max}) = 5.25$ V. Each input is at the minimum 1 condition, 2 V. The comparator will have a negative (no-go) output if the NAND gate output is more than 0.4 V under the test conditions, and a positive (go) output for a gate output of less than 0.4 V. Note that the comparator leads are reversed for a maximum test from a minimum test. The sink resistor R is chosen as

$$R = \frac{V_{cc}(\text{max}) - V_{out}(\text{max})}{I_{sink}} = 303\Omega$$

for our 7400 series gate.

11-1.4 Digital-to-Analog Conversion

Digital-to-analog converters (D/A) convert digital code to analog output voltages. In this fashion, digital codes can be used to drive devices that require an analog input such as motors. D/A converters are available in monolithic IC form up to about 10 digital bits in and hybrid IC form for extreme precision with 16 or more digital bits input. Op-amps are used primarily as buffers and amplifiers in D/A converters, so we include D/A converters here for completeness. Several types of D/A converters are available. We will look briefly at two types.

WEIGHTED-RESISTOR D/A

Figure 11.6 shows a four-bit weighted resistor D/A converter and the D/A output for a given binary input. The circuit is basically a summer with a very precision-fixed input. The binary inputs are switch closures. The switches are transistors or analog switches (to be covered soon) that are closed by the digital input bits. If the digital input is 1010 (decimal 10), switches 1 and 3 are closed. The amplifier is then a weighted summer with an output of

$$V_{out} = -V_{ref}\left(\frac{R_f}{R + 4\,R}\right)$$

Since the resistors are weighted in a binary fashion, each switch closure and resistor provides the proper gain for the value of the binary bit chosen. A close study of the table in Figure 11.6 will show the output versus the switch closures. Since there are four switches, 16 binary numbers can be converted to 16 different output voltage levels.

The major problem with the weighted resistor D/A converter is that a different value of resistor must be used for each bit. An 8-bit converter requires resistors ranging from R to 128 R. If R = 10 kΩ, 128 R = 1.28 MΩ. It is difficult to construct resistors with that great a size variance that are both very precise and have the same rate of change in resistance with temperature.

R-2*R* DIGITAL-TO-ANALOG CONVERTER

A more popular D/A converter type that solves the resistor problem is the *R*-2*R* ladder D/A converter. This converter switches only two resistor values (one if the 2*R* values are two-series *R* resistors) to provide any number of output levels (i.e., input bits) desired. Many variations of the basic *R*-2*R* ladder are the heart of both monolithic and hybrid IC D/A converters. A four-bit *R*-2*R* ladder D/A is shown in Figure 11.7.

As before, each digital bit is a switch closure. As before, a digital 1010 (decimal 10) would be entered into the circuit by closing switches S_1 and S_3. The op-amp is simply a buffer. The output of the amplifier will be

$$V_{out} = V_{pt.\ d}\left(\frac{R_1 + R_f}{R_1}\right)$$

The *R*-2*R* ladder simply functions as a binary weighted-voltage divider or current divider. The use of same-value resistors of the same type allow looser absolute resistor tolerances, as long as all the resistors are the same value, and closer temperature tracking. A table of ladder voltage divider ratios at pt. *d* versus binary input is shown in Figure 11.7*b*.

To see how the divider ratios arise, refer to Figure 11.8. For four bits in, 16 distinct voltage levels are required. If S_1 is at V_{ref} while switches S_2, S_3, and S_4 are at ground, the digital word in is half its maximum value so $V_{pt.\ d}$ should be 1/2 V_{ref}. Starting at the bottom of the *R*-2*R* ladder in Figure 11.7*a*,

$$\begin{aligned} R \text{ at pt. } a &= 2\,R \parallel 2\,R = R \\ R \text{ at pt. } b &= (R_{pt.\ a} + R) \parallel 2\,R = 2\,R \parallel 2\,R = R \\ R \text{ at pt. } c &= (R_{pt.\ b} + R) \parallel 2\,R = 2\,R \parallel 2\,R = R \\ R_{pt.\ d} \text{ to ground} &= R_{pt.\ c} + R = 2\,R \end{aligned}$$

The voltage at point *d* is then $V_{ref}/2$ as shown in Figure 11.8*a*.

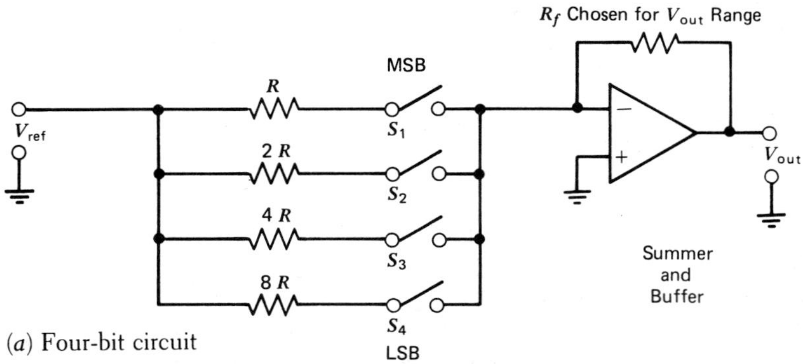

(*a*) Four-bit circuit

Decimal Number	Binary Number Most significant bit MSB			Least significant bit LSB	Summer Gain	Output Voltage
0	0	0	0	0	0	0
1	0	0	0	1	0.125	−0.625 V
2	0	0	1	0	0.25	−1.25 V
3	0	0	1	1	0.375	−1.875 V
4	0	1	0	0	0.5	−2.5 V
5	0	1	0	1	0.625	−3.125 V
6	0	1	1	0	0.75	−3.75 V
7	0	1	1	1	0.875	−4.375 V
8	1	0	0	0	1	−5 V
9	1	0	0	1	1.125	−5.625 V
10	1	0	1	0	1.25	−6.25 V
11	1	0	1	1	1.375	−6.875 V
12	1	1	0	0	1.5	−7.5 V
13	1	1	0	1	1.625	−8.125 V
14	1	1	1	0	1.75	−8.75 V
15	1	1	1	1	1.875	−9.375 V

(*b*) Output versus binary input for $V_{ref} = 5$ V, $R_f = 10$ kΩ, $R = 10$ kΩ, $2R = 20$ kΩ, $4R = 40$ kΩ, $8R = 80$ kΩ

FIGURE 11.6 Weighted resistor digital-to-analog converter.

If S_2 is at V_{ref} and S_1, S_3, and S_4 in the ground position, the resistance at point *b* is *R* as before and the resistance from point *c* to ground is 2 *R* in parallel with *R* and 2 *R* in series as shown in Figure 11.8*b*. The voltage at point *c* is

$$V_{pt.\ c} = V_{ref}\left(\frac{2\ R \parallel 3\ R}{2\ R \parallel 3\ R + 2\ R}\right) = V_{ref}\left(\frac{1.2\ R}{3.2\ R}\right) = 0.375\ V_{ref}$$

The voltage at point *c* is then divided by the series *R* and 2 *R* to yield

$$V_{pt.\ d} = 0.375\ V_{ref}\left(\frac{2\ R}{3\ R}\right) = 0.25\ V_{ref}$$

Every time S_2 is at V_{ref} a voltage of 0.25 V_{ref} is added to the voltage at point *d*.

If S_3 is at V_{ref} while S_1, S_2, and S_4 are at ground, the circuit of Figure 11.8*c* results. The circuit is simplified by the same method shown with switch S_2 in the V_{ref} position. Each time S_3 is in the V_{ref} position, 0.125 V_{ref} is added to the voltage at point *d*. Similarly, if S_4 is switched to V_{ref} with S_1, S_2 and S_3 at ground, the voltage at point *d* is 0.0625 V_{ref}.

The voltage at any set of switch positions can quickly be determined since each switch at V_{ref} adds a voltage equal to its relative binary weight. For example, if the input digital number is 1011 (decimal 13), the output at point *d* is

$$\begin{aligned} V_{pt.\ d} &= 0.5\ V_{ref} + 0.125\ V_{ref} + 0.0625\ V_{ref} \\ &= 0.6875\ V_{ref} \end{aligned}$$

All of the four bit values are shown in Figure 11.8*b*.

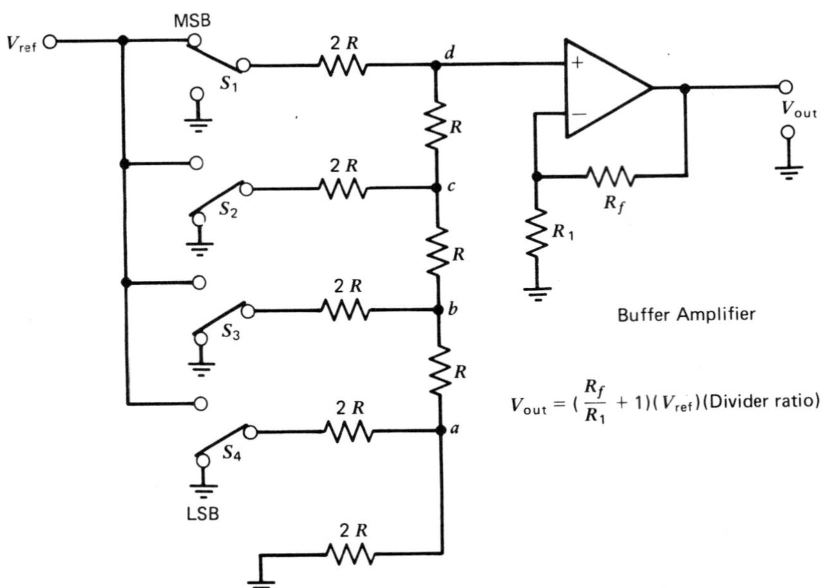

(*a*) *R* – 2*R* A/D circuit

Decimal Number	Binary Number MSB			LSB	Voltage Divider Ratio at Point d $\left(\frac{V_d}{V_{ref}}\right)$
0	0	0	0	0	0
1	0	0	0	1	0.0625
2	0	0	1	0	0.125
3	0	0	1	1	0.1875
4	0	1	0	0	0.25
5	0	1	0	1	0.3125
6	0	1	1	0	0.375
7	0	1	1	1	0.4375
8	1	0	0	0	0.5
9	1	0	0	1	0.5625
10	1	0	1	0	0.625
11	1	0	1	1	0.6875
12	1	1	0	0	0.75
13	1	1	0	1	0.8125
14	1	1	1	0	0.875
15	1	1	1	1	0.9375
Digital Bit Value	2^3	2^2	2^1	2^0	

(*b*) Voltage divider ratio at point *d* versus binary code

FIGURE 11.7 *R*-2*R* digital-to-analog converter.

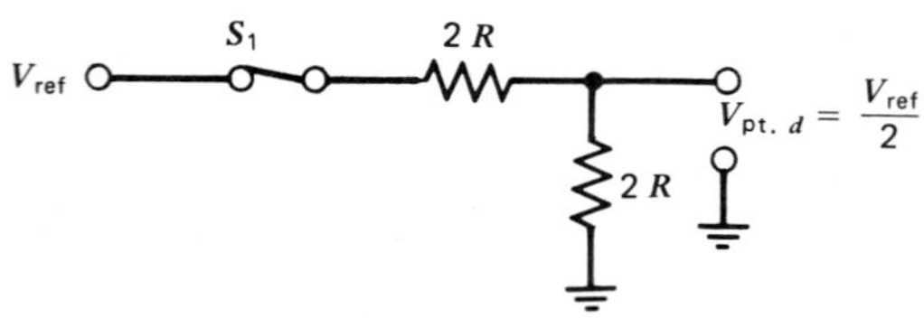

(*a*) S_1 to V_{ref}

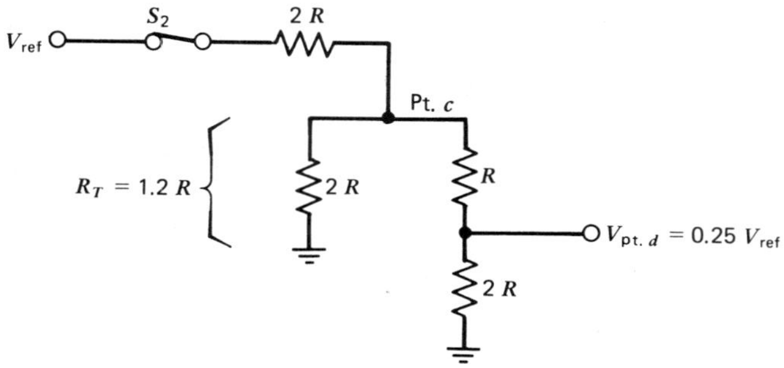

$$V_{pt.c} = V_{ref}\ \frac{1.2R}{3.2R} = 0.375\ V_{ref}$$

$$V_{pt.d} = 0.375\ V_{ref}\ \frac{2R}{3R} = 0.25\ V_{ref}$$

(b) S_2 to V_{ref}

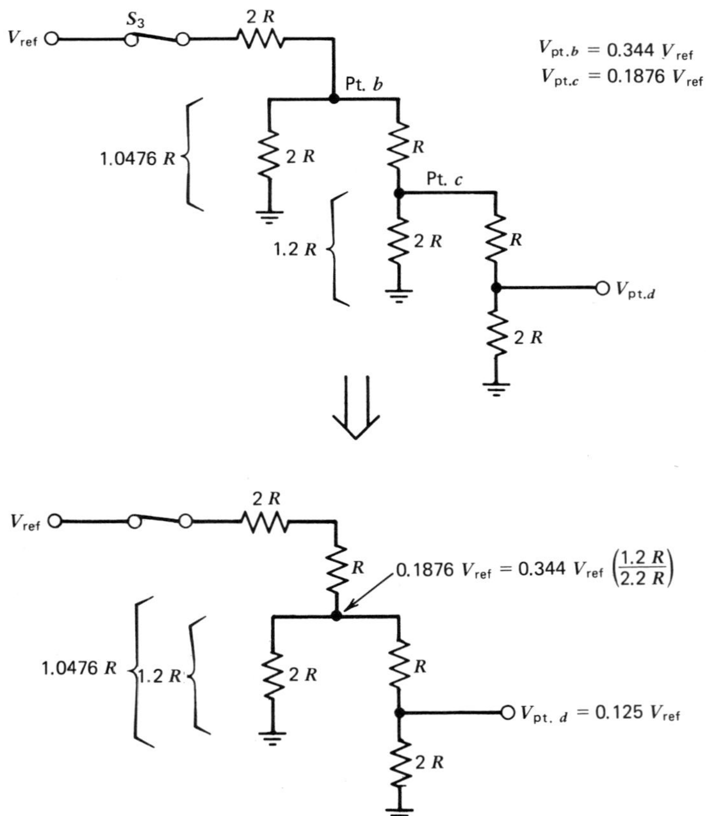

(c) S_3 to V_{ref}

FIGURE 11.8 *R*-2*R* ladder solutions.

11-1.5 Analog-to-Digital Converters

Analog-to-digital (A/D) converters change an analog input to a digital code. A/D converters are used when a circuit or sensor that has an analog output, such as a temperature-sensing bridge, must be converted to a digital code for convenient and economical data logging or calculations. A/D converters are used extensively in industrial process control, digital communications, and testing. The list of uses for A/D converters is very long and growing rapidly. There are many types of A/D converters, but most are variations on three main themes. These three will be discussed briefly here. They are:

1. Parallel or "flash" converters.
2. Integrating converters.
3. Successive approximation converters.

A number of monolithic eight- and ten-bit A/D converters are available, and hybrid A/D converters of 16 and more bits are produced. One or more comparators are used in all A/D converters and are essential to their operation.

PARALLEL A/D CONVERTER

Parallel A/D converters are essentially a group of parallel comparators as shown in Figure 11.9*a*. The analog input is applied to several comparators simultaneously. Each comparator has one input connected to a different reference voltage. The reference voltage of each comparator is the analog value of one LSB apart. All comparators for which $V_{in} > V_{ref}$ change output state when V_{in} is applied. All comparators for which $V_{in} < V_{ref}$ do not change states. The comparator outputs go to a decoding circuit that changes the output states with V_{in} applied to a digital word.

Parallel A/D converters are fast with conversion times as short as 30 ns, since the digital output is available immediately after the settling time of the comparators and the propagation time of the decoding logic. However, one comparator is needed for every possible output bit configuration [$(2^n - 1)$ where n = number of bits] so an eight-bit parallel comparator requires $2^8 - 1 = 256 - 1 = 255$ comparators. The parallel converter becomes very costly as the number of bits are increased.

A two-bit parallel A/D converter with decoding is shown in Figure 11.9*b*. The reference voltages start at 1/2 LSB × V_{ref} and continue a whole LSB × V_{ref} apart to V_{ref} − 1/2 LSB × V_{ref}. Thus, for a two-bit parallel comparator the reference voltages for the three $(2^2 - 1)$ comparators with $V_{ref} = 3$ V are 0.5 V, 1.5 V, and 2.5 V. If $V_{in} < 0.5$ V, all comparator outputs are low and the digital code is 00. If 0.5 V $\leq V_{in} \leq$ 1.5 V, comparator 1 output goes high, comparator 2 output is low. The comparator 2 output is inverted and fed to the AND gate so the AND gate output is high. Thus a high appears at the output of the OR gate and the digital code is 01.

If 1.5 V $\leq V_{in} \leq$ 2.5 V, the output of comparator 2 is high, disabling the AND gate so its output goes low. The digital code is now 10. If $V_{in} > 2.5$ the output of comparators 2 and 3 are high. The output of comparator 3 sets the LSB OR gate output high so the digital code is 11.

DUAL-SLOPE A/D CONVERTER

The dual-slope A/D converter is one of several types of integrating converter circuits. Integrating A/D converters are slow, with conversion times from 20 to 40 ms, generally very accurate, linear, and low cost. They are used for digital voltmeters, digital panel meters, and data acquisition where speed is not a factor.

The dual-slope A/D converter, shown in Figure 11.10, is one of the most popular integrating A/D converters. It provides high accuracy and requires only short-term clock stability. It also rejects noise on the input well.

Refer to Figure 11.10*a* and *b* for the operation. Conversion begins with the control logic setting the counter to zero. Switch S_1 is set to couple V_{in} to the integrator and the counter counts the clock. As the counter counts, V_{in} (if positive) causes the integrator output to swing negatively at a rate proportional to V_{in}. At the counter overflow signal the integrator output voltage

$$V_{\text{out integrator}} = \frac{(V_{in}/R)\,t_1}{C} = \frac{V_{in}\,t_1}{RC}$$

At the overflow S_1 is set to $-V_{ref}$, the counter reset to zero, the counter begins counting again, and the capacitor begins to discharge at a fixed rate proportional

FIGURE 11.9 Parallel analog-to-digital converter.

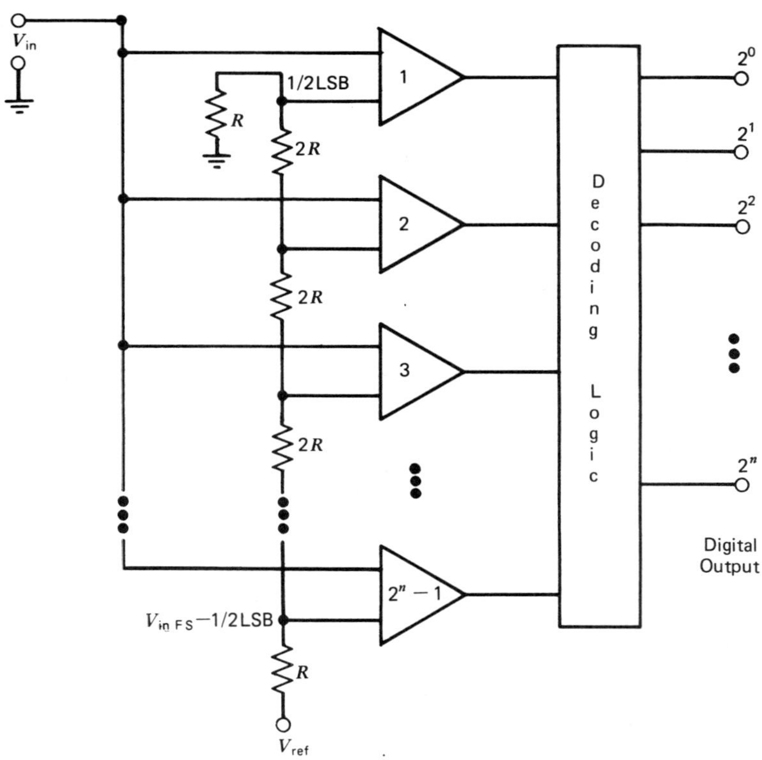

(*a*) General parallel conversion circuit

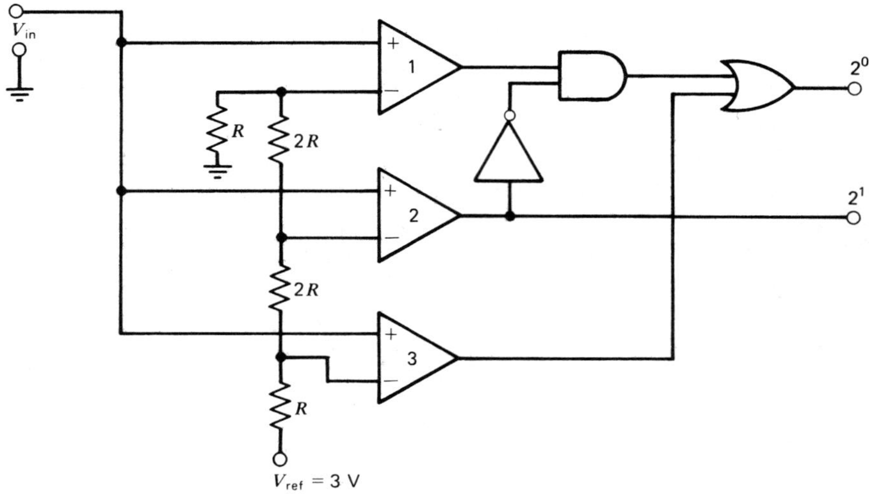

V_{IN}	Digital code
0 V 0.5 V	00
1 V 1.5 V	01
2 V 2.5 V	10
3 V	11

(b) Two-bit parallel A/D with decoding logic

to V_{ref}/RC. When the integrator capacitor has discharged to zero, a comparator stops the counter. The counter contents are decoded into the voltage level. The counter contents represent the ratio of V_{in} to V_{ref}. Since the integrator capacitor discharges a voltage in t_2 equal to its charge during t_1

$$\frac{V_{in}\, t_1}{RC} = \frac{V_{ref}\, t_2}{RC}$$

$$\frac{V_{in}}{V_{ref}} = \frac{t_2}{t_1}$$

Since the output is a ratio of the counts during t_1 and t_2, the clock must be stable only during times $t_1 + t_2$. The noise rejection results from the average noise voltage being zero over a long period of time. If $t_1 = 16.67$ ms, as is often the case, the ever-present 60 Hz noise is very well rejected.

Dual slope, and other integrating A/D converters, are available in both monolithic and hybrid IC packages.

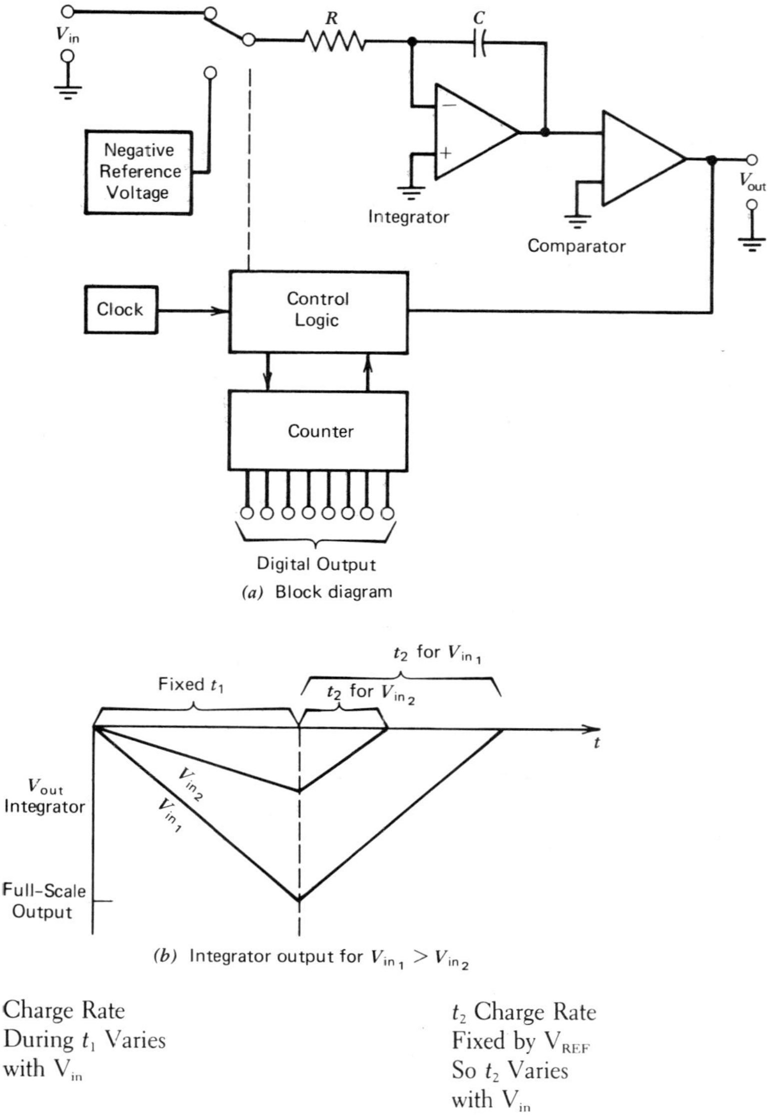

(a) Block diagram

(b) Integrator output for $V_{in_1} > V_{in_2}$

FIGURE 11.10 Dual-slope analog-to-digital converter.

SUCCESSIVE-APPROXIMATION A/D CONVERTER

The most popular A/D conversion technique is successive approximation. The method provides moderate cost, moderate-to-high conversion speed, and good accuracy. An n-bit converter requires n clock periods to convert an analog voltage to a digital number.

The converter consists of a reference, a clock, a D/A converter, a comparator, and a successive-approximation register (available as an IC). The successive-ap-

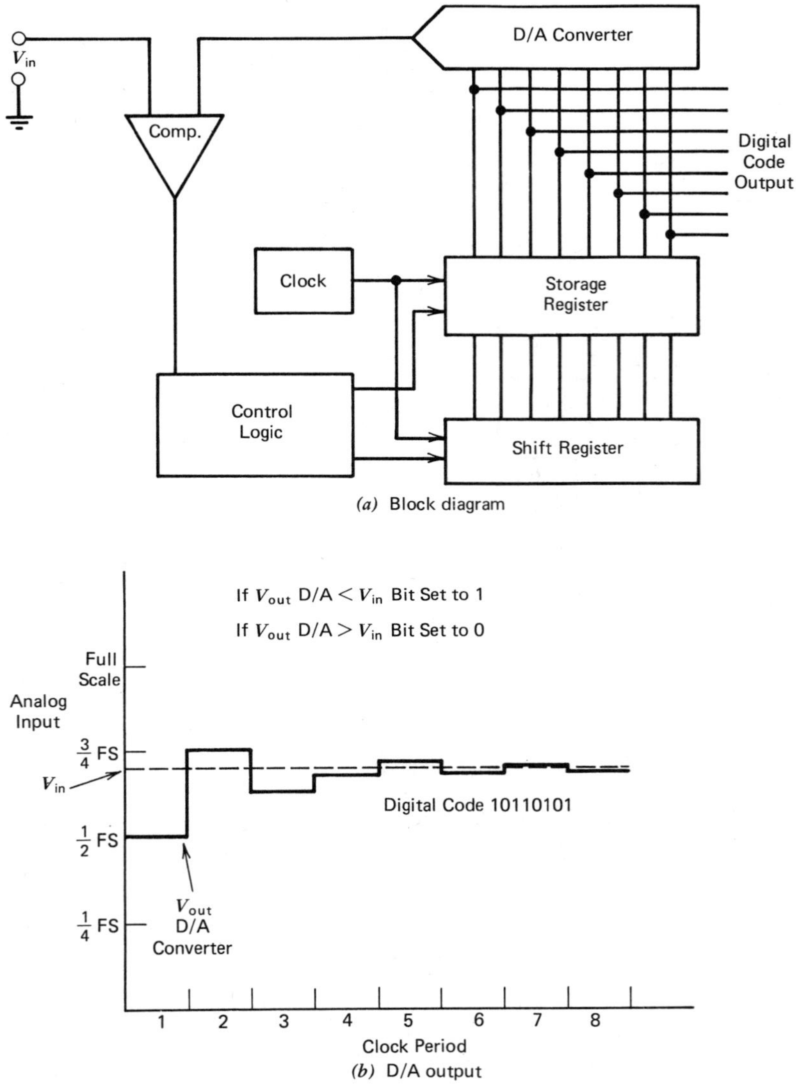

FIGURE 11.11 Successive-approximation analog-to-digital converter.

proximation register (SAR) consists of a storage register, a shift register, and the necessary control logic.

A block diagram of the successive-approximation register is shown in Figure 11.11*a* and the D/A converter output in Figure 11-11*b*. The conversion starts by setting the MSB of the storage register to 1 and all other bits to 0. The D/A converter converts the MSB to an analog output of 1/2 V_{FS} (one-half full-scale voltage). The comparator compares the D/A output to the analog input. If the input voltage is greater than the D/A output, the MSB is left at 1, if not the MSB is set to 0. The shift register shifts its 1 to the next bit position at the beginning of the second clock cycle. If the D/A output of the first two bits is less than V_{in}, the second bit is set to 1, if not, the second bit is set to 0. This process continues until all of the bits are tested.

The accuracy of the successive approximation A/D converter is only as good as the D/A accuracy and can be much worse. Once again this converter is available in IC form.

11-1.6 Analog Switches

The need to perform switching functions on IC converter chips and the need for switches compatible with IC technology led to the development of integrated circuit switches. The IC analog switches are available as switches in IC packages and are incorporated directly into many A/D and D/A integrated circuits. The switching elements are JFETs or MOSFETs. These switches are much faster than their electromechanical counterparts, but they cannot handle anywhere near the voltage or current.

A typical JFET analog switch, the National Semiconductor AHO140, is shown in Figure 11.12. The entire series—AHO120, AHO130, AHO140, AHO150, and AH0160—specifications are in Appendix C. The series features turn on times of 4 μs and turn off times of 1 μs. On resistance is available in five steps between 10 and 80 Ω. A high input voltage level (±10 V) and low input voltage level (±7.5 V) series is available. The switches operate from ±15 V but can be switched by TTL logic. The switch control circuitry needs only 2.5 V at 0.1 mA to switch. The switch contacts can switch up to 30 mA. The JFET switch contacts must have the drain to the higher voltage to be switched, which can be a disadvantage.

The circuit of Figure 11.12*a* operates as follows: The switch is enabled by grounding the enable pin. If this pin is high in potential as an input logic 1 level, the input transistors Q_1 cannot turn on and the switch is disabled. If the enable pin is low, a $V_{in} > 2.5$ V turns Q_1 on, which in turn switches Q_2 on, Q_3 on, and Q_4 off. The JFET gate voltage is now about zero volts and will remain above the source voltage. When $V_{in} \leqslant 0.8$ V the switch turns off since Q_1 turns off turning Q_2 off, Q_3 off, and Q_4 on. Q_4 connects the gate of the JFET to $-V$ holding it off.

MOSFET IC switches can be constructed so that it does not matter which contact is connected to the higher voltage. As shown in Figure 11.13, the MOSFET analog switch consists of a paralleled enhancement *p* channel and *n* channel device. Two amplifiers provide the proper interface between the control voltage and the gate drive. In the circuit of Figure 11.13, when $V_{in} < V_{ref}$ the gate of the *n* channel MOSFET, Q_1, is at $-V$ and the *p* channel device, Q_2, at $+V$; the switch is off for any contact voltage between the supply voltages. When $V_{in} > V_{ref}$ the gate of the *n* channel MOSFET goes to $+V$ and the *p* channel gate to $-V$. The switch is on for any contact voltage between the positive supply voltage less the *n* channel MOSFET threshold and the negative supply voltage less the *p* channel MOSFET threshold. MOSFET IC switches are available from several semiconductor manufacturers.

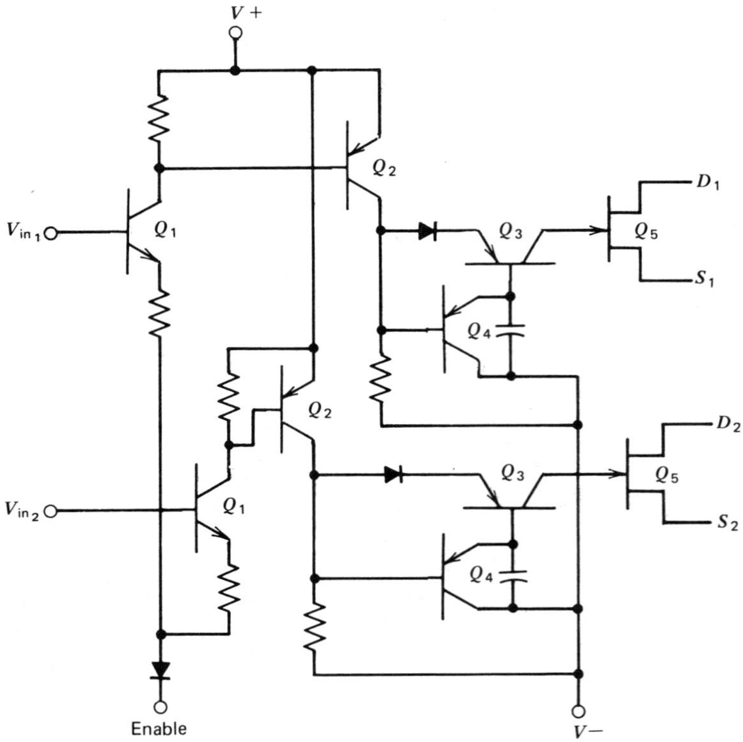

(a) Equivalent circuit, dual SPST

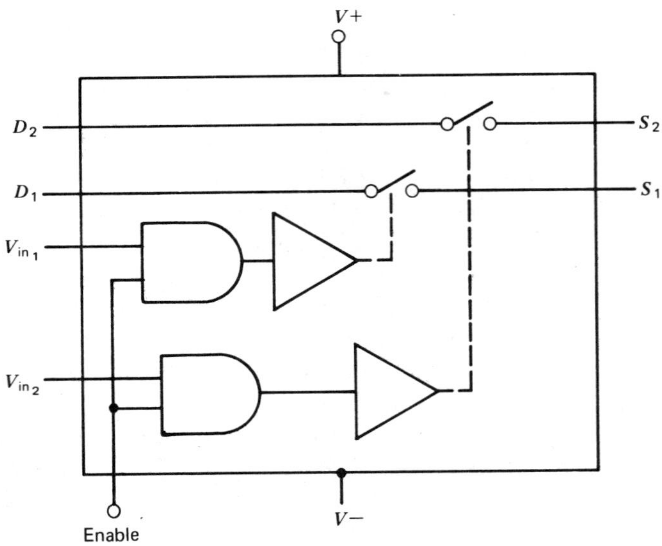

(b) IC block diagram

FIGURE 11.12 National Semiconductor AH0140 JFET analog switch. (*Courtesy of National Semiconductor Corporation.*)

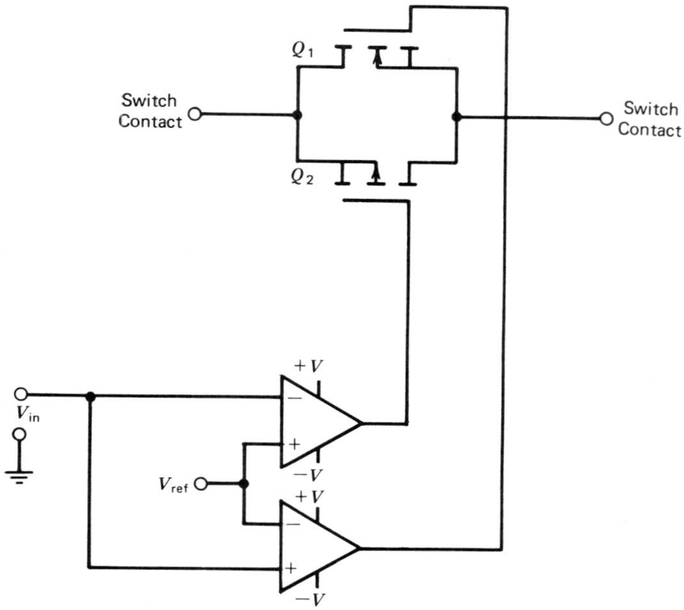

$V_{in} > V_{ref}$ Switch On $V_{GQ1} \cong +V,\ V_{GQ2} \cong -V$

$V_{in} < V_{ref}$ Switch Off $V_{GQ1} \cong -V,\ V_{GQ2} \cong +V$

FIGURE 11.13 MOSFET analog switch equivalent circuit.

11-1.7 Switched Capacitors

The advent of reliable MOSFET analog switches suitable for integration has allowed IC designers to use small, accurate ratio MOS capacitors and analog switches to perform functions for which resistors were formerly used. The MOS capacitors are both smaller and more accurately controllable than resistors on monolithic integrated circuits. Two uses of switched capacitors are comparators and gain setting.

SWITCHED CAPACITOR COMPARATOR

The switched capacitor, also called charge balancing, comparator is shown in Figure 11.14. It is used by National Semiconductor in monolithic IC A/D converters.

The circuit operates by sampling the reference and input voltage in turn. When S_1 is closed, C_1 and C_2 are charged to V_{ref}. The amplifier is shunted during the capacitor charge. S_1 is opened and S_2 is closed. If $V_{in} = V_{ref}$, no change in charge occurs on the capacitors. If $V_{in} < V_{ref}$, the amplifier output swings positive, and if $V_{in} > V_{ref}$, the amplifier output swings negative. An output swing is ac coupled through C_c to a latch.

The output voltage swing is the amplifier voltage gain multiplied by the voltage change on C_2. Since C_1 and C_2 are in series

$$V_{cs} = \frac{Q_T}{C_2} = \frac{C_T V_T}{C_2} = \frac{\left(\dfrac{C_1\,C_s}{C_1 + C_2}\right) V_T}{C_2} = \frac{C_1\,V_T}{C_1 + C_2}$$

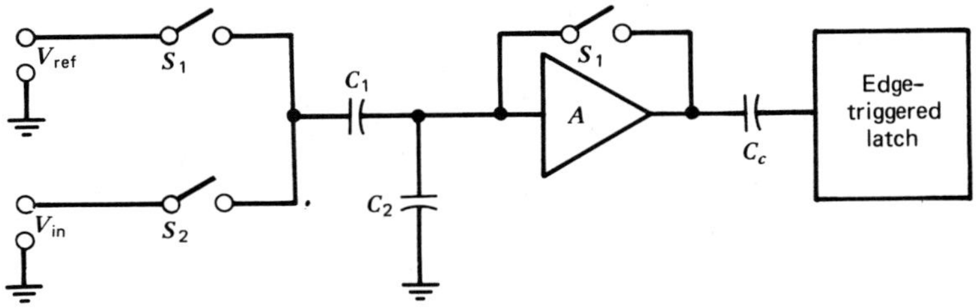

$$\Delta V_{out} = \frac{-A\, C_1(V_{in} - V_{ref})}{C_1 + C_2}$$

FIGURE 11.14 Switched capacitor comparator.

where

Q_T = total charge
C_T = C_1 in series with C_2
V_T = voltage applied to the series C_1, C_2 circuit

From this, we see

$$\Delta V_{in} = \frac{C_1\, \Delta V_T}{C_1 + C_2} = \frac{C_1(V_{in} - V_{ref})}{C_1 + C_2}$$

so

$$\Delta V_{out} = \frac{-A\, C_1(V_{in} - V_{ref})}{C_1 + C_2}$$

Switches S_1 and S_2 are, of course, MOSFET analog switches.

SWITCHED CAPACITOR GAIN SETTING

Capacitors can be used to set op-amp gain as shown in Figure 11.15*a*. As long as V_{in} is an ac voltage

$$A_v = \frac{Z_f}{Z_1} = \frac{\left(\dfrac{1}{2\pi f C_f}\right)}{\left(\dfrac{1}{2\pi f C_1}\right)} = \frac{C_1}{C_f}$$

The gain remains the same as the frequency changes. It is, however, easier to build MOS capacitors with one plate grounded. The circuit of Figure 11.15*a* cannot amplify a dc voltage.

A capacitor to ground, if switched at a rate (f_s) higher than the highest frequency applied, can act as a resistor as shown in Figure 11.15*b*. If the switch is open for

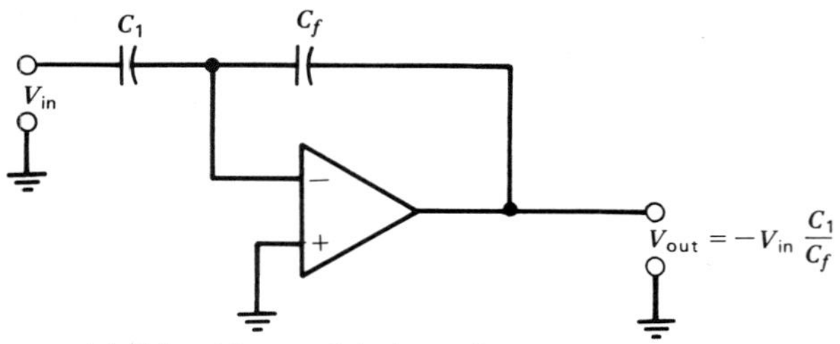

(a) Gain set by unswitched capacitors

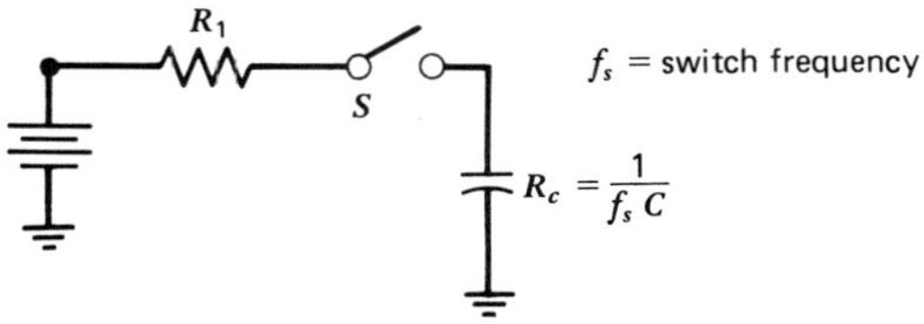

(b) Switched capacitor as a resistor

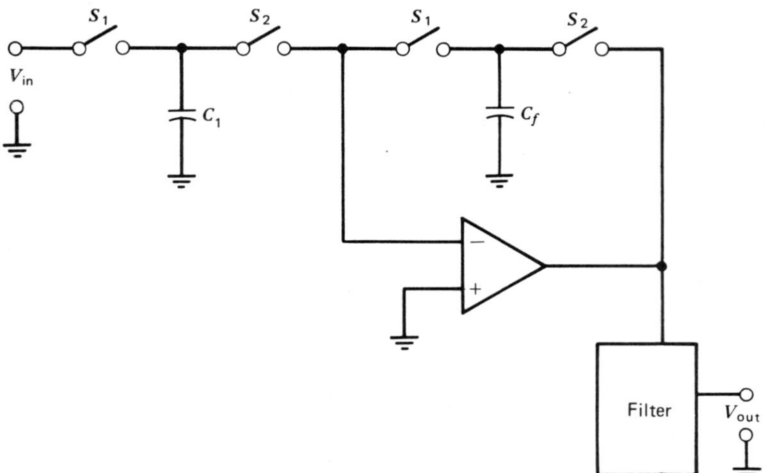

(c) Switched capacitors used to set gain

S_1 closed 50% of the Time
S_2 Closed other 50% of the Time

$$A_{fb} = -\frac{\left(\frac{1}{C_f f_s}\right)}{\left(\frac{1}{C_1 f_s}\right)} = -\frac{C_1}{C_f}$$

f_s = Switch Frequency
$f_s >> f_{in}(\max)$

FIGURE 11.15 Capacitor gain setting.

the same amount of time it is closed, we can find the effective resistance of the capacitor.

$$R_{eff} = \frac{V_c}{I_c}$$

but

$$V_c = Q/C$$

and

$$I_{c\ (ave)} = \frac{CV}{2\,t}$$

from

$$C = \frac{It}{V}$$

thus

$$R_{eff} = \frac{V_c}{I_c} = \frac{\frac{Q}{C}}{\frac{CV}{2\,t}} = \left(\frac{Q}{C}\right)\left(\frac{2\,t}{CV}\right)$$

since

$$CV = Q$$

$$R_{eff} = \frac{2\,t}{C}$$

but

$$f_s = \frac{1}{2\,t}$$

therefore

$$R_{eff} = \frac{1}{f_s C}$$

Two switched capacitors can be used to set the gain of an op-amp as shown in Figure 11.15*c*. The gain is

$$A_{fb} = -\frac{R_f}{R_1} = -\frac{\frac{1}{f_s C_f}}{\frac{1}{f_s C_1}} = -\frac{C_1}{C_f}$$

The output of the amplifier is an ac square wave whose average value is

$$V_{out} = -\left(\frac{C_1}{C_f}\right) V_{in}$$

and must be filtered. Note that the input to the circuit can be a dc voltage. Once again f_s must be much higher in frequency than the highest input frequency.

Switched capacitors can also replace resistors in active filters.

11-2 CURRENT-MODE AMPLIFIERS

Current-mode amplifiers, also called Norton amplifiers and current differencing amplifiers, amplify input current differences rather than input voltage differences as more conventional op-amps do. The major advantage of current differencing amplifiers is that they operate from a single power supply. Conventional op-amps can also be operated from a single supply, but sometimes lose performance. Current differencing amplifiers do not, however, offer the precision performance of conventional IC op-amps. The two most popular symbols for current-mode amplifiers are shown in Figures 11.16*a* and *b*. The most popular symbol has a current source

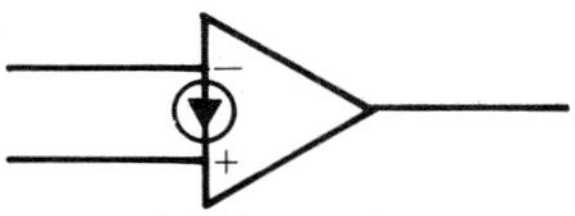

(*a*) Most used symbol

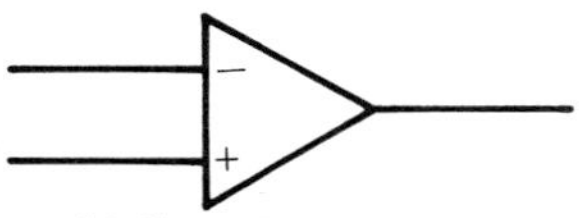

(*b*) Next most used symbol

FIGURE 11.16 Current differencing amplifier symbols.

drawn between the inverting and noninverting terminals. The other symbol has no distinguishing features, but the circuitry usually helps one identify the amplifier as current differencing.

The two most popular current-mode monolithic IC amplifiers are the quad Motorola MC3401 and the quad National Semiconductor LM3900. Although very similar, they are not exact replacements for each other. We will look at the MC3401.

11-2.1 MC3401

The Motorola MC3401 quad IC current differencing op-amp operates from a supply voltage from 5 to 18 V. It is internally frequency compensated and has an A_{ol} = 2000 typically. The typical value for bias current is 50 nA, input resistance is 1 M Ω, and linear output current is 1 mA.

The circuitry of current-mode amplifiers is distinctly different from more conventional op-amps, as can be seen in Figure 11.17*a*. Q_5 and Q_{10} are constant current sources that act as active loads for common emitter amplifier Q_1 and emitter follower Q_2, respectively. Q_5 provides high gain for Q_1, and Q_{10} biases Q_2 for class A, linear operation. Q_4 acts as an emitter follower for Q_1 to buffer Q_1 from Q_2. Q_4 provides current gain to drive Q_2, so that Q_1 can operate at a lower collector current and, therefore, will require less input current. The base of Q_1 is the amplifier inverting input. The capacitor provides frequency compensation.

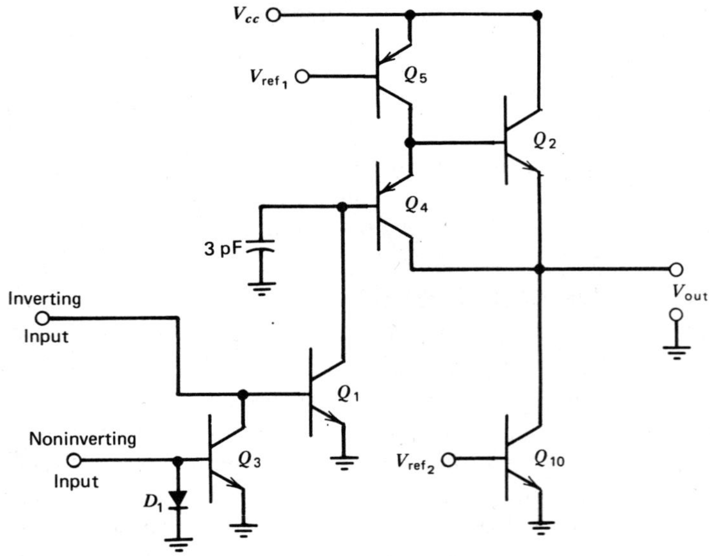

(a) Amplifier circuit

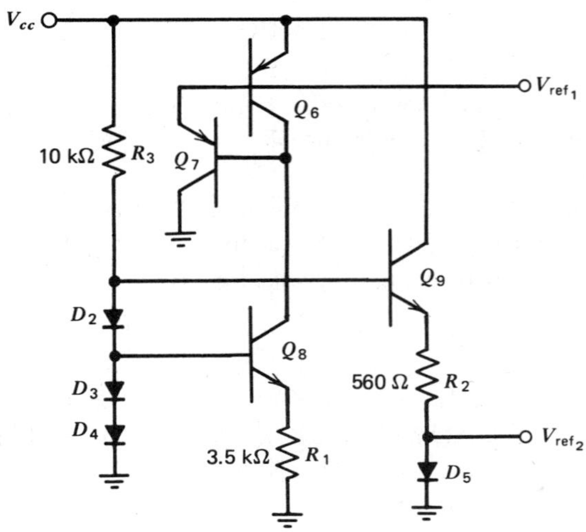

(b) Biasing circuit

FIGURE 11.17 Motorola MC 3401 current-mode operational amplifier. (*Courtesy of Motorola Semiconductor.*)

Q_3 and D_1 make up the *current mirror* for the noninverting input. Q_3 and D_1 are matched so that I of Q_3 is equal to the input current of the noninverting terminal. The noninverting terminal current is called the *mirror current*, I_m.

The circuit in Figure 11.17*b* provides the reference voltages for constant current sources Q_5 and Q_{10}. The voltage drops across D_2, D_3, and D_4 provide the reference voltages. The voltage across R_1 is

$$V_{R_1} = V_{D_3} + V_{D_4} - V_{BE_{Q8}}$$

The *pnp* current sources are set at a current equal to V_{BE}/R_1 by Q_6, whose collector current is equal to that of Q_8. Q_7 provides additional base current to the *pnp* current sources at a fixed potential to reduce loading. The current through D_5 is

$$I_{D_5} = \frac{V_{R_2}}{R_2} = \frac{V_{D_2} + V_{D_3} + V_{D_4} - V_{BE_{Q9}} - V_{D_5}}{R_2}$$

The base voltage of Q_{10} is the forward voltage of D_5. Since the characteristics of D_5 are similar to those of Q_{10}, $I_{Q10} = I_{D5}$.

If, as usually occurs, I_m is fixed as V_{in} on the inverting terminal increases, the base current of Q_1 increases causing V_{CQ_1} to decrease. If V_{CQ_1} decreases, so does the emitter voltage of Q_2. If V_{in} decreases, the base current of Q_1 decreases and V_{out} increases. Thus the inverting terminal does invert.

If the inverting input current is held constant and the noninverting input current is decreased, then I_{Q_3} decreases causing $I_{BQ1} = I_{in} - I_{CQ_3}$ to increase. As I_{BQ1} increases, the output voltage decreases.

11-2.2 Basic Current-Mode Amplifier Circuits

Figure 11.18*a* shows a current-mode inverting amplifier. Note it is connected as an ac coupled amplifier. As long as the dc output voltage of one stage is equal to the dc input voltage of the next, dc coupling can be used. The amplifier is biased by choosing R_m, which sets the mirror current, and R_f. The gain is set by the choice of R_1. Since the mirror current is usually constant, R_m is tied to V_{cc}. The resistors are chosen as follows:

$$R_m = \frac{V_{cc} - V_{D1}}{I_m} \tag{11-5}$$

where

I_m = noninverting input current (mirror current)
V_{D1} = noninverting input diode forward voltage drop (0.7 V typically)

$$R_f = \frac{V_{out\ Q} - V_{BEQ1}}{I_m} \tag{11-6}$$

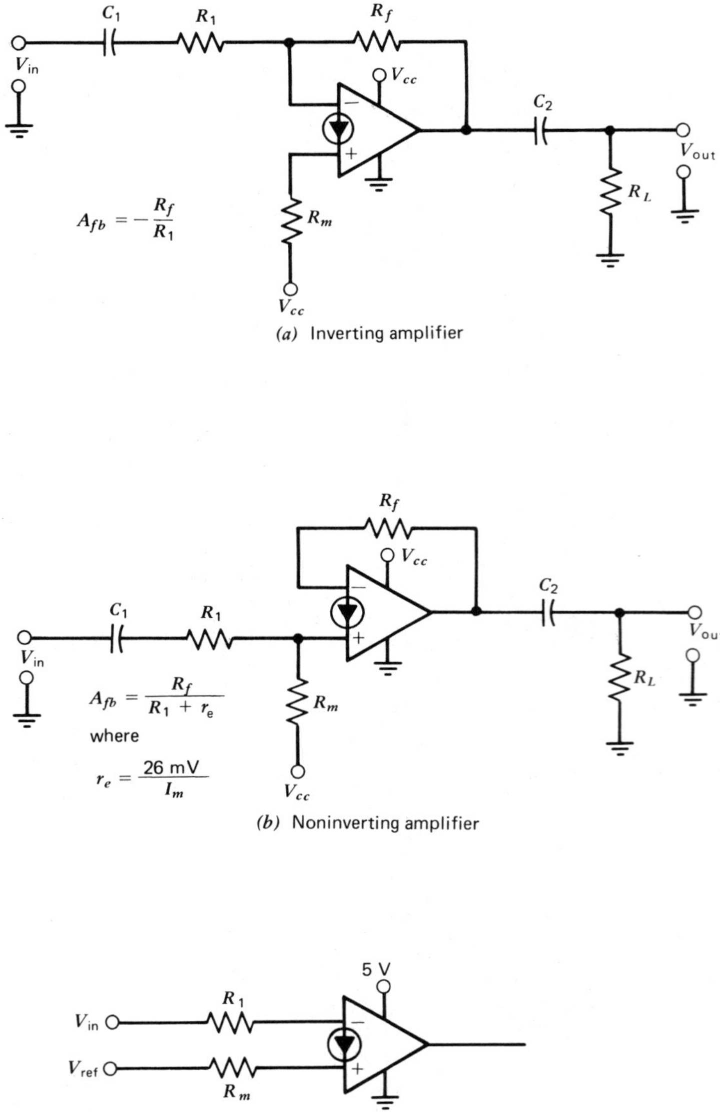

FIGURE 11.18 Current differencing amplifier configurations.

where

$V_{out\ Q}$ = quiescent output voltage
V_{BEQ1} = emitter-base voltage drop of *npn* inverting input transistor (0.7 V typically)

Since $I_{R_f} = I_m$, any change in I_{R_f} is from the current through R_1. If I_{R_1} decreases, I_{R_f} will increase by the same amount.

$$\Delta I_{R1} = -\Delta I_{Rf}$$

$$\frac{\Delta V_{in}}{R_1} = -\frac{\Delta V_{out}}{R_f}$$

$$\frac{\Delta V_{out}}{\Delta V_{in}} = -\frac{R_f}{R_1}$$

but

$$A_{fb} = \frac{\Delta V_{out}}{\Delta V_{in}}$$

so

$$A_{fb} = -\frac{R_f}{R_1} \tag{11-7}$$

EXAMPLE 11-4

Calculate R_1, R_f, R_m, C_1, and C_2 for the inverting amplifier of Figure 11.18*a*. Use an MC3401. $V_{cc} = 15$ V, $R_L = 10$ kΩ, $f_L = 100$ Hz, and $A_{f_b} = 20$.

SOLUTION

From the specification sheet in Appendix C, I_m can arbitrarily be chosen between 5 and 100 μA, so let $I_m = 50$ μA.

$$R_m = \frac{V_{cc} - V_{D_1}}{I_m} = \frac{15\text{ V} - 0.7\text{ V}}{50\ \mu\text{A}} = 286\text{ k}\Omega$$

$$R_f = \frac{V_{out\ Q} - V_{BEQ_1}}{I_m} = \frac{7.5\text{ V} - 0.7\text{ V}}{50\ \mu\text{A}} = 136\text{ k}\Omega$$

Note $R_f \cong \frac{1}{2} R_m$.

$$R_1 = \frac{R_f}{A_{fb}} = \frac{286\text{ k}\Omega}{20} = 6.8\text{ k}\Omega$$

$$C_1 = \frac{1}{2\ \pi\ f_L\ R_1} = \frac{1}{2\ \pi\ (100\text{ Hz})\ 6.8\text{ k}\Omega} = 0.234\ \mu\text{F}$$

$$C_2 = \frac{1}{2\ \pi\ f_L\ R_L} = \frac{1}{2\ \pi\ (100\text{ Hz})\ 10\text{ k}\Omega} = 0.16\ \mu\text{F}$$

Figure 11.18*b* shows a noninverting current mode amplifier. The mirror current must be provided for biasing so the input current subtracts from or adds to the mirror current. Since any change of the mirror current causes an equal current change in the current of Q_3 in Figure 11.17*a*, the current change in R_f is equal to the current change in R_1.

$$\Delta I_{R_f} = \Delta I_{R_1}$$

$$\frac{\Delta V_{out}}{R_f} = \frac{\Delta V_{in}}{R_1 + r_e}$$

where $r_e \cong \dfrac{26\text{ mV}}{I_m}$, the ac resistance of the noninverting input diode.

$$\frac{\Delta V_{out}}{\Delta V_{in}} = \frac{R_f}{R_1 + r_e}$$

thus

$$A_{fb} = \frac{R_f}{R_1 + r_e} \tag{11-8}$$

Since the diode ac noise voltage is only approximately 26mV, the gain equation is not exact. R_m and R_f are found from Equations 11-5 and 11-6.

EXAMPLE 11-5

Calculate the components for the noninverting amplifier of Figure 11.18*b*. Use an MC3401 op-amp where $I_m = 10\ \mu A$, $V_{cc} = 5\ V$, $A_{fb} = 10$, $f_L = 100\ Hz$, and $R_L = 20\ k\Omega$.

SOLUTION

$$R_m = \frac{V_{cc} - V_{D_1}}{I_m} = \frac{5\text{ V} - 0.7\text{ V}}{10\ \mu\text{A}} = 430\text{ k}\Omega$$

$$R_f = \frac{V_{out\ Q} - V_{BE_{Q_1}}}{I_m} = \frac{2.5\text{ V} - 0.7\text{ V}}{10\ \mu\text{A}} = 180\text{ k}\Omega$$

$$R_1 + r_e = \frac{R_f}{A_{fb}} = \frac{180\text{ k}\Omega}{10} = 18\text{ k}\Omega$$

$$r_e = \frac{26\text{ mV}}{I_m} = \frac{26\text{ mV}}{10\ \mu\text{A}} = 2.6\text{ k}\Omega$$

$$R_1 = 18\text{ k}\Omega - 2.6\text{ k}\Omega = 15.4\text{ k}\Omega$$

$$C_1 = \frac{1}{2\pi(R_1 + r_e)f_L} = \frac{1}{2\pi\ 18\text{ k}\Omega\ 100\text{ Hz}} = 0.089\ \mu\text{F}$$

$$C_2 = \frac{1}{2\pi R_L f_L} = \frac{1}{2\pi\ 20\text{ k}\Omega\ 100\text{ Hz}} = 0.08\ \mu\text{F}$$

Figure 11.18*c* shows a current-mode amplifier comparator. Since the common-mode input voltage is held at 0.7 V by Q_1 and D_1 (See Figure 11.17*a*), any voltage level can be compared if R_1 and R_m prevent excessive input current flow. Simply select

$$R_m = \frac{V_{ref} - V_{D1}}{I_m}$$

and set

$$R_1 = R_m$$

and the circuit is done. The voltage gain will be the op-amp open loop gain.

EXAMPLE 11-6

Set up an MC3401 as a TTL compatible comparator to detect when an input voltage exceeds 50 V. Let $I_m = 20\ \mu A$

SOLUTION

Set $V_{ref} = 50$ V and $V_{cc} = 5$ V

$$R_1 = R_m = \frac{V_{ref} - V_{D1}}{I_m} \cong \frac{50\text{ V}}{10\ \mu\text{A}} \doteq 2.5\ \text{M}\Omega$$

Current-mode amplifiers can be used in many applications just as a regular op-amp except for the selection of R_m for biasing. The quad current-mode amplifiers are very useful in low-cost active filters, amplifiers, summers, and comparators. The list of possible applications is nearly endless.

11-3 555 TIMERS

The 555 timer is a very useful medium-precision oscillator and timing circuit produced by several IC manufacturers. Its two modes of operation are as an astable oscillator and a monostable circuit. The circuit can oscillate up to about 100 kHz and there is a new CMOS version to 250 kHz. We will look at the National Semiconductor LM555 timer. The circuit can operate from 4.5 to 18 V. The low V_{out} is from 0.1 V at $I_{sink} = 10$ mA (V_{out} supplying current to a load connected to V_{cc}) to 2.5 V at $I_{sink} = 200$ mA. V_{out} high is about 1.3 V less than V_{cc} at $I_{source} = 1$ mA (I_{R_L} to ground from high V_{out}) to 1.8 V less than V_{cc} at $I_{source} = 100$ mA. The device may operate from TTL power supplies and will drive TTL logic. The output rise and fall times are 100 ns each. The LM555 is supplied in an operating temperature range of −55 to +125°C, and the LM555C in an operating temperature range from 0 to 70°C. The device is normally supplied in an eight-pin DIP or an eight-pin metal can. A dual 555 timer, the 556, is supplied in a 14-pin DIP.

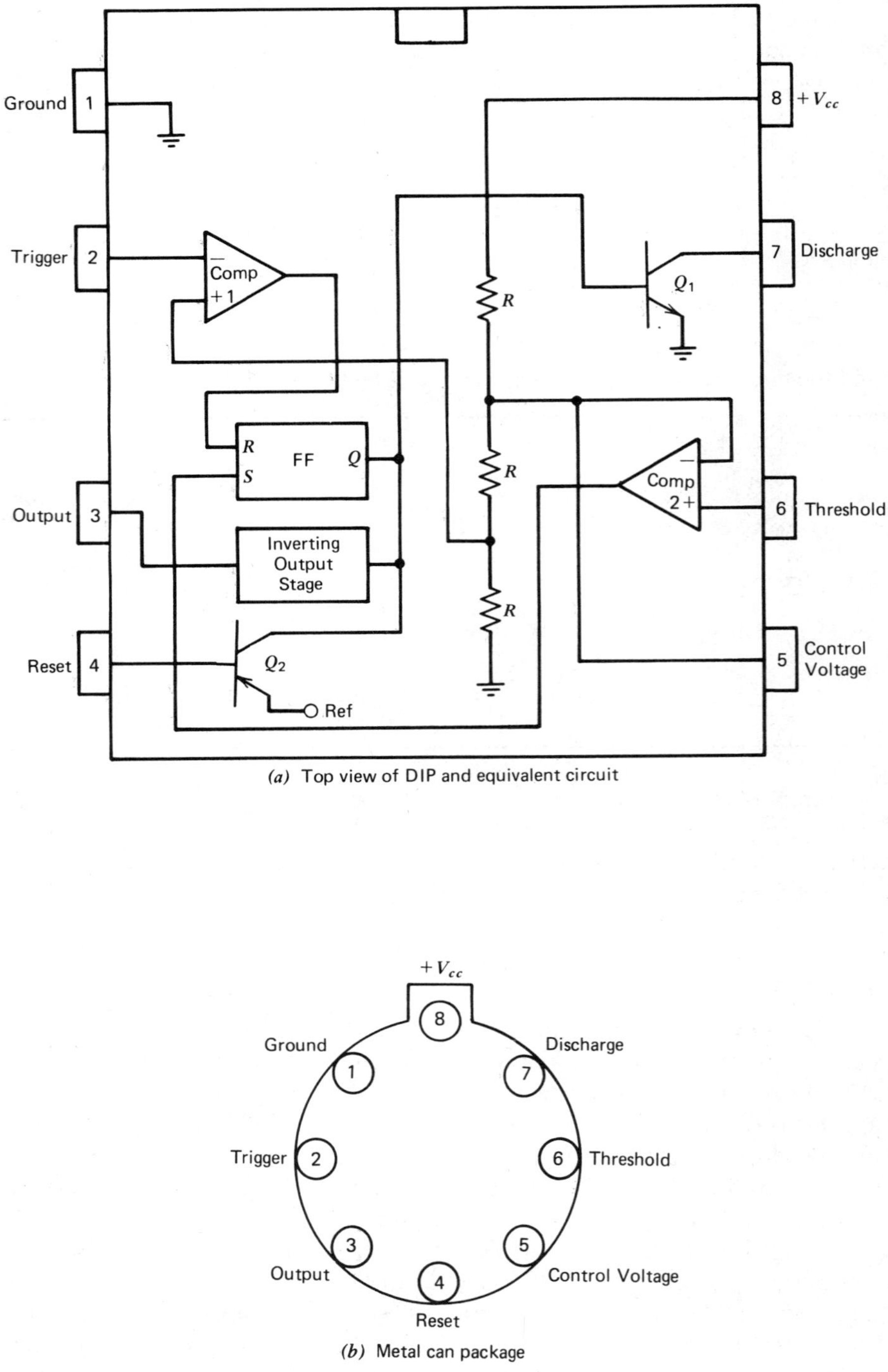

(a) Top view of DIP and equivalent circuit

(b) Metal can package

FIGURE 11.19 LM 555 timer.

11-3.1 555 Timer Circuit

The 555 timer block diagram is shown in Figure 11.19*a*. The reference for the comparators is provided by the *R-R-R* voltage divider. The reference voltages are 2/3 V_{cc} for the threshold comparator, comparator 2, and 1/3 V_{cc} for the trigger comparator, comparator 1. The flip-flop (FF) is set (driven high) by the threshold compartor when $V_{threshold} > 2/3\ V_{cc}$. The high FF output causes the complementary output stage to go to low and the discharge transitor (Q_1) to turn on.

The FF is reset (driven low) by the trigger comparator whenever $V_{trigger} < 1/3\ V_{cc}$. The low FF output causes the output to go high and Q_1 to turn off.

A low on the reset pin causes Q_1 to turn on and the output to go low immediately. The control pin allows the threshold reference voltage to be varied. A control voltage changes the frequency of a 555 astable and the pulse width of a 555 monostable by changing the reference voltage of comparator 2. When not in use, a 0.01 μF capacitor is connected from the control pin to ground to prevent unwanted modulation from noise.

The complete specifications are in Appendix C.

11-3.2 555 Astable

Astable operation is obtained from the circuit shown in Figure 11.20*a*. When V_{cc} is applied, V_{out} is high, C_T charges toward V_{cc} at a rate set by R_A and R_B. When $V_{C_T} = 2/3\ V_{cc}$, the threshold comparator sets the FF output high, V_{out} low, and turns Q_1 on. The timing capacitor now discharges through Q_1 and R_B. When $V_{C_T} = 1/3\ V_{cc}$, the trigger comparator resets the FF output low, V_{out} high, and turns Q_1 off. The process repeats. Note that the first pulse will be longer because C_T charges from 0 V to 2/3 V_{cc} rather than from 1/3 V_{cc} to 2/3 V_{cc} as it does once operation is in progress. This is normally no problem. Oscillation can be stopped at any time by bringing the reset pin low. Oscillation resumes when the reset pin goes high again.

The specification sheet provides equations for the timing components R_A, R_B, and C_T. They are

$$t_1 = 0.693\,(R_A + R_B)C$$

$$t_2 = 0.693\,R_B C$$

$$\tau = t_1 + t_2 = 0.693\,(R_A + 2R_B)C$$

$$f = \frac{1}{\tau} = \frac{1.44}{(R_A + 2\,R_B)C}$$

and duty cycle *D*

$$D = \frac{t_1}{t_1 + t_2} = \frac{t_1}{\tau} = \frac{R_B}{R_A + 2R_B}$$

It is very difficult to get an exact 50% duty cycle from a 555 timer, although if $R_B = 100\ R_A$, it can be within 1%. If an exact 50% duty cycle is required, a 555 can drive a flip-flop very well.

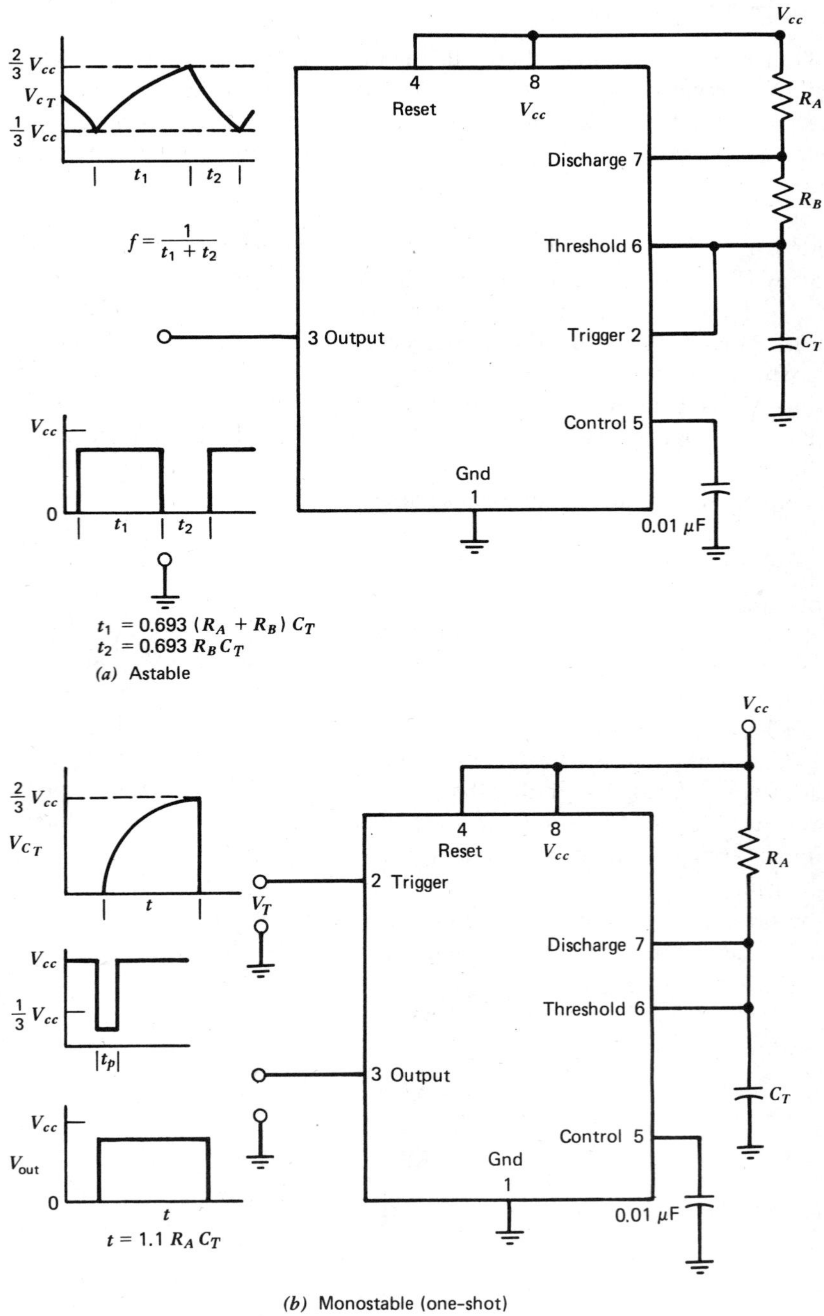

FIGURE 11.20 555 timer circuits.

EXAMPLE 11-7

Calculate the components for the 555 astable of Figure 11.20*a* for an oscillating frequency of 2 kHz, a duty cycle (D) of 75%. Let $V_{cc} = 5$ V.

SOLUTION

From the specifications we note that maximum temperature stability is obtained when

$$1\ \text{k}\Omega \leq R_A \leq 100\ \text{k}\Omega$$

$$1\ \text{k}\Omega \leq R_B \leq 100\ \text{k}\Omega$$

We will choose R_A and R_B between these limits even though the 555 will operate with $R_A + R_B = 20$ MΩ.

Now we find the times t_1 and t_2

$$\tau = \frac{1}{f_{osc}} = \frac{1}{2\ \text{kHz}} = 0.5\ \text{ms}$$

$$t_1 = D\tau = 0.75(0.5\ \text{ms}) = 375\ \mu\text{s}$$

$$t_2 = \tau - t_1 = 500\ \mu\text{s} - 375\ \mu\text{s} = 125\ \mu\text{s}$$

Choosing $R_A + R_B = 10$ kΩ

$$C_T = \frac{t_1}{0.693\ (R_A + R_B)} = \frac{375\ \mu\text{s}}{0.693\ (10\ \text{k}\Omega)} = 0.054\ \mu\text{F}$$

$$R_B = \frac{t_2}{0.693C} = \frac{125\ \mu\text{s}}{0.693\ (0.054\ \mu\text{F})} = 3.34\ \text{k}\Omega$$

$$R_A = (R_A + R_B) - R_B = 10\ \text{k}\Omega - 3.34\ \text{k}\Omega = 6.66\ \text{k}\Omega$$

and from the specifications the control pin capacitor is 0.01 μF.

11-3.3 555 Monostable

The 555 monostable is shown in Figure 11.20*b*. Pin 2, the trigger input, will be high so that the trigger comparator holds the FF output high, Q_1 on, and V_{out} low. Since Q_1 is on, the timing capacitor cannot charge. A momentary negative trigger pulse ($0.1\ \mu\text{s} < t_p < t$) whose negative peak is less than 1/3 V_{cc} will reset the FF output low, turn Q_1 off, and set V_{out} high. C_T now charges through R_A until $V_{C_T} = 2/3\ V_{cc}$, at which time the threshold comparator sets the FF output high, V_{out} low, and turns on Q_1. C_T is rapidly discharged through Q_1. The circuit will now set until the next trigger pulse.

A negative on the reset pin will immediately set the output low and turn on Q_1. The circuit will not produce another output pulse after reset until the next trigger pulse. The minimum output pulse time is the trigger pulse width. The reset is inoperative until the trigger pulse is removed. From the specification sheet

$$t = 1.1\ R_A C$$

EXAMPLE 11-8

Calculate R_A and C_T in Figure 11.20*b* so the 555 operates as an monostable with $t = 50\ \mu s$.

SOLUTION

Choose R_A between 1 and 100 kΩ. Let $R_A = 47\ k\Omega$.

$$C = \frac{t}{1.1\ R_A} = \frac{50\ \mu s}{1.1\ (47\ k\Omega)} = 967\ pF$$

11-3.4 555 Timer PWM Switching Power Supply with Current Limiting

The circuit in Figure 11.21 shows a 555 timer PWM step-down switching power supply with foldback current limiting. In practice, a dual 556 timer would be used, but for clarity two 555 timers are shown. The basic circuit can be used to control a step-up or a voltage-inverting power supply.

Q_1 is the switching transistor, and Q_2 is the switching transistor driver. The monostable (one shot) turns on Q_2 during t_c and turns Q_2 off when its output is low. C_s is a speed-up capacitor to aid in turning on Q_1, and D_2 removes the positive voltage spikes from the base of Q_1 when the collector of Q_2 goes high.

The op-amp compares the output voltage to the reference voltage developed across Z_2. Almost any op-amp, including a current differencing amplifier, will work, although for our example a 741 is chosen. As the output voltage drops, the control voltage of the 555 one-shot increases. C_T now must charge to a higher voltage before timing out, so the pulse width, t_c, is lengthened. The increase in t_c causes the output voltage to rise back to its original value. If V_{out} increases, $V_{control}$ decreases, t_c decreases, and V_{out} drops back to its original value. Zener diode, Z_1, prevents $V_{control}$ from rising to V_{in} so that the one shot will turn off under start-up conditions.

The astable triggers the one shot every cycle. The trigger pulse width sets the minimum one-shot pulse width for current limiting. The minimum pulse width of the trigger is 1 μs for a 555 timer astable.

Q_3 is the current limiting transistor. It turns on whenever $V_{Rsc} > V_{BEQ_3}$. When Q_3 turns on, Q_4 turns on and pulls the reset pin of the one shot low. This terminates the output pulse until the next trigger. Since the average output current depends on t_c, the average output current can be limited to a value less than I_o(max) when overload conditions exist.

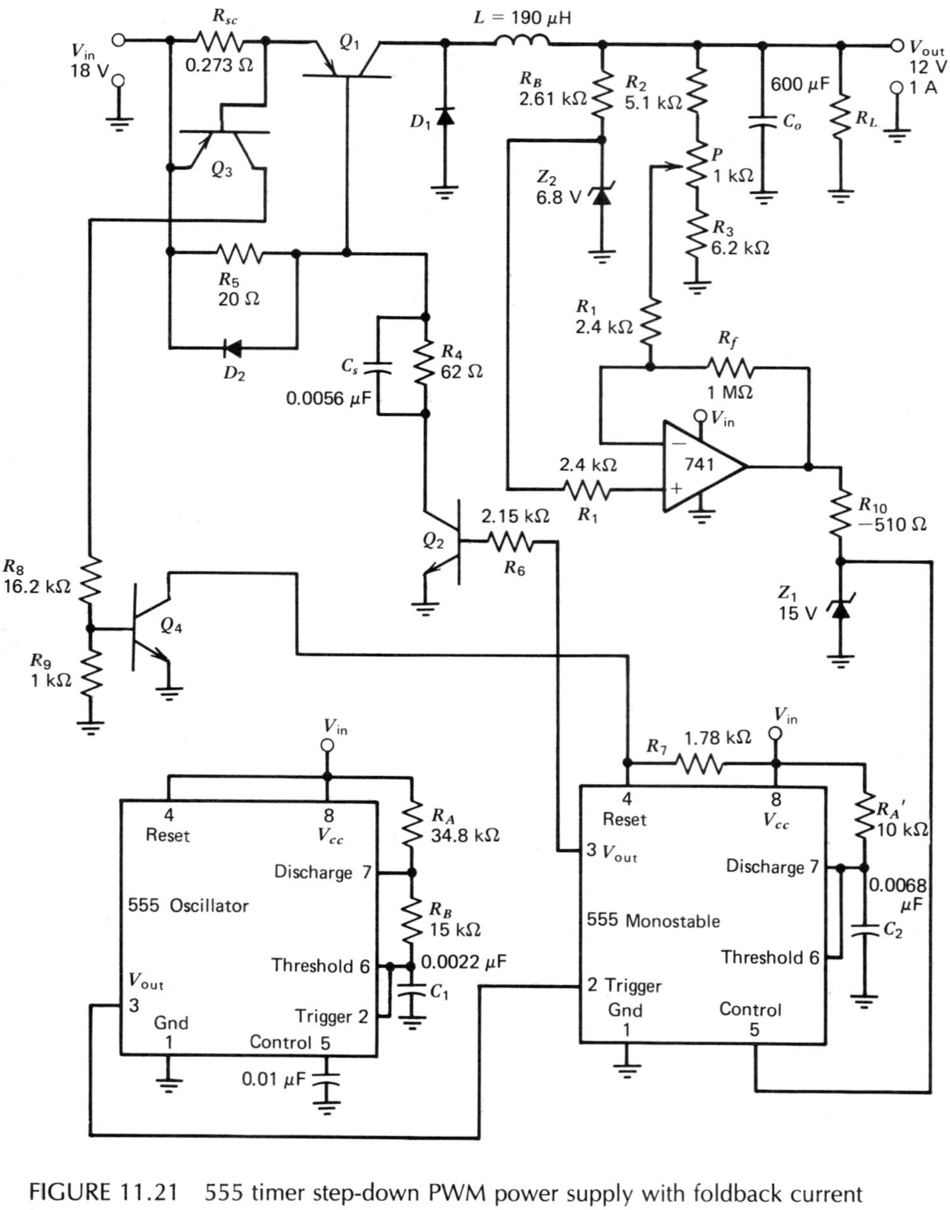

FIGURE 11.21 555 timer step-down PWM power supply with foldback current limiting.

When the output is short circuited, all three basic inverters, step down, step up, and voltage inverting, apply the full input voltage to the inductor. Thus the average input current is

$$I_{in\,(ave)} = (I_p/2)\,\frac{t_c}{\tau}$$

The peak output current is difficult to predict under short-circuit conditions, so the average input current, thus the *average* output current, is limited by detecting I_p. The minimum t_c is equal to the time the reset pin is low; in other words, $t_c(\text{min}) = t_L$ of the 555 oscillator as shown in Figure 11.22. The foldback current (I_{FB}) is chosen, and from the average input current equation of the voltage inverter, $t_L = t_c(\text{min})$ is chosen. Thus

$$t_L = t_c(\text{min}) = \frac{2\, I_{FB}\, \tau}{I_{cl}}$$

where

$I_{cl} \cong 1.1\, I_p$ so maximum I_L can be supplied
I_{FB} = foldback current, usually $0.25\, I_o(\text{max})$.

Switching inverters cannot be operated with no load. At least about 5 to 10% of the full load current must be drawn or the output voltage rises to V_{in} because no inductor current flows.

EXAMPLE 11-9

Calculate the components for the circuit of Figure 11.21 so that with $V_{in} = 18$ V: $V_{out} = 12$ V at 1 A, $V_{ripple} = 30\ \text{mV}_{pp}$, $I_{FB} = 0.25$ A, and $f_{osc} = 10$ kHz.

SOLUTION

With $I_o(\text{max}) = 1$ A, $I_p = 2$ A and $I_{cl} = 2.2$ A. $V_{in} = 18$ V, thus $2/3\, V_{in} < V_{Z_1} < V_{in}$, so V_{Z_1} is chosen to be 15 V. V_{Z_2} is chosen to be 6.8 V at $I_B = 2$ mA. The op-amp is a 741.

The peak current through D_1 can be very large, although of short duration, under short-circuit conditions. D_1 is chosen with the following specifications:

$$PIV = 40\ \text{V},\ V_F = 1\ \text{V at 2 A},\ I_{F(\text{ave})} = 4\ \text{A}$$

Q_1 specifications are:

$$BV_{CEO} = 40\ \text{V},\ \beta_{min} = 10 \text{ at } I_C = 2\ \text{A}$$
$$V_{BE} = 1\ \text{V at } I_C = 2\ \text{A},\ P_D = 10\ \text{W},\ V_{CE}(\text{sat}) = 1\ \text{V at } I_C = 2\ \text{A}$$
$$t_{on} = 0.2\ \mu\text{s}$$

Q_2 specifications are:

$$BV_{CEO} = 40\ \text{V},\ \beta_{min} = 35 \text{ at } I_C = 0.2\ \text{A}$$
$$V_{BE} = 0.8\ \text{V at } I_C = 0.2\ \text{A},\ P_D = 1\ \text{W},\ V_{CE}(\text{sat}) = 0.5\ \text{V at } I_C = 0.2\ \text{A}$$

Q_3 and Q_4 specifications are:

$$BV_{CEO} = 40\ \text{V},\ V_{CE}(\text{sat}) = 0.3\ \text{V at } I_C = 10\ \text{mA}$$
$$\beta_{(min)} = 50,\ V_{BE} = 0.6\ \text{V at } I_C = 0.1\ \text{mA}$$

Calculations

Inverter

$$\tau = \frac{1}{f_{osc}} = 100\ \mu s$$

$$\frac{t_c}{t_o} = \frac{V_{out} + V_D}{V_{in} - [V_{out} + V_{CE}(sat)_{Q1}]}$$

$$= \frac{13\ V}{18\ V - 13\ V} = 2.6$$

$$t_o = \frac{\tau}{3.6} = \frac{100\ \mu s}{3.6} = 27.8\ \mu s$$

$$t_c = \tau - t_o = 100\ \mu s - 27.8\ \mu s = 72.2\ \mu s$$

$$L = \left(\frac{V_{out} + V_D}{I_p}\right) t_o = \left(\frac{13\ V}{2\ A}\right) 27.8\ \mu s = 181\ \mu H$$

$$C = \frac{I_o(\max) t_c}{4 V_{ripple\,pp}} = \frac{1\ A\ (72.2\ \mu s)}{4\ (30\ mV)} = 602\ \mu F$$

$$t_L = t_c(\min) = \frac{2\ I_{FB}\ \tau}{I_{cl}}$$

$$= \frac{2(0.25\ A)(100\ \mu s)}{2.2\ A} = 22.7\ \mu s$$

555 Timer

The 555 oscillator has $t_2 = t_c = 22.7\ \mu s$, so $t_1 = \tau - t_2 = 100\ \mu s - 22.7\ \mu s = 77.3\ \mu s$. Selecting $R_A + R_B = 50\ k\Omega$

$$C_1 = \frac{t_1}{0.693\ (R_A + R_B)} = \frac{77.3\ \mu s}{0.693\ (50\ k\Omega)}$$

$$= 0.0022\ \mu F$$

$$R_B = \frac{t_2}{0.693\ C_1} = \frac{22.7\ \mu s}{0.693\ (2.2\ nF)}$$

$$= 14.9\ k\Omega$$

$$R_A = 50\ k\Omega - 14.9\ k\Omega = 35.1\ k\Omega$$

As usual, pin 5 has a 0.01 μF capacitor to ground.

The 555 monostable is triggered by the oscillator. Its output is t_c. Selecting $R'_A = 10\ k\Omega$

$$C_2 = \frac{t_c}{1.1\ R_A} = \frac{72.2\ \mu s}{1.1\ (10\ k\Omega)} = 0.0066\ \mu F$$

The control pin is driven by the op-amp output.

Op-amp circuit

The op-amp needs a gain of about 400 to provide load regulation of 0.5% and line regulation of 0.1%/V. If $R_f = 1\ M\Omega$ then

$$R_1 = \frac{R_f}{A_{fb}} = \frac{1\ M\Omega}{400} = 2.5\ k\Omega$$

$$R_s = R_f \| R_1 \cong R_1 = 2.5\ k\Omega$$

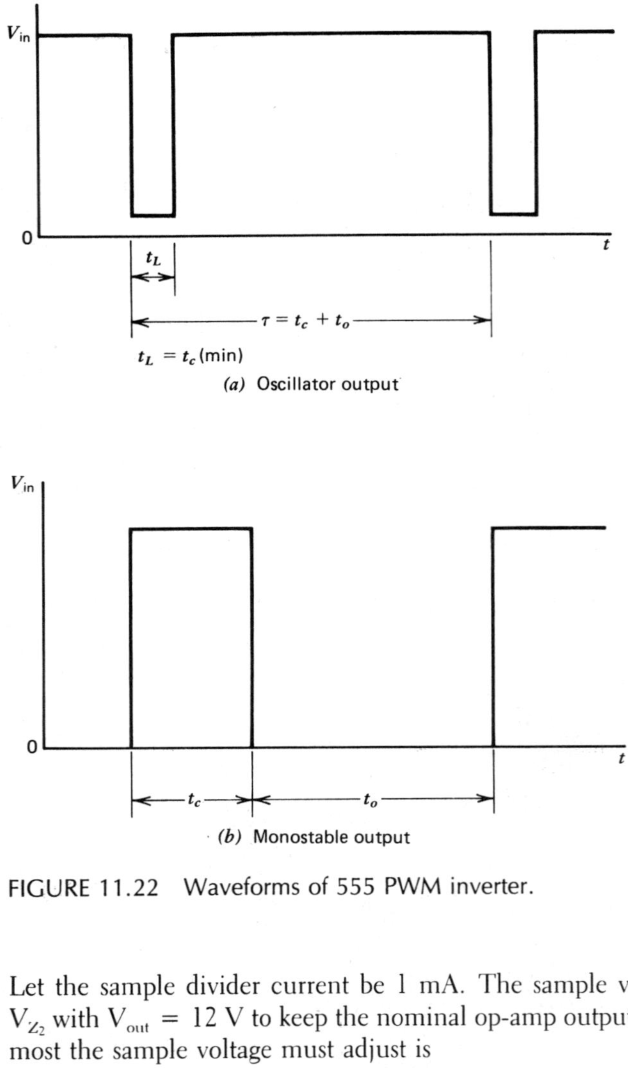

FIGURE 11.22 Waveforms of 555 PWM inverter.

Let the sample divider current be 1 mA. The sample voltage must be less than V_{Z_2} with $V_{out} = 12$ V to keep the nominal op-amp output voltage at 0.67 V_{in}. The most the sample voltage must adjust is

$$\Delta V_{sample} = \frac{0.67\ V_{in}}{A_{fb}} = 30\ \text{mV}$$

If P drops 1 V at $V_{out} = 12$ V, the output can be adjusted to 12 V precisely. Let one end of the potentiometer be 7 V and the other 6 V for full adjustment. Now

$$R_2 = \frac{V_{out} - 7\ \text{V}}{I_D} = \frac{12\ \text{V} - 7\ \text{V}}{1\ \text{mA}} = 5\ \text{k}\Omega$$

$$P = \frac{7\ \text{V} - 6\ \text{V}}{I_D} = \frac{1\ \text{V}}{1\ \text{mA}} = 1\ \text{k}\Omega$$

$$R_3 = \frac{6\ \text{V}}{I_D} = \frac{6\ \text{V}}{1\ \text{mA}} = 6\ \text{k}\Omega$$

R_B provides 2 mA of bias current to Z_2.

$$R_B = \frac{V_{out} - V_{Z2}}{I_{Z_2}} = \frac{12\text{ V} - 6.8\text{ V}}{2\text{ mA}} = 2.6\text{ k}\Omega$$

R_{10} provides bias current to Z_1 whenever the op-amp output is greater than V_{Z_1}. Let $I_{B_1} = 5\text{ mA} < I_{out}$ of the 741

$$R_{10} = \frac{+V_{sat} - V_{Z1}}{I_{Z1}} = \frac{17\text{ V} - 15\text{ V}}{2\text{ mA}} = 500\ \Omega$$

Q_1 drive circuit

$$I_{BQ1} = \frac{I_p}{\beta_{Q_1}} = \frac{2\text{ A}}{10} = 200\text{ mA}$$

Choose $I_{CQ_2} > I_{BQ_1}$ for variations in V_{BE} of Q_1. Let $I_{CQ_2} = 250\text{ mA}$

$$R_4 = \frac{V_{in} - V_{BEQ1} - V_{CE}(\text{sat})_{Q2}}{I_{CQ2}}$$
$$= \frac{18\text{ V} - 1\text{ V} - 0.5\text{ V}}{250\text{ mA}} = 66\ \Omega$$
$$P_{R4} \geqslant V_{in}I_{CQ2} = 4.5\text{ W}$$
$$R_5 = \frac{V_{BEQ1}}{I_{CQ2} - I_{BQ1}} = \frac{1\text{ V}}{50\text{ mA}} = 20\ \Omega$$
$$R_6 = \frac{V_{out\ 555} - V_{BEQ2}}{I_{CQ_2}/\beta_{Q_2}}$$
$$= \frac{16\text{ V} - 0.8\text{ V}}{250\text{ mA}/35} = 2.13\text{ k}\Omega$$
$$C_s = \frac{2\ I_{BQ1}\ t_{on}}{V_{in} - V_{CEQ2} - V_{BEQ1}}$$
$$= \frac{0.5\text{ A}\ (0.2\ \mu\text{s})}{16.5\text{ V}} = 0.006\ \mu\text{F}$$

Current limit

$$R_{sc} = \frac{V_{BEQ3}}{1.1\ I_p} = 0.273\ \Omega$$

Let $I_{CQ4} = 10\text{ mA}$

$$R_7 = \frac{V_{in} - V_{CE}(\text{sat})_{Q4}}{I_{CQ4}} = \frac{18\text{ V} - 0.3\text{ V}}{10\text{ mA}} = 1.77\text{ k}\Omega$$
$$I_{BQ4} = \frac{I_{CQ4}}{\beta_{min\ Q4}} = \frac{10\text{ mA}}{50} = 0.2\text{ mA}$$

Set $I_{CQ_3} = 1$ mA

$$R_8 = \frac{V_{in} - V_{BEQ4} - V_{CE}(sat)_{Q3}}{I_{CQ3}} = 17.1\ k\Omega$$

$$R_9 = \frac{V_{BEQ4}}{I_{CQ3} - I_{BQ4}} = \frac{0.6\ V}{0.8\ mA} = 1\ k\Omega$$

The components are calculated. The components on the schematic are the nearest 5% values to the calculated values.

11-4 INSTRUMENTATION AMPLIFIERS

The instrumentation amplifier is an op-amp or op-amp circuit used to measure small differential voltages riding on a common-mode voltage that is frequently larger than the differential voltage. Other common terms for this type of amplifier are transducer amplifier, difference amplifier, error amplifier, and bridge amplifier. Instrumentation amplifiers require differential inputs, high gain, low offset, and very high (usually greater than 80 dB) CMRR. The high CMRR is necessary because the voltage to be amplified is often riding on a higher common-mode voltage, as already mentioned.

The differential signal input is often the output voltage of a bridge circuit such as that shown in Figure 11.23*a*. The transducer will detect whatever change the circuit is designed to measure. A *transducer* is a device that converts a physical parameter and its variations, such as temperature or pressure, into an electrical parameter, such as resistance or voltage change. Examples of transducers are:

1. Photoresistors, for changing light levels into resistance levels.
2. Thermistors, for changing temperature variations into resistance variations.
3. Strain gauges, specially configured resistors that produce a resistance change proportional to the mechanical pressure applied to them.
4. Thermocouples, dissimilar metal junctions that have a voltage output proportional to the temperature applied to them.

Referring to Figure 11.23*a*, if all four resistors of the bridge are equal in value, then $e_1 = e_2 = V_R/2$. If a physical variable causes a change in the transducer resistance (ΔR), e_1 no longer equals e_2 and the polarity of e_1 with respect to e_2 depends on whether ΔR is an increase or decrease in the transducer resistance. The output voltage of the bridge ($e_1 - e_2$) is:

$$e_1 - e_2 = V_R\left(\frac{\Delta R}{4\ R + 2\ \Delta R}\right) \tag{11-9}$$

or, if ΔR is small.

$$e_1 - e_2 \cong V_R \frac{\Delta R}{4\ R} \tag{11-10}$$

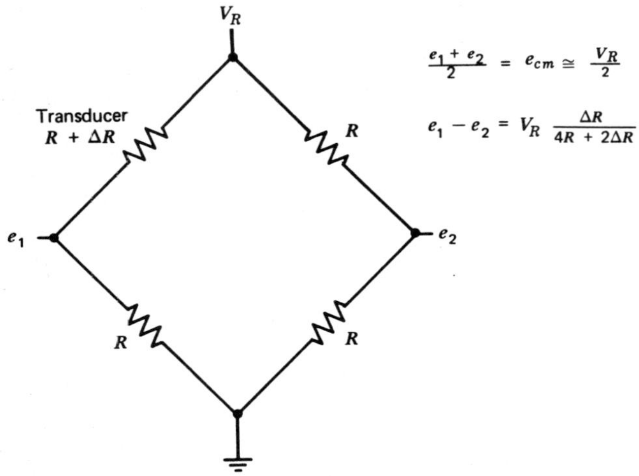

(a) Bridge circuit

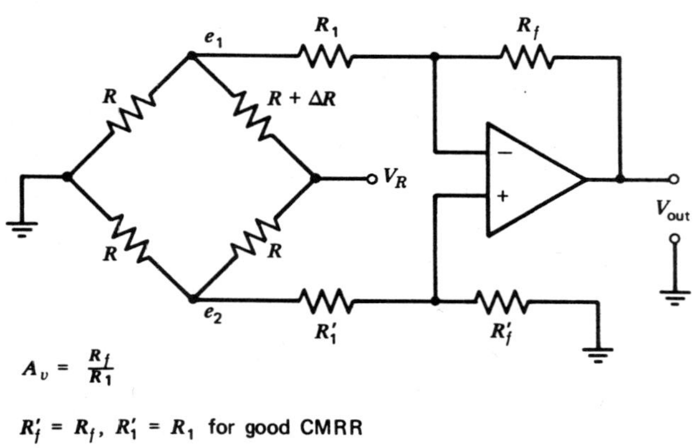

(b) Adder–subtracter as bridge output amplifier

FIGURE 11.23 Bridge circuit and amplifier.

where

$$\Delta R << 4\,R$$

The bridge output voltage will normally be quite small.

11-4.1 Instrumentation Amplifier Circuits

An op-amp circuit that amplifies the differential bridge output $(e_1 - e_2)$, and rejects or attenuates the common-mode voltage (V_{cm}) that the $(e_1 - e_2)$ is superimposed on, is the adder-subtracter or differential input circuit. The output voltge of this circuit, shown in Figure 11.23 *b* is

$$V_{out} = \frac{R_f}{R_1}(e_2 - e_1)$$

where

$$R_f = R_f' \qquad \text{and} \qquad R_1 = R_1'$$

The feedback resistors R_f and R_f' and the input resistors R_1 and R_1' must be very closely matched if the common-mode voltage on each input is to be accurately subtracted. If the feedback and input resistors are matched and the op-amp has a high CMRR, the common-mode signal is rejected and the differential signal $(e_2 - e_1)$ amplified, thus the circuit of Figure 11.23*b* has a high CMRR.

The circuit of Figure 11.23*b* has two drawbacks that limit its application. The first is its relatively low input resistance, and the second is the difficulty of varying the circuit gain.

If the input resistors of the circuit (Figure 11.23*b*) are increased to increase input resistance, offset because of bias currents will also be increased, but a low value of R_1 and R_1' cause bridge loading. If each input to the amplifier is fed by a voltage follower as shown in the instrumentation amplifier circuit of Figure 11.24*a*, the input resistance problem is overcome. If op-amps A_1 and A_2 are FET input op-amps, the input resistance will be very high.

The circuit of Figure 11.24*a* works very well for fixed-gain applications, but to vary gain both R_f and R_f' must be varied. The feedback resistors, R_f and R_f' are matched and must remain matched as they are varied to vary gain, or CMRR of the circuit will drop. This is a very difficult condition to meet. Figure 11.24*b* shows a circuit with adjustable gain in which the gain adjustment does not degrade the circuit CMRR. The output voltage of Figure 11.24*b* is

$$V_{out} = 2\left(1 + \frac{R_f}{R_2}\right)\frac{R_f}{R_1}(e_2 - e_1) \qquad (11\text{-}11)$$

Note that R_1 and R_1' must be matched and four resistors of value R_f must be matched to avoid degrading CMRR. Also note from Equation 11-11 that the gain is not linear. The gain is adjusted by varying the amount of feedback from V_{out} with the voltage divider R_f and R_2.

A circuit in which mismatch of the gain setting resistors causes only gain error and does not effect circuit CMRR is shown in Figure 11.25. Op-amps A_2 and A_2 are parallel-connected noninverting amplifiers. The common-mode voltage passes

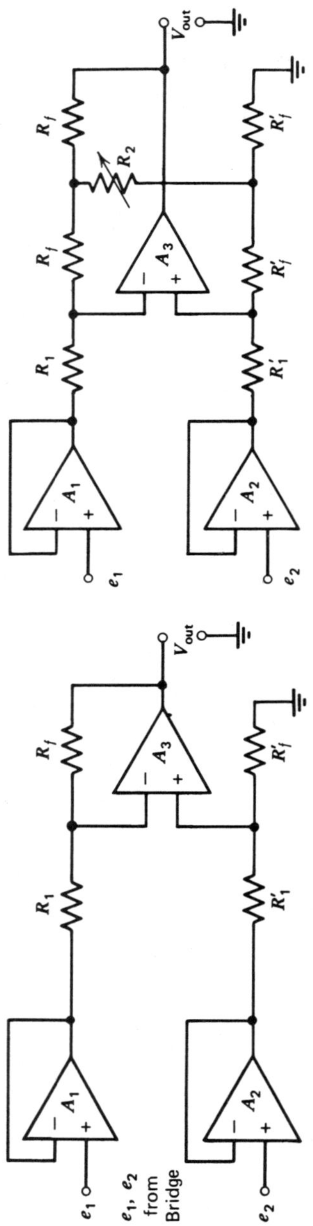

$A_v = \frac{R_f}{R_1}$, $R_f = R'_f, R_1 = R'_1$

(a) Followers A_1 and A_2 increase input resistance of differential input (adder–subtracter) amplifier A

$A_v = 2\left(1 + \frac{R_f}{R_2}\right)\frac{R_f}{R_1}$, $R_1 = R'_1$, $R_f = R'_f$

(b) Adjustable gain added to circuit of Figure 11–24 a

FIGURE 11.24 Instrumentation amplifier circuit.

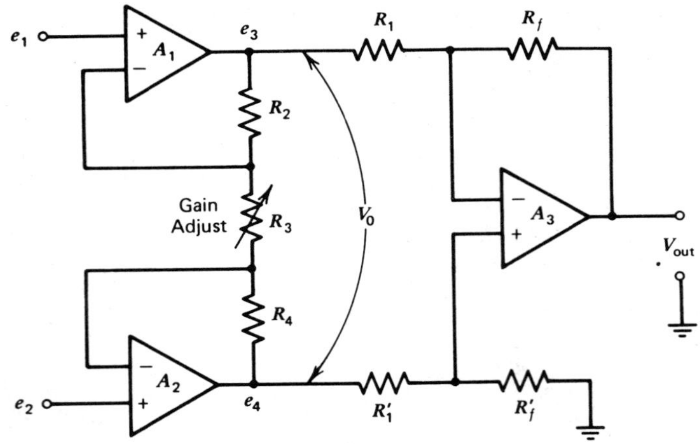

$R_2 = R_4,\ R_1 = R_1',\ R_f = R_f'$

$V_o = \left(1 + 2\,\dfrac{R_2}{R_3}\right)(e_1 - e_2) \qquad V_{out} = \left(\dfrac{R_f}{R_1}\right)\left(1 + 2\,\dfrac{R_2}{R_3}\right)(e_1 - e_2)$

FIGURE 11.25 Gain adjust added to circuit of Figure 11.23 using parallel connected noninverting amplifiers A_1 and A_2.

through A_1 and A_2 with a gain of 1, but is rejected by adder-subtracter A_3. The output voltage of $A_1(e_3)$ is

$$e_3 = \left(1 + \frac{R_2}{R_3}\right)e_1 - \frac{R_2}{R_3}e_2 + V_{cm} \tag{11-12}$$

and the output voltage of A_2 (e_4) is

$$e_4 = \left(1 + \frac{R_4}{R_3}\right)e_2 - \frac{R_4}{R_3}e_1 + V_{cm} \tag{11-12a}$$

The first term of Equations 11-12 and 11-12a is the noninverting amplifier gain term to the circuit inputs e_1 and e_2. The second term of each equation is the inverting gain to the other input. Since the voltage on the inverting and noninverting terminal of an op-amp is approximately the same (because of the high A_{ol}), e_2 is an input to amplifier A_1, where A_1, R_3, and R_2 make up an inverter. The last term is the common-mode voltage passed through each amplifier A_1 and A_2 with a gain of 1. The output voltage (V_{out}) of the parallel-connected noninverting amplifiers A_1 and A_2 is

$$V_{out} = e_4 - e_3$$

which, if $R_2 = R_4$ (normally they are equal), is

$$V_{out} = \left(1 + \frac{2\,R_2}{R_3}\right)(e_2 - e_1) \tag{11-13}$$

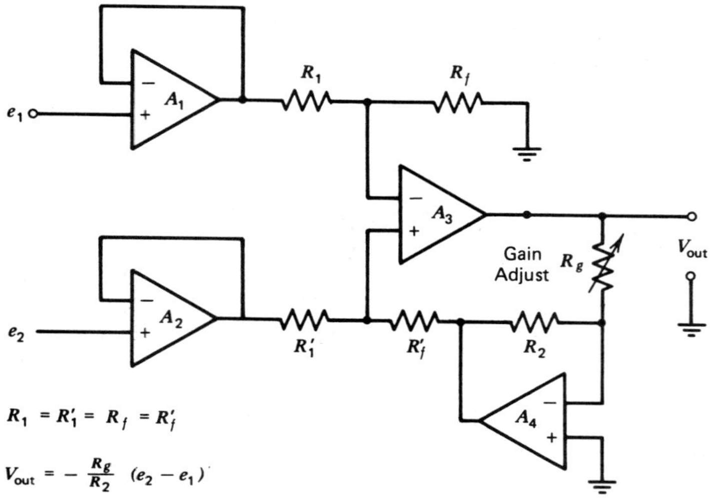

FIGURE 11.26 Linear gain adjust added with amplifier A_4.

As mentioned before, a mismatch between R_2 and R_4 leads to gain error, but not a reduced CMRR. The disadvantage of the instrumentation amplifier of Figure 11.25 is the circuit gain (the total gain of both stages).

$$A_v = \left(1 + \frac{2\,R_2}{R_3}\right)\left(\frac{R_f}{R_1}\right) \tag{11-14}$$

is still not completely linear. It is fairly linear over small ranges of gain, however. Normally most of the circuit gain will be obtained from A_1 and A_2 with A_3 supplying only unity gain or slightly more. By keeping A_3 gain low, output offset voltage is kept low. The resistors of this circuit can be kept low to minimize offset due to bias current differences.

Linear gain adjustment can be obtained with the circuit of Figure 11.26. Op-amps A_1 and A_2 are unity gain followers, and differential input amplifier A_3 is set for unit gain. The circuit gain is adjusted by varying the linear potentiometer R_g, the input resistor of A_4. A_4 acts as an attenuator in the feedback loop, varying the amount of V_{out} fed back to A_3. The voltage fed back to A_3 (the output of A_4) must equal $e_2 - e_1$. For this to be true, the output voltage is

$$V_{out} = -\frac{R_g}{R_2}(e_2 - e_1) \tag{11-15}$$

which is linear.

11-4.2 Input Guarding

In applications requiring high CMRR at frequencies above dc, stray capacitances at the input can degrade performance. If, in Figure 11.27*a*, $R_{source_1}C_1$ and R_{source_2}

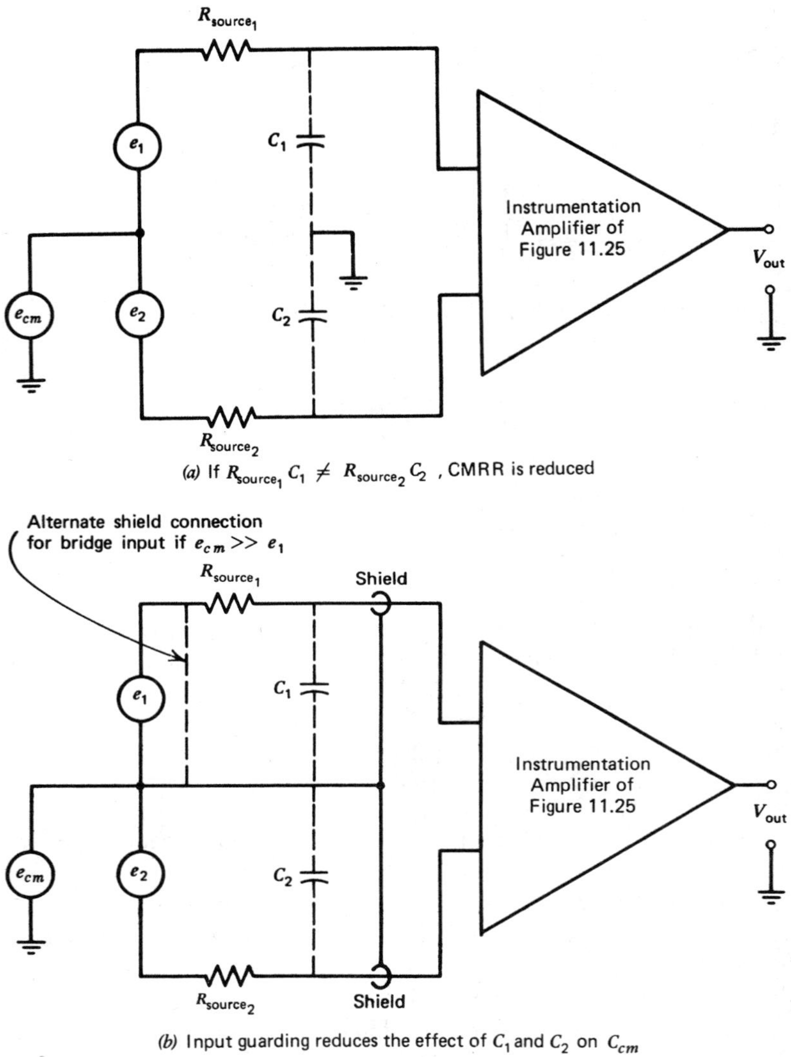

FIGURE 11.27 Input guarding.

C_2 are not equal, the signals into the instrumentation amplifier are not equally attenuated by these *RC* networks. If the input $R_{source}C_{stray}$ networks do not attenuate ac signals (especially V_{cm}) equally, CMRR is reduced. The capacitances C_1 and C_2 are op-amp input capacitances, stray capacitances, and shield capacitances. Shielding of the input leads is routine in instrumentation applications where the input signals must be routed any great distance. If the shielded leads are long, the shield capacitance can easily be the bulk of C_1 and C_2. If the shield conductors are driven by a signal equal to the common-mode voltage as shown in Figure 11.27*b*, then the common-mode voltage is not attenuated by the input *RC* networks and, more

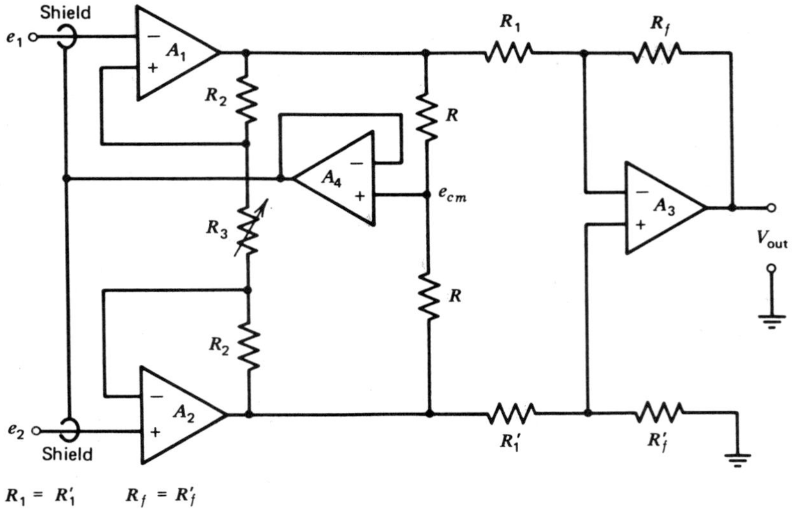

FIGURE 11.28 Instrumentation amplifier with input guard circuit.

importantly, not attenuated more on one input than the other. Recall that unequal attenuation of the common-mode signal degrades circuit CMRR. In the circuit shown in Figure 11.27*b*, only the differential input voltage is seen by the C_1 and C_2.

Connecting the shield as shown in Figure 11.27*b* may lead to unacceptable loading on the input source. By connecting the shield as shown in Figure 11.28, this loading is avoided. The two resistors (R) are equal in value and chosen so as not to load the outputs of A_1 and A_2. The voltage at the junction of the two resistors is approximately equal to the common-mode voltage (V_{cm}). The follower A_4 acts as a buffer between the circuit and the shield. If the shield capacitance is low and amplifiers A_1 and A_2 can provide sufficient driving current, the buffer can be eliminated and the shield connected directly to the voltage divider midpoint.

11-4.3 National Semiconductor LH0036 Instrumentation Amplifier

The National Semiconductor LH0036, whose block diagram is shown in Figure 11.29, is like the one shown in Figure 11.25 and operates in the same way. R_1, R_1', R_f, and R_f' are all equal in the monolithic IC. The amplifier features: R_{in} = 300 MΩ, CMRR = 80 dB (trimmed to 100 dB), gain adjust from 1 to 1000 with a single resistor, and operation from supply voltages of ±1 to ±18 V. The LH0036 provides a guard drive output, but an auxiliary voltage follower is needed for best performance. The bias current and bandwidth can be adjusted as outlined in the specification sheet in Appendix C. From the specifications

$$A_{fb} = 1 + \frac{50\ \text{k}\Omega}{R_g}$$

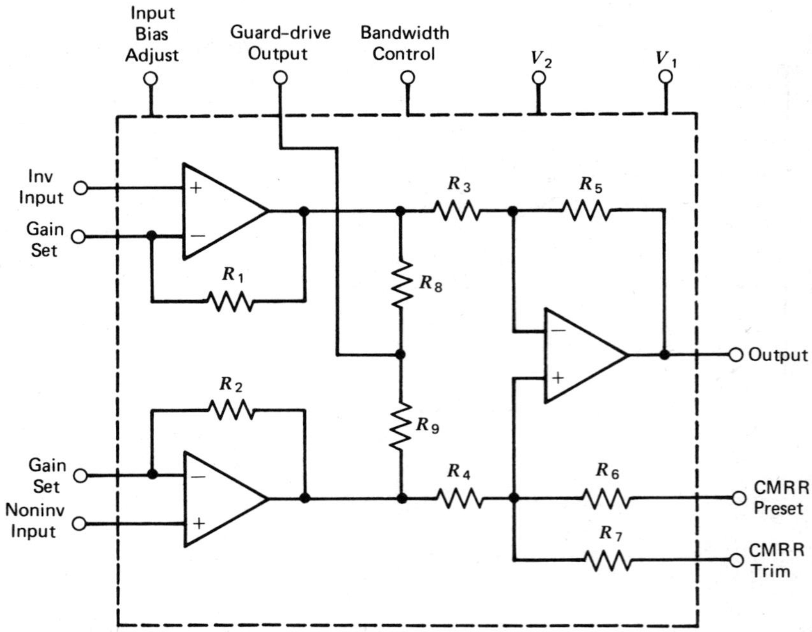

To 8 Metal Can Package

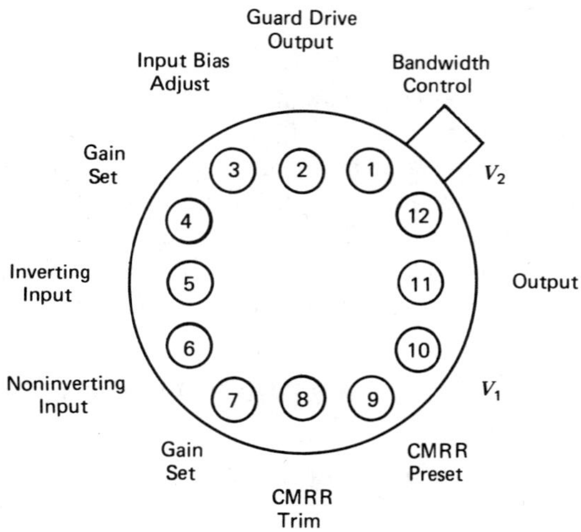

Top View

Order Number LH003G or LH0036C

Features

☐ High input impedance	300 MΩ
☐ High CMRR	100 dB
☐ Single resistor gain adjust	1 to 1000

□ Low power	90 μW
□ Wide supply range	±1 to 18 V
□ Adjustable input bias current	
□ Adjustable output bandwidth	
□ Guard drive output	

FIGURE 11.29 Equivalent circuit and connection diagrams for LH 0036.

where R_g = gain setting resistor. Thus

$$R_g = \frac{50\ \text{k}\Omega}{A_{fb} - 1}$$

The LH0036 is very easy to use.

EXAMPLE 11-10

Connect the LH0036 as shown in Figure 11.30 to measure the output of the strain gauge. The strain gauge resistance, $R = 350\ \Omega$, $\Delta R = 1\ \Omega$ and the bridge bias $V_R = 10$ V. Set $A_{fb} = 100$. Check to see that the output from the strain gauge is larger than from the common-mode voltage. Assume A_{ol} of $A_1 = 20{,}000$.

SOLUTION

An auxiliary op-amp is used to drive the shield for optimum performance. The CMRR adjust circuit from the specification sheet in Appendix C is used.

$$R_g = \frac{50\ \text{k}\Omega}{A_v - 1} = \frac{50\ \text{k}\Omega}{99} = 505\Omega$$

The circuit is complete.

From Equation 11-9 the strain-gauge output is

$$e_1 - e_2 = V_R\left(\frac{\Delta R}{4R + 2\ \Delta r}\right) = 10\ \text{V}\left(\frac{1\Omega}{4(350\ \Omega) + 2\ \Omega}\right) = 7.13\ \text{mV}$$

The differential output is

$$V_{out} = A_{fb}(e_1 - e_2) = 100\ (7.13\ \text{mV}) = 0.713\ \text{V}$$

The output voltage from the common-mode input voltage is

$$V_{o\ (cm)} = V_{cm}\left(\frac{A_{ol}/\text{CMRR}}{1 + A_{ol}\beta}\right) = 5\ \text{V}\left(\frac{20000/10000}{1 + 20000(0.01)}\right)$$

$$= 4.9\ \text{mV}$$

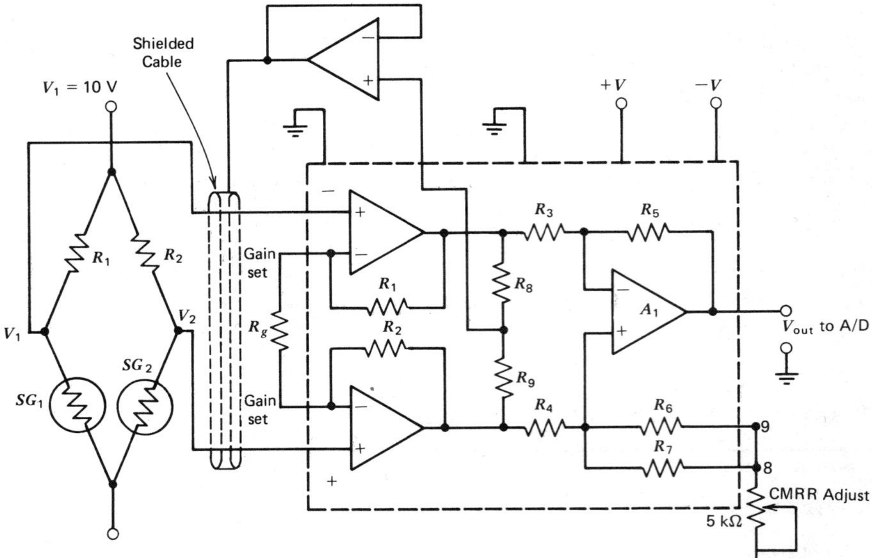

SG_1 = Active strain gauge
SG_2 = Temperature-compensating strain gauge

$$R_1 = R_2 = R_{SG_1} = R_{SG_2} = 350\Omega$$

$\Delta R = 1\Omega$
$V_{out} = 0.714$ V from $V_1 - V_2$
$V_{out} = 4.9$ mV from V_{cm}

FIGURE 11.30 LH 0036 application, strain-gauge bridge.

IC INSTRUMENTATION AMPLIFIERS

Instrumentation amplifier modules, either specially designed monolithic IC circuits, such as the LM0036, or hybrid IC circuits, are available from a number of op-amp manufacturers. These IC instrumentation amplifiers will have internal circuitry similar to that shown in Figures 11.24*b*, 11.25, 11.26, or 11.28. Instrumentation amplifier IC modules will normally have provisions for connecting a single external resistor or potentiometer to adjust gain, an output offset voltage nulling provision (using an external potentiometer), and a reference terminal where a set-point voltage may be applied. IC instrumentation amplifiers will be designed to have high R_{in}, low offset voltage errors, and a very high CMRR at all gain settings. Some instrumentation amplifiers now feature digital gain control.

11-5 ISOLATION AMPLIFIERS

Isolation amplifiers are a special type of amplifier available only in hybrid IC packaging, although monolithic IC op-amps and instrumentation amplifiers may be used in them. They provide almost complete isolation between the input and output. There are many different designs, but all use transformer coupling or optical coupling to provide the input-to-output isolation.

Isolation amplifiers provide dc isolation between the input and output from typically 2000 to 7500 V, CMRRs normally over 100 dB, and input impedances typically between 10^8 Ω and 10^{11} Ω. Isolation amplifiers normally have rather narrow bandwidths varying from dc-500 Hz to dc-3 kHz for transformer-coupled types and up to dc-15 kHz in optically coupled types. Optically coupled isolation transformers are generally not quite as linear as transformer-coupled isolation amplifiers.

The major uses of isolation amplifiers are medical monitoring, where small signals are hidden on much larger signals and dc leakage can be fatal, and isolating hazardous equipment with high voltages from sensitive A/D and computer equipment, where long two-wire monitoring of transducers is essential but common grounds would cause ground loops, and where electrically noisy equipment, such as motors, must be monitored.

11-5.1 Operation

Referring to Figure 11.31*b*, we see that a typical isolation amplifier is a rather complex communication system. The main amplifier power is supplied to the output section.

An oscillator operating at a much higher frequency than the maximum amplifier input frequency provides a reference signal to the demodulator, a carrier signal to the modulator, and ac to the input side rectifier and power supply.

The input power supply provides a floating voltage source for the amplifier input circuitry and often supplies auxiliary input power for additional input circuitry.

The input amplifier can be a single op-amp, but is normally a complete high performance instrumentation amplifier. One or two gain-setting terminals and offset adjustment terminals are usually supplied on the input side. The entire input circuit is usually shielded.

The input amplifier output modulates the modulator. The modulation can be AM, PWM, or LED light intensity. The demodulator extracts the information from the modulated carrier. The signal is then filtered and fed to a buffer amplifier and the output.

Isolation amplifiers can be used as high performance instrumentation amplifiers or for isolation. Figure 11.32 shows a biomedical application of isolation amplifiers. The high CMRR of the amplifier separates the small voltage fetal heart signal from the larger amplitude mother's heartbeat and the 60 Hz noise picked up by the mother's skin. The isolation assures that none of the well-grounded stainless steel medical furniture will provide a hazard to the patient in the event of equipment failure.

SUMMARY

1. Special purpose linear-integrated circuits exist for any circuit function that can be provided by IC technology if a sufficient market exists for that circuit function.
2. Comparators are special-purpose integrated circuits designed for use in any application

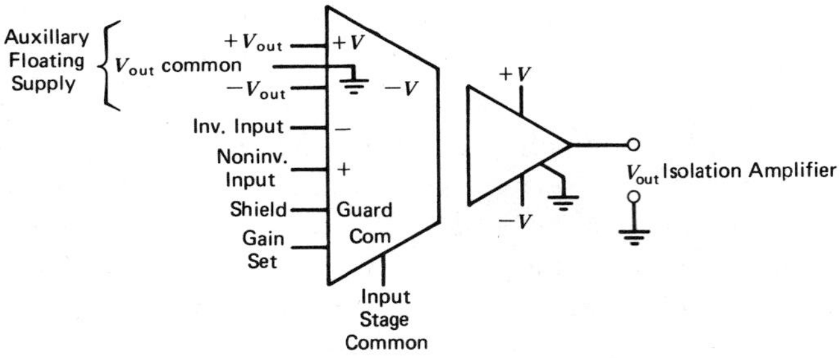

(a) Symbol

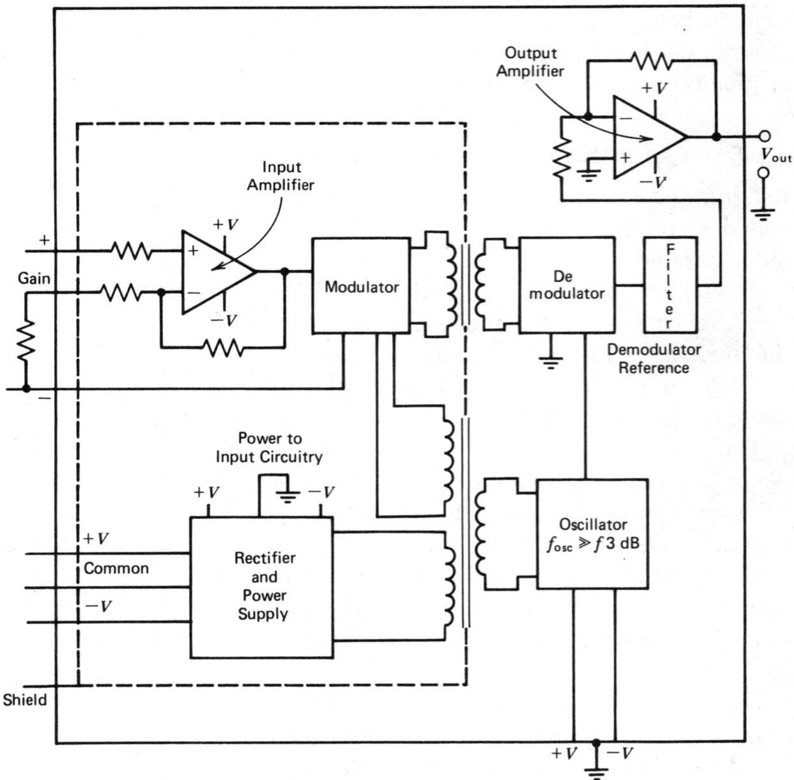

(b) Block diagram, transformer coupled isolation amplifier

FIGURE 11.31 Isolation amplifier.

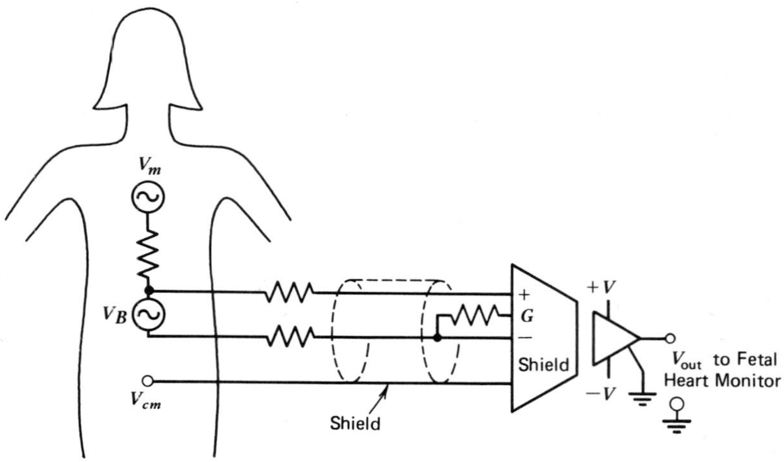

(a) Connection

V_m = mother's heart signal (typ. 1 mV)
V_B = fetal heart signal (typ. 50 μV)
V_{cm} = common-mode pickup (1 to 100 mV)

Signal separation provided by high CMRR. V_{cm} cannot leak across isolation boundary. Patient safe from ground loops and equipment voltages.

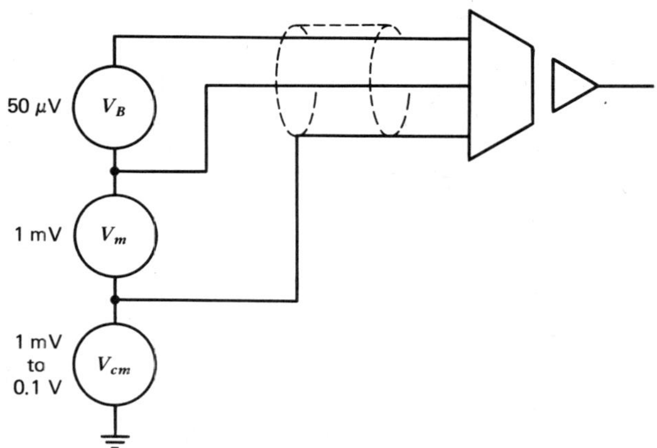

(b) Equivalent circuit

FIGURE 11.32 Fetal heart monitoring, an isolation amplifier application.

that requires a switching output when one input exceeds another. Even though regular op-amps can be used, comparators are more easily used for this function.
3. Current-mode amplifiers operate differently from more conventional op-amps yet can perform many of the same functions if less precision is acceptable. Current-mode amplifiers are designed for single-supply operation.
4. 555 timers are very useful, medium precision oscillators and pulse generators. Although not strictly a linear circuit, the 555 demonstrates the use of comparators as level detectors. As seen in the switching power supply example, 555 timers have many uses in larger circuits.
5. The instrumentation amplifier is a very high performance special-purpose amplifier that normally consists of several op-amps. Instrumentation amplifiers must have very high input resistance, CMRR, and stability. Ease of use is an important factor in instrumentation amplifier marketing.
6. Isolation amplifiers may be thought of as super instrumentation amplifiers. They offer extremely high CMRR, input resistance, and stability as well as dc isolation of input and output. Isolation amplifiers are rather complicated systems involving modulation by the input signal and demodulation in the output section. Isolation amplifiers achieve isolation by transformer-coupling or optically coupling input signals to the output. Optically coupled isolation amplifiers have wider bandwidths than transformer-coupled types, but lack their linearity. In the field of biomedical instrumentation, isolation amplifiers can be real lifesavers.

SELF-TEST QUESTIONS

11-1 State the two conditions that must be met for a special-purpose IC to be developed.
11-2 State the major function of a comparator.
11-3 State the purpose for hysteresis in a comparator.
11-4 State the disadvantage of hysteresis in a comparator.
11-5 Calculate R_1 and R_2 for the circuit in Figure 11.2 so that UTP $= +4$ V and LTP $= -4$ V. $\pm V = \pm 15$ V and $I_{R1} = I_{R2} = 0.1$ mA.
11-6 Calculate R_1, R_2, R_3, R_B, and R_4 for the circuit of Figure 11.3. $\pm V = \pm 15$ V, $V_{ref} = 15$ V, $V_{Z1} = 4.7$ V at $I_B = 5$ mA, $V_{CE}(\text{sat})_{Q1} = 0.2$ V, $V_{BEQ1} = 0.8$ V, $\beta_{Q1} = 40$. Let $I_{divider} = 0.1$ mA, UTP $= 8$ V, LTP $= 4$ V, $I_L = 3$ mA.
11-7 Calculate the components for the window comparator of Figure 11.4 so that UTP $= 4.8$ V and LTP $= 2.4$ V. Set $V_{ref} = 5$ V and $I_{divider} = 0.2$ mA.
11-8 For the circuit in Figure 11.5, what must be done for $V_{out} = +V$ to be a no-go and $V_{out} = -V$ to be a go condition.
11-9 State the major advantage of a *R*-2*R* D/A converter over a weighted resistor D/A converter.
11-10 State the major advantage and disadvantage of a parallel A/D converter.
11-11 The dual slope A/D converter need not have a highly stable clock oscillator. Why not?
11-12 State briefly in your own words the operation of a dual-slope A/D converter.
11-13 The successive approximation A/D converter uses a D/A converter. State the function of the D/A converter.
11-14 List four uses for analog switches.
11-15 State the major disadvantage and advantage of the capacitor gain set circuit of Figure 11.15*c* with respect to the one in Figure 11.15*a*.
11-16 State the major advantage and disadvantage of current-mode amplifiers with respect to "regular" op-amps.
11-17 Calculate the components for the inverting amplifier of Figure 11.18*a* if $V_{cc} = 15$ V, $A_{fb} = 20$, $I_m = 30\ \mu A$, $R_L = 10\ k\Omega$, $f_L = 300$ Hz, and $V_{out\ dc} = 1/2\ V_{cc}$.
11-18 Calculate the components for the noninverting amplifier of Figure 11.18*b* if $V_{cc} = 10$ V, $I_m = 50\ \mu A$, $A_{fb} = 12$, $R_L = 5\ k\Omega$, $V_{out} = ½\ V_{cc}$, and $f_L = 60$ Hz.

11-19 Calculate R_A and R_B for a 555 oscillator. Let f_{osc} = 20 kHz, duty cycle = 80%, and C_T = 0.001 μF.
11-20 Calculate C_T for a 555 monostable if R_A = 5.6 kΩ, and t_p = 3 ms.
11-21 For the circuit of Figure 11.21, what must $t_L = t_c(\text{min})$ of the oscillator be if I_{FB} in Example 11-9 is chosen to be 0.4 I_p.
11-22 Calculate $e_1 - e_2$ for the bridge of Figure 11.23 if R = 2 kΩ, ΔR = 10 Ω, and V_R = 10 V.
11-23 In the circuit of Figure 11-25 $R_1 = R_1' = R_f = R_f'$ = 100 kΩ and $R_2 = R_4$ = 50 kΩ. Calculate R_3 for A_{fb} = 40.
11-24 Let $R_1 = R_1' = R_f = R_f'$ = 100 kΩ in Figure 11.26. If R_2 = 20 kΩ, calculate R_g for A_{fb} = 20.
11-25 State the purpose of input guarding.
11-26 State the four major uses of isolation amplifiers.
11-27 How are the input signals coupled to the output of an isolation amplifier?

If you cannot answer certain questions, place a check next to them and review appropriate parts of the text to find the answers.

LABORATORY EXERCISE

Objective

After completing this laboratory, the student will be able to set up an op-amp as a comparator, a Schmitt trigger, and a window detector. The student will be able to set up a current-mode amplifier as an inverting and noninverting amplifier and a 555 timer as an astable and monostable.

Equipment

1. Two μA741 operational amplifiers or equivalent. (Optional: LM311 comparators.)
2. General-purpose oscilloscope.
3. Motorola MC3401 or equivalent.
4. Switching transistor, *npn*, 2N2369A or equivalent.
5. Breadboard such as EL Instruments SK-10 mounted on vectorboard.
6. Power supply, ±15 V dc.
7. Adjustable 5 V power supply.
8. Audio signal generator.
9. Two 10 kΩ potentiometers, one 25 kΩ potentiometer.
10. ±2% resistor assortment.
11. Metalized mylar capacitor assortment.
12. A TTL NAND gate of your (or your instructor's) choosing.
13. Zener diode, V_Z = 4.7 V.
14. LED
15. Two diodes, 1N914 or equivalent.

Procedure

1. TTTL gate test.
   (a) Connect the 741 as a comparator with a gain of 200 (±0.5% trip point accuracy) as shown in Figure 11.33. Let R_f = 1 MΩ and I_Z = 1 mA.

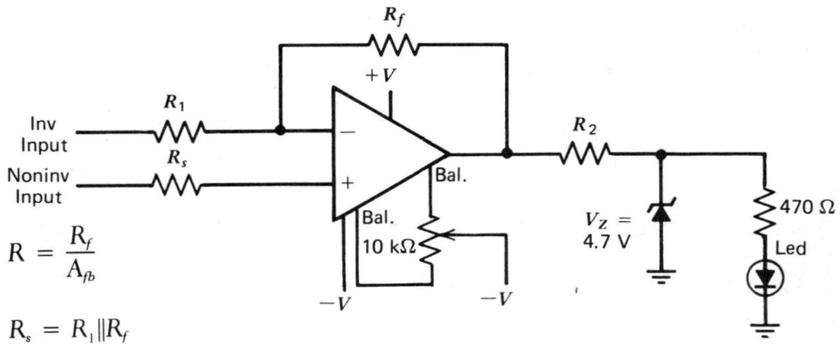

$$R = \frac{R_f}{A_{fb}}$$

$$R_s = R_1 \| R_f$$

$$R_2 = \frac{V_{SAT} - V_Z}{I_{LED} + I_Z}$$

FIGURE 11.33 μA 741 connected as a low-gain comparator.

(b) Set both inputs to ground and null the offset.

(c) Let the led on be a no-go indication and the led off be a go indication. Connect the comparator to the gate as shown in Figure 11.34*a* and *b*. This will test for V_{out} high minimum and V_{out} low maximum.

Note. Do not change the offset balance potentiometer for any of these measurements.

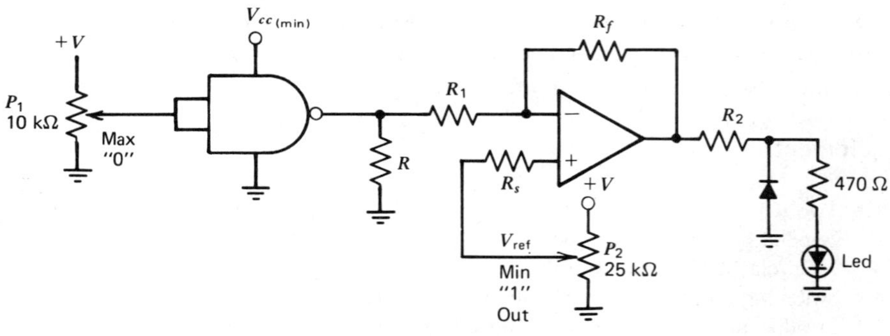

(*a*) V_{out} high minimum test

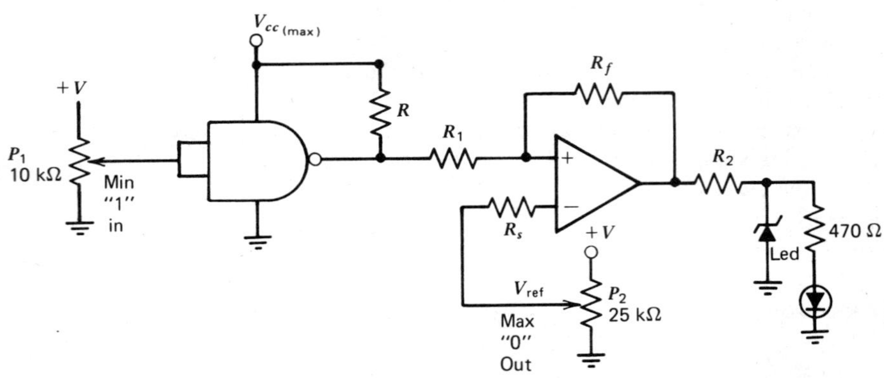

(*b*) V_{out} low maximum test

FIGURE 11.34 TTL gate-test circuit.

2. Schmitt trigger.
   (a) Connect the Schmitt trigger shown in Figure 11.35. Calculate R_1, R_2, R_3, and R_B so that UTP = 6 V and LTP = 2 V. Let $I_{R1} = I_{R2} = I_{R3} = 1$ mA. Use the R_f, R_1, and R_s values calculated in part 1. Do not remove the offset balance potentiometer.
   (b) Connect a dc voltage from a potentiometer wiper. Observe the output as V_{in} passes the UTP and LTP.
   (c) Connect a signal generator with $V_{out} = 10\ V_p$ to the Schmitt trigger. Observe the output at $f \cong 1$ kHz.
3. Window comparator.
   (a) Connect two comparators, set up as in part 1, as a window comparator as shown in Figure 11.36.
   (b) Using the adjustable dc voltage from part 2b, vary the input voltage.
   Note that the output changes as V_{in} passes through the LTP and UTP.
4. Current differencing amplifier.
   (a) Connect the MC3401 as an inverting amplifier as shown in Figure 11.37*a*. Set $I_m = 50\ \mu A$, $A_{fb} = 10$, and $V_{out\ Q} = 0.5\ V_{cc}$.
   (b) Measure A_{fb} at $f_{in} = 1$ kHz and measure V_{out} quiescent.

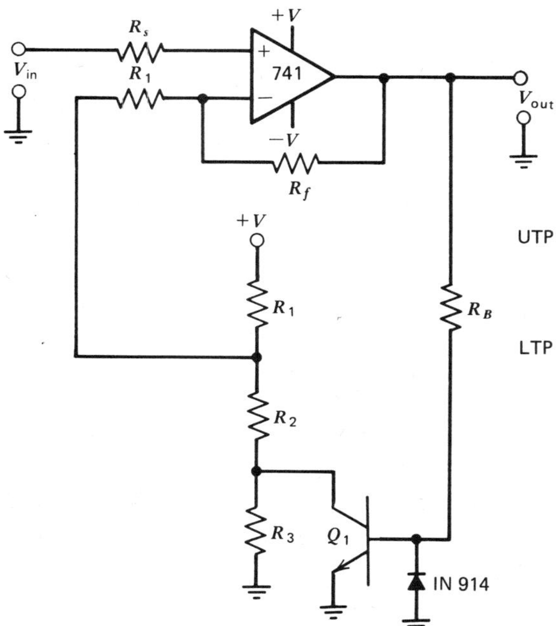

$$\text{UTP} = +V\left(\frac{R_2 + R_3}{R_1 + R_2 + R_3}\right)$$

$$\text{LTP} = +V\left(\frac{R_2}{R_1 + R_2}\right) + V_{CE\,(sat)}\ Q_1$$

(a) Schmitt trigger connection

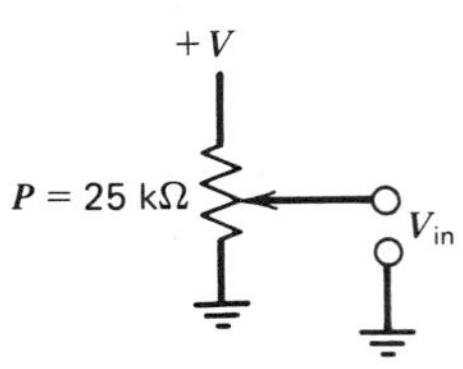

(b) Adjustable dc voltage for part 2b

FIGURE 11.35 Schmitt trigger.

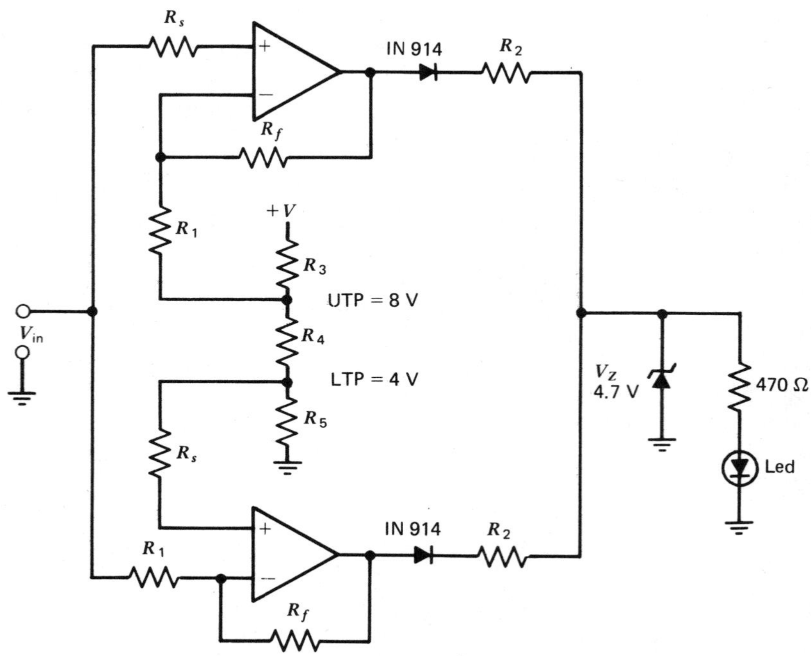

FIGURE 11.36 Window comparator.

(c) Connect the MC3401 as a noninverting amplifier as shown in Figure 11.37*b*. Set $I_m = 50\ \mu A$, $A_{fb} = 20$, and $V_{out\,Q} = 0.5\ V_{cc}$.
(d) Measure V_{out} quiescent and measure A_{fb} at $f_{in} = 1$ kHz.

5. 555 Timer.
(a) Connect the 555 timer as an oscillator as shown in Figure 11.38*a*. Set $f_{osc} = 1$ kHz, duty cycle = 60%. Measure t_1, t_2, and f_{osc}.
(b) Remove V_{cc} from the Reset (pin 4) and ground pin 4. Observe the oscillator operation as pin 4 is grounded.
(c) Connect a potentiometer wiper to the control pin as shown in Figure 11.38*a* (inset). Observe the change in oscillation as the control voltage is varied.
(d) Connect the 555 timer as a monostable as shown in Figure 11.38*b*. Q_1 is simply a convenient trigger. Let $t_p = 0.6$ ms.

FIGURE 11.37 Current-mode amplifier.

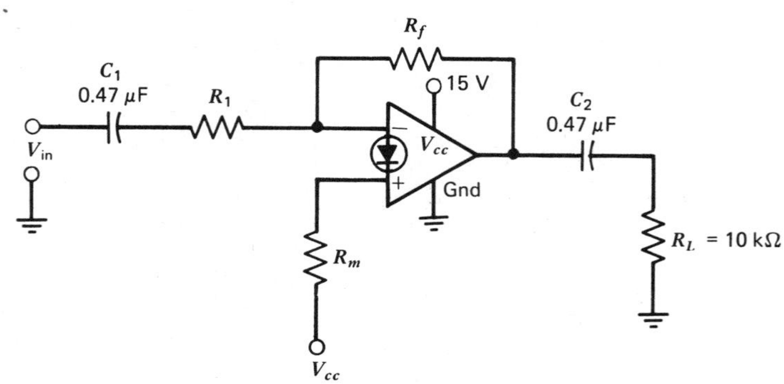

(a) Inverting amplifier

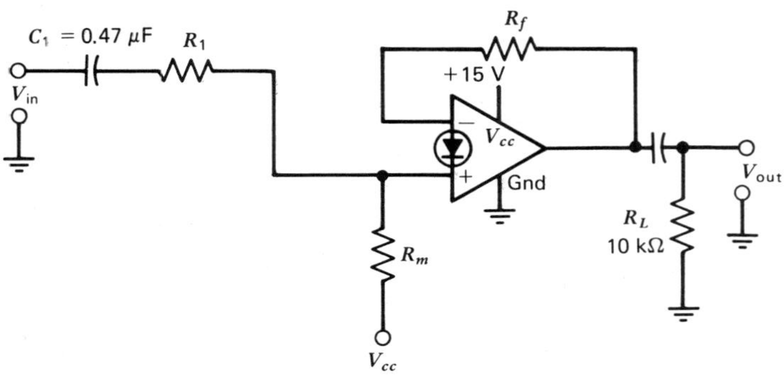

(b) Noninverting amplifier

Inverting

$$R_m = \frac{V_{cc} - 0.7\text{ V}}{I_m}$$

$$R_f = \frac{0.5\ V_{cc} - 0.7\text{ V}}{I_m}$$

$$R_1 = \frac{R_f}{A_{fb}}$$

Noninverting

$$R_m = \frac{V_{cc} - 0.7\text{ V}}{I_m}$$

$$R_f = \frac{0.5\ V_{cc} - 0.7\text{ V}}{I_m}$$

$$r_e = \frac{26\text{mV}}{I_m}$$

$$R_1 = \frac{R_t}{A_{fb}} - r_e$$

(e) Connect an ac signal source to the input of Q_1. Let $V_{in} = 1.5\ V_p$ and $f = 1$ kHz. Observe the output and measure t_p.

(f) Vary the control voltage as in step (c). Observe t_p as the control voltage is varied.

FIGURE 11.38 555 timer.

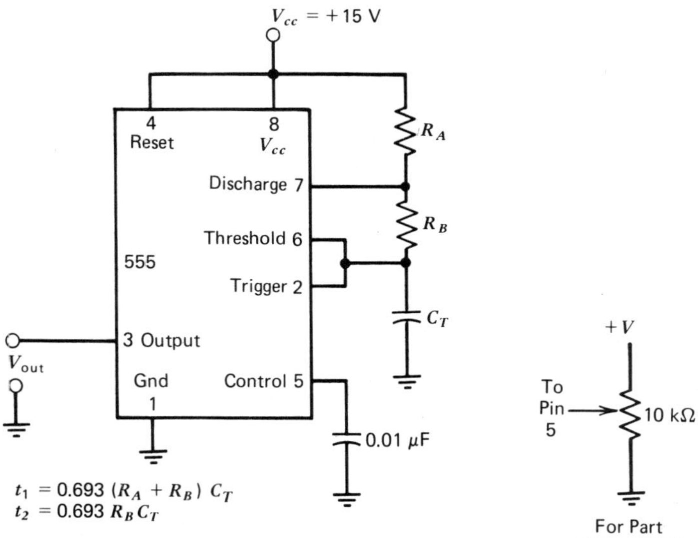

$t_1 = 0.693\ (R_A + R_B)\ C_T$
$t_2 = 0.693\ R_B C_T$

(a) Astable

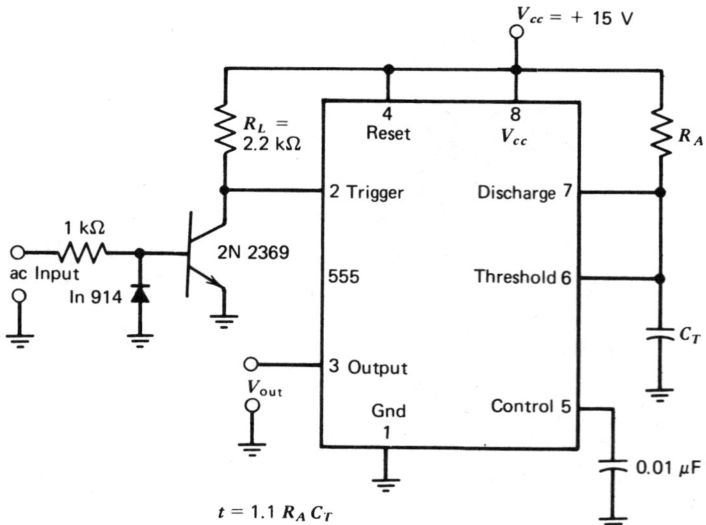

(b) Monostable

CHAPTER

12

NOISE

No discussion of operational amplifiers is complete without mention of noise. A broad definition of noise is: alternating current signals generated by random electric charge motion. A better understanding of noise can be obtained by looking at the various types of noise that plague the op-amp user. The three types of noise that can occur within all amplifying devices are Johnson noise (also called thermal noise), Schottky noise (usually called shot noise), and flicker (or 1/f) noise.

OBJECTIVES

After completing the study of this chapter and the self-test questions, the student should be able to:

1. Define Johnson, Shottky, and flicker noise.
2. State three basic principles of shielding.
3. State the basic principle of proper grounding.
4. Perform the laboratory for Chapter 12.

12-1 JOHNSON NOISE (THERMAL NOISE)

Thermal noise is caused by the random motion of charges due to the thermal energy they receive from their surroundings. This noise is random in frequency with an amplitude proportional to the square root of temperature. The wide-band frequency content is due to the random motion of the charges, and the higher the temperature, the higher the amplitude of the random motion. All materials (conductors, semiconductors) with free charge exhibit thermal noise. In an ohmic resistance the average square (usually called mean-square) open circuit thermal noise voltage across the ends of the resistance is

$$\overline{e_n^2} = 4\,k\,T\,R\,\Delta f \tag{12-1}$$

where

$\overline{e_n^2}$ = mean-square noise voltage
k = Boltzmann's constant, 1.38×10^{-23} joules/°Kelvin
T = temperature in degrees Kelvin (°C + 273)
Δf = frequency range of interest (i.e., amplifier bandwidth)

Note that the noise voltage depends on bandwidth, not on the frequency value. Also note that the noise voltage is proportional to the resistance. These facts imply that low impedance levels and narrow bandwidths reduce the effect of noise. In other words, if a 100 kΩ feedback resistor will do as well as a 1 MΩ, we would be wise to use the 100 kΩ. Also, we should use only as much bandwidth as is needed where noise is critical. Since all the charge-carrying components in the amplifier are generating this noise, and the amplifier amplifies all the noise at its input, the noise riding on the output signal may be significant.

EXAMPLE 12.1

Calculate the rms noise voltage generated by a 100 kΩ wire-wound resistor if it is to be connected to a device with a 10 kHz bandwidth at room temperature.

SOLUTION

A wire-wound resistor will have close to the minimum theoretical noise. The square root of the mean-squared value gives us the rms value.

$$e_n\ (\text{rms}) = \sqrt{4k\,TR\,\Delta f}$$

$$= \sqrt{4\,(1.38 \times 10^{-23})\,(298)\,(100\text{ k}\Omega)\,(10\text{ kHz})}$$

$$e_n = 4.0558\ \mu\text{V (rms)}$$

This noise, should the resistor be hooked between the input of an op-amp to ground, will be amplified along with any signal. The peak value of the noise voltage will be about five times the rms value.

12-2 SHOT NOISE

Shot noise is a result of current carriers being particles, that is, electrons. Shot noise is associated with semiconductor materials (which include many resistor compounds) and vacuum tubes. To see how the particular nature of current causes shot noise, consider a steady-state dc current. Even though the average current is steady, the exact number of electrons per unit time passing a particular point varies randomly. This variation causes a noise current to be impressed on the steady-state current measured at a point. The mean-squared value of this noise current in a semiconductor is

$$\overline{i_n^2} = 2\,q\,I_{dc}\,\Delta f \qquad 12\text{-}2$$

where

$\overline{i_n^2}$ = mean-squared noise current
q = charge of electron, 1.6×10^{-19} coulombs
Δf = bandwidth of interest
I_{dc} = average dc current in device or circuit of interest

Note that shot noise, like thermal noise, is frequency independent. Observe also that as dc current is reduced, shot noise in a semiconductor is reduced.

EXAMPLE 12-2

The bias current of a bipolar input op-amp is 200 nA. Calculate the input noise current and the total input noise voltage if the resistance from Example 12-1 is the source resistance of the op-amp. $\Delta f = 10$ kHz

SOLUTION

The noise current is simply

$$\begin{aligned} i_n &= \sqrt{2q\, I_{DC}\, \Delta f} \\ &= \sqrt{2\,(1.6 \times 10^{-19})\; 200\text{ nA}\; (10\text{ kHz})} \\ &= 2.53 \times 10^{-11}\text{ A} = 25.3\text{ pA rms} \end{aligned}$$

The total equivalent noise voltage of Examples 12-1 and 12-2 combine as follows

$$e_{n_t} = \sqrt{e_n^2 + 2i_n^2\, R_g^2}$$

where

e_{nt} = total input noise
e_n^2 = thermal mean-squared input noise
i_n^2 = mean-squared shot noise current
R_g = source resistance into the op-amp, i.e., resistance from the input to ground.

solving

$$e_{nt} = \sqrt{1.64 \times 10^{-11}\text{V} + 2(25.3\text{ pA})^2(100\text{ k}\Omega)^2}$$

$$e_{nt} = 5.4\ \mu\text{V (RMS)}$$

As with thermal noise, keeping the circuit resistances as low as is practical reduces noise problems.

The first stage of a multistage amplifier contributes most to the noise, since for each succeeding stage the signal level is higher. For low-noise amplifier systems it is very important that the first stage be very low in noise.

The noise bandwidth, Δf, of a device or amplifier is wider than the difference between the upper and lower 3 dB points ($f_{cu} - f_{cl}$). If the roll-off slopes of an amplifier are 6 dB/octave, the noise upper and lower cutoff frequencies, f_{cnu} and f_{cnl}, are

$$f_{cnu} = 1.57\, f_{cu}$$

$$f_{cnl} = f_{cu}/1.57$$

This is because noise power is produced in the roll-off frequency band from f_c to f crossing.

12-3 FLICKER NOISE (1/f) NOISE

The cause of flicker noise in semiconductors is believed to be the variations of electron (or hole) velocity effected by flaws in the semiconducting material. Flicker noise is characteristic of semiconducting material (which is used to make many kinds of resistors) but is absent in wire-wound and metal alloy resistors. The reason flicker noise is called 1/f noise is that it increases as frequency decreases. Although flicker noise varies from device to device, it does not usually vary widely within the same device type. Flicker noise can then be measured for a given semiconductor. A graph of flicker noise versus frequency is shown in Figure 12.1. The only way to avoid flicker noise is to use devices that have very little or that operate at a high enough frequency that flicker noise is not significant. This is often impossible to do.

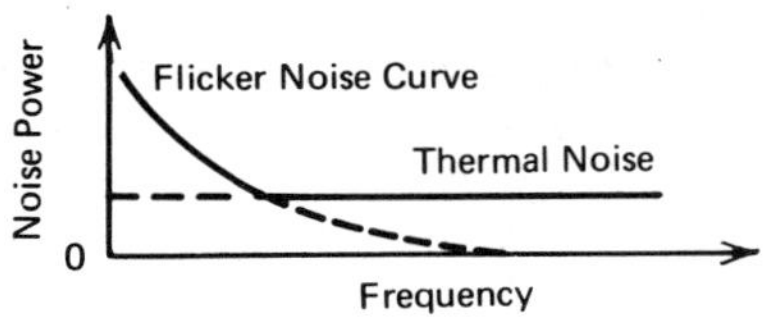

FIGURE 12.1 Flicker noise vs. frequency.

12-4 SIGNAL-TO-NOISE RATIO

The signal-to-noise ratio is the ratio of overall signal to existing noise. If N = noise, and S = signal, then

$$\frac{Si}{Ni} = \text{signal-to-noise ratio at input} \qquad (12\text{-}3)$$

and

$$\frac{S_o}{N_o} = \text{signal-to-noise ratio at output} \qquad (12\text{-}4)$$

12-5 NOISE FIGURE

The noise figure is defined as

$$NF \text{ dB} = 10 \log \frac{(S_i/N_i)_{\text{power}}}{(S_o/N_o)_{\text{power}}} \qquad (12\text{-}5)$$

Observe that if $S_i/N_i = S_o/N_o$ the noise figure is O dB or 1. The noise figure is a measure of how much noise the device or amplifier adds. A low noise figure is usually desired, but designing for the lowest noise figure may not result in the lowest signal-to-noise ratio at the output.

EXAMPLE 12-3

An inverting operational amplifier is fed from a voltage source with a 5 kΩ internal impedance. The op-amp gain is 20 and its load resistance in parallel with R_f is 9.09 kΩ. The amplifier bandwidth is 1 MHz. The equivalent input noise of the op-amp from the specification sheet is 24 μV RMS. The total generator noise voltage is 11.3 μV rms, and the signal voltage is 200 μV RMS. Calculate the noise figure.

SOLUTION

First we will calculate the input signal-to-noise ratio in power.

$$\frac{S_i}{N_i} = \frac{V_{\text{in}}^2/R_g}{e_{ng}^2/R_g}$$

where $R_g = 5$ kΩ, the generator resistance. Thus

$$\frac{S_i}{N_i} = \frac{(200\ \mu\text{V})^2/5\ \text{k}\Omega}{(11.3\ \mu\text{V})^2/5\ \text{k}\Omega} = \frac{8\ \text{pW}}{0.026\ \text{pW}}$$

$$= 313.26$$

Now we will calculate the output signal-to-noise ratio.

$$V_{\text{out}} = 0.2\ \text{mV}\ A_{fb}$$

$$= 4\ \text{mV}$$

The gain of a noninverting amplifier to noise is just A_{fb}, but the gain of an inverting amplifier to noise (A_{fn}) is

$$A_{fn} = A_{fb}\left(\frac{R_1 + R_f}{R_f}\right)$$

If $R_f = 100$ kΩ and $R_1 = 5$ kΩ

$$A_{fn} = 20\left(\frac{105\ \text{k}\Omega}{100\ \text{k}\Omega}\right) = 21$$

so

$$e_{no} = A_{fn}\, e_{ni_t}$$

but

$$e_{ni_t} = \sqrt{e_{ni}^2 + e_{ng}^2}$$

$$= \sqrt{(24\ \mu V)^2 + (11.3\ \mu V)^2}$$

$$= \sqrt{7.04 \times 10^{-10}} = 26.5\ \mu V$$

$$e_{on} = 21\ (26.5\ \mu V) = 557.1\ \mu V$$

and the output power signal-to-noise ratio is

$$\frac{S_o}{N_o} = \frac{V_{out}^2/(R_L \parallel R_f)}{e_{no}^2/(R_L \parallel R_f)}$$

$$= \frac{1.76\ nW}{34.1\ pW}$$

$$= 51.6$$

Finally, the noise figure is

$$NF\ dB = 10 \log \frac{S_i/N_i}{S_o/N_o}$$

$$= 10 \log \frac{331.26}{51.6}$$

$$= 8.08\ dB$$

12-6 INTERFERENCE

Since interference is a source of unwanted signals, it can be classified as noise. Protection against interference can be accomplished by proper grounding, careful circuit layout, and shielding. The shielding problem can be quite complex but, reduced to three simple principles (which may not be simple to follow), we can say:

1. The shield conductor should be connected to the zero reference of the signal (ground) only once.
2. The shield and signal reference ground should be grounded at the same point physically.
3. To be effective, all signal-carrying conductors should be enclosed within the shield.

The circuit layout should avoid excessive crowding, long adjacent signal paths with or without shielding, and unnecessary wire crossings. In other words, follow good construction practice.

12-7 GROUNDING

A sure way to pick up unwanted signals is to ground carelessly. The principle of proper grounding is simple to state, but sometimes hard to follow. The principle is:

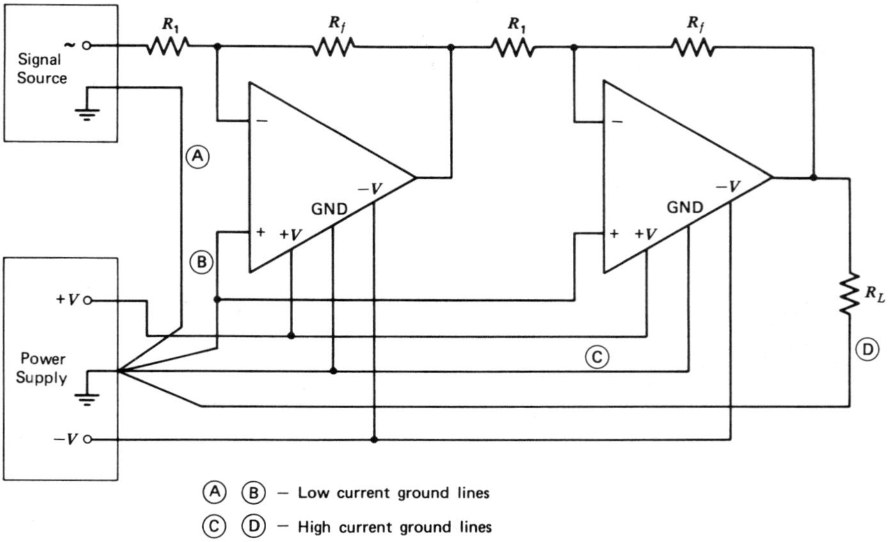

FIGURE 12.2 Proper grounding.

Grounds that must carry load current should be returned to the power supply common through a different wire from the signal ground.

This principle is shown in Figure 12.2. The reason for this is that load current is often many times greater than the signal current. The load current, even when flowing through rather large conductors, can cause an *IR* drop along the ground line. This would cause the reference line of each op-amp connected to that line to be at a slightly different potential. In extreme cases this potential difference could be several millivolts. This is a source of considerable error.

SUMMARY

1. Noise is an unwanted signal. Most noise is broadband—that is, it contains many frequencies—and is due to random charge motion within semiconductors and resistors, or is from interfering signals. Whatever the source of noise, it will mask small input signals and cause uncertainty in output signals, so we must try to minimize its effect on our circuitry.
2. The three major sources of noise within circuitry are Johnson (also called thermal noise), flicker noise, and shot noise. These are minimized by using low noise components, keeping circuit resistance as low as practical, and using low current and reduced bandwidth where possible.
3. Interference is the major source of external noise; it can be minimized by careful shielding.
4. If circuits are not grounded carefully, the output signal of a circuit will interfere with its input. Grounds carrying large currents should be conductors different from signal grounds.

SELF-TEST QUESTIONS

12-1 Name the source of Johnson noise.
12-2 Name the source of shot noise.
12-3 State the frequency dependence of flicker noise.

12-4 Define the signal-to-noise ratio.
12-5 Define noise figure.
12-6 Name three steps that can be taken to minimize interference.
12-7 State the prime principle of grounding.
12-8 A 10 kΩ resistor at room temperature (25°C) has a bandwidth of 100 kHz. Calculate the rms noise voltage.
12-9 An op-amp has a bias current of 50 nA and a bandwidth of 100 kHz. Calculate the rms noise current, i_n.
12-10 If the source resistance is 10 kΩ, calculate the total noise voltage from questions 12-8 and 12-9.
12-11 State three steps that can be taken to minimize noise.

If you cannot answer certain questions, place a check next to them and review appropriate parts of the text to find the answers.

LABORATORY EXERCISE

Objective

The student will measure the noise figure of an operational amplifier.

Equipment

1. Two Fairchild μA741 operational amplifiers or equivalent.
2. ±2% resistor assortment.
3. Power supply, ±15 V.
4. Audio signal generator.
5. One 10 kΩ potentiometer.
6. Oscilloscope, dual sweep with alternate triggering mode.
7. Breadboard such as EL Instruments SK-10.

Procedure

There are several ways to measure the noise figure of an amplifier circuit. Many methods use a calibrated noise source for an input signal so that the input contains all of the frequencies in the amplifier's bandwidth. Since noise is broadband, a noise generator meets this criterion.

The noise voltage is usually measured with a true RMS voltmeter of extreme accuracy and high cost. We will use an oscilloscope and the "tangential method" to obtain medium accuracy. This method will enable us to measure noise to approximately 1 dB accuracy. A direct reading of noise from the oscilloscope trace is not accurate because the random peaks are much higher than the visible noise on the trace. The tangential method is as follows (refer to Figure 12.3):

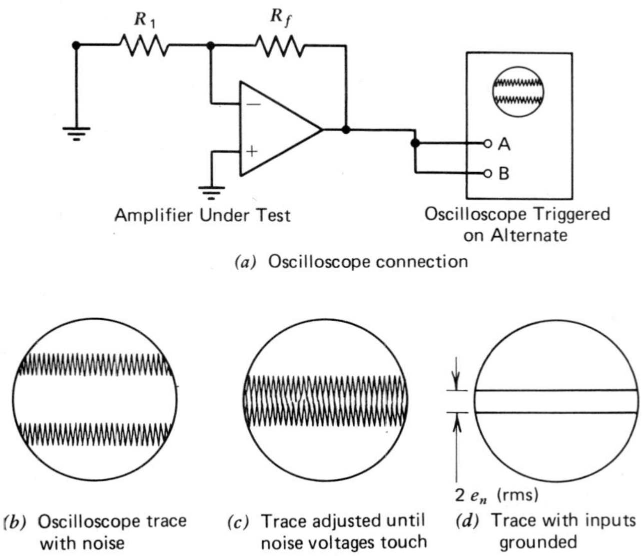

FIGURE 12.3 Tangential method of noise voltage measurement.

1. Connect both inputs of a dual-trace oscilloscope to the noise voltage to be measured. A 1:1 oscilloscope probe should be used. Trigger the oscilloscope on alternate. Both channels should be on the same voltage setting and identically calibrated. Use the vertical amplifier magnifier if available and needed.
2. Adjust the traces with the noise voltage until the voltages just touch and there is no clear CRT face between them, as shown in Figure 12.3*c*.
3. Ground the oscilloscope probes to obtain a noise-free trace. The distance between the two traces will be twice the rms noise voltage. For example, if the traces are 1.6 divisions apart and the oscilloscope is on 20 mV/div, the rms noise voltage is

$$e_n \text{ rms} = \frac{1.6\ (20\ \text{mV/div})}{2} = 16\ \text{mV rms}$$

Many noise figure measuring methods use a calibrated noise source. This provides a signal source with frequency components throughout the amplfier's bandwidth. We will use a signal generator with a frequency well within the amplifier's band pass since few schools have a calibrated noise generator.

1. Input signal-to-noise ratio.
   (a) Connect the signal generator as shown in Figure 12.4*a*.
   (b) Measure the noise voltage with the signal generator output voltage at zero volts if the oscilloscope is sensitive enough.
   (c) If you measured a noise voltage in step (b), skip this step. If not, calculate the thermal noise of the generator resistance.

$$e_{ng} = \sqrt{4\,kT\,\Delta f\,R_g}$$

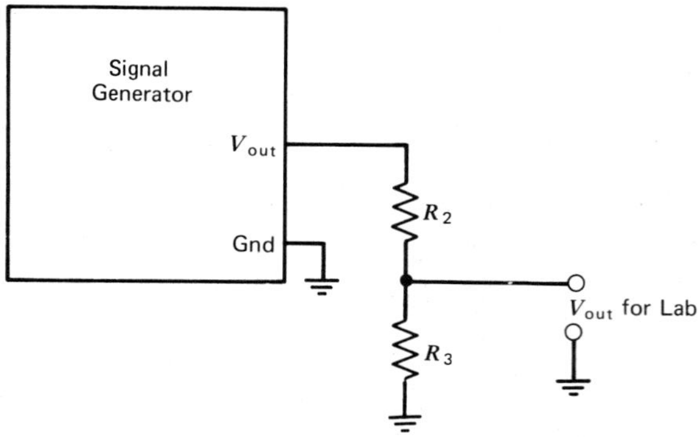

(a) Signal generator

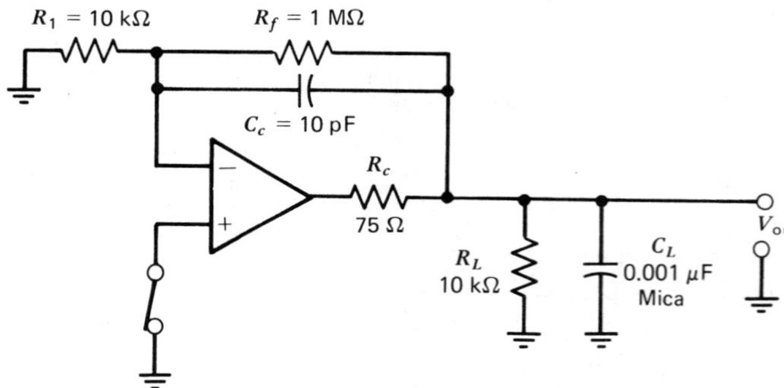

(b) Amplifier circuit for part 2 (a) and (b)

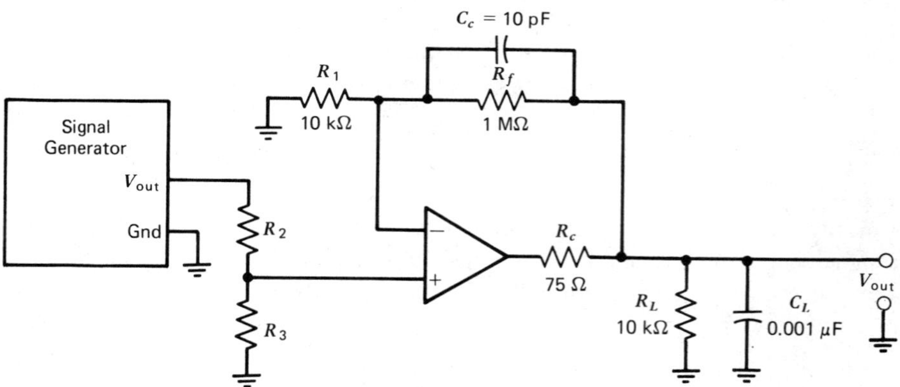

(c) Set up for part 2 (c)

FIGURE 12.4 Noise figure measurement.

where

k = Boltzman's constant 1.38×10^{-23}, joules/°K
T = absolute room temperature, about 300°K
R_g = $R_2 \parallel R_3$
f = op-amp bandwidth × 1.57. This is set at 15 kHz by R_L and C so Δf = 23.55 kHz.

(d) Set the ouptut voltage (V_s) of the generator to 1 mV rms (2.8 mV peak to peak) at about 5 kHz. If this is too difficult to measure, set the voltage across voltage divider R_2 and R_3 to 28 mV peak to peak.
(e) Calculate the input signal to noise ratio

$$\frac{S_i}{N_i} = \frac{V_s^2/R_g}{e_{ng}^2/R_g}$$

2. Output signal-to-noise ratio.
(a) Ground the op-amp amplifier input as shown in Figure 12.4*b*.
(b) Measure the output noise (e_{no}).
(c) Connect the generator to the op-amp input and measure the output signal (V_o). Convert the peak-to-peak output signal to an rms value, $(V_{out\,pp})(0.3535) = V_o$.
(d) Calculate the ouptut signal-to-noise ratio.

$$\frac{S_o}{N_o} = \frac{V_o^2/[R_L \parallel (R_1 + R_f)]}{e_{no}^2/[R_L \parallel (R_1 + R_f)]}$$

(e) Calculate the noise figure.

$$NF\ dB = 10 \log \left(\frac{S_i/N_i}{S_o/N_o}\right)$$

Your value will normally be between 7 and 15 dB.

APPENDIX

A

DIFFERENTIAL AMPLIFIERS

A-1 DIFFERENTIAL AMPLIFIER REVIEW

The first stage of virtually all op-amps is a differential amplifier, so a brief look at the properties of differential amplifiers is useful. The differential amplifier should amplify the difference between two input signals [$(V_2 - V_1)$ in Figure A.1], but not amplify when the two inputs are the same, $V_2 = V_1$.

Note that R_E determines the emitter current. When $R_{L_1} = R_{L_2}$, $\beta_{Q_1} = \beta_{Q_2}$ and the V_{BE} of Q_1 and Q_2 are the same, the current flowing through each transistor is half of that flowing through R_E.

Suppose the voltage on the base of $Q_1(V_1)$ goes positive as the voltage on the base of $Q_2(V_2)$ goes negative. The forward bias on Q_1 will be increased and I_{C_1} will

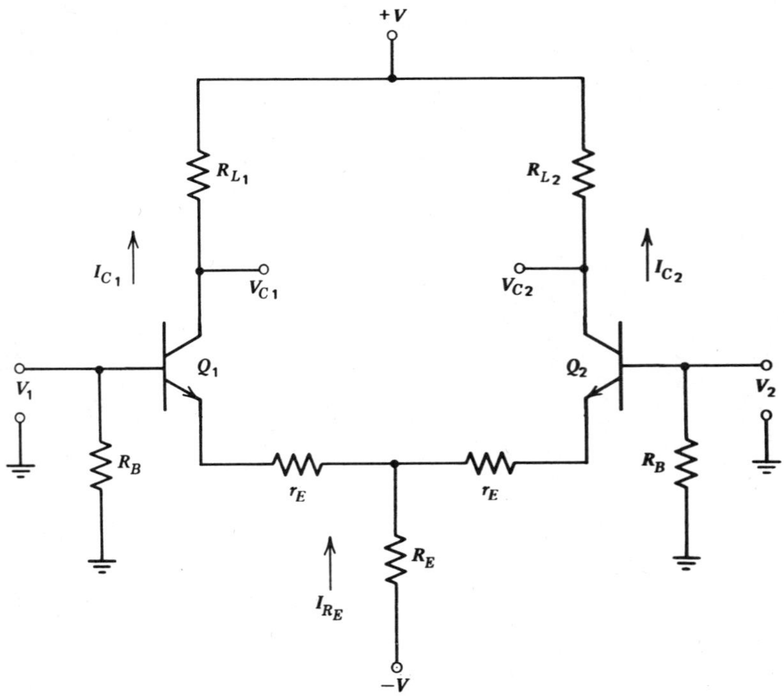

FIGURE A.1 Differential amplifier.

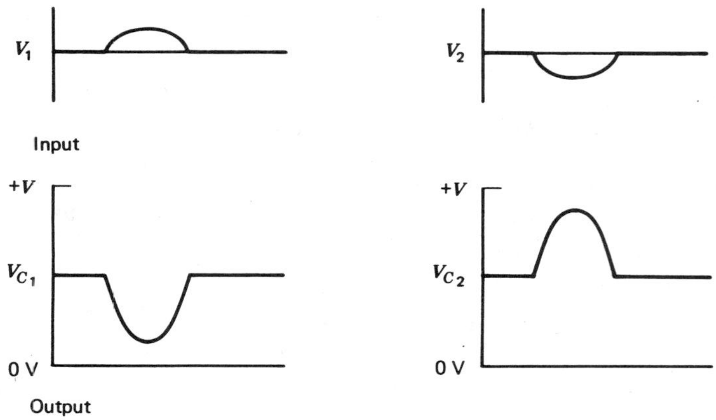

FIGURE A.2 Input and output waveforms, V_1 out of phase with V_2.

increase causing the voltage on the collector of Q_1 to go negative from its quiescent value. Similarly, Q_2 sees a decrease in its forward bias causing its collector voltage to rise above the quiescent value as illustrated in Figure A.2.

If the output voltage is taken between the collectors of Q_1 and Q_2, the output $V_{C_2} - V_{C_1}$ will be proportional to the difference in V_2 and V_1, but opposite in phase. Suppose V_1 and V_2 both go positive, but V_1 goes more positive than V_2. Then both Q_1 and Q_2 have increased forward bias so I_{C_1} and I_{C_2} both increase, but I_{C_1} will increase more than I_{C_2}. This causes V_{C_1} to go more negative than V_{C_2}. The output voltage is the difference in V_{C_1} and V_{C_2} as illustrated in Figure A.3.

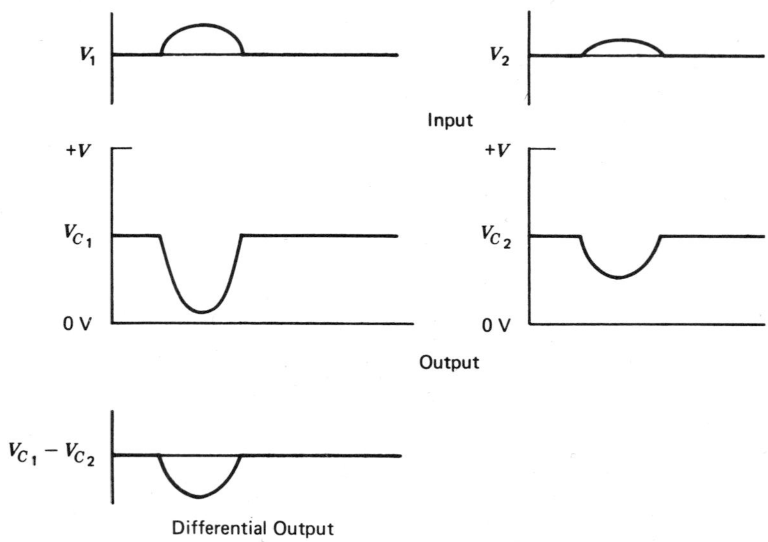

FIGURE A.3 Input and output waveforms, V_1 in phase with V_2.

If the same voltage is applied to Q_1 and Q_2 simultaneously, the output voltage is zero. For example, if V_1 and V_2 both go negative by the same amount, the forward bias on Q_1 and Q_2 will decrease by the same amount. This causes the voltage on the collectors of Q_1 and Q_2 to go positive, but since they both go positive by the same amount the output voltage ($V_{C_2} - V_{C_1}$) is zero.

The preceding remarks apply to an ideal amplifier. In reality there is always some small output since the beta and V_{BE} of the transistors and the collector resistors of the amplifier cannot be matched perfectly. In a good differential amplifier the output voltage is small whenever the input voltages are the same.

Since the differential amplifier is essentially an emitter bias stage with two supplies, the emitter current is given by

$$I_E = \frac{|-V| - V_{BE}}{R_E + \dfrac{R_B}{2(\beta + 1)}} \tag{A-1}$$

$R_B/2$ is used in Equation A-1 since two base resistors are present instead of one. Of course the collector current of each transistor is one-half the emitter current. The voltage gain of the circuit is developed as follows. Referring to Figure A.4, a signal into the base of Q_1 is coupled to the emitter of Q_2 as though Q_1 were an emitter follower. Note that $(r_{E_2} + r_{D_2})\|R_E \cong (r_{E_2} + r_{D_2})$ since R_E is usually much larger than r_E or r_D.

The emitter resistance of Q_1 is then approximately $r_{D_1} + r_{E_1} + r_{E_2} + r_{D_2}$. Since the voltage is applied to the amplifying circuit of Q_2 between r_{E_1} and r_{E_2} the voltage into Q_2 is ½ V_1. Q_2 is essentially a common-base amplifier with r_{E_2} as an emitter swamping resistor, having a gain of $A_V = R_L/(r_E + r_D)$, therefore $V_{C_2} = ½(V_1 - V_2)\, R_L/(r_{E_2} + r_{D_2})$ whenever we set $V_2 = 0$ in the preceding example (Figure A.4). A similar analysis can be applied to a voltage on the base of Q_2 to find V_{C_1}.

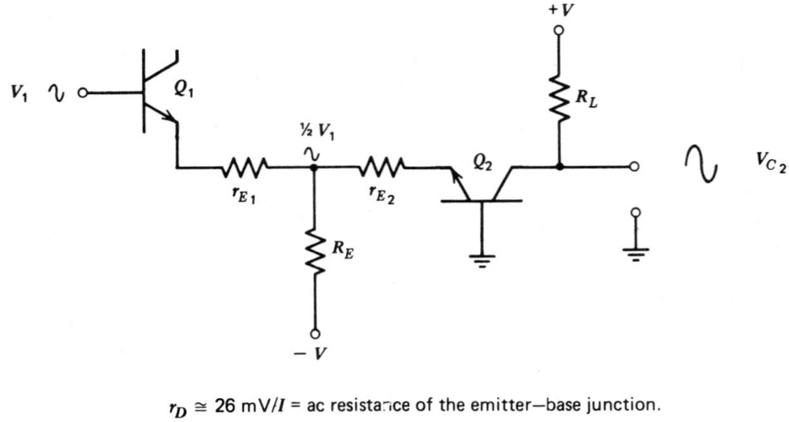

FIGURE A.4 Equivalent circuit for gain analysis.

Since $r_{E_1} = r_{E_2}$ and $r_{D_1} \cong r_{D_2}$ in most cases, we will just write r_E and r_D henceforth. The output voltage is $V_{C_2} - V_{C_1}$, so we can write

$$V_{C2} - V_{C1} = \frac{1}{2}(V_1 - V_2)\frac{R_L}{r_D + r_E} - \frac{1}{2}(V_2 - V_1)\frac{R_L}{r_D + r_E}$$

Thus

$$\begin{aligned} V_{C2} - V_{C1} &= \frac{1}{2}\frac{R_L}{r_D + r_E}(V_1 - V_2 - V_2 + V_1) \\ &= \frac{1}{2}\frac{R_L}{r_D + r_E}[2(V_1 - V_2)] \\ &= \frac{R_L}{r_D + r_E}(V_1 - V_2) \end{aligned}$$

If r_E is omitted,

$$V_{C2} - V_{C1} = \frac{R_L}{r_D}(V_1 - V_2) \tag{A-2}$$

If the output is taken from either collector to ground, we are using the amplifier single-ended. This results in one half the gain that we would have if we take the output from V_{C_1} to V_{C_2}. However, by choosing the collector we use for an output of a single-ended stage, we can have an output that is either in phase or out of phase with the input.

The input resistance of the differential amplifier can be increased and the input current reduced by the use of field-effect transistors (JFET or depletion MOSFET usually) or Darlington pairs for Q_1 and Q_2. When FETs are used, the voltage gain will be lower than that of a bipolar transistor.

By similar analyses the balanced output voltage gain of a JFET differential amplifier is

$$A_{VD} = g_m R_D$$

where

g_m = transconductance
R_D = drain load resistance

and the single-ended gain $A_{V(SE)}$ is

$$A_{V(SE)} = \frac{g_m R_D}{2}$$

if no source resistance is used other than for biasing.

A-2 COMMON-MODE GAIN

If a voltage of the same magnitude and phase is applied to both inputs of a differential amplifier, the applied voltage is called a common-mode voltage. If the resistors and transistors of a differential amplifier are not matched, there will be a difference in the collector voltages and some differential output voltage ($V_{C_1} - V_{C_2}$) will result. This is minimized (but never completely eliminated) by carefully matching all components of the amplifier.

We noted earlier that if a common-mode voltage is applied to a perfectly balanced differential amplifier, the collector voltages shift in level. If the input stage (or a subsequent stage) is used single-ended (the output taken from one collector to ground), a change in the output voltage can be expected. Even with a balanced output (from the collector of Q_1 to the collector of Q_2), a shift in the level of the collectors may unbalance a later stage. The change in the collector voltages with a common-mode input is called common-mode gain and should be as small as possible.

If we redraw Figure A.1, as in Figure A.5, we see that the source of common-mode gain as the differential amplifier acts like a common emitter stage with two paralleled transistors to a common-mode input. The circuit in Figure A.5*b* has a common-mode gain of

$$A_{cm} = \frac{R_{L/2}}{R_E + r_{E/2} + r_{D/2}} \tag{A-3}$$

Thus we can see that the higher we can make R_E, the lower the common-mode gain. Ordinarily, $R_L \cong R_E$ and $A_{cm} \cong ½$. A common method to reduce A_{cm} is to replace R_E with a constant current generator. A constant current generator has a very high ac resistance, so its use in place of R_E causes the common-mode gain to be very low. This is illustrated in Figure A.6. The common-mode signal sensitivity of a differential amplifier is often expressed as the common-mode rejection ratio (CMRR). CMRR is defined as the differential gain divided by the common-mode gain, CMRR = A/A_{cm}. The higher this ratio, the better the amplifier performs.

The CMRR of an FET differential amplifier is usually lower than that of a bipolar transistor differential amplifier.

A-3 OFFSET VOLTAGE

If both inputs of the differential amplifier are set at zero volts, both collector voltages are not equal. This is due primarily to the fact that the emitter-to-base voltage drops of the transistors are not equal. This V_{BE} difference is defined as the input offset voltage (V_{os}).

$$V_{os} = |V_{BE_1} - V_{BE_2}| \tag{A-4}$$

The offset voltage acts exactly like a differential input signal to the amplifier, causing an output equal to $A_v V_{os}$. If the collector resistors are not matched perfectly, they will contribute to the difference in the collector voltages with zero input

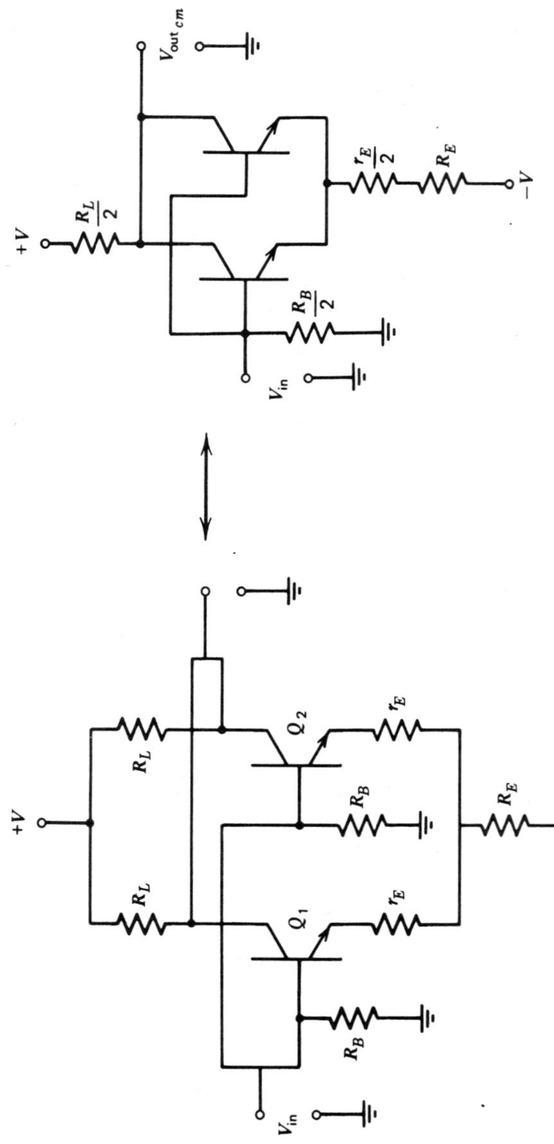

(a) Common–mode circuit

(b) Equivalent common–mode circuit

FIGURE A.5 Common-mode circuits.

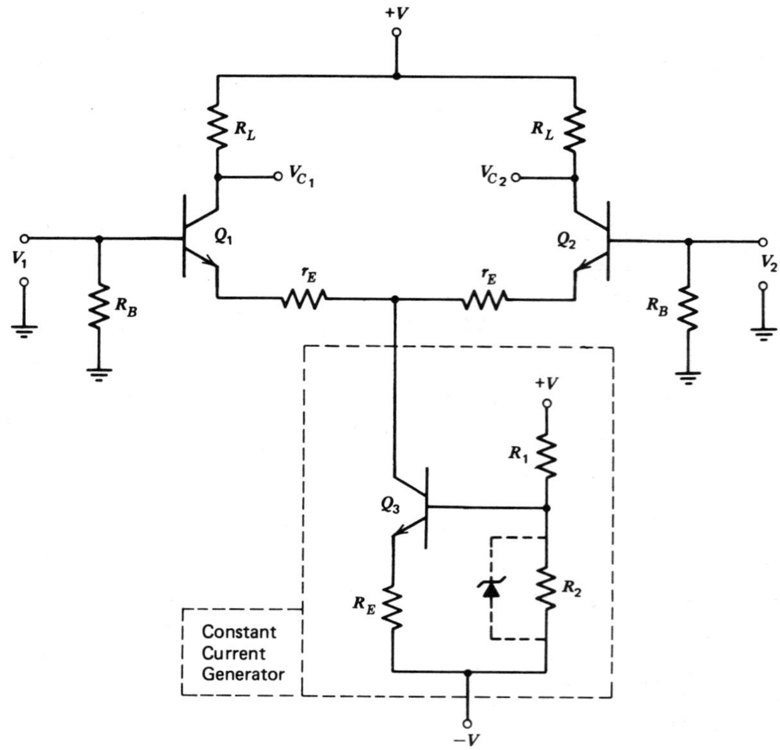

$$I_E = \frac{V_{R_2} - V_{BE_{Q3}}}{R_E + \dfrac{R_1 \| R_2}{(\beta_{Q3} + 1)}}$$

Note: R_2 can be replaced with a zener diode in which case

$$I = \frac{V_Z - V_{BE_{Q3}}}{R_E}$$

FIGURE A.6 Use of a constant current generator to reduce A_{cm}.

voltages. Another factor that can cause $V_{C_1} \neq V_{C_2}$ with the input voltage to both transistors equal to zero volts is beta mismatch. If the betas of the transistors are not equal, the collector currents of the transistors are not equal. These effects must be compensated for to obtain good performance. Compensating techniques are discussed in Chapters 1 and 2.

APPENDIX

B

FAIRCHILD μA741 OPERATION

Figure B.1 shows the very popular Fairchild μA741 operational amplifier. This appendix describes its operation.

Diode connected Q_8 biases constant current source Q_9. Q_8 and Q_9 are matched transistors so that the collector currents of Q_8 and Q_9 are identical if their emitter-base voltages are equal. Therefore, the current through Q_8 sets its voltage drop, and the voltage drop across Q_8 sets the collector current of Q_9 equal to the current through Q_8. Q_9 is called the *current mirror* of Q_8.

Diode connected Q_{12} biases constant current source Q_{13}, and diode connected Q_{11} biases constant current source Q_{10}. The emitter resistor, R_4, reduces the current of current source Q_{10} to about the same value as that of current source Q_9.

Transistors Q_1 through Q_7 and resistors R_1, R_2, and R_3 make up the input differential amplifier. Input transistors Q_1 and Q_2 are emitter followers that feed common-base transistors Q_3 and Q_4. The bases of Q_3 and Q_4 are biased by constant

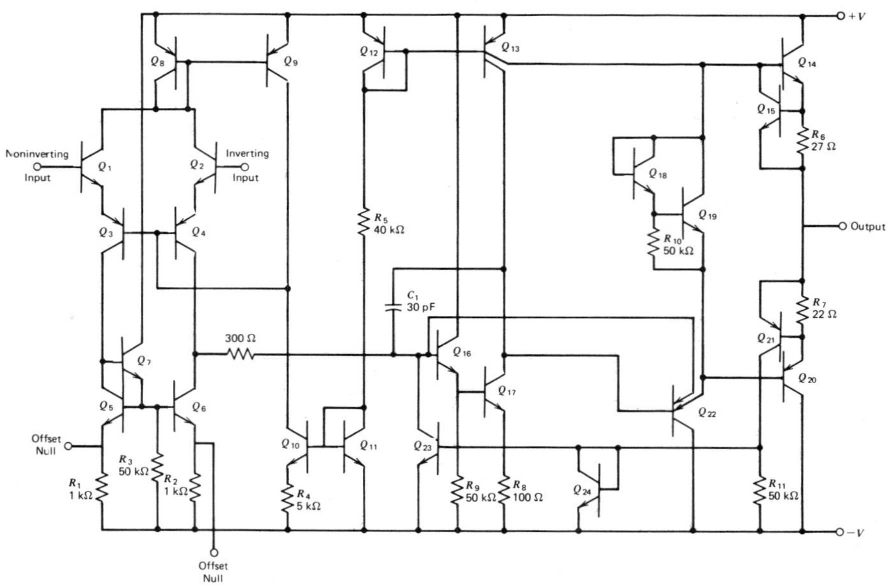

FIGURE B.1 Fairchild μA 741 operational amplifier. (*Reprinted courtesy of Fairchild Camera and Instrument Corporation.*)

current sources Q_9 and Q_{10}. The high impedance of the constant current sources keeps the input resistance of Q_3 and Q_4 high. Q_5 and Q_6 are constant current sources that set the collector currents of Q_1, Q_3, Q_2, and Q_4. Constant current sources Q_5 and Q_6 also provide high resistance loads to Q_3 and Q_4 so that the voltage gain of the common-base transistors is high. Q_7 provides the dc base voltage for Q_5 and Q_6. However, for an ac input signal, Q_7 is an emitter follower that feeds the collector signal of Q_3 to the base of Q_6 in phase. Q_6 inverts and amplifies the signal. The collector of Q_6 swings negative when the collector of Q_4 swings negative and vice versa. Thus the collector voltage change of Q_6 is added to that of Q_4 providing a larger single-ended voltage gain than the differential amplifier would otherwise be capable of (about 1000).

The offset null connections are connected one to each end of a 10 kΩ potentiometer whose wiper goes to $-V$. As the wiper is moved, the emitter resistances of Q_5 and Q_6 vary, so their collector currents vary. The connection of the potentiometer is shown in the laboratory of Chapter 1 and in the 741 specifications in Appendix C. If the collector current of Q_1 and Q_2 is varied, their emitter-base voltage varies, so input offset voltage can be nulled out.

If the IC temperature rises, the V_{BE} of Q_1 and Q_2 will decrease. This will cause their collector currents to increase and the voltage drop across Q_8 to increase. An increase in the voltage drop of Q_8 will increase the collector current of Q_9, decreasing the base current of Q_3 and Q_4. The base current decrease of Q_3 and Q_4 will decrease their collector current. Since Q_3 and Q_4 are in series with Q_1 and Q_2, the collector current of Q_1 and Q_2 will decrease also. In this fashion the differential amplifier is temperature stabilized.

Q_{16} is an emitter follower driving common emitter amplifier Q_{17}. Q_{16} reduces the loading on the output of the differential amplifier. The voltage gain of Q_{17} is very high because constant current source Q_{13} acts as a very high resistance load resistor. Q_{22} is an emitter follower that provides current gain to drive the complementary output stage, Q_{14} and Q_{20}, and also reduces loading on the collector of Q_{17}. The upper emitter-base diode of Q_{22} is reverse biased under normal operation. If the collector of Q_{17} goes excessively negative from Q_{16} being overdriven positive, the upper emitter of Q_{22} clamps the collector of Q_{17}. This prevents Q_{17} from saturating and improves the recovery time from excessive input voltage to the op-amp. The constant current source provided by the upper collector of Q_{13} provides bias current to Q_{22} and keeps its gain at almost 1.

Diode connected Q_{18}, Q_{19}, and R_{10} provide a dc level shifter. Q_{18} and R_{10} bias Q_{19} so that its $V_{CE} \cong 1.2$ V. The function of Q_{19} is to bias Q_{14} and Q_{20} in class AB operation to prevent *crossover distortion*, which is explained in the next paragraph.

Q_{14} and Q_{20} are a complementary connected *npn-pnp* pair. They are the output stage. The emitter follower connection provides the amplifier with a low output resistance. If the collector voltage of Q_{17} goes positive, Q_{14} turns on, and if the collector voltage of Q_{17} goes negative, Q_{20} turns on. Without Q_{19} the collector voltage of Q_{17} would have to go about 0.6 V positive before Q_{14} turned on and about 0.6 V negative before Q_{20} turned on, resulting in crossover distortion on the

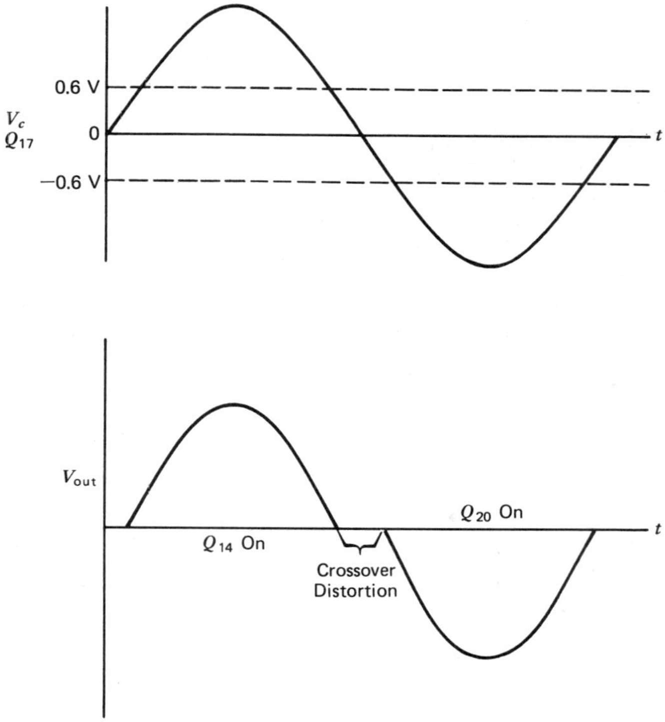

FIGURE B.2 Crossover distortion.

output as shown in Figure B.2. The $V_{CE} = 1.2$ V of Q_{19} holds Q_{14} and Q_{20} barely on, so they respond even to very small changes in the emitter voltage of Q_{22}.

Q_{15} is a current-limiting transistor for Q_{14}. When the emitter current of Q_{15} is high enough (about 20 mA) so that the voltage across R_9 is equal to the V_{BE} of Q_{15}, Q_{15} turns on and removes base drive from Q_{14}, causing its current to remain constant. The emitter current of Q_{14} can never rise above

$$I_E(Q_{14})\text{max} = \frac{V_{BE(Q15)}}{R_9}$$

Q_{21} provides current limiting for Q_{20}. When the emitter current of Q_{21} is more than about 20 mA, Q_{21} turns on causing Q_{23} to conduct, which removes base drive from Q_{16}. Q_{16} then causes the collector of Q_{17}, thus the emitter of Q_{22}, to go positive. This reduces the base drive of Q_{20}, limiting its emitter current.

The 30 pF capacitor keeps the op-amp from oscillating. This is explained in Chapter 4.

APPENDIX

C

INTEGRATED CIRCUIT AND OPERATIONAL AMPLIFIER SPECIFICATIONS

- □ Fairchild μA741
- □ Texas Instruments TL080/081/082/084
- □ Fairchild μA791
- □ National Semiconductor LH0036
- □ National Semiconductor LM311
- □ Motorola MC3401
- □ National Semiconductor LM105/205/305A
- □ National Semiconductor LM309
- □ Fairchild μA723
- □ Fairchild μA78S40
- □ Motorola MC3420/3520
- □ National Semiconductor AH0120/130/140/150/160
- □ National Semiconductor LM555

The author gratefully acknowledges Fairchild Camera and Instrument Corporation, Texas Instruments Incorporated, National Semiconductor Corporation, and Motorola Semiconductor Products Incorporated for their permission to reprint the following specifications.

μA741

FREQUENCY-COMPENSATED OPERATIONAL AMPLIFIER

FAIRCHILD LINEAR INTEGRATED CIRCUITS

GENERAL DESCRIPTION – The μA741 is a high performance monolithic Operational Amplifier constructed using the Fairchild Planar* epitaxial process. It is intended for a wide range of analog applications. High common mode voltage range and absence of latch-up tendencies make the μA741 ideal for use as a voltage follower. The high gain and wide range of operating voltage provides superior performance in integrator, summing amplifier, and general feedback applications.

- **NO FREQUENCY COMPENSATION REQUIRED**
- **SHORT CIRCUIT PROTECTION**
- **OFFSET VOLTAGE NULL CAPABILITY**
- **LARGE COMMON MODE AND DIFFERENTIAL VOLTAGE RANGES**
- **LOW POWER CONSUMPTION**
- **NO LATCH-UP**

ABSOLUTE MAXIMUM RATINGS

Supply Voltage	
μA741A, μA741, μA741E	±22 V
μA741C	±18 V
Internal Power Dissipation (Note 1)	
Metal Can	500 mW
Molded and Hermetic DIP	670 mW
Mini DIP	310 mW
Flatpak	570 mW
Differential Input Voltage	±30 V
Input Voltage (Note 2)	±15 V
Storage Temperature Range	
Metal Can, Hermetic DIP, and Flatpak	−65°C to +150°C
Mini DIP, Molded DIP	−55°C to +125°C
Operating Temperature Range	
Military (μA741A, μA741)	−55°C to +125°C
Commercial (μA741E, μA741C)	0°C to +70°C
Pin Temperature (Soldering)	
Metal Can, Hermetic DIPs, and Flatpak (60 s)	300°C
Molded DIPs (10 s)	260°C
Output Short Circuit Duration (Note 3)	Indefinite

CONNECTION DIAGRAMS

8-PIN METAL CAN
(TOP VIEW)
PACKAGE OUTLINE 5B
PACKAGE CODE H

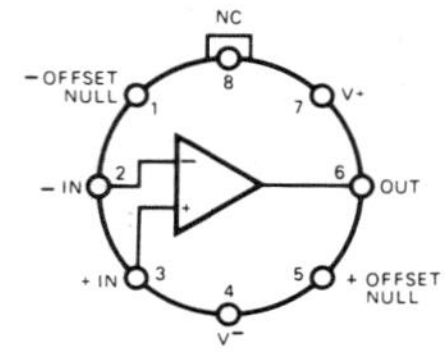

Note: Pin 4 connected to case

ORDER INFORMATION

TYPE	PART NO.
μA741A	μA741AHM
μA741	μA741HM
μA741E	μA741EHC
μA741C	μA741HC

14-PIN DIP
(TOP VIEW)
PACKAGE OUTLINES 6A, 9A
PACKAGE CODES D P

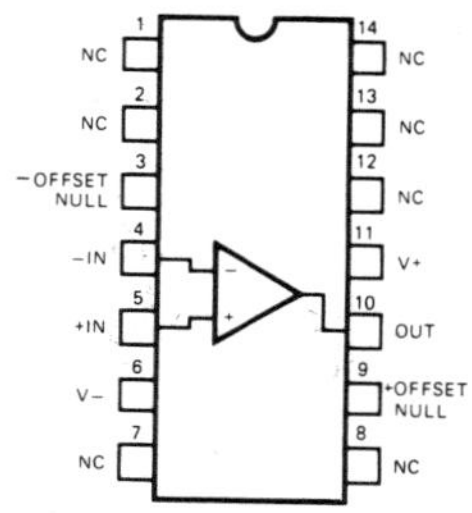

ORDER INFORMATION

TYPE	PART NO.
μA741A	μA741ADM
μA741	μA741DM
μA741E	μA741EDC
μA741C	μA741DC
μA741C	μA741PC

8-PIN MINI DIP
(TOP VIEW)
PACKAGE OUTLINES 6T 9T
PACKAGE CODES R T

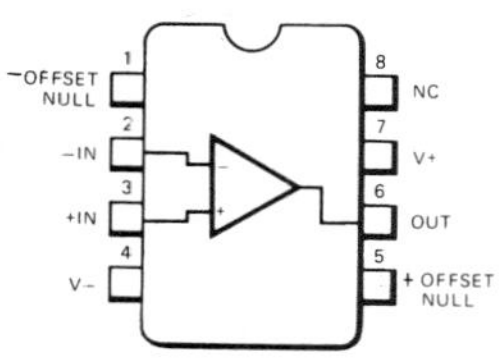

ORDER INFORMATION

TYPE	PART NO.
μA741C	μA741TC
μA741C	μA741RC

10-PIN FLATPAK
(TOP VIEW)
PACKAGE OUTLINE 3F
PACKAGE CODE F

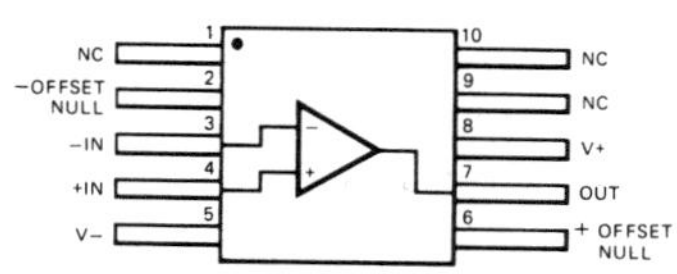

ORDER INFORMATION

TYPE	PART NO.
μA741A	μA741AFM
μA741	μA741FM

*Planar is a patented Fairchild process.

Courtesy of Fairchild Camera and Instruments Corporation.

μA741

ELECTRICAL CHARACTERISTICS: $V_S = \pm 15$ V, $T_A = 25°C$ unless otherwise specified.

CHARACTERISTICS (see definitions)		CONDITIONS	MIN	TYP	MAX	UNITS
Input Offset Voltage		$R_S \leq 10$ kΩ		1.0	5.0	mV
Input Offset Current				20	200	nA
Input Bias Current				80	500	nA
Input Resistance			0.3	2.0		MΩ
Input Capacitance				1.4		pF
Offset Voltage Adjustment Range				±15		mV
Large Signal Voltage Gain		$R_L \geq 2$ kΩ, $V_{OUT} = \pm 10$ V	50,000	200,000		
Output Resistance				75		Ω
Output Short Circuit Current				25		mA
Supply Current				1.7	2.8	mA
Power Consumption				50	85	mW
Transient Response (Unity Gain)	Rise time	$V_{IN} = 20$ mV, $R_L = 2$ kΩ, $C_L \leq 100$ pF		0.3		μs
	Overshoot			5.0		%
Slew Rate		$R_L \geq 2$ kΩ		0.5		V/μs

The following specifications apply for $-55°C \leq T_A \leq +125°C$:

CHARACTERISTICS	CONDITIONS	MIN	TYP	MAX	UNITS
Input Offset Voltage	$R_S \leq 10$ kΩ		1.0	6.0	mV
Input Offset Current	$T_A = +125°C$		7.0	200	nA
	$T_A = -55°C$		85	500	nA
Input Bias Current	$T_A = +125°C$		0.03	0.5	μA
	$T_A = -55°C$		0.3	1.5	μA
Input Voltage Range		±12	±13		V
Common Mode Rejection Ratio	$R_S \leq 10$ kΩ	70	90		dB
Supply Voltage Rejection Ratio	$R_S \leq 10$ kΩ		30	150	μV/V
Large Signal Voltage Gain	$R_L \geq 2$ kΩ, $V_{OUT} = \pm 10$ V	25,000			
Output Voltage Swing	$R_L \geq 10$ kΩ	±12	±14		V
	$R_L \geq 2$ kΩ	±10	±13		V
Supply Current	$T_A = +125°C$		1.5	2.5	mA
	$T_A = -55°C$		2.0	3.3	mA
Power Consumption	$T_A = +125°C$		45	75	mW
	$T_A = -55°C$		60	100	mW

TYPICAL PERFORMANCE CURVES FOR μA741A AND μA741

OPEN LOOP VOLTAGE GAIN AS A FUNCTION OF SUPPLY VOLTAGE

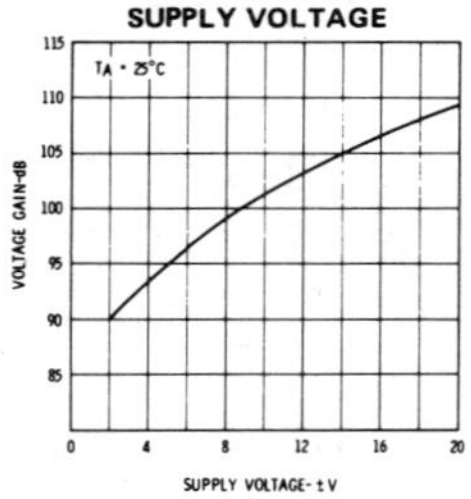

OUTPUT VOLTAGE SWING AS A FUNCTION OF SUPPLY VOLTAGE

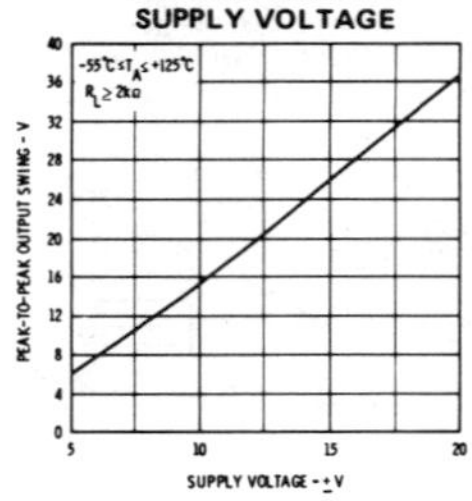

INPUT COMMON MODE VOLTAGE RANGE AS A FUNCTION OF SUPPLY VOLTAGE

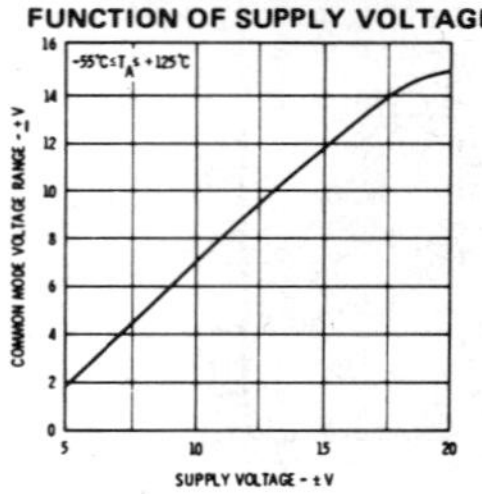

μA741E

ELECTRICAL CHARACTERISTICS: $V_S = \pm 15$ V, $T_A = 25°C$ unless otherwise specified.

CHARACTERISTICS (see definitions)		CONDITIONS	MIN	TYP	MAX	UNITS
Input Offset Voltage		$R_S < 50\Omega$		0.8	3.0	mV
Average Input Offset Voltage Drift					15	μV/°C
Input Offset Current				3.0	30	nA
Average Input Offset Current Drift					0.5	nA/°C
Input Bias Current				30	80	nA
Power Supply Rejection Ratio		$V_S = +10, -20$; $V_S = +20, -10$V, $R_S = 50\Omega$		15	50	μV/V
Output Short Circuit Current			10	25	40	mA
Power Dissipation		$V_S = \pm 20$V		80	150	mW
Input Impedance		$V_S = \pm 20$V	1.0	6.0		MΩ
Large Signal Voltage Gain		$V_S = \pm 20$V, $R_L = 2k\Omega$, $V_{OUT} = \pm 15$V	50			V/mV
Transient Response	Rise Time			0.25	0.8	μs
(Unity Gain)	Overshoot			6.0	20	%
Bandwidth (Note 4)			.437	1.5		MHz
Slew Rate (Unity Gain)		$V_{IN} = \pm 10$V	0.3	0.7		V/μs
The following specifications apply for $0°C < T_A < 70°C$						
Input Offset Voltage					4.0	mV
Input Offset Current					70	nA
Input Bias Current					210	nA
Common Mode Rejection Ratio		$V_S = \pm 20$V, $V_{IN} = \pm 15$V, $R_S = 50\Omega$	80	95		dB
Adjustment For Input Offset Voltage		$V_S = \pm 20$V	10			mV
Output Short Circuit Current			10		40	mA
Power Dissipation		$V_S = \pm 20$V			150	mW
Input Impedance		$V_S = \pm 20$V	0.5			MΩ
Output Voltage Swing		$V_S = \pm 20$V, $R_L = 10k\Omega$	±16			V
		$V_S = \pm 20$V, $R_L = 2k\Omega$	±15			V
Large Signal Voltage Gain		$V_S = \pm 20$V, $R_L = 2k\Omega$, $V_{OUT} = \pm 15$V	32			V/mV
		$V_S = \pm 5$V, $R_L = 2k\Omega$, $V_{OUT} = \pm 2$ V	10			V/mV

EQUIVALENT CIRCUIT

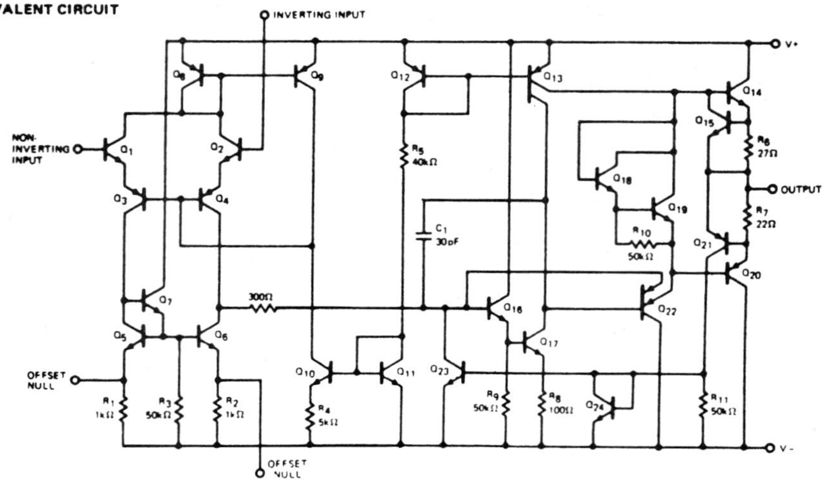

μA741C

ELECTRICAL CHARACTERISTICS: $V_S = \pm 15$ V, $T_A = 25°C$ unless otherwise specified.

CHARACTERISTICS (see definitions)		CONDITIONS	MIN	TYP	MAX	UNITS
Input Offset Voltage		$R_S \leq 10$ kΩ		2.0	6.0	mV
Input Offset Current				20	200	nA
Input Bias Current				80	500	nA
Input Resistance			0.3	2.0		MΩ
Input Capacitance				1.4		pF
Offset Voltage Adjustment Range				±15		mV
Input Voltage Range			±12	±13		V
Common Mode Rejection Ratio		$R_S \leq 10$ kΩ	70	90		dB
Supply Voltage Rejection Ratio		$R_S \leq 10$ kΩ		30	150	μV/V
Large Signal Voltage Gain		$R_L \geq 2$ kΩ, $V_{OUT} = \pm 10$ V	20,000	200,000		
Output Voltage Swing		$R_L \geq 10$ kΩ	±12	±14		V
		$R_L \geq 2$ kΩ	±10	±13		V
Output Resistance				75		Ω
Output Short Circuit Current				25		mA
Supply Current				1.7	2.8	mA
Power Consumption				50	85	mW
Transient Response (Unity Gain)	Rise time	$V_{IN} = 20$ mV, $R_L = 2$ kΩ, $C_L \leq 100$ pF		0.3		μs
	Overshoot			5.0		%
Slew Rate		$R_L \geq 2$ kΩ		0.5		V/μs
The following specifications apply for $0°C \leq T_A \leq +70°C$:						
Input Offset Voltage					7.5	mV
Input Offset Current					300	nA
Input Bias Current					800	nA
Large Signal Voltage Gain		$R_L \geq 2$ kΩ, $V_{OUT} = \pm 10$ V	15,000			
Output Voltage Swing		$R_L \geq 2$ kΩ	±10	±13		V

TYPICAL PERFORMANCE CURVES FOR μA741E AND μA741C

OPEN LOOP VOLTAGE GAIN AS A FUNCTION OF SUPPLY VOLTAGE

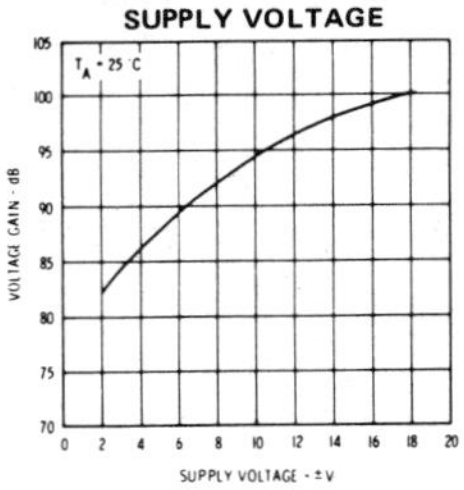

OUTPUT VOLTAGE SWING AS A FUNCTION OF SUPPLY VOLTAGE

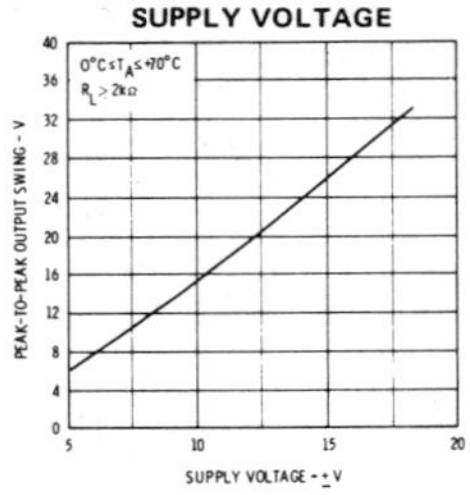

INPUT COMMON MODE VOLTAGE RANGE AS A FUNCTION OF SUPPLY VOLTAGE

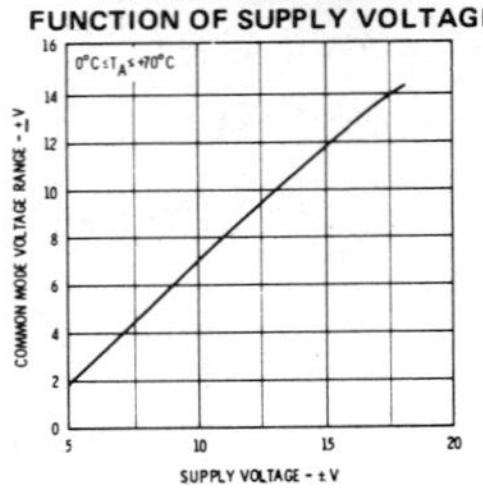

TYPICAL PERFORMANCE CURVES FOR μA741A, μA741, μA741E AND μA741C

POWER CONSUMPTION AS A FUNCTION OF SUPPLY VOLTAGE

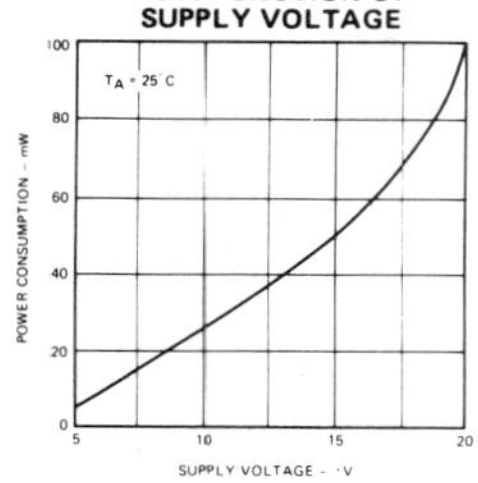

OPEN LOOP VOLTAGE GAIN AS A FUNCTION OF FREQUENCY

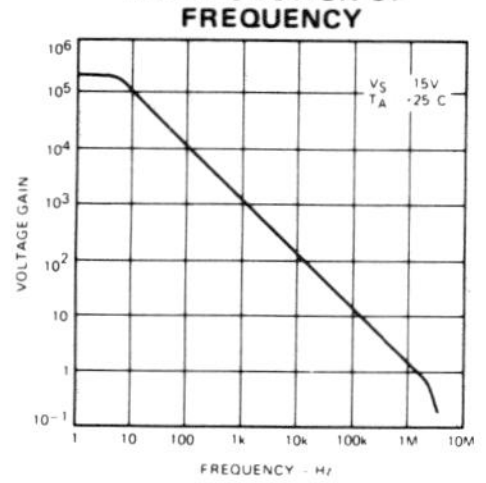

OPEN LOOP PHASE RESPONSE AS A FUNCTION OF FREQUENCY

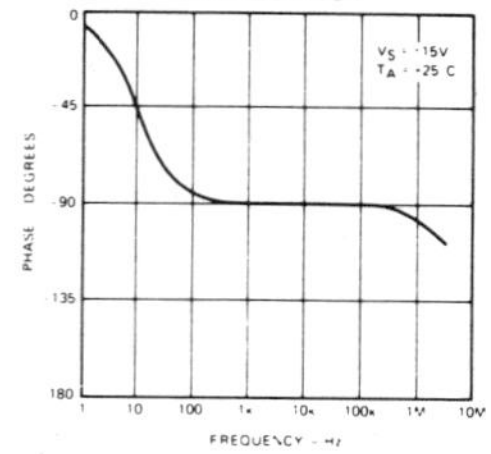

INPUT OFFSET CURRENT AS A FUNCTION OF SUPPLY VOLTAGE

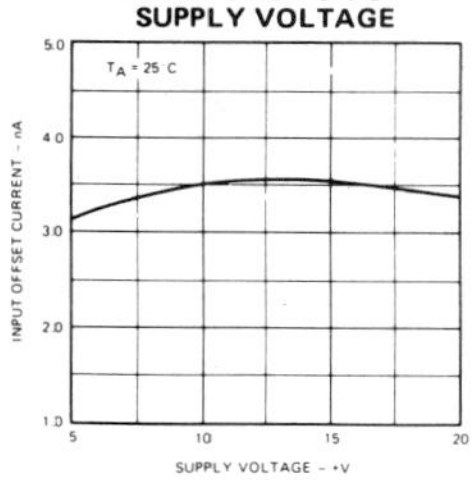

INPUT RESISTANCE AND INPUT CAPACITANCE AS A FUNCTION OF FREQUENCY

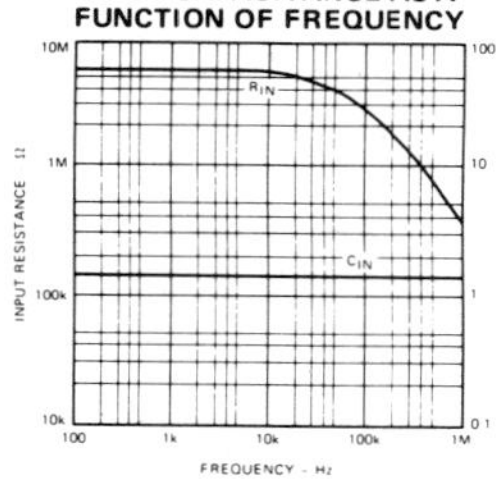

OUTPUT RESISTANCE AS A FUNCTION OF FREQUENCY

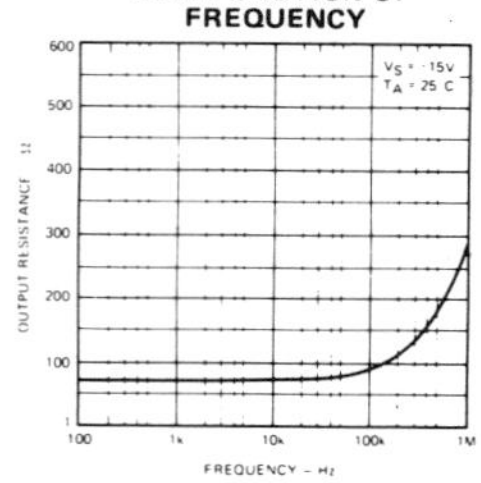

OUTPUT VOLTAGE SWING AS A FUNCTION OF LOAD RESISTANCE

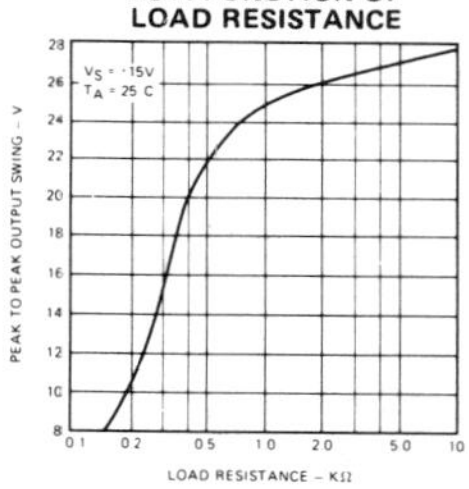

OUTPUT VOLTAGE SWING AS A FUNCTION OF FREQUENCY

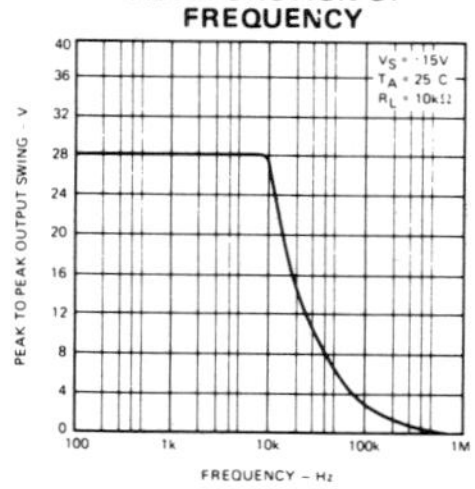

ABSOLUTE MAXIMUM POWER DISSIPATION AS A FUNCTION OF AMBIENT TEMPERATURE

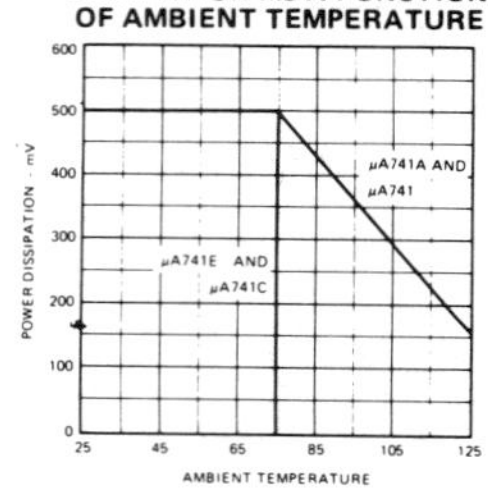

INPUT NOISE VOLTAGE AS A FUNCTION OF FREQUENCY

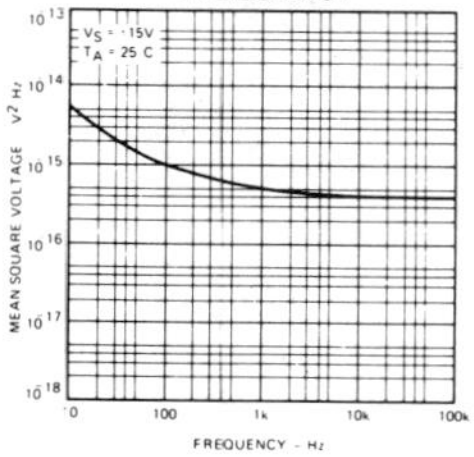

INPUT NOISE CURRENT AS A FUNCTION OF FREQUENCY

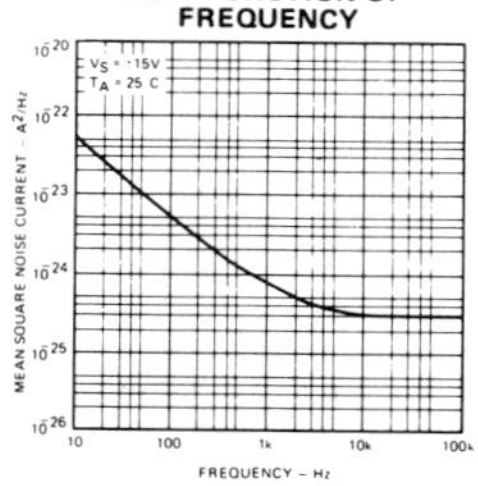

BROADBAND NOISE FOR VARIOUS BANDWIDTHS

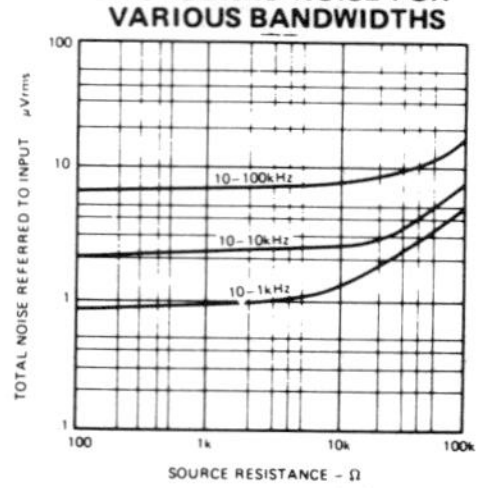

TYPICAL PERFORMANCE CURVES FOR μA741A AND μA741

INPUT BIAS CURRENT AS A FUNCTION OF AMBIENT TEMPERATURE

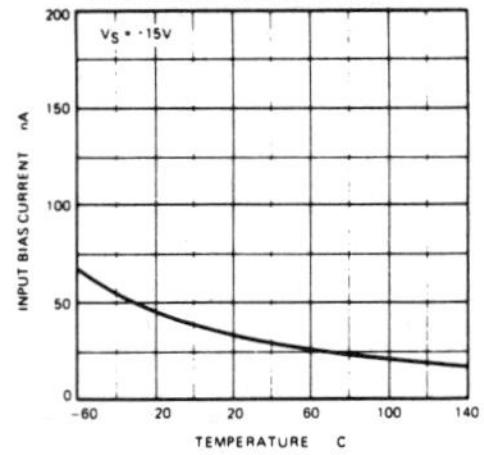

INPUT RESISTANCE AS A FUNCTION OF AMBIENT TEMPERATURE

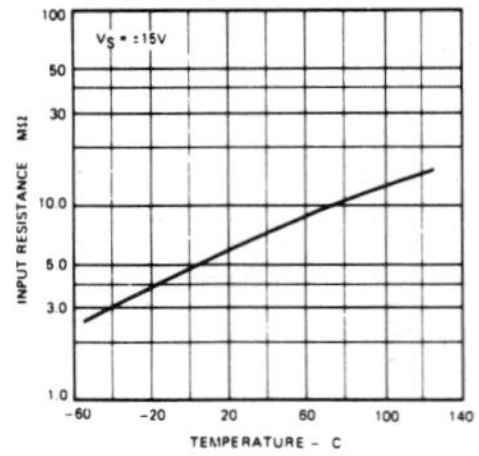

OUTPUT SHORT-CIRCUIT CURRENT AS A FUNCTION OF AMBIENT TEMPERATURE

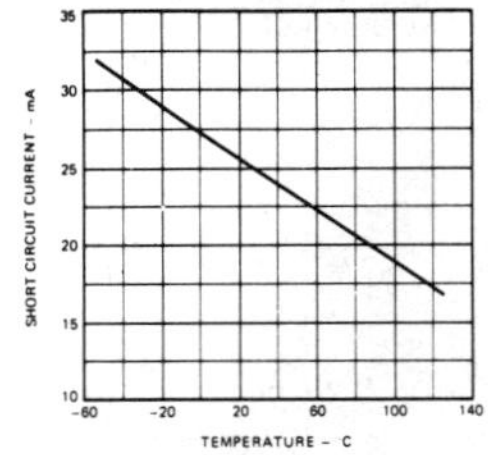

INPUT OFFSET CURRENT AS A FUNCTION OF AMBIENT TEMPERATURE

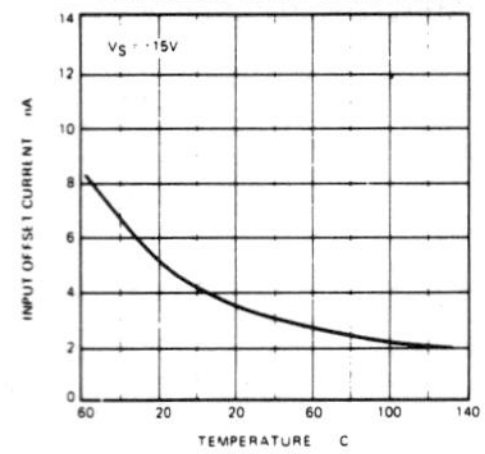

POWER CONSUMPTION AS A FUNCTION OF AMBIENT TEMPERATURE

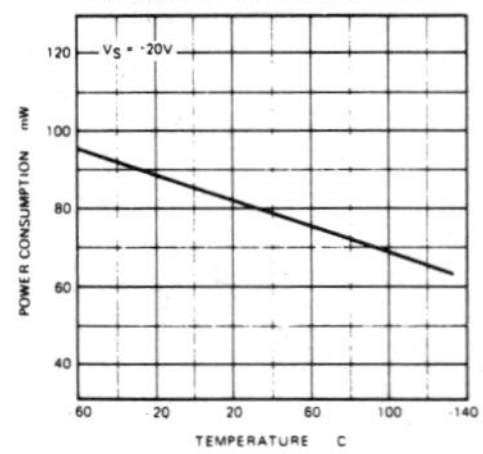

FREQUENCY CHARACTERISTICS AS A FUNCTION OF AMBIENT TEMPERATURE

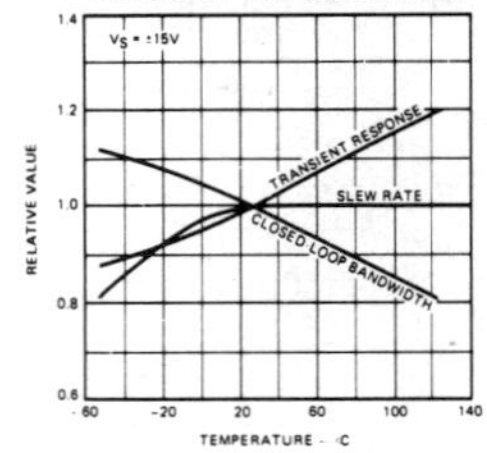

TYPICAL PERFORMANCE CURVES FOR μA741E AND μA741C

INPUT BIAS CURRENT AS A FUNCTION OF AMBIENT TEMPERATURE

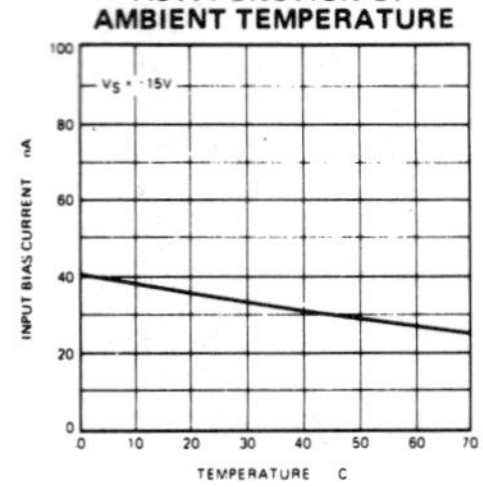

INPUT RESISTANCE AS A FUNCTION OF AMBIENT TEMPERATURE

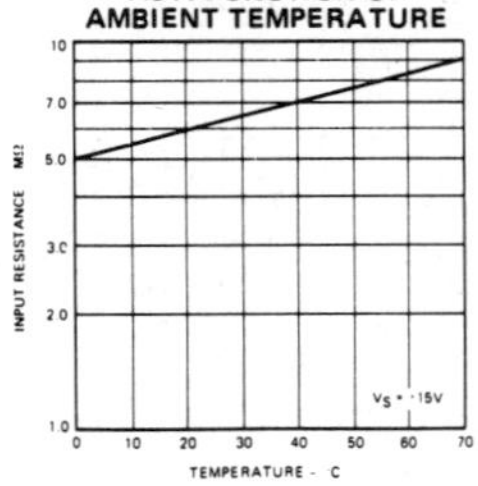

INPUT OFFSET CURRENT AS A FUNCTION OF AMBIENT TEMPERATURE

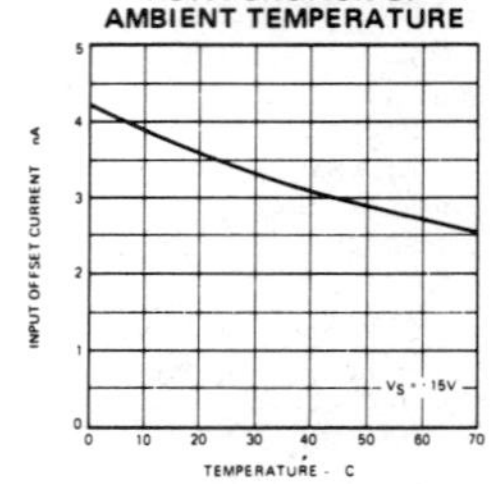

POWER CONSUMPTION AS A FUNCTION OF AMBIENT TEMPERATURE

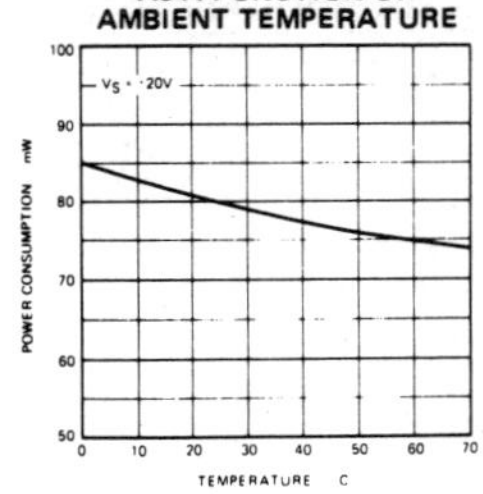

OUTPUT SHORT CIRCUIT CURRENT AS A FUNCTION OF AMBIENT TEMPERATURE

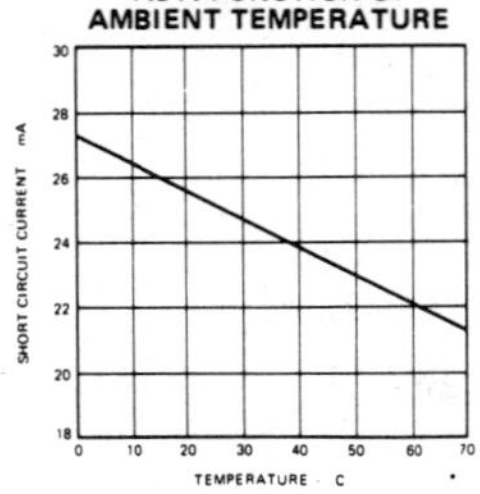

FREQUENCY CHARACTERISTICS AS A FUNCTION OF AMBIENT TEMPERATURE

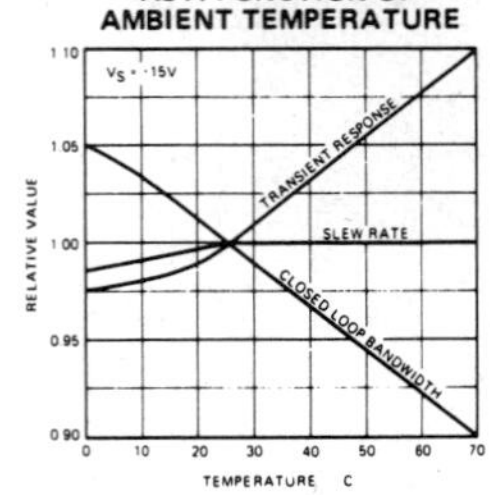

TRANSIENT RESPONSE

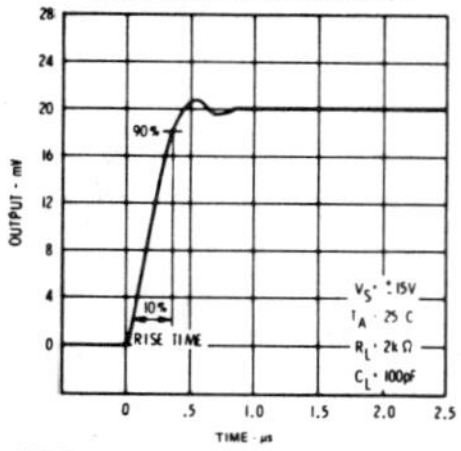

TRANSIENT RESPONSE TEST CIRCUIT

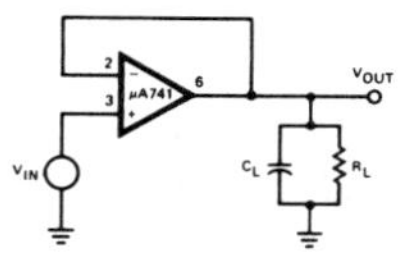

COMMON MODE REJECTION RATIO AS A FUNCTION OF FREQUENCY

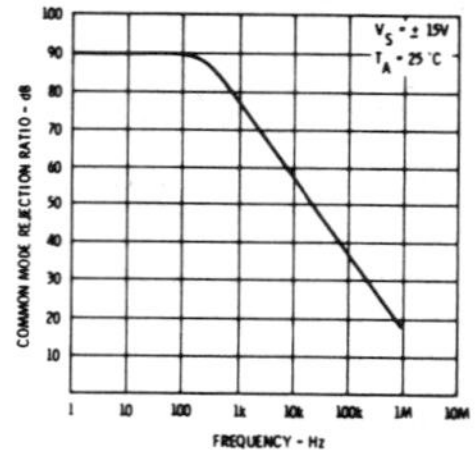

FREQUENCY CHARACTERISTICS AS A FUNCTION OF SUPPLY VOLTAGE

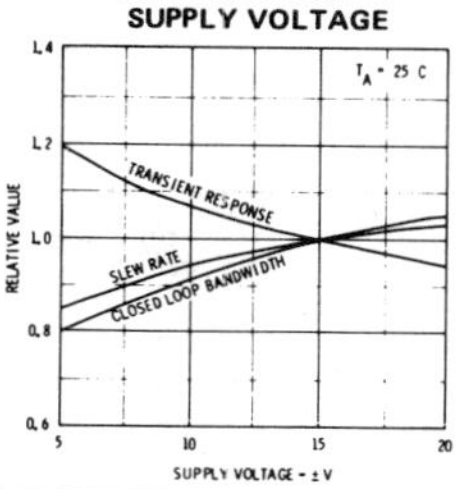

VOLTAGE OFFSET NULL CIRCUIT

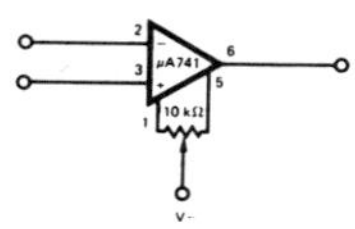

VOLTAGE FOLLOWER LARGE SIGNAL PULSE RESPONSE

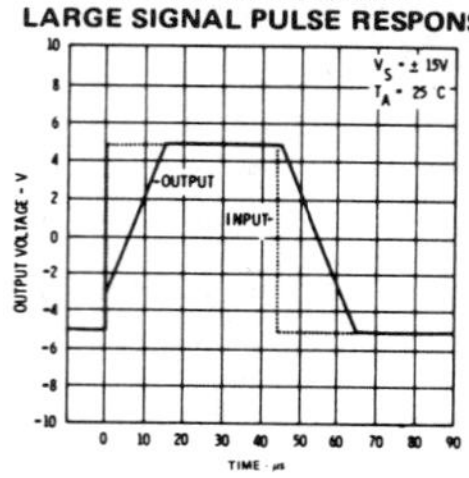

TYPICAL APPLICATIONS

UNITY-GAIN VOLTAGE FOLLOWER

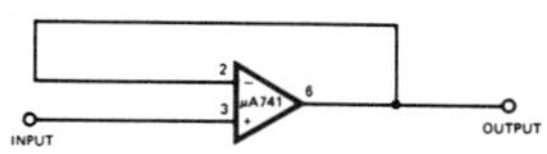

R_{IN} = 400 MΩ

C_{IN} = 1 pF

R_{OUT} << 1 Ω

B.W. = 1 MHz

NON-INVERTING AMPLIFIER

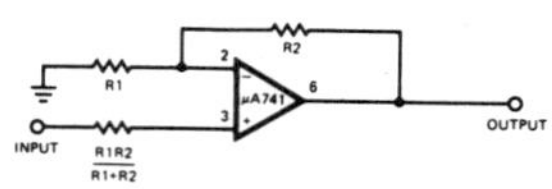

GAIN	R1	R2	B W	R_{IN}
10	1 kΩ	9 kΩ	100 kHz	400 MΩ
100	100 Ω	9.9 kΩ	10 kHz	280 MΩ
1000	100 Ω	99.9 kΩ	1 kHz	80 MΩ

INVERTING AMPLIFIER

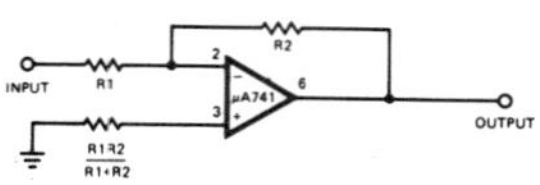

GAIN	R1	R2	B W	R_{IN}
1	10 kΩ	10 kΩ	1 MHz	10 kΩ
10	1 kΩ	10 kΩ	100 kHz	1 kΩ
100	1 kΩ	100 kΩ	10 kHz	1 kΩ
1000	100 Ω	100 kΩ	1 kHz	100 Ω

CLIPPING AMPLIFIER

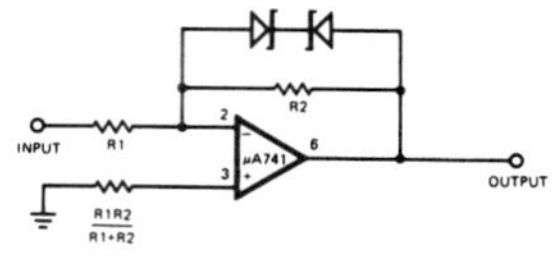

$$\frac{E_{OUT}}{E_{IN}} = \frac{R2}{R1} \text{ if } |E_{OUT}| \leq V_Z + 0.7\text{ V}$$

where V_Z = Zener breakdown voltage

TYPICAL APPLICATIONS (Cont'd)

SIMPLE INTEGRATOR

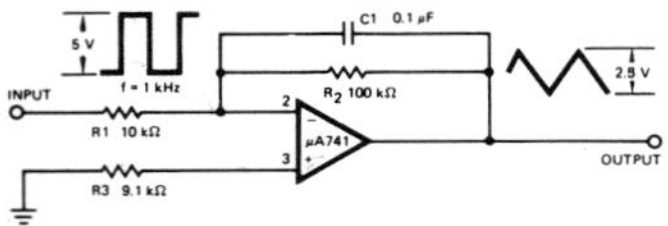

$$E_{OUT} = -\frac{1}{R_1C_1}\int E_{IN}dt$$

SIMPLE DIFFERENTIATOR

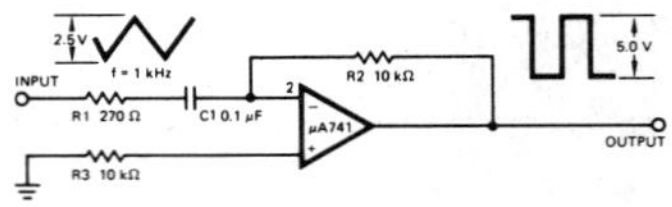

$$E_{OUT} = -R2C\frac{dE_{IN}}{dt}$$

LOW DRIFT LOW NOISE AMPLIFIER

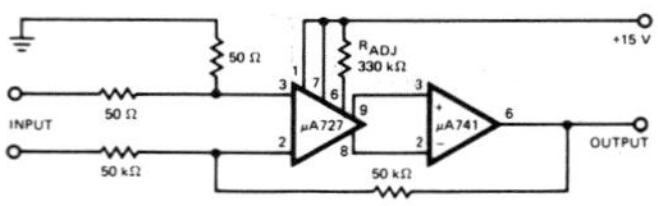

Voltage Gain = 10^3
Input Offset Voltage Drift = 0.6 μV/°C
Input Offset Current Drift = 2.0 pA/°C

HIGH SLEW RATE POWER AMPLIFIER

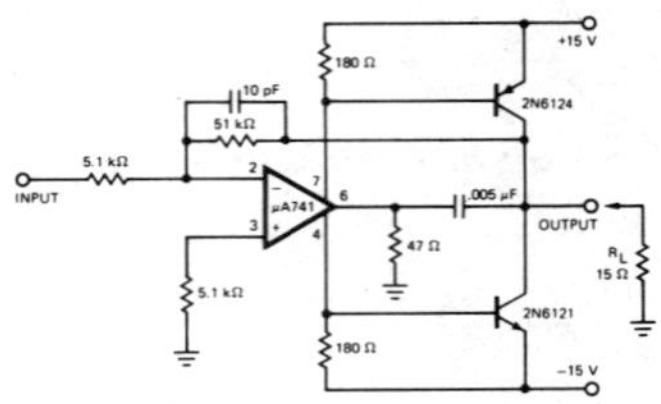

NOTCH FILTER USING THE μA741 AS A GYRATOR

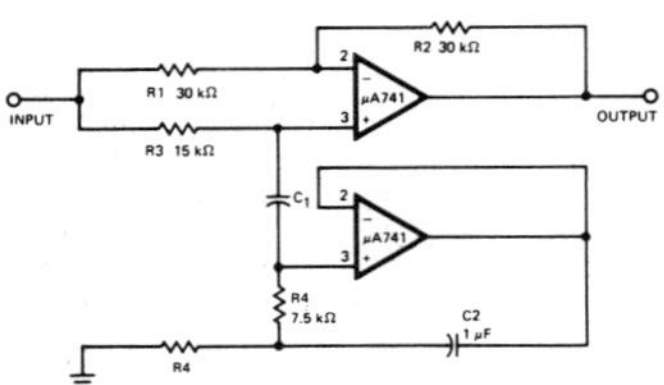

Trim R3 such that

$$\frac{R1}{R2} = \frac{R3}{2\,R4}$$

NOTCH FREQUENCY AS A FUNCTION OF C_1

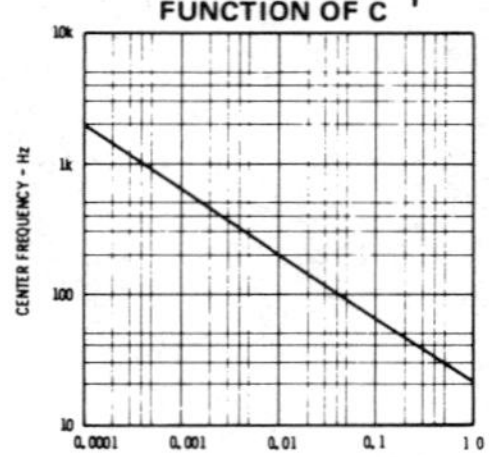

LINEAR INTEGRATED CIRCUITS

TYPES TL080 THRU TL085, TL080A THRU TL084A, TL081B, TL082B, TL084B JFET-INPUT OPERATIONAL AMPLIFIERS

BULLETIN NO. DL-S 12484, FEBRUARY 1977–REVISED OCTOBER 1979

24 DEVICES COVER COMMERCIAL, INDUSTRIAL, AND MILITARY TEMPERATURE RANGES

- **Low Power Consumption**
- **Wide Common-Mode and Differential Voltage Ranges**
- **Low Input Bias and Offset Currents**
- **Output Short-Circuit Protection**
- **High Input Impedance . . . JFET-Input Stage**
- **Internal Frequency Compensation (Except TL080, TL080A)**
- **Latch-Up-Free Operation**
- **High Slew Rate . . . 13 V/μs Typ**

description

The TL081 JFET-input operational amplifier family is designed to offer a wider selection than any previously developed operational amplifier family. Each of these JFET-input operational amplifiers incorporates well-matched, high-voltage JFET and bipolar transistors in a monolithic integrated circuit. The devices feature high slew rates, low input bias and offset currents, and low offset voltage temperature coefficient. Offset adjustment and external compensation options are available within the TL081 Family.

Device types with an "M" suffix are characterized for operation over the full military temperature range of –55°C to 125°C, those with an "I" suffix are characterized for operation from –25°C to 85°C, and those with a "C" suffix are characterized for operation from 0°C to 70°C.

TL080, TL080A
JG OR P DUAL-IN-LINE
PACKAGE (TOP VIEW)

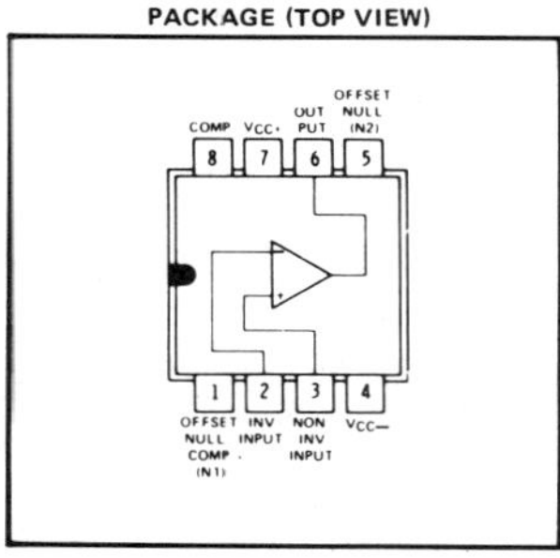

TL081, TL081A, TL081B
JG OR P DUAL-IN-LINE
PACKAGE (TOP VIEW)

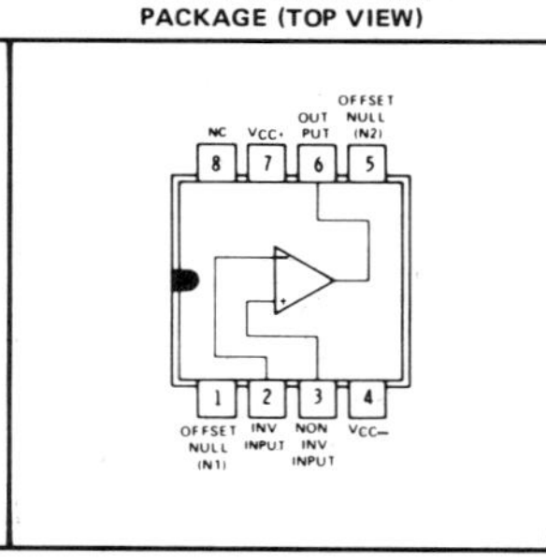

TL082, TL082A, TL082B
JG OR P DUAL-IN-LINE
PACKAGE (TOP VIEW)

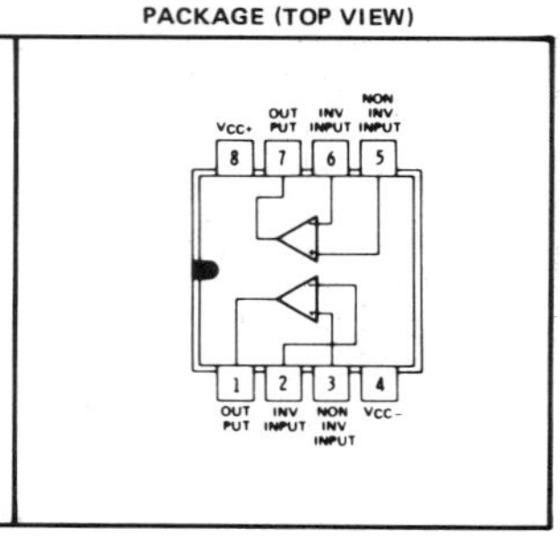

TL083, TL083A
J OR N DUAL-IN-LINE
PACKAGE (TOP VIEW)

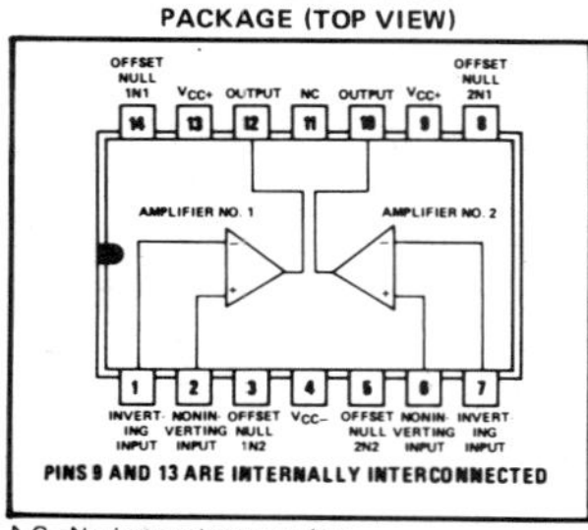

TL084, TL084A, TL084B
J OR N DUAL-IN-LINE
OR W FLAT PACKAGE
(TOP VIEW)

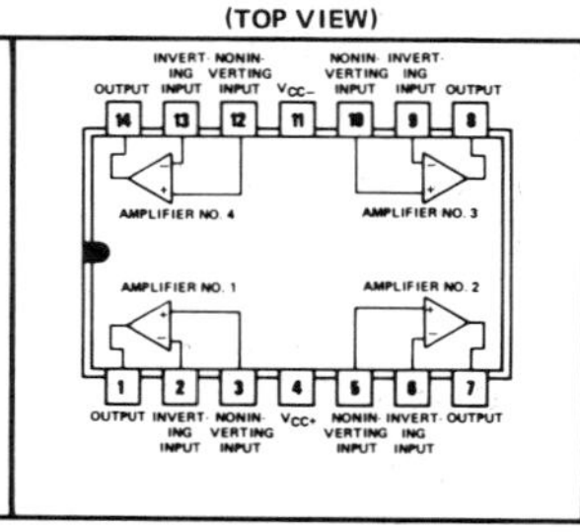

TL085
N DUAL-IN-LINE
PACKAGE (TOP VIEW)

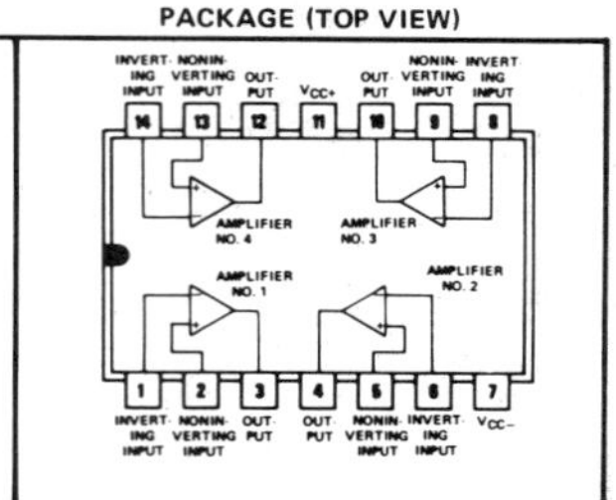

NC–No internal connection

Courtesy of Texas Instruments Incorporated.

TYPES TL080 THRU TL085, TL080A THRU TL084A, TL081B, TL082B, TL084B JFET-INPUT OPERATIONAL AMPLIFIERS

schematic (each amplifier)

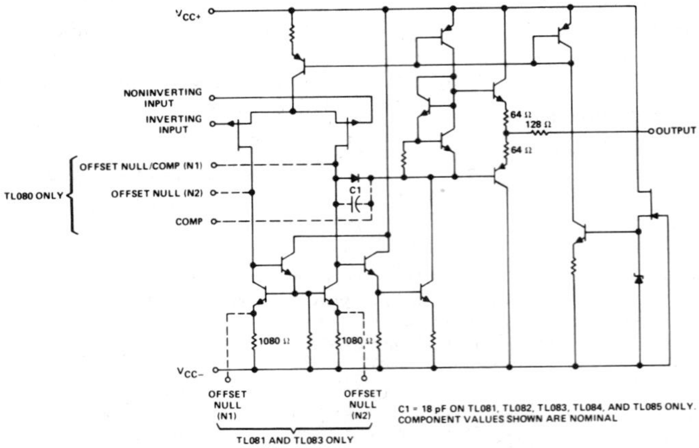

absolute maximum ratings over operating free-air temperature range (unless otherwise noted)

		TL08_M	TL08_I	TL08_C TL08_AC TL08_BC	UNIT
Supply voltage, V_{CC+} (see Note 1)		18	18	18	V
Supply voltage, V_{CC-} (see Note 1)		−18	−18	−18	V
Differential input voltage (see Note 2)		±30	±30	±30	V
Input voltage (see Notes 1 and 3)		±15	±15	±15	V
Duration of output short circuit (see Note 4)		Unlimited	Unlimited	Unlimited	
Continuous total dissipation at (or below) 25°C free-air temperature (See Note 5)		680	680	680	mW
Operating free-air temperature range		−55 to 125	−25 to 85	0 to 70	°C
Storage temperature range		−65 to 150	−65 to 150	−65 to 150	°C
Lead temperature 1/16 inch (1,6 mm) from case for 60 seconds	J, JG, or W package	300	300	300	°C
Lead temperature 1/16 inch (1,6 mm) from case for 10 seconds	N or P package		260	260	°C

NOTES:
1. All voltage values, except differential voltages, are with respect to the midpoint between V_{CC+} and V_{CC-}.
2. Differential voltages are at the noninverting input terminal with respect to the inverting input terminal.
3. The magnitude of the input voltage must never exceed the magnitude of the supply voltage or 15 volts, whichever is less.
4. The output may be shorted to ground or to either supply. Temperature and/or supply voltages must be limited to ensure that the dissipation rating is not exceeded.
5. For operation above 25°C free-air temperature, refer to Dissipation Derating Table. In the J and JG packages, TL08_M chips are alloy-mounted; TL08_I, TL08_C, TL08_AC, and TL08_BC chips are glass-mounted.

DISSIPATION DERATING TABLE

PACKAGE	POWER RATING	DERATING FACTOR	ABOVE T_A
J (Alloy-Mounted Chip)	680 mW	11.0 mW/°C	88°C
J (Glass-Mounted Chip)	680 mW	8.2 mW/°C	67°C
JG (Alloy-Mounted Chip)	680 mW	8.4 mW/°C	69°C
JG (Glass-Mounted Chip)	680 mW	6.6 mW/°C	47°C
N	680 mW	9.2 mW/°C	76°C
P	680 mW	8.0 mW/°C	65°C
W	680 mW	8.0 mW/°C	65°C

Also see Dissipation Derating Curves, Section 2.

DEVICE TYPES, SUFFIX VERSIONS, AND PACKAGES

	TL080	TL081	TL082	TL083	TL084	TL085
TL08_M	JG	JG	JG	J	J, W	•
TL08_I	JG, P	JG, P	JG, P	J, N	J, N	•
TL08_C	JG, P	JG, P	JG, P	J, N	J, N	N
TL08_AC	JG, P	JG, P	JG, P	J, N	J, N	•
TL08_BC	•	JG, P	JG, P	•	J, N	•

•These combinations are not defined by this data sheet.

TYPES TL080 THRU TL085, TL080A THRU TL084A, TL081B, TL082B, TL084B JFET-INPUT OPERATIONAL AMPLIFIERS

electrical characteristics, $V_{CC\pm} = \pm 15$ V

PARAMETER		TEST CONDITIONS†		TL08_M MIN	TL08_M TYP	TL08_M MAX	TL08_I MIN	TL08_I TYP	TL08_I MAX	TL08_C, TL08_AC, TL08_BC MIN	TL08_C, TL08_AC, TL08_BC TYP	TL08_C, TL08_AC, TL08_BC MAX	UNIT
V_{IO}	Input offset voltage	$R_S = 50\ \Omega$, $T_A = 25°C$	'80,'81,'82,'83,'85‡		3	6		3	6		5	15	mV
			TL084		3	9		3	6		5	15	
			TL08_A								3	6	
			'81B,'82B,'84B								2	3	
		$R_S = 50\ \Omega$, T_A = full range	'80,'81,'82,'83,'85‡			9			9			20	
			TL084			15			9			20	
			TL08_A									7.5	
			'81B,'82B,'84B									5	
α_{VIO}	Temperature coefficient of input offset voltage	$R_S = 50\ \Omega$,	T_A = full range		10			10			10		μV/°C
I_{IO}	Input offset current§	$T_A = 25°C$	TL08_‡		5	100		5	100		5	200	pA
			TL08_A								5	100	
			'81B,'82B,'84B								5	100	
		T_A = full range	TL08_‡			20			10			5	nA
			TL08_A									3	
			'81B,'82B,'84B									3	
I_{IB}	Input bias current§	$T_A = 25°C$	TL08_‡		30	200		30	200		30	400	pA
			TL08_A								30	200	
			'81B,'82B,'84B								30	200	
		T_A = full range	TL08_‡			50			20			10	nA
			TL08_A									7	
			'81B,'82B,'84B									7	
V_{ICR}	Common-mode input voltage range	$T_A = 25°C$	TL08_‡	±11	±12		±11	±12		±10	±11		V
			TL08_A							±11	±12		
			'81B,'82B,'84B							±11	±12		
V_{OPP}	Maximum peak-to-peak output voltage swing	$T_A = 25°C$	$R_L = 10\ k\Omega$	24	27		24	27		24	27		V
		T_A = full range	$R_L \geq 10\ k\Omega$	24			24			24			
			$R_L \geq 2\ k\Omega$	20	24		20	24		20	24		
A_{VD}	Large-signal differential voltage amplification	$R_L \geq 2\ k\Omega$, $V_O = \pm 10$ V, $T_A = 25°C$	TL08_‡	25	200		50	200		25	200		V/mV
			TL08_A							50	200		
			'81B,'82B,'84B							50	200		
		$R_L \geq 2\ k\Omega$, $V_O = \pm 10$ V, T_A = full range	TL08_‡	15			25			15			
			TL08_A							25			
			'81B,'82B,'84B							25			
B_1	Unity-gain bandwidth	$T_A = 25°C$			3			3			3		MHz
r_i	Input resistance	$T_A = 25°C$			10^{12}			10^{12}			10^{12}		Ω
CMRR	Common-mode rejection ratio	$R_S \geq 10\ k\Omega$, $T_A = 25°C$	TL08_‡	80	86		80	86		70	76		dB
			TL08_A							80	86		
			'81B,'82B,'84B							80	86		
k_{SVR}	Supply voltage rejection ratio ($\Delta V_{CC\pm}/\Delta V_{IO}$)	$R_S \geq 10\ k\Omega$, $T_A = 25°C$	TL08_‡	80	86		80	86		70	76		dB
			TL08_A							80	86		
			'81B,'82B,'84B							80	86		
I_{CC}	Supply current (per amplifier)	No load, $T_A = 25°C$	No signal,		1.4	2.8		1.4	2.8		1.4	2.8	mA
V_{o1}/V_{o2}	Channel separation	$A_{VD} = 100$,	$T_A = 25°C$		120			120			120		dB

† All characteristics are specified under open-loop conditions unless otherwise noted. Full range for T_A is −55°C to 125°C for TL08_M; −25°C to 85°C for TL08_I; and 0°C to 70°C for TL08_C, TL08_AC, and TL08_BC.

‡ Types TL085I and TL085M are not defined by this data sheet.

§ Input bias currents of a FET-input operational amplifier are normal junction reverse currents, which are temperature sensitive as shown in Figure 18. Pulse techniques must be used that will maintain the junction temperature as close to the ambient temperature as is possible.

TYPES TL080 THRU TL085, TL080A THRU TL084A, TL081B, TL082B, TL084B JFET-INPUT OPERATIONAL AMPLIFIERS

operating characteristics, $V_{CC\pm} = \pm 15$ V, $T_A = 25°$ C

PARAMETER		TEST CONDITIONS		TL08_M			ALL OTHERS			UNIT
				MIN	TYP	MAX	MIN	TYP	MAX	
SR	Slew rate at unity gain	V_I = 10 V, C_L = 100 pF,	R_L = 2 kΩ, See Figure 1	8	13			13		V/μs
t_r	Rise time	V_I = 20 mV,	R_L = 2 kΩ,		0.1			0.1		μs
	Overshoot factor	C_L = 100 pF,	See Figure 1		10%			10%		
V_n	Equivalent input noise voltage	R_S = 100 Ω,	f = 1 kHz		25			25		nV/$\sqrt{Hz}$

PARAMETER MEASUREMENT INFORMATION

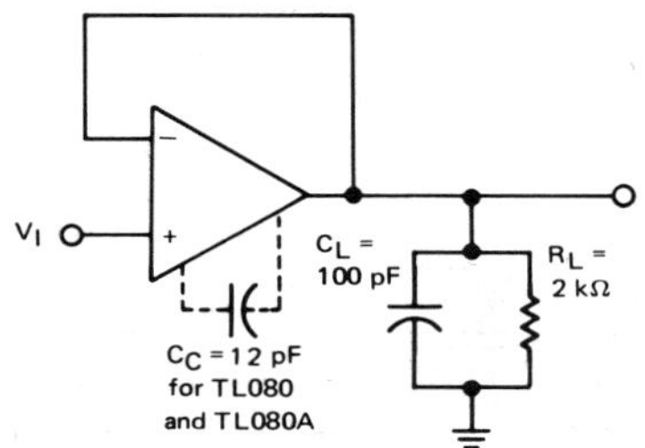

FIGURE 1—UNITY-GAIN AMPLIFIER

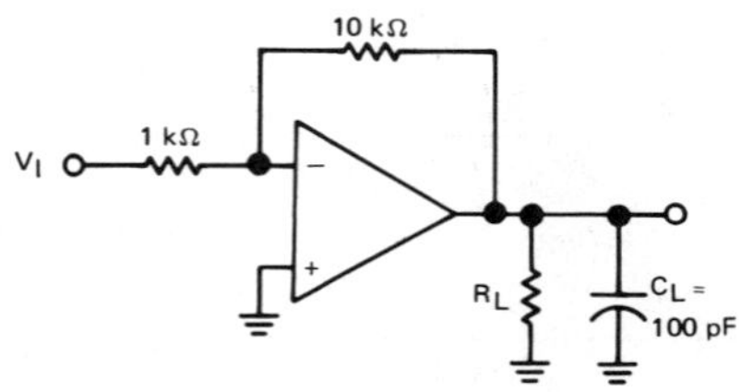

FIGURE 2—GAIN-OF-10 INVERTING AMPLIFIER

INPUT OFFSET VOLTAGE NULL CIRCUITS

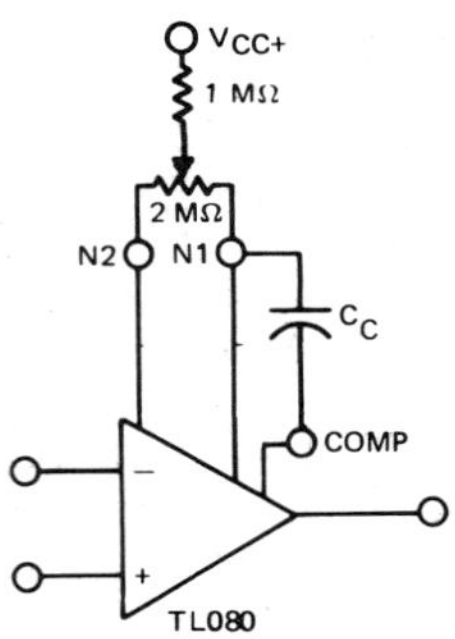

FIGURE 3

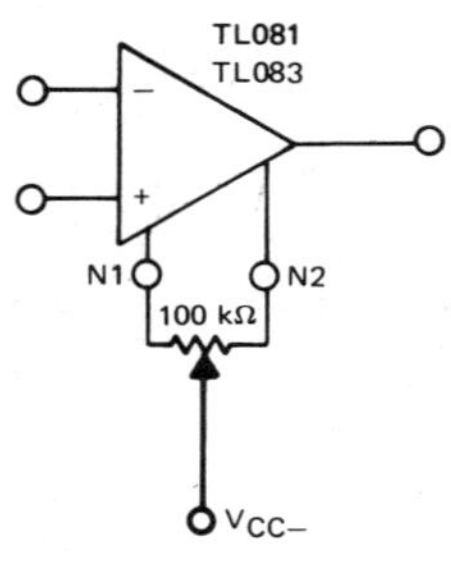

FIGURE 4

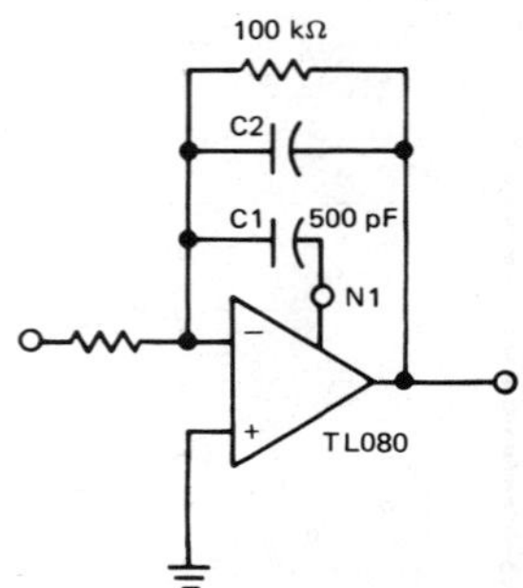

FEED-FORWARD COMPENSATION

FIGURE 5

TYPES TL080 THRU TL085, TL080A THRU TL084A, TL081B, TL082B, TL084B JFET-INPUT OPERATIONAL AMPLIFIERS

TYPICAL CHARACTERISTICS†

MAXIMUM PEAK-TO-PEAK OUTPUT VOLTAGE vs FREQUENCY

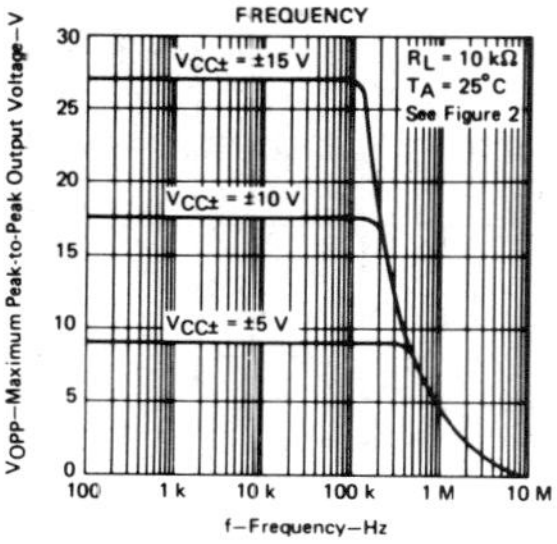

FIGURE 6

MAXIMUM PEAK-TO-PEAK OUTPUT VOLTAGE vs FREQUENCY

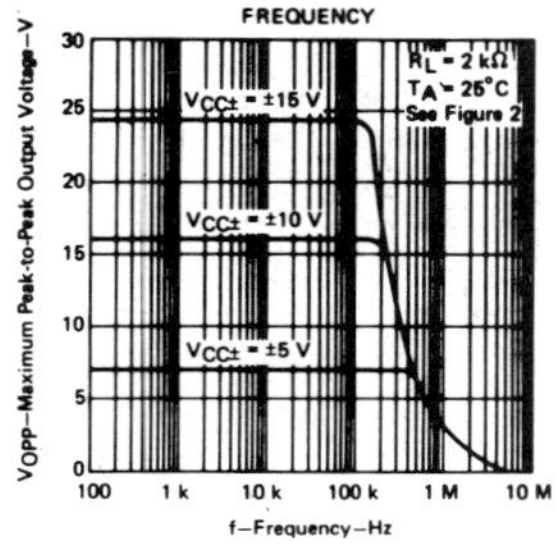

FIGURE 7

MAXIMUM PEAK-TO-PEAK OUTPUT VOLTAGE vs FREQUENCY

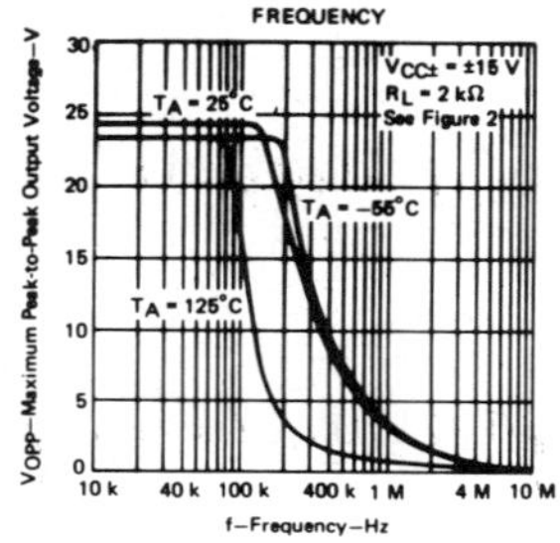

FIGURE 8

MAXIMUM PEAK-TO-PEAK OUTPUT VOLTAGE vs FREE-AIR TEMPERATURE

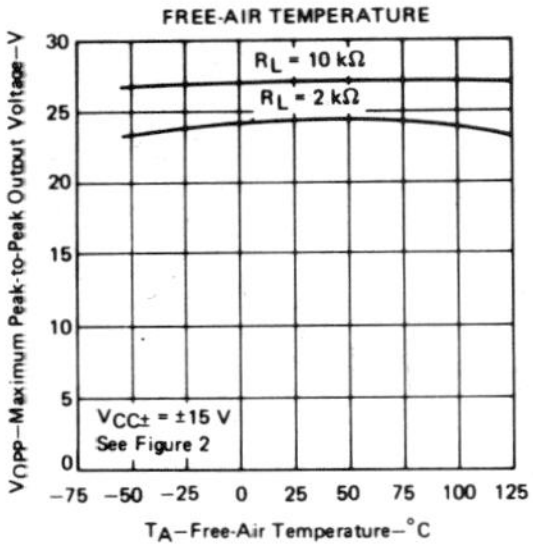

FIGURE 9

MAXIMUM PEAK-TO-PEAK OUTPUT VOLTAGE vs LOAD RESISTANCE

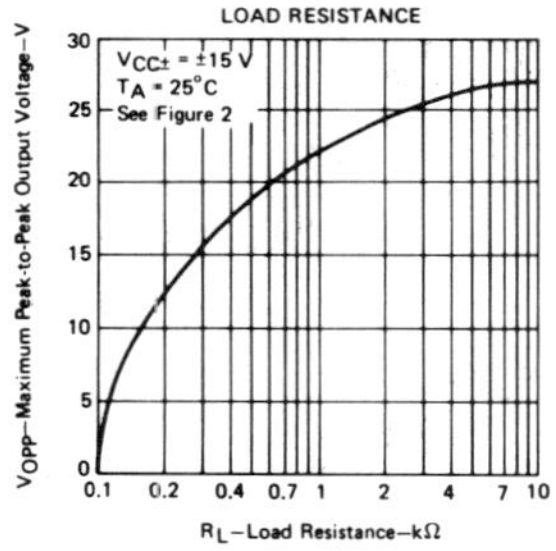

FIGURE 10

MAXIMUM PEAK-TO-PEAK OUTPUT VOLTAGE vs SUPPLY VOLTAGE

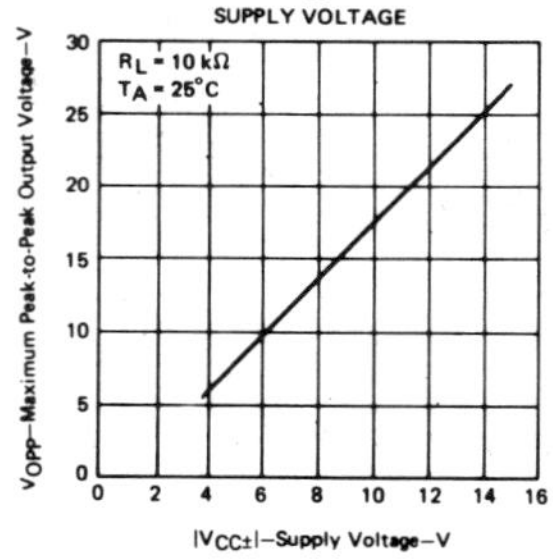

FIGURE 11

LARGE-SIGNAL DIFFERENTIAL VOLTAGE AMPLIFICATION vs FREE-AIR TEMPERATURE

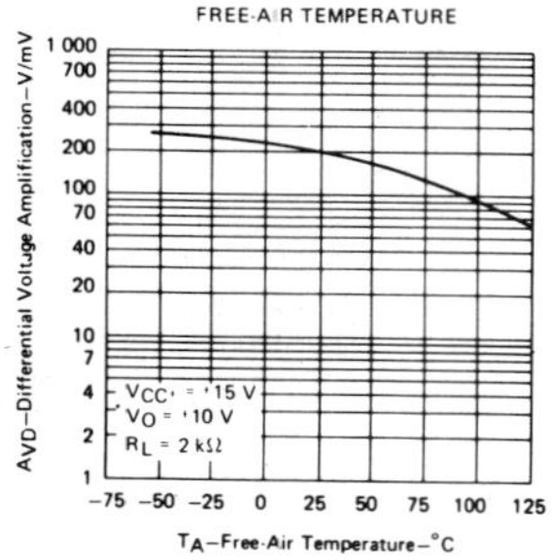

FIGURE 12

LARGE-SIGNAL DIFFERENTIAL VOLTAGE AMPLIFICATION and PHASE SHIFT vs FREQUENCY

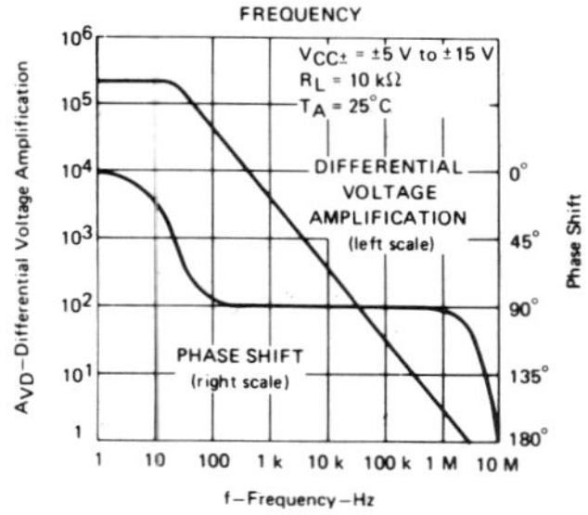

FIGURE 13

TL080, TL080A DIFFERENTIAL VOLTAGE AMPLIFICATION vs FREQUENCY WITH FEED-FORWARD COMPENSATION

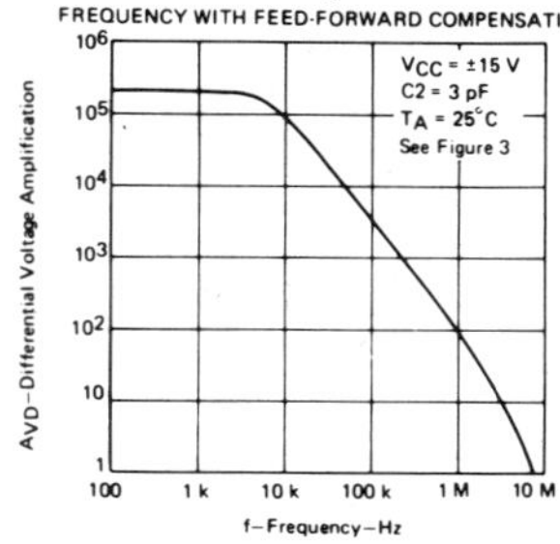

FIGURE 14

†Data at high and low temperatures are applicable only within the rated operating free-air temperature ranges of the various devices. A 12-pF compensation capacitor is used with TL080 and TL080A.

TYPES TL080 THRU TL085, TL080A THRU TL084A, TL081B, TL082B, TL084B JFET-INPUT OPERATIONAL AMPLIFIERS

TYPICAL CHARACTERISTICS†

TOTAL POWER DISSIPATED vs FREE-AIR TEMPERATURE

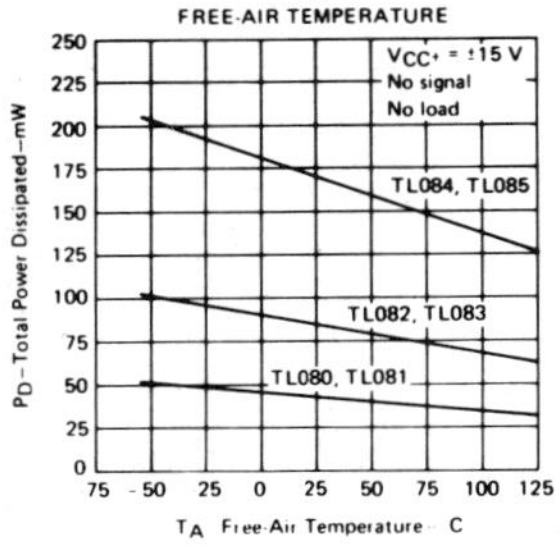

FIGURE 15

SUPPLY CURRENT PER AMPLIFIER vs FREE-AIR TEMPERATURE

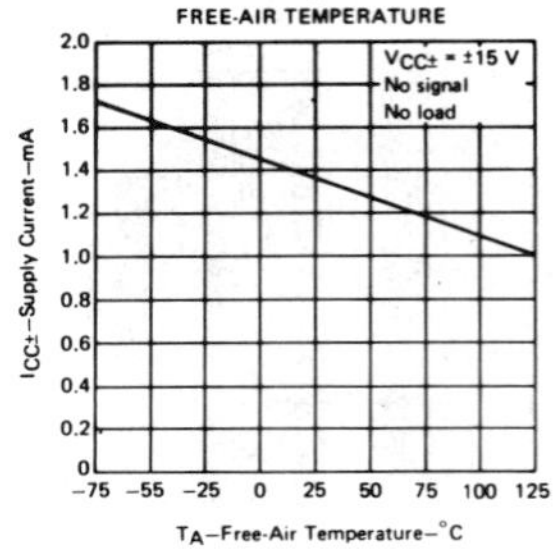

FIGURE 16

SUPPLY CURRENT vs SUPPLY VOLTAGE

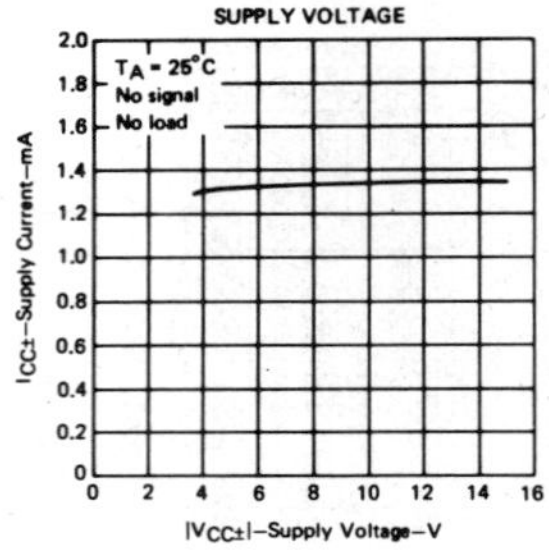

FIGURE 17

INPUT BIAS CURRENT vs FREE-AIR TEMPERATURE

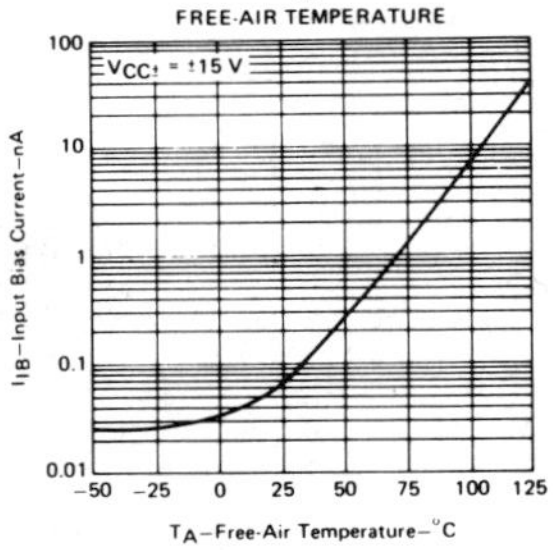

FIGURE 18

VOLTAGE-FOLLOWER LARGE-SIGNAL PULSE RESPONSE

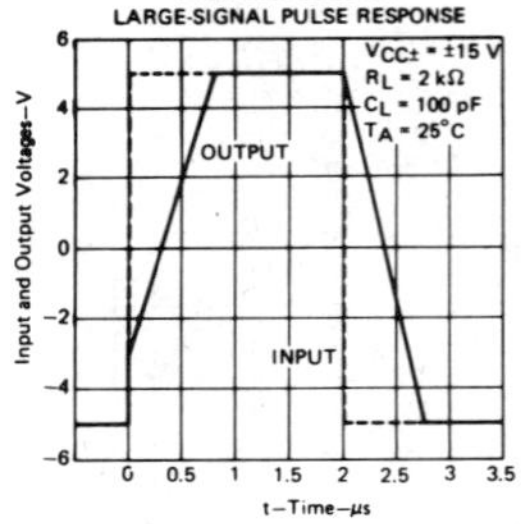

FIGURE 19

OUTPUT VOLTAGE vs ELAPSED TIME

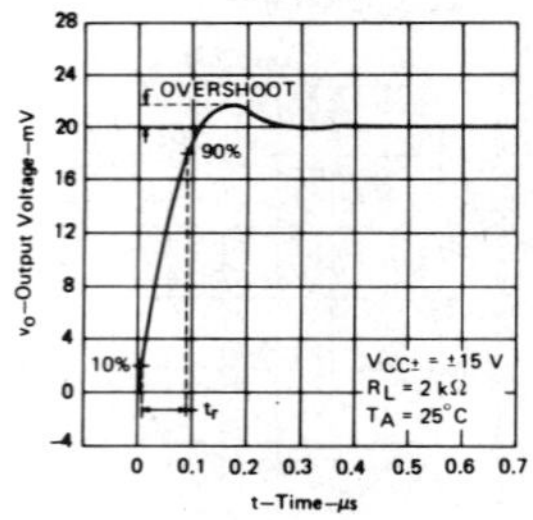

FIGURE 20

ALL EXCEPT TL08_C
COMMON-MODE REJECTION RATIO vs FREE-AIR TEMPERATURE

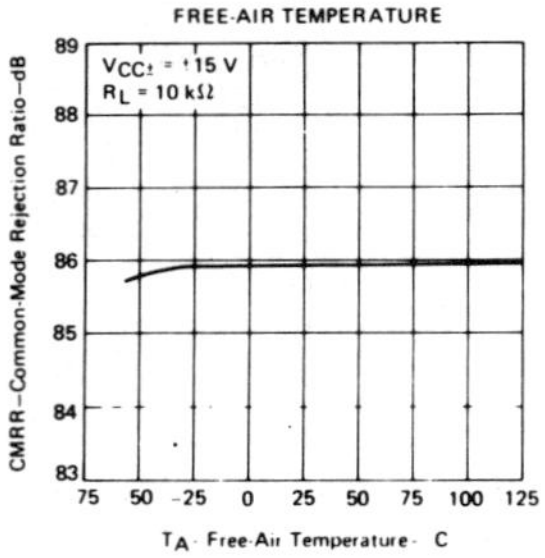

FIGURE 21

EQUIVALENT INPUT NOISE VOLTAGE vs FREQUENCY

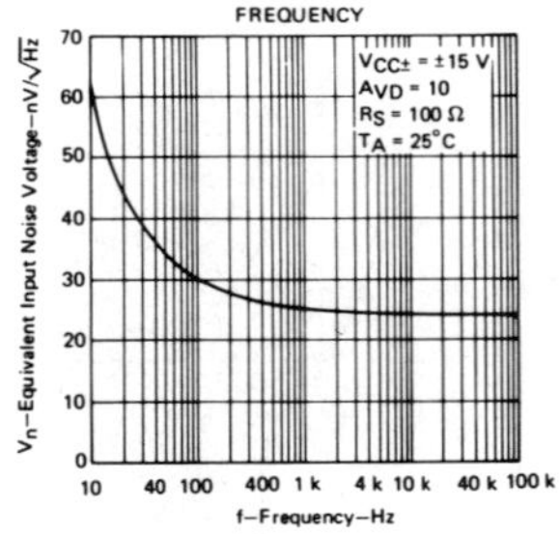

FIGURE 22

TOTAL HARMONIC DISTORTION vs FREQUENCY

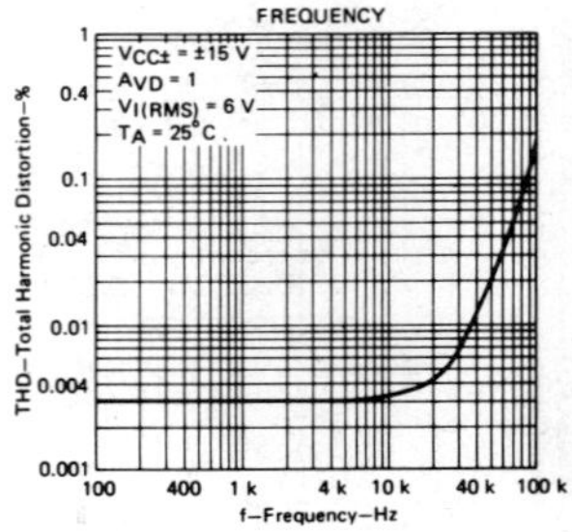

FIGURE 23

†Data at high and low temperatures are applicable only within the rated operating free-air temperature ranges of the various devices. A 12-pF compensation capacitor is used with TL080 and TL080A.

μA791
POWER OPERATIONAL AMPLIFIER
FAIRCHILD LINEAR INTEGRATED CIRCUIT

GENERAL DESCRIPTION – The μA791 is a high performance monolithic Operational Amplifier constructed using the Fairchild Planar* Epitaxial process with input characteristics similar to the μA741 operational amplifier and 1A available output current. It is intended for use in a wide variety of applications including audio amplifiers, servo amplifiers, and power supplies. The high gain and high output power capability provide superior performance wherever an operational amplifier/power booster combination is required. The μA791 is thermal overload and short circuit protected.

- **CURRENT OUTPUT TO 1 A**
- **SHORT CIRCUIT PROTECTION**
- **OFFSET VOLTAGE NULL CAPABILITY**
- **NO LATCH UP**
- **LARGE COMMON MODE AND DIFFERENTIAL MODE RANGES**
- **THERMAL OVERLOAD PROTECTION**

EQUIVALENT CIRCUIT

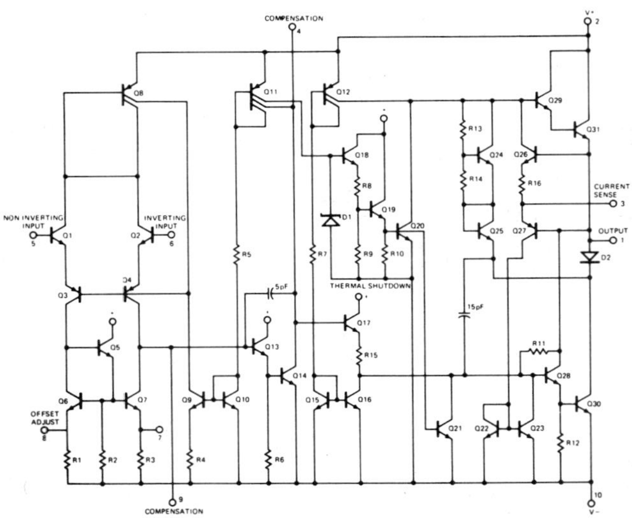

NOTE: Pin connections shown are for metal can.

CONNECTION DIAGRAMS
10-PIN METAL CAN
(TOP VIEW)
PACKAGE OUTLINE 5H
PACKAGE CODE K

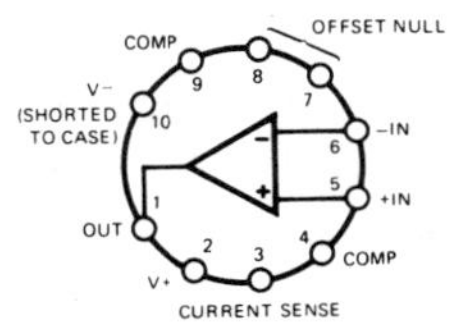

ORDER INFORMATION

TYPE	PART NO.
μA791C	μA791KC
μA791	μA791KM

12-PIN DIP
(TOP VIEW)
PACKAGE OUTLINE 9W
PACKAGE CODE P5

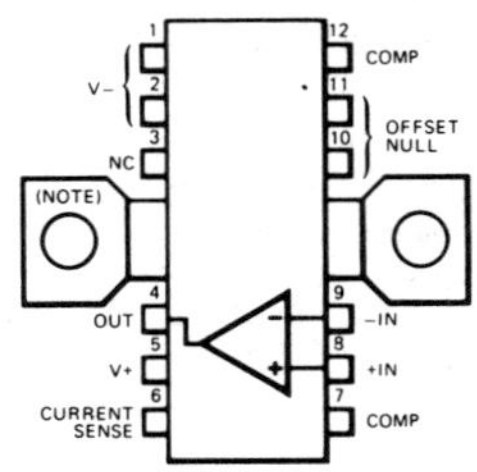

ORDER INFORMATION

TYPE	PART NO.
μA791C	μA791P5

NOTES:
The heat sink wings on the P-package are internally connected to V−.
Both pin 1 and pin 2 must be connected externally to V−.

*Planar is a patented Fairchild process.

Courtesy of Fairchild Camera and Instrument Corporation.

ABSOLUTE MAXIMUM RATINGS

Supply Voltage	
Military (μA791)	±22 V
Commercial (μA791C)	±18 V
Peak Output Current	1.25 A
Continuous Internal Power Dissipation (Total Package) (Note 1)	Internally Limited
Peak Internal Power Dissipation (Per Output Transistor for t ≤ 5 s, Note 2)	15 W
Differential Input Voltage	±30 V
Input Voltage (Note 3)	±15 V
Voltages between offset Null and V−	±0.5 V
Operating Junction Temperature	
Military (μA791)	-55° C to +150° C
Commercial (μA791C)	0° C to +125° C
Storage Temperature Range	
Metal Can	-65° C to +150° C
Molded Power DIP	-55° C to +125° C
Pin Temperatures	
Metal Can (Soldering, 60 s max.)	280° C
Molded Power DIP (Soldering, 10 s max.)	260° C

NOTES:
1. Thermal resistance of the packages (without a heat sink)

Package	Junction to Case		Junction to Ambient		Unit
	Typ	Max	Typ	Max	
TO-3 Type (5H)	4	6	35	40	°C/W
Dual In-Line Power (9W)	8	12	50	55	°C/W

2. Under short circuit conditions, the safe operating area and dc power dissipation limitations must be observed.
3. For supply voltages less than ±15V, the absolute maximum input voltage is equal to the supply voltage.

μA791C

ELECTRICAL CHARACTERISTICS: V_S = ±15 V, T_J = 25°C unless otherwise specified.

CHARACTERISTICS	CONDITIONS	MIN	TYP	MAX	UNITS
Input Offset Voltage	$R_S \le 10\ k\Omega$		2.0	6.0	mV
Input Offset Current			20	200	nA
Input Bias Current			80	500	nA
Input Resistance		0.3	1.0		MΩ
Offset Voltage Adjustment Range			±15		mV
Input Voltage Range		±12	±13		V
Common Mode Rejection Ratio		70			dB
Power Supply Rejection Ratio				150	μV/V
Large Signal Voltage Gain	$R_L = 1\ k\Omega, V_{OUT} = \pm 10\ V$	20k			V/V
	$R_L = 10\ \Omega, V_{OUT} = \pm 10\ V$	20k			V/V
Output Voltage Swing	$R_{SC} = 0, R_L = 1k\Omega$	±11.5	±14		V
	$R_{SC} = 0, R_L = 10\ \Omega$	±10	±12.2		V
Output Short Circuit Current	$R_{SC} = 0.7\Omega$		1000		mA
	$R_{SC} = 1.5\Omega$		500		mA
Supply Current (Zero Signal)				30	mA
The following specifications apply for $0°C \le T_J \le 125°C$					
Input Offset Voltage	$R_S \le 10\ k\Omega$			7.5	mV
Input Offset Current				300	nA
Input Bias Current				800	nA
Common Mode Rejection Ratio		70			dB
Power Supply Rejection Ratio				150	μV/V
Large Signal Voltage Gain	$R_L = 1\ k\Omega, V_{OUT} = \pm 10\ V$	15k			V/V
	$R_L = 10\ \Omega, V_{OUT} = \pm 10\ V$	15k			V/V
Output Voltage Swing	$R_{SC} = 0, R_L = 1\ k\Omega$	±10			V
	$R_{SC} = 0, R_L = 10\ \Omega$	±10			V
Supply Current (Zero signal)				30	mA

FREQUENCY COMPENSATION

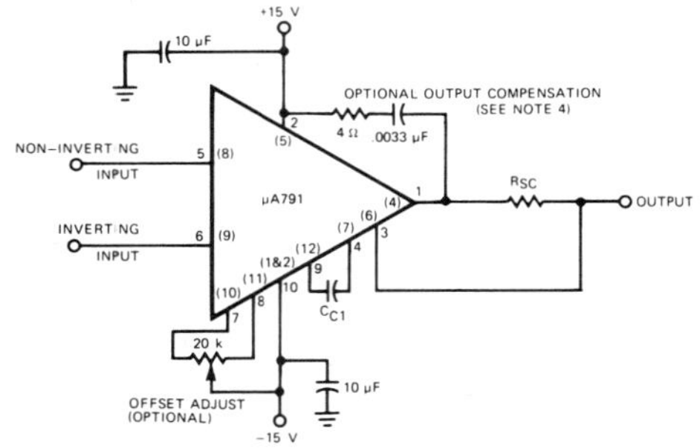

GAIN	C_C
1	100 pF
10	5 pF
100	Not Req.

R_{SC}	I_{SC}
0.6Ω	1.0 A
1.5Ω	500 mA
3.0Ω	250 mA

NOTES
1. Power supply decoupling capacitors and compensation network components must have short leads and they must be located at the amplifier pins.
2. When short circuit limiting is not required, connect terminals one and three together.
3. Pin connections in parentheses are for plastic packages.
4. Output compensation may be required for some loads.

μA791

ELECTRICAL CHARACTERISTICS: $V_S = \pm 15$ V, $T_J = 25°C$ **unless otherwise specified.**

CHARACTERISTICS	CONDITIONS	MIN	TYP	MAX	UNITS
Input Offset Voltage	$R_S \leq 10$ kΩ		1.0	5.0	mV
Input Offset Current			20	200	nA
Input Bias Current			80	500	nA
Input Resistance		0.3	2.0		MΩ
Offset Voltage Adjustment Range			±15		mV
Input Voltage Range		±12	±13		V
Common Mode Rejection Ratio		70			dB
Power Supply Rejection Ratio				150	μV/V
Large Signal Voltage Gain	$R_L = 1$ kΩ	50,000			V/V
	$R_L = 10$ Ω	50,000			V/V
Output Voltage Swing	$R_{SC} = 0$, $R_L = 1$kΩ	±12	±14		V
	$R_{SC} = 0$, $R_L = 10$Ω	±10	±12.2		V
Output Short Circuit Current	$R_{SC} = 0.7$Ω		1000		mA
	$R_{SC} = 1.5$Ω		500		mA
Supply Current (Zero Signal)				25	mA
The following specifications apply for $-55°C \leq T_J \leq 150°C$					
Input Offset Voltage	$R_S \leq 10$ kΩ			6	mV
Input Offset Current				500	nA
Input Bias Current				1.5	μA
Common Mode Rejection Ratio		70			dB
Power Supply Rejection Ratio				150	μV/V
Large Signal Voltage Gain	$R_L = 1$ kΩ	25,000			V/V
	$R_L = 10$Ω	25,000			V/V
Output Voltage Swing	$R_{SC} = 0$, $R_L = 1$ kΩ	±10			V
	$R_{SC} = 0$, $R_L = 10$Ω	±10			V
Supply Current (Zero Signal)				30	mA

TYPICAL PERFORMANCE CURVES FOR μA791 AND μA791C

INPUT BIAS CURRENT AS A FUNCTION OF JUNCTION TEMPERATURE

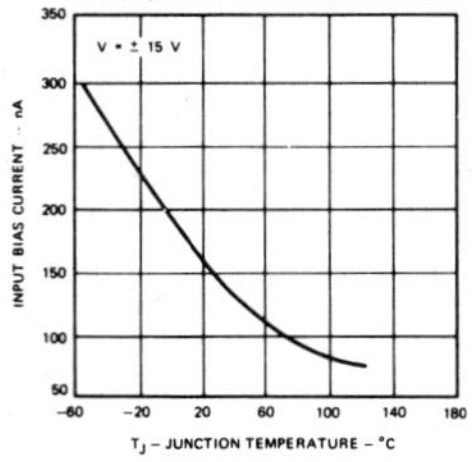

INPUT OFFSET CURRENT AS A FUNCTION OF JUNCTION TEMPERATURE

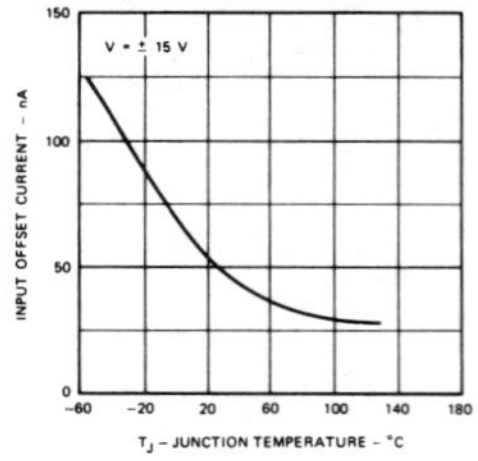

OUTPUT RESISTANCE AS A FUNCTION OF FREQUENCY (OPEN LOOP)

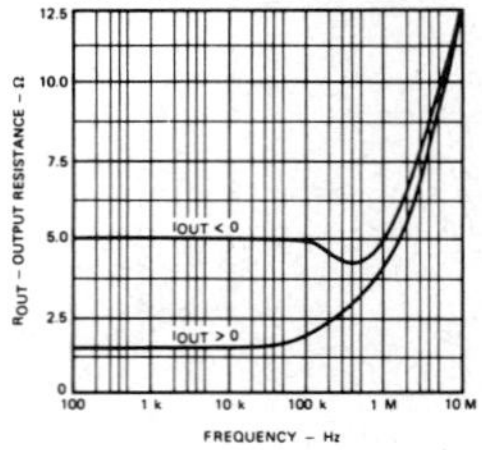

VOLTAGE GAIN AS A FUNCTION OF POWER SUPPLY VOLTAGE

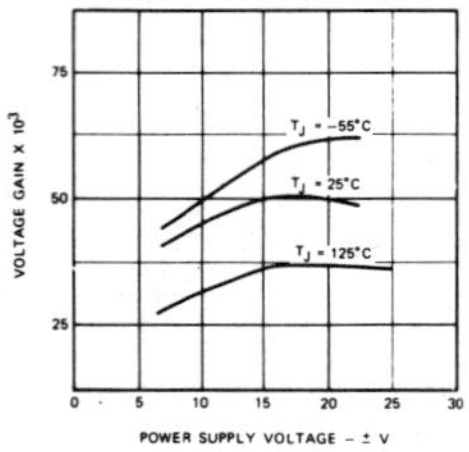

POWER SUPPLY REJECTION RATIO AS A FUNCTION OF FREQUENCY

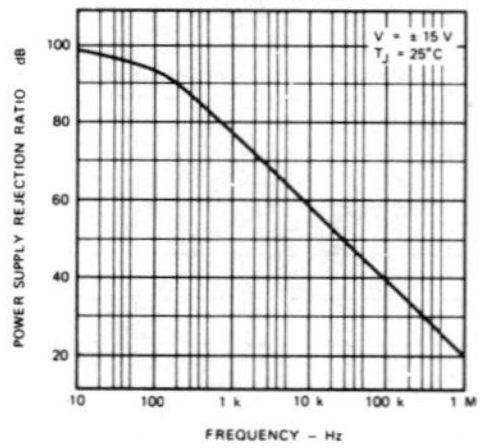

INPUT NOISE VOLTAGE AS A FUNCTION OF FREQUENCY

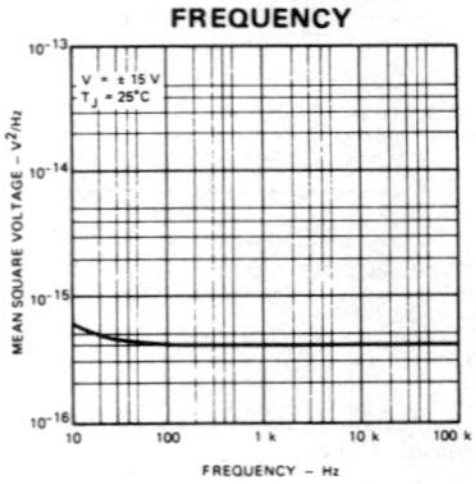

INPUT NOISE CURRENT AS A FUNCTION OF FREQUENCY

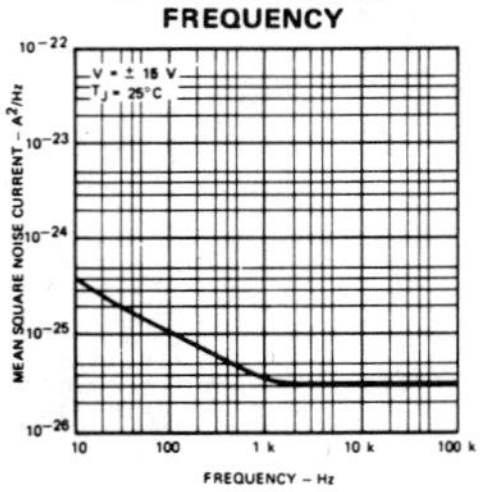

TOTAL NOISE (20 Hz-20 kHz) AS A FUNCTION OF JUNCTION TEMPERATURE

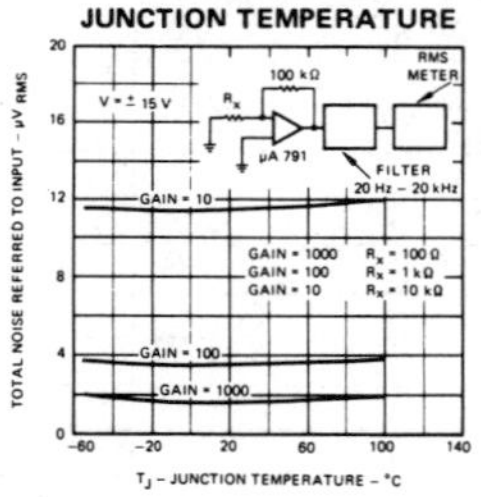

OUTPUT VOLTAGE SWING AS A FUNCTION OF OUTPUT CURRENT

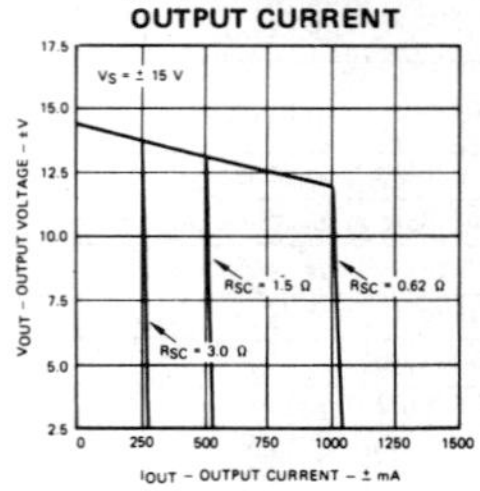

TYPICAL PERFORMANCE CURVES FOR μA791 and μA791C (Cont'd)

OUTPUT SAFE OPERATING AREA PER OUTPUT TRANSISTOR

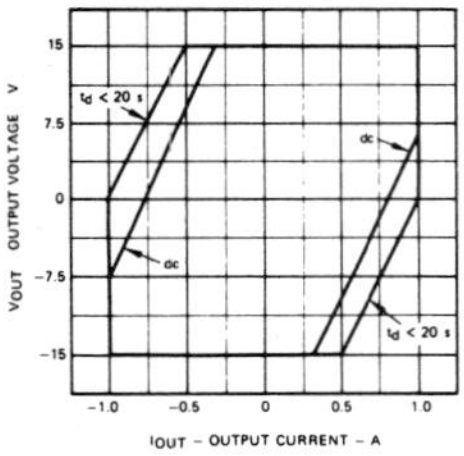

SHORT CIRCUIT CURRENT AS A FUNCTION OF CURRENT SENSE RESISTOR, R_{SC}

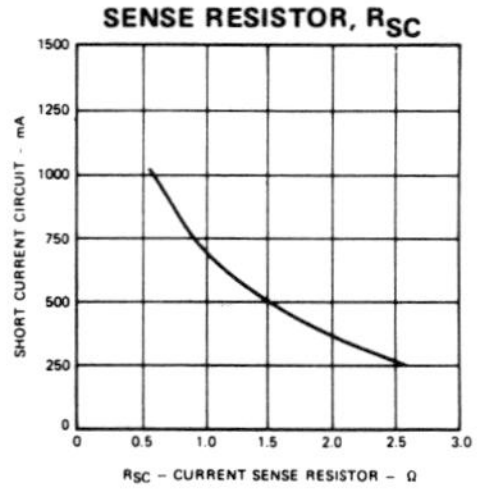

SHORT CIRCUIT CURRENT AS A FUNCTION OF CASE TEMPERATURE

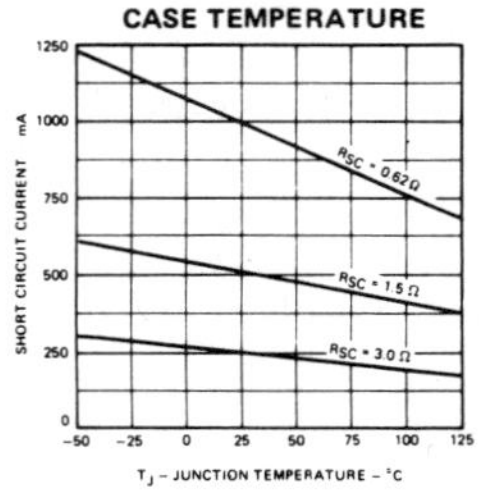

POWER BANDWIDTH AS A FUNCTION OF CLOSED LOOP GAIN

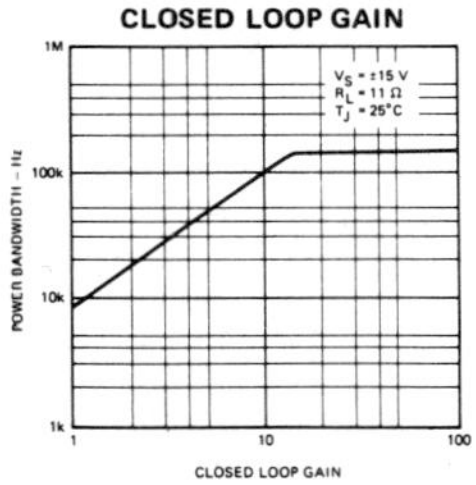

SLEW RATE AS A FUNCTION OF CLOSED LOOP GAIN

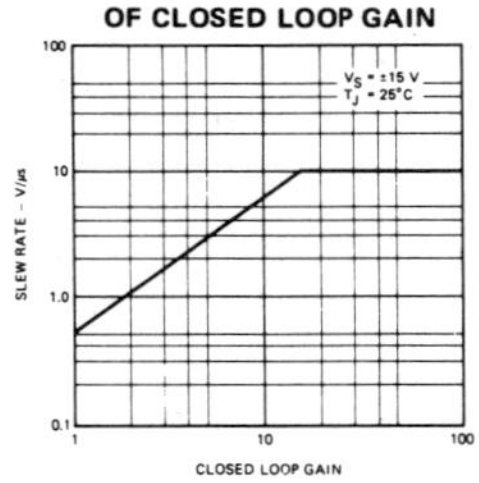

VOLTAGE GAIN AS A FUNCTION OF OPEN LOOP FREQUENCY RESPONSE

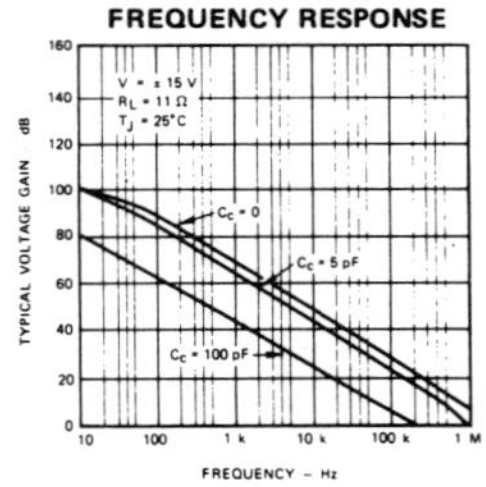

OPEN LOOP PHASE RESPONSE AS A FUNCTION OF FREQUENCY

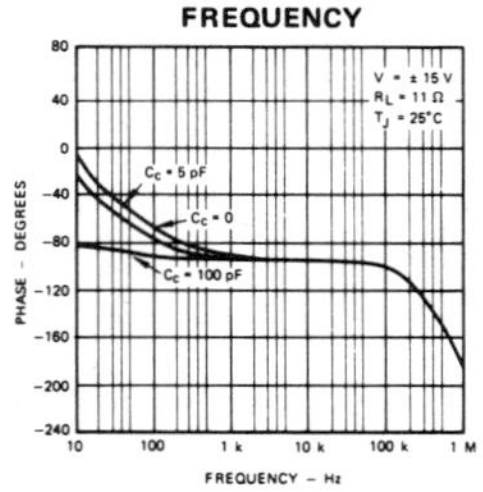

FREQUENCY RESPONSE FOR CLOSED LOOP GAINS

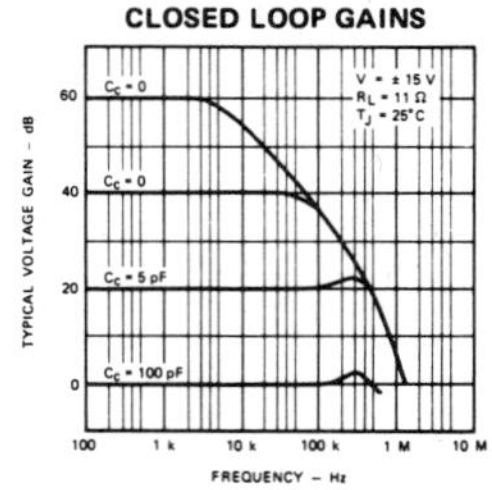

National Semiconductor

Amplifiers

LH0036/LH0036C Instrumentation Amplifier

General Description

The LH0036/LH0036C is a true micro power instrumentation amplifier designed for precision differential signal processing. Extremely high accuracy can be obtained due to the 300 MΩ input impedance and excellent 100 dB common mode rejection ratio. It is packaged in a hermetic TO-8 package. Gain is programmable with one external resistor from 1 to 1000. Power supply operating range is between ±1V and ±18V. Input bias current and output bandwidth are both externally adjustable or can be set by internally set values. The LH0036 is specified for operation over the −55°C to +125°C temperature range and the LH0036C is specified for operation over the −25°C to +85°C temperature range.

Features

- High input impedance 300 MΩ
- High CMRR 100 dB
- Single resistor gain adjust 1 to 1000
- Low power 90μW
- Wide supply range ±1V to ±18V
- Adjustable input bias current
- Adjustable output bandwidth
- Guard drive output

Equivalent Circuits and Connection Diagrams

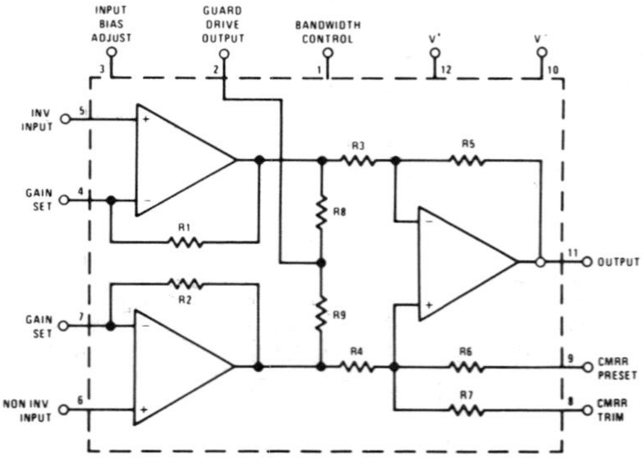

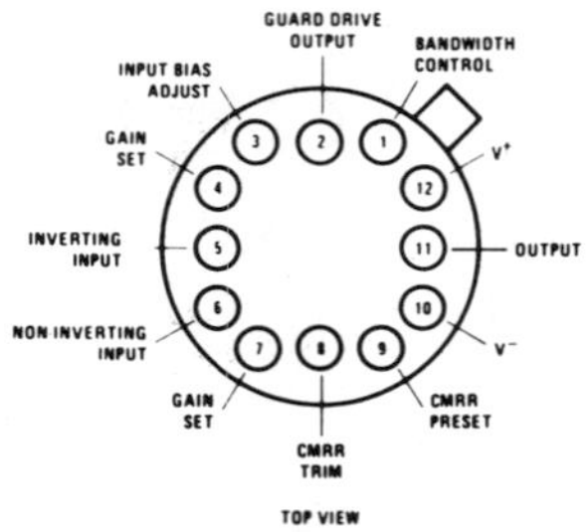

TOP VIEW

Order Number LH0036G or LH0036CG
See NS Package H12B

Courtesy of National Semiconductor Corporation.

Absolute Maximum Ratings

Supply Voltage	±18V	Short Circuit Duration	Continuous
Differential Input Voltage	±30V	Operating Temperature Range	
Input Voltage Range	$\pm V_S$	LH0036	-55°C to +125°C
Shield Drive Voltage	$\pm V_S$	LH0036C	-25°C to +85°C
CMRR Preset Voltage	$\pm V_S$	Storage Temperature Range	-65°C to +150°C
CMRR Trim Voltage	$\pm V_S$	Lead Temperature, Soldering 10 seconds	300°C
Power Dissipation (Note 3)	1.5W		

Electrical Characteristics (Notes 1 and 2)

PARAMETER	CONDITIONS	LIMITS						UNITS
		LH0036			LH0036C			
		MIN	TYP	MAX	MIN	TYP	MAX	
Input Offset Voltage (V_{IOS})	$R_S = 1.0k\Omega$, $T_A = 25°C$		0.5	1.0		1.0	2.0	mV
	$R_S = 1.0k\Omega$			2.0			3.0	mV
Output Offset Voltage (V_{OOS})	$R_S = 1.0k\Omega$, $T_A = 25°C$		2.0	5.0		5.0	10	mV
	$R_S = 1.0k\Omega$			6.0			12	mV
Input Offset Voltage Tempco ($\Delta V_{IOS}/\Delta T$)	$R_S \leq 1.0k\Omega$		10			10		μV/°C
Output Offset Voltage Tempco ($\Delta V_{OOS}/\Delta T$)			15			15		μV/°C
Overall Offset Referred to Input (V_{OS})	$A_V = 1.0$		2.5			6.0		mV
	$A_V = 10$		0.7			1.5		mV
	$A_V = 100$		0.52			1.05		mV
	$A_V = 1000$		0.502			1.005		mV
Input Bias Current (I_B)	$T_A = 25°C$		40	100		50	125	nA
				150			200	nA
Input Offset Current (I_{OS})	$T_A = 25°C$		10	40		20	50	nA
				80			100	nA
Small Signal Bandwidth	$A_V = 1.0$, $R_L = 10k\Omega$		350			350		kHz
	$A_V = 10$, $R_L = 10k\Omega$		35			35		kHz
	$A_V = 100$, $R_L = 10k\Omega$		3.5			3.5		kHz
	$A_V = 1000$, $R_L = 10k\Omega$		350			350		Hz
Full Power Bandwidth	$V_{IN} = \pm 10V$, $R_L = 10k$, $A_V = 1$		5.0			5.0		kHz
Input Voltage Range	Differential	±10	±12		±10	±12		V
	Common Mode	±10	±12		±10	±12		V
Gain Nonlinearity			0.03			0.03		%
Deviation From Gain Equation Formula	$A_V = 1$ to 1000		±0.3	±1.0		±1.0	±3.0	%
PSRR	$\pm 5.0V \leq V_S \leq \pm 15V$, $A_V = 1.0$		1.0	2.5		1.0	5.0	mV/V
	$\pm 5.0V \leq V_S \leq \pm 15V$, $A_V = 100$		0.05	0.25		0.10	0.50	mV/V
CMRR	$A_V = 1.0$ DC to		1.0	2.5		2.5	5.0	mV/V
	$A_V = 10$ 100 Hz		0.1	0.25		0.25	0.50	mV/V
	$A_V = 100$ $\Delta R_S = 1.0k$		50	100		50	100	μV/V
Output Voltage	$V_S = \pm 15V$, $R_L = 10k\Omega$,	±10	±13.5		±10	±13.5		V
	$V_S = \pm 1.5V$, $R_L = 100k\Omega$	±0.6	±0.8		±0.6	±0.8		V
Output Resistance			0.5			0.5		Ω
Supply Current			300	400		400	600	μA
Equivalent Input Noise Voltage	$0.1 Hz < f < 10 kHz$, $R_S < 50\Omega$		20			20		μV p-p
Slew Rate	$\Delta V_{IN} = \pm 10V$, $R_L = 10k\Omega$, $A_V = 1.0$		0.3			0.3		V/μs
Settling Time	To ±10 mV, $R_L = 10k\Omega$, $\Delta V_{OUT} = 1.0V$							
	$A_V = 1.0$		3.8			3.8		μs
	$A_V = 100$		180			180		μs

Note 1: Unless otherwise specified, all specifications apply for $V_S = \pm 15V$, Pins 1, 3, and 9 grounded, -25°C to +85°C for the LH0036C and -55°C to +125°C for the LH0036.

Note 2: All typical values are for $T_A = 25°C$.

Note 3: The maximum junction temperature is 150°C. For operation at elevated temperature derate the G package on a thermal resistance of 90°C/W, above 25°C.

Typical Performance Characteristics

Supply Current vs Temperature

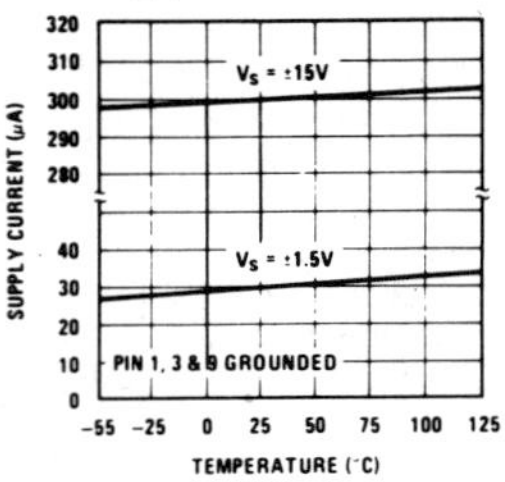

Supply Current vs Supply Voltage

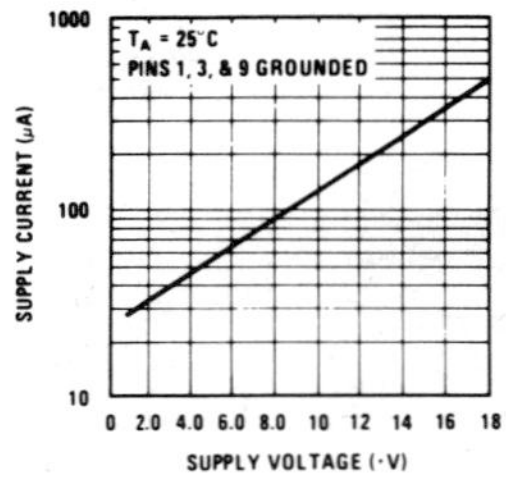

Output Voltage Swing vs Supply Voltage

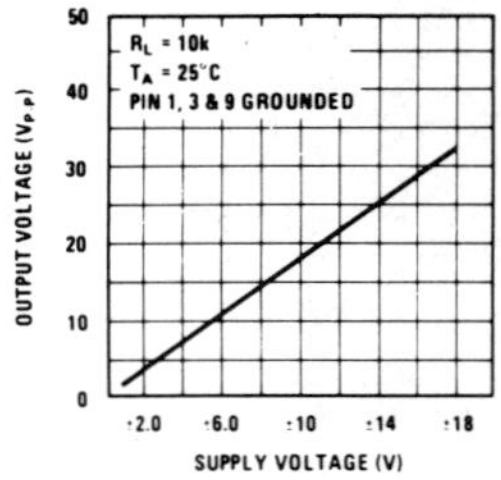

Peak to Peak Output Voltage Swing vs R_L

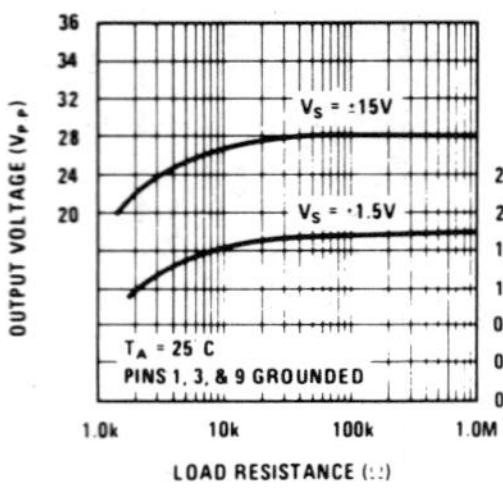

Total Input Noise Voltage* vs Frequency

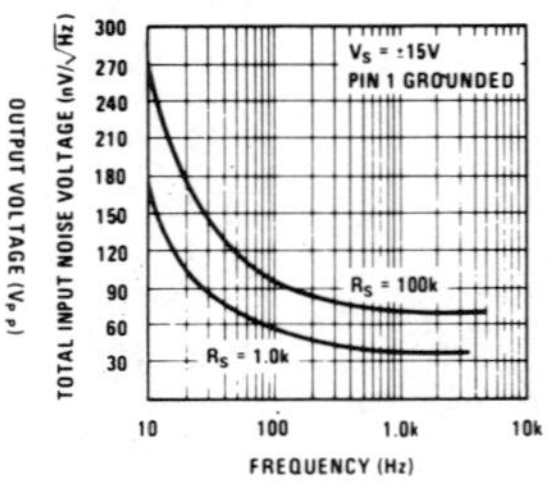

*Noise voltage includes contribution from source resistance

Closed Loop Voltage Gain vs Frequency

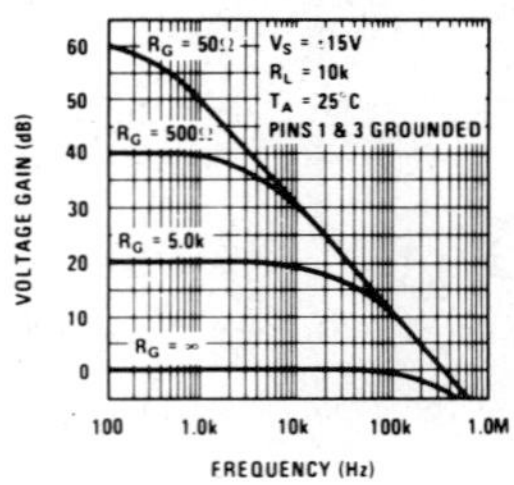

Closed Loop Voltage Gain

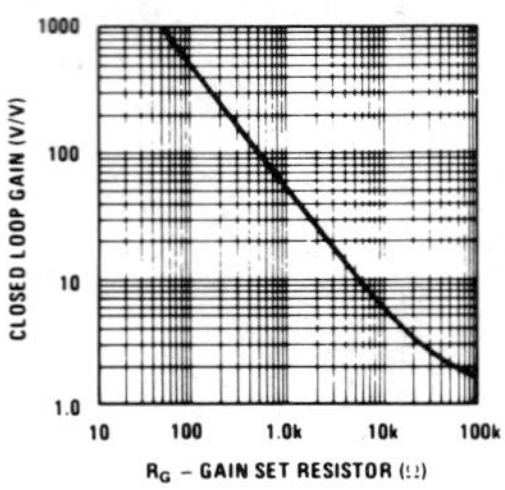

CMRR vs Frequency

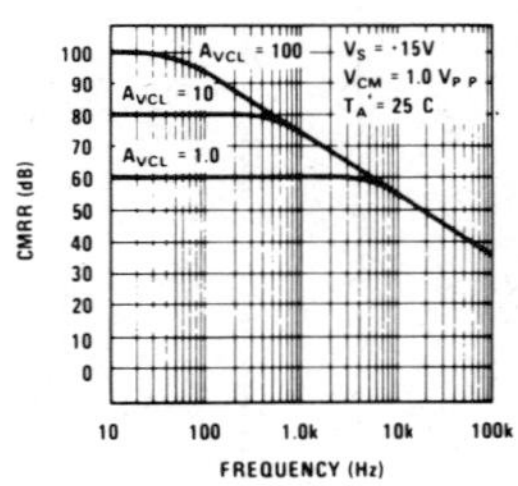

Common Mode Voltage vs Supply Voltage

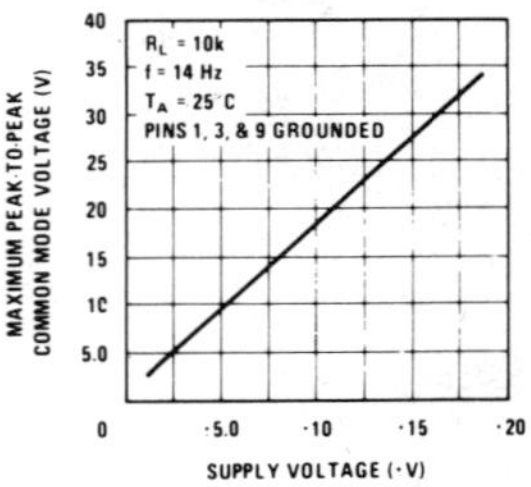

Input Bias Current

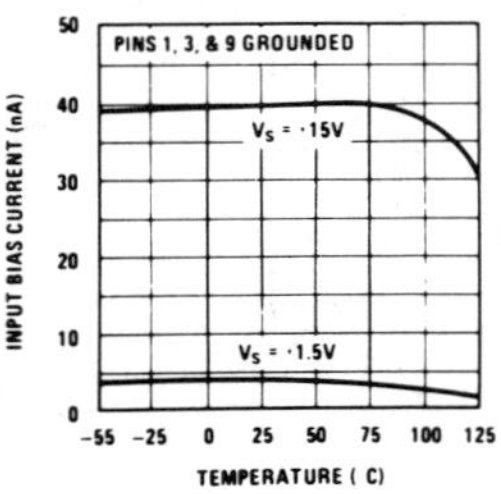

Output Voltage Swing vs Frequency

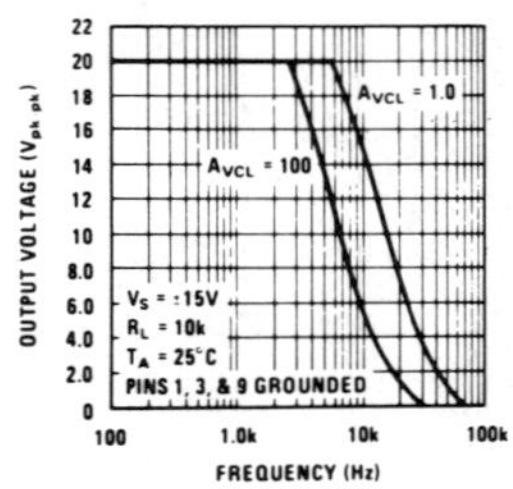

Large Signal Pulse Response

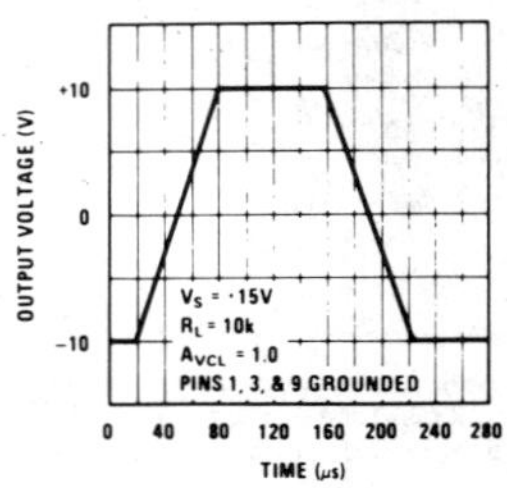

Typical Applications

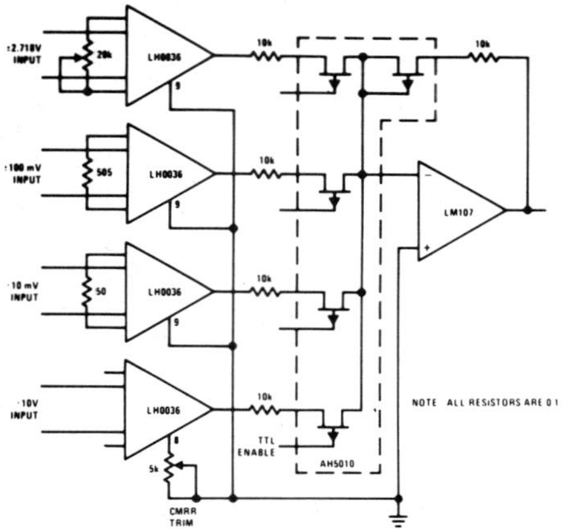

Pre MUX Signal Conditioning

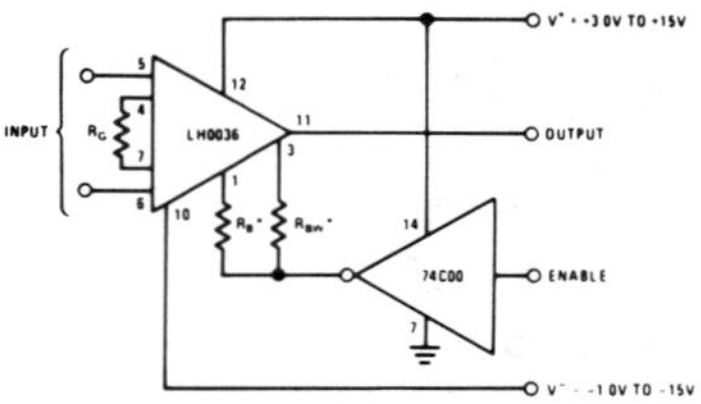

*R_{BW} AND R_B ARE OPTIONAL BANDWIDTH AND INPUT BIAS CURRENT CONTROLLING RESISTORS

Instrumentation Amplifier with Logic Controlled Shut-Down

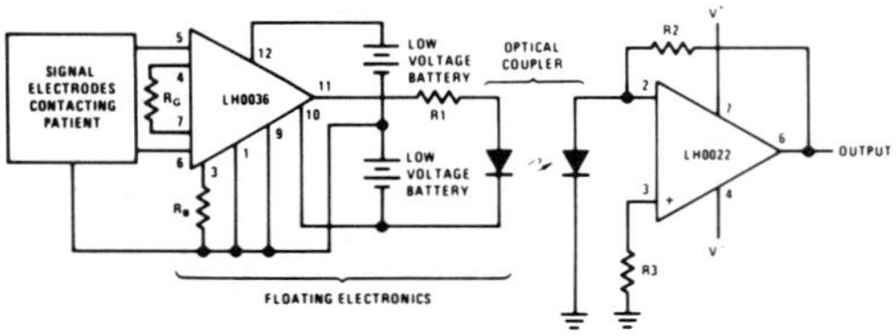

Isolation Amplifier for Medical Telemetry

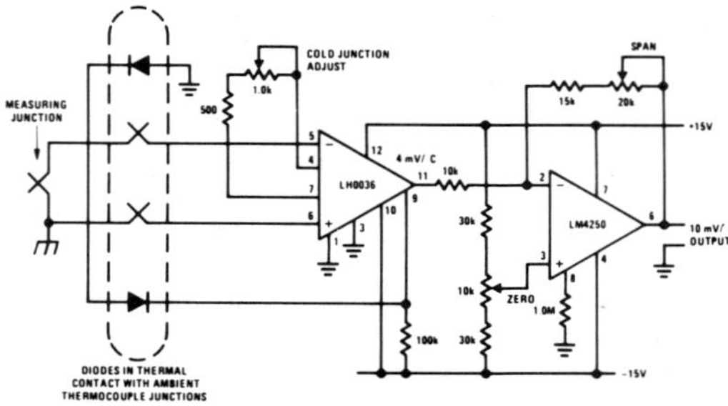

Thermocouple Amplifier with Cold Junction Compensation

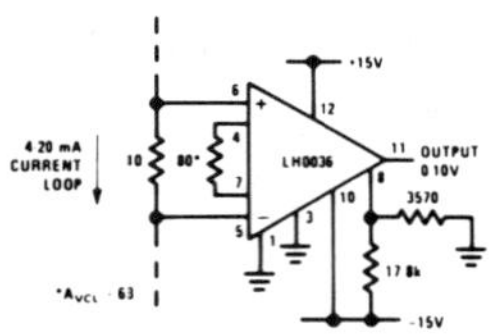

Process Control Interface

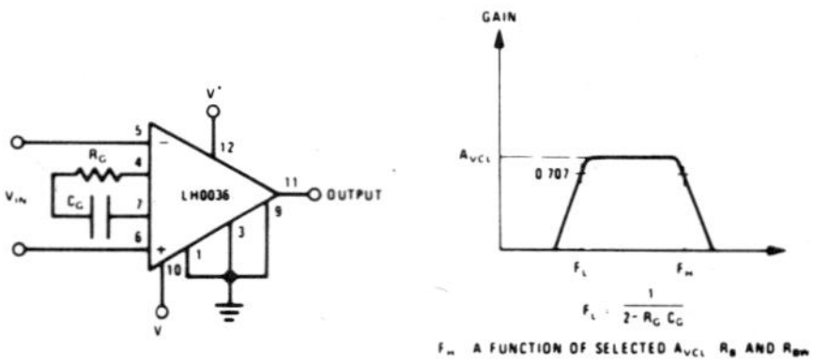

High Pass Filter

Applications Information

THEORY OF OPERATION

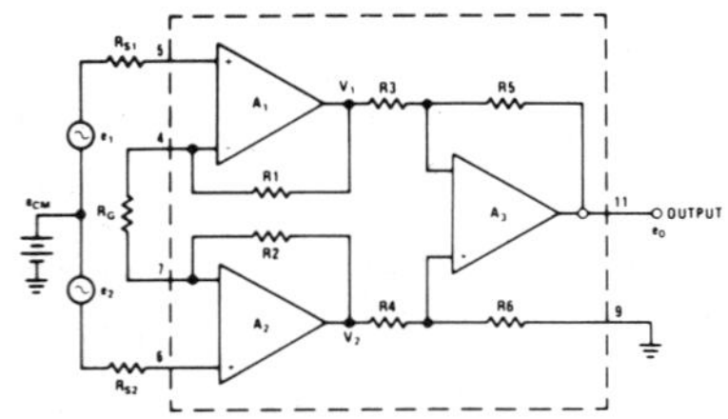

FIGURE 1. Simplified LH0036

The LH0036 is a 2 stage amplifier with a high input impedance gain stage comprised of A_1 and A_2 and a differential to single-ended unity gain stage, A_3. Operational amplifier, A_1, receives differential input signal, e_1, and amplifies it by a factor equal to $(R1 + R_G)/R_G$.

A_1 also receives input e_2 via A_2 and R2. e_2 is seen as an inverting signal with a gain of $R1/R_G$. A_1 also receives the common mode signal e_{CM} and processes it with a gain of +1.

Hence:

$$V_1 = \frac{R1 + R_G}{R_G} e_1 - \frac{R1}{R_G} e_2 + e_{CM} \quad (1)$$

By similar analysis V_2 is seen to be:

$$V_2 = \frac{R2 + R_G}{R_G} e_2 - \frac{R2}{R_G} e_1 + e_{CM} \quad (2)$$

For R1 = R2:

$$V_2 - V_1 = \left[\left(\frac{2R1}{R_G}\right) + 1\right](e_2 - e_1) \quad (3)$$

Also, for R3 = R5 = R4 = R6, the gain of $A_3 = 1$, and:

$$e_o = (1)(V_2 - V_1) = (e_2 - e_1)\left[1 + \left(\frac{2R1}{R_G}\right)\right] \quad (4)$$

As can be seen for identically matched resistors, e_{CM} is cancelled out, and the differential gain is dictated by equation (4).

For the LH0036, equation (4) reduces to:

$$A_{VCL} = \frac{e_o}{e_2 - e_1} = 1 + \frac{50k}{R_G} \quad (5a)$$

The closed loop gain may be set to any value from 1 ($R_G = \infty$) to 1000 ($R_G \cong 50\Omega$). Equation (5a) re-arranged in more convenient form may be used to select R_G for a desired gain:

$$R_G = \frac{50k}{A_{VCL} - 1} \quad (5b)$$

USE OF BANDWIDTH CONTROL (pin 1)

In the standard configuration, pin 1 of the LH0036 is simply grounded. The amplifier's slew rate in this configuration is typically 0.3V/μs and small signal bandwidth 350 kHz for $A_{VCL} = 1$. In some applications, particularly at low frequency, it may be desirable to limit bandwidth in order to minimize the overall noise bandwidth of the device. A resistor R_{BW} may be placed between pin 1 and ground to accomplish this purpose. Figure 2 shows typical small signal bandwidth versus R_{BW}.

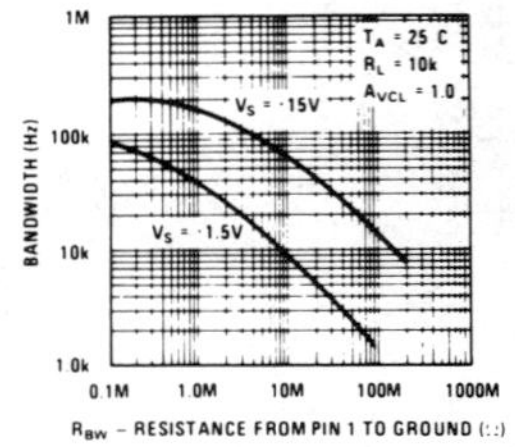

FIGURE 2. Bandwidth vs R_{BW}

It also should be noted that large signal bandwidth and slew rate may be adjusted down by use of R_{BW}. Figure 3 is plot of slew rate versus R_{BW}.

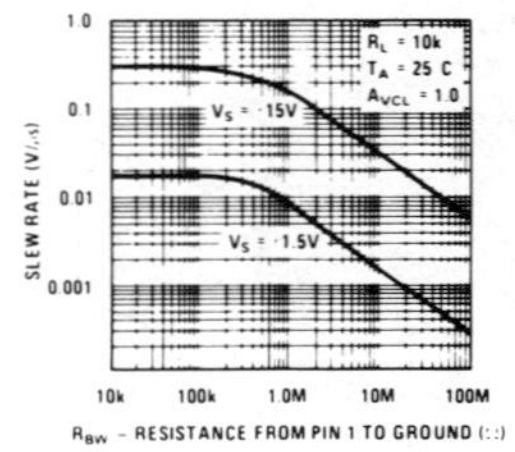

FIGURE 3. Output Slew Rate vs R_{BW}

CMRR CONSIDERATIONS

Use of Pin 9, CMRR Preset

Pin 9 should be grounded for nominal operation. An internal factory trimmed resistor, R6, will yield a CMRR in excess of 80 dB (for $A_{VCL} = 100$). Should a higher CMRR be desired, pin 9 should be left open and the procedure, in this section followed.

DC Off-set Voltage and Common Mode Rejection Adjustments

Off-set may be nulled using the circuit shown in Figure 4.

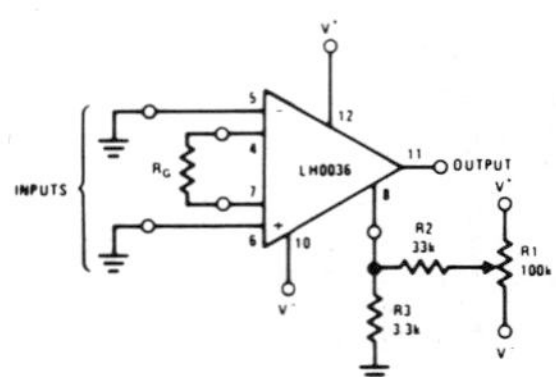

FIGURE 4. V_{OS} Adjustment Circuit

Pin 8 is also used to improve the common mode rejection ratio as shown in Figure 5. Null is

Applications Information (Continued)

achieved by alternately applying ±10V (for V^+ & V^- = 15V) to the inputs and adjusting R1 for minimum change at the output.

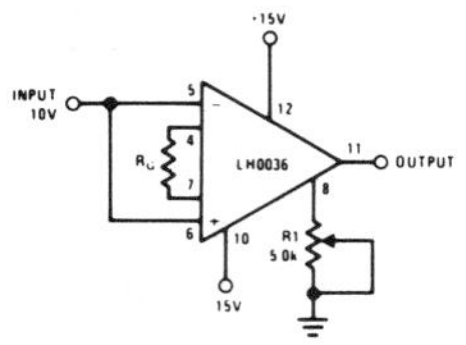

FIGURE 5. CMRR Adjustment Circuit

The circuits of Figure 4 and 5 may be combined as shown in Figure 6 to accomplish both V_{OS} and CMRR null. However, the V_{OS} and CMRR adjustment are interactive and several iterations are required. The procedure for null should start with the inputs grounded.

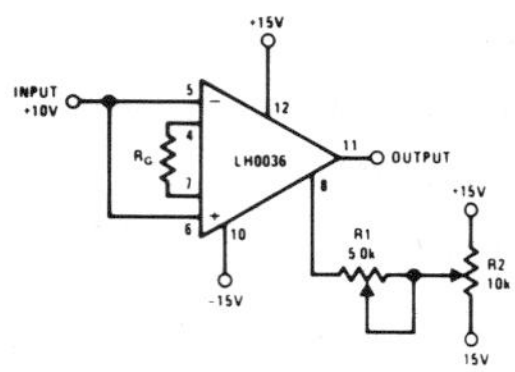

FIGURE 6. Combined CMRR, V_{OS} Adjustment Circuit

R2 is adjusted for V_{OS} null. An input of +10V is then applied and R1 is adjusted for CMRR null. The procedure is then repeated until the optimum is achieved.

A circuit which overcomes adjustment interaction is shown in Figure 7. In this case, R2 is adjusted first for output null of the LH0036. R1 is then adjusted for output null with +10V input. It is always a good idea to check CMRR null with a −10V input. The optimum null achievable will yield the highest CMRR over the amplifiers common mode range.

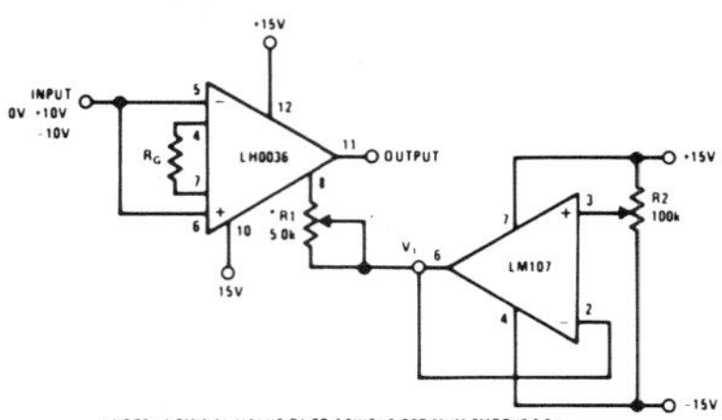

* NOTE: NOMINAL VALUE R1 TO ACHIEVE OPTIMUM CMRR IS 3.0k

FIGURE 7. Improved V_{OS}, CMRR Nulling Circuit

AC CMRR Considerations

The ac CMRR may be improved using the circuit of Figure 8.

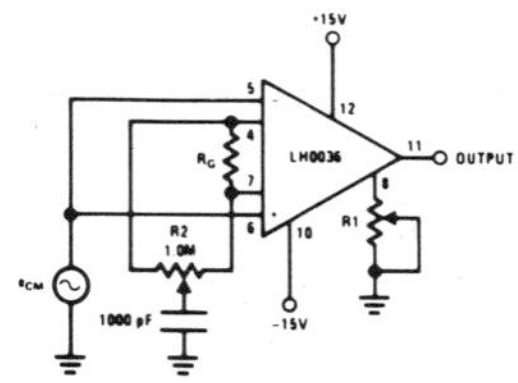

FIGURE 8. Improved AC CMRR Circuit

After adjusting R1 for best dc CMRR as before, R2 should be adjusted for minimum peak-to-peak voltage at the output while applying an ac common mode signal of the maximum amplitude and frequency of interest.

INPUT BIAS CURRENT CONTROL

Under nominal operating conditions (pin 3 grounded), the LH0036 requires input currents of 40 nA. The input current may be reduced by inserting a resistor (R_B) between 3 and ground or, alternatively, between 3 and V^-. For R_B returned to ground, the input bias current may be predicted by:

$$I_{BIAS} \cong \frac{V^+ - 0.5}{4 \times 10^8 + 800\ R_B} \tag{6a}$$

or

$$R_B = \frac{V^+ - 0.5 - (4 \times 10^8)\,(I_{BIAS})}{800\ I_{BIAS}} \tag{6b}$$

Where:

I_{BIAS} = Input Bias Current (nA)

R_B = External Resistor connected between pin 3 and ground (Ohms)

V^+ = Positive Supply Voltage (Volts)

Figure 9 is a plot of input bias current versus R_B.

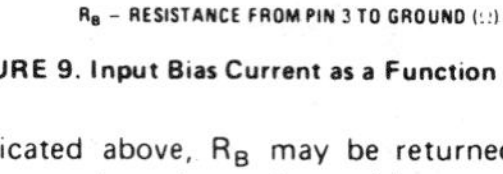

FIGURE 9. Input Bias Current as a Function of R_B

As indicated above, R_B may be returned to the negative supply voltage. Input bias current may then be predicted by:

$$I_{BIAS} \cong \frac{(V^+ - V^-) - 0.5}{4 \times 10^8 + 800\ R_B}$$

Applications Information (Continued)

or

$$R_B \cong \frac{(V^+ - V^-) - 0.5 - (4 \times 10^8)(I_{BIAS})}{800\ I_{BIAS}} \quad (8)$$

Where:

I_{BIAS} = Input Bias Current (nA)

R_B = External resistor connected between pin 3 and V^- (Ohms)

V^+ = Positive Supply Voltage (Volts)

V^- = Negative Supply Voltage (Volts)

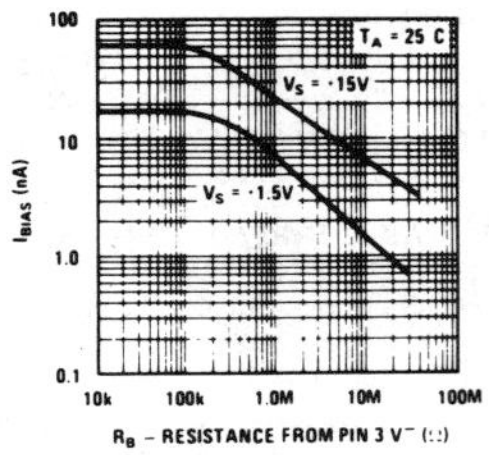

FIGURE 10. Input Bias Current as a Function of R_B

Figure 10 is a plot of input bias current versus R_B returned to V^- it should be noted that bandwidth is affected by changes in R_B. Figure 11 is a plot of bandwidth versus R_B.

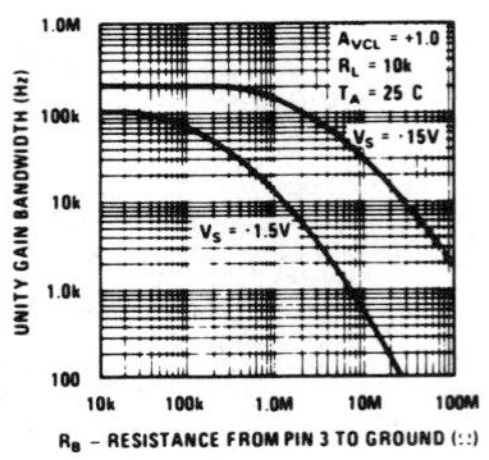

FIGURE 11. Unity Gain Bandwidth as a Function of R_B

BIAS CURRENT RETURN PATH CONSIDERATIONS

The LH0036 exhibits input bias currents typically in the 40 nA region in each input. This current must flow through R_{ISO} as shown in Figure 12.

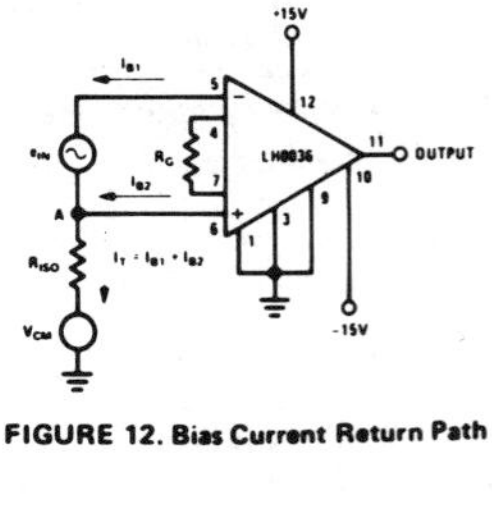

FIGURE 12. Bias Current Return Path

In a typical application, $V_S = \pm 15V$, $I_{B1} \cong I_{B2} \cong 40$ nA, the total current, I_T, would flow through R_{ISO} causing a voltage rise at point A. For values of $R_{ISO} \geq 150\ M\Omega$, the voltage at point A exceeds the +12V common range of the device. Clearly, for $R_{ISO} = \infty$, the LH0036 would be driven to positive saturation.

The implication is that a finite impedance must be supplied between the input and power supply ground. The value of the resistor is dictated by the maximum input bias current, and the common mode voltage. Under worst case conditions:

$$R_{ISO} \leq \frac{V_{CMR} - V_{CM}}{I_T} \quad (9)$$

Where:

V_{CMR} = Common Mode Range (10V for the LH0036)

V_{CM} = Common Mode Voltage

$I_T = I_{B1} + I_{B2}$

In applications in which the signal source is floating, such as a thermocouple, one end of the source may be grounded directly or through a resistor.

GUARD OUTPUT

Pin 2 of the LH0036 is provided as a guard drive pin in those stringent applications which require very low leakage and minimum input capacitance. Pin 2 will always be biased at the input common mode voltage. The source impedance looking into pin 2 is approximately 15 kΩ. Proper use of the guard/shield pin is shown in Figure 13.

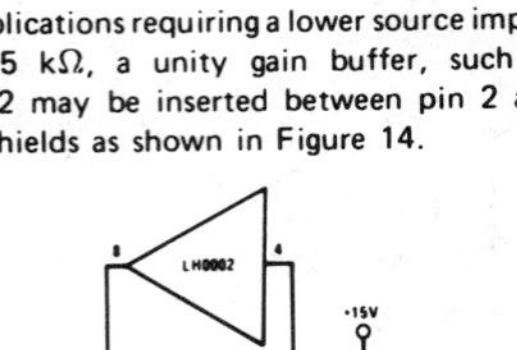

FIGURE 13. Use of Guard

For applications requiring a lower source impedance than 15 kΩ, a unity gain buffer, such as the LH0002 may be inserted between pin 2 and the input shields as shown in Figure 14.

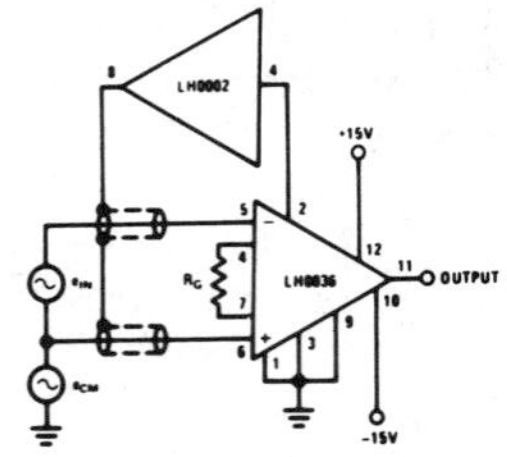

FIGURE 14. Guard Pin With Buffer

National Semiconductor

Successive Approximation Registers/Comparators

LM311 Voltage Comparator

General Description

The LM311 is a voltage comparator that has input currents more than a hundred times lower than devices like the LM306 or LM710C. It is also designed to operate over a wider range of supply voltages: from standard ±15V op amp supplies down to the single 5V supply used for IC logic. Its output is compatible with RTL, DTL and TTL as well as MOS circuits. Further, it can drive lamps or relays, switching voltages up to 40V at currents as high as 50 mA.

Features

- Operates from single 5V supply
- Maximum input current: 250 nA
- Maximum offset current: 50 nA
- Differential input voltage range: ±30V
- Power consumption: 135 mW at ±15V

Both the input and the output of the LM311 can be isolated from system ground, and the output can drive loads referred to ground, the positive supply or the negative supply. Offset balancing and strobe capability are provided and outputs can be wire OR'ed. Although slower than the LM306 and LM710C (200 ns response time vs 40 ns) the device is also much less prone to spurious oscillations. The LM311 has the same pin configuration as the LM306 and LM710C. See the "application hints" of the LM311 for application help.

Auxiliary Circuits**

**Note: Pin connections shown on schematic diagram and typical applications are for TO-5 package.

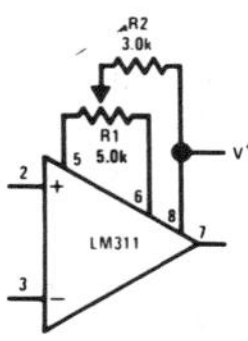

Offset Balancing

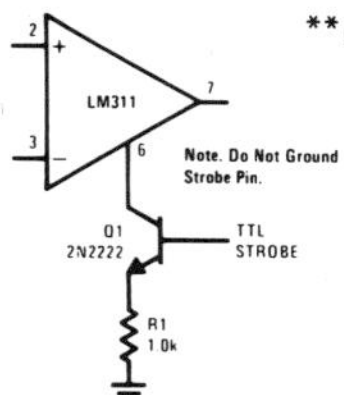

Strobing

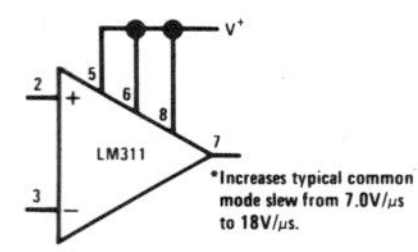

Increasing Input Stage Current*

Typical Applications**

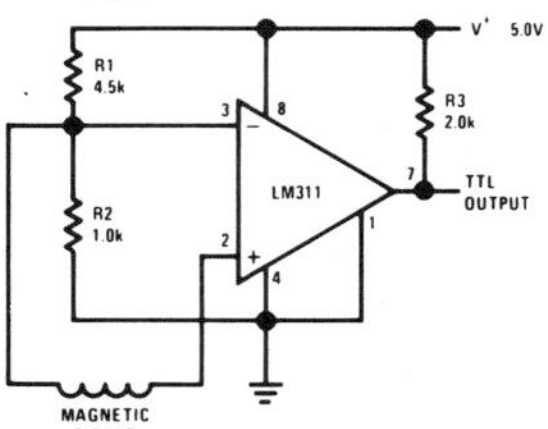

Detector for Magnetic Transducer

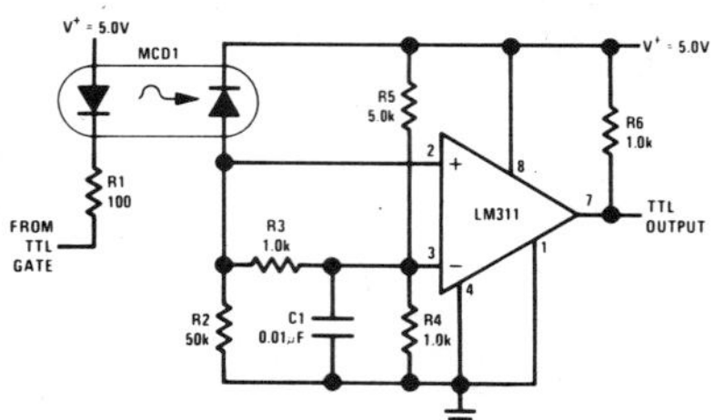

Digital Transmission Isolator

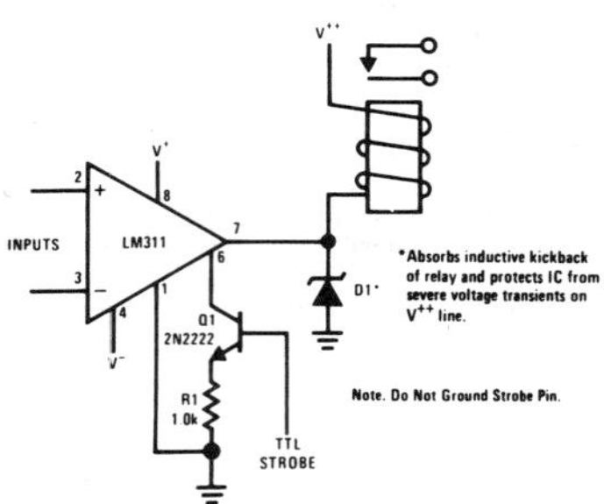

Relay Driver with Strobe

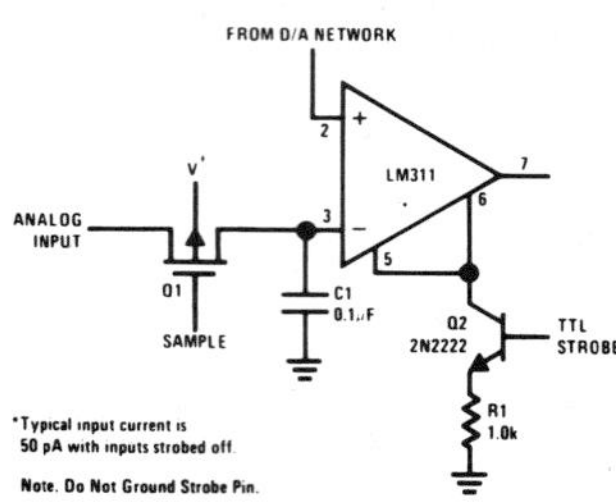

Strobing off Both Input* and Output Stages

Courtesy of National Semiconductor Corporation.

Absolute Maximum Ratings

Total Supply Voltage (V_{84})	36V
Output to Negative Supply Voltage (V_{74})	40V
Ground to Negative Supply Voltage (V_{14})	30V
Differential Input Voltage	±30V
Input Voltage (Note 1)	±15V
Power Dissipation (Note 2)	500 mW
Output Short Circuit Duration	10 sec
Operating Temperature Range	0°C to 70°C
Storage Temperature Range	-65°C to 150°C
Lead Temperature (soldering, 10 sec)	300°C
Voltage at Strobe Pin	$V^+ - 5V$

Electrical Characteristics (Note 3)

PARAMETER	CONDITIONS	MIN	TYP	MAX	UNITS
Input Offset Voltage (Note 4)	$T_A = 25°C$, $R_S \leq 50k$		2.0	7.5	mV
Input Offset Current (Note 4)	$T_A = 25°C$		6.0	50	nA
Input Bias Current	$T_A = 25°C$		100	250	nA
Voltage Gain	$T_A = 25°C$	40	200		V/mV
Response Time (Note 5)	$T_A = 25°C$		200		ns
Saturation Voltage	$V_{IN} \leq -10$ mV, $I_{OUT} = 50$ mA $T_A = 25°C$		0.75	1.5	V
Strobe ON Current	$T_A = 25°C$		3.0		mA
Output Leakage Current	$V_{IN} \geq 10$ mV, $V_{OUT} = 35V$ $T_A = 25°C$, $I_{STROBE} = 3$ mA		0.2	50	nA
Input Offset Voltage (Note 4)	$R_S \leq 50k$			10	mV
Input Offset Current (Note 4)				70	nA
Input Bias Current				300	nA
Input Voltage Range		−14.5	13.8,−14.7	13.0	V
Saturation Voltage	$V^+ \geq 4.5V$, $V^- = 0$ $V_{IN} \leq -10$ mV, $I_{SINK} \leq 8$ mA		0.23	0.4	V
Positive Supply Current	$T_A = 25°C$		5.1	7.5	mA
Negative Supply Current	$T_A = 25°C$		4.1	5.0	mA

Note 1: This rating applies for ±15V supplies. The positive input voltage limit is 30V above the negative supply. The negative input voltage limit is equal to the negative supply voltage or 30V below the positive supply, whichever is less.

Note 2: The maximum junction temperature of the LM311 is 110°C. For operating at elevated temperatures, devices in the TO-5 package must be derated based on a thermal resistance of 150°C/W, junction to ambient, or 45°C/W, junction to case. The thermal resistance of the dual-in-line package is 100°C/W, junction to ambient.

Note 3: These specifications apply for $V_S = \pm15V$ and the Ground pin at ground, and $0°C < T_A < +70°C$, unless otherwise specified. The offset voltage, offset current and bias current specifications apply for any supply voltage from a single 5V supply up to ±15V supplies.

Note 4: The offset voltages and offset currents given are the maximum values required to drive the output within a volt of either supply with 1 mA load. Thus, these parameters define an error band and take into account the worst-case effects of voltage gain and input impedance.

Note 5: The response time specified (see definitions) is for a 100 mV input step with 5 mV overdrive.

Note 6: Do not short the strobe pin to ground; it should be current driven at 3 to 5 mA.

Typical Performance Characteristics

Input Bias Current

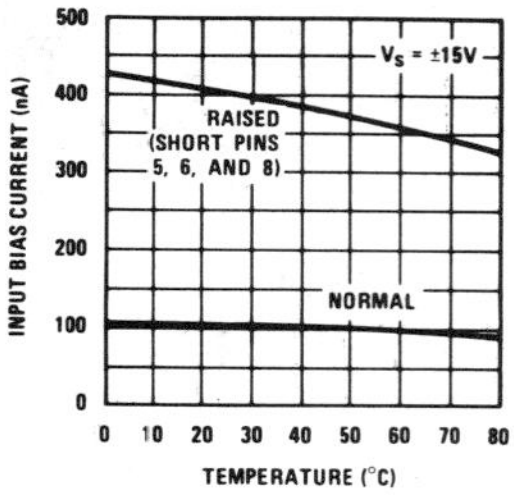

Input Offset Current

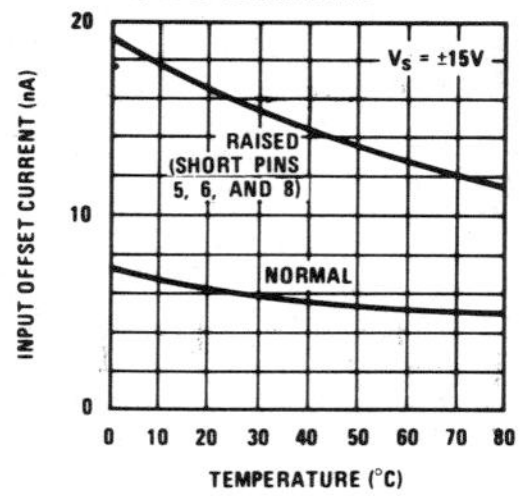

Offset Error

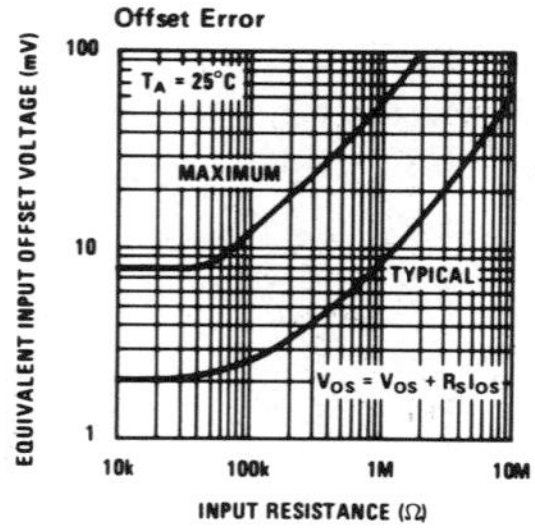

Input Characteristics

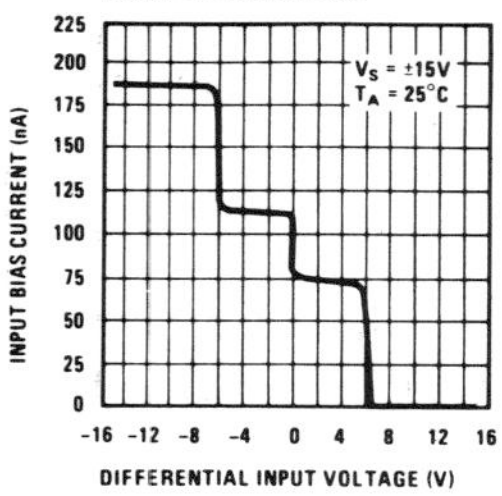

Common Mode Limits

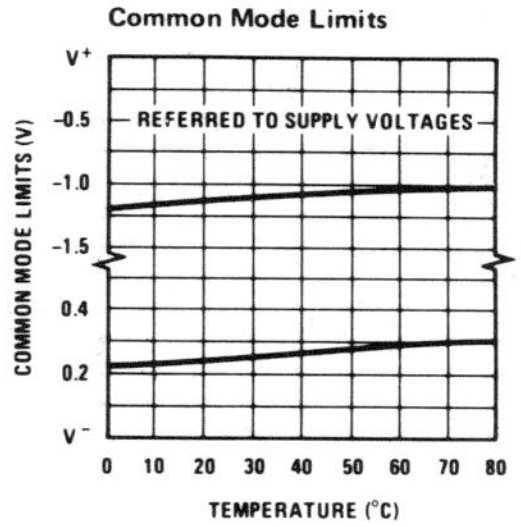

Transfer Function

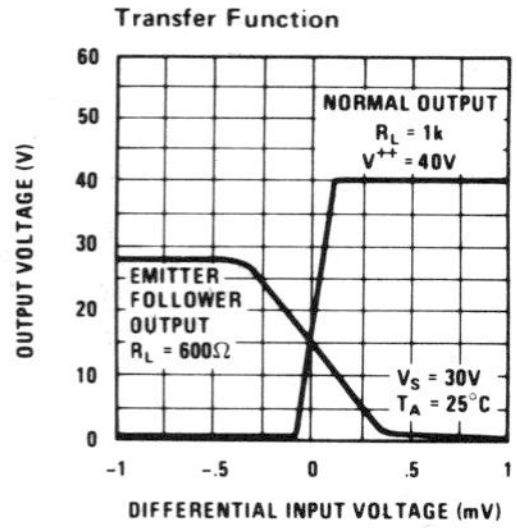

Response Time for Various Input Overdrives

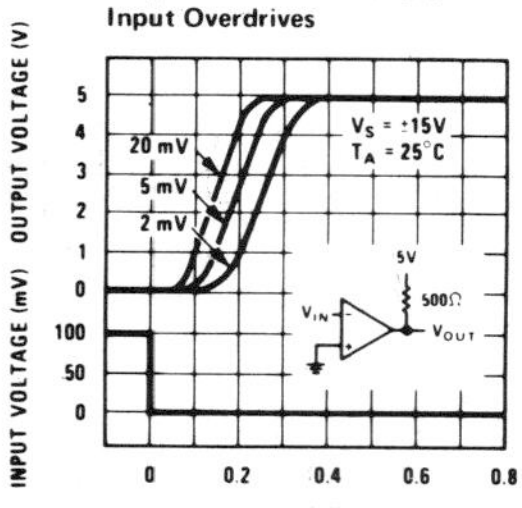

Response Time for Various Input Overdrives

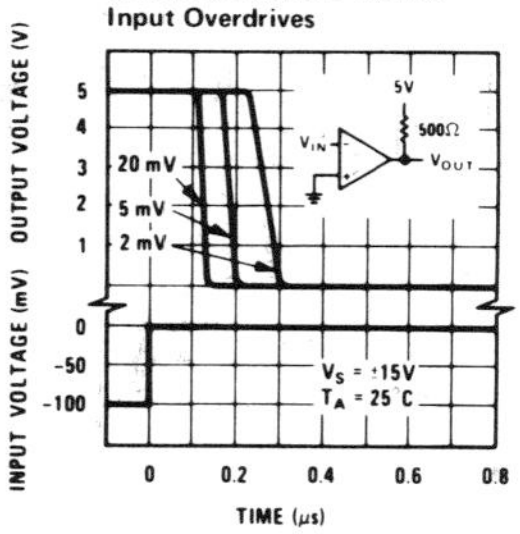

Output Saturation Voltage

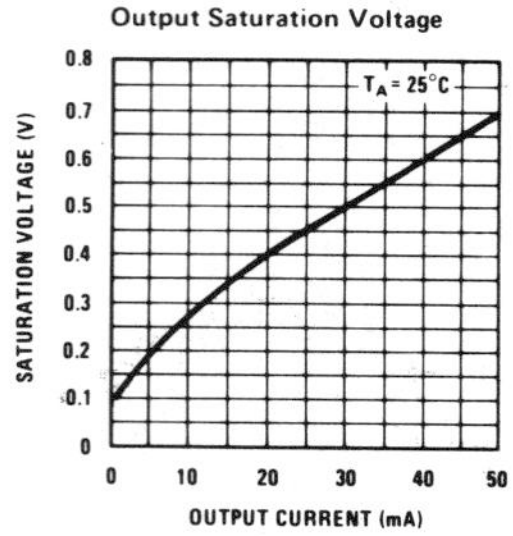

Response Time for Various Input Overdrives

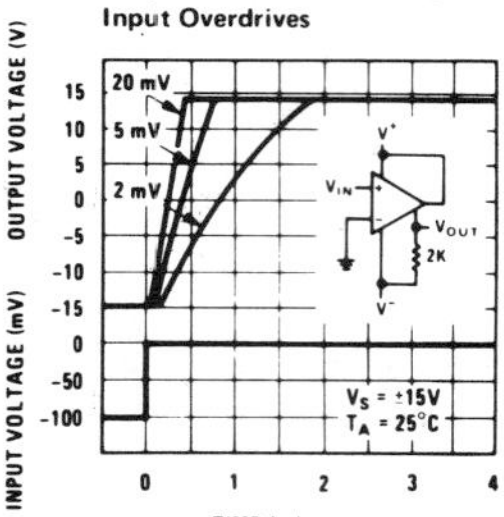

Response Time for Various Input Overdrives

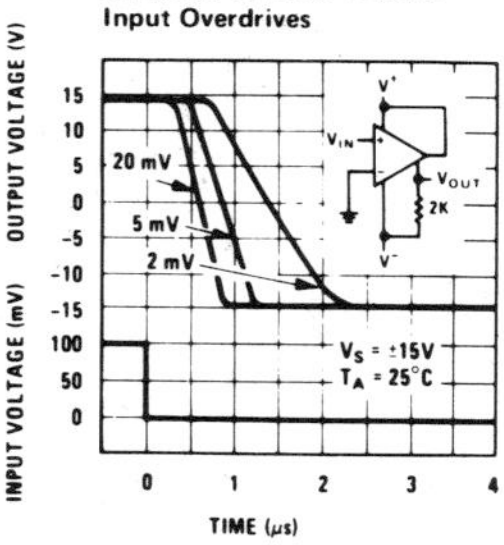

Output Limiting Characteristics

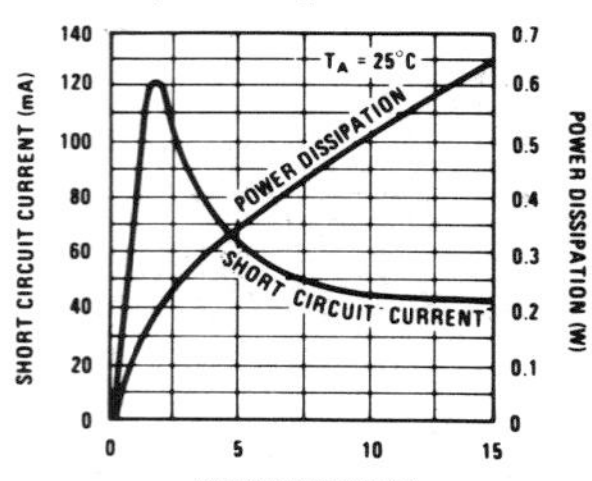

Typical Performance Characteristics (Continued)

Supply Current

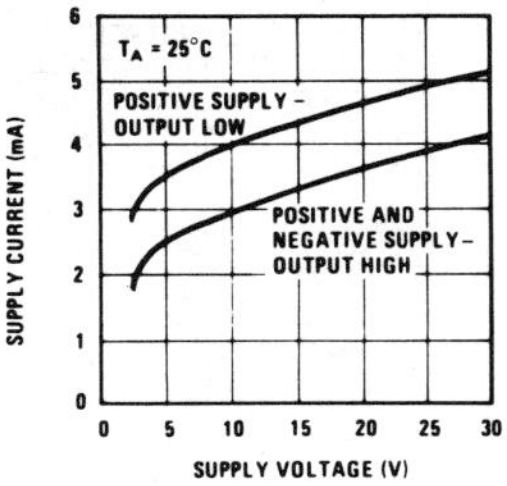

Supply Current

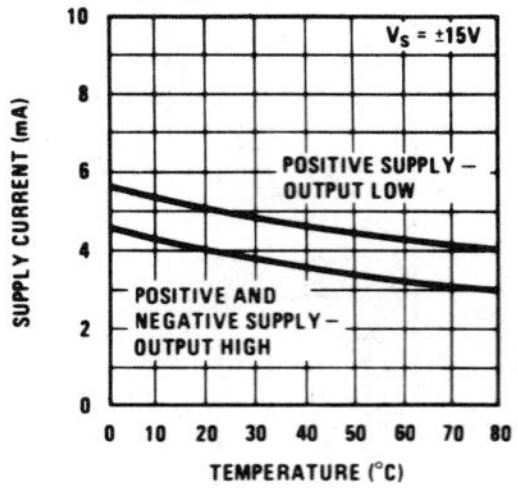

Leakage Currents

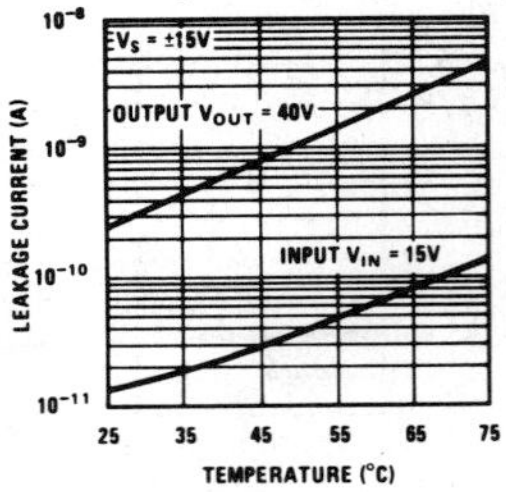

Typical Applications

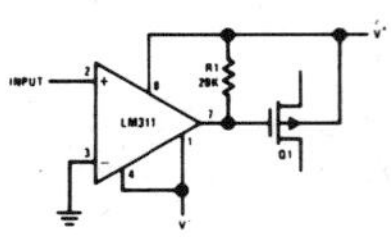

Zero Crossing Detector Driving MOS Switch

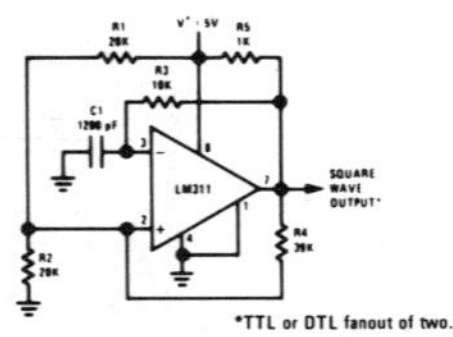

*TTL or DTL fanout of two.

100 kHz Free Running Multivibrator

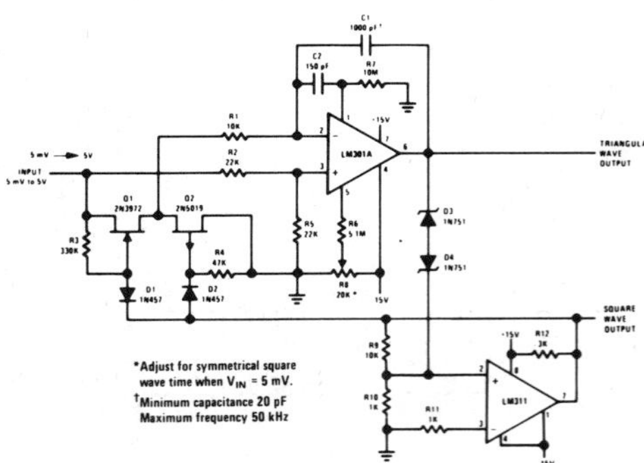

*Adjust for symmetrical square wave time when $V_{IN} = 5$ mV.
†Minimum capacitance 20 pF Maximum frequency 50 kHz

10 Hz to 10 kHz Voltage Controlled Oscillator

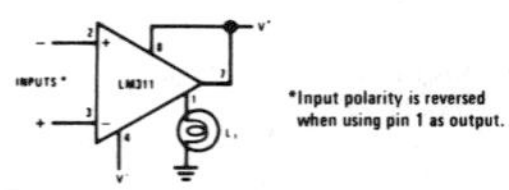

*Input polarity is reversed when using pin 1 as output.

Driving Ground-Referred Load

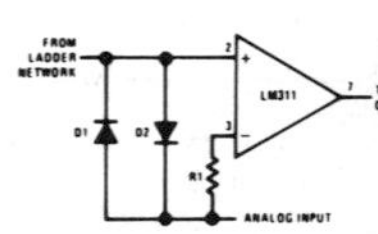

Using Clamp Diodes to Improve Response

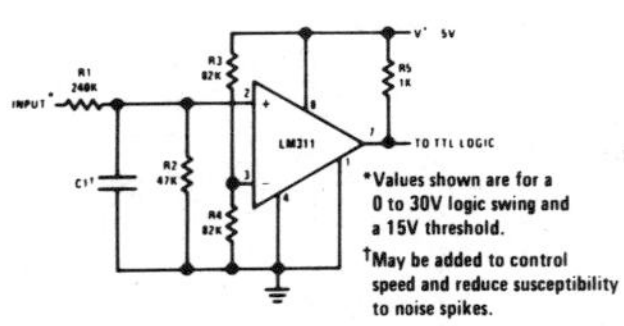

*Values shown are for a 0 to 30V logic swing and a 15V threshold.
†May be added to control speed and reduce susceptibility to noise spikes.

TTL Interface with High Level Logic

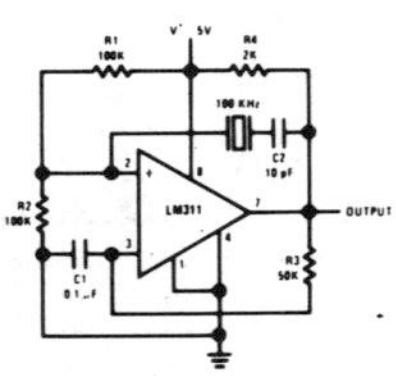

Crystal Oscillator

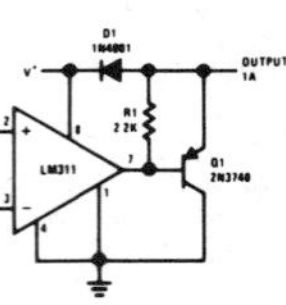

Comparator and Solenoid Driver

Schematic Diagram

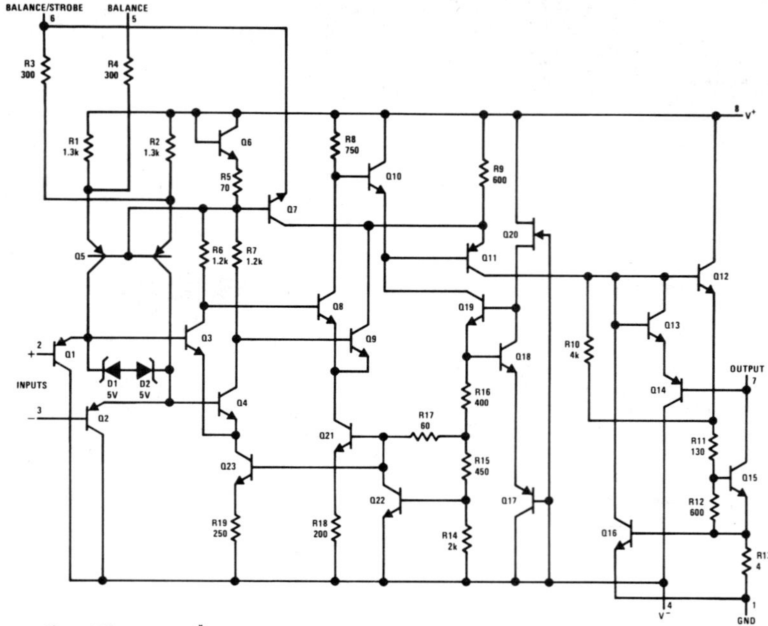

Connection Diagrams *

Metal Can Package

Order Number LM311H
See NS Package H08C

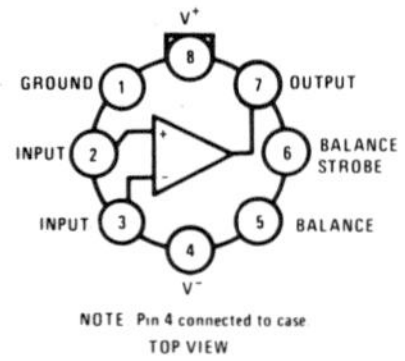

Dual-In-Line Package

Order Number LM311N
See NS Package N08B
Order Number LM311J-8
See NS Package J08A

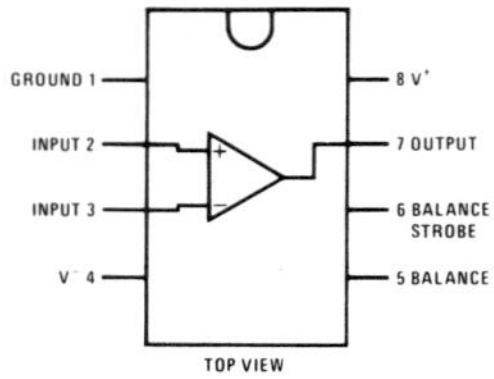

Dual-In-Line Package

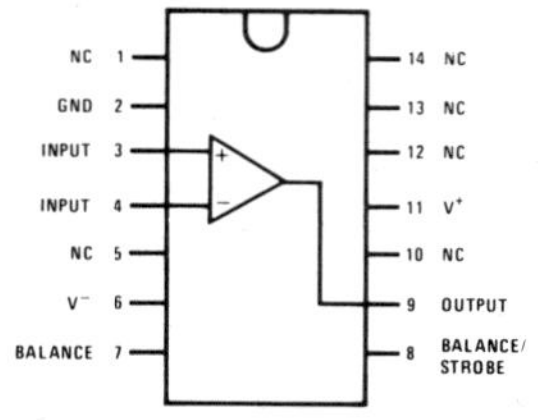

Order Number LM311N-14
See NS Package N14A
Order Number LM311J
See NS Package J14A

Note: Pin 6 connected to bottom of package.

*Pin connections shown on schematic diagram and typical applications are for TO-5 package.

MOTOROLA Semiconductors
BOX 20912 • PHOENIX, ARIZONA 85036

MC3401P

Specifications and Applications Information

QUAD
OPERATIONAL AMPLIFIER
SILICON MONOLITHIC
INTEGRATED CIRCUIT

QUAD SINGLE-SUPPLY OPERATIONAL AMPLIFIER

These internally compensated operational amplifiers are designed specifically for single positive power supply applications found in industrial control systems and automotive electronics. Each MC3401P device contains four independent amplifiers – making it ideal for applications such as active filters, multi-channel amplifiers, tachometer, oscillator and other similar usages.

- Single-Supply Operation – +5.0 Vdc to +18 Vdc
- Internally Compensated
- Wide Unity Gain Bandwidth – 5.0 MHz typical
- Low Input Bias Current – 50 nA typical
- High Open-Loop Gain – 1000 V/V minimum

PLASTIC PACKAGE
CASE 646

FIGURE 1 – EQUIVALENT CIRCUIT

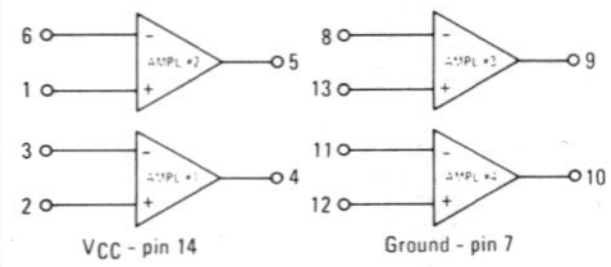

FIGURE 2 – SMALL-SIGNAL TRANSIENT RESPONSE

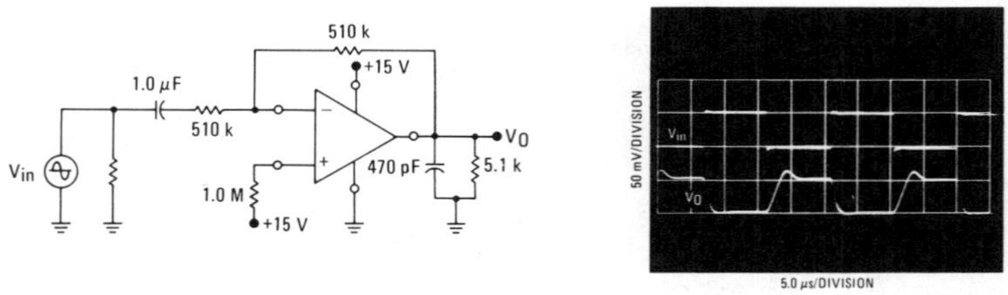

FIGURE 3 – INVERTING AMPLIFIER

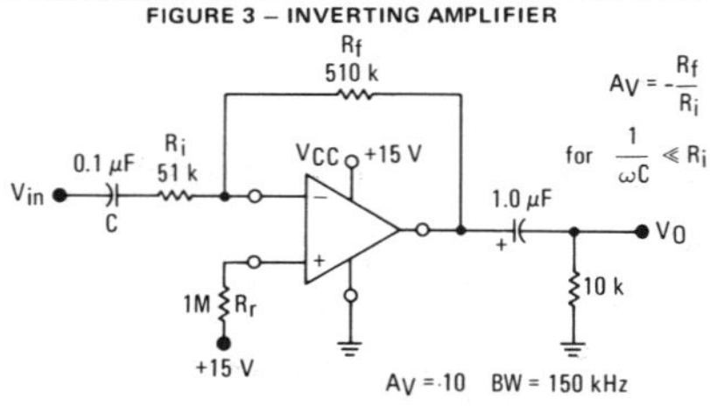

FIGURE 4 – NONINVERTING AMPLIFIER

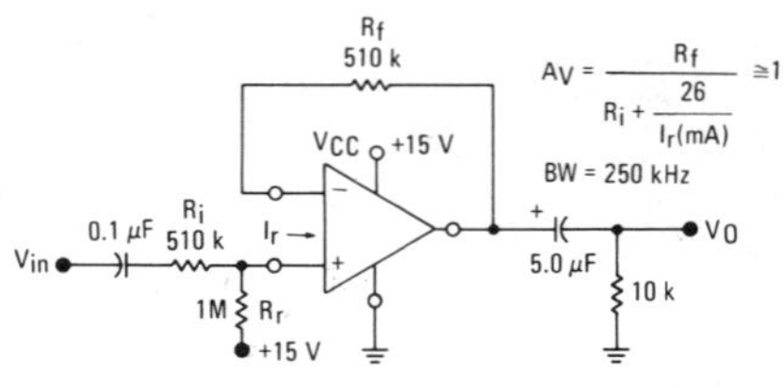

Courtesy of National Semiconductor Corporation.

MAXIMUM RATINGS (T_A = +25°C unless otherwise noted)

Rating	Symbol	Value	Unit
Power Supply Voltage	V_{CC}	+18	Vdc
Non-inverting Input Current	I_{in}	5.0	mA
Power Dissipation Derate above T_A = +25°C	P_D	625 5.0	mW mW/°C
Operating Temperature Range	T_A	0 to +75	°C
Storage Temperature Range	T_{stg}	-65 to +150	°C

ELECTRICAL CHARACTERISTICS [V_{CC} = +15 Vdc, R_L = 5.0 kΩ, T_A = +25°C (each amplifier) unless otherwise noted]

Characteristic	Fig. No.	Note	Symbol	Min	Typ	Max	Unit
Open-Loop Voltage Gain T_A = +25°C 0°C ⩽ T_A ⩽ +75°C	5,9,10	1	A_{vol}	 1000 800	 2000 –	 – –	V/V
Quiescent Power Supply Current (Total for four amplifiers) Noninverting inputs open Noninverting inputs grounded	6,12	2	 I_{DO} I_{DG}	 – –	 6.9 7.8	 10 14	mAdc
Input Bias Current, R_L = ∞ T_A = +25°C 0°C ⩽ T_A ⩽ +75°C	5	3	I_{IB}	 – –	 50 –	 300 500	nAdc
Output Current Source Capability Sink Capability	5 13 14	4	 I_{source} I_{sink}	 5.0 0.5	 10 1.0	 – –	mAdc
Output Voltage High Voltage Low Voltage Undistorted Output Swing (0°C < T_A < +75°C)	 7 7 8	 5 5 6	 V_{OH} V_{OL} $V_{O(p-p)}$	 13.5 – 10	 14.2 0.03 13.5	 – 0.1 –	Vdc $V_{(p-p)}$
Input Resistance	5		R_{in}	0.1	1.0	–	MEG Ω
Slew Rate (C_L = 100 pF, R_L = 5.0 k)			SR	–	0.6	–	V/μs
Unity Gain Bandwidth			BW	–	5.0	–	MHz
Phase Margin			ϕ_m	–	70	–	Degrees
Power Supply Rejection (f = 100 Hz)		7	PSSR	–	55	–	dB
Channel Separation (f = 1.0 kHz)			e_{o1}/e_{o2}	–	65	–	dB

NOTES

1. Open loop voltage gain is defined as the voltage gain from the inverting input to the output.
2. The quiescent current will increase approximately 0.3 mA for each noninverting input which is grounded. Leaving the noninverting input open causes the apparent input bias current to increase slightly (100 nA) at high temperatures.
3. Input bias current can be defined only for the inverting input. The noninverting input is not a true "differential input" – as with a conventional IC operational amplifier. As such this input does not have a requirement for input bias current.
4. Sink current is specified for linear operation. When the device is used as a gate or a comparator (non-linear operation), the sink capability of the device is approximately 5.0 milliamperes.
5. When used as a noninverting amplifier, the minimum output voltage is the V_{BE} of the inverting input transistor.
6. Peak-to-peak restrictions are due to the variations of the quiescent dc output voltage in the standard configuration (Figure 8).
7. Power supply rejection is specified at closed loop unity gain, and therefore indicates the supply rejection of both the biasing circuitry and the feedback amplifier.

TYPICAL CHARACTERISTICS

(V_{CC} = +15 Vdc, R_L = 5.0 kΩ, T_A = +25°C
[each amplifier] unless otherwise noted.)

FIGURE 9 – OPEN-LOOP VOLTAGE GAIN versus FREQUENCY

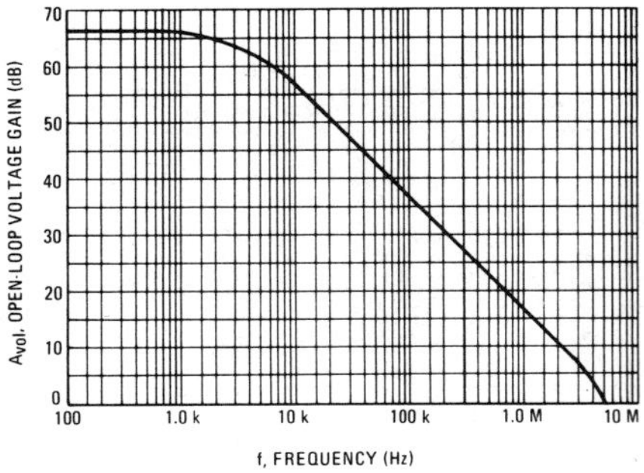

FIGURE 10 – OPEN-LOOP VOLTAGE GAIN versus SUPPLY VOLTAGE

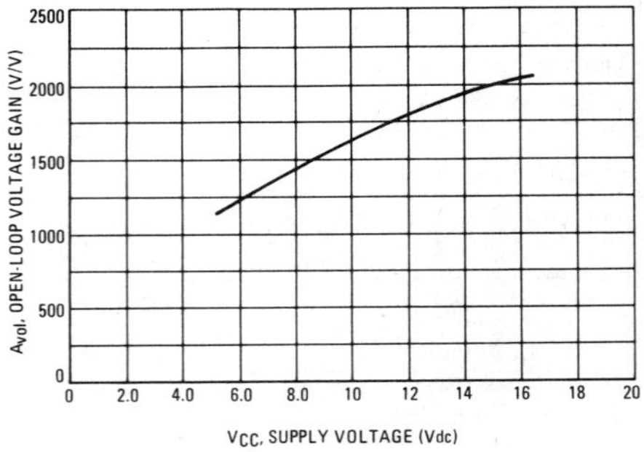

FIGURE 11 – OUTPUT RESISTANCE versus FREQUENCY

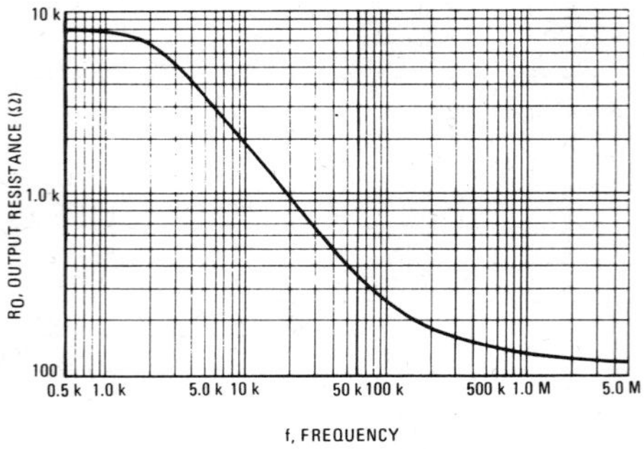

FIGURE 12 – SUPPLY CURRENT versus SUPPLY VOLTAGE

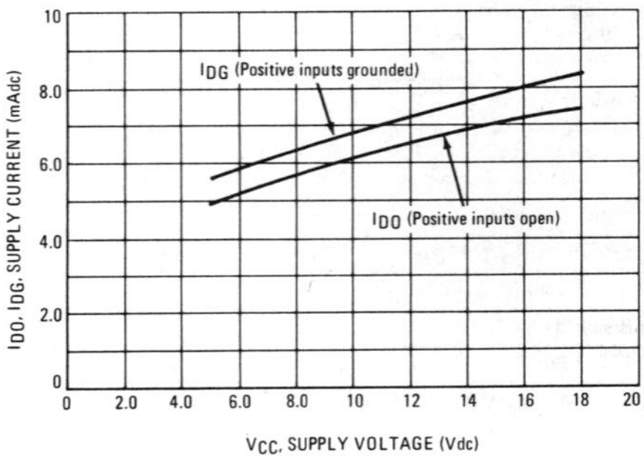

FIGURE 13 – LINEAR SOURCE CURRENT versus SUPPLY VOLTAGE

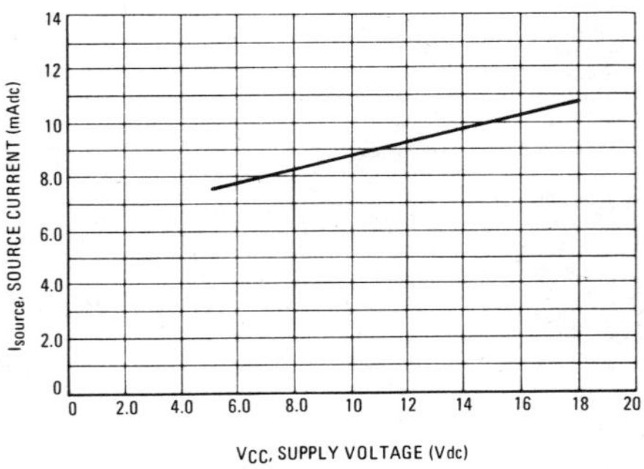

FIGURE 14 – LINEAR SINK CURRENT versus SUPPLY VOLTAGE

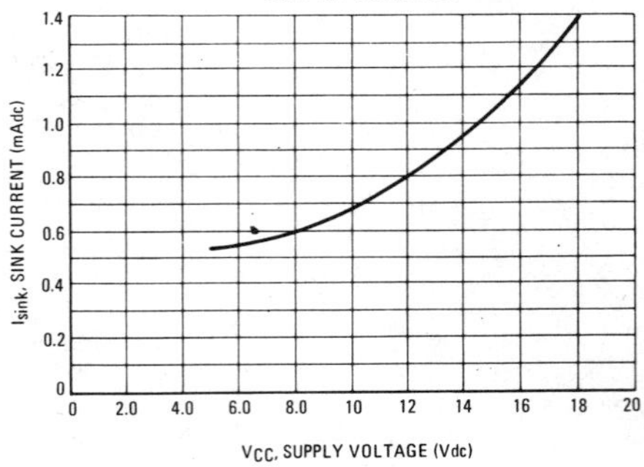

OPERATION AND APPLICATIONS

Basic Amplifier

The basic amplifier is the common emitter stage shown in Figures 15 and 16. The active load I_1 is buffered from the input transistor by a PNP transistor, Q4, and from the output by an NPN transistor, Q2. Q2 is biased class A by the current source I_2. The magnitude of I_2 (specified I_{sink}) is a limiting factor in capacitively coupled linear operation at the output. The sink current of the device can be forced to exceed the specified level with an increase in the distortion appearing at the output. Closed loop stability is maintained by an on-the-chip 3-pF capacitor shown in Figure 18. No external compensation is required.

FIGURE 15

BLOCK DIAGRAM

CIRCUIT SCHEMATIC

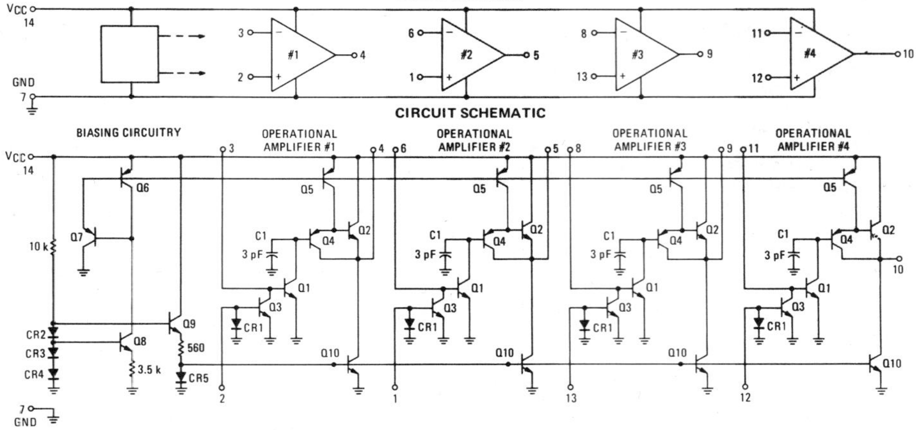

A noninverting input is obtained by adding a current mirror as shown in Figure 17. Essentially all current which enters the noninverting input, I_{in2}, flows through the diode CR1. The voltage drop across CR1 corresponds to this input current magnitude and this same voltage is applied to a matched device, Q3. Thus Q3 is biased to conduct an emitter current equal to I_{in2}. Since the alpha current gain of Q3 $\approx$ 1, its collector current $\approx I_{in2}$ also. In operation this current flows through an external feedback resistor which generates the output voltage signal. For inverting applications, the noninverting input is often used to set the dc quiescent level at the output. Techniques for doing this are discussed in the "Normal Design Procedure" section.

FIGURE 16 – A BASIC GAIN STAGE

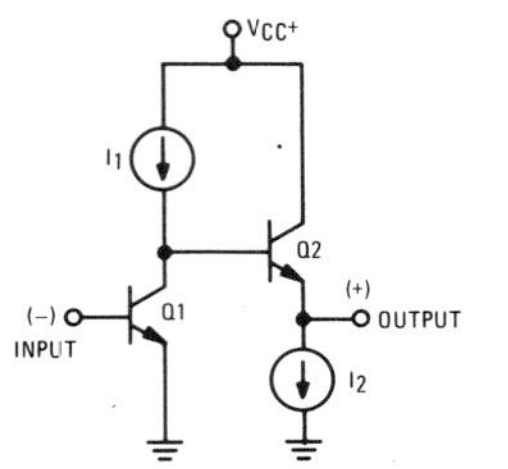

FIGURE 17 – OBTAINING A NONINVERTING INPUT

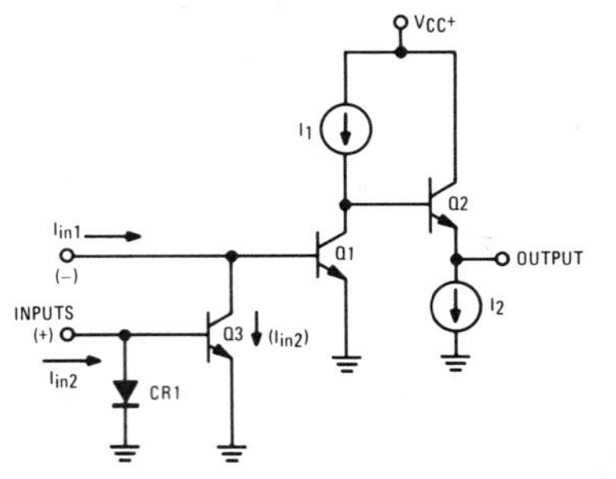

Biasing Circuitry

The circuitry common to all four amplifiers is shown in Figure 19. The purpose of this circuitry is to provide biasing voltage for the PNP and NPN current sources used in the amplifiers.

The voltage drops across diodes CR2, CR3 and CR4 are used as references. The voltage across resistor R1 is the sum of the drops across CR4 and CR3 minus the V_{BE} of Q8. The PNP current sources (Q5, etc.) are set to the magnitude $V_{BE}/R1$ by transistor Q6. Transistor Q7 reduces base current loading. The voltage across resistor R2 is the sum of the voltage drops across CR2, CR3 and CR4, minus the V_{BE} drops of transistor Q9 and diode CR5. The current thus set is established by CR5 in all the NPN current sources (Q10, etc.). This technique results in current source magnitudes which are relatively independent of the supply voltage.

National Semiconductor

Voltage Regulators

LM105/LM205/LM305/LM305A, LM376 Voltage Regulators

General Description

The LM105 series are positive voltage regulators similar to the LM100, except that an extra gain stage has been added for improved regulation. A redesign of the biasing circuitry removes any minimum load current requirement and at the same time reduces standby current drain, permitting higher voltage operation. They are direct, plug-in replacements for the LM100 in both linear and switching regulator circuits with output voltages greater than 4.5V. Important characteristics of the circuits are:

- Output voltage adjustable from 4.5V to 40V
- Output currents in excess of 10A possible by adding external transistors
- Load regulation better than 0.1%, full load with current limiting
- DC line regulation guaranteed at 0.03%/V
- Ripple rejection of 0.01%/V
- 45 mA output current without external pass transistor (LM305A)

Like the LM100, they also feature fast response to both load and line transients, freedom from oscillations with varying resistive and reactive loads and the ability to start reliably on any load within rating. The circuits are built on a single silicon chip and are supplied in either an 8-lead, TO-5 header or a 1/4" x 1/4" metal flat package.

The LM105 is specified for operation for $-55^\circ C \leq T_A \leq +125^\circ C$, the LM205 is specified for $-25^\circ C \leq T_A \leq +85^\circ C$, and the LM305/LM305A, LM376 is specified for $0^\circ C \leq T_A \leq +70^\circ C$.

Schematic and Connection Diagrams

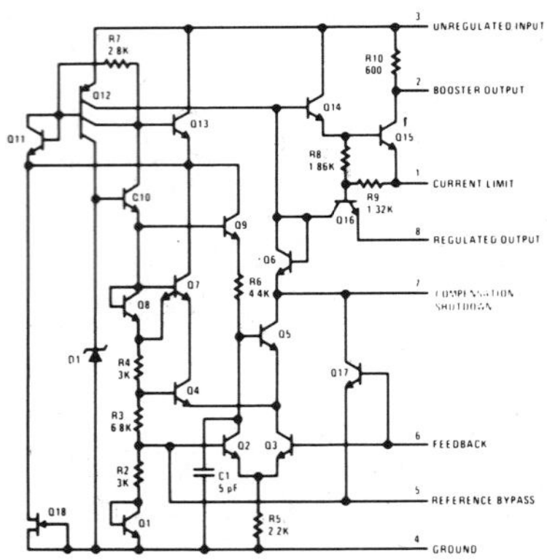

Pin connections shown are for metal can.

Dual-In-Line Package

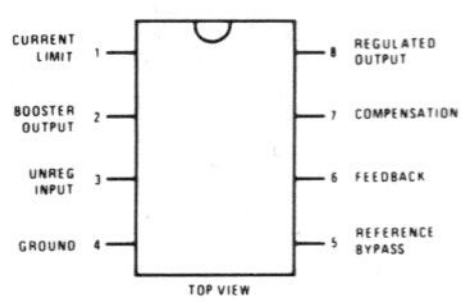

Order Number LM376N
See NS Package N08B

Metal Can Package

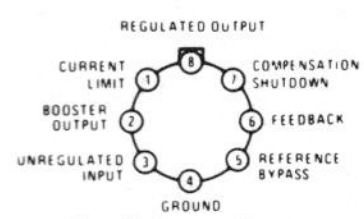

Note: Pin 4 connected to case.
TOP VIEW

Order Number LM105H, LM205H, LM305H or LM305AH
See NS Package H08C

Typical Applications

10A Regulator with Foldback Current Limiting

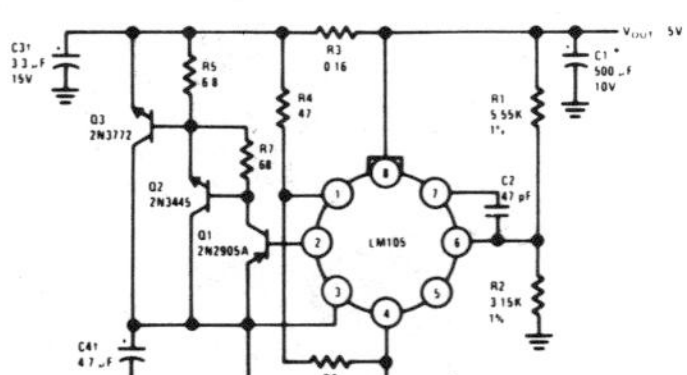

†Solid tantalum.
*Electrolytic.

1.0A Regulator with Protective Diodes

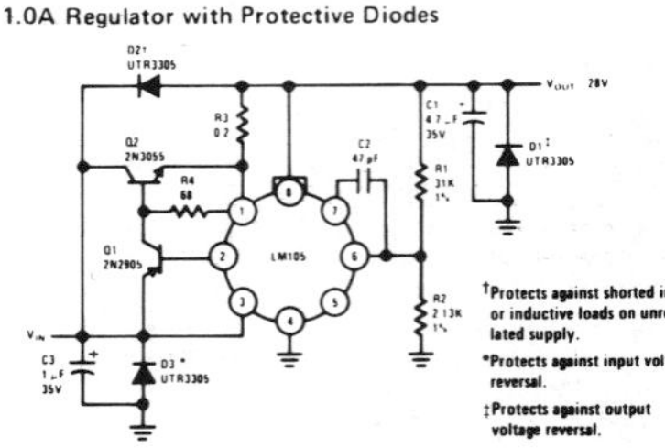

†Protects against shorted input or inductive loads on unregulated supply.
*Protects against input voltage reversal.
‡Protects against output voltage reversal.

Courtesy of National Semiconductor Corporation.

Absolute Maximum Ratings

	LM105	LM205	LM305	LM305A	LM376
Input Voltage	50V	50V	40V	50V	40V
Input-Output Differential	40V	40V	40V	40V	40V
Power Dissipation (Note 1)	800 mW	800 mW	800 mW	800 mW	400 mW
Operating Temperature Range	−55°C to +125°C	−25°C to +85°C	0°C to +70°C	0°C to +70°C	0°C to +70°C
Storage Temperature Range	−65°C to +150°C	−65°C to +150°C	−65°C to +150°C	−65°C to +150°C	−65°C to +150°C
Lead Temperature (Soldering, 10 seconds)	300°C	300°C	300°C	300°C	300°C

Electrical Characteristics (Note 2)

PARAMETER	CONDITIONS	LM105			LM205			LM305			LM305A			LM376			UNITS
		MIN	TYP	MAX	MIN	TYP	MAX	MIN	TYP	MAX	MIN	TYP	MAX	MIN	TYP	MAX	
Input Voltage Range		8.5		50	8.5		50	8.5		40	8.5		50	9.0		40	V
Output Voltage Range		4.5		40	4.5		40	4.5		30	4.5		40	5.0		37	V
Input-Output Voltage Differential		3.0		30	3.0		30	3.0		30	3.0		30	3.0		30	V
Load Regulation (Note 3)	R_{SC} = 10Ω, T_A = 25°C		0.02	0.05		0.02	0.05		0.02	0.05							%
	R_{SC} = 10Ω, $T_A = T_{A(MAX)}$		0.03	0.1		0.03	0.1		0.03	0.1							%
	R_{SC} = 10Ω, $T_A = T_{A(MIN)}$		0.03	0.1		0.03	0.1		0.03	0.1							%
			$0 \le I_O \le 12$ mA			$0 \le I_O \le 12$ mA			$0 \le I_O \le 12$ mA								
	R_{SC} = 0Ω, T_A = 25°C											0.02	0.2			0.2	%
	R_{SC} = 0Ω, T_A = 70°C											0.03	0.4			0.5	%
	R_{SC} = 0Ω, T_A = 0°C											0.03	0.4			0.5	%
												$0 \le I_O \le 45$ mA			$0 \le I_O \le 25$ mA		
Line Regulation	T_A = 25°C															0.03	%/V
	0°C ≤ T_A ≤ +70°C															0.1	%/V
	$V_{IN} - V_{OUT} \le 5V$, T_A = 25°C		0.025	0.06		0.025	0.06		0.025	0.06		0.025	0.06				%/V
	$V_{IN} - V_{OUT} \ge 5V$, T_A = 25°C		0.015	0.03		0.015	0.03		0.015	0.03		0.015	0.03				%/V
Temperature Stability	$T_{A(MIN)} \le T_A \le T_{A(MAX)}$		0.3	1.0		0.3	1.0		0.3	1.0		0.3	1.0				%
Feedback Sense Voltage		1.63	1.7	1.81	1.63	1.7	1.81	1.63	1.7	1.81	1.55	1.7	1.85	1.60	1.72	1.80	V
Output Noise Voltage	10 Hz ≤ f ≤ 10 kHz																
	C_{REF} = 0		0.005			0.005			0.005			0.005					%
	C_{REF} = 0.1μF		0.002			0.002			0.002			0.002					%
Standby Current Drain	V_{IN} = 30V, T_A = 25°C															2.5	mA
	V_{IN} = 40V								0.8	2.0							mA
	V_{IN} = 50V		0.8	2.0		0.8	2.0					0.8	2.0				mA
Current Limit Sense Voltage	T_A = 25°C, R_{SC} = 10Ω, V_{OUT} = 0V, (Note 4)	225	300	375	225	300	375	225	300	375	225	300	375		300		mV
Long Term Stability			0.1	1.0		0.1	1.0		0.1	1.0		0.1	1.0				%
Ripple Rejection	C_{REF} = 10μF, f = 120 Hz		0.003	0.01		0.003	0.01		0.003	0.01		0.003				0.1	%/V

Note 1: The maximum junction temperature of the LM105 and LM305A is 150°C, the LM205 and LM376 is 100°C, and the LM305 is 85°C. For operation at elevated temperatures, devices in the TO-5 package must be derated based on a thermal resistance of 150°C/W junction to ambient, or 45°C/W junction to case. For the epoxy dual-in-line package, derating is based on a thermal resistance of 187°C/W junction to ambient. Peak dissipations to 1W are allowable providing the dissipation rating is not exceeded with the power averaged over a five second interval for the LM105 and LM205, and averaged over a two second interval for the LM305.

Note 2: Unless otherwise specified, these specifications apply for temperatures within the operating temperature range, for input and output voltages within the range given, and for a divider impedance seen by the feedback terminal of 2 kΩ. Load and line regulation specifications are for a constant junction temperature. Temperature drift effects must be taken into account separately when the unit is operating under conditions of high dissipation.

Note 3: The output currents given, as well as the load regulation, can be increased by the addition of external transistors. The improvement factor will be roughly equal to the composite current gain of the added transistors.

Note 4: With no external pass transistor.

Typical Performance Characteristics LM105/LM205/LM305/LM305A

Load Regulation

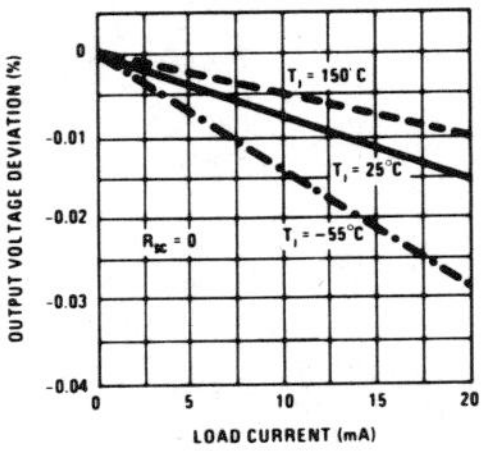

Load Regulation

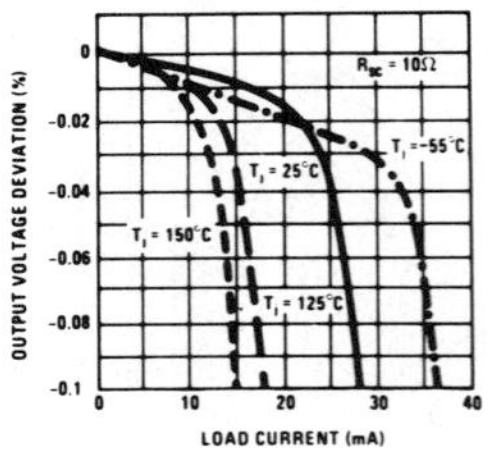

Current Limiting Characteristics

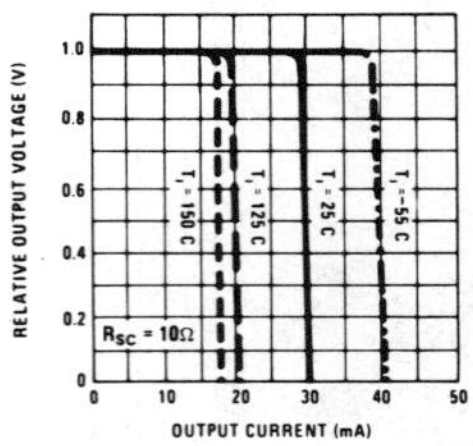

Current Limit Sense Voltage

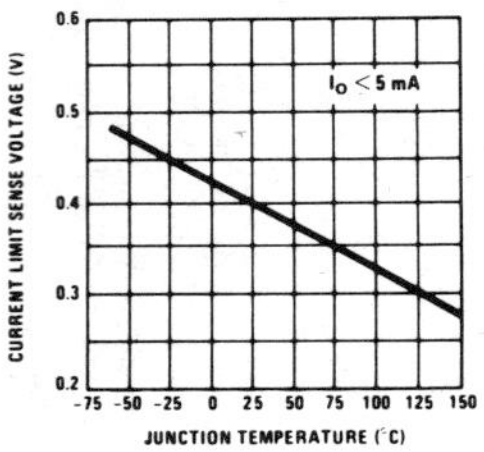

Short Circuit Current

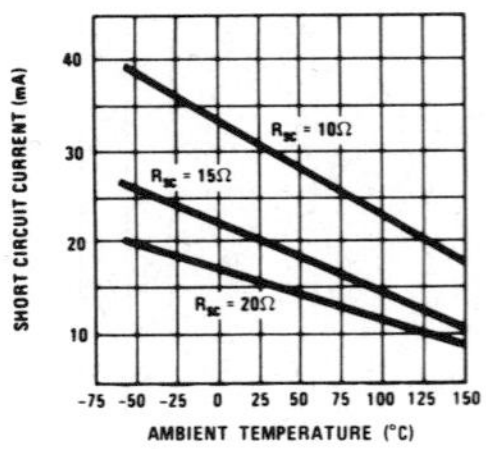

Optimum Divider Resistance Values

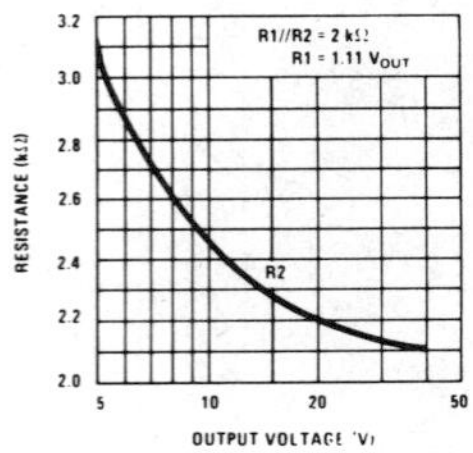

Minimum Input Voltage

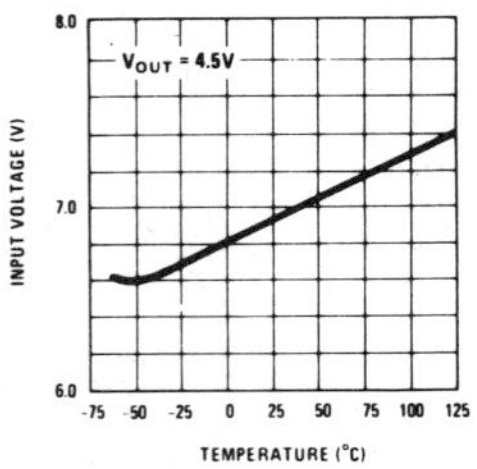

Regulator Dropout Voltage

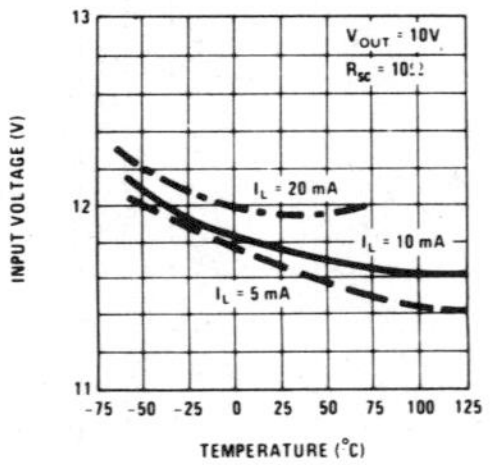

Supply Voltage Rejection

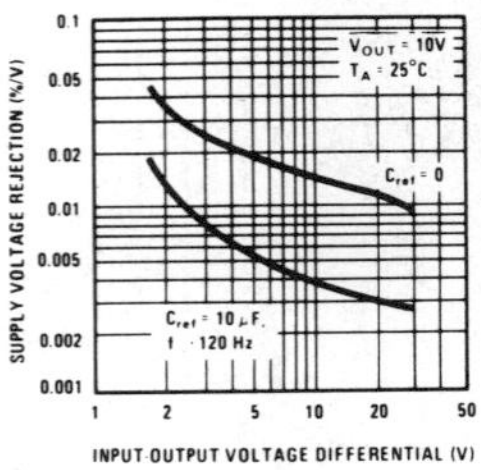

Minimum Output Voltage

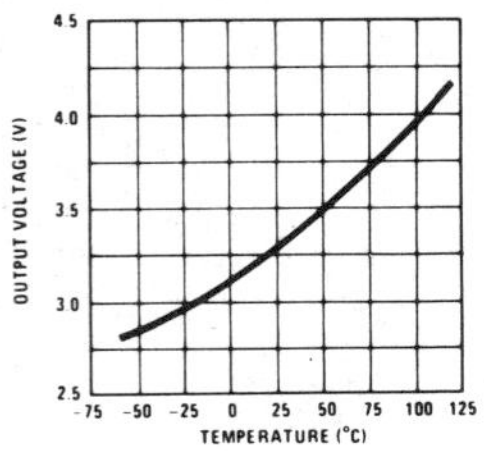

Standby Current Drain

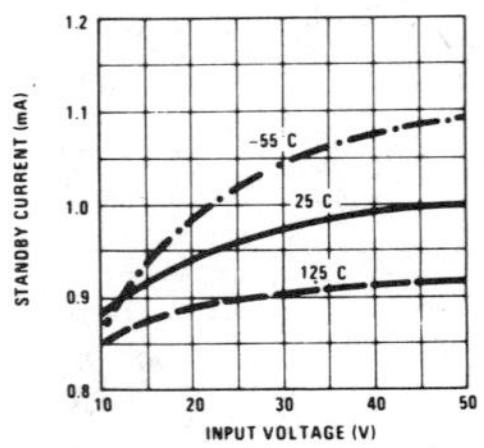

Transient Response

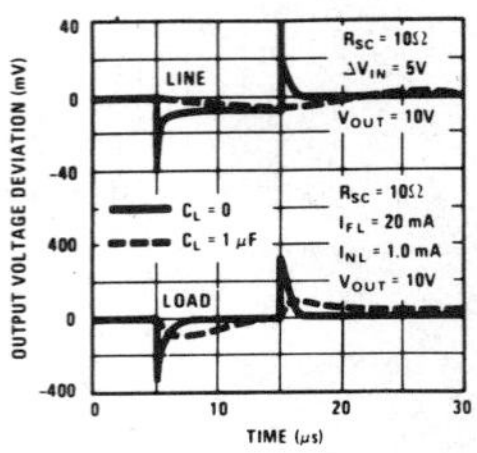

Typical Performance Characteristics LM376

Load Regulation

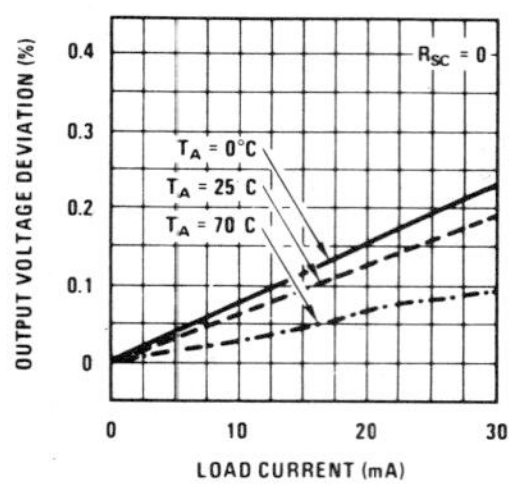

Current Limiting Characteristics

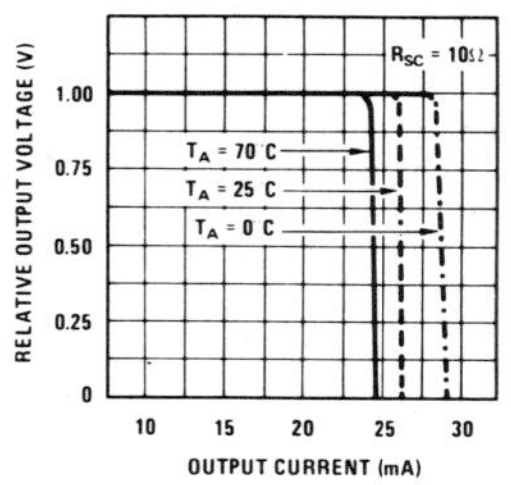

Current Limit Sense Voltage

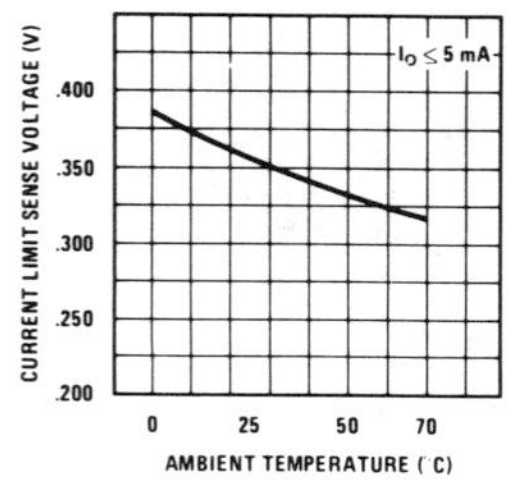

Minimum Input Voltage

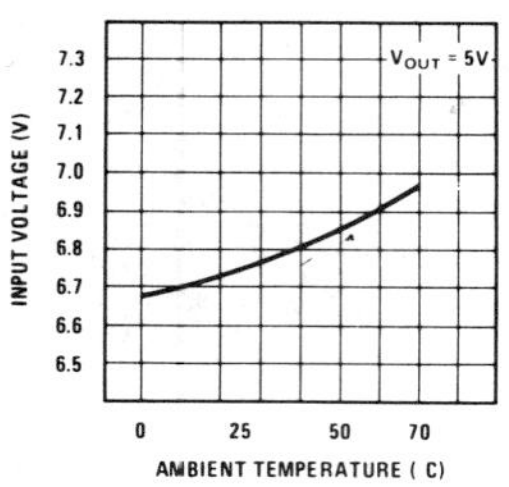

Regulator Dropout Voltage

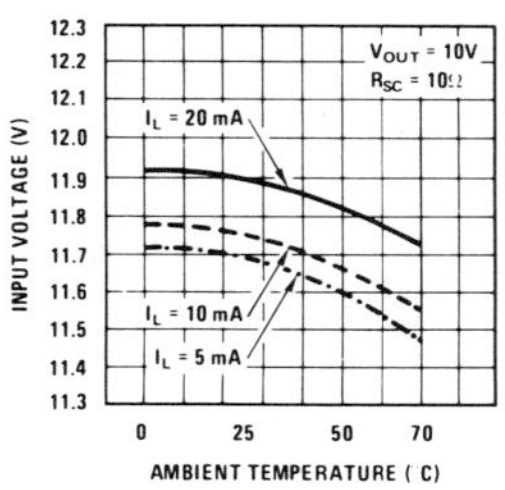

Optimum Divider Resistance

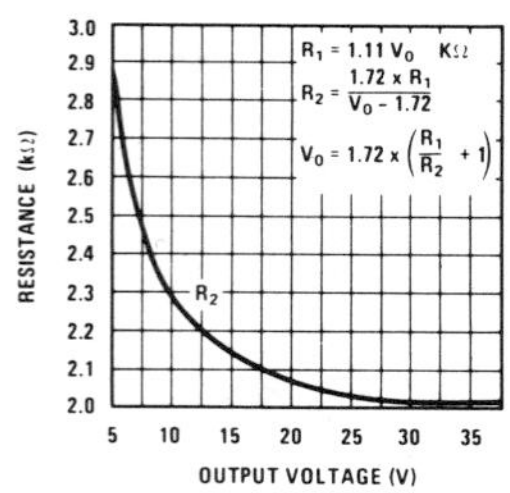

Standby Current Drain
T_A = 25°C

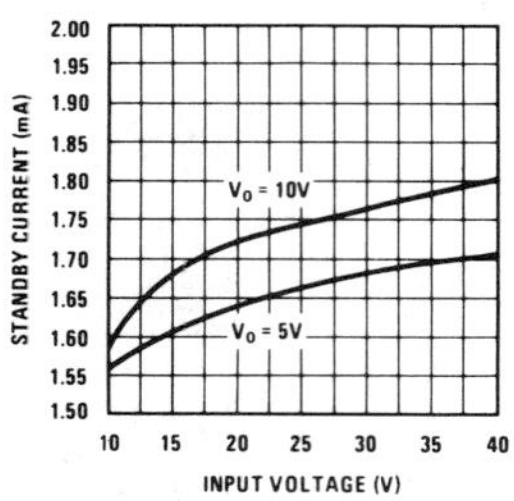

Supply Voltage Rejection

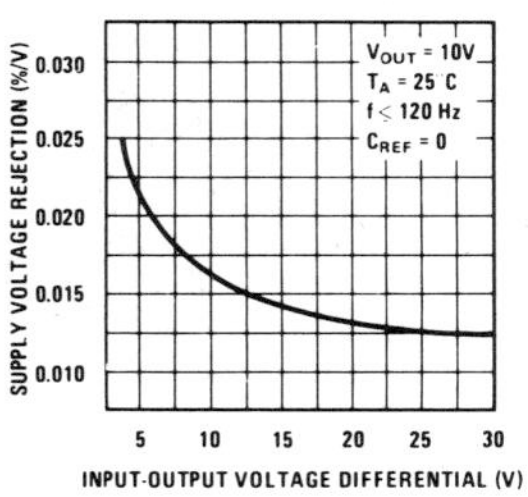

Transient Response

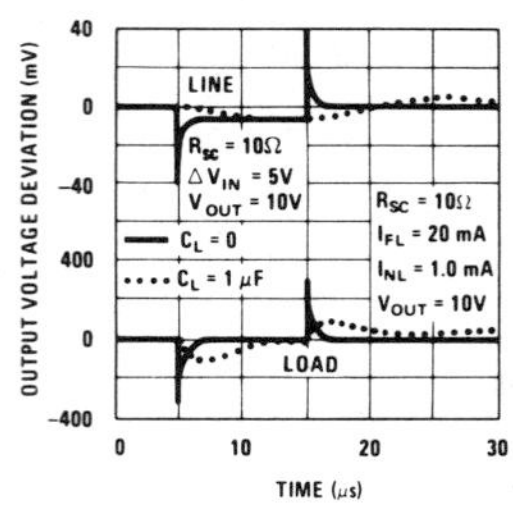

Typical Applications (Continued)

Linear Regulator with Foldback Current Limiting

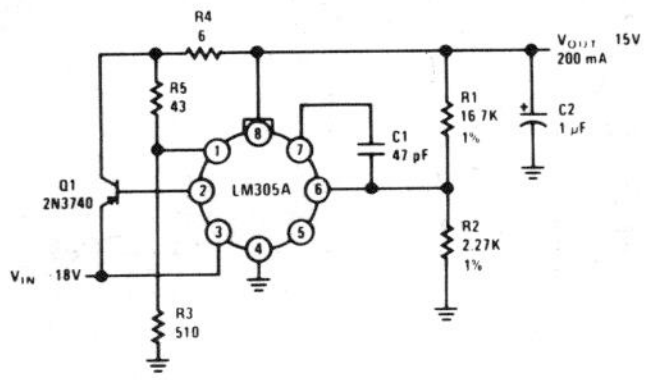

Current Regulator

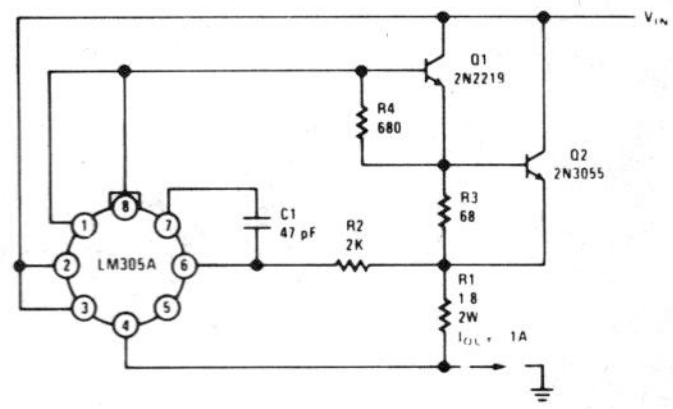

Shunt Regulator

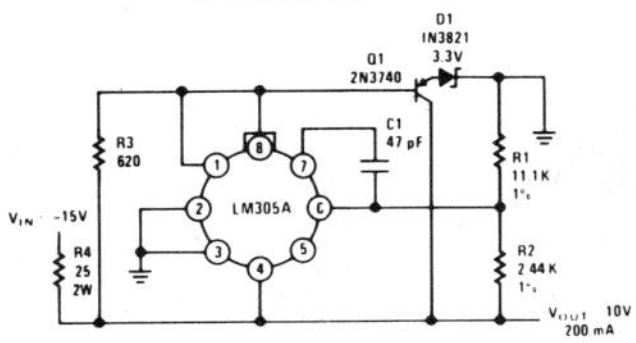

Switching Regulator

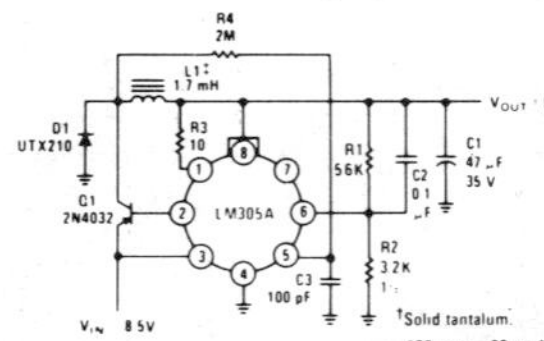

†Solid tantalum.

125 turns #22 on Arnold Engineering A262123-2 molybdenum permally core.

Basic Positive Regulator with Current Limiting

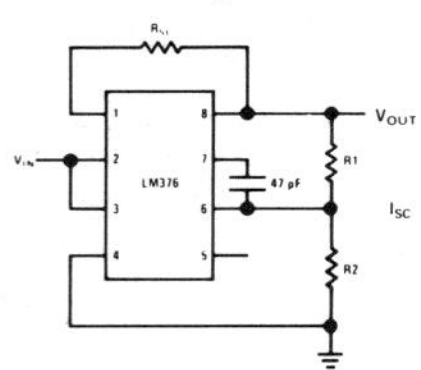

$$V_{OUT} = 1.72 \frac{R1 + R2}{R2} \text{ V}$$

$$I_{SC} = \frac{325}{R_{SC}} \text{ mA}$$

1.0A Regulator with Protective Diodes

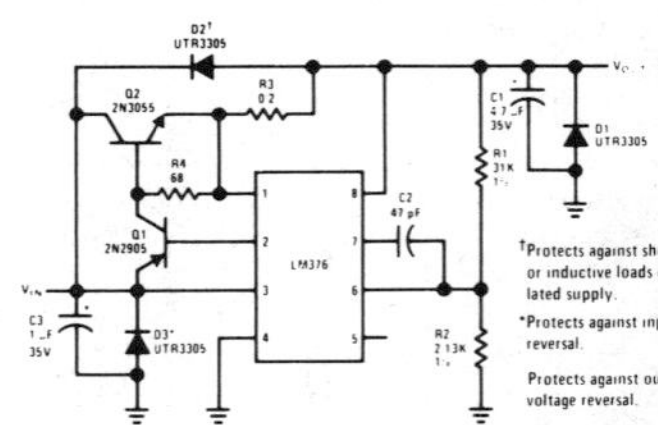

†Protects against shorted input or inductive loads on unregulated supply.

*Protects against input voltage reversal.

Protects against output voltage reversal.

Linear Regulator with Foldback Current Limiting

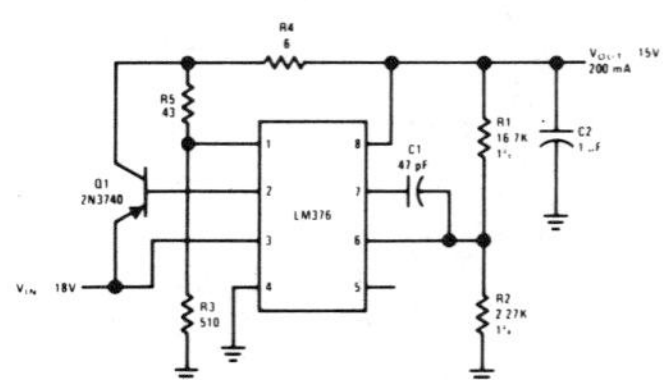

National Semiconductor

Voltage Regulators

LM109/LM209/LM309 5-Volt Regulator

General Description

The LM109 series are complete 5 V regulators fabricated on a single silicon chip. They are designed for local regulation on digital logic cards, eliminating the distribution problems associated with single-point regulation. The devices are available in two standard transistor packages. In the solid-kovar TO-5 header, it can deliver output currents in excess of 200 mA, if adequate heat sinking is provided. With the TO-3 power package, the available output current is greater than 1 A.

The regulators are essentially blowout proof. Current limiting is included to limit the peak output current to a safe value. In addition, thermal shutdown is provided to keep the IC from overheating. If internal dissipation becomes too great, the regulator will shut down to prevent excessive heating.

Considerable effort was expended to make these devices easy to use and to minimize the number of external components. It is not necessary to bypass the output, although this does improve transient response somewhat. Input bypassing is needed, however, if the regulator is located very far from the filter capacitor of the power supply. Stability is also achieved by methods that provide very good rejection of load or line transients as are usually seen with TTL logic.

Although designed primarily as a fixed-voltage regulator, the output of the LM109 series can be set to voltages above 5 V, as shown below. It is also possible to use the circuits as the control element in precision regulators, taking advantage of the good current-handling capability and the thermal overload protection.

Features

- Specified to be compatible, worst case, with TTL and DTL
- Output current in excess of 1 A
- Internal thermal overload protection
- No external components required

Schematic Diagram

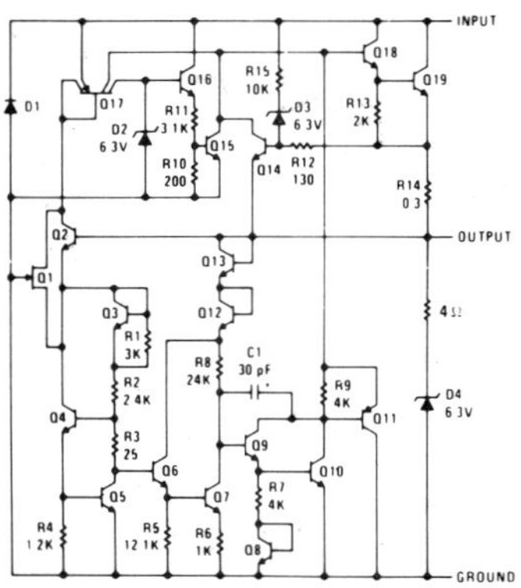

Connection Diagrams

Metal Can Packages

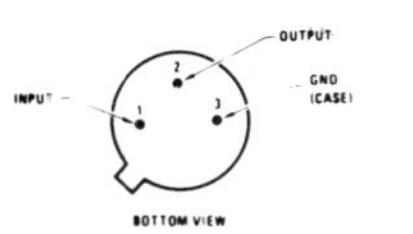

Order Number LM109H, LM209H or LM309H
See Package H03A

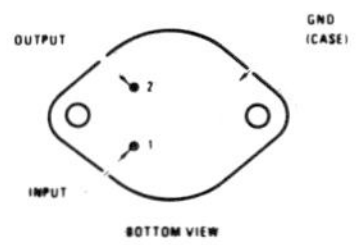

Order Number LM109K STEEL, LM209K STEEL, LM309K STEEL or LM309K (Aluminum)
See Package K02A

Typical Application

Fixed 5V Regulator

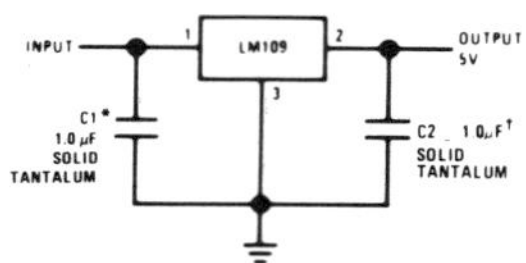

*Required if regulator is located more than 4" from power supply filter capacitor.

†Although no output capacitor is needed for stability, it does improve transient response.
C2 should be used whenever long wires are used to connect to the load, or when transient response is critical.

NOTE: Pin 3 electrically connected to case.

Adjustable Output Regulator

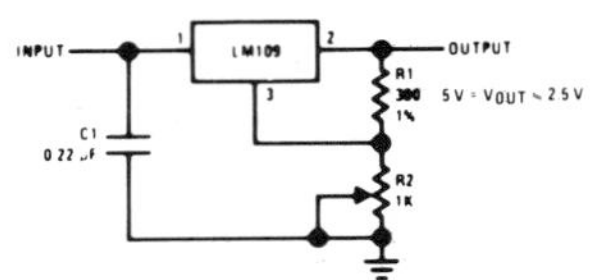

Courtesy of National Semiconductor Corporation.

Absolute Maximum Ratings

Input Voltage	35 V
Power Dissipation	Internally Limited
Operating Junction Temperature Range	
LM109	−55°C to +150°C
LM209	−25°C to +150°C
LM309	0°C to +125°C
Storage Temperature Range	−65°C to +150°C
Lead Temperature (Soldering, 10 seconds)	300°C

Electrical Characteristics

PARAMETER	CONDITIONS	LM109/LM209			LM309			UNITS
		MIN	TYP	MAX	MIN	TYP	MAX	
Output Voltage	$T_j = 25°C$	4.7	5.05	5.3	4.8	5.05	5.2	V
Line Regulation	$T_j = 25°C$, $7V \leq V_{IN} \leq 25V$		4.0	50		4.0	50	mV
Load Regulation	$T_j = 25°C$							
TO-5 Package	$5mA \leq I_{OUT} \leq 0.5A$		15	50		15	50	mV
TO-3 Package	$5mA \leq I_{OUT} \leq 1.5A$		15	100		15	100	mV
Output Voltage	$7V \leq V_{IN} \leq 25V$, $5mA \leq I_{OUT} \leq I_{MAX}$, $P < P_{MAX}$	4.6		5.4	4.75		5.25	V
Quiescent Current	$7V \leq V_{IN} \leq 25V$		5.2	10		5.2	10	mA
Quiescent Current Change	$7V \leq V_{IN} \leq 25V$			0.5			0.5	mA
	$5mA \leq I_{OUT} \leq I_{MAX}$			0.8			0.8	mA
Output Noise Voltage	$T_A = 25°C$, $10Hz \leq f \leq 100kHz$		40			40		μV
Long Term Stability				10			20	mV
Ripple Rejection	$T_j = 25°C$	50			50			dB
Thermal Resistance, Junction to Case	(Note 2)							
TO-5 Package			15			15		°C/W
TO-3 Package			2.5			2.5		°C/W

Note 1: Unless otherwise specified, these specifications apply for $-55°C \leq T_j \leq +150°C$ for the LM109, $-25°C \leq T_j \leq +150°C$ for the LM209, and $0°C \leq T_j \leq +125°C$ for the LM309; $V_{IN} = 10V$ and $I_{OUT} = 0.1A$ for the TO-5 package or $I_{OUT} = 0.5A$ for the TO-3 package. For the TO-5 package, $I_{MAX} = 0.2A$ and $P_{MAX} = 2.0W$. For the TO-3 package, $I_{MAX} = 1.0A$ and $P_{MAX} = 20W$.

Note 2: Without a heat sink, the thermal resistance of the TO-5 package is about 150°C/W, while that of the TO-3 package is approximately 35°C/W. With a heat sink, the effective thermal resistance can only approach the values specified, depending on the efficiency of the sink.

Typical Applications (Continued)

High Stability Regulator*

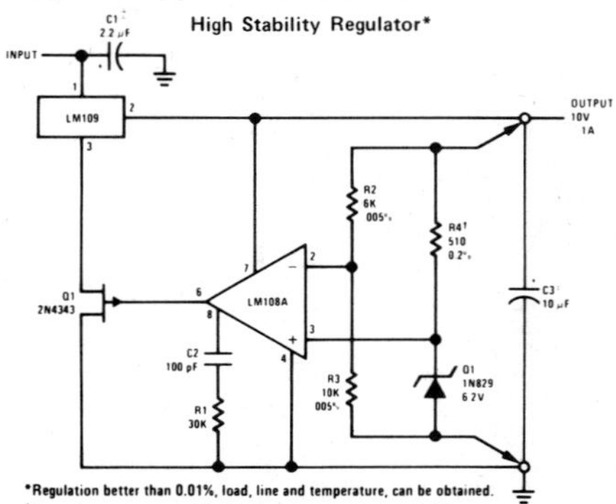

*Regulation better than 0.01%, load, line and temperature, can be obtained.

†Determines zener current. May be adjusted to minimize thermal drift.

‡Solid tantalum.

Current Regulator

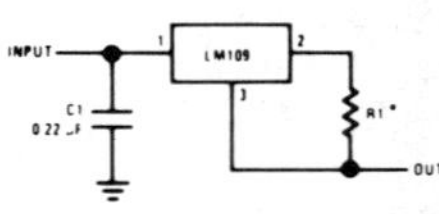

*Determines output current. If wirewound resistor is used, bypass with 0.1 μF.

Typical Performance Characteristics

MAXIMUM AVERAGE POWER DISSIPATION (LM109K)

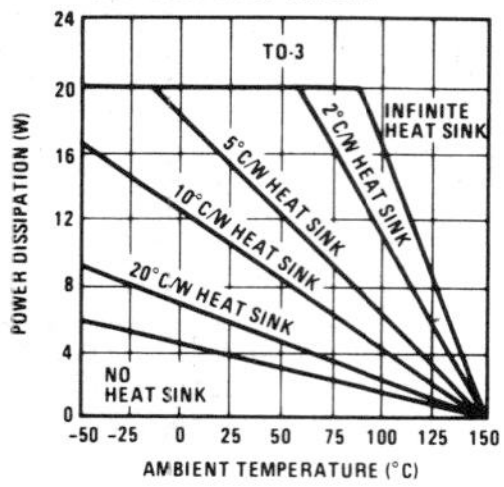

MAXIMUM AVERAGE POWER DISSIPATION (LM309K)

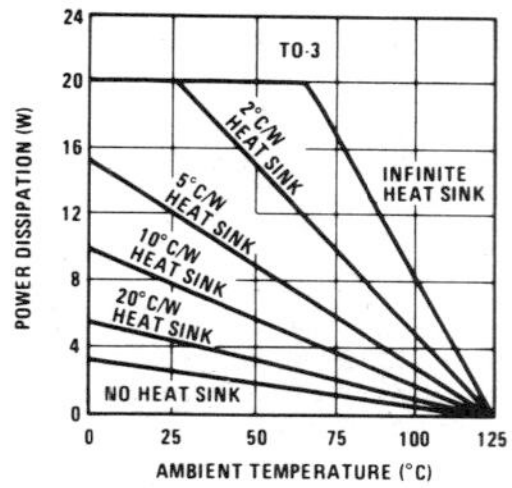

OUTPUT IMPEDANCE

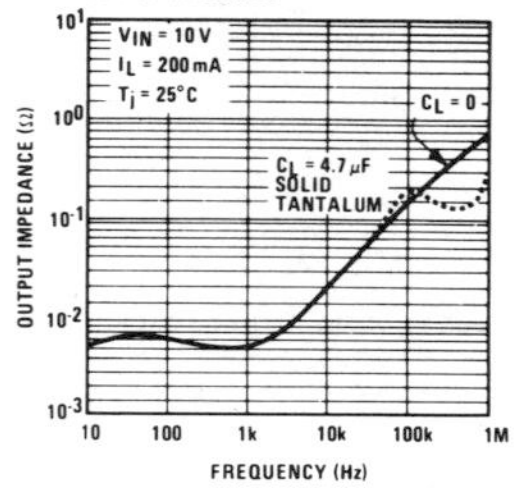

MAXIMUM AVERAGE POWER DISSIPATION (LM109H)

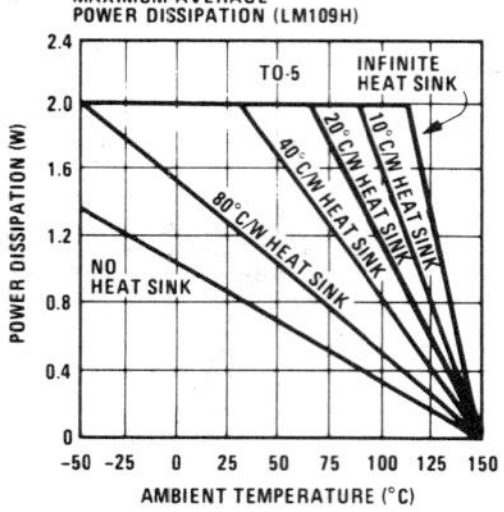

MAXIMUM AVERAGE POWER DISSIPATION (LM309H)

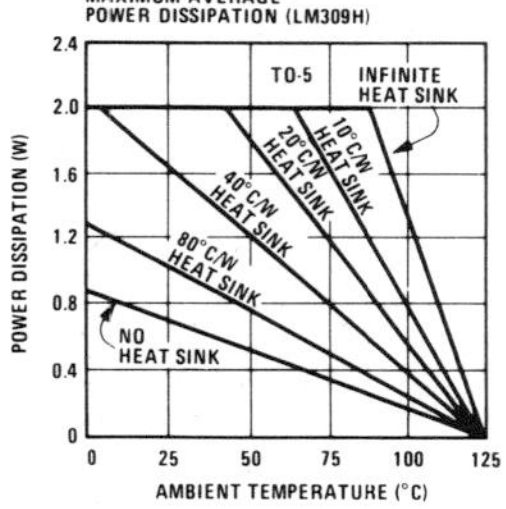

RIPPLE REJECTION

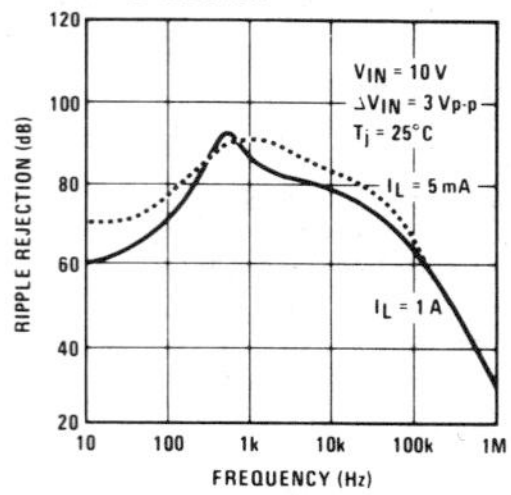

CURRENT LIMIT CHARACTERISTICS (NOTE 1)

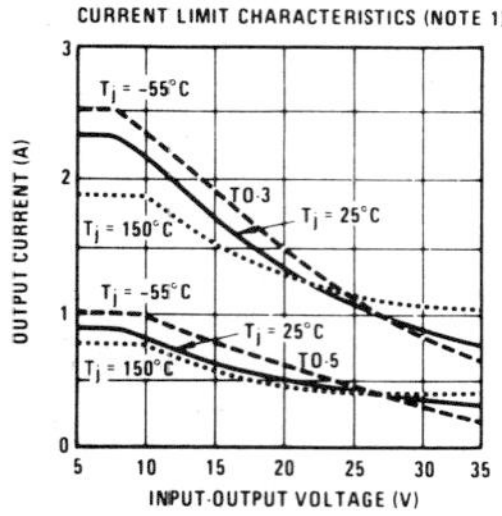

Note 1: Current limiting foldback characteristics are determined by input-output differential, not by output voltage.

THERMALLY INDUCED OUTPUT VOLTAGE VARIATION

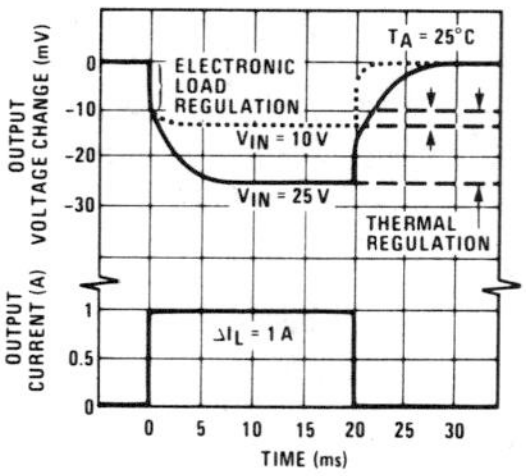

RIPPLE REJECTION

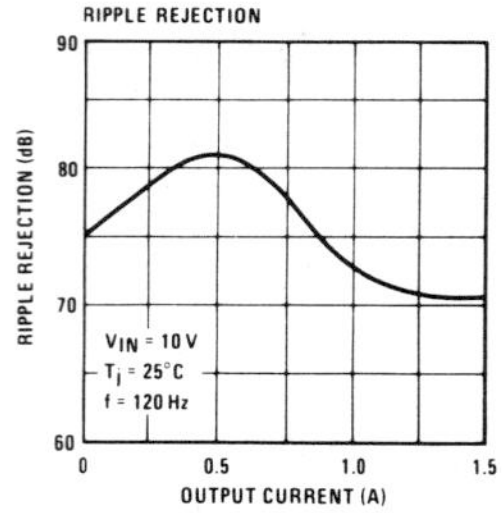

μA723

PRECISION VOLTAGE REGULATOR

FAIRCHILD LINEAR INTEGRATED CIRCUITS

GENERAL DESCRIPTION – The μA723 is a monolithic Voltage Regulator constructed using the Fairchild Planar* epitaxial process. The device consists of a temperature compensated reference amplifier, error amplifier, power series pass transistor and current limit circuitry. Additional NPN or PNP pass elements may be used when output currents exceeding 150 mA are required. Provisions are made for adjustable current limiting and remote shutdown. In addition to the above, the device features low standby current drain, low temperature drift and high ripple rejection. The μA723 is intended for use with positive or negative supplies as a series, shunt, switching or floating regulator. Applications include laboratory power supplies, isolation regulators for low level data amplifiers, logic card regulators, small instrument power supplies, airborne systems and other power supplies for digital and linear circuits.

- **POSITIVE OR NEGATIVE SUPPLY OPERATION**
- **SERIES, SHUNT, SWITCHING OR FLOATING OPERATION**
- **0.01% LINE AND LOAD REGULATION**
- **OUTPUT VOLTAGE ADJUSTABLE FROM 2 TO 37 VOLTS**
- **OUTPUT CURRENT TO 150 mA WITHOUT EXTERNAL PASS TRANSISTOR**

ABSOLUTE MAXIMUM RATINGS

Pulse Voltage from V+ to V–, (50 ms) (μA723)	50 V
Continuous Voltage from V_+ to V_-	40 V
Input/Output Voltage Differential	40 V
Differential Input Voltage	±5 V
Voltage Between Non-Inverting Input and V_-	+8 V
Current from V_Z	25 mA
Current from V_{REF}	15 mA
Internal Power Dissipation (Note 1)	
Metal Can	800 mW
DIP	1000 mW
Storage Temperature Range	–65°C to +150°C
Operating Temperature Range	
Military (μA723)	–55°C to +125°C
Commercial (μA723C)	0°C to +70°C
Lead Temperature (Soldering, 60 s)	300°C

EQUIVALENT CIRCUIT

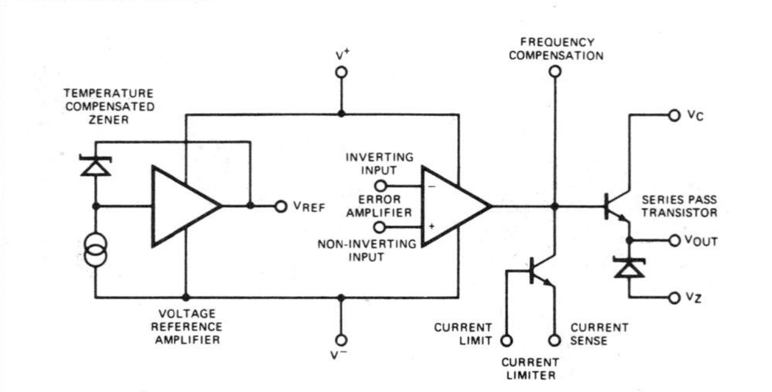

Notes on following pages.

CONNECTION DIAGRAMS
10-LEAD METAL CAN
(TOP VIEW)

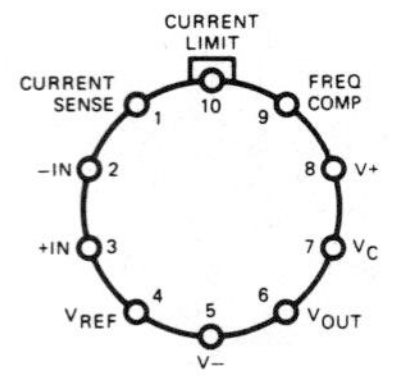

Note: Pin 5 connected to case.

ORDER INFORMATION

TYPE	PART NO.
μA723	μA723HM
μA723C	μA723HC

14-LEAD DIP
(TOP VIEW)

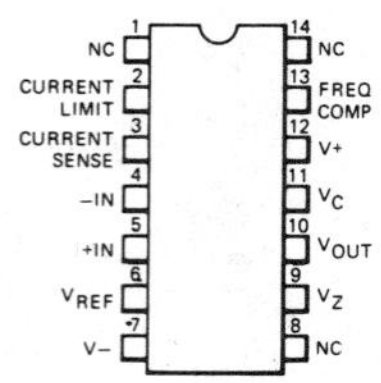

ORDER INFORMATION

TYPE	PART NO.
μA723	μA723DM
μA723C	μA723DC
μA723C	μA723PC

*Planar is a patented Fairchild process.

Courtesy of Fairchild Camera and Instrument Corporation.

μA723

ELECTRICAL CHARACTERISTICS: $T_A = 25°C$, $V_{IN} = V+ = V_C = 12$ V, $V- = 0$, $V_{OUT} = 5$ V, $I_L = 1$ mA, $R_{SC} = 0$, C1 = pF, $C_{ref} = 0$, unless otherwise specified. Divider impedance as seen by error amplifier ≤ 10 kΩ connected shown in Fig. 1. Line and load regulation specifications are given for the condition of constant chip temperature Temperature drifts must be taken into account separately for high dissipation conditions.

CHARACTERISTICS	CONDITIONS	MIN	TYP	MAX	UNITS
Line Regulation	$V_{IN} = 12V$ to $V_{IN} = 15V$		0.01	0.1	$\%V_O$
	$V_{IN} = 12V$ to $V_{IN} = 40V$		0.02	0.2	$\%V_O$
	$-55°C \leq T_A \leq +125°C$, $V_{IN} = 12V$ to $V_{IN} = 15V$			0.3	$\%V_O$
Load Regulation	$I_L = 1$ mA to $I_L = 50$ mA		0.03	0.15	$\%V_O$
	$-55°C \leq T_A \leq +125°C$, $I_L = 1$ mA to $I_L = 50$ mA			0.6	$\%V_O$
Ripple Rejection	f = 50 Hz to 10 kHz		74		dB
	f = 50 Hz to 10 kHz, $C_{REF} = 5\ \mu F$		86		dB
Average Temperature Coefficient of Output Voltage	$-55°C \leq T_A \leq +125°C$		0.002	0.015	%/°C
Short Circuit Current Limit	$R_{SC} = 10\ \Omega$, $V_O = 0$		65		mA
Reference Voltage		6.95	7.15	7.35	V
Output Noise Voltage	BW = 100 Hz to 10 kHz, $C_{REF} = 0$		20		μV_{rms}
	BW = 100 Hz to 10 kHz, $C_{REF} = 5\ \mu F$		2.5		μV_{rms}
Long Term Stability			0.1		%/1000 hrs
Standby Current Drain	$I_L = 0$, $V_{IN} = 30V$		2.3	3.5	mA
Input Voltage Range		9.5		40	V
Output Voltage Range		2.0		37	V
Input/Output Voltage Differential		3.0		38	V

μA723C

ELECTRICAL CHARACTERISTICS: $T_A = 25°C$, $V_{IN} = V+ = V_C = 12$ V, $V- = 0$, $V_{OUT} = 5$ V, $I_L = 1$ mA, $R_{SC} = 0$, C1 = 100 pF, $C_{ref} = 0$, unless otherwise specified. Divider impedance as seen by error amplifier ≤ 10 kΩ connected as shown in Fig. 1. Line and load regulation specifications are given for the condition of constant chip temperature. Temperature drifts must be taken into account separately for high dissipation conditions.

CHARACTERISTICS	CONDITIONS	MIN	TYP	MAX	UNITS
Line Regulation	$V_{IN} = 12V$ to $V_{IN} = 15V$		0.01	0.1	$\%V_O$
	$V_{IN} = 12V$ to $V_{IN} = 40V$		0.1	0.5	$\%V_O$
	$0°C \leq T_A \leq 70°C$, $V_{IN} = 12V$ to $V_{IN} = 15V$			0.3	$\%V_O$
Load Regulation	$I_L = 1$ mA to $I_L = 50$ mA		0.03	0.2	$\%V_O$
	$0°C \leq T_A \leq 70°C$, $I_L = 1$ mA to $I_L = 50$ mA			0.6	$\%V_O$
Ripple Rejection	f = 50 Hz to 10 kHz		74		dB
	f = 50 Hz to 10 kHz, $C_{REF} = 5\ \mu F$		86		dB
Average Temperature Coefficient of Output Voltage	$0°C \leq T_A \leq 70°C$		0.003	0.015	%/°C
Short Circuit Current Limit	$R_{SC} = 10\ \Omega$, $V_O = 0$		65		mA
Reference Voltage		6.80	7.15	7.50	V
Output Noise Voltage	BW = 100 Hz to 10 kHz, $C_{REF} = 0$		20		μV_{rms}
	BW = 100 Hz to 10 kHz, $C_{REF} = 5\ \mu F$		2.5		μV_{rms}
Long Term Stability			0.1		%/1000 hrs
Standby Current Drain	$I_L = 0$, $V_{IN} = 30V$		2.3	4.0	mA
Input Voltage Range		9.5		40	V
Output Voltage Range		2.0		37	V
Input/Output Voltage Differential		3.0		38	V

TYPICAL PERFORMANCE CURVES FOR μA723 AND μA723C

LINE REGULATION AS A FUNCTION OF INPUT/OUTPUT VOLTAGE DIFFERENTIAL

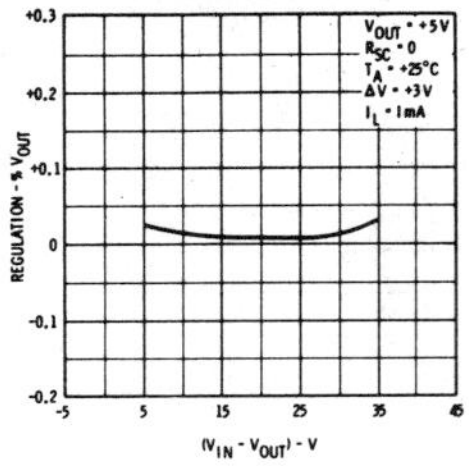

LOAD REGULATION AS A FUNCTION OF INPUT/OUTPUT VOLTAGE DIFFERENTIAL

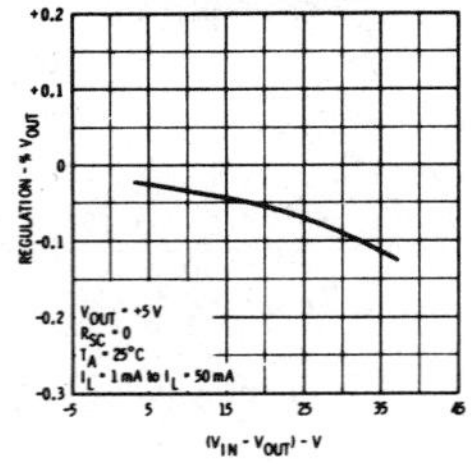

CURRENT LIMITING CHARACTERISTICS AS A FUNCTION OF JUNCTION TEMPERATURE

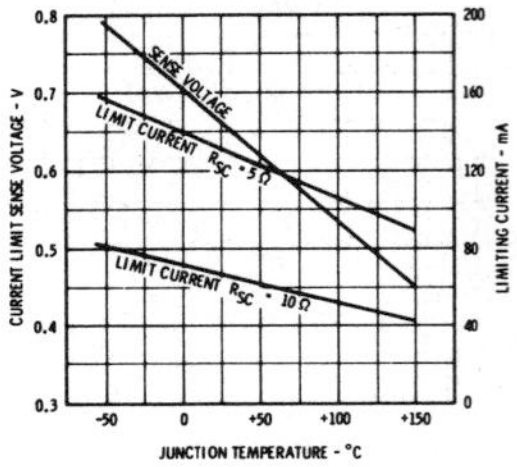

LINE TRANSIENT RESPONSE

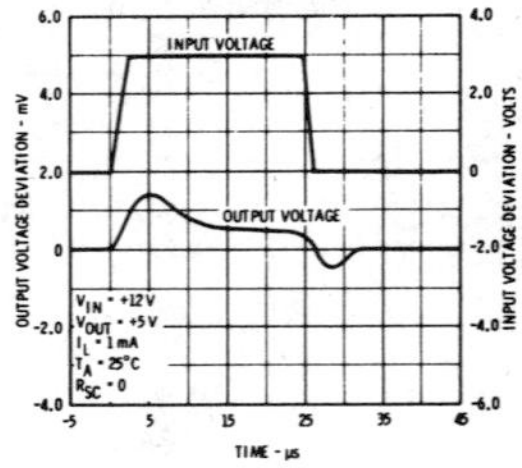

LOAD TRANSIENT RESPONSE

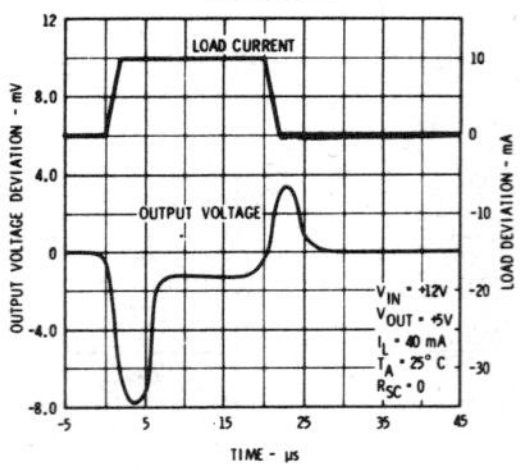

OUTPUT IMPEDANCE AS A FUNCTION OF FREQUENCY

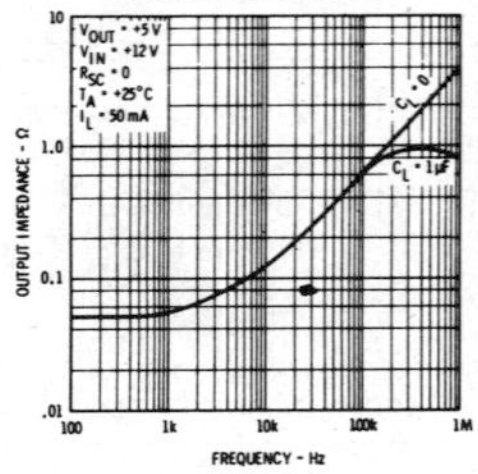

NOTES:

1. Rating applies to ambient temperatures up to 25°C. Above 25°C ambient derate based on the following thermal resistance values:

	θ_{JA} TYP	θ_{JA} MAX
TO-5	150	190
Plastic DIP	150	190
Ceramic DIP	125	160

2. L_1 is 40 turns of No. 20 enameled copper wire wound on Ferroxcube P36/22-3B7 pot core or equivalent with 0.009" air gap.
3. Figures in parentheses may be used if R_1/R_2 divider is placed on opposite side of error amp.
4. Replace R_1/R_2 in figures with divider shown in figure 13.
5. V^+ must be connected to a +3 V or greater supply.
6. For metal can applications where V_Z is required, an external 6.2 volt zener diode should be connected in series with V_{OUT}.

TYPICAL PERFORMANCE CURVES FOR μA723

MAXIMUM LOAD CURRENT AS A FUNCTION OF INPUT-OUTPUT VOLTAGE DIFFERENTIAL

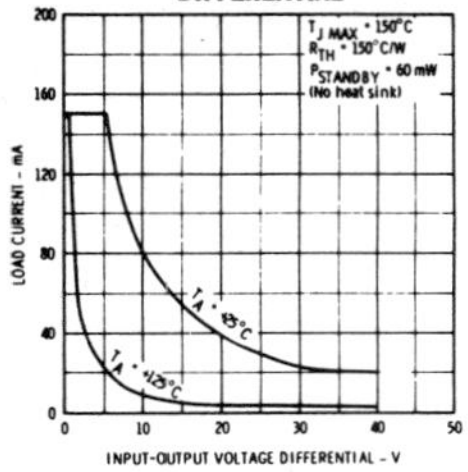

LOAD REGULATION CHARACTERISTICS WITHOUT CURRENT LIMITING

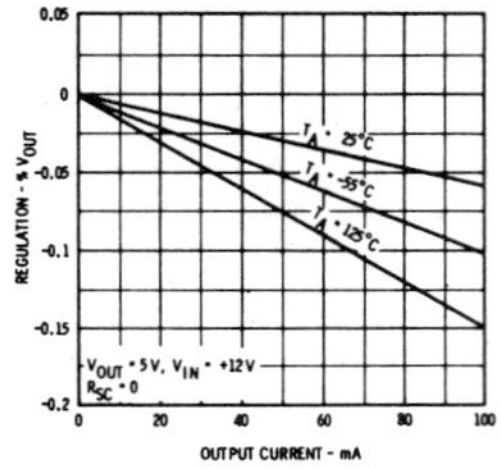

LOAD REGULATION CHARACTERISTICS WITH CURRENT LIMITING

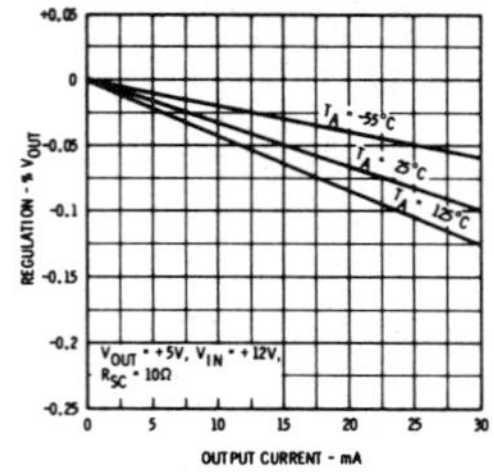

LOAD REGULATION CHARACTERISTICS WITH CURRENT LIMITING

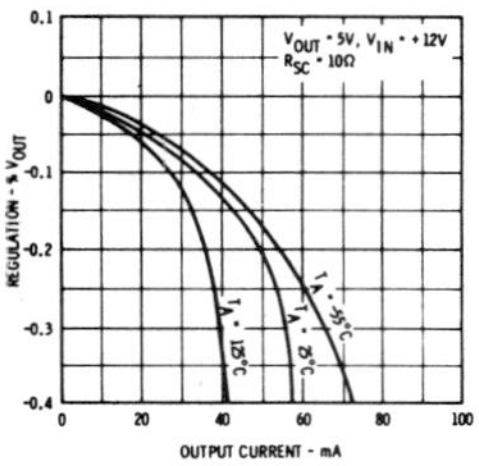

CURRENT LIMITING CHARACTERISTICS

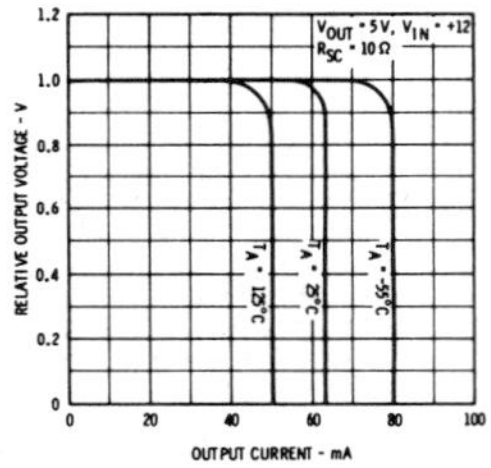

STANDBY CURRENT DRAIN AS A FUNCTION OF INPUT VOLTAGE

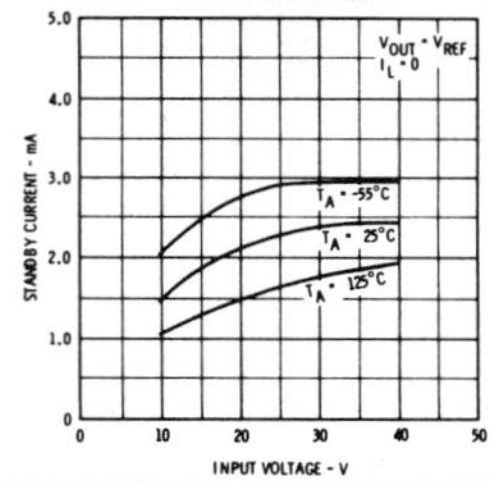

TYPICAL PERFORMANCE CURVES FOR μA723C

MAXIMUM LOAD CURRENT AS A FUNCTION OF INPUT/OUTPUT VOLTAGE DIFFERENTIAL

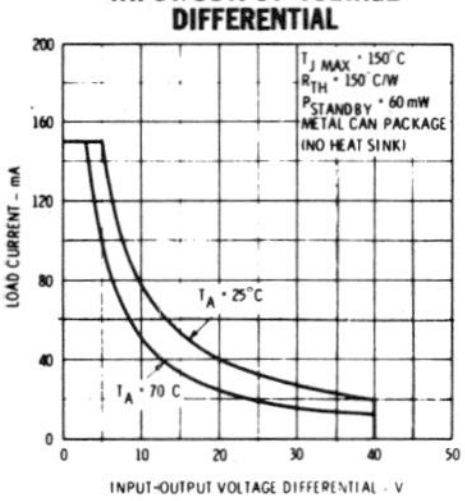

MAXIMUM LOAD CURRENT AS A FUNCTION OF INPUT/OUTPUT VOLTAGE DIFFERENTIAL

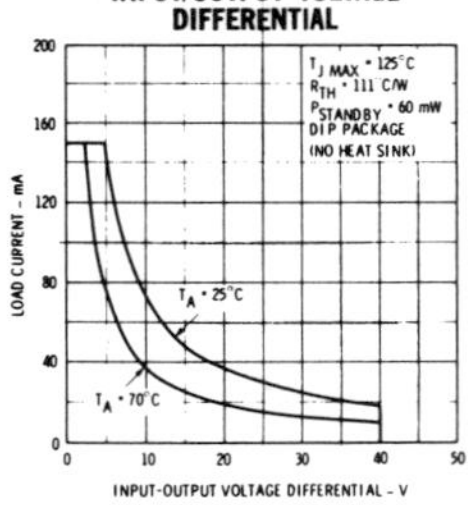

LOAD REGULATION CHARACTERISTICS WITHOUT CURRENT LIMITING

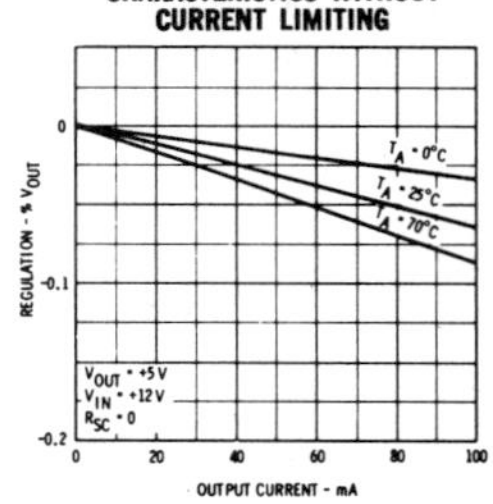

LOAD REGULATION CHARACTERISTICS WITH CURRENT LIMITING

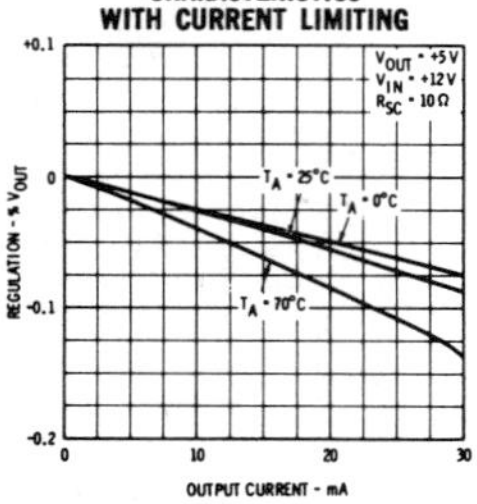

CURRENT LIMITING CHARACTERISTICS

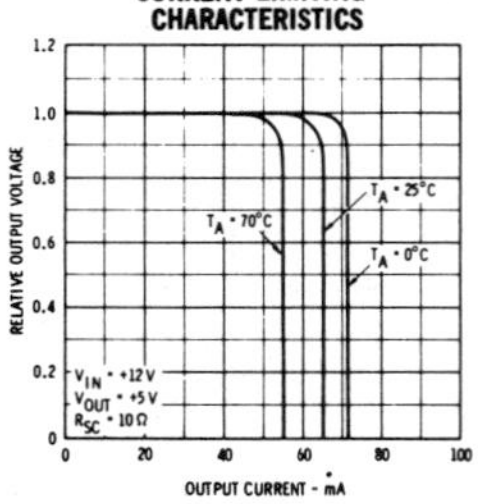

STANDBY CURRENT DRAIN AS A FUNCTION OF INPUT VOLTAGE

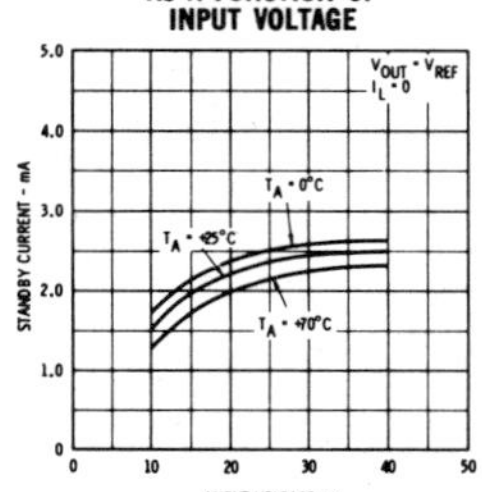

TABLE I
RESISTOR VALUES (kΩ) FOR STANDARD OUTPUT VOLTAGES

POSITIVE OUTPUT VOLTAGE	APPLICABLE FIGURES	FIXED OUTPUT ±5%		OUTPUT ADJUSTABLE ±10% (Note 4)			NEGATIVE OUTPUT VOLTAGE	APPLICABLE FIGURES	FIXED OUTPUT ±5%		5% OUTPUT ADJUSTABLE ±10%		
	(Note 3)	R_1	R_2	R_1	P_1	R_2			R_1	R_2	R_1	P_1	R_2
+3.0	1, 5, 6, 9, 12 (4)	4.12	3.01	1.8	0.5	1.2	+100	7	3.57	102	2.2	10	91
+3.6	1, 5, 6, 9, 12 (4)	3.57	3.65	1.5	0.5	1.5	+250	7	3.57	255	2.2	10	240
+5.0	1, 5, 6, 9, 12 (4)	2.15	4.99	.75	0.5	2.2	−6 (Note 5)	3, (10)	3.57	2.43	1.2	0.5	.75
+6.0	1, 5, 6, 9, 12 (4)	1.15	6.04	0.5	0.5	2.7	−9	3, 10	3.48	5.36	1.2	0.5	2.0
+9.0	2, 4, (5, 6, 12, 9)	1.87	7.15	.75	1.0	2.7	−12	3, 10	3.57	8.45	1.2	0.5	3.3
+12	2, 4, (5, 6, 9, 12)	4.87	7.15	2.0	1.0	3.0	−15	3, 10	3.65	11.5	1.2	0.5	4.3
+15	2, 4, (5, 6, 9, 12)	7.87	7.15	3.3	1.0	3.0	−28	3, 10	3.57	24.3	1.2	0.5	10
+28	2, 4, (5, 6, 9, 12)	21.0	7.15	5.6	1.0	2.0	−45	8	3.57	41.2	2.2	10	33
+45	7	3.57	48.7	2.2	10	39	−100	8	3.57	97.6	2.2	10	91
+75	7	3.57	78.7	2.2	10	68	−250	8	3.57	249	2.2	10	240

TABLE II
FORMULAE FOR INTERMEDIATE OUTPUT VOLTAGES

Outputs from +2 to +7 volts [Figures 1, 5, 6, 9, 12, (4)]	Outputs from +4 to +250 volts [Figure 7]	Current Limiting
$V_{OUT} = [V_{REF} \times \frac{R_2}{R_1 + R_2}]$	$V_{OUT} = [\frac{V_{REF}}{2} \times \frac{R_2 - R_1}{R_1}]$; $R_3 = R_4$	$I_{LIMIT} = \frac{V_{SENSE}}{R_{SC}}$
Outputs from +7 to +37 volts [Figures 2, 4, (5, 6, 9, 12)]	**Outputs from −6 to −250 volts [Figures 3, 8, 10]**	**Foldback Current Limiting**
$V_{OUT} = [V_{REF} \times \frac{R_1 + R_2}{R_2}]$	$V_{OUT} = [\frac{V_{REF}}{2} \times \frac{R_1 + R_2}{R_1}]$; $R_3 = R_4$	$I_{KNEE} = [\frac{V_{OUT} R_3}{R_{SC} R_4} + \frac{V_{SENSE}(R_3 + R_4)}{R_{SC} R_4}]$ $I_{SHORT\ CKT} = [\frac{V_{SENSE}}{R_{SC}} \times \frac{R_3 + R_4}{R_4}]$

BASIC LOW VOLTAGE REGULATOR
(V_{OUT} = 2 to 7 V)

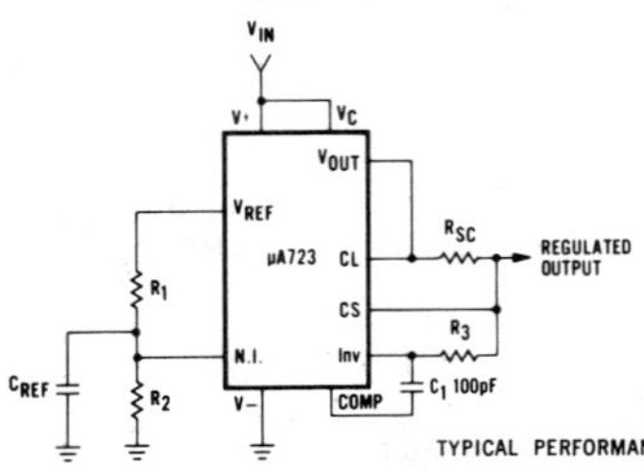

TYPICAL PERFORMANCE

Regulated Output Voltage	5 V
Line Regulation (ΔV_{IN} = 3 V)	0.5 mV
Load Regulation (ΔI_L=50 mA)	1.5 mV

Note: $R_3 = \frac{R_1 R_2}{R_1 + R_2}$ for minimum temperature drift.

Fig. 1

BASIC HIGH VOLTAGE REGULATOR
(V_{OUT} = 7 to 37 V)

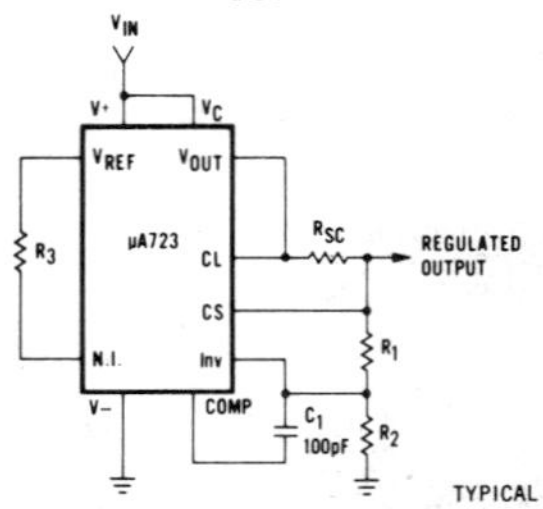

TYPICAL PERFORMANCE

Regulated Output Voltage	15 V
Line Regulation (ΔV_{IN} = 3 V)	1.5 mV
Load Regulation (ΔI_L=50 mA)	4.5 mV

Note: $R_3 = \frac{R_1 R_2}{R_1 + R_2}$ for minimum temperature drift.

R_3 may be eliminated for minimum component count.

Fig. 2

NEGATIVE VOLTAGE REGULATOR

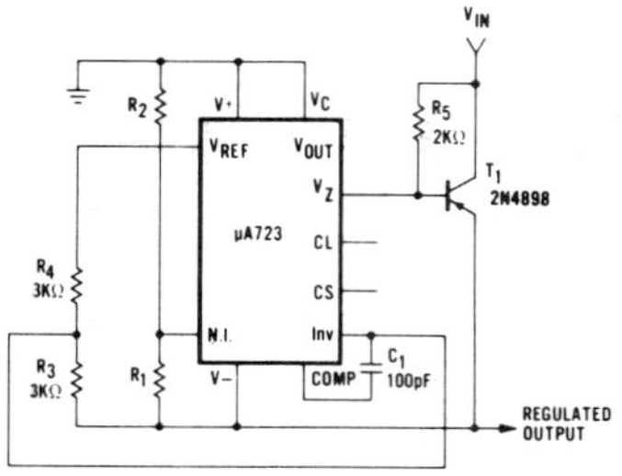

TYPICAL PERFORMANCE

Regulated Output Voltage	−15 V
Line Regulation (ΔV_{IN} = 3 V)	1 mV
Load Regulation (ΔI_L = 100 mA)	2 mV

Note 6

Fig. 3

POSITIVE VOLTAGE REGULATOR
(External NPN Pass Transistor)

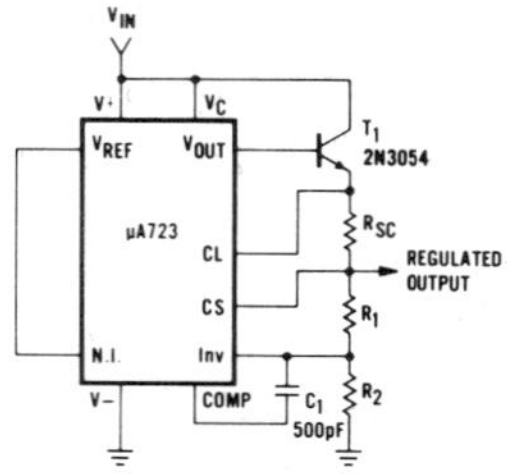

TYPICAL PERFORMANCE

Regulated Output Voltage	+15 V
Line Regulation (ΔV_{IN} = 3 V)	1.5 mV
Load Regulation (ΔI_L = 1 A)	15 mV

Fig. 4

POSITIVE VOLTAGE REGULATOR
(External PNP Pass Transistor)

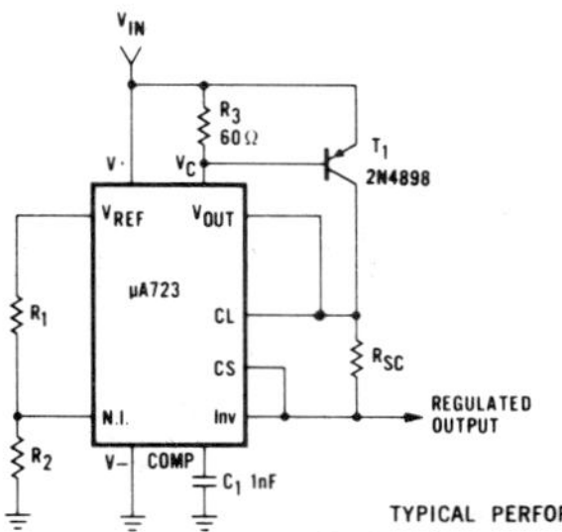

TYPICAL PERFORMANCE

Regulated Output Voltage	+5 V
Line Regulation (ΔV_{IN} = 3 V)	0.5 mV
Load Regulation (ΔI_L = 1 A)	5 mV

Fig. 5

FOLDBACK CURRENT LIMITING

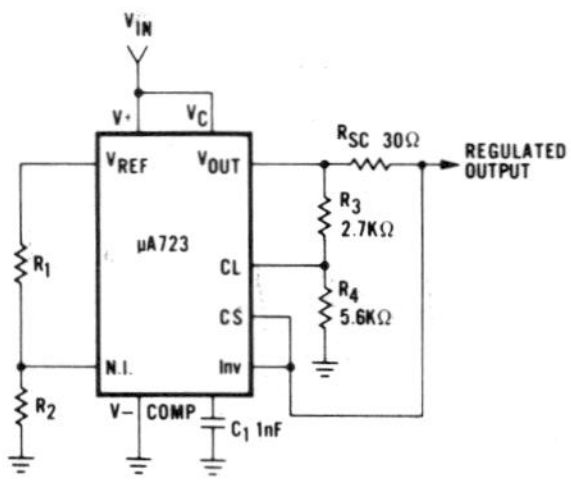

TYPICAL PERFORMANCE

Regulated Output Voltage	+5 V
Line Regulation (ΔV_{IN} = 3 V)	0.5 mV
Load Regulation (ΔI_L = 10 mA)	1 mV
Short Circuit Current	20 mA

Fig. 6

POSITIVE FLOATING REGULATOR

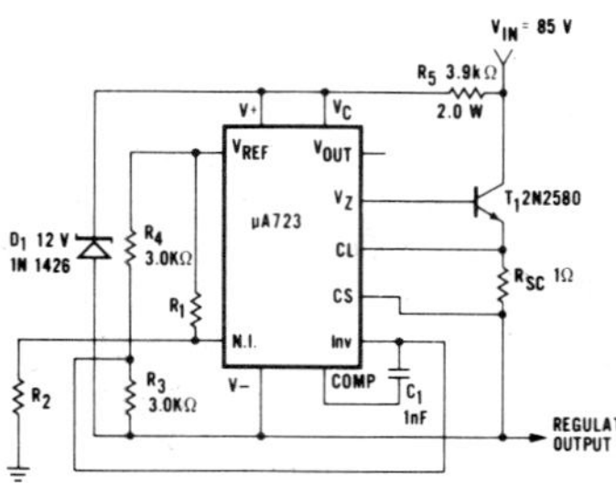

TYPICAL PERFORMANCE

Regulated Output Voltage	+50 V
Line Regulation (ΔV_{IN} = 20 V)	15 mV
Load Regulation (ΔI_L = 50 mA)	20 mV

Note 6

Fig. 7

NEGATIVE FLOATING REGULATOR

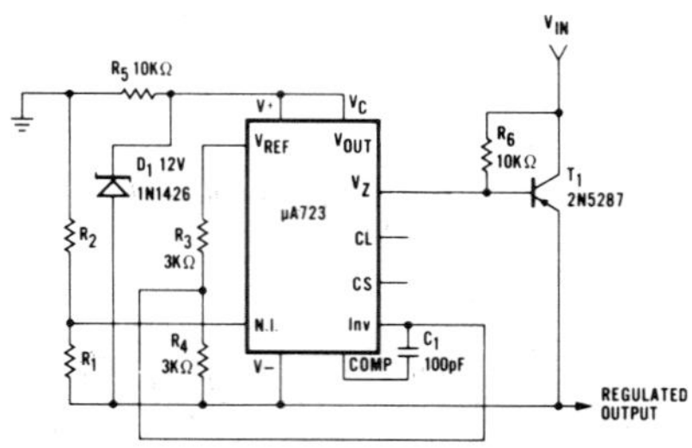

TYPICAL PERFORMANCE

Regulated Output Voltage	−100 V
Line Regulation (ΔV_{IN} = 20 V)	30 mV
Load Regulation (ΔI_L=100 mA)	20 mV

Note 6

Fig. 8

POSITIVE SWITCHING REGULATOR

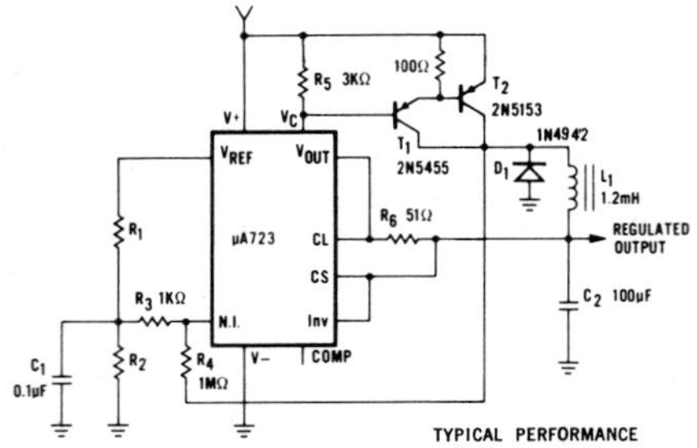

TYPICAL PERFORMANCE

Regulated Output Voltage	+5 V
Line Regulation (ΔV_{IN} = 30 V)	10 mV
Load Regulation (ΔI_L = 2 A)	80 mV

Note 2

Fig. 9

NEGATIVE SWITCHING REGULATOR

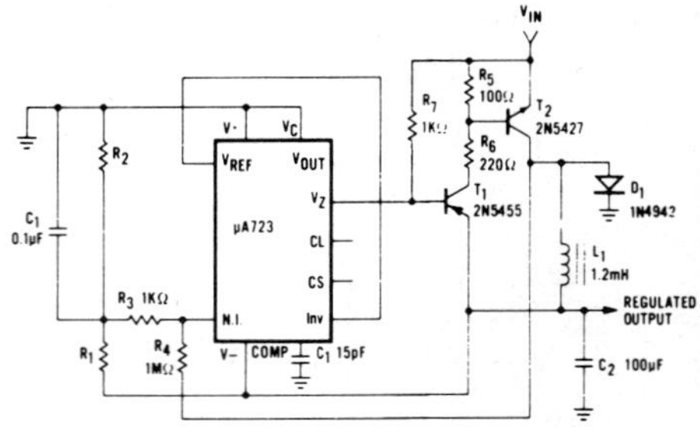

TYPICAL PERFORMANCE

Regulated Output Voltage	−15 V
Line Regulation (ΔV_{IN} = 20 V)	8 mV
Load Regulation (ΔI_L = 2 A)	6 mV

Notes 2, 6

Fig. 10

REMOTE SHUTDOWN REGULATOR WITH CURRENT LIMITING

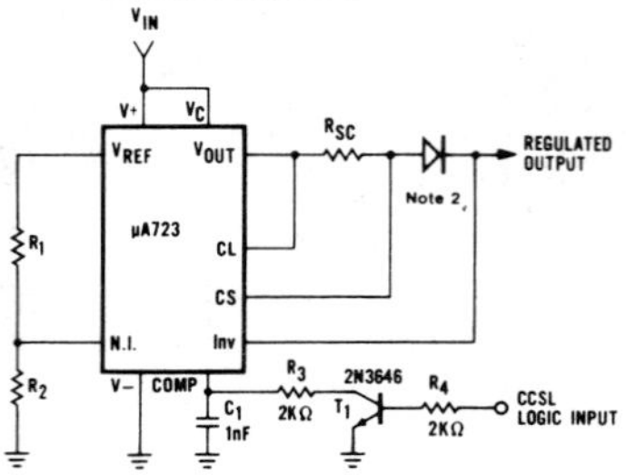

Note 1: Current limit transistor may be used for shutdown if current limiting is not required.

2: Add if Vout > 10V

TYPICAL PERFORMANCE

Regulated Output Voltage	+5 V
Line Regulation (ΔV_{IN} = 3 V)	0.5 mV
Load Regulation (ΔI_L=50 mA)	1.5 mV

Fig. 11

SHUNT REGULATOR

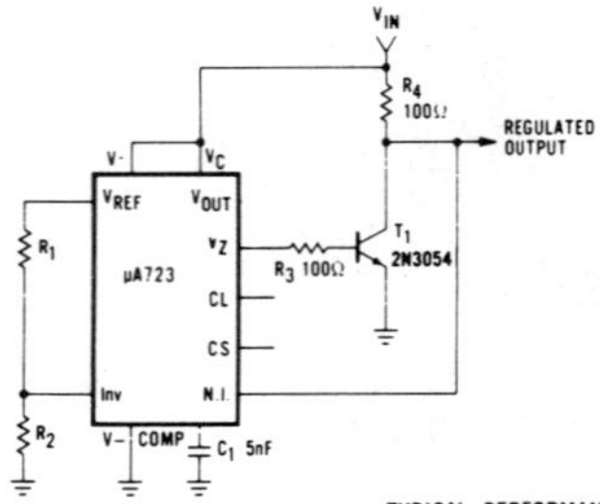

TYPICAL PERFORMANCE

Regulated Output Voltage	+5 V
Line Regulation (ΔV_{IN} = 10 V)	0.5 mV
Load Regulation (ΔI_L = 100 mA)	1.5 mV

Note 6

Fig. 12

OUTPUT VOLTAGE ADJUST

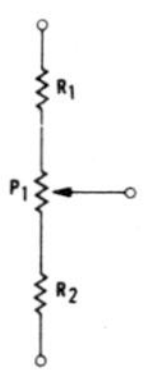

Fig. 13

EQUIVALENT CIRCUIT

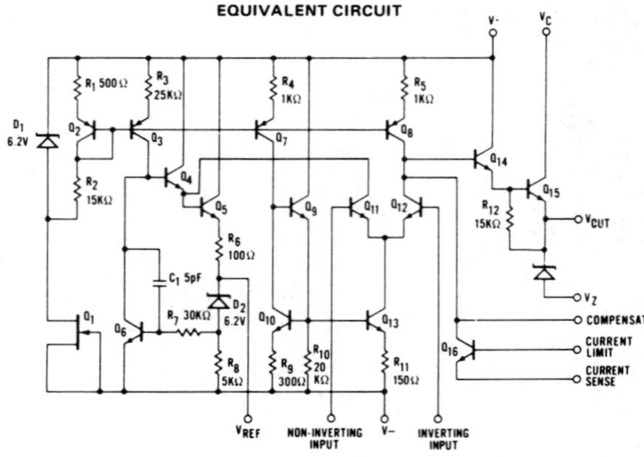

μA78S40

UNIVERSAL SWITCHING REGULATOR SUBSYSTEM

FAIRCHILD LINEAR INTEGRATED CIRCUITS

GENERAL DESCRIPTION — The μA78S40 is a Monolithic Regulator Subsystem consisting of all the active building blocks necessary for switching regulator systems. The device consists of a temperature-compensated voltage reference, a duty-cycle controllable oscillator with an active current limit circuit, an error amplifier, high-current, high-voltage output switch, a power diode and an uncommitted operational amplifier. The device can drive external npn or pnp transistors when currents in excess of 1.5 A or voltages in excess of 40 V are required. The device can be used for step down, step up or inverting switching regulators as well as for series pass regulators. It features wide supply voltage range, low standby power dissipation, high efficiency and low drift. It is useful for any stand-alone, low part-count switching system and works extremely well in battery operated systems.

- **STEP UP, STEP DOWN OR INVERTING SWITCHING REGULATORS**
- **OUTPUT ADJUSTABLE FROM 1.3 to 40 V**
- **OUTPUT CURRENTS TO 1.5 A WITHOUT EXTERNAL TRANSISTORS**
- **OPERATION FROM 2.5 TO 40 V INPUT**
- **LOW STANDBY CURRENT DRAIN**
- **80 dB LINE AND LOAD REGULATION**
- **HIGH GAIN, HIGH CURRENT, INDEPENDENT OP AMP**

CONNECTION DIAGRAM
16-PIN DIP
(TOP VIEW)

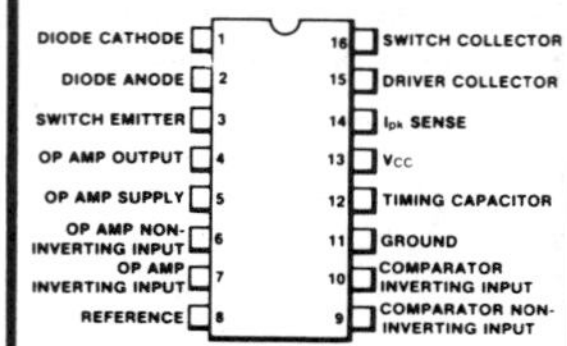

ORDER INFORMATION

TYPE	PART NO.
μA78S40	μA78S40DM
μA78S40	μA78S40DC
μA78S40	μA78S40PC

BLOCK DIAGRAM

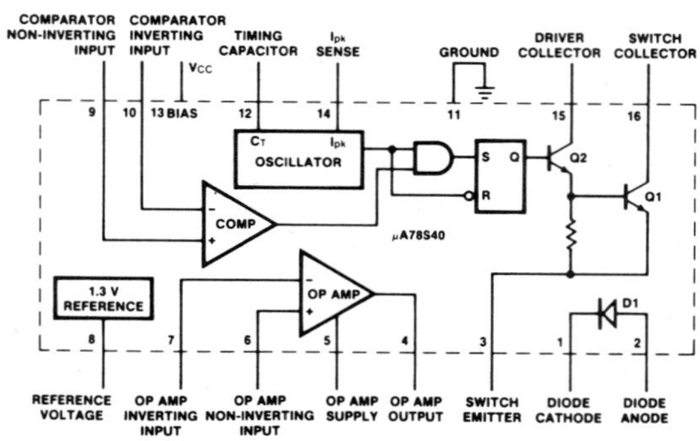

Courtesy of Fairchild Camera and Instruments Corporation.

ABSOLUTE MAXIMUM RATINGS

Input voltage from V^+ to V^-	40 V
Input voltage from V^+ op amp to V^-	40 V
Common mode input range (Error Amplifier and Op Amp)	-0.3 to V +
Differential input voltage (Note 1)	±30 V
Output Short Circuit Duration (Op Amp)	continuous
Current from V_{REF}	10 mA
Voltage from Switch Collectors to GND	40 V
Voltage from Switch Emitters to GND	40 V
Voltage from Switch Collectors to Emitter	40 V
Voltage from Power Diode to GND	40 V
Reverse Power Diode Voltage	40 V
Current through Power Switch	1.5 A
Current through Power Diode	1.5 A
Internal Power Dissipation (Note 2)	
Plastic DIP	1500 mW
Hermetic DIP	1000 mW
Storage Temperature Range	−65°C to + 150°C
Operating Temperature Range	
Military (μA78S40M)	−55°C to 125°C
Commercial (μA78S40C)	0°C to 70°C
Lead Temperature	
Hermetic DIP (Soldering, 60 s)	300°C
Plastic DIP (Soldering, 10 s)	260°C

NOTES:

1. For supply voltages less than 30 V, the absolute maximum voltage is equal to the supply voltage.
2. Ratings apply to 25° C. Above 25° C ambient, derate hermetic DIP at 8 mW/° C and plastic DIP at 14 mW/° C.

ELECTRICAL CHARACTERISTICS: V_{IN} = 5.0 V, $V_{Op\ Amp}$ = 5.0 V, T_A = 25° C unless otherwise specified.

CHARACTERISTICS	CONDITIONS	MIN	TYP	MAX	UNITS
GENERAL CHARACTERISTICS					
Supply Voltage		2.5		40	V
Supply Current (Op Amp Disconnected)	V_{IN} = 5.0 V		1.8	2.5	mA
	V_{IN} = 40 V		2.3	3.5	mA
Supply Current Op Amp	V_{IN} = 5.0 V		0.4	1.0	
	V_{IN} = 40 V		0.5	1.5	
REFERENCE SECTION					
Reference Voltage	I_{REF} = 1.0 mA	1.180	1.245	1.310	V
Reference Voltage Temperature Coefficient	I_{REF} = 1.0 mA		100		ppm/°C
Reference Voltage Line Regulation	V_{IN} = 3.0 V to V_{IN} = 40 V, I_{REF} = 1.0 mA		0.04	0.2	mV/V
Reference Voltage Load Regulation	I_{REF} = 1.0 mA to I_{REF} = 10 mA		0.2	0.2	mV/mA
OSCILLATOR SECTION					
Charging Current			25		μA
ON Time	C_T = 0.01 μF		200		μs
Discharge Current			225		μA
OFF Time	C_T = 0.01 μF		22		μs
Oscillator Voltage Swing			0.5		V
CURRENT LIMIT SECTION					
Current Limit Sense Voltage			330		mV
OUTPUT SWITCH SECTION					
Output Saturation Voltage 1	I_{SW} = 1.0 A		1.1	1.3	V
Output Saturation Voltage 2	I_{SW} = 1.0 A		0.45	0.7	V
Output Transistor h_{FE}	I_C = 1.0 A, V_{CE} = 5.0 V		70		
Output Leakage Current	V_{OUT} = 40 V		10		nA

ELECTRICAL CHARACTERISTICS: V_{IN} = 5.0 V, $V_{Op\ Amp}$ = 5.0 V, T_A = Operating temperature range, unless otherwise specified.**

COMPARATOR					
Input Offset Voltage	$V_{CM} = V_{REF}$		1.5	15	mV
Input Bias Current	$V_{CM} = V_{REF}$		35	200	nA
Input Offset Current	$V_{CM} = V_{REF}$		5.0	75	nA
Common Mode Voltage Range	T_A = 25°C	0		V+ -2	V
Power Supply Rejection Ratio	V_{IN} 3.0 V to 40 V, T_A = 25°C	70	96		dB
OUTPUT OPERATIONAL AMPLIFIER					
Input Offset Voltage	V_{CM} = 2.5 V		4.0	15	mV
Input Bias Current	V_{CM} = 2.5 V		30	200	nA
Input Offset Current	V_{CM} = 2.5 V		5.0	75	nA
Voltage Gain +	R_L = 2.0 k to GND; V_O = 1.0 to 2.5 V, T_A = 25°C	25 k	250 k		V/V
Voltage Gain -	R_L = 2.0 k to V+ Op Amp; V_O = 1.0 to 2.5 V, T_A = 25°C	25 k	250 k		V/V
Common Mode Voltage Range	T_A = 25°C	0		V+ -2	V
Common Mode Rejection Ratio	V_{CM} = 0 to 3.0 V, T_A = 25°C	76	100		dB
Power Supply Rejection Ratio	V+ Op Amp = 3.0 to 40 V, T_A = 25°C	76	100		dB
Output Source Current	T_A = 25°C	75	150		mA
Output Sink Current	T_A = 25°C	10	35		mA
Slew Rate	T_A = 25°C		0.6		V/μs
Output LOW Voltage	I_L = −5.0 mA, T_A = 25°C			1.0	V
Output HIGH Voltage	I_L = 50 mA, T_A = 25°C	V+Op Amp −3.0 V			V

DESIGN FORMULAS

CHARACTERISTIC	STEP DOWN	STEP UP	INVERTING	UNITS
I_{pk}	$2\,I_{OUT(Max)}$	$2\,I_{OUT(Max)} \bullet \dfrac{V_{OUT}+V_D-V_S}{V_{IN}-V_S}$	$2\,I_{OUT(Max)} \bullet \dfrac{V_{IN}+\lvert V_{OUT}\rvert+V_D-V_S}{V_{IN}-V_S}$	A
R_{SC} *	$0.33/I_{pk}$	$0.33\ V/I_{pk}$	$0.33\ V/I_{pk}$	Ω
$\dfrac{t_{on}}{t_{off}}$	$\dfrac{V_{OUT}+V_D}{V_{IN}-V_S-V_{OUT}}$	$\dfrac{V_{OUT}+V_D-V_{IN}}{V_{IN}-V_S}$	$\dfrac{\lvert V_{OUT}\rvert+V_D}{V_{IN}-V_S}$	–
L *	$\dfrac{V_{OUT}+V_D}{I_{pk}} \bullet t_{off}$	$\dfrac{V_{OUT}+V_D-V_{IN}}{I_{pk}} \bullet t_{off}$	$\dfrac{\lvert V_{OUT}\rvert+V_D}{I_{pk}} \bullet t_{off}$	μh
t_{off}	$\dfrac{I_{pk}\cdot L}{V_{OUT}+V_D}$	$\dfrac{I_{pk}\cdot L}{V_{OUT}+V_D-V_{IN}}$	$\dfrac{I_{pk}\cdot L}{\lvert V_{OUT}\rvert+V_D}$	μs
C_T*(μF)	$45 \times 10^{-5}\ t_{off}(\mu s)$	$45 \times 10^{-5}\ t_{off}(\mu s)$	$45 \times 10^{-5}\ t_{off}(\mu s)$	μF
C_O *	$\dfrac{I_{pk}\cdot(t_{on}+t_{off})}{8\,V_{ripple}}$	$\dfrac{(I_{pk}-I_{OUT})^2\cdot t_{off}}{2\,I_{pk}\cdot V_{ripple}}$	$\dfrac{(I_{pk}-I_{OUT})^2\cdot t_{off}}{2\,I_{pk}\cdot V_{ripple}}$	μF
Efficiency	$\dfrac{V_{IN}-V_S+V_D}{V_{IN}} \bullet \dfrac{V_{OUT}}{V_{OUT}+V_D}$	$\dfrac{V_{IN}-V_S}{V_{IN}} \bullet \dfrac{V_{OUT}}{V_{OUT}+V_D-V_S}$	$\dfrac{V_{IN}-V_S}{V_{IN}} \bullet \dfrac{\lvert V_{OUT}\rvert}{\lvert V_{OUT}\rvert+V_D}$	–
$I_{IN(Avg)}$ (Max load condition)	$\dfrac{I_{pk}}{2} \bullet \dfrac{V_{OUT}+V_D}{V_{IN}-V_S+V_D}$	$\dfrac{I_{pk}}{2}$	$\dfrac{I_{pk}}{2} \bullet \dfrac{\lvert V_{OUT}\rvert+V_D}{V_{IN}+\lvert V_{OUT}\rvert+V_D-V_S}$	A

MOTOROLA Semiconductors
BOX 20912 • PHOENIX, ARIZONA 85036

MC3420
MC3520

SWITCHMODE REGULATOR CONTROL CIRCUIT

The MC3520/3420 is an inverter control unit which provides all the control circuitry for PWM push-pull, bridge and series type switchmode power supplies.

These devices are designed to supply the pulse width modulated drive to the base of two external power transistors. Other applications where these devices can be used are in transformerless voltage doublers, transformer coupled dc-to-dc converters and other power control functions.

The MC3520 is specified over the military operating range of $-55^{\circ}C$ to $+125^{\circ}C$. The MC3420 is specified from $0^{\circ}C$ to $+70^{\circ}C$.

- Includes Symmetrical Oscillator
- On Chip Pulse Width Modulator, Voltage Reference, Dead Time Comparator, and Phase Splitter
- Output Frequency Adjustable (2 kHz to 100 kHz)
- Inhibit and Symmetry Correction Inputs Available
- Controlled Start-Up
- Frequency and Dead Time are Independently Adjustable (0% to 100%)
- Can be Slaved to Other MC3420s
- Open Collector Outputs
- Output Capability 50 mA (Max.)
- On Chip Protection Against Double Pulsing of Same Output During Load Transient Condition

FIGURE 1—TYPICAL APPLICATION

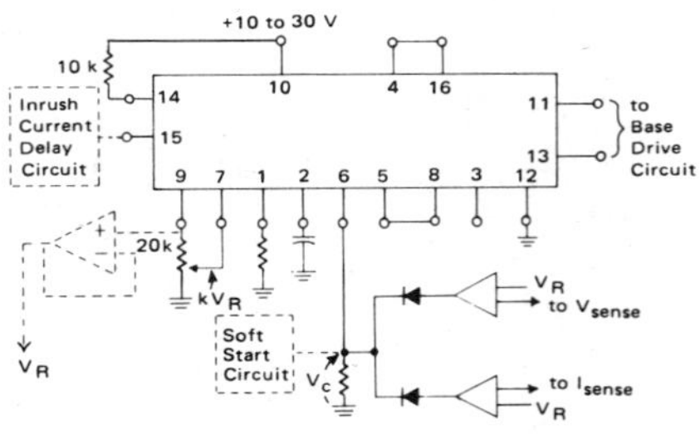

SWITCHMODE REGULATOR CONTROL CIRCUIT

SILICON MONOLITHIC INTEGRATED CIRCUITS

P SUFFIX
PLASTIC PACKAGE
CASE 648

L SUFFIX
CERAMIC PACKAGE
CASE 620

PIN CONNECTIONS

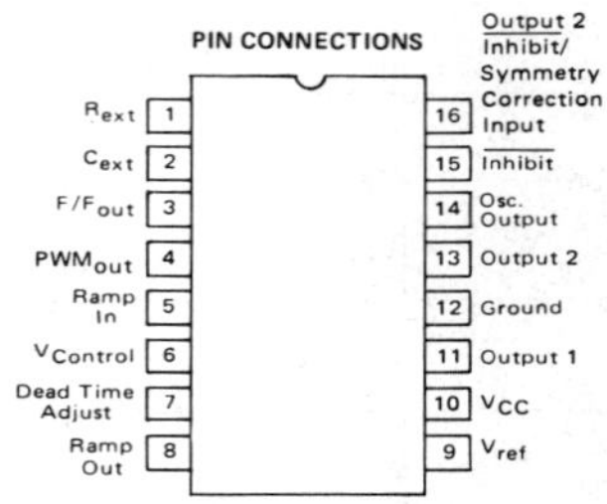

ORDERING INFORMATION

DEVICE	TEMPERATURE RANGE	PACKAGE
MC3420P	0 to +70°C	Plastic DIP
MC3420L	0 to +70°C	Ceramic DIP
MC3520L	-55 to +125°C	Ceramic DIP

Courtesy of Motorola Semiconductor Products Incorporated.

MAXIMUM RATINGS

Rating	Symbol	MC3520	MC3420	Unit
Power Supply Voltage	V_{CC}	30		V
Output Voltage (pins 11 and 13)	V_{out}	40		V
Oscillator Output Voltage (pin 14)	V_{14}	30		V
Voltage at pin 4	V_4	2.0		V
Voltage at pins 3 and 8	V_3, V_8	5.0		V
Voltage at pin 5	V_5	7.0		V
Power Dissipation	P_D	See Thermal Information		
Operating Junction Temperature	T_J			°C
Plastic Package		–	125	
Ceramic Package		150	150	
Operating Ambient Temperature Range	T_A	-55 to +125	0 to +70	°C
Storage Temperature Range	T_{stg}	-65 to +150	-65 to +150	°C

ELECTRICAL CHARACTERISTICS (V_{CC} = 10 to 30 V, T_A = 25°C unless otherwise noted.)

Characteristic	Figure	Symbol	MC3520 Min	MC3520 Typ	MC3520 Max	MC3420 Min	MC3420 Typ	MC3420 Max	Unit
REFERENCE SECTION									
Reference Voltage (I_{ref} = 400 μA)	5	V_{ref}	7.6	7.8	8.0	7.4	7.8	8.2	V
Temperature Coefficient of Reference Voltage (V_{CC} = 15 V, I_{ref} = 400 μA)	5	TCV_{ref}	–	0.008	0.03	–	0.008	0.03	%/°C
Input Regulation of Reference Voltage	5	$Reg_{(in)}$							mV/V
(I_{ref} = 400 μA)			–	3.0	7.5	–	4.0	7.5	
(I_{ref} = 1.0 mA)			–	5.0	–	–	5.0	–	
DC SUPPLY SECTION									
Supply Voltage	5	V_{in}	10	–	30	10	–	30	V
Supply Current (R_{ext} = 10 kΩ, excluding load and current and reference current)	5	I_D	–	–	16	–	–	22	mA
OSCILLATOR SECTION									
Line Frequency Stability	5								
(f = 20 kHz)		Δf	–	–	3.0	–	–	5.0	%
(f = 20 kHz, V_{CC} = 15 V, T_{low} to T_{high})		Δf	–	0.03	–	–	0.04	–	%/°C
Maximum Output Frequency (V_{CC} = 15 V)	6	f_{max}	100	200	–	100	200	–	kHz
Minimum Output Frequency (V_{CC} = 15 V)	6	f_{min}	–	2.0	5.0	–	2.0	5.0	kHz
Oscillator Output Saturation Voltage ($I_{14\ sink}$ = 5.0 mA)	11	$V_{osc(sat)}$	–	0.2	0.5	–	0.2	0.5	V
OUTPUT SECTION									
Output Saturation Voltage	7	$V_{CE(sat)}$							V
(I_L = 40 mA, T_{high} to T_{low})			–	0.33	0.5	–	0.33	0.5	
(I_L = 25 mA, T_{high} to T_{low})			–	0.22	–	–	0.22	–	
Output Leakage Current (V_{CE} = 40 V, pins 11 and 13)	8	I_{CE}	–	–	50	–	–	50	μA
COMPARATOR SECTION									
Pulse Width Adjustment Range	9	ΔPW	0	–	100	0	–	100	%
Dead Time Adjustment Range	9	ΔDT	0	–	100	0	–	100	%
Temperature Coefficient of Dead Time	–	TCDT	–	0.1	–	–	0.1	–	%/°C
Comparator Bias Currents	12, 13	I_{IB}	–	5.0	15	–	5.0	15	μA
	14	I_{IB}	–	10	30	–	10	30	μA

ELECTRICAL CHARACTERISTICS (continued)

Characteristic	Figure	Symbol	MC3520 Min	MC3520 Typ	MC3520 Max	MC3420 Min	MC3420 Typ	MC3420 Max	Unit
AUXILIARY INPUTS/OUTPUTS									
Ramp Voltage	5								V
Peak High		$V_{ramp(Hi)}$	5.5	6.0	6.5	5.5	6.0	6.5	
Peak Low		$V_{ramp(Low)}$	2.0	2.4	2.8	2.0	2.4	2.8	
Ramp Voltage Change ($V_{ramp\ Hi} - V_{ramp\ Low}$)	5	ΔV_{ramp}	3.0	3.5	4.0	3.0	3.5	4.0	V
Ramp Out Sink Current	5	I_{sink}	–	400	–	–	400	–	μA
Ramp Out Source Current	5	I_{source}	–	3.0	–	–	3.0	–	mA
$\overline{\text{Inhibit}}$ Input Current – High (V_{IH} = 2.0 V)	10	I_{IH}	–	–	40	–	–	40	μA
$\overline{\text{Inhibit}}$ Input Current – Low (V_{IL} = 0.8 V)	10	I_{IL}	–	−25	−180	–	−25	−180	μA
Symmetry Correction Input/Output 2 Inhibit Current – High (V_{SY} = 2.0 V, pin 16)	10	$I_{SY/H}$	–	–	40	–	–	40	μA
Symmetry Correction Input/Output 2 Inhibit Current – Low (V_{SY} = 0.8 V, pin 16)	10	$I_{SY/L}$	–	−10	−180	–	−10	−180	μA
F/F_{out} Source Current	–	I_{source}	–	2.0	–	–	2.0	–	mA
OUTPUT AC CHARACTERISTICS ($T_A = T_{high}$, V_{CC} = +15 V, f = 20 kHz)									
Rise Time	15	t_r	–	40	–	–	40	–	ns
Fall Time	15	t_f	–	150	–	–	150	–	ns
Overlap Time	15	t_{ov}	–	275	–	–	275	–	ns
Assymmetry (Duty Cycle = 50%)	15	$\frac{t_{on1} - t_{on2}}{t_{on1}}$	–	±1.0	–	–	±1.0	–	%

NOTE:

T_{high} = +125°C for MC3520
+70°C for MC3420

T_{low} = −55°C for MC3520
0°C for MC3420

FIGURE 2–EQUIVALENT CIRCUIT

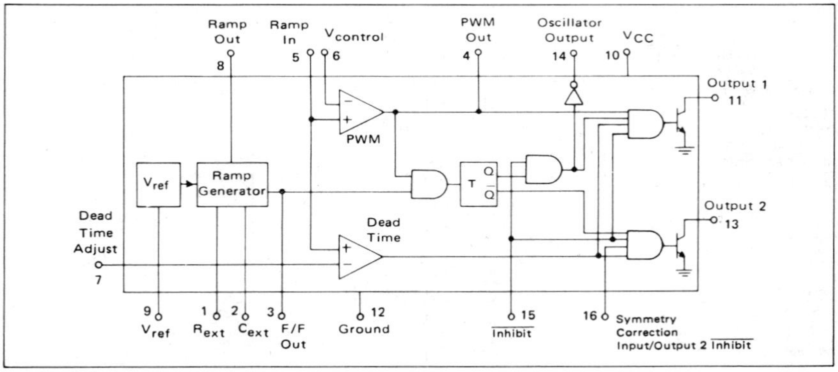

GENERAL INFORMATION

The internal block diagram of the MC3420 is shown in Figure 2, and consists of the following sections:

Voltage Reference

A stable reference voltage is generated by the MC3420 primarily for internal use. However, it is also available externally at Pin 9 (V_{ref}) for use in setting the dead time (Pin 7) and for use as a reference for the external control loop error amplifiers.

Ramp Generator

The ramp generator section produces a symmetrical triangular waveform ramping between 2.4 V and 6.0 V, with frequency determined by an external resistor (R_{ext}) and capacitor (C_{ext}) tied from Pins 1 and 2, respectively, to ground.

PWM Comparator

The output of the ramp generator at pin 8 is normally connected to Pin 5, RAMP IN. The PWM (pulse width modulation) comparator compares the voltage at Pin 6 ($V_{control}$) to the ramp generator output. The level of $V_{control}$ determines the outputs' pulse width or duty cycle. The duty cycle of each output can vary, exclusive of dead time, from 50% (when $V_{control}$ is at approximately 2.4 V) to 0% ($V_{control}$ approximately 6.0 V).

Dead Time Comparator

An additional comparator has been included in MC3420 to allow independent adjustment of system dead time or maximum duty cycle. By dividing down V_{ref} at Pin 9 with a resistive divider or potentiometer, and applying this voltage to Pin 7, a stable dead time is obtained for prevention of inverter switching transistor cross conduction at high duty cycles due to storage time delays.

Phase Splitter

A phase splitter is included to obtain two 180° out of phase outputs for use in multiple transistor inverter systems. It consists of a toggle flip-flop whose clock signal is derived by "ANDing" the output of the PWM comparator and a signal from the ramp generator section. This "AND" gate ensures that the outputs truly alternate under control loop transient conditions. Better understanding of this feature and MC3420 operation may be gained by studying the circuit waveforms, shown in Figure 4.

FIGURE 4 – INTERNAL WAVEFORMS

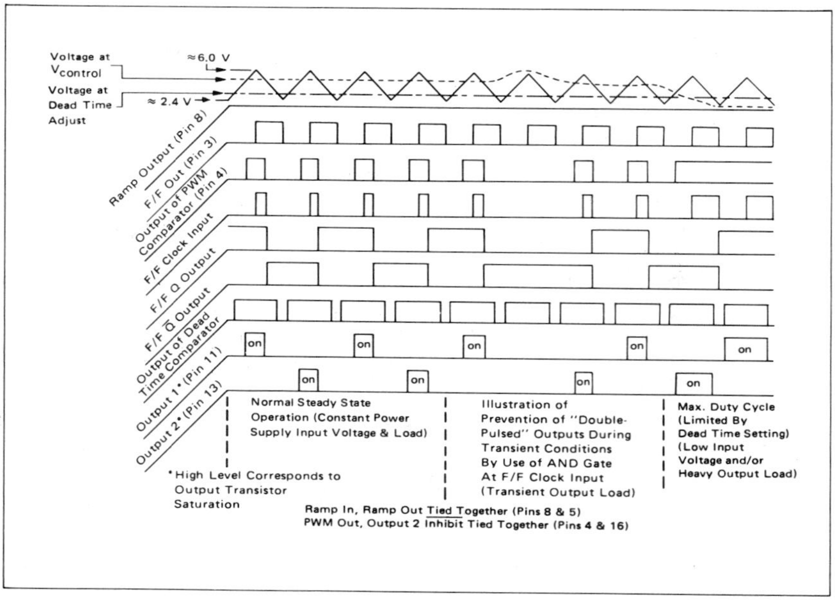

FIGURE 8 – OUTPUT LEAKAGE TEST CIRCUIT

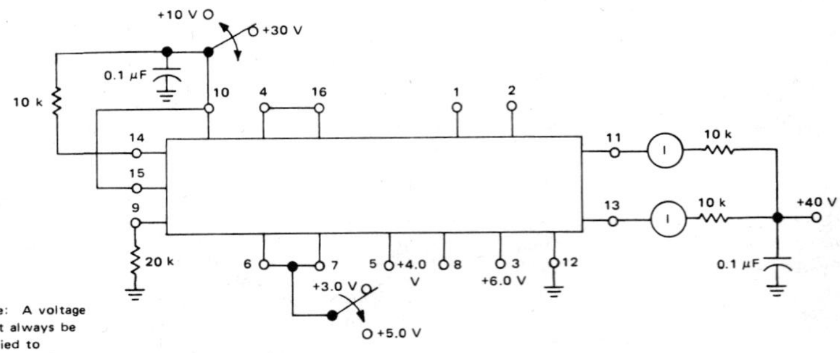

Note: A voltage must always be applied to pins 6 and 7.

FIGURE 9 – OUTPUT DUTY CYCLE TEST CIRCUIT

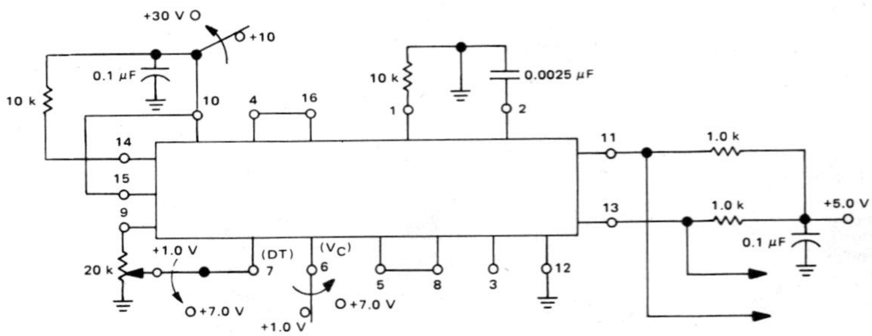

TYPICAL DUTY CYCLE versus DEAD TIME VOLTAGE	
PIN 7. DEAD TIME VOLTAGE (V) ($V_{control}$ = 2.0 V)	% DUTY CYCLE (FOR EACH OUTPUT)
2.0	50
2.5	46
3.0	40
3.5	33
4.0	26
4.5	18
5.0	11
5.5	4.0
6.0	0

TYPICAL DUTY CYCLE versus PWM VOLTAGE ($V_{control}$)	
PIN 6. $V_{control}$ (V) (DEAD TIME VOLTAGE = 1.0 V)	% DUTY CYCLE (FOR EACH OUTPUT)
2.0	50
2.5	46
3.0	40
3.5	33
4.0	26
4.5	18
5.0	11
5.5	4.0
6.0	0

	V_6	V_7	
	Volts		
100% Adjust			
Dead Time	1.0	1.0	(Pin 11 + Pin 13 = Logic "1")
Pulse Width	1.0	1.0	
0% Adjust			
Dead Time	7.0	1.0	(Pin 11)(Pin 13) = Logic "1"
Pulse Width	1.0	7.0	

NOTE: Logic "1" is TTL-Compatible V_{OH}.

OPERATION AND APPLICATIONS INFORMATION

The Voltage Reference

The temperature coefficient of V_{ref} has been optimized for a 400 μA (≅20 kΩ) load. If increased current capability is required, an op amp buffer may be used, as shown in Figure 22.

FIGURE 22

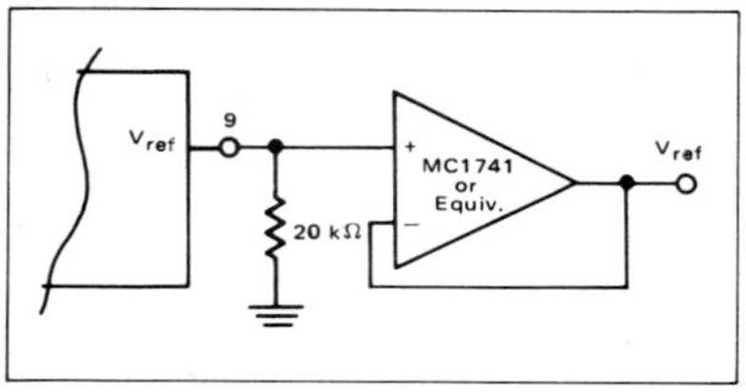

Output Frequency

The values of R_{ext} and C_{ext} for a given output frequency, f_o, can be found from:

$$f_o \cong \frac{0.55}{R_{ext}\,C_{ext}}\,;\ 5.0\ k\Omega \leq R_{ext} \leq 20\ k\Omega \quad \text{(Eq. 1)}$$

or from the graph shown in Figure 23.

Note that f_o refers to the frequency of Output 1 (Pin 11) or Output 2 (Pin 13). The frequency of the ramp generator output waveform at Pin 8 will be twice f_o.

FIGURE 23

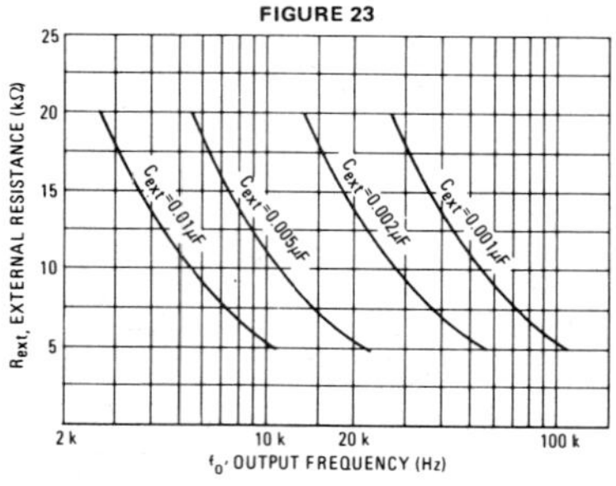

Dead Time

Figure 24 illustrates how to set or adjust the MC3420 outputs' dead time or maximum duty cycle. For minimum dead time drift with temperature or supply voltage, $V_{D.T.}$ should be derived from V_{ref} as shown.

FIGURE 24

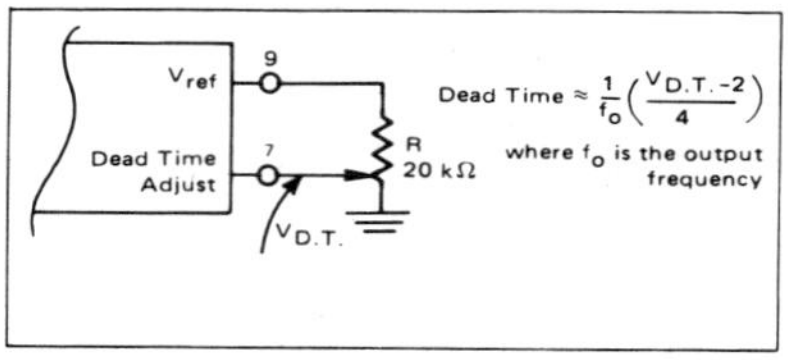

Connections to the $V_{control}$ Pin

In many systems, it is necessary to make multiple connections to the $V_{control}$ Pin in order to implement features in addition to voltage regulation such as current limiting, soft start, etc. These can be made by the use of a simple "diode-OR" connection, as shown in Figure 25. This allows whichever control element is seeking the lowest PWM duty cycle to dominate. Note that a resistor, R1, whose value is ≤ 50 kΩ is placed from the $V_{control}$ Pin to ground. This is necessary to provide a dc path for the PWM comparator input bias current under all conditions.

The system duty cycle is given by:

$$\text{D.C. (\%)} \cong \frac{V_{Control} - 2}{4} \times 100 \quad \text{(Eq. 2)}$$

FIGURE 25

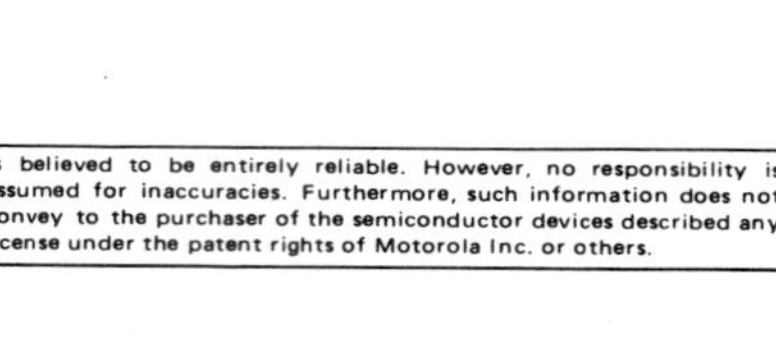

Soft Start

In most PWM switching supplies, a soft start feature is desired to prevent output voltage overshoots and magnetizing current imbalances in the power transformer primary. This feature forces the duty cycle of the switching elements to gradually increase from zero to their normal operating point during initial system power-up or after an inhibit. This feature can be easily implemented with the MC3420. One method is shown in Figure 26.

FIGURE 26

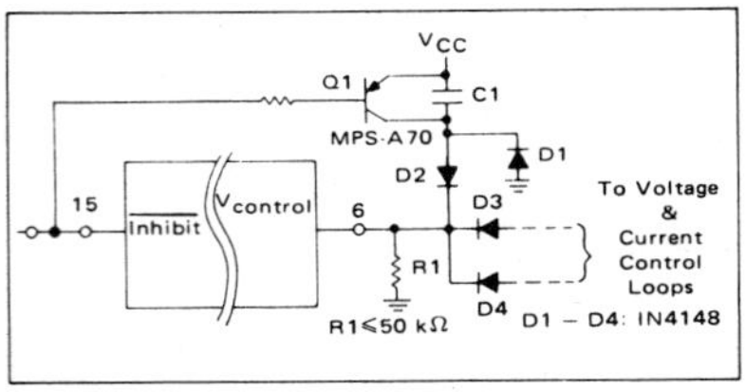

After an inhibit command or during power-up, the voltage on R1 and Pin 6 exponentially decays from V_{CC} toward ground with a time constant of R1C1, allowing a gradual increase in duty cycle. Diodes D2 – D4 provide a diode-or function at the $V_{control}$ Pin, while Q1 serves to reset the timing capacitor, C1, when an inhibit command is received thereby reinitializing the soft-start feature. D1 allows C1 to reset when power (V_{CC}) is turned off.

Inrush Current Limiting

Since many PWM switching supplies are operated directly off the rectified 110 Vac line with capacitive input filters, some means of preventing rectifier failure due to inrush surge currents is usually necessary. One method which can be used is shown in Figure 27.

In this circuit, a series resistor, R_S, is used to provide inrush surge current limiting. After the filter capacitor, C1, is charged, Q1 receives a trigger signal from the control circuitry through T1 and shorts R_S out of the circuit, eliminating its otherwise larger power dissipation. The trigger signal for Q1 may be derived from either the oscillator output (Pin 14) or one of the MC3420's outputs. If the oscillator output is used, it will be necessary

FIGURE 27

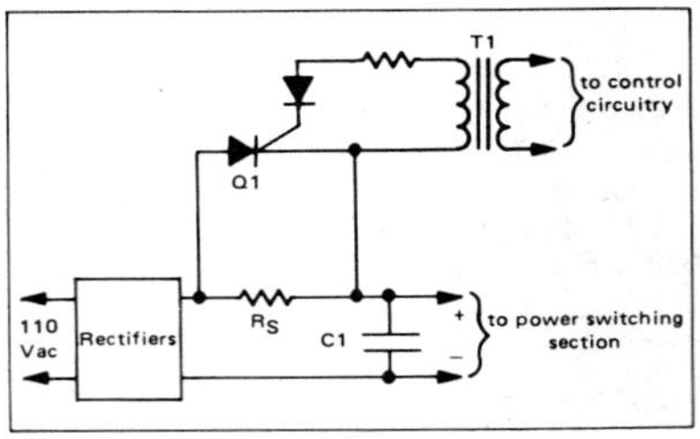

to provide a time delay on the inhibit pin to keep it low until the input filter capacitor, C1, has had time to charge, whereas the initial portion of the soft start timing cycle can be used for this delay if this signal is derived from one of the output pins. However, using the Oscillator Output Pin does offer the advantage that its waveform has a constant 50% duty cycle, independent of the outputs' duty cycle which can simplify the design of a drive circuit for T1.

Slaving

In some applications, as when one PWM inverter/converter is used to feed another, it may be desired that their frequencies be synchronized. This can be done with multiple MC3420s as shown in Figure 28. By omitting their R_{ext} and C_{ext}, up to two MC3420s may be slaved to a master MC3420.

FIGURE 28 – SLAVING THE MC3420

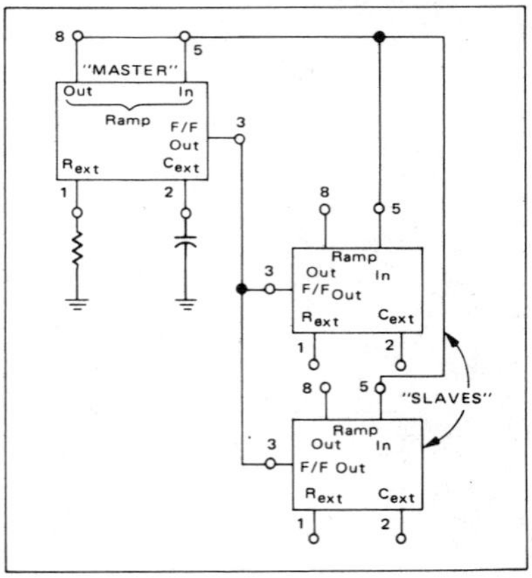

FIGURE 29 – 15 V, 2A DC-TO-DC CONVERTER

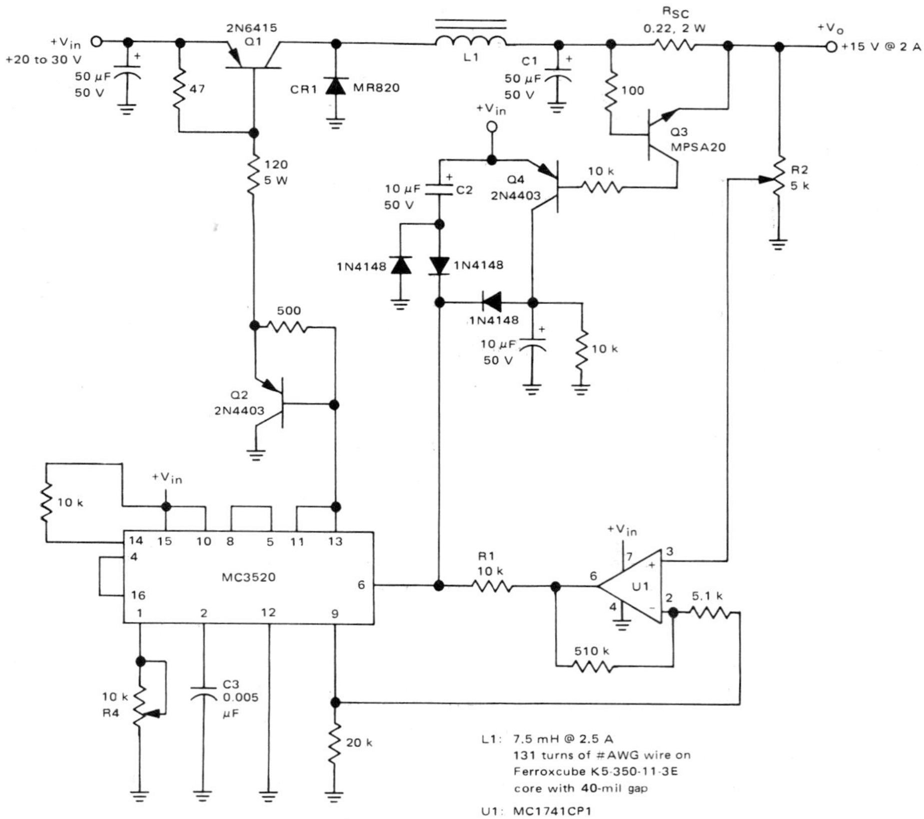

15 V, 2 A DC-to-DC Converter

Figure 29 illustrates the use of the MC3520 in a PWM switching power supply utilizing a single series switching element (see Appendix for description of PWM switching supply configurations). The series switching transistor, Q1, chops the dc input voltage, V_{in}, at a frequency of ≅ 25 kHz, and the resulting waveform is filtered by L1 and C1 to provide the dc output voltage. The frequency is set by R4 and C3, and since the outputs of the MC3520 are wire-ORed together, fo is twice that given by Equation 1 and Figure 23. V_o is regulated by comparing its value to the MC3520's reference voltage and amplifying the error voltage with U1. The output of U1 is fed into the MC3520 to provide PWM to Q1, thereby controlling its duty cycle and thus the value of V_o.

C2 provides a soft-start feature during power up to prevent output voltage overshoots and excessive start up currents through Q1.

Short circuit protection is provided by R_{SC}, Q3 and Q4. When an overcurrent condition occurs, Q3 is turned on by the voltage across R_{SC}; Q3 drives Q4 on, which raises the voltage at pin 6 ($V_{control}$) of the MC3520, reducing Q1's duty cycle and maintaining a constant output current of ≅ 2.5 A.

National Semiconductor

Analog Switches

AH0120/AH0130/AH0140/AH0150/AH0160 Series Analog Switches

General Description

The AH0100 series represents a complete family of junction FET analog switches. The inherent flexibility of the family allows the designer to tailor the device selection to the particular application. Switch configurations available include dual DPST, dual SPST, DPDT, and SPDT. $r_{ds(ON)}$ ranges from 10 ohms through 100 ohms. The series is available in both 14 lead flat pack and 14 lead cavity DIP. Important design features include:

- TTL/DTL and RTL compatible logic inputs
- Up to 20V p-p analog input signal
- $r_{ds(ON)}$ less than 10Ω (AH0140, AH0141, AH0145, AH0146)
- Analog signals in excess of 1 MHz
- "OFF" power less than 1 mW
- Gate to drain bleed resistors eliminated
- Fast switching, t_{ON} is typically 0.4 μs, t_{OFF} is 1.0 μs
- Operation from standard op amp supply voltages, ±15V, available (AH0150/AH0160 series)
- Pin compatible with the popular DG 100 series

The AH0100 series is designed to fulfill a wide variety of analog switching applications including commutators, multiplexers, D/A converters, sample and hold circuits, and modulators/demodulators. The AH0100 series is guaranteed over the temperature range -55°C to +125°C; whereas, the AH0100C series is guaranteed over the temperature range -25°C to +85°C

Schematic Diagrams

DUAL DPST and DUAL SPST

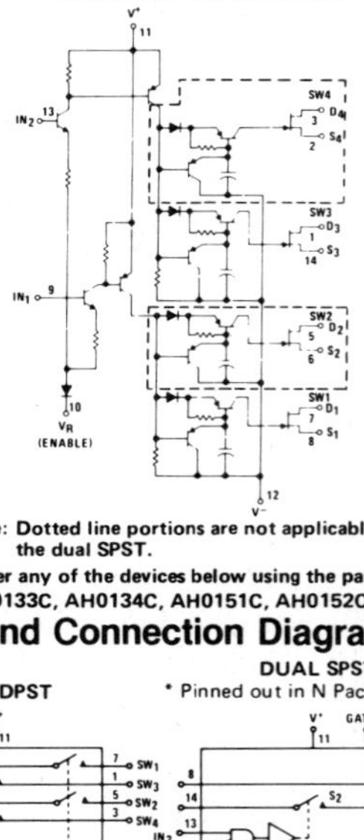

Note: Dotted line portions are not applicable to the dual SPST.

DPDT (diff.) and SPDT (diff.)

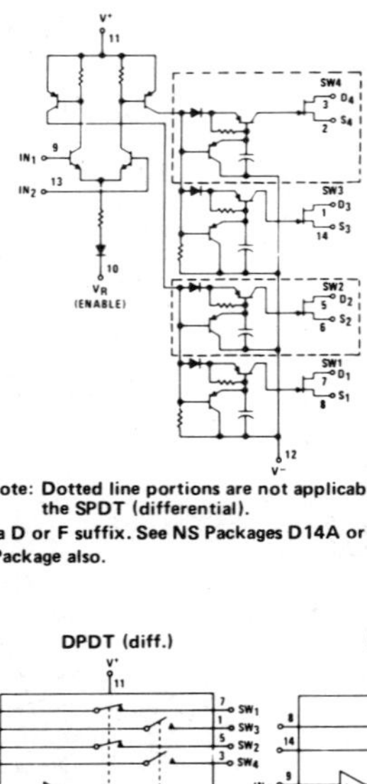

Note: Dotted line portions are not applicable to the SPDT (differential).

Order any of the devices below using the part number with a D or F suffix. See NS Packages D14A or F14A. AH0133C, AH0134C, AH0151C, AH0152C available in N Package also.

Logic and Connection Diagrams

DUAL DPST

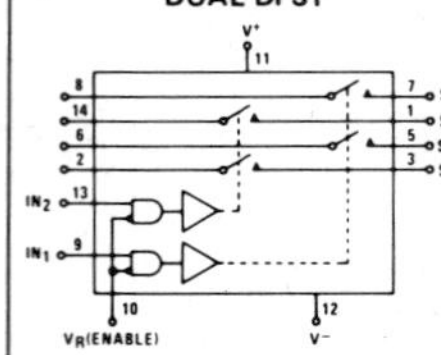

HIGH LEVEL (±10V)
AH0140 (10Ω)
AH0129 (30Ω)
AH0126 (80Ω)
MEDIUM LEVEL (±7.5V)
AH0153 (15Ω)
AH0154 (50Ω)

DUAL SPST

* Pinned out in N Package only.

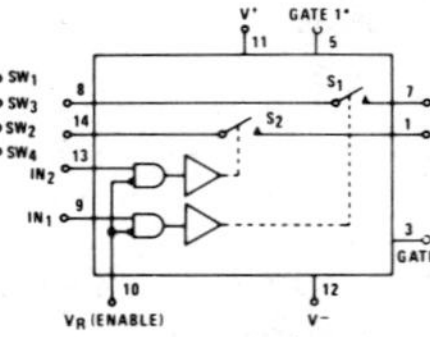

HIGH LEVEL (±10V)
AH0141 (10Ω)
AH0133 (30Ω)
AH0134 (80Ω)
MEDIUM LEVEL (±7.5V)
AH0151 (15Ω)
AH0152 (50Ω)

DPDT (diff.)

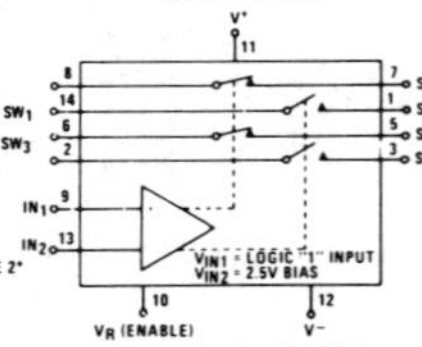

HIGH LEVEL (±10V)
AH0145 (10Ω)
AH0139 (30Ω)
AH0142 (80Ω)
MEDIUM LEVEL (±7.5V)
AH0163 (15Ω)
AH0164 (50Ω)

SPDT (diff.)

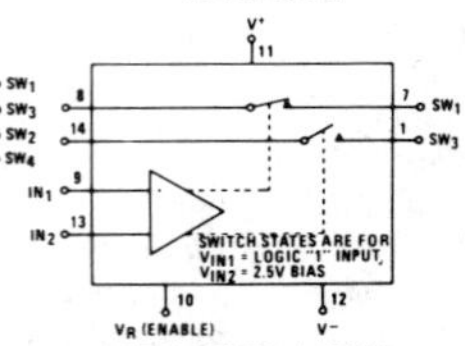

HIGH LEVEL (±10V)
AH0146 (10Ω)
AH0144 (30Ω)
AH0143 (80Ω)
MEDIUM LEVEL (±7.5V)
AH0161 (15Ω)
AH0162 (50Ω)

Courtesy of National Semiconductor Corporation.

Absolute Maximum Ratings

		High Level	Medium Level
Total Supply Voltage ($V^+ - V^-$)		36V	34V
Analog Signal Voltage ($V^+ - V_A$ or $V_A - V^-$)		30V	25V
Positive Supply Voltage to Reference ($V^+ - V_R$)		25V	25V
Negative Supply Voltage to Reference ($V_R - V^-$)		22V	22V
Positive Supply Voltage to Input ($V^+ - V_{IN}$)		25V	25V
Input Voltage to Reference ($V_{IN} - V_R$)		±6V	±6V
Differential Input Voltage ($V_{IN} - V_{IN2}$)		±6V	±6V
Input Current, Any Terminal		30 mA	30 mA
Power Dissipation			See Curve
Operating Temperature Range	AH0100 Series	−55°C to +125°C	
	AH0100C Series	−25°C to +85°C	
Storage Temperature Range		−65°C to +150°C	
Lead Temperature (Soldering, 10 sec)			300°C

Electrical Characteristics for "HIGH LEVEL" Switches (Note 1)

PARAMETER	SYMBOL	DEVICE TYPE: DUAL DPST	DUAL SPST	DPDT (DIFF)	SPDT (DIFF)	CONDITIONS: $V^+ = 12.0V$, $V^- = -18.0V$, $V_R = 0.0V$		LIMITS: TYP	MAX	UNITS
Logic "1" Input Current	$I_{IN(ON)}$	All Circuits				Note 2	T_A = 25°C	2.0	60	μA
							Over Temp. Range		120	μA
Logic "0" Input Current	$I_{IN(OFF)}$	All Circuits				Note 2	T_A = 25°C	.01	1	μA
							Over Temp. Range		2.0	μA
Positive Supply Current Switch ON	$I^+_{(ON)}$	All Circuits				One Driver ON Note 2	T_A = 25°C	2.2	3.0	mA
							Over Temp. Range		3.3	mA
Negative Supply Current Switch ON	$I^-_{(ON)}$	All Circuits				One Driver ON Note 2	T_A = 25°C	−1.0	−1.8	mA
							Over Temp. Range		−2.0	mA
Reference Input (Enable) ON Current	$I_{R(ON)}$	All Circuits				One Driver ON Note 2	T_A = 25°C	−1.0	−1.4	mA
							Over Temp. Range		−1.6	mA
Positive Supply Current Switch OFF	$I^+_{(OFF)}$	All Circuits				$V_{IN1} = V_{IN2} = 0.8V$	T_A = 25°C	1.0	10	μA
							Over Temp. Range		25	μA
Negative Supply Current Switch OFF	$I^-_{(OFF)}$	All Circuits				$V_{IN1} = V_{IN2} = 0.8V$	T_A = 25°C	−1.0	−10	μA
							Over Temp. Range		−25	μA
Reference Input (Enable) OFF Current	$I_{R(OFF)}$	All Circuits				$V_{IN1} = V_{IN2} = 0.8V$	T_A = 25°C	−1.0	−10	μA
							Over Temp. Range		−25	μA
Switch ON Resistance	$r_{ds(ON)}$	AH0126	AH0134	AH0142	AH0143	V_D = 10V, I_D = 1 mA	T_A = 25°C	45	80	Ω
							Over Temp. Range		150	Ω
Switch ON Resistance	$r_{ds(ON)}$	AH0129	AH0133	AH0139	AH0144	V_D = 10V, I_D = 1 mA	T_A = 25°C	25	30	Ω
							Over Temp. Range		60	Ω
Switch ON Resistance	$r_{ds(ON)}$	AH0140	AH0141	AH0145	AH0146	V_D = 10V, I_F = 1 mA	T_A = 25°C	8	10	Ω
							Over Temp. Range		20	Ω
Driver Leakage Current	$(I_D + I_S)_{ON}$	All Circuits				$V_D = V_S = -10V$	T_A = 25°C	.01	1	nA
							Over Temp. Range		100	nA
Switch Leakage Current	$I_{S(OFF)}$ OR $I_{D(OFF)}$	AH0126 AH0129	AH0134 AH0133	AH0142 AH0139	AH0143 AH0144	V_{DS} = ±20V	T_A = 25°C	0.8	1	nA
							Over Temp. Range		100	nA
Switch Leakage Current	$I_{S(OFF)}$ OR $I_{D(OFF)}$	AH0140	AH0141	AH0145	AH0146	V_{DS} = ±20V	T_A = 25°C	4	10	nA
							Over Temp. Range		1.0	μA
Switch Turn-ON Time	t_{ON}	AH0126 AH0129	AH0134 AH0133	AH0142 AH0139	AH0143 AH0144	See Test Circuit V_A = ±10V T_A = 25°C		0.5	0.8	μs
Switch Turn-ON Time	t_{ON}	AH0140	AH0141	AH0145	AH0146	See Test Circuit V_A = ±10V T_A = 25°C		0.8	1.0	μs
Switch Turn-OFF Time	t_{OFF}	AH0126 AH0129	AH0134 AH0133	AH0142 AH0139	AH0143 AH0144	See Test Circuit V_A = ±10V T_A = 25°C		0.9	1.6	μs
Switch Turn-OFF Time	t_{OFF}	AH0140	AH0141	AH0145	AH0146	See Test Circuit V_A = ±10V T_A = 25°C		1.1	2.5	μs

Note 1: Unless otherwise specified these limits apply for −55°C to +125°C for the AH0100 series and −25°C to +85°C for the AH0100C series. All typical values are for T_A = 25°C.

Note 2: For the DPST and Dual DPST, the ON condition is for V_{IN} = 2.5V; the OFF condition is for V_{IN} = 0.8V. For the differential switches and SW1 and 2 ON, V_{IN2} = 2.5V, V_{IN1} = 3.0V. For SW3 and 4 ON, V_{IN2} = 2.5V, V_{IN1} = 2.0V.

Electrical Characteristics for "MEDIUM LEVEL" Switches (Note 1)

PARAMETER	SYMBOL	DEVICE TYPE: DUAL DPST	DUAL SPST	DUAL DPDT	SPDT (DIFF)	CONDITIONS: $V^+ = +15.0V$, $V^- = -15V$, $V_R = 0V$		LIMITS: TYP	MAX	UNITS
Logic "1" Input Current	$I_{IN(ON)}$	All Circuits				Note 2	$T_A = 25°C$	20	60	μA
							Over Temp. Range		120	μA
Logic "0" Input Current	$I_{IN(OFF)}$	All Circuits				Note 2	$T_A = 25°C$	.01	0.1	μA
							Over Temp. Range		2	μA
Positive Supply Current Switch ON	$I^+_{(ON)}$	All Circuits				One Driver ON Note 2	$T_A = 25°C$	2.2	3.0	mA
							Over Temp. Range		3.3	mA
Negative Supply Current Switch ON	$I^-_{(ON)}$	All Circuits				One Driver ON Note 2	$T_A = 25°C$	-1.0	-1.8	mA
							Over Temp. Range		-2.0	mA
Reference Input (Enable) ON Current	$I_{R(ON)}$	All Circuits				One Driver ON Note 2	$T_A = 25°C$	-1.0	-1.4	mA
							Over Temp. Range		-1.6	mA
Positive Supply Current Switch OFF	$I^+_{(OFF)}$	All Circuits				$V_{IN1} = V_{IN2} = 0.8V$	$T_A = 25°C$	1.0	10	μA
							Over Temp. Range		25	μA
Negative Supply Current Switch OFF	$I^-_{(OFF)}$	All Circuits				$V_{IN1} = V_{IN2} = 0.8V$	$T_A = 25°C$	-1.0	-10	μA
							Over Temp. Range		-25	μA
Reference Input (Enable) OFF Current	$I_{R(OFF)}$	All Circuits				$V_{IN1} = V_{IN2} = 0.8V$	$T_A = 25°C$	-1.0	-10	μA
							Over Temp. Range		-25	μA
Switch ON Resistance	$r_{ds(ON)}$	AH0153	AH0151	AH0163	AH0161	$V_D = 7.5V$, $I_D = 1 mA$	$T_A = 25°C$	10	15	Ω
							Over Temp. Range		30	Ω
Switch ON Resistance	$r_{ds(ON)}$	AH0154	AH0152	AH0164	AH0162	$V_D = 7.5V$, $I_D = 1 mA$	$T_A = 25°C$	45	50	Ω
							Over Temp. Range		100	Ω
Driver Leakage Current	$(I_D + I_S)_{ON}$	All Circuits				$V_D = V_S = -7.5V$	$T_A = 25°C$	.01	2	nA
							Over Temp. Range		500	nA
Switch Leakage Current	$I_{D(OFF)}$ OR $I_{S(OFF)}$	AH0153	AH0151	AH0163	AH0161	$V_{DS} = \pm 15V$	$T_A = 25°C$	5	10	nA
							Over Temp. Range		1.0	μA
Switch Leakage Current	$I_{D(OFF)}$ OR $I_{S(OFF)}$	AH0154	AH0152	AH0164	AH0162	$V_{DS} = \pm 15.0V$	$T_A = 25°C$	1.0	2.0	nA
							Over Temp. Range		200	nA
Switch Turn-ON Time	t_{ON}	AH0153	AH0151	AH0163	AH0161	See Test Circuit, $V_A = \pm 7.5V$, $T_A = 25°C$		0.8	1.0	μs
Switch Turn-ON Time	t_{ON}	AH0154	AH0152	AH0164	AH0162	See Test Circuit, $V_A = \pm 7.5V$, $T_A = 25°C$		0.5	0.8	μs
Switch Turn-OFF Time	t_{OFF}	AH0153	AH0151	AH0163	AH0161	See Test Circuit, $V_A = \pm 7.5V$, $T_A = 25°C$		1.1	2.5	μs
Switch Turn-OFF Time	t_{OFF}	AH0154	AH0152	AH0164	AH0162	See Test Circuit, $V_A = \pm 7.5V$, $T_A = 25°C$		0.9	1.5	μs

Note 1: Unless otherwise specified, these limits apply for −55°C to +125°C for the AH0100 series and −25°C to +85°C for the AH0100C series. All typical values are for $T_A = 25°C$.

Note 2: For the DPST and Dual DPST, the ON condition is for $V_{IN} = 2.5V$; the OFF condition is for $V_{IN} = 0.8V$. For the differential switches and SW1 and 2 ON, $V_{IN2} = 2.5V$, $V_{IN1} = 3.0V$. For SW3 and 4 ON, $V_{IN2} = 2.5V$, $V_{IN1} = 2.0V$.

Applications Information

1. INPUT LOGIC COMPATIBILITY

A. Voltage Considerations

In general, the AH0100 series is compatible with most DTL, TTL, and RTL logic families. The ON-input threshold is determined by the V_{BE} of the input transistor plus the V_f of the diode in the emitter leg, plus I x R_1, plus V_R. At room temperature and V_R = 0V, the nominal ON threshold is: 0.7V + 0.7V + 0.2V, = 1.6V. Over temperature and manufacturing tolerances, the threshold may be as high as 2.5V and as low as 0.8V. The rules for proper operation are:

$V_{IN} - V_R \geq 2.5V$ All switches ON

$V_{IN} - V_R \leq 0.8V$ All switches OFF

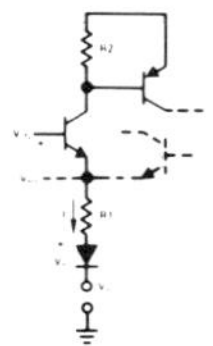

B. Input Current Considerations

$I_{IN(ON)}$, the current drawn by the driver with V_{IN} = 2.5V is typically 20 μA at 25°C and is guaranteed less than 120 μA over temperature. DTL, such as the DM930 series can supply 180 μA at logic "1" voltages in excess of 2.5V. TTL output levels are comparable at 400 μA. The DTL and TTL can drive the AH0100 series directly. However, at low temperature, DC noise margin in the logic "1" state is eroded with DTL. A pull-up resistor of 10 kΩ is recommended when using DTL over military temperature range.

If more than one driver is to be driven by a DM930 series (6K) gate, an external pull-up resistor should be added. The value is given by:

$$R_P = \frac{11}{N - 1} \text{ for } N > 2$$

where:

R_P = value of the pull-up resistor in kΩ

N = number of drivers.

C. Input Slew Rate

The slew rate of the logic input must be in excess of 0.3V/μs in order to assure proper operation of the analog switch. DTL, TTL, and RTL output rise times are far in excess of the minimum slew rate requirements. Discrete logic designs, however, should include consideration of input rise time.

2. ENABLE CONTROL

The application of a positive signal at the V_R terminal will open all switches. The V_R (ENABLE) signal must be capable of rising to within 0.8V of $V_{IN(ON)}$ in the OFF state and of sinking $I_{R(ON)}$ milliamps in the ON state (at $V_{IN(ON)} - V_R >$ 2.5V). The V_R terminal can be driven from most TTL and DTL gates.

3. DIFFERENTIAL INPUT CONSIDERATIONS

The differential switch driver is essentially a differential amplifier. The input requirements for proper operation are:

$$|V_{IN1} - V_{IN2}| \geq 0.3V$$

$$2.5 \leq (V_{IN1} \text{ or } V_{IN2}) - V_R \leq 5V$$

The differential driver may be furnished by a DC level as shown below. The level may be derived from a voltage divider to V^+ or the 5V V_{CC} of the DTL logic. In order to assure proper operation, the divider should be "stiff" with respect to I_{IN2}. Bypassing R1 with a 0.1 μF disc capacitor will prevent degradation of t_{ON} and t_{OFF}.

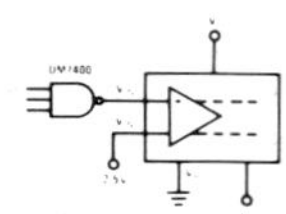

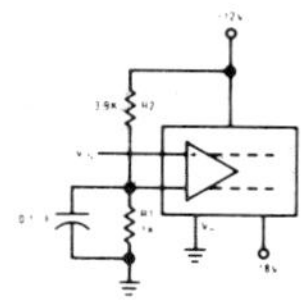

Alternatively, the differential driver may be driven from a TTL flip-flop or inverter.

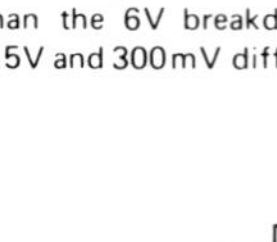

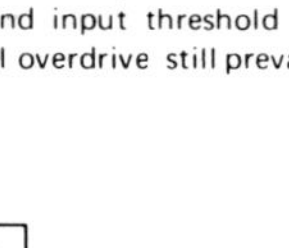

Connection of a 1 mA current source between V_R and V^- will allow operation over a ±10V common mode range. Differential input voltage must be less than the 6V breakdown, and input threshold of 2.5V and 300mV differential overdrive still prevail.

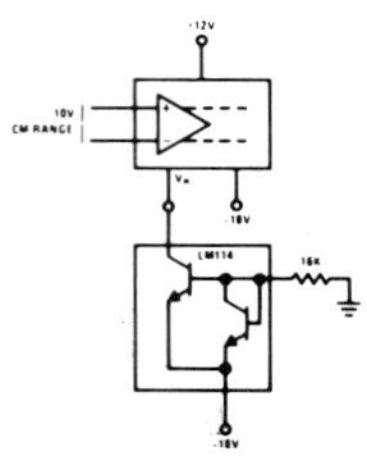

4. ANALOG VOLTAGE CONSIDERATIONS

The rules for operating the AH0100 series at supply voltages other than those specified essentially breakdown into OFF and ON considerations. The OFF considerations are dictated by the maximum negative swing of the analog signal and the pinch off of the JFET switch. In the OFF state, the gate of the FET is at $V^- + V_{BE} + V_{SAT}$ or about 1.0V above the V^- potential. The maximum V_P of the FET switches is 7V. The most negative analog voltage, V_A, swing which can be accommodated for any given supply voltage is:

$$|V_A| \leq |V^-| - V_P - V_{BE} - V_{SAT} \text{ or}$$

$$|V_A| \leq |V^-| - 8.0 \text{ or } |V^-| \geq |V_A| + 8.0V$$

For the standard high level switches, $V_A \leq |-18|+8 = -10V$. The value for V^+ is dictated by the maximum positive swing of the analog input voltage. Essentially the collector to base junction of the turn-on PNP must remain reversed biased for all positive value of analog input voltage. The base of the PNP is at $V^+ - V_{SAT} - V_{BE}$ or $V^+ - 1.0V$. The PNP's collector base junction should have at least 1.0V reverse bias. Hence, the most positive analog voltage swing which may be accommodated for a given value of V^+ is:

$$V_A \leq V^+ - V_{SAT} - V_{BE} - 1.0V \text{ or}$$

$$V_A \leq V^+ - 2.0V \text{ or } V^+ \geq V_A + 2.0V$$

For the standard high level switches, $V_A = 12 - 2.0V = +10V$.

5. SWITCHING TRANSIENTS

Due to charge stored in the gate-to-source and gate-to-drain capacitances of the FET switch, transients may appear in the output during switching. This is particularly true during the OFF to ON transition. The magnitude and duration of the transient may be minimized by making source and load impedance levels as small as practical.

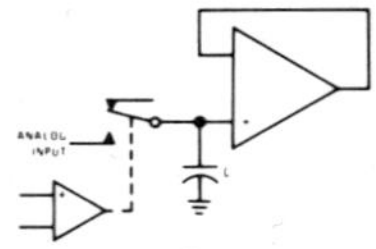

Furthermore, transients may be minimized by operating the switches in the differential mode; i.e., the charge delivered to the load during the ON to OFF transition is, to a large extent, cancelled by the OFF to ON transition.

Typical Applications

Programmable One Amp Power Supply

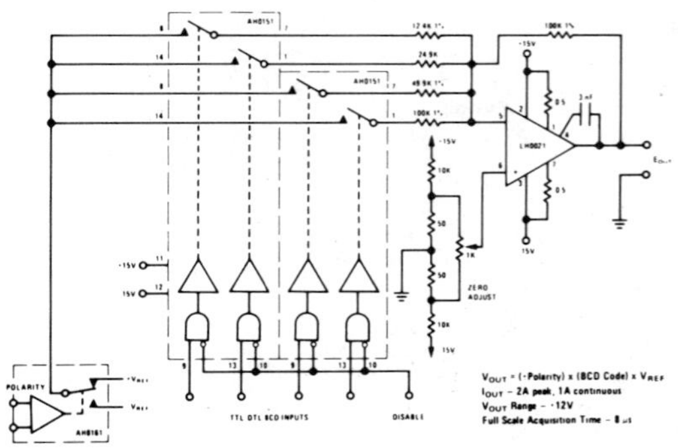

Four to Ten Bit D to A Converter (4 Bits Shown)

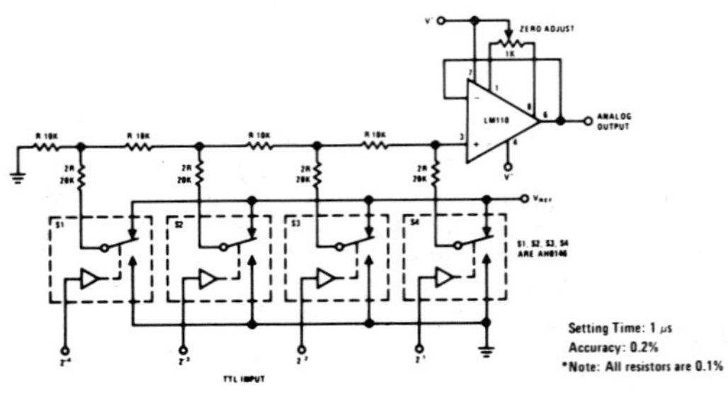

National Semiconductor

Industrial/Automotive/Functional Blocks/Telecommunications

LM555/LM555C Timer

General Description

The LM555 is a highly stable device for generating accurate time delays or oscillation. Additional terminals are provided for triggering or resetting if desired. In the time delay mode of operation, the time is precisely controlled by one external resistor and capacitor. For astable operation as an oscillator, the free running frequency and duty cycle are accurately controlled with two external resistors and one capacitor. The circuit may be triggered and reset on falling waveforms, and the output circuit can source or sink up to 200 mA or drive TTL circuits.

Features

- Direct replacement for SE555/NE555
- Timing from microseconds through hours
- Operates in both astable and monostable modes
- Adjustable duty cycle
- Output can source or sink 200 mA
- Output and supply TTL compatible
- Temperature stability better than 0.005% per °C
- Normally on and normally off output

Applications

- Precision timing
- Pulse generation
- Sequential timing
- Time delay generation
- Pulse width modulation
- Pulse position modulation
- Linear ramp generator

Schematic Diagram

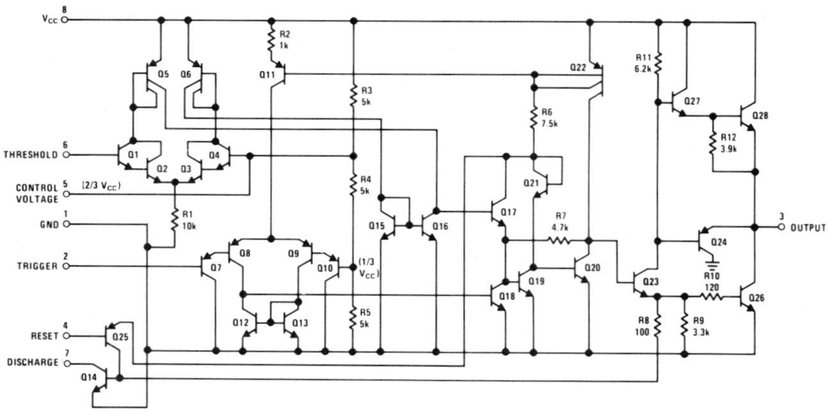

Connection Diagrams

Metal Can Package

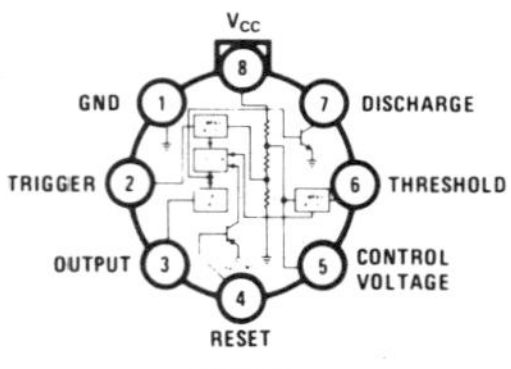

TOP VIEW

Order Number LM555H, LM555CH
See NS Package H08C

Dual-In-Line Package

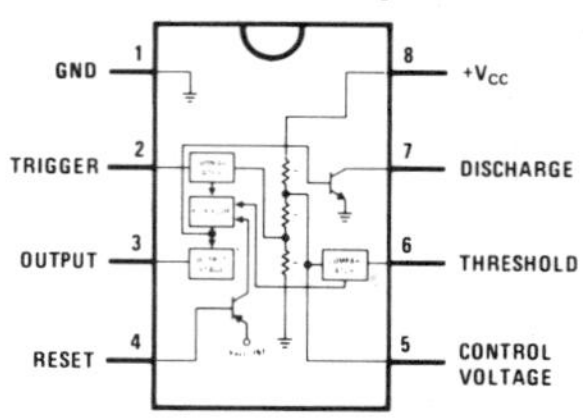

TOP VIEW

Order Number LM555CN
See NS Package N08B
Order Number LM555J or LM555CJ
See NS Package J08A

Absolute Maximum Ratings

Supply Voltage	+18V
Power Dissipation (Note 1)	600 mW
Operating Temperature Ranges	
LM555C	0°C to +70°C
LM555	−55°C to +125°C
Storage Temperature Range	−65°C to +150°C
Lead Temperature (Soldering, 10 seconds)	300°C

Electrical Characteristics (T_A = 25°C, V_{CC} = +5V to +15V, unless otherwise specified)

PARAMETER	CONDITIONS	LIMITS						UNITS
		LM555			LM555C			
		MIN	TYP	MAX	MIN	TYP	MAX	
Supply Voltage		4.5		18	4.5		16	V
Supply Current	V_{CC} = 5V, R_L = ∞		3	5		3	6	mA
	V_{CC} = 15V, R_L = ∞		10	12		10	15	mA
	(Low State) (Note 2)							
Timing Error, Monostable								
Initial Accuracy			0.5	2		1		%
Drift with Temperature	R_A, R_B = 1k to 100 k,		30			50		ppm/°C
	C = 0.1µF, (Note 3)							
Accuracy over Temperature			1.5	3.0		1.5		%
Drift with Supply			0.05	0.2		0.1		%/V
Timing Error, Astable								
Initial Accuracy			1.5	5		2.25	7	%
Drift with Temperature			90			150		ppm/°C
Accuracy over Temperature			2.5			3.0		%
Drift with Supply			0.15	0.2		0.30	0.5	%/V
Threshold Voltage			0.667			0.667		x V_{CC}
Trigger Voltage	V_{CC} = 15V	4.8	5	5.2		5		V
	V_{CC} = 5V	1.45	1.67	1.9		1.67		V
Trigger Current			0.01	0.5		0.5	0.9	µA
Reset Voltage		0.4	0.5	1	0.4	0.5	1	V
Reset Current			0.1	0.4		0.1	0.4	mA
Threshold Current	(Note 4)		0.1	0.25		0.1	0.25	µA
Control Voltage Level	V_{CC} = 15V	9.6	10	10.4	9	10	11	V
	V_{CC} = 5V	2.9	3.33	3.8	2.6	3.33	4	V
Pin 7 Leakage Output High			1	100		1	100	nA
Pin 7 Sat (Note 5)								
Output Low	V_{CC} = 15V, I_7 = 15 mA		150			180		mV
Output Low	V_{CC} = 4.5V, I_7 = 4.5 mA		70	100		80	200	mV
Output Voltage Drop (Low)	V_{CC} = 15V							
	I_{SINK} = 10 mA		0.1	0.15		0.1	0.25	V
	I_{SINK} = 50 mA		0.4	0.5		0.4	0.75	V
	I_{SINK} = 100 mA		2	2.2		2	2.5	V
	I_{SINK} = 200 mA		2.5			2.5		V
	V_{CC} = 5V							
	I_{SINK} = 8 mA		0.1	0.25				V
	I_{SINK} = 5 mA					0.25	0.35	V
Output Voltage Drop (High)	I_{SOURCE} = 200 mA, V_{CC} = 15V		12.5			12.5		V
	I_{SOURCE} = 100 mA, V_{CC} = 15V	13	13.3		12.75	13.3		V
	V_{CC} = 5V	3	3.3		2.75	3.3		V
Rise Time of Output			100			100		ns
Fall Time of Output			100			100		ns

Note 1: For operating at elevated temperatures the device must be derated based on a +150°C maximum junction temperature and a thermal resistance of +45°C/W junction to case for TO-5 and +150°C/W junction to ambient for both packages.

Note 2: Supply current when output high typically 1 mA less at V_{CC} = 5V.

Note 3: Tested at V_{CC} = 5V and V_{CC} = 15V.

Note 4: This will determine the maximum value of $R_A + R_B$ for 15V operation. The maximum total $(R_A + R_B)$ is 20 MΩ.

Note 5: No protection against excessive pin 7 current is necessary providing the package dissipation rating will not be exceeded.

Typical Performance Characteristics

Minimum Pulse Width Required for Triggering

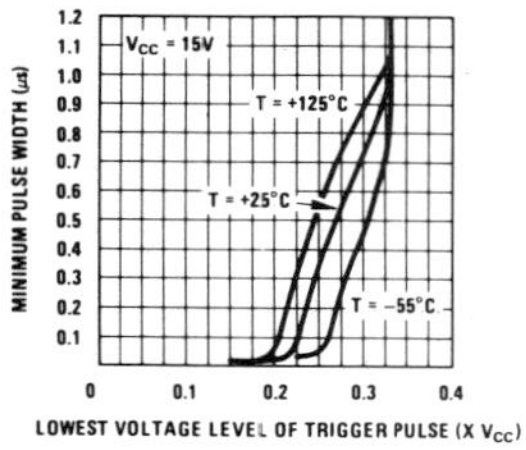

Supply Current vs Supply Voltage

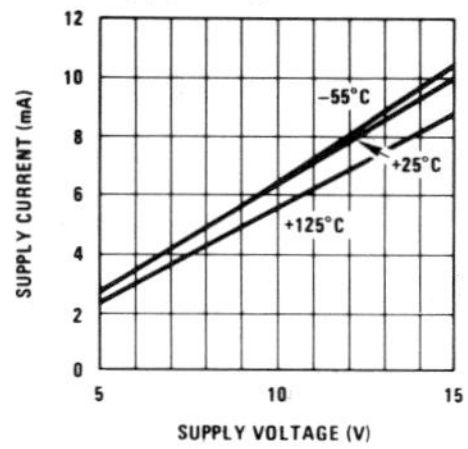

High Output Voltage vs Output Source Current

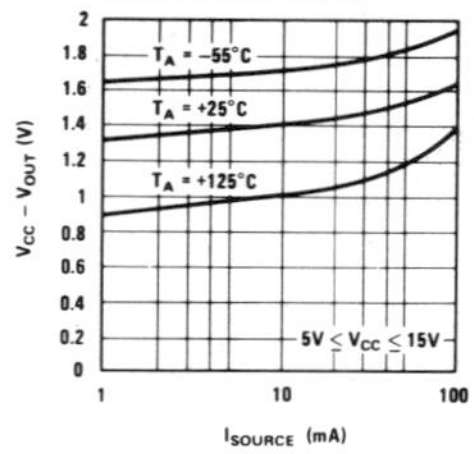

Low Output Voltage vs Output Sink Current

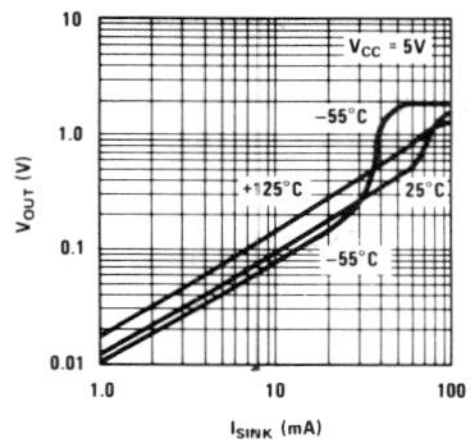

Low Output Voltage vs Output Sink Current

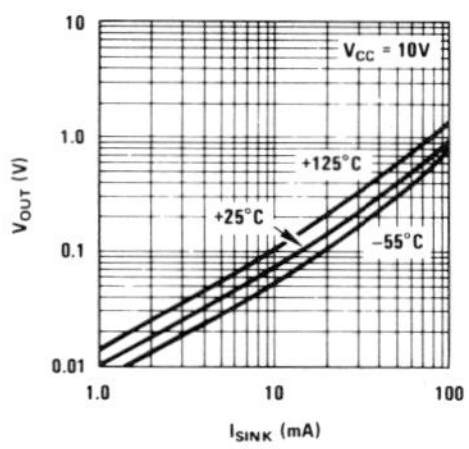

Low Output Voltage vs Output Sink Current

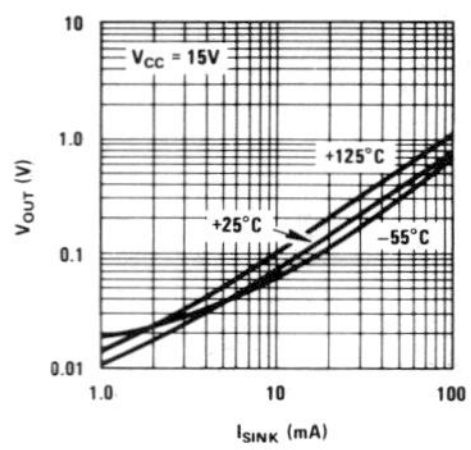

Output Propagation Delay vs Voltage Level of Trigger Pulse

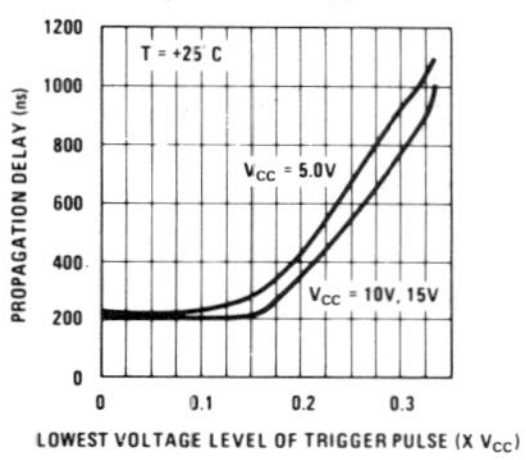

Output Propagation Delay vs Voltage Level of Trigger Pulse

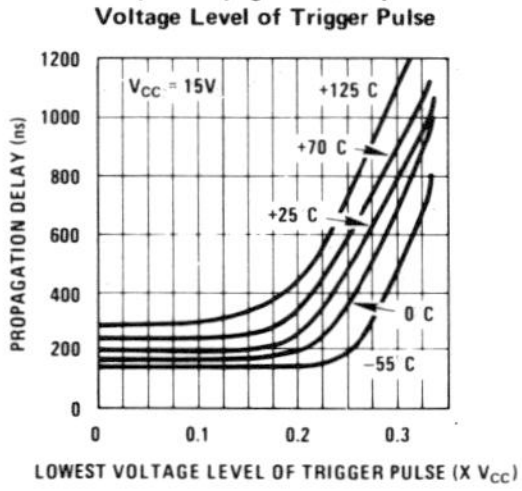

Discharge Transistor (Pin 7) Voltage vs Sink Current

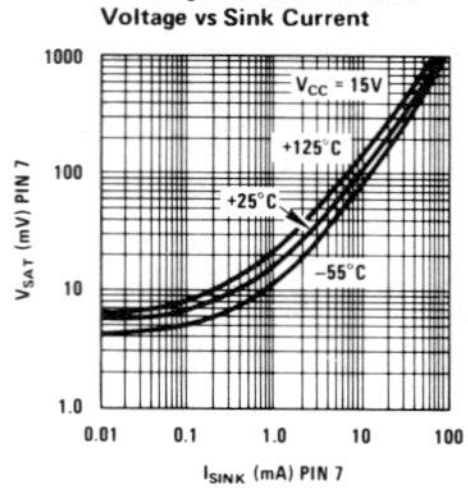

Discharge Transistor (Pin 7) Voltage vs Sink Current

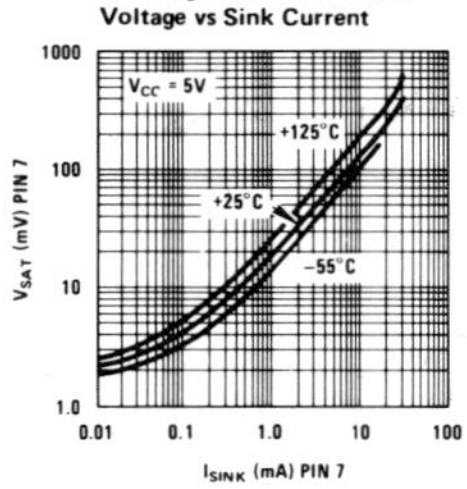

Applications Information

MONOSTABLE OPERATION

In this mode of operation, the timer functions as a one-shot (*Figure 1*). The external capacitor is initially held discharged by a transistor inside the timer. Upon application of a negative trigger pulse of less than 1/3 V_{CC} to pin 2, the flip-flop is set which both releases the short circuit across the capacitor and drives the output high.

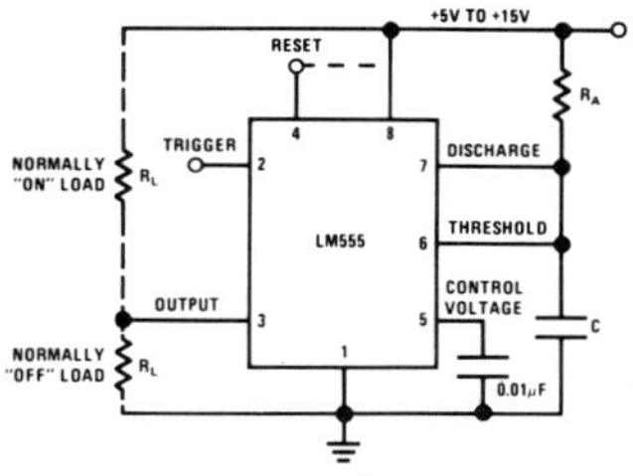

FIGURE 1. Monostable

The voltage across the capacitor then increases exponentially for a period of $t = 1.1\ R_A C$, at the end of which time the voltage equals 2/3 V_{CC}. The comparator then resets the flip-flop which in turn discharges the capacitor and drives the output to its low state. *Figure 2* shows the waveforms generated in this mode of operation. Since the charge and the threshold level of the comparator are both directly proportional to supply voltage, the timing internal is independent of supply.

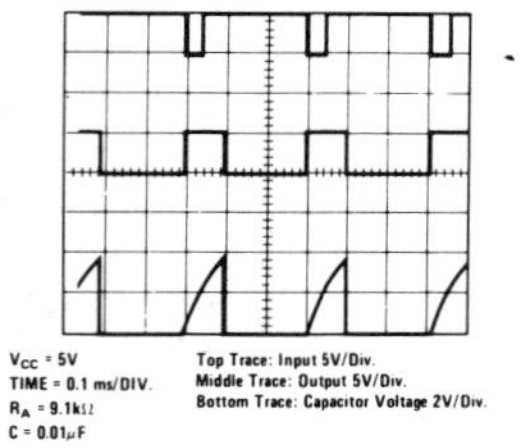

FIGURE 2. Monostable Waveforms

During the timing cycle when the output is high, the further application of a trigger pulse will not effect the circuit. However the circuit can be reset during this time by the application of a negative pulse to the reset terminal (pin 4). The output will then remain in the low state until a trigger pulse is again applied.

When the reset function is not in use, it is recommended that it be connected to V_{CC} to avoid any possibility of false triggering.

Figure 3 is a nomograph for easy determination of R, C values for various time delays.

NOTE: In monostable operation, the trigger should be driven high before the end of timing cycle.

ASTABLE OPERATION

If the circuit is connected as shown in *Figure 4* (pins 2 and 6 connected) it will trigger itself and free run as a multivibrator. The external capacitor charges through $R_A + R_B$ and discharges through R_B. Thus the duty cycle may be precisely set by the ratio of these two resistors.

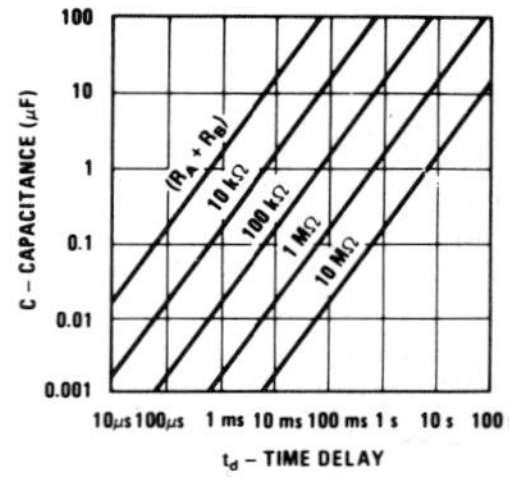

FIGURE 3. Time Delay

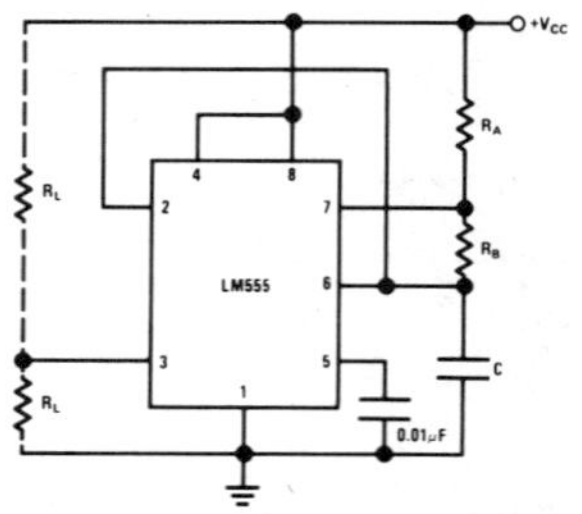

FIGURE 4. Astable

In this mode of operation, the capacitor charges and discharges between 1/3 V_{CC} and 2/3 V_{CC}. As in the triggered mode, the charge and discharge times, and therefore the frequency are independent of the supply voltage.

Figure 5 shows the waveforms generated in this mode of operation.

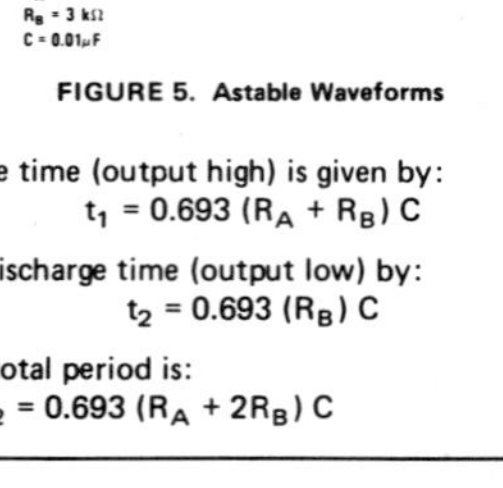

FIGURE 5. Astable Waveforms

The charge time (output high) is given by:

$$t_1 = 0.693\ (R_A + R_B)\ C$$

And the discharge time (output low) by:

$$t_2 = 0.693\ (R_B)\ C$$

Thus the total period is:

$$T = t_1 + t_2 = 0.693\ (R_A + 2R_B)\ C$$

Applications Information (Continued)

The frequency of oscillation is:

$$f = \frac{1}{T} = \frac{1.44}{(R_A + 2R_B)C}$$

Figure 6 may be used for quick determination of these RC values.

The duty cycle is: $D = \frac{R_B}{R_A + 2R_B}$

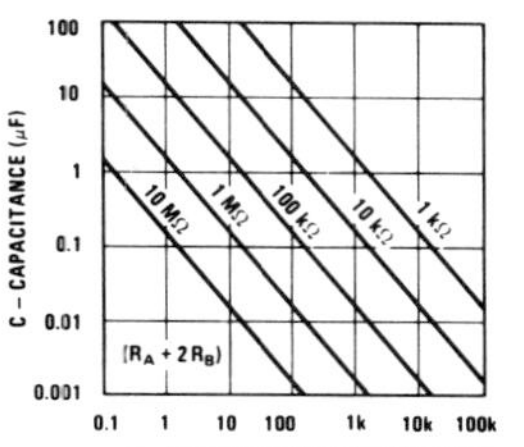

FIGURE 6. Free Running Frequency

FREQUENCY DIVIDER

The monostable circuit of *Figure 1* can be used as a frequency divider by adjusting the length of the timing cycle. *Figure 7* shows the waveforms generated in a divide by three circuit.

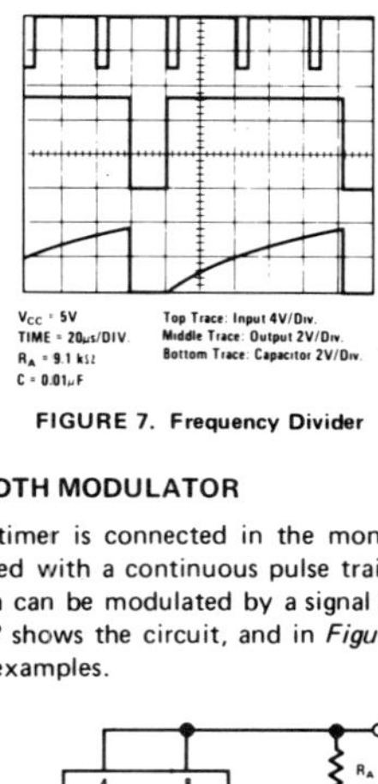

FIGURE 7. Frequency Divider

PULSE WIDTH MODULATOR

When the timer is connected in the monostable mode and triggered with a continuous pulse train, the output pulse width can be modulated by a signal applied to pin 5. *Figure 8* shows the circuit, and in *Figure 9* are some waveform examples.

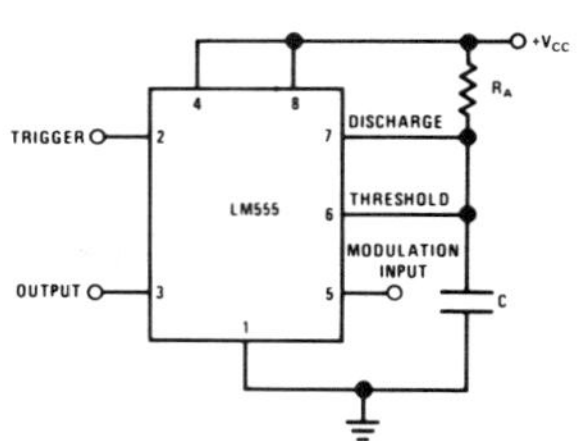

FIGURE 8. Pulse Width Modulator

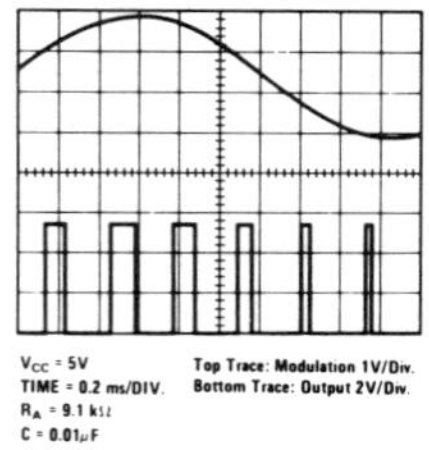

FIGURE 9. Pulse Width Modulator

PULSE POSITION MODULATOR

This application uses the timer connected for astable operation, as in *Figure 10*, with a modulating signal again applied to the control voltage terminal. The pulse position varies with the modulating signal, since the threshold voltage and hence the time delay is varied. *Figure 11* shows the waveforms generated for a triangle wave modulation signal.

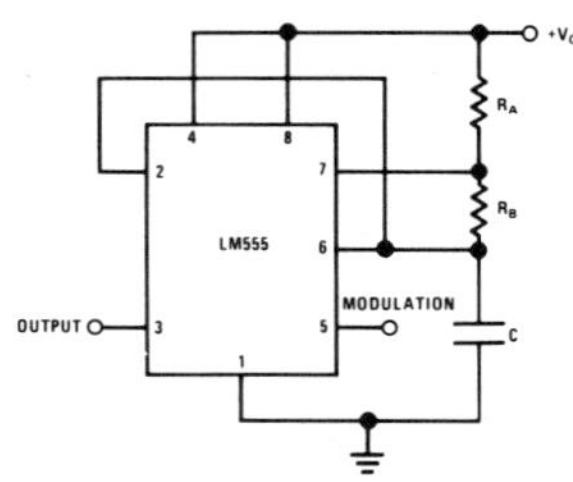

FIGURE 10. Pulse Position Modulator

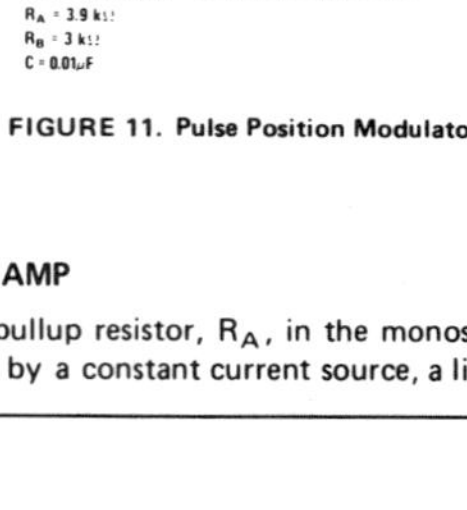

FIGURE 11. Pulse Position Modulator

LINEAR RAMP

When the pullup resistor, R_A, in the monostable circuit is replaced by a constant current source, a linear ramp is

Applications Information (Continued)

generated. *Figure 12* shows a circuit configuration that will perform this function.

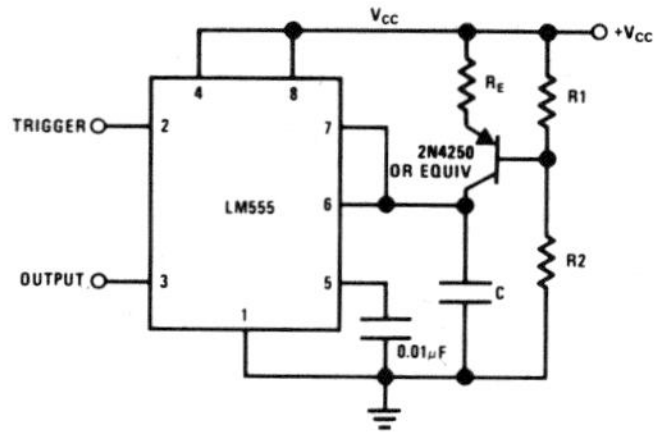

FIGURE 12.

Figure 13 shows waveforms generated by the linear ramp.

The time interval is given by:

$$T = \frac{2/3\ V_{CC}\ R_E\ (R_1 + R_2)\ C}{R_1\ V_{CC} - V_{BE}\ (R_1 + R_2)}$$

$$V_{BE} \simeq 0.6V$$

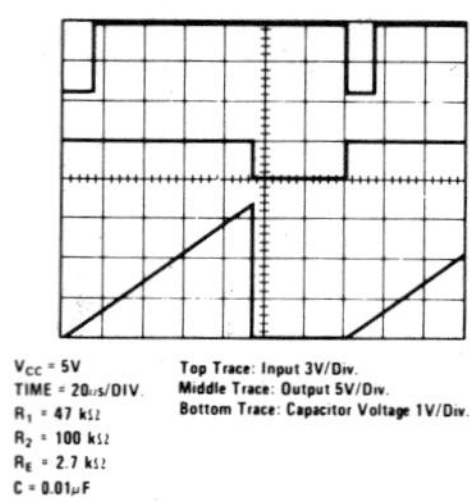

FIGURE 13. Linear Ramp

50% DUTY CYCLE OSCILLATOR

For a 50% duty cycle, the resistors R_A and R_B may be connected as in *Figure 14*. The time period for the output high is the same as previous, $t_1 = 0.693\ R_A\ C$. For the output low it is $t_2 =$

$$[(R_A\ R_B)/(R_A + R_B)]\ \ CLn \left[\frac{R_B - 2R_A}{2R_B - R_A}\right]$$

Thus the frequency of oscillation is $f = \dfrac{1}{t_1 + t_2}$

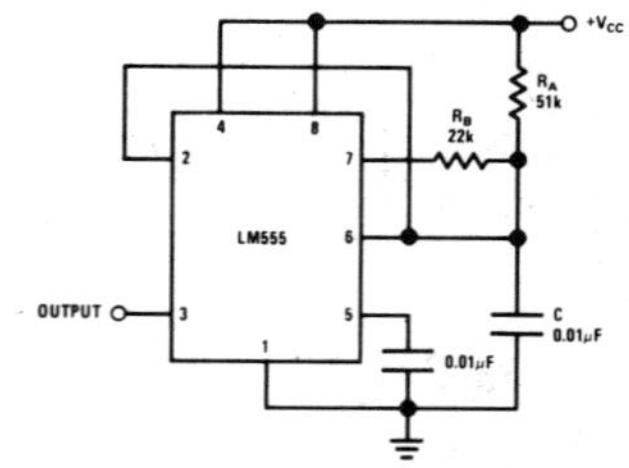

FIGURE 14. 50% Duty Cycle Oscillator

Note that this circuit will not oscillate if R_B is greater than 1/2 R_A because the junction of R_A and R_B cannot bring pin 2 down to 1/3 V_{CC} and trigger the lower comparator.

ADDITIONAL INFORMATION

Adequate power supply bypassing is necessary to protect associated circuitry. Minimum recommended is 0.1μF in parallel with 1μF electrolytic.

Lower comparator storage time can be as long as 10μs when pin 2 is driven fully to ground for triggering. This limits the monostable pulse width to 10μs minimum.

Delay time reset to output is 0.47μs typical. Minimum reset pulse width must be 0.3μs, typical.

Pin 7 current switches within 30 ns of the output (pin 3) voltage.

APPENDIX

D

DERIVATION OF EQUATION 4-1—THE FREQUENCY DEPENDENT OPEN LOOP GAIN

Let us develop an expression for the voltage gain of an amplifier stage at any frequency. Gain can be expressed as follows:

$$A = \frac{V_{out}}{V_{in}}$$

Referring to Figure D.1, we can see that V_{out} is the voltage across the parallel

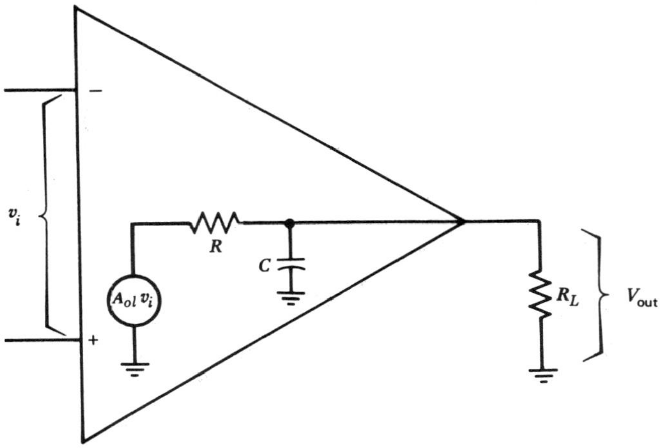

A_{ol} = low frequency voltage gain
$A_{ol}\,v_i$ = equivalent voltage source
R = R_{out}
C = stray and junction capacitance

FIGURE D.1 Equivalent circuit for frequency response.

combination of R_L and C, so we can rewrite the gain as:

$$A = \frac{A_{ol}v_i}{v_i} \qquad \frac{R_L \parallel \frac{1}{j\omega C}}{R + \left(R_L \parallel \frac{1}{j\omega C}\right)}$$

Doing the necessary algebra:

$$A = \frac{A_{ol}\left(\frac{R_L\left(\frac{1}{j\omega C}\right)}{\left(R_L + \frac{1}{j\omega C}\right)}\right)}{R + \left(\frac{R_L\left(\frac{1}{j\omega C}\right)}{R_L + \left(\frac{1}{j\omega C}\right)}\right)}$$

Multiplying numerator and denominator by $R_L + \frac{1}{j\omega C}$ we obtain:

$$A = A_{ol}\frac{R_L\left(\frac{1}{j\omega C}\right)}{R\left(R_L + \frac{1}{j\omega C}\right) + R_L\left(\frac{1}{j\omega C}\right)}$$

We expand this equation to get

$$A = \frac{A_{ol}R_L\left(\frac{1}{j\omega C}\right)}{R\,R_L + \left(\frac{1}{j\omega C}\right)R + R_L\left(\frac{1}{j\omega C}\right)}$$

and

$$A = \frac{A_{ol}R_L\left(\frac{1}{j\omega C}\right)}{R\,R_L + \left(\frac{1}{j\omega C}\right)(R + R_L)}$$

Now let us multiply the numerator and denominator by $j\omega C$ and $\dfrac{1}{(R + R_L)}$ to get

$$A = A_{ol}\left[\frac{\left(\dfrac{R_L}{R + R_L}\right)}{\left(\dfrac{R\,R_L}{R + R_L}\right)j\omega C + 1}\right]$$

So

$$A = A_{ol}\left(\frac{R_L}{R + R_L}\right)\left[\frac{1}{\left(\dfrac{R\,R_L}{R + R_L}\right)j\omega C + 1}\right]$$

We now let $\dfrac{R\,R_L}{R + R_L} = R_{th}$, a Thevenin equivalent resistance and write

$$A = A_{ol}\frac{R_L}{R + R_L}\left(\frac{1}{R_{th}j\omega C + 1}\right)$$

If $R_L >> R_{out}$, which is frequently the case, then

$$\frac{R_L}{R + R_L} \simeq 1 \qquad \text{and} \qquad R_{th} = \frac{R\,R_L}{R + R_L} \simeq R$$

so we can write

$$A \simeq \frac{A_{ol}}{Rj\omega C + 1}$$

Since $\omega = 2\pi f$,

$$A = \frac{A_{ol}}{1 + j\,2\pi f\,RC}$$

We now define a frequency $f_1 = \dfrac{1}{2\pi RC}$, the upper cutoff frequency so that

$$A = \frac{A_{ol}}{1 + j\dfrac{f}{f_1}}$$

and, finally,

$$|A| = \frac{A_{ol}}{\left[1 + \left(\frac{f}{f_1}\right)^2\right]^{1/2}} \angle - \arctan\left(\frac{f}{f_1}\right) \qquad (4\text{-}1)$$

APPENDIX

E

DERIVATION OF EQUATION FOR R_C OF LAG-COMPENSATION CIRCUIT

To find R_c, we make use of the voltage divider equation to write a formula for V_o of Figure E.1.

$$V_o = v_i \frac{R_c + \dfrac{1}{j2\pi f C_c}}{R + R_c + \dfrac{1}{j2\pi f C_c}}$$

We can now express the ratio as

$$\frac{V_o}{v_i} = \frac{1 + j2\pi f\, R_c C_c}{1 + j2\pi f C_c(R + R_c)} = \frac{1 + j\left(\dfrac{f}{f_y}\right)}{1 + j\left(\dfrac{f}{f_x}\right)}$$

Since

$$f_x = \frac{1}{2\pi C_c(R + R_c)}$$

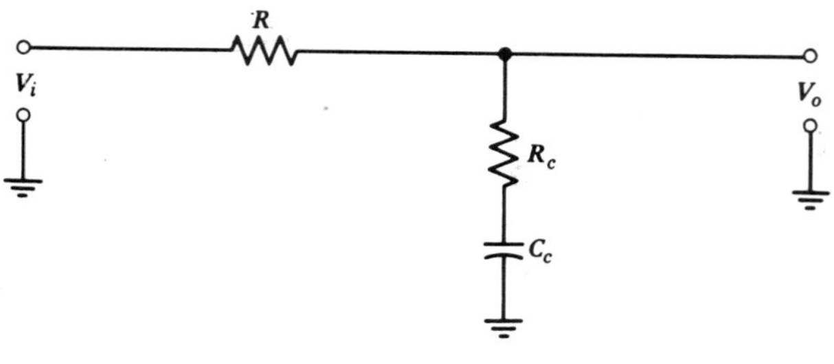

V_i = voltage of amplifier stage to which compensating network is attached

R = R_{out} of amplifier stage to which compensating network is attached

R_c = compensating resistor

C_c = compensating capacitor

FIGURE E.1 Lag compensation network.

and

$$f_y = \frac{1}{2\pi C_c R_c}$$

We now write

$$\frac{V_o}{v_i} = \frac{\left[1 + \left(\frac{f}{f_y}\right)^2\right]^{1/2}}{\left[1 + \left(\frac{f}{f_x}\right)^2\right]^{1/2}} \angle \arctan\left(\frac{f}{f_y}\right) - \arctan\left(\frac{f}{f_x}\right)$$

Converting to decibels we have

$$\frac{V_o}{v_i}\,(\text{dB}) = 20 \log \sqrt{1 + \left(\frac{f}{f_y}\right)^2} - 20 \log \sqrt{1 + \left(\frac{f}{f_x}\right)^2}$$

Where $f >> f_y > f_x$ we can write

$$\frac{V_o}{v_i}\,(\text{dB}) = 20 \log \frac{f}{f_y} - 20 \log \frac{f}{f_x} = 20 \frac{f_x}{f_y}$$

Substituting in f_x and f_y we see

$$\frac{V_o}{v_i}\,(\text{dB}) = 20 \log \left(\frac{\frac{1}{2\pi C(R + R_c)}}{\frac{1}{2\pi C_c R_c}}\right) = -20 \log\left(\frac{R + R_c}{R_c}\right)$$

which is the attenuation introduced by R_c and C_c at frequencies above f_y. The attenuation introduced must be equal to the reduction in gain necessary to have a smooth roll-off curve at f_1 in the compensated op-amp. This attenuation is marked M in Figure 4.16. From the preceding equation we see that

$$M(\text{dB}) = -20 \log \frac{R + R_c}{R_c}$$

We now have sufficient information to find R_c. Solving the preceding equation for R_c:

$$\frac{M}{20} = \log \frac{R + R_c}{R_c}$$

$$\text{antilog} \frac{M}{20} = \frac{R + R_c}{R_c}$$

$$R_c \text{ antilog} \frac{M}{20} - R_c = R$$

and

$$R_c = \frac{R}{\left[\text{antilog}\left(\frac{M}{20}\right)\right] - 1} \tag{4-15}$$

We neglected the negative sign of M since we are aware that is an attenuation.

APPENDIX

F

DERIVATION OF $\Delta V_{out}/\Delta V_{in}$, $\Delta V_{out}/\Delta I_O$, AND STEP-UP SWITCHING SUPPLY EFFICIENCY

$\Delta V_{out}/\Delta V_{in}$

If V_{in} changes, $\Delta V_{out}/\Delta V_{in}$ can be found as follows (refer to Figure F.1):

$$\Delta V_{out} = V_{R_L} - V_{out} \begin{pmatrix}\text{from amplifier with error}\\ \text{voltage from } \Delta V_{in}\end{pmatrix}$$

$$= \frac{V_{in}R_L}{R_L + R_c} - A(V_S - V_R)$$

where

R_L = load resistance

R_c = ac collector-to-emitter resistance of Q_1, $\left.\dfrac{\Delta V_{CE}}{\Delta I_C}\right|_{I_B}$

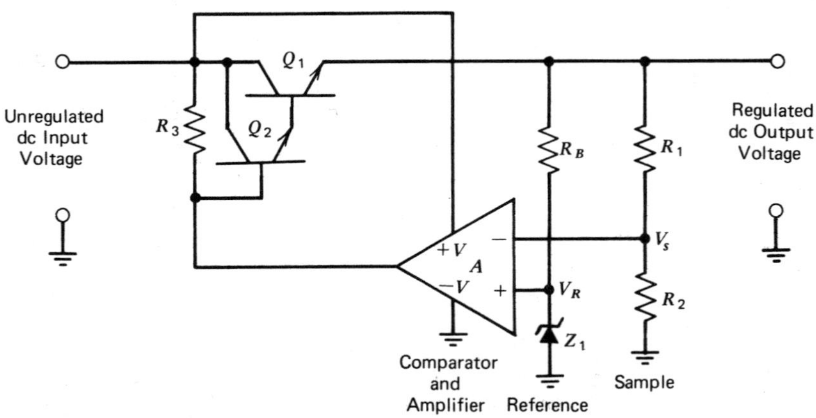

FIGURE F.1 Basic series-pass voltage regulator.

$V_R = V_{Z1}$
A = op-amp gain

$$\Delta V_{out} = \frac{V_{in}R_L}{R_L + R_C} - \frac{A\Delta V_{out}R_2}{R_1 + R_2} + A\,V_R$$

$$\Delta V_{out} + \frac{A\Delta V_{out}R_2}{R_1 + R_2} = \frac{V_{in}R_L}{R_L + R_c} + A\,V_R$$

$$\Delta V_{out}\left(1 + \frac{A\,R_2}{R_1 + R_2}\right) = \frac{V_{in}R_L}{R_L + R_c} + A\,V_R$$

$$\Delta V_{out} = \frac{V_{in}\left(\frac{R_L}{R_L + R_c}\right)}{1 + A\left(\frac{R_2}{R_1 + R_2}\right)} + \frac{A\,V_R}{1 + A\left(\frac{R_2}{R_1 + R_2}\right)}$$

It is worth noting at this time that the second term is equal to the drift in output if the reference voltage, V_{Z1}, changes from temperature or aging. In this discussion V_{Z1} is considered fixed, so any change in the output will come from the first term, thus:

$$\Delta V_{out} = \frac{\Delta V_{in}\left(\frac{R_L}{R_L + R_c}\right)}{1 + A\left(\frac{R_2}{R_1 + R_2}\right)} = \frac{\Delta V_{in}\left(\frac{\frac{R_L}{R_c}}{1 + R_L/R_c}\right)}{1 + A\left(\frac{R_2}{R_1 + R_2}\right)}$$

Since $R_L/R_c \ll 1$ and $A\,R_2/(R_1 + R_2) \gg 1$

$$\Delta V_{out} = \frac{\Delta V_{in}(R_L/R_c)}{A\,R_2/(R_1 + R_2)} = \frac{\Delta V_{in}}{A}\left(\frac{R_L}{R_c}\right)\left(\frac{R_2 + R_1}{R_2}\right)$$

so,

$$\frac{\Delta V_{out}}{\Delta V_{in}} = \frac{R_L}{R_c}\left(\frac{R_2 + R_1}{A\,R_2}\right) \tag{10-3}$$

With a Darlington driver, the effective R_c of Q_1 is increased by a factor of β_{Q2}

$$R_c(\text{eff})_{Q1} \cong R_{cQ1}\beta_{Q2}$$

Note. R_{cQ2} is usually about 10 times R_c of Q_1 since Q_2 is a lower current transistor. β_{Q2}, the current gain of Q_2, is usually from 40 to 60 at the lower I_C of Q_2. Recall $I_{CQ2} = I_{BQ1} = I_{out}/\beta_{Q1}$.

R_c(eff) is substituted into Equation 10-3 for a Darlington output configuration. R_c of both transistors is taken from the specification sheet or a transistor curve tracer.

$\Delta V_{out}/\Delta I_L$

The output voltage change with respect to a change in load current (I_o) can be found as follows:

Recall for any op-amp, $V_{out} = Av_i$. The power supply regulator op-amp $v_i = V_R - V_s$, but V_R is fixed, so any change in v_i is ΔV_s. An output voltage change can be expressed as:

$$A\Delta V_s = \Delta V_{out}$$

ΔV_s is caused by ΔV_{out}, but the attenuation of the voltage divider R_1 and R_2 causes:

$$\Delta V_s = \frac{\Delta V_{out} R_2}{R_1 + R_2}$$

therefore,

$$A\Delta V_{out}\left(\frac{R_2}{R_1 + R_2}\right) = \Delta V_{out}$$

The output voltage change ΔV_{out} is from $\Delta I_o R_L$ so,

$$A\Delta V_{out}\left(\frac{R_2}{R_1 + R_2}\right) = \Delta I_o R_L$$

Thus

$$\frac{\Delta V_{out}}{\Delta I_o} = \frac{R_L}{A}\left(\frac{R_1 + R_2}{R_2}\right) \tag{10-4}$$

EFFICIENCY OF STEP-UP SWITCHING INVERTER

$$P_{out} = V_{out} I_o$$

$$P_{in} = V_{out} I_o + P_{Q1} + P_D$$

$$P_{in} = V_{out} I_o + I_m V_{CE}(\text{sat}) + I_o V_D$$

But

$$I_{in} = I_o\left(\frac{t_c + t_o}{t_o}\right)$$

$$P_{in} = V_{out} I_o + I_o\left(\frac{t_c + t_o}{t_o}\right) V_{CE}(\text{sat}) + V_D I_o$$

$$\text{Eff} = \frac{P_{out}}{P_{in}} = \frac{V_{out} I_o}{V_{out} I_o + V_{CE}(\text{sat}) I_o\left(\frac{t_c + t_o}{t_o}\right) + V_D I_o}$$

Factoring out I_o and multiplying through by $t_o/(t_c + t_o)$ leaves

$$\text{Eff} = \frac{V_{out}\left(\dfrac{t_o}{t_c + t_o}\right)}{V_{out}\left(\dfrac{t_o}{t_c + t_o}\right) + V_{CE}(\text{sat}) + V_D\left(\dfrac{t_o}{t_c + t_o}\right)}$$

but

$$V_{in} = V_{out}\left(\frac{t_o}{t_c + t_o}\right)$$

therefore

$$\text{Eff} = \frac{V_{in}}{V_{in} + V_{CE}(\text{sat}) + V_D\left(\dfrac{t_o}{t_c + t_o}\right)}$$

APPENDIX

G

ANSWERS TO SELF-TEST QUESTIONS

Chapter 1

1. $A_{ol} = \infty$, $R_{in} = \infty$, $R_{out} = 0$, $V_{os} = 0$; 2. See page 0; 3. V_{os} – mismatched V_{BE}, I_{os} – mismatched β of input transistor; 4. FET input amplifier has higher R_{in}, lower I_B, higher V_{os} than bipolar transistor input amp; 5. High R_{in}, low R_{out}; 6. (a) $R_f = 300$ kΩ, (b) $A_{fb} = 150$, (c) $R_1 = 90$ kΩ; 7. (a) $A_{fb} = 21$, (b) $R_f = 380$ kΩ, (c) $R_1 = 222$ kΩ; 8. 4 V; 9. −12 V; 10. $V_{out} = -2.4$ V, $v_i = 0.24$ mV; 11. Constant current source; 12. See pages 7, 9, 12, and 13.

Chapter 2

1. $A_{fb} = 100.5$, $R_{outfb} = 5.025$ Ω, $R_{infb} = 39.8$ MΩ; 2. $A_{fb} = 99.66$, $A_{oe} =$ 29,703, $\beta = 0.01$, $R_{infb} = 10.033$ kΩ, $R_{outfb} = 1.68$ Ω; 3. A_{fb} decreases, R_{infb} increases, R_{outfb} decreases; 4. V_{os} due to $I_b = I_b(R_1 \| R_f)$; 5. Canceling V_{os} with equal but opposite voltage; 6. $R_2 = 8958$ Ω, $R_A = 133$ Ω; 7. $R_s = 9.524$ kΩ, $R_B =$ 9.907 kΩ, $R_A = 93.3$ Ω; 8. $I_{R_4} > I_{fb} >> I_b$; 9. $A_{ol} >> A_{fb}$; 10. $A_{ol} =$ 49,950.

Chapter 3

1. V due to $I_b = I_{b_1}R_1$ or $I_{b_2}R_2$. I_{out} of follower>>I_b and is therefore easy to measure, and V_{out} of follower $= I_{b_1}R_1$ or $I_{b_2}R_2$; 2. $R_1 I_{b_1}$ and $I_{b_2}R_2$ easy to measure accurately; 3. $I_{b_1} = 60$ nA, $I_{b_2} = 100$ nA, $I_{os} = 40$ nA; 4. See page 47; 5. In a noninverting configuration, the absolute as well as the differential signal will be amplified causing error; 6. $A_{fb} = 49.955$; 7. The noninverting input is grounded so $V_{cm} = 0$; 8. CMRR = 24,024, CMRR(dB) = 87.6 dB; 9. $\Delta V_{os}/\Delta T$, $\Delta I_{os}/\Delta T$; 10. $E = 1.593$ V; 11. $E_i = 79.65$ mV; 12. Low drift; 13. See page 55; 14. To detect small voltage changes at the summing point.

Chapter 4

1. Stray capacitance, junction capacitance; 2. A at 10 kHz = 1492; 3. $A_{ol} =$ 10,000; 5. Loop gain (dB) = 45 dB; 6. (a) $A_v = 1.5$, (b) $BW = 100$ kHz; 7. $\theta > 180°$, loop gain >1; 8. −6 dB/octave-stable, −12 dB/octave-conditional stability (may or may not be stable), −18 dB/octave-unstable; 9. $\theta_{CL} = -176.1°$; 10. (a) $\theta_{CL} = -188.6°$, (b) $\theta_{CL} = -139.4°$; 11. $V_p = 3.979$ V; 12. $f = 42.4$ kHz; 13. $R_c = 203.6$ Ω, $C_1 = 0.041$ μF; 14. Higher slew rate; 15. See pages 87, 93–97.

Chapter 5

1. -3 V; 2. -5 V; 3. $R_1 = 250$ kΩ, $R_2 = 333$ kΩ, $R_3 = 167$ kΩ; 4. $R_x = 15.4$ kΩ, $V_{out} = 2.9$ V; 5. $R_1' = 333$ kΩ, $R_2' = 250$ kΩ, $R_1 = 500$ kΩ, $R_2 = 1$ MΩ, $R_x = 250$ kΩ; 6. $R_1' = 50$ kΩ, $R_2' = 33.3$ kΩ, $R_1 = 20$ kΩ; 7. See pages 114–118. 8. Advantage—no common-mode error, disadvantage—may require more amplifiers; 9. Error amplifier; 10. provides feedback signal; 11. $R_2 = R_3 = R_4 = 200$ kΩ, $R_1 = 100$ kΩ, $P = 90$ kΩ.

Chapter 6

1. Averages voltage over time; 2. $R = 100$ kΩ; 3. (a) $V_{out} = -0.6$ μV, (b) $V_{out} = -0.053$ μV, (c) $V_{out} = -0.1$ V; 4. Bias current, capacitor leakage current, V_{os}; 5. Periodic reset, high-quality low-leakage capacitors, very low offset op-amp such as chopper-stabilized; 6. $R_c = 200$ kΩ, $C = 796$ pF; 7. Avoid amplifier saturation; 8. $R_f = 500$ kΩ, $C = 1$ μF; 9. See page 140; 10. $R = 159$ Ω; 11. 0.1 μF; 12. (a) $V_{out} = -0.004$ V. (b) $V_{out} = -0.004$ V; 13. Reduce noise, improve stability; 14. $R_c = 79.6$ Ω, $R = 79.6$ kΩ, $C_c = 100$ pF; 15. $R_c = 3.18$ kΩ, $R = 63.7$ kΩ, $C = 0.00125$ μF; 16. 10 kHz.

Chapter 7

1. Normally, diode junction or emitter-base junction of transistor; 2. $\Delta V/\Delta T$ and $\Delta I/\Delta T$ of semiconductor junction; 3. Placement of logarithmic component so as to produce exponential output in an antilog amp; 4. See pages 174–176; 5. Increase the range of amplitudes a circuit can process; 6. $V_{out} = -0.192$ V; 7. $V_{out} = 18.72$ mV; 8. X is fed to a log amp. The log amp output is fed to one input of a multiplier and n into the other. The multiplier output is then $n \ln X$. The antilog of the multiplier is then taken providing X^n; 9. Place the synthesizing network across R_f; 10. Take the antilog of the output with a bipolar antilog circuit; 11. Prevent loading the op-amp output.

Chapter 8

1. See p. 185; 2. Need a power supply, frequency response limited by the op-amp; 3. See p. 184; 4. 6 dB/octave/pole transition region attenuation; 5. Butterworth: maximally flat passband; Chebyshev: maximum rate of attenuation change in the transition region; Bessel: linear phase delay; 6. As α decreases, peaking increases; 7. $R_1 = R_2 = 2.122$ kΩ, if $R_A = 10$ kΩ, $R_B = 5.86$ kΩ, $A_p = 1.586$; 8. $R_1 = R_2 = 1.895$ kΩ, if $R_A = 10$ kΩ, $R_B = 5.86$ kΩ; 9. $f_1 = 1$ kHz, $f_2 = 1.2$ kHz, $C = 0.0033$ μF, $f_o = 1.095$ kHz, $Q = 5.47$, $R_1 = 24.114$ kΩ; $R_2 = 4.822$ kΩ, $R_3 = 482.3$ kΩ; 10. $f_{(3\text{ dB})}/f_c = 1.218$, $\alpha = 1.059$, $f_c = 9.744$ kHz, $R_1 = R_2 = R_3 = R_4 = R_f' = R_f = 16.33$ kΩ, $R_5 = 29.94$ kΩ; 11. $f_o = 1.125$ kHz, $Q = 22.5$, $R_1 = R_2 = R_3 = R_4 = R_f = R_f' = 6.43$ kΩ, $R_5 = 427.6$ kΩ; 12. $f_{(3\text{ dB})}/f_c = 1.074$, $\alpha = 0.886$, $f_c = 2.328$ kHz, $R_1 = R_2 = R_3 = R_5 = R_f = R_A = 20.7$ kΩ, $R_4 = 2.589$ kΩ, $R_B = 18.36$ kΩ; 13. $f_o = 512.35$ Hz, $Q = 20.5$, $G = 0.4879$, $R_1 = R_2 = R_3 = R_5 = R_f = R_A = 31.06$ kΩ, $R_B = R_A/Q =$

1.516 kΩ, R_4 = 63.66 kΩ, R_5 = 8.91 kΩ; 14. f_o = 749.9 Hz, Q = 37.5, G = 3.749, R_2 = R_3 = R_4 = R_5 = 45.15 kΩ, R_1 = 169.3 kΩ, R_c = 1.693 MΩ, R_5 = 34.91 kΩ; 15. 1st stage: α = 1.076, f_c ratio = 0.471, f_c = 1.413 kHz, 2nd stage: α = 0.218, f_c ratio = 0.964, f_c = 2.892 kHz, stage 1: R_A = R_1 = R_2 = 34.13 kΩ, R_B = 31.54 kΩ, stage 2: R_A = R_1 = R_2 = 16.68 kΩ, R_B = 29.72 kΩ, A_p = 5.35; 16. Stage 1: α = 1.328, f_c ratio = 0.265, stage 2: α = 0.511, f_c ratio = 0.584, stage 3: α = 0.234, f_c ratio = 0.851, stage 4: α = 0.702, f_c ratio = 0.997; 17. Simulate inductance; 18. Low noise, TC, wide frequency range.

Chapter 9

1. A capacitor is charged to V_p through a diode so it cannot discharge. A high R_{in} buffer is used to detect the capacitor voltage; 2. The high R_{in} holds the capacitor voltage longer; 3. Sum the peak detector outputs. 4. If the booster is within the feedback loop, any output voltage offset produced by the current booster is reduced by A_{ol}; 5. The high R_{in} causes very little loading of the zener, and the high A_{ol} with feedback causes the output voltage to be independent of output current. 6. Reduce distortion; 7. Comparator that automatically resets; 8. A_1 − comparator, A_2 − integrator. 9. So the reset voltage is reached quickly; 10. The charging current for C varies; 11. A_1 − switched input buffer, A_2 − integrator, A_3 − comparator and square-wave generator.

Chapter 10

1. Linear, Adv: Good regulation, inexpensive. Disadv: Inefficient. Switching. Adv: Light, efficient. Disadv: Complex; 2. Preregulators. Reduce ripple on series-pass transistor base; 3. R_B = 700 Ω; if I_D = 1 mA, R_1 = 2.8 kΩ, R_2 = 6.8 kΩ, R_3 < 600 Ω, but R_3 > 300 Ω; 4. 75%; 5. R_{cl} = 0.19 Ω; 6. R_A = 320 Ω, R_B = 11.2 kΩ; 7. R_5 = 10.6 Ω, R_1 = 25.9 kΩ, R_2 = 2.4 kΩ, R_{sc} = 1 Ω, R_b = 930 Ω, R_5 = 70 Ω; 8. The LM 309 cannot current limit an external pass transistor because the current-limit transistor does not have external connection; 9. R_1 = 10.85 kΩ, R_2 = 7.15 kΩ, R_3 = 4.3 kΩ, R_{sc} = 30.95 Ω; 10. R_1 = 7.85 kΩ, R_2 = 7.15 kΩ, R_{sc} = 10.8 Ω, R_A = 2.6 kΩ, R_B = 12.4 kΩ; 11. If V_{out} < V_{ref}, V_{ref} must have voltage divider, If V_{out} > V_{ref}, V_{out} must have voltage divider; 12. A drop in voltage from a negative supply causes a positive going voltage on the comparator input. Unless the Inv and Noninv inputs are switched, no regulation occurs; 13. P_{ave} = 30.3 mW; 14. See pages 307–312; 15. t_o = 48.54 μs, t_c = 51.46 μs, L = 303 μH, C_o = 312 μF, C_T = 0.022 μF, R_2 = 12.45 kΩ, R_1 = 37.6 kΩ, R_{sc} = 0.33 Ω, I_p = 1 A; 16. t_o = 22.56 μs, t_c = 27.44 μs, I_p = 0.886 A, L = 337 μH, C_o = 275 μF, C_T = 0.01 μF, R_2 = 12.45 kΩ, R_1 = 227.6 kΩ, R_{sc} = 0.372 Ω; 17. t_o = 20 μs, t_c = 30 μs, I_p = 5 A, L = 240 μH, C_o = 750 μF, C_T = 0.009 μF, If R_1 = 10 kΩ, R_f = 2.49 kΩ, and R_s = 2 kΩ, R_{sc} = 0.066 Ω; 18. The frequency varies as load current varies; 19. Dead time is the time between turning the "on" transistor off and the "off" transistor on. It prevents high switching dissipation; 20. Provides same power with transistors that have V_{CE}(sus) = V_{in} and I_c one-half half-bridge rating. 21. Yes. See Figure G.1.

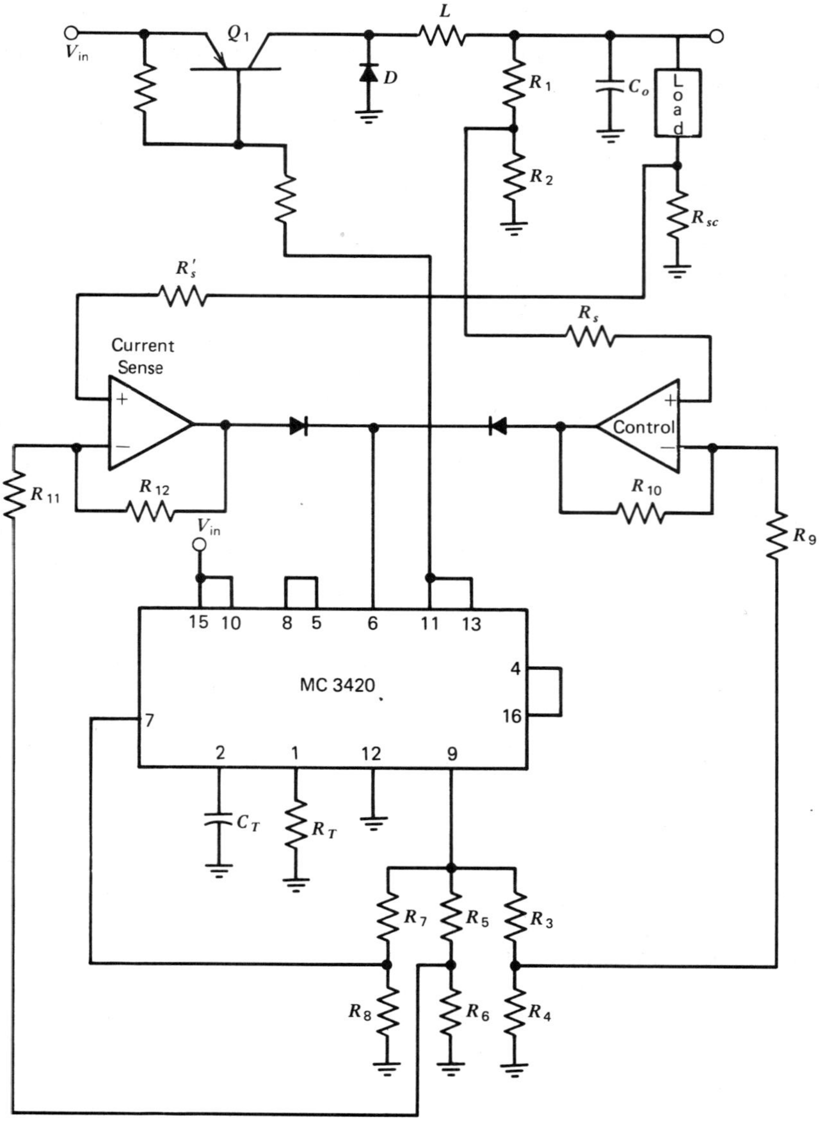

$$\text{Let} - f_{osc} = (1/2)\,\frac{0.55}{R_{EXT}\,C_{EXT}}$$

FIGURE G.1 Answer to Question 10-21.

Chapter 11

1. Sufficient market and necessary technology available; 2. Detect when an input voltage is greater than or less than a reference voltage applied to the other input; 3. Hysterisis (where UTP ≠ LTP) provides immunity against output voltage changes from noise on the input; 4. Hysterisis prevents the comparator from being able to detect an exact input voltage level independent of the direction of approach of the input to that level; 5. $R_2 = 40$ kΩ, $R_1 = 100$ kΩ; 6. $R_1 = 70$ kΩ, $R_2 = 24.6$ kΩ, $R_3 = 55.4$ kΩ, $R_4 = 1.16$ kΩ, $R_B = 5.28$ MΩ; 7. $R_1 = 1$ kΩ, $R_2 = R_3 = 12$ kΩ; 8. Reverse the inverting and noninverting input for each test; 9. Only two (or one if 2 *R* is two series *R* resistors) resistor values must be used; 10. Adv: speed, disadv: complex if many bits are to be converted; 11. The voltage readout is a ratio of counts taken during the conversion period, which is fairly short; 12. See pp. 361–363; 13. Convert iterated digital bits to an analog voltage to be compared to V_{in}; 14. Switching in A/D and D/A converters, automatic testing, switched capacitor gain setting, switched capacitor comparators, and signal routing; 15. Adv: amplifies dc, disadv: switching noise; 16. Adv: single-supply operation, disadv: less precision operation; 17. $R_m = 476.7$ kΩ, $R_f = 226.7$ kΩ, $R_1 = 11.33$ kΩ, $C_1 = 0.047$ μF, $C_2 = 0.053$ μF; 18. $R_m = 186$ kΩ, $R_f = 86$ kΩ, $R_1 = 6.65$ kΩ, $r_e = 520$ Ω, $C_1 = 0.37$ μF, $C_2 = 0.53$ μF; 19. $\tau = 50$ μs, $t_1 = 40$ μs, $t_2 = 10$ μs, $R_A = 43.27$ kΩ, $R_B = 14.43$ kΩ; 20. $C_T = 0.487$ μF; 21. $t_c = 36.36$ μs; 22. $e_1 - e_2 = 12.47$ mV; 23. $R_3 = 2.5$ kΩ; 24. $R_g = 400$ kΩ; 25. Input guarding is holding the input wire's shield at V_{cm}. This prevents unequal capacitive attenuation of the input signal because the wire-to-shield capacitance is already charged; 26. See pp. 398–399; 27. Transformer coupling or optical coupling.

Chapter 12

1. Thermal agitation of current carriers; 2. Discrete particle nature of current carriers causes number variation per unit time in any stream of current carriers. This number variation produces noise; 3. Noise amplitude ∝ $1/f$; 4. (Signal amplitude/noise amplitude) or (signal power/noise power); 5. Logarithm of output signal-to-noise ratio divided by input signal-to-noise ratio; 6. See page 414; 7. Return high currents through a different conductor than the signal reference; 8. 4.06 μV; 9. 40 pA; 10. 4.1 μV; 11. Use low noise op-amps, minimize resistances, use only the bandwidth necessary.

APPENDIX

H

SYMBOLS GLOSSARY

Symbol	Meaning
A	Voltage gain
A_{cl}	Closed loop gain (gain with feedback)
A_{cm}	Common-mode gain
A_{fb}	Closed loop gain (gain with feedback)
A_{fb_i}	Ideal closed loop gain
A_{fb_l}	Low frequency closed loop gain
A_l	Low frequency open loop gain
A_{ol}	Open loop gain
A_{sa}	Stabilizing amplifier gain
BV_{CEO}	Breakdown voltage collector to emitter with the base open
C	Capacitance
C_c	Compensating capacitance
CMRR	Common-mode rejection ratio
dB	Decibel
E	Error
e_{cm}	Common-mode voltage
E_I	Error referred to the input
e	Voltage
f	Frequency
f_1	Corner frequency
$f_{1_{fb}}$	Corner frequency with feedback
f_c	Crossing frequency
f_{cl}	Crossing frequency with feedback
i	Current
I_B	Bias current
I_{ES}	Emitter-base leakage at low voltage
I_m	Maximum specified load current
I_m	Mirror current
I_{os}	Input offset current
I_s	Leakage current of diode at low voltage
k	Boltzmann's constant
log	Base 10 logarithm
ln	Base e logarithm
M	Attenuation
NF	Noise figure
N_i	Input noise

N_o	Output noise
prf	pulse repetition frequency
Q	Charge or the quality factor of an active filter
R	Resistance
R_c	Compensating resistance
r_D, r_e	ac junction resistance
R_f	Feedback resistor
r_{oc}	Output resistance of constant current source
S	Slew rate
S_i	Input signal
S_n $n = 1, 2, 3 \ldots$	Slope
S_o	Output signal
t	Time
T	Temperature
V_{BE}	Emitter-base voltage
V_{CEO}(sus)	Maximum collector to emitter voltage with the base open
V_D	Diode voltage
V_i	Voltage from inverting input to noninverting input
V_{IC}	Integrator reset voltage
V_{os}	Input offset voltage
V_z	Zener breakdown voltage
α	Alpha of transistor or the damping factor of an active filter
β	Feedback factor or transistor beta
θ_{cl}	Phase angle when $f = f_{1fb}$
θ_{pm}	Phase margin

APPENDIX

I

GLOSSARY OF TERMS USED IN TEXT

Active filter: Op-amp circuits, usually consisting of an op-amp with various *RC* feedback networks, which are used as low-pass, high-pass, band-pass, band-reject, and notch filters.

Adder: An operational amplifier circuit whose output voltage is the algebraic sum of two or more input voltages.

Adder-subtracter: An op-amp circuit whose output is the algebraic difference in two sums.

Antilog amplifier: A circuit whose output is proportional to the exponential of its input voltage.

Augmenting differentiator: An op-amp circuit that is a combination amplifier and differentiator.

Augmenting integrator: An op-amp circuit that is a combination amplifier and integrator.

Averager: An inverting adder arranged so that its output is the average of its inputs.

Bias current: The input current of an operational amplifier; base current in a biopolar amplifier and leakage current in a FET input amplifier.

Bias current (I_B): The input current of an operational amplifier.

Bode plot: A straight-line approximation of a frequency response curve, usually drawn on a semilog graph paper. Although easily drawn, a Bode plot will have a maximum 3 dB error at the corner frequency.

Bound: A circuit used across the feedback network of an op-amp circuit that prevents the op-amp output voltage from exceeding a chosen value, usually less than the supply voltage.

Chopper channel: The circuitry in a chopper-stabilized op-amp that contains the chopper circuitry and the stabilizing amplifier.

Chopper stabilization: The process by which temperature drift in the offset voltage of an op-amp is detected, changed to ac (chopped), amplified, changed back to dc, and applied back to the main amplifier in such a manner as to cancel the original drift in offset voltage.

Closed loop gain (A_{fb} or A_{cl}): The voltage gain of an operational amplifier with feedback.

Common-mode gain (A_{cm}): The gain of an operational amplifier to signals of the same voltage and phase connected to both the inverting and noninverting inputs simultaneously.

Common-mode rejection ratio (CMRR): A method used to express operational amplifier common-mode gain. $CMRR = A_{ol}/A_{cm}$

Common-mode signal: A voltage of the same phase and amplitude applied to both the inverting and noninverting inputs of an operational amplifier at the same time.

Comparator: A circuit used to detect whether the voltage on one input is higher or lower than the voltage on another input.

Corner frequency: The frequency at which an amplifier's gain has decreased 3 dB from its low frequency value.

Crossing frequency: The frequency at which $A_{ol} = 1$.

Damping factor (α): The response at the 3 dB frequency of an active filter.

Decade: An increase or decrease by a factor of ten.

Difference differentiator: An op-amp circuit that differentiates the difference in two input signals.

Differential input amplifier: A method of connecting an operational amplifier so that the circuit responds to a difference in input signals.

Differential integrator: An op-amp circuit that integrates the difference in two input signals.

Differential (balanced) output amplifier: A type of operational amplifier that has both differential input terminals and differential output terminals.

Direct adder: An op-amp adder that does not have an inverted output or scaling of voltages.

Feedback: Applying a portion of the output of an operational amplifier to the input through a resistive voltage divider.

Feedback factor (β): The fraction of the output voltage fed back to the input by a negative feedback network. The feedback factor is the reciprocal of the ideal closed loop gain, $\beta = 1/A_{fbi}$.

Flicker noise: Noise in semiconductor materials believed to be caused by the variation in charge velocity because of flaws in the semiconducting material.

Frequency compensation: The addition of an *RC* network to an op-amp to keep it from oscillating.

Frequency response curve: A plot of voltage gain versus frequency.

Function synthesizer: A circuit that provides a precise nonlinear response to an input voltage.

Gain-bandwidth product: The gain multiplied by the corner frequency at that gain. If the roll-off rate is −6 dB/octave, the gain-bandwidth product is constant.

$\Delta I_{os}/\Delta T$: Change in input offset current per degree centigrade.

Ideal closed loop gain (A_{fbi}): The ratio of the feedback resistors that set the op-amp circuit gain. For example, for an inverter $A_{fbi} = R_f/R_1$.

Input guarding: Connecting the shield conductors of the inputs to the common-mode voltage of the inputs.

Input offset current (I_{os}): The difference in the bias current of the inverting and noninverting inputs of an operational amplifier.

Input offset voltage (V_{os}): Small, undesired signals generated internally by the amplifier, causing some voltage output with no input voltage. The primary source of V_{os} is emitter-base voltage mismatch of the differential input transistors in bipolar transistor amplifiers and gate-source voltage mismatch in FET input amplifiers.

Instrumentation amplifier: An op-amp circuit designed to function as a differential input amplifier that has a high R_{in}, low offset voltage errors, high gain, and a very high CMRR.

Integrator: A circuit whose output voltage is proportional to the average over time of its input voltage.

Inverting adder: An op-amp adder circuit that inverts the algebraic sum of two or more inputs.

Interference: Electromagnetic signals picked up in a circuit from sources external to that circuit. An example would be a television picture interferred with by the motor noise of a power tool

Inverting amplifier: An operational amplifier feedback circuit in which the input and output voltages are out of phase.

Isolation amplifier: An amplifier system that has no dc current path between the input and output sections.

Johnson (thermal) noise: Noise from the random motion of charges in a material because of the thermal energy the charges receive from their surroundings.

Logarithmic amplifier: A circuit whose output voltage is proportional to the logarithm of its input voltage.

Loop gain: The product of the open loop gain and the feedback factor. Loop gain $= A_{ol}\beta$.

Negative feedback: Feedback in which the portion of the output applied to the input is out of phase with the input.

Noise: A term generally used to describe the electromagnetic energy generated by the random motion of charges within a material.

Noise figure: A measure of the noise added to a system by a particular circuit: 10 log (input signal to noise ratio/output signal to noise ratio).

Nonideal operational amplifier: An op-amp with $R_{in} < \infty$, $A_{ol} < \infty$ and $R_{out} > 0$. All manufactured op-amps are nonideal.

Noninverting amplifier: An op-amp circuit using feedback in which the input voltage and output voltage of the circuit are in phase.

Null: A condition in which the set point and state voltages cancel each other.

Octave: An increase or decrease by a factor to two.

Offset compensation: The use of external or internal adjustable circuits to cancel out an amplifier's inherent offset voltage.

Open loop gain (A_{ol}): The voltage gain of the operational amplifier package. The voltage gain obtained from an operational amplifier without feedback.

Operational amplifier (op-amp): A multistage, differential input amplifier that approximates the characteristics of an ideal amplifier.

Peak detector: A circuit that detects and stores the maximum input voltage fed into it.

Phase compensation: The addition of an *RC* network to an amplifier to keep it from oscillating.

Phase detector: A circuit that measures the difference in phase between two input signals of the same frequency.

Phase margin: The difference in degrees or radians between the phase shift of a signal through a system and the phase shift that will cause the system to oscillate.

Proportional control: A control in which the output voltage is proportional to the difference between the set point (control) voltage and a sample of the output (state) voltage. Also called a closed loop *servo system*.

Roll-off rate: The rate at which the gain of an amplifier decreases with increasing frequency, usually given in decibels per octave.

Scaling adder: An adder that amplifies as well as adds some of its inputs.

Set point: The voltage used as a reference in proportional control. This voltage tells the system what to do.

Servo amplifier: An amplifier used in a servo system that provides the proper current and voltage to drive the load.

Signal compressor: A circuit whose response to small input signals is larger than its response to large input signals.

Signal-to-noise ratio (S/N): The ratio of available signal to existing noise, that is, signal power/noise power.

Slew rate (S): The slew rate of an op-amp is the maximum change in output voltage per unit time that an op-amp can deliver.

Stabilizing amplifier: A low offset ac amplifier in the chopper channel used to amplify the chopped error signal.

State voltage: The feedback voltage that indicates what the load is doing.

Summing circuit: Any adder or linear signal combining circuit.

Summing differentiator: An op-amp circuit that differentiates and sums the derivative of two or more input signals.

Summing integrator: An op-amp circuit that integrates and sums two or more input signals.

Three-mode integrator: An integrator circuit used to integrate very low frequency signals.

Transducer: A device that converts a physical parameter or parameter variation, such as pressure or temperature, to an electrical signal or signal variation.

$\Delta V_{os}/\Delta$T: Change in input offset voltage per degree centigrade.

Voltage follower: An operational amlplifier circuit in which all the output voltage is fed back to the inverting terminal. The input voltage is applied to the noninverting input. The circuit has a gain of one.

Zero-crossing detector: A circuit that has a change in output every time its input goes through zero volts.

INDEX